Consciousness is the Universe

Conscious Awareness
Neuroscience
Quantum Physics
Evolution, Development
Unconscious Mind

Rhawn Gabriel Joseph, Ph.D.

I0084564

Consciousness is the Universe

Conscious Awareness
Neuroscience
Quantum Physics
Evolution, Development
Unconscious Mind

Consciousness is the Universe

Conscious Awareness
Neuroscience
Quantum Physics
Evolution, Development
Unconscious Mind

Rhawn Gabriel Joseph, Ph.D.
Cosmology.com / BrainMind.com

Copyright © 2009, 2010, 2014, 2016, 2017,, 2018 Cosmology Science Publishers, Cosmology.com

Published by: Cosmology Science Publishers, Cambridge, MA

ISBN: 978-1-938024-32-0
1-938024-32-X

Contents

1. Freud, Jung, and the Collective Quantum Continuum of Cosmic Consciousness

(From below) cried ID: "If thou be Christ, save thyself and us."
But Superego rebuked ID, saying: "Dost not thou fear God?"

THE DUALITY OF MIND

For many thousands of years humans have speculated as to the possible existence of a non-conscious psychic or spiritual domain from which intuitive leaps, inspirations, dreams and forbidden impulses originate. Some of the ancients believed the ultimate source for these mysterious forms of experience were the various gods who populated the heavens. For their own unknown reasons and purposes, the gods transmitted these thoughts, dreams and ideas into human hearts and heads. There was no unconscious mind. There were unconscious gods. For a human to truly know himself, and his destiny he had to supplicate these gods and study the dreams they sent.

Not all of the ancients, however, looked to impersonal, external gods for self understanding. Five thousand years ago the Sumerians (of ancient Sumer, Iraq) worshiped and believed in a pantheon of gods, as well as a personal god whose source and abode was within themselves, and which served almost like a conscience. It was to this personal deity that the individual sufferer bared his heart and soul as to personal short comings, sins and misdeeds, and to which he admitted responsibility for his own bad behavior. This personal god in turn acted as a mediator with the pantheon of gods that ruled the world at that time.

However, insofar as the Sumerians (as well as many other ancient cultures) were concerned, the mind and soul were to be found in the blood colored liver. Hence, if sufficient blood were lost, life was lost and the soul and spirit departed.

By contrast, the ancient Egyptians, concerned about "judgment day" and their well being after death, conceived of a personal soul and spirit which represented the Self in the Hereafter. The ancient Egyptians referred to these psychic entities as the "Ba " and the "Ka " which they thought existed within one's self, and later after death, within the realm of the supernatural either as a spirit like self-image or as a god-like creative life force. Like modern humans, the Egyptians also believed the soul to be affected by sin and to contain all those aspects of the Self which were considered both debase and noble. However, it was not the brain, but the bowels and the heart in which these entities were to be found. The brain, in fact was discarded after death, whereas these other organs were preserved with the mummified remains.

By contrast the people of the Indus Valley of ancient India began to worship Brahma. Brahma was described as an all pervasive, sustaining, and creative force which exists in ourself as the most essential aspect of our being, our soul.

In addition to this all encompassing God Within, they also postulated the existence of a surface layer of the Self which engaged in the seeking of knowledge; referred to as the atman. It was this psychic cleavage and thus the duality of the mind and spirit (atman and Brahman), which prevented as well as made possible the seeking of one's hidden self and essential being, and thus the attainment of unity and enlightenment.

It is this same dualistic philosophical system, first written in Sanskrit and in the ancient Vedic literature, which eventually gave rise to modern Hinduism and which also influenced Plato and Kant and thus the very foundations of Western philosophical thought. Hence, the duality of the mind has been recognized and debated for at least 4 thousand years if not longer. Nevertheless, with the advent of Greek Philosophical thought, as advocated by Plato, and with the establishment of the Hippocratic school of physicians, around 350 BC, the brain was finally posited as the abode of the mind and the source from which emotions arose.

Over the ensuing centuries, Western and middle Eastern religious and spiritual thought continued to be influenced by the ancient Sumerians,

Egyptians as well as the Greeks. In consequence, a marriage of these different philosophical and religious belief systems resulted, at least in part. Hence, by the 19th Century, the duality of the mind (including both conscious and non-conscious elements), the presence of a personal soul or spirit, the existence of a hidden unknown self, and the possibility that seat of the mind was to be found within the brain, as well as other body organs, were widely debated beliefs.

Moreover, the "unconscious" (as well as the pineal gland of the brain) began to be viewed by some as the source within which the soul could be found. In fact, at the beginning of the last century the distinguished American psychologist William James argued that the unconscious mind, particularly during sleep, acted as a doorway or portal through which God could exert his influences. "If there be higher spiritual agencies that can directly touch us," it is via "our possession of a subconscious region which yields access to them."

The belief in a linkage between the spiritual and supernatural, and a hidden, unknown, unconscious self continues unabated even today. Indeed, the very roots of modern day psychology are completely entangled in such beliefs for even the term "psychology," as many people know, originally meant the "study of the soul."

Unfortunately, this same presumed link between the unconscious and the spiritual has so traumatized modern scientists that many of them refuse to consider the unconscious as a serious construct; they fear the taint of association and the ridicule of their colleagues for toying with what they believe to be mysticism. Indeed, this has been a major complaint regarding Sigmund Freud and Carl Jung in particular. Presumably, mystical experiences are not worthy of the attentions of a true scientist.

Nonetheless, it is noteworthy that extremely heightened feelings of the mystical, and hyper religiousness, hyper morality, and spirituality sometimes result when certain areas of the brain (i.e. the limbic system and temporal lobe) are injured, abnormally activated or subject to seizure disorders. Fantastic verbal and visual hallucinations (often filled with dream-like, or religious imagery) may be experienced, and the erroneous assignment of religious and emotional significance to commonplace, mundane events may occur, particularly when brain structures such as the amygdala and hippocampus (which are buried in the depth of the temporal lobe) are hyper-activated.

In fact, individuals who were never very religious may suddenly "find religion" when this area of the limbic brain has been injured. Some scholars have even postulated that Moses, who may have been learning disabled, had a speech disorder, and suffered from murderous rages and tremendous headaches (all signs of possible brain damage and temporal lobe, limbic seizures), may have also suffered from a temporal lobe epileptic seizure disorder. This in turn would also explain his supposed hallucinations (for example, seeing a burning bush and hearing the voice of God), hyper religious fervor, hyper morality, and a need to write that was so pronounced that although lacking pen and ink he spent months and maybe even years carving his thoughts and God's laws in stone; a condition resembling hypergraphia (the urge to write). These conditions are all associated with an irritative lesion that abnormally activates the temporal lobe and limbic system of the brain. However, even if he did suffer from this condition (and I am not saying that he did), does this invalidate his experiences?

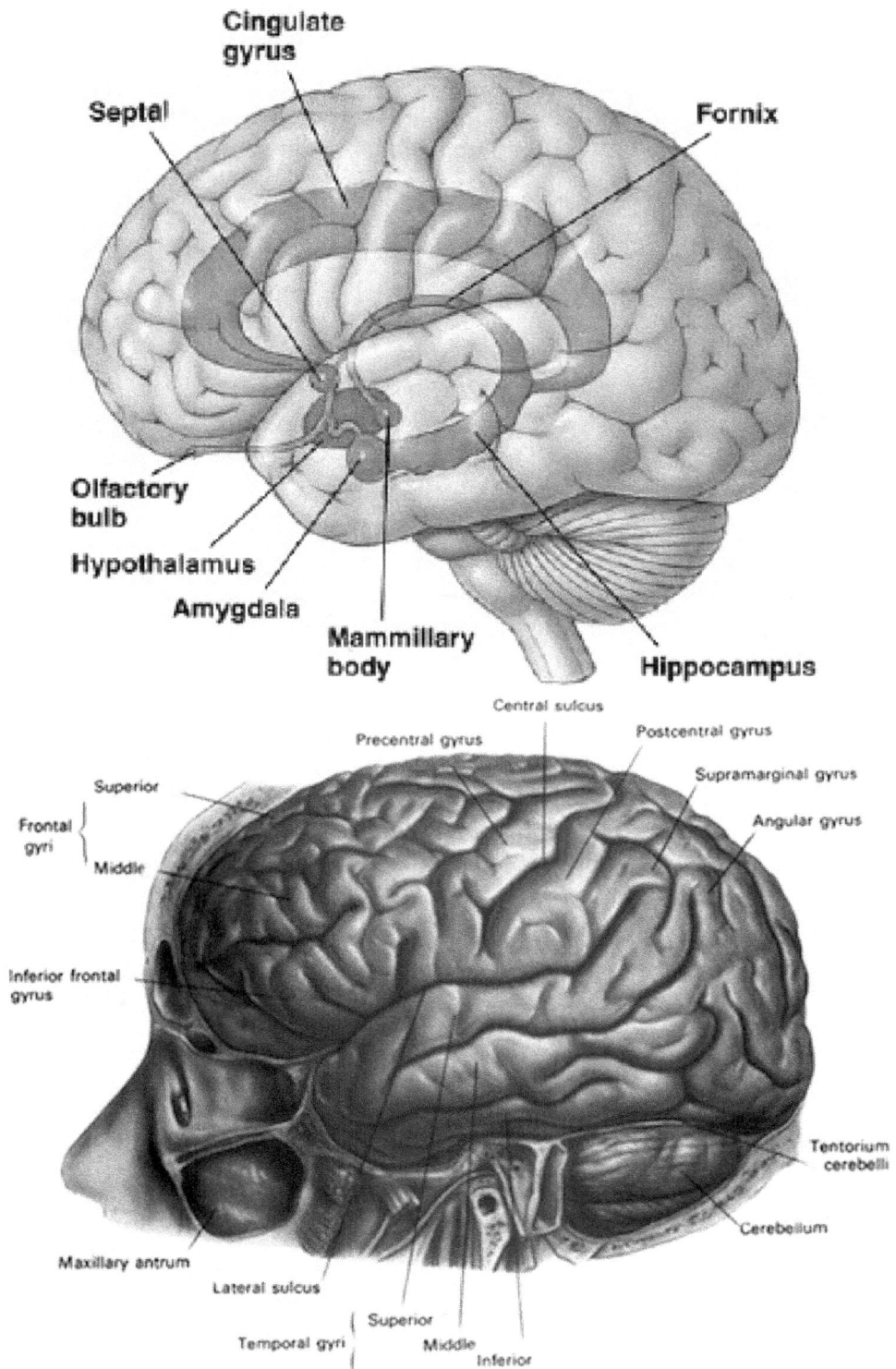

Cingulate gyrus

Septal

Fornix

Olfactory bulb

Hypothalamus

Amygdala

Mammillary body

Hippocampus

Central sulcus

Precentral gyrus

Postcentral gyrus

Supramarginal gyrus

Angular gyrus

Superior

Frontal gyri

Middle

Inferior frontal gyrus

Tentorium cerebelli

Cerebellum

Maxillary antrum

Lateral sulcus

Temporal gyri

Superior

Middle

Inferior

THE ABODE OF THE MIND

Although the Egyptians recognized that speech disorders may result from left cerebral injuries and the ancient Greeks acknowledged the importance of the brain in regard to emotions, mind, and even epilepsy, the brain has only recently been seriously thought by mainstream scholars as an organ of the mind. Indeed, as noted, the soul, spirit, or mind was thought by various ancient and even recent cultures and philosophers to reside in the liver, heart, lungs or diaphragm, and it was from these organs that inspiration, love, emotion and desire were believed to have their source and origin; thus, the changing in the rhythm of the heart or the intensity of one's breathing when beset by strong feelings and emotions. In fact, God and the Holy Spirit were long associated with the wind and the "breath of life." However, it was also thought that these spirit like forces may be transmitted from these various internal organs to the fluid filled ventricular cavities of the brain where they could then be expressed.

Ancient and not so ancient beliefs still influence modern scientists, popular literature and psychiatry. For example, the term "schizophrenia" originally and long ago referred to a splitting of the soul (or mind). This was thought to be caused by damage or a splitting of the phrenic nerve which serves the diaphragm and thus the ability to breathe; within which was the abode of the mind. Hence: schizo-phrenic.

By the dawn of the 19th century, neither the lungs, the heart, the pineal gland, nor the fluid filled ventricular cavities of the brain, were thought by most scientists to have much psychic significance and were viewed as exerting little if any influence on mental or emotional functioning except indirectly. Increasingly, the brain itself (and not it's cavities) was being viewed as the abode of the mind. Although seriously challenged and debated, the duality of the mind and the presence of a non-conscious psychic domain continued to be viewed and recognized by many as exerting significant influences on mental activity, creativity, and emotional functioning.

THE UNCONSCIOUS MIND

Many modern theorists have referred to this non-conscious portion of the mind as the "subconscious," the "pre-conscious," the "unconscious" or the "Collective Unconscious." Indeed, 100 years ago Sigmund Freud and Carl Gustav Jung began to develop elaborate models of the mind and posited the existence of an unconscious psychic realm from which dreams, intuitive understanding, emotional conflicts, and embarrassing and unpleasant impulses have their source and origin.

They also believed that the unconscious contained many ancient and archaic elements and was much more primitive than the more recently evolved conscious mind. Freud and Jung also believed that the unconscious in fact consisted of two levels. Freud classified these two levels as the unconscious and pre-conscious, whereas Jung posited the existence of a personal unconscious and a collective unconscious.

FREUD'S THEORY OF MENTAL FUNCTIONING: THE CONSCIOUS, PRE-CONSCIOUS & UNCONSCIOUS MIND

According to Sigmund Freud, the psyche consists of three mental domains; the unconscious, the pre-conscious, and the conscious mind. The unconscious, argued Freud, is the deepest, least accessible region of the mind, which contains hidden and seemingly forgotten memories and unacceptable thoughts, feelings, and ideas.

The Pre-conscious contains information and memories that exist just below the surface of consciousness and in this respect it is part of the unconscious. Once information reaches the pre-conscious it becomes relatively accessible to the conscious mind. However, the pre-conscious also contains information that is pushed out of consciousness.

Initially all psychic material, thoughts, feelings, and desires, exist as unconscious impulses which attempt to discharge into consciousness. To reach the conscious mind, however, these unconscious impulses must first pass through the pre-conscious. The pre-conscious mind serves in some respects like a psychic corridor which link yet separate these two mental realms.

All impulses must also be approved by two censors which lurk at opposite ends of the pre-conscious corridor. One censor guards the door which links the unconscious to the pre-conscious, the other guards the door that links the pre-conscious with the conscious mind. If an impulse is deemed undesirable by either censor, entry into consciousness is met with resistance and prevented. The information is either transformed into something less threatening or is repressed and remains unconscious. Although repressed, it may continue to influence the conscious psyche, however.

Freud argues that the conscious mind has a very minor role in psychic functioning, being only the "outermost superficial portion of the mental apparatus." Rather, all behavior is determined by the interaction of unconscious mental forces which he termed the Id, Superego and Ego (the latter of which is both conscious and unconscious). Ultimately all behavior, even slips of the tongue or forgetfulness, are due to impulses which initially exist totally beyond conscious awareness; there is no such thing as an accident as all behavior is purposeful and serves unconscious desires.

Thus, at a conscious level we have no control and no choices over our actions and thoughts and are instead pushed about by unconscious forces which literally have a mind of their own.

THE EGO, ID & SUPEREGO

In developing his theory Freud posited the existence of an Ego, which simultaneously occupies the conscious, pre-conscious and unconscious; a Superego which could be found in the preconscious and unconscious mind; and an Id, which was wholly unconscious. The Id (a term possibly coined by Nietzsche) was posited as the seat of the pleasure principle and instincts. It acts to maximize pleasant feelings and minimize unpleasant tensions.

The Superego acts to instill guilt and was thought to represent an internalized parental voice. It serves in an inhibitory fashion so as to promote high standards of behavior while quashing all that which had been taught to be undesirable. At a very basic level, the Id states figuratively: "I want it," whereas the parental Superego states: "No! You can't have it. Don't do that. That's wrong!"

Essentially, the Ego is identified with our conscious sense of Self. However, it is also unconscious and acts as a mediator between the Id and the Superego, so that compromises can be fashioned. The Ego has the additional duties of censorship as it is the responsibility of the unconscious portion of the Ego to protect the conscious aspect of the mind by standing guard at either end of the interlinking psychic corridor; the preconscious.

The Ego, thus serves as both mediator and censor, and can enable an impulse which the Superego finds objectionable to be resisted and repressed, or refashioned and transformed into something more acceptable to the conscious self image. This reduces tension by permitting the original impulse to be partially expressed, albeit in disguise or indirectly.

For example, the Id may spy something that catches its fancy and proclaim (albeit non-verbally): "I want it," and then try to obtain it regardless of the consequences. The Superego may admonish the Id for these desires and in its own unique language proclaim, "You can't have it." The Ego, acting as mediator may then try to meet the wishes of the Id while attending to the standards of the Superego. Instead of just taking the item (which the Id urges but the Superego finds objectionable) the Ego pays cash and rationalizes the purchase through the generation of numerous believable explanations.

THE MIND AND BRAIN

It is noteworthy that long before developing his psychoanalytical theories Sigmund Freud had spent many years studying the nervous system and the effects of brain damage on the mind. In 1895, he even wrote an almost forgotten treatise "Project for a Scientific Psychology" where he proposed a neurological theory of the psyche. However, he soon abandoned his mind brain proposals and in fact later argued against attempts to draw any parallels between psychoanalytic theory and brain functioning as he believed that simply not enough was then known about the brain.

Fortunately, much has been learned about the brain in the last 100 years. Moreover, parallels between certain aspects of brain functioning and specific features of Freud's and even Jung's theory of the mind in fact exist, and some scholars have expended considerable effort exploring these and similar relationships as limited as they may be. Indeed, the so called unconscious mind may in fact be a manifestation of right brain and limbic system mental activity. Conversely what we regard as the conscious mind appears to be localized within and is maintained and supported via the functional integrity of the left half of the brain in most people.all human memory and experience ends and where the ancestral and spiritual memories of our non-human ancestors begin.

THE COLLECTIVE UNCONSCIOUS

According to Jung, for a good part of our ancestral history all mental functioning was governed by the Original Mind. Over time consciousness appeared and became increasingly differentiated and separated from the original unconscious psyche. Consciousness is thus a relatively recent evolutionary development. This separation was not complete, however, and both remain linked.

According to Jung the difference between the conscious and unconscious mind is in part a matter of distinctness, organization and differentiation as both tend to process and respond to information in a different fashion. If we descend from the surface levels of consciousness into the unconscious, the material we encounter becomes increasingly less distinct or logical and more disorganized, and the capacity to specifically focus the minds eye quickly wanes. Delving deeper into the swirling mists of the unconscious we encounter forgotten and painful memories, and a variety of significant impulses, and ideas, some of which are about to emerge into or have just been pushed out of consciousness.

Jung noted, however, that the unconscious contained a very ancient psychic dimension, the contents of which formed the core of what he referred to as the "Collective Unconscious."

Jung argues that if we continue to descend we would soon enter a level of the mind where the personal layer ends. It is here that the residues of ancestral life and the earliest memories and most profound experiences of our ancient ancestors begin to be encountered. We have thus entered the collective unconscious.

The collective unconscious is an aspect of the mind and brain which functions in accordance with inherited tendencies to respond to certain experiences with emotional, spiritual, mythical, or religious feelings, images, thoughts and ideas. Collectively these primordial feelings and images are called archetypes. Archetypes are often encountered in dreams (what Freud and others have referred to as the "royal road to the unconscious.") and mythology. However, these archetypes can also ascend into consciousness and are represented in art, film, literature, and architecture.

When archetypes reveal themselves as images it is usually as emotionally invested symbols or concept which tend to transcend meaning or precise definition. God is such an archetype, as is the ideal of motherhood, and the cross and the Christmas tree, which, one might argue are presumably symbolically descended from or related to the Egyptian Key of Life, the Hebrew tree of life and the tree of good and evil (as described in Genesis) and all of which are probably related in some unknown fashion to the fact that our extremely remote ancestors at one time spent a good deal of their lives in trees for several millions years. Hence, even when people from completely different cultures encounter these and other archetypical symbols, they often view them with similar emotional, spiritual, or mystical awe.

For instance, in the 16th century when the Spanish conquistadors first arrived in Mexico they were astonished not only by the highly advanced culture and grand architecture they found everywhere but by what appeared to them to be the sign of the cross (as well as the swastika) which repeatedly appeared on various buildings and temples. The Aztecs and the many other tribes that inhabited these regions had never heard of Christ, however, nor were they familiar with the ancient Romans who also employed the swastika (as did the Nazi's several hundred years later). Although they held the cross with reverence, it was in no way associated with a crucified god. Rather, the sign of the cross is a celestial symbol associated with the four seasons, the four corners of the universe, and the winter/summer and Spring/Autumn solstices and equinox.

Since large masses of humanity respond similarly to certain symbols such as the cross both across cultures, continents and time, and since many people have similar dreams and myths with similar archaic and often mysterious or spiritual imagery, then these elements must be the residues of a collective memory of the same ancestral heritage, argues Jung. Because these experiences and dreams contain symbols or actions which have no bearing on one's personal experience they can represent only something that is inherited. However, it is not the image per se which is inherited, it is the feelings and emotions associated with certain images and ancestral experiences, at least, so argues Jung.

In so far as these ancestral experiences had significantly impacted many generations and large masses of ancient humanity, they eventually became figuratively engraved into the minds, memories and even the genetic structure of Homo sapiens. These memories have passed down succeeding generations and are forever recalled in the dreams and myths of the masses. Hence, ten thousand years from now our descendants will probably be dreaming of mushroom clouds and nuclear explosions.

Interestingly, Jung was convinced that if one continued their journey into the farthest reaches of the collective unconscious they would reach a point where all human memory and experience ends and where the ancestral and spiritual memories of our non-human ancestors begin. That ancestry, however, would lead backward in time when life first took root on Earth... and if life on Earth came from other planets, then that journey would therefore continue and incorporate memories and experiences from those who may have evolved on other worlds. If true, then the collective unconscious, therefore, would encompass the collective experiences of life forms which hail from throughout the cosmos and include those who long ago evolved before Earth became a "twinkle in god's eye."

COSMIC CONSCIOUSNESS AND THE QUANTUM CONTINUUM

The Universe may be conscious--that is, the electromagnetic quantum continuum that forms the fabric of the cosmos may serve as a continuum of consciousness which contains all the information of the universe.

Perhaps human consciousness could be likened to a television or radio receiver--we tune into certain stations... or into an aspect of the conscious continuum and "hear" or "see" what already exists within the collective cosmic mind. Human consciousness may be only a fragment of the totality which is already "out there."

As will be explained in later chapters, that quantum continuum encompasses not only all of existence, but time. Einstein had said: "The distinctions between past present and future are an illusion." Experiments in quantum physics have demonstrated, through entanglement, that events in the future can affect the present, and thus, the past. If the quantum continuum encompasses all of time and what we perceive as the future or remember as the past, then that collective cosmic consciousness would also encompass and have consciousness of what will happen, or may happen, or which has already happened in the future and all possible futures. If true, then the quantum continuum of consciousness would be omnipresent and omniscient--characteristics which some have attributed to their "gods."

2. How Consciousness Became the Universe

ATOMISTS VS CREATIONISTS

For thousands of years humans have gazed into the heavens pondering the nature of universe, with some arguing the Universe was created, and others championing an eternal, infinite cosmos, in continual flux and change, and with no beginning and no end. This latter view, developed by the Greek "atomists", and the ancient Hindus and Buddhists, was the prevailing theory accepted by most scientists until the 20th century, when the creationists, with the backing of the Catholic Church, came up with a scientific explanation for "the creation" which was eventually accepted by the majority of scientists. Ironically, it was Fred Hoyle, a leading proponent of an infinite universe, who dismissively coined the phrase which popularized the creationist theory. Hoyle called it: "the big bang."

It was in 1927 that Monsignor Georges Lemaître, published what became known as the theory of the big bang. His paper was titled "A homogeneous Universe of constant mass and growing radius accounting for the radial velocity of extragalactic nebulae." Lemaître designed and based his big bang theory on the Biblical story of Genesis. Lemaître hoped to make the Bible scientific and in accord with testable observations. The universe, he said, was created by a creator, as detailed in Genesis, chapter 1: "In the beginning God created the heavens...And God said, Let there be light: and there was light. And God saw the light, that it was good: and God divided the light from the darkness."

Lemaître initially called his theory the 'hypothesis of the primeval atom" and described it as "the Cosmic Egg exploding at the moment of the creation." Today is is known as "The Big Bang."

Lemaître's physics, as Lemaître admitted, had a spiritual foundation. Monsignor Lemaître firmly believed Jesus Christ was God and God created the universe, as advocated by the Catholic Church and as described in Genesis 1. Lemaître was in fact an honorary prelate with the rank of Bishop in the Catholic Church, a professor at the Catholic University of Leuven. and president of the Pontifical Academy of Sciences which is under the direction and authority of the Pope of the Catholic Church. Lemaître's "big bang" theory was in accordance with the teachings of the Church, was supported by the Pope, and backed by the authority of the Bible: The Universe was created. God created the Universe.

For thousands of years, however, other scientists, philosophers and sages, have argued the universe is infinite and eternal, and that the very foundation of existence, undergoes constant change. For example the Greek "atomists" Democritus, Anaxagoras and Epicurus argued that the universe was infinite;

that permanence was an illusion--a manifestation of the act of observing and the limitation of the senses.

As their name suggests, the "atomists" based their reasoning on atomic theory which they originated. Specifically, in the fifth century B.C. Democritus and Anaxagoras proposed that all matter is made up of tiny indestructible units, called atoms. According to Democritus, these atoms move about in space and combine in various ways to give shape and form to all macroscopic objects.

Anaxagoras argued that the fundamental unit of matter, the atom, could neither be created nor destroyed—perhaps the earliest formulation of the law of conservation of mass. All matter was comprised of smaller elements called "atoms"which have always existed and which could not be created. Therefore the essence of existence, including the universe which is made of atoms, has no beginning but has existed for all of time. There was thus no need for a creator god.

The atomists were repeatedly assailed and attacked by Jewish philosophers and religious authorities who described these theories as "heresy." Attacks against the "atomists" continued with the establishment of the Catholic Church who labeled all atomists "heretics" and "blasphemers."

Supporters of the atomists countered: if there is a god, then who created god? The theologians replied: God the creator, became the creator at the moment of creation, just as a carpenter becomes a carpenter at the moment he first builds something. Thus it was argued that god existed before the creation, and it was only at the moment of creation, that god became "god the creator."

Hence, the "creation-event" which gave rise to the universe was created by an all powerful omnipotent Lord God who existed prior to and is responsible for the creation event; exactly as described in Genesis 1. God is self-creating. In fact, similar arguments have been put forth by those supporting the theory of the big bang: the universe did not exist until it was created, and existed as a singularity or as a nothingness prior to the creation event. The universe created itself.

The theory of the big bang creation event, and the argument in favor of god the creator, rely upon similar philosophical arguments. Prior to the big bang there was nothing, or a singularity which for unknown reasons exploded and became the universe. God was the singularity which created the universe, and thus god and the universe are self-creating: the universe became the universe at the moment of creation, and god the creator became "god the creator" at the moment of creating the universe.

Be it God or a "big bang" creation event, both theories rest upon the same belief: something always existed, i.e. "god" and/or the "singularity." And both god and the universe are self-creating. God became god by creating the universe and thus God created God by the act of creation. The Universe became the universe by creating itself and thus the universe created the universe by the act of creation. Be it the Jewish/Christian religion, or the "big bang" theorists, all assume something existed which gave rise to the universe; i.e. god or a singularity.

According to Lemaître, God is that singularity. The universe, therefore, is not just a manifestation of the creative act, but is also god. The universe is god since all things come from God. Every galaxy, stars, planet, human, dog, cat, molecule, atom--all come from God, and in the beginning were one with god--this singularity. And in the end, all will return to god, becoming one with god. So in the end, is, as the beginning: A oneness from which all existence emanates and returns. The parallels with Hinduism, Taoism, atomic theory, and quantum physics--as will be explained--are striking.

THE HINDU UNIVERSE: THE UPANISHADS / THE BHAGAVAD-GI-TA

In HinduVedic literature, it is said that Visnu is the primary creator who creates the basic universal shell and provides all the raw materials for what is perceived as the material world.

"In the beginning of the creation, the Lord first expanded Himself in the universal form of the puruṣa incarnation and manifested all the ingredients for the material creation. And thus at first there was the creation of the sixteen principles of material action. This was for the purpose of creating the material universe." -The Bhagavad-gītā

"This form [the manifestation of the puruṣa] is the source and indestructible seed of multifarious incarnations within the universe. From the particles and portions of this form, different living entities, like demigods, men and others, are created." -The Bhagavad-gītā

"Not-being was this in the beginning; From it arose. Self-fashioned indeed out of itself." -Upanishads

"Invisible, incomprehensible, without genealogy, colorless, Without eye or ear, without hands or feet, Eternal, pervading all and over all, scarce knowable, That unchanging one Whom the wise regard as being's womb." -Upanishads

THE TAO

"All things begin from nothing and end in nothing." -Chang Tzu

"This is the Tao--it may be apprehended by the mind, but it cannot be seen. It has Its root and ground of existence in Itself. Before there was heaven and Earth, It was securely existing. From It came the mysterious existences of spirits, from It the mysterious existence of God. It produced heaven. It produced Earth." -Chang Tzu

"From nonexistence we proceed to existence."-Chang Tzu

"Heaven, Earth, and I were produced together; and all things and I are one." -Chang Tzu

"It is from the nameless that Heaven and Earth sprang." -Tao Te Ching

"The nameless Tao was the beginning of Heaven and Earth and was before God." -Tao Te Ching

"There was something undefined and complete, coming into existence before Heaven and Earth; formless, reaching everywhere and in no danger of being exhausted. It may be regarded as the Mother of all things." -Tao Te Ching

THE JEWISH / CHRISTIAN GOD
"Lift up your eyes and look to the heavens:
  Who created all these?
He who brings out the starry host one by one
  and calls forth each of them by name." -Isaiah 40:26

"...when I was being made in secret,
intricately woven in the depths...
Your eyes saw my unformed substance...the days that were formed for me, when as yet there was none of them" - Psalm 139

"He reveals deep and hidden things;
he knows what lies in darkness,
and light dwells with him" - Daniel 2:22

"The secret things belong to God, but the things revealed belong to us" - Deuteronomy 29:29...
"I am God... declaring the end from the beginning" -Isaiah 46

CONSCIOUSNESS CREATED THE UNIVERSE
But what exactly was that "creative act"? How did non-being produce being? If there is a god, how did "god" create? How did "god the creator" create "god the creator"? The same questions have been asked about the "big bang." What caused it and why?

How do we know something exists?

One answer might be: Consciousness. Consciousness, exists. Consciousness has energy. Energy may become matter and matter, energy. Consciousness is always a consciousness of something. Consciousness requires a duality in order to exist as a consciousness. Self-consciousness, consciousness of consciousness, is also a duality.

As summed up the Heisenberg, one of the founders of quantum mechanics: "the transition from the possible to the actual takes place during the act of observation... and the interaction of the object with the measuring device, and thereby with the rest of the world...Since through the observation our knowledge of the system has changed discontinuously, its mathematical representation has also undergone the discontinuous change and we speak of a quantum jump" (Heisenberg, 1958). In other words, something comes into existence, by becoming conscious of it.

Heisenberg (1958), cautioned, however, that the observer is not the creator of reality, but instead merely registers, at a particular moment, certain isolated fragments of activity within the continuum, the nature of which is determined by our senses which perceives only parts and not the continuum which the part is, part of: "Quantum theory does not introduce the mind of the physicist as part of the atomic event. But it starts from the division of the world into the object and the rest of the world. What we observe is not nature in itself but nature exposed to our method of questioning."

The act of knowing, of observing, or measuring, that is, interacting with the environment in any way, creates an entangled state and a knot in the quantum continuum described as a "collapse of the wave function;" a knot of energy that is a kind of blemish in the continuum of the quantum field. This quantum knot bunches up at the point of observation, at the assigned value of measurement and can be entangled. Consciousness perceives a blemish in the continuum, but which is still part of the continuum, even though it is perceived as distinct.

The universe exists, because there is consciousness of the universe. This also means: consciousness must have come first, creating a quantum jump from singularity to duality. The universe may have become conscious of itself.

If the universe is a conscious universe, then it could be said that the universe came into being when the universe became conscious of itself. Likewise, if there is a "god" then it could be said that God created the universe and became god the creator, by becoming conscious of itself; that is, god became god the creator at the moment god achieved self-consciousness. If there is a a "God" and "god" and the universe are "one" then the god-universe became conscious of itself and thus the universe and god came into existence. Consciousness was the creative act. Consciousness creates by becoming conscious.

In the Upanishads, it is stated: "In the beginning the atman was this universe. He gazed around, he saw nothing there but himself. Thereupon he cried out at the beginning: It is I."

Thus, in the beginning, atman and the universe were a singularity, a oneness. However, upon becoming conscious of its existence, the universe came into existence, but as a duality:

"Not-being was this in the beginning; From it arose. Self-fashioned indeed out of itself." -Upanishads

The act of becoming conscious, created a duality also known as Brahman and Atman:

"Brahman...After he had created it, he entered into it; after he had entered into it, he was: The being and the beyond, Expressible and inexpressible, Founded and foundationless. Consciousness and unconsciousness; Reality and unreality... In truth, there are two forms of Brahman, The formed and the unformed, The mortal and immortal. The abiding and the fleeting, The being and the beyond....Truly the Brahman is this Atman." -Upanishads

It is the Atman, consciousness/self-consciousness which creates the universe, thereby creating a duality of Brahman and Atman--a universe which is conscious of itself. Brahman is the unknown that needs to be explained, atman is the known through which the unknown finds its explanation. Brahman and Atman became a duality, upon achieving self-consciousness; and this is how "god" and the universe came into being, according to the Upanishads.

Genesis, chapter 1: "In the beginning God created the heavens...And God said, Let there be light: and there was light. And God saw the light, that it was good: and God divided the light from the darkness." What is this light? It is the illumination of consciousness.

THE QUANTUM CONTINUUM: BEING AND NON-BEING

As pointed out by Neils Bohr and Werner Heisenberg, the founders of quantum theory, there are direct parallels between quantum mechanics and Taoism, Buddhism, and Hinduism: "The great scientific contributions in theoretical physics... has.. a relationship between the philosophical ideas in the tradition of the Far East and the philosophical substance of quantum theory." -Werner Heisenberg

Being become nonbeing, and nonbeing becoming being, is a major principle of quantum physics. Form and substance are manifestations of dynamic patterns of energy and electromagnetic radiation that have no material reality. Form and substance, that is, the "particles" they are comprised of, exist only as probabilities. Form and substance have probable realities and can become something else entirely, or cease to exist, returning to the basic oneness of the continuum--the most obvious example of which, is, death and decay. What is, can undergo a transformation from being into nonbeing

Consider the theoretical molecular foundations of matter, e.g., electrons, protons, neutrons, photons, neutrinos, etc., Each of these "particles" consist of yet smaller particles and all are balanced by the existence of anti-particles which are in all respects opposite in charge. However, in some instances, these elements also comprise their own anti-particle. Thus, there is being, and anti-being.

However, according to quantum mechanics, these particles do not really exist, except as probabilities. These "subatomic" particles have probable existences and display tendencies to assume certain patterns of activity that we perceive as shape and form by a conscious mind wielding an measurement apparatus. Yet, they may also begin to display a different pattern of activity such that being can become nonbeing or something else altogether.

Shape and form are a function of our perception of these dynamic interactions within the frenzy of activity which is the quantum continuum. However, we can only perceive what our senses can detect, and what we detect as form and shape are really a mass of frenzied subatomic electromagnetic activity that is amenable to detection by our senses and conscious mind at a particular moment in time. The perception and our consciousness of certain aspects of these oscillating patterns of activity are dependent on our sensory capabilities, which give rise to the impressions of shape and form. If we possessed additional senses, or an increased sensory channel capacity, we would perceive yet other patterns and other realities.

More specifically, this electromagnetic activity is so frenzied, that a particle can exist here, and now there. Although it is neither here nor there, it is perceived by consciousness as a solid object that has shape, form, length, mass, weight, and so on. This is because the rapidity of movement obscures the fact that much of what we perceive are particular patterns of electromagnetic activity that are in frenzied motion.

Mass, as Einstein explained in his famous equation, is energy. That is, mass (M) is equal to energy (E) divided by the speed of light squared ($M=E/C^2$). However, whereas mass can be destroyed, energy cannot. Rather, as mass is destroyed, an equal amount of energy is released back into the electromagnetic continuum. The energy released as a function of destruction, is equal to the mass of the object times the square of the speed of light ($E=MC^2$). Conversely, if we increase the amount of energy, we increase the amount of apparent mass.

Since energy cannot be created, then energy was never created and has existed always.

Mass does not consist of tiny particles that are packed tightly together.

Rather, what we perceive as mass (shape, form, length, weight) are dynamic patterns of energy which we selectively attend to and then perceive as stable and static. And, we are perceiving only fragments of the quantum continuum.

This energy that makes up the object of our conscious perceptions, is but an aspect of the electromagnetic continuum which has assumed a specific pattern that may be sensed and processed by our brain. As dictated by quantum mechanics, the universe exists, because we are conscious of it.

Hence, consciousness created the universe. If the universe is self-creating, then the universe became conscious of itself.

THE TAO OF QUANTUM CONSCIOUSNESS

If the universe came into being by becoming conscious if itself, then what is, or was, the singularity, this oneness, before it became conscious of itself?

"The nameless Tao was the beginning of Heaven and Earth and was before God." -Tao Te Ching

According to the Taoists, the basic oneness is not compacted into a single point, but is all pervasive, everywhere, at the same time; and not only do all things emerge from it, but all things return to it, becoming one:

"Tao is forever...All pervading is the Great Tao. It may be found on the left hand and on the right. All things depend on it for their production.. All things return to their root and disappear and do not know that it is it which presides over their doing." -Tao Te Ching

"The Tao passes on in constant flow. Passing on it becomes remote. Having become remote, it returns." -Tao Te Ching

"The Tao produces all things and nourishes them; it produces them and does not claim them as its own; it does all; it presides over all, yet does not control. This is what is called the Mystery of Tao." -Tao Te Ching

"In Tao the only motion is returning. Although Heaven and Earth were produced by being, being was produced by nonbeing." -Tao Te Ching

"Endless the series of things without a name on the way way to where there is nothing. They are called shapeless shapes; Forms without form; are called vague semblances." -Tao Te Ching

"This endless cycle of being emerging from non-being, and returning to its source, is consistent with quantum theory, i.e. the quantum continuum of electromagnetic activity, from which all existence arises and returns; and similar ideas are repeated in the Upanishads:

"In Him in whom this universe is interwoven; Whatever moves or is motionless, In Brahman everything is lost; Like bubbles in the ocean. In whim in whom the living creatures of the universe Emptying themselves become invisible. They disappear and come to light again. As bubbles rise to the surface." -Upanishads

"After he had created it, he entered into it; after he had entered into it, he was: The being and the beyond... Reality and unreality... The formed and the unformed, The mortal and immortal. The abiding and the fleeting, The being and the beyond. -Upanishads

The central view of Upanishads and the Tao, and which is consistent with quantum mechanics, is the cyclic nature of non-being becoming being, being becoming non-being--ceaseless motion and constant change: "Returning is the motion of Tao... and going far means returning." (Lao Tzu)

This cyclic change of continual motion is also represented by the Chinese concepts of yin and yang: "The yang having reached its end become yin; the yin having reached its end becomes the yang." -I Ching. Book of Changes

Its tao is forever changing -alteration, movement without rest, Flowing through the six empty places, Rising and sinking without fixed law; Firm and yielding transform each other. They cannot be confined within a rule. It is only change that is at work here." -I Ching. Book of Changes

What was the singularity which existed before the universe or the god/universe became the god/universe? The quantum continuum, which is forever changing. At the moment that quantum continuum became conscious of itself, the universe was created: being emerged from non-being.

QUANTUM CONSCIOUSNESS vs EINSTEIN

Einstein ridiculed the implications of quantum theory and its emphasis on consciousness as having any role in the nature of existence: "Do you really think the moon isn't there if you aren't looking at it?"

As theorized by Einstein (1961), and unlike the Copenhagen model of quantum physic, space-time is relative to but independent of any observer. Consciousness and the act of measurement is relative but irrelevant having no effect on the passage of time or events. In relativity, each event, which occurs at certain moments of time, in a given region of space, and are relative to those observers in different regions of space. Each observer chooses a convenient metrical coordinate system in which these events are specified by four real numbers.

In relativity, consciousness is merely relative. In quantum physics, consciousness and the act of observation and measurement constitute a separate reference frame which can collapse the wave function of the quantum continuum and register entangled interactions within the environment. Consciousness by the act of observation or measurement takes a static or series of pictures-in-time which then becomes discontinuous from the quantum continuum (Heisenberg 1958; Planck 1931; von Neumann 2001). These entanglements (Francis 2012; Juan et al. 2013; Plenio 2007), or blemishes in the quantum continuum, may be observed as shape, form, cause, effect, past, present, future, the passage of time, and thus reality; the result of a decoupling of quanta from the quantum (coherent) continuum which leaks out and then couples together in a knot of activity which is observed as a wave form collapse.

As based on the Copenhagen theory of quantum mechanics (Bohr, 1958, 1963; Heisenberg 1955, 1958), what we perceive as reality are a manifestation of wave functions and alterations in patterns of activity within the quantum continuum which are perceived by consciousness as discontinuous. Wave form collapse is always a matter of probability, and is non-local, indeterministic and a consequence of conscious observation, measurement, and entanglement. Consciousness and the act of measurement, therefore, are entangled with the quantum continuum and can after the continuum and the space-time manifold.

In summary, form and substance are comprised of particles which take on specific patterns and which are the result of concentrations of energy that emerge and disappear back into the electromagnetic quantum field; a knot of energy that is a kind of blemish in the continuum of the quantum field which is perceived by a conscious mind.

This reality, therefore, is a manifestation of consciousness and consciousness of alterations in the patterns of activity within the electromagnetic field. The electromagnetic field, this energy, is therefore the fundamental entity, the continuum that is constitutes the basic oneness and unity of all things.

QUANTUM PHYSICS AND PROBABILITY OF EXISTENCE

Einstein and his followers waged war against quantum theory which posed a direct challenge to his theories and those of Newton. For example, the equations of Newton and Einstein were replaced by equations between matrices representing the position and momentum of electrons which were found to be unpredictable. In fact, whereas Newtonian and Einsteinian theories do an admirable job of explaining the macro-world, they completely break down when applied to the sub-atomic world. This is because the patterns that resemble particles, display tendencies to exist and tendencies to no longer exist. Thus measurement becomes uncertain and can only be based on probabilities.

Because particles have only a probable existence it is impossible to predict an atomic event with 100% certainty. This principle, in fact, is called the uncertainty principle, because one can never predict when a particle may exist, and where it may exist when it does exist. Again, these particles exist and do not exist, and are always in motion--which gives rise to the illusion of permanent shape and form when perceived or measured by a conscious mind. It is also these principles which form the crux of quantum electrodynamics, or field theory.

Broadly considered, atoms consist of empty space at the center of which is a positively charged nucleus and which is orbited by electrons. The positive charge of the atom's nucleus determines the number of surrounding electrons, making the atom electrically neutral. However, it was determined that it was impossible to make precise predictions about the position and momentum of electrons based on Newtonian or Einsteinian physics, and this led to the Copenhagen interpretation (Heisenberg 1925, 1927) which Einstein repeatedly attacked because of all the inherent paradoxes. Matrix mechanics is referred to now as quantum mechanics whereas the "statistical matrix" is known as the "probability function;" all of which are central to quantum theory.

As summed up by Heisenberg (1958) "the probability function represents our deficiency of knowledge... it does not represent a course of events, but a tendency for events to take a certain course or assume certain patters. The probability function also requires that new measurements be made to determine the properties of a system, and to calculate the probable result of the new measurement; i.e. a new probability function."

Quantum physics, as exemplified by the Copenhagen school (Bohr, 1934, 1958, 1963; Heisenberg, 1925, 1927, 1930), like Einsteinian physics, makes assumptions about the nature of reality as related to an observer, the "knower" who is conceptualized as a singularity. As summed up by Heisenberg (1958), "the concepts of Newtonian or Einsteinian physics can be used to describe events in nature." However, because the physical world is relative to being known by a "knower" (the observing consciousness), then the "knower" can influence the nature of the reality which is being observed through the act of measurement and registration at a particular moment in time.

And yet, what is observed or measured at one moment can never include all the properties of the object under observation. In consequence, what is known vs what is not known, becomes relatively imprecise (Bohr, 1934, 1958, 1963; Heisenberg, 1925, 1927).

As expressed by the Heisenberg uncertainty principle (Heisenberg, 1927), the more precisely one physical property is known the more unknowable become other properties. The more precisely one property is known, the less precisely the other can be known and this is true at the molecular and atomic levels of reality. Therefore it is impossible to precisely determine, simultaneously, for example, both the position and velocity of an electron (Bohr, 1934, 1958, 1963).

Einstein objected to quantum mechanics and Heisenberg's formulations of potentiality and indeterminacy by proclaiming "god does not play dice."

In Einstein's and Newton's physics, the state of any isolated mechanical system at a given moment of time is given precisely. Numbers specifying the position and momentum of each mass in the system are empirically determined at that moment of time of the measurement. Probability never enters into the equation. Therefore, the position and momentum of objects including subatomic particles are precisely located in space and time as designated by a single pair of numbers, all of which can be determined causally and deterministically. However, quantum physics proved that Einstein and Newton's formulation are not true at the atomic and subatomic level (Bohr, 1934, Born et al. 1925; Heisenberg 1925, 1927).

According to Heisenberg (1925, 1927, 1930), chance and probability enters into the state and the definition of a physical system because the very act of measurement can effect the system; and this is because the observing consciousness and measurement apparatus are also entangled with the quantum continuum and effect it, and vice-verse. No system is truly in isolation, but is contiguous with the quantum continuum. Moreover, no system can be viewed from all perspectives in totality simultaneously which would require a god's eye view; and "god's eye" would also have to observe itself observing. Only if the entire universe is included can one apply the qualifying condition of "an isolated system." By including the observer, his eye, the measuring apparatus and the object, creates uncertainty because every action is entangled with the continuum.

"This crucial point...implies the impossibility of any sharp separation between the behaviour of atomic objects and the interaction with the measuring instruments which serve to define the conditions under which the phenomena appear...." -Bohr 1949.

As determined by Niels Bohr (1949), the properties of physical entities exist only as complementary or conjugate pairs: duality. A profound aspect of complementarity is that it not only applies to measurability or knowability of some property of a physical entity, but more importantly it applies to the limitations of that physical entity's very manifestation of the property in the physical world. Physical reality is defined by manifestations of properties which are limited by the interactions and trade-offs between these complementary pairs. For example, the accuracy in measuring the position of an electron requires a complementary loss of accuracy in determining its momentum. Precision in measuring one pair is complimented by a corresponding loss of precision in measuring the other pair (Bohr, 1949, 1958, 1963). The ultimate limitations in precision of property manifestations are quantified by Heisenberg's uncertainty principle and matrix mechanics. Complementarity and Uncertainty dictate that all properties and actions in the physical world are therefore non-deterministic to some degree.

Bohr (1949) called this the principle of complementarity, a concept fundamental to quantum mechanics and closely associated with the Uncertainty Principle. "The knowledge of the position of a particle is complementary to the knowledge of its velocity or momentum." If we know the one with high accuracy we cannot know the other with high accuracy (Bohr, 1949, 1958, 1963; Heisenberg, 1927, 1955, 1958).

THE WAVE FUNCTION OF REALITY

In quantum physics, nature and reality are represented by the quantum state. The electromagnetic field of the quantum state is the fundamental entity, the continuum that constitutes the basic oneness and unity of all things. The physical nature of this state can be "known" by assigning it mathematical properties and probabilities (Bohr, 1958, 1963; Heisenberg, 1927). Therefore, abstractions, i.e., numbers and probabilities become representational of a hypothetical physical state. Because these are abstractions, the physical state is also an abstraction and does not possess the material consistency, continuity, and hard, tangible, physical substance as is assumed by Classical (Newtonian) physics. Instead, reality, the physical world, is a process of observing, measuring, and knowing and is based on probabilities and the wave function (Heisenberg, 1955).

Central to quantum mechanics is the wave function (Bohr, 1963; Heisenberg, 1958). All of existence has a wave function, including light. Every aspect of existence can be described as sharing particle-like properties and wave-like properties. The wave function is the particle spread out over space and describes all the various possible states of the particle. According to quantum theory the probability of findings a particle in time or space is determined by the probability wave which obeys the Schrodinger equation. Everything is reduced to probabilities. Moreover, these particle/waves and these probabilities are entangled.

The act of conscious observations, by perception, by using a measuring device, interacts with the quantum continuum, because it is part of and entangled with the continuum. By becoming conscious, by focusing the conscious mind (or measuring device) at a particular area of the quantum continuum, triggers a "wave form collapse"--and it is this collapse of the wave function which gives rise to the perception of galaxies, stars, planets, moons, Earth, people, dogs, cats, and molecules.

Wave function collapse has also been described as "decoherence." The wave function describes all the various possible states of the particle. Rocks, trees, cats, dogs, humans, planets, stars, galaxies, the universe, the cosmos, past, present, future, as a collective, all have wave functions. The universe as a whole, has a wave function. If there is a god, then god has a wave function.

Waves can also be particles, thereby giving rise to a particle-wave duality and the Uncertainty Principle. Particle-waves interact with other particle-waves. The wave function of a person sitting on their rocking chair would, within the immediate vicinity of the person and the chair, resemble a seething quantum cloud of frenzied quantum activity in the general shape of the body and rocking chair. This quantum cloud of activity, when perceived by a conscious mind, gives shape and form to the man in his chair, even though the man/chair is part of the quantum continuum. The man/chair is a blemish in the continuum which is still part of the continuum and interacts with other knots of activity thus giving rise to cause and effect as well as violations of causality: "spooky action at a distance."

Reality is a manifestation of wave functions and alterations in patterns of activity within the quantum continuum which are entangled and perceived by conscious observations and measurement. When perceived, the wave function collapses and becomes discontinuous from the continuum, even though it is still part of it. The perception of a structural unit of information is not just perceived, but is inserted into the quantum state which causes the reduction of the wave-packet and the collapse of the wave function. It is this collapse which describes shape, form, length, width, and future and past events and locations within space-time (Bohr, 1963; Heisenberg, 1958).

In quantum physics, the wave function describes all possible states of the particle and larger objects, thereby giving rise to probabilities, and this leads to the "Many Worlds" interpretation of quantum mechanics (Dewitt, 1971; Everett 1956, 1957). That is, since there are numerous if not infinite probable outcomes, each outcome and probable outcome represents a different "world" with some worlds being more probable than others.

In the Copenhagen model, objects are viewed as quantum mechanical systems which are best described by the wave function and the probability function. "The reduction of wave packets occurs when the transition is completed from the possible to the actual" (Heisenberg, 1958).

Likewise, it could be said that the universe underwent a transition from the possible to the actual, at the moment of conscious registration which triggered a wave form collapse.

The measuring apparatus, the observer, and the conscious mind of the observer, also have a wave function and therefore interact with what is being measured. The effect of this is obvious when its a macro-structure measuring a micro-structure vs a macro-structure measuring a macro-structure. According to the Copenhagen interpretation (Bohr, 1949, 1963; Heisenberg, 1958), it is the act of measurement which collapses the wave function. It is also the measurement and observation of one event which triggers the instantaneous alteration in behavior of another object at faster than light speeds; i.e. entanglement. However, that which takes place faster than the speed of light means that an event in the future can travel from the future into the present and then into the past. Therefore, the future can effect the present and the past.

For example, an electron may collide with and bounce to the left of a proton on one trial, then to the right on the next, and then at a different angle on the third trial, and another angle on the fourth and so on, even though conditions are seemingly identical. This gives rise to the Uncertainty Principle and this is why the rules of quantum mechanics are indeterministic and based on probabilities. The state of a system one moment cannot determine what will happen next. Instead, we have probabilities which are based on the wave function. The wave function describes all the various possible states of the particle (Bohr, 1963; Heisenberg, 1958).

THE WAVE FUNCTION OF GOD/UNIVERSE CONSCIOUSNESS

From a singularity, out of nothingness, emerged the universe which emerged upon becoming conscious of itself. By becoming conscious, this caused a "collapse of the wave function" and the universe underwent a transition "from the possible to the actual" (Heisenberg, 1958).

Thus, if the universe became conscious of itself, it collapsed the wave function and the universe was transformed into the actual. If there is a god, and if the universe is god, then god is that quantum consciousness which created itself and the universe, by becoming conscious.

A conscious universe, or rather, a quantum consciousness which is identical with and yet distinct from the quantum continuum, is predicted by quantum physics, the theories of the "atomists" and the "Eastern" philosophies of Hinduism, Buddhism, and Taoism: the act of perceiving, of being conscious, creates an impression of distinction, of unique individuality; when in fact what is perceived are only fragments of the totality which is all.

Some Big Bang theorist like to believe the hypothetical "singularity" that gave rise to the universe, may have initially existed as a solid mass, no larger than a Plank length; with all matter compacted into the smallest of all measurable space. And then it exploded outward. However, as predicted by quantum mechanics, and Newtonian and Einsteinian physics, the gravity would be so immense that instead of blowing outward, it would have blown inward, imploding, creating a hole in the fabric of space-time; at the bottom of which would be a mirror universe. Thus, a universe of matter and a universe of anti-matter, a duality, would have been fashioned by the creative event. Two universes, being and non-being, matter and anti-matter, and each conscious of the others' existence.

This singularity, this basic oneness, from the perspective of quantum mechanics, need not have been compacted, however, but exists as a quantum continuum forever in flux, from which being emerges from non-being at the moment of consciousness of it. That is, when the singularity became conscious of itself, by becoming conscious of the "singularity", this created a fragmentation within the quantum continuum, the basic oneness which is everything. The quantum continuum became discontinuous by the act of becoming conscious. Again, consciousness, to be consciousness, must be consciousness of something; and that something is distinct from consciousness. Even consciousness of consciousness (self-consciousness) creates a duality which may then become a multiplicity (Joseph 2011).

By becoming a conscious observer, fragments of the quantum continuum were observed as distinct from the continuum, and perceived as galaxies, stars, planets, molecules, atoms, particles. And collections of those molecules and atoms also became conscious of themselves as things, and were perceived as having shape, form, with length, width, height, weight, duration, and individual identities. The quantum continuum, as a quantum consciousness, begets not just self-consciousness, but islands of consciousness: consciousness begets itself.

THE CONSCIOUS UNIVERSE

Since the universe, as a collective, must have a wave function, then this universal wave function would describe all the possible states of the universe and thus all possible universes. Hence, there must be multiple universes which exist simultaneously as probabilities (Dewitt, 1971; Everett 1956, 1957).

And if there is a god, the same would be true of "god" and the universe as a whole. "God" and the universe would have a wave function. Therefore, there would be more than one god--and this leads to the Hindu religion with its millions of gods, each of which is actually a manifestation of the one god which is the universe.

Consciousness, too, would have a wave function. In consequence, because the wave function of consciousness is entangled with the quantum continuum it can cause a collapse of the wave function.

If consciousness is energy, then the energy which is the quantum continuum also has the probability of becoming conscious. If the universe, as a whole, is a manifestation of the quantum continuum, as perceived by consciousness, then the continuum could have become conscious of itself, and in achieving self-consciousness, created the universe, which, created itself by becoming conscious.

And if the wave function is consciousness, and if there is a god and that god is the universe, then god/universe became god/universe, when the quantum continuum became conscious.

"In the beginning the atman was this universe. He gazed around, he saw nothing there but himself. Thereupon he cried out at the beginning: It is I."
-Upanishads

33

And from the quantum continuum, through quantum consciousness, the one became the many: galaxies, stars, planets, moons, people, dogs, cats, molecules....

"O Lord of the universe, O universal form, I see in Your body many, many arms, bellies, mouths, and eyes, expanded everywhere, without limit I see in You no end, no middle, and no beginning. You have numberless arms, and the sun and moon are Your eyes. I see You with blazing fire coming forth from Your mouth, burning this entire universe by Your own radiance." -The Bhagavad-gītā

Sūta said: In the beginning of the creation, the Lord first expanded Himself in the universal form of the puruṣa incarnation and manifested all the ingredients for the material creation. And thus at first there was the creation of the sixteen principles of material action. This was for the purpose of creating the material universe. -The Bhagavad-gītā

This form [the second manifestation of the puruṣa] is the source and indestructible seed of multifarious incarnations within the universe. From the particles and portions of this form, different living entities, like demigods, men and others, are created. -The Bhagavad-gītā

In Him in whom this universe is interwoven; Whatever moves or is motionless" -Upanishads

THE QUANTUM PHYSICS OF ALL KNOWING GOD

In a quantum universe all of existence consists of a frenzy of subatomic energetic activity which can be characterized as possessing pure potentiality, and all of which are linked and entangled as a basic oneness which extends in all directions and encompasses all dimensions including time (Bohr, 1958, 1963; Dirac, 1966a,b; Planck 1931, 1932, Heisenberg 1955, 1958; Joseph 2014; von Neumann 1937, 1955). If there is a "god" then that god is that basic oneness.

The quantum continuum is the ultimate reality, an all inclusive oneness which is all things, and which becomes all things at the moment of consciousness--and the same has been said of the Tao, Hindu, Christian, Buddhist, and Jewish god.

It has been said that "god" is all knowing, omnipotent, omnipresent, existing in the future and the past, and in all things and the same can be said of the quantum continuum. And because this frenzied electromagnetic continuum extends in all directions and dimensions, encompassing even what is experienced as time, then the continuum which could be described as "god" would encompass the future, present, and past, and thus, would be eternal and all knowing.

34

"The Supreme Truth exists both internally and externally, in the moving and nonmoving. He is beyond the power of the material senses to see or to know. Although far, far away, He is also near to all." --Bhagavad Gita, 13:16

The omniscience of God is to be all knowing, to know all things past, present and future, including what is hidden from human sight and knowledge--all is still known by God. God's knowledge originates in himself and is complete--so says the Tao, Apanishads, Bhagavad Gita, and the Christian and Jewish religious texts (see Isa 40:13-14, Job 21:22, Mt 10:30, Dt 29:29).

> "Who has directed the Spirit of the Lord
> or instructed him as his adviser?
> Whom did he consult?
> Who gave him understanding?
> Who taught him the right way?
> Who taught him knowledge?
> Who showed him the path to understanding?" -Isaiah 40:13-14

To be all-knowing is to be knowing of all time--time itself is but a manifestation of the oneness as experienced as a single moment, an eternal "now" by a conscious observer. According to Einstein, time is relative to the observer (Einstein 1905a,b,c, 1906, 1961). Since there are innumerable observers, there is no universal "past, present, future" which are infinite in number and all of which are in motion. However, as all observers are but blemishes in the electromagnetic continuum, knots of energy which perceive themselves as distinct but which are still entangled in the continuum, then the basic oneness of the continuum, consists of all observers and all pasts presents and futures--in which case, there is only one observer: the quantum consciousness which is the universe, and what some have described as Tao, and God. From a god's eye view, the eternal "now" consists of all pasts, presents, futures, which are one.

"....the days that were formed for me, when as yet there was none of them" - Psalm 139

"God" would thus encompass the past, present, and future, and would have consciousness of all pasts presents futures.

Because space is "isotropic" there is nothing in the law of physics indicating that a particular direction is preferred; down, up, sideways, backwards, its all the same. Why should space-time, or time, be any different? The laws of electromagnetism do not make a distinction between past and future (Pollack & Stump, 2001; Slater & Frank, 2011). Since the past, present and future overlap and are relative to observers and differ according to location, gravity, and speed of movement, then as Einstein stated, the distinctions between them are an illusion.

Like a flowing river, the "present," "past" and "future" are relative to an observer, just as "downstream" is relative to the location of an observer. In quantum physics, the river has no present or past or future, it just is; and this is why events which, from the perspective of an observer, occur in the future, can effect events which take place in the present: the eternal now.

As demonstrated by what is known as "entanglement" (Plenio 2007; Juan et al. 2013; Francis 2012) effects cannot always be traced to an earlier cause, for the cause may occur in the future (Megidish et al 2013). Effects may occur simultaneously with causes, and take place at faster than light speeds (Francis 2012; Juan et al. 2013; Lee et al. 2011; Matson 2012; Plenio 2007). As indicated by entanglement, the future may effect or take place before the past-present (Megidish et al 2013).

GOD AS ENTANGLMENT: PAST PRESENT FUTURE ARE ONE

According to Einstein's theorems of relativity (Einstein 1905a,b,c, 1907, 1910, 1961), the past, present and future overlap and exist simultaneously but in different distant locations in the dimension known as space-time, and as such "The distinction between past, present and future is only an illusion" (Einstein 1955). Quantum physics, the Uncertainty Principle, the "Many Worlds" interpretation of quantum physics, and what Einstein (1930) called "spooky action at a distance" all call into question the causal distinctions between past, present and future.

As predicted by quantum mechanics and relativity, time is a circle--a closed loop with no beginning and no end--time is a continuum, where past, present, future, are linked and entangled. If there is a god, and if that god is the quantum continuum, then from the perspective of the god-continuum, time and god are one and the same, except at the moment of consciousness, which triggers a wave form collapse which is experienced as the eternal "now." God, the quantum continuum, is entangled with the future and the past, as all a basic oneness.

These are not just thought experiments. There is considerable evidence of what Einstein (1955) called "spooky action at a distance" and faster than light "entanglement" (Plenio 2007; Juan et al. 2013; Francis 2012). It is well established that causes and effects can occur simultaneously and ever faster than light speed (Lee et al. 2011; Matson 2012; Olaf et al. 2003); a consequence of the connectedness of all things in the quantum continuum.

For example, entanglement, between photons, has been demonstrated even before the second photon even exists, such that the photon in the future effects the photon in the present; "a manifestation of the non-locality of quantum mechanics not only in space, but also in time (Megidish et al 2013). Time-space is interactional. Time, and time-space are embedded in the quantum continuum and can effect as well as be effected by other particle-waves even at great distances; a concept referred to as "entanglement." Time and space-time are entangled in the continuum which is all things.

Consider photons which are easily manipulated. They preserve their coherence for long times and can be entangled by projection measurements (Weinfurter 1994; Kwiat et al. 1995). A pump photon, for example, can split light into two lower- energy photons while preserving momentum and energy, and these photons remained maximally entangled although separated spatially (Goebel et al 2008; Pan et al. 1998). However, entanglement swapping protocols can entangle two remote photons without any interaction between them and even with a significant time-like separation (Ma et al., 2012; Megidish et al. 2013; Peres 2000). In one set of experiments entanglement was demonstrated following a delayed choice and even before there was a decision to make a choice. Specifically, four photons were created and two were measured and which became entangled. However, if a choice was then made to measure the remaining two photons, all four became entangled before it was decided to do a second measurement (Ma et al., 2012; Peres 2000). Entanglement can occur independent of and before the act of measurement. "The time at which quantum measurements are taken and their order, has no effect on the outcome of a quantum mechanical experiment" (Megidish et al. 2013).

Moreover, "two photons that exist at separate times can be entangled" (Megidish et al. 2013). As detailed by Megidish et al (2013): "In the scenario we present here, measuring the last photon affects the physical description of the first photon in the past, before it has even been measured. Thus, the "spooky action" is steering the system's past. Another point of view...is that the measurement of the first photon is immediately steering the future physical description of the last photon. In this case, the action is on the future of a part of the system that has not yet been created."

Hence, entanglement, between photons has been demonstrated even before the second photon even exists; "a manifestation of the non-locality of quantum mechanics not only in space, but also in time (Megidish et al 2013). In other words, a photon may become entangled with another photon even before that photon is created, before it even exists. Even after the first photon ceases to exist and before the second photon is created, both become entangled even though there is no overlap in time. Photons that do not exist can effect photons which do exist and photons which no longer exist and photons which will exist (Megidish et al. 2013); and presumably the same applies to all particles, atoms, molecules (Wiegner, et al 2011). As dictated by the "uncertainty principle" energy and mass can be time-independent (Heisenberg 1927, 1958).

However, if the future present past are a continuum and the distinctions between them illusions, then the photons in the future already exist, in the future. The future already exists, now.

KARMA: HOW GOD AND THE UNIVERSE CAN BE SELF-CREAT-ING

Since time is entangled with the electro-magnetic continuum; in a continuum, the future can overlap with and even come before the past and catch up with itself in the past, so that an event can be "simultaneous" with or occur before its cause. An event may be able to cause itself.

Thus, the universe-quantum-universe-continuum, could have caused itself--and if there is a god, then god could have created god--the universe could have created itself.

If time is considered from the perspective of space-like intervals and not time-like intervals, then causality can be forward, backward, or simultaneous (Bonor & Steadman, 2005; Buser et al. 2013; Carroll 2004; Gödel 1995). The future and the past become entangled as a continuity in space-time. If "God" is that continuum, made conscious, then the God-continuum can also have knowledge of the future before the future takes place. Likewise, effects in the future can affect the present.

For example, it is believed that "Karma" can affect a person's future, or reincarnated life. However, if there is Karma, then that future life can have a backwards-in-time effect. In other words, someone's "evil deeds" in the future, or in a future life, may cause suffering in the present. God may punish the sinner before they sin, because the sin has already taken place: in the future.

It is well established that objects respond to and can influence and affect distant objects at speeds faster than light. This "spooky action at a distance" has been attributed to "fields," "mediator particles," gravity, and "quantum entanglement" (Bokulich & Jaeger, 2010; Juan et al. 2013; Sonner 2013).

It is believed that an electric "field" may mediate "electrostatic" interactions between electromagnetic charges and currents separated by great distances across space. However, these changes can take place at faster than light speeds. Charged particles, for example, produce an electric field around them which creates a "force" that effects other charges even at a distance. Maxwell's theories and equations incorporate these electrostatic physical "fields" to account for all electromagnetic interactions including action at a distance.

Since mass can become energy and energy mass, the "field" is therefore a physical entity that contains energy and has momentum which can be transmitted across space. Therefore, "action at a distance" may be both distant and local, a consequence of the interactions of these charges within the force field they create. However, the effects can be simultaneous, even at great distances, and occur faster than the speed of light (Plenio 2007; Juan et al. 2013; Francis 2012; Schrödinger & Dirac 1936), effecting electrons, photons, atoms, molecules and even diamonds (Lee et al. 2011; Matson 2012; Olaf et al. 2003;

Schrödinger & Born 1935). The effect, therefore, may precede the cause since it takes place faster than light.

Correlation is not causation and it can't always be said with certainty which is the cause and which is the effect and this is because the cosmos is entangled as a basic oneness. According to the Tao, Gita, Upanishads, and so on: that oneness is god.

"Everywhere are His hands and legs, His eyes and faces, and He hears everything." --Bhagavad Gita, 13:14

"He is the source of light in all luminous objects. He is beyond the darkness of matter and is unmanifested. He is knowledge, He is the object of knowledge, and He is the goal of knowledge. He is situated in everyone's heart." --Bhagavad Gita, 13:18

GOD CONSCIOUSNESS AND THE QUANTUM CONTINUUM

The quantum continuum is without dimensions and encompasses space and time in its basic unity of oneness. Everything within the quantum continuum can be effected by local effects and distant effects simultaneously at and beyond light speeds; even the future can effect the present. Therefore, the future, and the "present" being part of this continuum can effect the past-present-future by effecting the wave function of the past, present, future, and thus, the space-time continuum, as all are entangled.

Quantum entanglement is a feature of time and the quantum continuum, this frenzied electromagnetic activity which is all things, encompassing even what some have called "god."

As based on quantum mechanics, it could be said that in space-time all things overlaps and coincide and exist side by side, including time, god, and consciousness. If consciousness is considered as spatial and different aspects of consciousness all all connected to the quantum continuum, then the future, past, and present coexists simultaneously, albeit continually in flux. Thus quantum consciousness, or god-consciousness, can be viewed as the union of all events in the same way that a line is the union of all of its points. God becomes all knowing, and what effects the future, can therefore effect the past as a continuity which is linked and not broken up and into isolated fragments which are separated by artificial time-like intervals. Likewise, when considered as phenomenon taking place in space, then the past can also effect the future as they are entangled as a unity.

If the world-line of cosmic consciousness is conceptualized as a string, then no matter where the string is plucked, the entire string will vibrate.

Since space-time includes "consciousness" then since space-time consists of energy which can become matter. Then consciousness is also interactional, which is why it can be experienced and perceived and why it can perceive. Consciousness can act on matter and time has energy which can become matter, a particle-wave duality which propagates through space. This implies that consciousness can also effect and warp the space-time continuum which includes multiple futures and multiple pasts which share world lines which can overlap and intersect one another. Again, if there is a god, or a quantum consciousness, then this god/consciousness is also a unity which creates singularities by becoming conscious and collapsing the wave function even of events which take place in the future; and this is because the future is entangled with the present, past, and with the continuum which is all things.

"And his mind was afterwards clear as tehe morning, and after this he was able to see beyond his own individuality. that individuality perceived, he was able to banish all thoughts of past and present...." --Chuang Tzu

THE LIMITATIONS OF THE SENSES

"Whatever you see in existence, both moving and unmoving, is only the combination of the field of activities and the knower of the field."--Bhagavad Gita, 13:27

In relativity consciousness is merely relative. In quantum physics, consciousness and the act of observation and measurement constitute a separate reference frame which can collapse the wave function and register entangled interactions within the environment. Consciousness by the act of observation or measurement takes a static or series of pictures-in-time which then becomes discontinuous from the quantum continuum (Heisenberg 1958; Planck 1931; von Neumann 2001). These entanglements (Francis 2012; Juan et al. 2013; Plenio 2007), or blemishes in the quantum continuum, may be observed as shape, form, cause, effect, past, present, future, the passage of time, and thus reality; the result of a decoupling of quanta from the quantum (coherent) continuum which leaks out and then couples together in a knot of activity which is observed as a wave form collapse.

As based on the Copenhagen theory of quantum mechanics (Bohr, 1958, 1963; Heisenberg 1955, 1958), consciousness and reality are a manifestation of wave functions and alterations in patterns of activity within the quantum continuum which are perceived by consciousness as discontinuous--and that ultimate consciousness, the collective consciousness, could be construed as "god."

Wave form collapse is always a matter of probability, and is non-local, indeterministic and a consequence of conscious observation, measurement, and entanglement. Consciousness is entangled with the quantum continuum and can after the continuum and the space-time manifold.

Most religious, in their conceptions of "god" share similar conceptions.

Consciousness, the act of observation be it visual, auditory, tactile, mechanical, digital, is entangled with the quantum continuum and creates a static impression of just a fragment of that quantum frenzy that is registered in the mind of the observer as length, width, height, first, second, and so on; like taking a single picture of something in continual motion, metamorphosis, and transformation. That is, the act of sensory registration, be it a function of a single cell, or the conscious mind of a woman or man, selects a fragment of the infinite quantum possibilities and experiences it as real, but only to that mind or that cell at the moment of registration (Heisenberg 1955, 1958).

As demonstrated in quantum physics, the act of observation, measurement, and registration of an event, can effect that event, causing a collapse of a the wave function (Dirac 1966a,b; Heisenberg 1955), thereby registering form, length, shape which emerges like a blemish on the face of the quantum continuum. Thus, it can be said, at the moment the quantum continuum became conscious, achieved consciousness, this triggered a collapse of the wave function, and the universe came into being.

The mind, however, is not a singularity, but a multiplicity. The mind is also limited by its senses. Consciousness cannot perceive what it cannot perceive. Only fragments of the quantum continuum may be perceived by consciousness, and different aspect of consciousness--the multiplicity of mind--may perceive aspects of reality that other regions of the mind cannot perceive. However, they are perceived, because a mind becomes conscious of them.

"These realms are not come from somewhere outside thyself. They come from Within... they exist from eternity within the faculties of thine own intellect... issuing from within thine own brain... reflection of thine own thought-forms." --Tibetan Book of the Dead

A radio or television receiver may be capable of receiving radio or television transmissions from hundreds or thousands of stations. These transmissions contain all manner of messages, voices, shapes, images, and information, and continue to be transmitted, even if the radio or television is off. There are other channels and other transmissions even if we have only one channel. This information is out there if we know it or not, and if we receive it or not.

However, we can only receive and perceive these messages, these images if the television, internet, or radio is on. The number of stations and messages we can receive depends on the channel capacity of the radio/television receiver. That we cannot see or hear it without the proper receiver, or if the radio/television is off, does not mean that this information, these images, these sounds, do not exist, or that those who do perceive these stimuli are hallucinating.

The same is true regarding other realities, including those which may contain spirits, souls, gods or demons. What some perceive as a "demon" could well be an entity living in one of the multiple (probable) worlds predicted by quantum mechanics. Or, this same "demon" may be created by the act of observation--collapsing the wave form of the continuum, created by an observing consciousness; even when that consciousness has been liberated from a body which has died.

"No terrible god punishes you. The shapes of frightening monsters who take hold of you, place a rope along your neck and drag you along, are just illusions which you create from the forces within you.... there are no gods, and no demons." --Tibetan Book of the Dead

We can only perceive what we are able to perceive. Perception is also reality. Our senses not only limit and shape our perceptual reality, but the very act of perceiving can alter that reality; a phenomenon known as entanglement. What is observed is effected by being observed, as has also been demonstrated in quantum physics. Likewise, by changing one's perceptions, by changing the observer, can also change the reality that is being perceived; just like changing the channel on the radio or television.

Each conscious mind is entangled with the quantum continuum. Each mind, and each brain housing that mind, come into existence via the collapse of the wave function and the creation of decoherence. The mind is not "one" with the continuum because it represents a collapse of the wave function and is entangled yet separate from the continuum.

In part, the mind is restrained by its senses and perceptual capabilities, and the inhibitory influences that filter out stimuli which may be overwhelming. Our intellectual and perceptual capabilities cannot process what they cannot perceive or comprehend. Language and belief also act as yoke, by shaping perceptions and labeling that which is beyond the norm, as abnormal, unnatural, crazy, and sinful.

However, trance states, isolation, fasting, prayer, meditation, dream states, and LSD, can free the mind of inhibitory restraint, producing not just dream-like hallucinations, but by opening the mind to a fuller range of experience, so that what is concealed may be revealed.

Yet, to enter into this other perceptual reality so that what is hidden is revealed, also entails the temporary annihilation of this world and this reality. Being becomes non-being, and non-being becomes being--which is also the philosophical view of the Tao.

IN THE BEGINNING THERE WAS LIFE

"Material nature and the living entities should be understood to be beginningless."---Bhagavad Gita, 13:20

According to Darwin (1871): "A belief in all-pervading spiritual agencies seems to be universal, and apparently follows from a considerable advance in man's reason, and from a still greater advance in his faculties of imagination, curiosity, and wonder. I am aware that the assumed instinctive belief in God has been used by many persons as an argument for His existence. But this is a rash argument, as we should be thus compelled to believe in the existence of many cruel and malignant spirits, only a little more powerful than man; for the belief in them is far more general than in a beneficent Deity."

Yet, despite, Darwin's rash claims to the otherwise, we are in fact spiritual beings, and there is scientific evidence to support these beliefs. We in fact possess the genetic and neurological capability to experience "gods" demons, spirits, souls, and other realities, for this reality is only one of many.

And, we possess a brain that not only shapes and filters this reality, but which, under certain conditions enables us to perceive at least some of these other probable realities. These same areas of the brain enable humans (and perhaps other animals) to experience the spirit and the soul as they transcend the body to traverse these myriad realms.

However, the spirit or soul, if comprised of energy, remain tethered to the body during life. But upon death, the energy is released to become one with the quantum continuum.

"All things begin from nothing and end in nothing."-Chuang Tzu

"There is no difference between life and death. The destruction of life is not dying and the beginning of new life is not living. There is no differences between life and death."-Chang Tzu

"There is a limit to our life, but to knowledge there is no limit." -Chuang Tzu

"All things return to their root and disappear and do not know that it is it which presides over their doing." -Tao Te Ching

"Tao is forever, and he that possesses it, though his body ceases, is not destroyed." -Tao Te Ching

"And his was mind was afterwards clear as the morning, and after this he was able to see beyond his own individuality. Freed of this he was able to penetrate to the truth-- how the destruction of life is not dying-- that there is no differences between life and death." -Chuang Tzu

Life, is common to life. Life has "life energy"--what some might call the "soul" or the essence or "spirit of god."

If life is energy, then life/energy cannot be created or destroyed. There has always been life. There will always be life. The universe itself, is alive.

What we perceive as "individuality" is a manifestation of the wave form collapse; an entangled blemish in the quantum continuum. At the moment of death, as being is transformed into nonbeing, the energy which was the mass which was the individual, returns to the quantum continuum, and this is because energy cannot be destroyed. Energy can neither be created nor destroyed.

If life is energy, then life can be neither created nor destroyed. Rather the structural organization that gives form to life, the body, is subject to disintegration and decay. What we call death.

Again, consider Einstein's theorem: i.e. $E=M/C^2$. Although energy cannot be destroyed or created, the destruction and transmutation of matter is something wholly different and is dependent on the organization and stability of the force field which is energy.

Although material forms may become unstable, disintegrate or assume new organization, the constituent fabric that gives rise to matter and its manifest structure, is pure energy -which cannot be destroyed. However, if matter decays, if the body dies, that energy is liberated, although its material form appears to disintegrate and to die.

Birth is being. Death is the return to non-being: a return to the quantum continuum which is all things.

"Tao is forever and he that possesses it, Though his body ceases, is not destroyed." -Lao Tzu

"Thine own consciousness, shining, void, and inseparable from the Great Body of Radiance, hath no birth, nor death, and is the Immutable Light--Buddha Amitabha, the source of life and boundless light." -Bardo Thodol / Tibetan Book of the Dead

Death is generally a gradual process, with some cells and tissues disintegrating in advance of others, and yet other tissues living for hours or even days before the body completely decays.

Presumably, even after death, so long as the body and its brain lives, one's sense of a personal soul and identity remains intact. This personal identity is perceived, after the death of the body, as an out-of-body experience. Moreover, this personal identity, the energy field associated with the dying body, may be perceived by itself, or by others, as a ghost, spirit, or departing soul. Presumably, this ethereal after-death existence and sense of personal identity remains tethered to the body/brain until the body completely dies and decays.

Indeed, the linkage of the personal soul and individual immortality to the body were widespread beliefs and practices among the ancient Egyptians which is why they expended so much effort to preserve the body via mummification. If the body could be preserved, so could one's personal soul and sense of individuality, leading to "immortality".

Others, including the Tibetan Buddhists sought just the opposite, to free the soul from the body and so as to escape the "illusion" of individuality and personal existence.

Therefore, what some experience as their personal soul upon death, may be but a gradual liberation of LIFE that at first retains its bodily links, thus preserving one's sense of individuality; the shadow of one's previous form as the body dies.

As the body is consumed, perhaps so too is the sense of individuality, freeing the soul, one's LIFE, to be embraced by the radiance of all LIFE thereby becoming One with the god/universe: the quantum continuum.

REFERNCES

Bell, J. S. (1964) On the Einstein Podolsky Rosen Paradox, Physics 1, 3, 195–200 (1964)

Bell,J. S. (1966) On the problem of hidden variables in quantum mechanics, Rev. Mod. Phys. 38, 447 Bohr, N. (1934/1987), Atomic Theory and the Description of Nature, reprinted as The Philosophical Writings of Niels Bohr, Vol. I, Woodbridge: Ox Bow Press.

Bohr. N. (1949). "Discussions with Einstein on Epistemological Problems in Atomic Physics". In P. Schilpp. Albert Einstein: Philosopher-Scientist. Open Court.

Bohr, N. (1958/1987), Essays 1932-1957 on Atomic Physics and Human Knowledge, reprinted as The Philosophical Writings of Niels Bohr, Vol. II, Woodbridge: Ox Bow Press.

Bohr, N. (1963/1987), Essays 1958-1962 on Atomic Physics and Human Knowledge, reprinted as The Philosophical Writings of Niels Bohr, Vol. III, Woodbridge: Ox Bow Press.

Bonnor, W. Steadman, B.R (2005). "Exact solutions of the Einstein-Maxwell equations with closed timelike curves". Gen. Rel. Grav. 37 (11): 1833.

Born, M. Heisenberg, W. & Jordan, P. (1925) Zur Quantenmechanik II, Zeitschrift für Physik, 35, 557-615, 1925

Buser, M. et al. (2013). Visualization of the Gödel universe. New Journal of Physics. Vol. 15.

Carroll, S (2004). Spacetime and Geometry. Addison Wesley.

DeWitt, B. S., (1971). The Many-Universes Interpretation of Quantum Mechanics, in B. D.'Espagnat (ed.), Foundations of Quantum Mechanics, New York: Academic Press. pp. 167–218.

DeWitt, B. S. and Graham, N., editors (1973). The Many-Worlds Interpretation of Quantum Mechanics. Princeton University Press, Princeton, New-Jersey.

Dirac, P. (1928). "The Quantum Theory of the Electron". Proceedings of the Royal Society of London. Series A, 117 (778): 610–24.

Dirac, P. (1930) Principles of Quantum Mechanics

Dirac, P. (1966a) Lectures on Quantum Mechanics

Dirac, P. (1966b). Lectures on Quantum Field Theory .

Einstein, A. (1905a). Does the Inertia of a Body Depend upon its Energy Content? Annalen der Physik 18, 639-641.

Einstein, A. (1905b). Concerning an Heuristic Point of View Toward the Emission and Transformation of Light. Annalen der Physik 17, 132-148.

Einstein, A. (1906a). On the Theory of Light Production and Light Absorption. Annalen der Physik 20, 199-206.

Einstein, A. (1906b). The Principle of Conservation of Motion of the Center of Gravity and the Inertia of Energy. Annalen der Physik 20, 627-633.

Einstein, A. (1926). Letter to Max Born. The Born-Einstein Letters (translated by Irene Born) Walker and Company, New York.

Everett , H (1956), Theory of the Universal Wavefunction",Thesis, Princeton University

Everett, H. (1957) Relative State Formulation of Quantum Mechanics, Reviews of Modern Physics vol 29, 454–462.

Einstein A, Podolsky B, Rosen N (1935). "Can Quantum-Mechanical Description of Physical Reality Be Considered Complete?". Phys. Rev. 47 (10): 777–780.

Gödel (1995) Lecture on rotating universes Kurt Gödel: Collected Works (Unpublished Essays and Lectures vol 3) ed S Feferman (Oxford: Oxford University Press)

Heisenberg, W. (1925) Über quantentheoretische Umdeutung kinematischer und mechanischer Beziehungen, ("Quantum-Theoretical Re-interpretation of Kinematic and Mechanical Relations") Zeitschrift für Physik, 33, 879-893, 1925

Heisenberg, W. (1927), "Über den anschaulichen Inhalt der quantentheoretischen Kinematik und Mechanik", Zeitschrift für Physik 43 (3–4): 172–198,

Heisenberg. W. (1930), Physikalische Prinzipien der Quantentheorie (Leipzig: Hirzel). English translation The Physical Principles of Quantum Theory, University of Chicago Press.

Heisenberg, W. (1955). The Development of the Interpretation of the Quantum Theory, in W. Pauli (ed), Niels Bohr and the Development of Physics, 35, London: Pergamon pp. 12-29.

Heisenberg, W. (1958), Physics and Philosophy: The Revolution in Modern Science, London: Goerge Allen & Unwin.

Joseph, R. (2014). The Quantum Physics of Time Travel. Cosmology Science Publishers

Juan Yin, et al. (2013). "Bounding the speed of `spooky action at a distance". Phys. Rev. Lett. 110, 260407.

Langevin, P. (1911), "The evolution of space and time", Scientia X: 31–54

Lee, K.C., et al. (2011). "Entangling macroscopic diamonds at room temperature". Science 334 (6060): 1253–1256.

Matson, J. (2012) Quantum teleportation achieved over record distances, Nature, 13 August

Matthew. F. (2012). Quantum entanglement shows that reality can't be local, Ars Technica, 30 October 2012

Nairz, O. et al. (2003) "Quantum interference experiments with large molecules", American Journal of Physics, 71 (April 2003) 319-325.

Olaf, N.. et al. (2003) "Quantum interference experiments with large molecules", American Journal of Physics, 71 (April 2003) 319-325.

Plenio, V. (2007). "An introduction to entanglement measures". Quant. Inf. Comp. 1: 1–51

Schrödinger, E. (1926). "An Undulatory Theory of the Mechanics of Atoms and Molecules". Physical Review 28 (6): 1049–1070. Bibcode:1926PhRv...28.1049S. doi:10.1103/PhysRev.28.1049.

Schrödinger E; Born, M. (1935). "Discussion of probability relations between separated systems". Mathematical Proceedings of the Cambridge Philosophical Society 31 (4): 555–563.

Schrödinger E; Dirac, P. A. M. (1936). "Probability relations between separated systems". Mathematical Proceedings of the Cambridge Philosophical Society 32 (3): 446–452.

Yin, J. et al. (2013). "Bounding the speed of `spooky action at a distance". Phys. Rev. Lett. 110, 260407.

3. Awareness, the Origin of Thought, and the Role of Conscious Self-Deception in Resistance and Repression.

Non-conscious processes involved in the formulation, organization, and expression of thought and consciousness are examined. It is argued that non-organized impulses and imageless non-linguistic knowledge exist prior to and result in the organization recognized as consciousness and thought. Hence consciousness is viewed as the developmental endpoint of a unitary process originating as non-activated sensations and knowledge, i.e., awareness. Because one can be aware of and have knowledge of "things" prior to being conscious of them, it is possible to know, yet not know, or rather not think about certain objectionable feelings or tacit ideas. These imageless, non-linguistic forms of knowledge are not unconscious; rather they are non-labeled and non-descriptive. Within this framework the Freudian conception of unconscious processes is examined and shown to support a model of self-deception, defined as a conscious refusal to attend to unstructured tacit knowledge of which one is simultaneously aware. Awareness is posited to be bilateral and the domain of the right hemisphere and limbic system, the verbal aspects of consciousness are associated with the left hemisphere, super-ego functions are linked to the frontal lobes, whereas the purely unconscious aspects of the mind are associated with the limbic system and brainstem.

> From below cried Id,
> If thou be Christ,
> save theyself and us!"
> But Superego rebuked Id, saying:
> "Dost not thou fear God?"

The conception of unconscious mental processes finds part of its popular appeal in the shifting of responsibility for one's actions from the individual to an independent casual factor arising within a primitive Id and unconscious ego. In particular, we find acceptance of the notion that thoughts, ideas, and desires originate within and are dictated by unconscious processes while consciousness is relegated to a position of mediator between environment, Id and Super-ego. Necessarily, the belief in the unconscious is largely based on inference and intuition, whereas consciousness is experientially undeniable.

The concern of this paper is with the origin of thought and consciousness and the non-conscious processes subserving their development and the expression of knowledge. Included in this discussion is a reexamination of the Freudian position concerning the so-called unconscious domain with particular emphasis on conscious self-deception as revealed by resistance and repression. In addition, a preliminary stage to conscious activity is examined, i.e., awareness, and then discussed in relation to self-deception, that is, conscious attempts to disavow knowledge of which one is simultaneously aware.

Thought

Thought is an activity which involves both conscious and explicit, descriptive organization. Its process is an evolution conceived in anticipation and expectation, unfolding before an observer. The process of thought is evoked by as well as aimed at questioning and explaining. Systemic symbolizing in the form of imagery, symbols, and language is inherent in its production.

Thinking is a conscious enterprise. It is a means of consciously contacting and apprehending both the world and Self. Although thought and consciousness are not synonymous, they are interwoven in a mutually dependent process.

Consciousness must by definition be conscious of something. The process of thought serves consciousness as a vehicle for both thinking about Self, which is posited as an object for consciousness, and for thinking about the world, within which consciousness is posited as a subject.

It is through thought that consciousness comes to better understand and know the world. Thought is not, however, a perceptual mechanism. Thinking is merely a means of explaining something, be it an idea, plan, thing-in-the-world, or desire. Yet the need to explain things to oneself seems paradoxical. It might be asked, "who is explaining what to whom?" Apparently the "I" that I am explains these "things" to the "I" that I am.

Assuming that the subject of thought originates in me, then I should know something of its aim and content prior to symbolizing the substance of the subject into the organization that my thinking process generates. Thinking, however, seems to act as a conscious inner language which organizes and objectifies our "not thought out" ideas, which are not fully conscious, into an organization that may be understood by consciousness. We must conclude, therefore, that the thought expressed is not conscious before it is expressed, and the idea or substance of thought, before it is expressed in the organization which characterizes thought, is pre-thought, a state we shall equate with awareness.

Imageless Knowledge

This indicates that there is a preliminary stage to thought and consciousness (Luria, 1973; Schilder, 1951; Stekel, 1951) but, being without explicit organization, it lacks distinctiveness (Schilder, 1951). That is, prior to the conscious and symbolic organization which characterizes thought, there exists an imageless knowing (Ach, 1951), which underlies and forms the basis of imaginal experience, conscious thought, language usage, and overt responses (Franks, 1974).

According to Narciss Ach (1951): Analysis of the contents of conscious-ness have shown a variety of experiences in which all of a complex content is simultaneously present in the form of a "knowledge." This knowledge exists in imageless form, that is, no phenomenological components are demonstrable--neither visual, acoustic, nor kinesthetic sensations, nor their memory images--which would qualitatively define the content of this knowledge (p.24).

Though these forms of imageless knowing are indeed not-conscious, this does not imply that they exist unconsciously (Stekel, 1949). However, image-less knowing is distinguished from consciousness in that it exists tacitly as non-activated knowledge which may become conscious (Ach, 1951; Franks, 1974; Stekel, 1949). When activated by consciousness (Luria, 1973), the re-sult is a realization which takes the form of phenomenal experience and overt responding (Franks, 1974). Hence, thought may be considered the endpoint which develops from imageless knowing (Schilder, 1951; Turvey, 1974).

In summary, it can be said that images and impulses pass through vari-ous transitory stages which evolve from an indefinite non-organized state that is prior to consciousness to one of increasing distinctiveness which achieves form and organization characteristic of thought and consciousness (cf. Schil-der, 1951).

Awareness and Consciousness

It is important to note that thought imageless forms of knowing are not im-mediately amenable to conscious analysis we are intrinsically aware of their existence as formless knowledge (Ach, 1951; Turvey, 1974). Hence, we know and are aware of the probable content and direction of thought prior to and while we are thinking. It seems, however, that the verbal aspects of our con-scious attention are directed to and dependent upon the symbolic-abstracting-thinking-language processes which organize, describe, and explain the indis-tinct and imageless, as well as that which is directly sensed and known. In this regard, the verbal aspects of consciousness are associated with the left hemisphere in the majority of the adult population, whereas the environmental, visual-spatial, and melodic and emotional aspects of non-verbal mental pro-cesses are linked to the right hemisphere.

In contrast to left hemisphere, verbal-related aspects of consciousness which is limited to understanding verbal information, the non-verbal aspects of awareness are bilateral and include awareness of environmental sounds and even the emotion in one's voice--and these sounds and related events may not be processed consciously, though we are aware of their presence.

For example, our conscious waking life is composed of events in which we may engage without consciously or verbally attending to their function or pro-cess, that is, we do not think about them--at least in words.

When I wind my watch, drive my car, eat my breakfast, or urinate, I may or may not be thinking about what I am doing or attending consciously to the many perceptual-motor components comprising my actions or their motivation. However, I am aware, at the moment, of their taking place.

Moreover, if I am questioned concerning these actions at a later date, my memory will be at best fragmentary, corresponding only to those aspects consciously contemplated. These processes, however, are not limited to my own actions. For instance, I may be sitting on my porch aware of the trees swaying slightly in the wind, my dog rolling in the grass, or a car going by. I may or may not consciously attend or think about them, though I know that they are there and occurring.

Nevertheless, if I am questioned about a certain vehicle that has passed but moments before, although not conscious of it at the time, I may reflect upon its occurrence which remains as it appeared at the level of awareness. However, hours or days later this "iconic" remainder of my experience will have dissipated and all queries will draw blanks. Thus, not all knowledge within the realm of awareness is without form. However, the form of that which is sensed, of which one is aware, always remains as it appears, an experience that is not descriptive. Imageless knowledge is the experience. Once one becomes conscious of this knowledge, the experience becomes descriptive, an abstraction of the original appearance.

The verbal and temporal sequential aspects of conscious perceptual experience are thus a process that may be said to coincide secondarily with the awareness of an event's occurrence or an object's existence. We may be aware of "something" but may or may not consciously attend, represent, think about, or express it--at a verbal level. Turvey (1974) has cited much published evidence in support of this view, which he believes indicates that one may "know something about the identity of an event before one knows the event's identity" (p.174).

Awareness and Self-consciousness

Because all consciousness is consciousness of something, it always requires an object (Sartre, 1956). However, although one is conscious of an object and has knowledge of it, consciousness, the object, and the knowledge of the object remain separate and distinct. This is because knowledge is an abstraction and serves to represent the object in a form necessitated by the nature of consciousness for understanding. Moreover, because conscious knowledge is a descriptive representation of something other than consciousness--at least in respect to the verbal aspects of consciousness-- knowledge cannot be identified with consciousness, even knowledge of consciousness.

Awareness, as opposed to consciousness, is non-verbal, bilateral, and pre-reflective and is not always subject to the more abstract forms of understanding that characterize conscious knowledge. Though we are aware of our awareness, once we consciously scrutinize this tacit dimension it becomes transformed and abstracted and all information gained is an indirect verbal representation of that which is without verbal representation. Since awareness is pre-language and pre-verbal thought, and may encompass emotions, sounds, spatial relations, or other variables that are non-verbal, it is thus difficult for the verbal aspects of consciousness to perceive the existence of this dimension without transforming it so that it may be understood consciously. Therefore, though consciousness can posit itself as an object in order to know itself, awareness is pre-objective and cannot be an object for consciousness except in the abstract.

Awareness, then, as defined in this paper, is distinct from consciousness. Awareness is a non-organization, existing prior to consciousness, thought, language, or other information processes that utilize organized temporal-sequential processes for expression and communication. Awareness as a mental process, is associated with the right hemisphere and limbic system.

Consciousness, for the purposes of this paper, is defined as being dependent on language, and temporal sequential modes of organization and perception. The verbal aspects of consciousness are linked to the left hemisphere.

Consciousness cannot think about or organize that which is without organization, without organizing and transforming this information so that it may be understood by consciousness. However, once information is altered, it becomes an abstraction and thus becomes something else.

Hence, awareness always remains non-explicit and cannot be made conscious, except as an abstraction. Hence, when consciousness thinks about itself, the "itself" of which it is conscious is its own organization, an abstraction.

Although such attempts at self-knowledge form the bases of self-consciousness, the "I" that is thought about can never be the "I" that is thinking.

However, consciousness and awareness cannot be posited as a duality. That is because consciousness is also an abstraction, the attempt to organize that which one is aware of, as well as the result of this attempt.

Consciousness is thus always a response, a result or rather the consequence of acting in the world, the developmental endpoint of awareness. Nevertheless, consciousness is distinct from awareness in that consciousness is descriptively representational, relating as either a passive force manipulating representations of the world through thought, or actively transforming reality through organized goal directed behavior. It can be stated that awareness is often represented by consciousness which acts to test and manipulate reality.

Hence, consciousness and awareness are processes of acting and being in the world; acting as an organization of thought and behavior, and being that which it is: aware. Consciousness, through organized actions, symbolically represents that which is without organization, or descriptive representation and therefore appears to be separate from that which has no appearance, i.e., awareness.

However, it seems we have nevertheless created a duality in our distinction between awareness and consciousness. It might be asked, if consciousness is distinct yet the same as awareness, what causes this appearance of separation?

We must place the blame on consciousness, which separates--when attempting to know and think about itself--into both observer and observed. Thus the distinguishing characteristic of consciousness is that it does not coincide with itself (Sartre, 1956). Consciousness points itself as a duality--a reflection which is its own reflecting.

Hence, we have the consciousness that we are, actively representing and reflecting upon the awareness that we are, thereby creating a separation of itself through itself by attempting to witness itself.

Self-consciousness, or consciousness of self, is thus a consciousness of awareness, for self-consciousness appears only when consciousness attempts to reflect upon the self-image. However, consciousness is no more separate from itself than a man is separate from the self he ponders in a mirror. The image is not the man, nor is the reflection of consciousness the consciousness. The "I" is neither image nor reflection, it simply is.

Because consciousness is always relational, it appears separate yet remains identical with itself in that the separation appears to occur when consciousness seeks to be conscious of itself as a consciousness. Nevertheless, consciousness cannot apprehend itself because it is already that which it attempts to apprehend.

THE TOPOLOGICAL AND STRUCTURAL VIEW

Freud, in both his structural and topological view of psychological processes is completely at odds with this position, presenting us with a mental realm which is hidden from, yet identical with, itself, i.e., the unconscious and conscious. It is the position of this paper that the magical separation Freud imposes on the psychic apparatus unnecessarily forces us to resort to supposition and groundless hypotheses concerning the origin and true meaning of our thoughts and ideas.

According to Sartre (1956), "the very essence of the reflexive idea of hiding something from oneself implies the unity of one and the same psychic mechanism and consequently a double activity in the heart of unity" (p. 53). That is, the psychic mechanism Freud proposed to us as split is actually one and the same thing, a unity that can only be described as consciousness. Simply, there is no unconscious; the name only serves as a metaphor for conscious processes of self-deception: a conscious attempt to be other than one is (Stekel, 1949). In fact, a careful reading of Freud's voluminous works admits of no other interpretation.

The Unconscious

According to Freud, all mental processes (except those arising in direct response to external events) originate in the unconscious, and at this level exist totally beyond conscious awareness (Freud, 1900, 1915b, 1917, 1940). That is, all psychic material, thoughts, feelings, desires, etc., exist first as unconscious impulses which attempt to discharge into consciousness.

Furthermore, the unconscious ego has total jurisdiction over what may become conscious. This is because all impulses must pass through a double censorship, the unconscious and preconscious censors (Freud, 1915b, 1917) which restrict and limit conscious knowledge of what occurs in psychic life, including, to a limited degree, that which occurs in the external environment. Consequently, the censors keep harmful or distasteful information from ever reaching consciousness.

Hence, we find within the Freudian framework two types of unconscious. One is the preconscious which is latent and can easily become conscious, and the other is the unconscious, whose contents may become conscious only with extreme difficulty, or not at all (Freud, 1915b, 1917, 1933, 1940). According to our neurodynamic model, these two types of unconscious are associated with the limbic system and right hemisphere, with the limbic system forming the more inaccessible regions of the psyche and the right hemisphere being associated with the "pre-conscious."

Freud's distinction between these unconscious mental processes lies within Freud's assumptions concerning their respective psychic energies, which in turn explains censorship. The unconscious has at is disposal mobile psychic energy which presses toward conscious recognition, i.e., the primary processes.

The preconscious, however, uses bound energy which is associated with the secondary process and the inhibition of cathected ideas belonging to the primary process.

Consciousness, therefore, has a very minor role being the "outermost superficial portion of the mental apparatus" acting as a medium for the perception of events occurring in the outside world and thus as a sensor of external events, and to a limited extent a sensor of internal events (Freud, 1933, p. 72). Thus, according to Freud, the psyche can be separated into three realms: the unconscious (Id and Ego), the preconscious (Ego and Superego), and the conscious (Ego and Superego).

The Ego: A Conscious Unconscious

As noted, the ego is both conscious and unconscious. However, though the Ego is unconscious, there are aspects of the unconscious which are wholly foreign to it, i.e., the Id seat of the pleasure principle and instinctual aims, which is expressed by the mobile energy of the primary process.

In our NeuroDynamic conception, the Id is linked to the limbic system. The unconscious aspects of the Ego are linked to the limbic system and right hemi-sphere. The conscious aspects of the Ego are linked to the left hemisphere.

According to Freud's view, the Ego also includes yet another subdomain: the SuperEgo--which we link to the frontal lobes.

According to Frued, the Ego, and the Superego originate from the Id and are identical with it, while simultaneously being a differentiated aspect of it. That is, the Ego is the organized portion of the Id that which comes in contact with the outside world (Freud, 1926a, 1926b, 1933, 1940). This is similarly true of the Superego and the Ego; both are indistinguishable unless a conflict arises between them (Freud, 1926b, 1933, 1940).

Moreover, in that the Ego transcends all three realms, the unconscious as-pect of the Ego has knowledge of what occurs consciously, whereas the con-scious aspect of the Ego is ignorant of the activities occurring within the un-conscious mind (Freud, 1913, 1915b, 1933, 1940). Thus the Ego is required to be both separate and yet identical with itself, particularly in that it acts as an unconscious instrument of repression--the censor (Freud, 1915a, 1917).

The Ego is therefore required to be a unity which is nevertheless divided into unorganized (Id) and organized aspects (Ego and Superego). It is sepa-rated into both conscious and unconscious and is required to hide information form itself. Hence, the Ego simultaneously knows what it does not know, act-ing in a dual capacity of both deceiver (censor) and deceived (censee) and is actually an "unconscious-consciousness."

This indicates that the unconscious-preconscious ego is in fact conscious and self-reflective in that it is simultaneously conscious though separate from and indirectly in control of what becomes conscious and what must remain unconscious. Because the unconscious Ego must be completely knowledge-able of what the conscious Ego knows and what it does not and should not know, so as to selectively conceal thoughts, desires, etc., the conscious self is allowed only a superfluous existence, unconscious of the fact that it is pro-tected by an independent and elusive structure. Whereas we have described the unconscious Ego as an unconscious-conscious, the conscious is unconscious, because (in the Freudian framework) it is not conscious of its own process. The conscious is a subordinate phantasm to this all powerful governor of the mind, the unconscious Ego which in turn is linked to the limbic system, and, in sofar as it is unconscious and aware, the right hemisphere (the domain of awareness and the "preconscious."

Resistance and Repression

The cornerstone upon which the whole conception of unconscious as well as psychoanalysis rests is the theory of repression (Freud, 1914). The concept of repression, however, would be meaningless without the existence of an uncon-scious. Freud (1914, 1915a) considers repression to be a concept demonstrated by psychoanalysis rather than a premise on which psychoanalysis is based. Freud posits two forms or phases of repression; primal repression is the first phase, acting to bar psychical representation of instincts from consciousness. Repression proper is the second stage, affecting ideas associated with material that has been primally repressed (Freud, 1914, 1915a). According to our NeuroDynamic model, "primal repression" is due to the frontal lobe inhibitory influences exerted on the limbic system and right hemisphere. However, primal repression may also be due to severe trauma which damaged the memory centers, thus preventing this information from every reaching consciousness.

Repression proper occurs usually after-the-fact--that is, after information has been consciously perceived. Repression proper occurs consciously, though according to Freud (1915a, 1915b) it also appears in conjunction with primary repression processes. In the Freudian model, however, material that is repressed requires a constant expenditure of bound energy on the part of the unconscious Ego to counter continually the repressed material's persistent attempts to gain access to consciousness.

It might asked, if that which has been repressed exists as an unconscious process, how is it that psychoanalysis is able to know of its presence? Repressions are revealed in the course of psychoanalysis when the patient begins to offer resistances which thwart the continuing of therapeutic work (Freud, 1914).

Resistances emerge when an attempt is made to trace neurotic symptoms to their source--that which has been repressed. For example, during the course of psychoanalysis, the patient is instructed to "say everything" and yet may resist these instructions by lapsing into silence, feigning a loss of memory, or finding critical exceptions to this provision. As illustrated in an early case history cited by Freud (1893-1895):

She often made such assertions as that there was nothing, after which she allowed a long interval to pass during which her tense and preoccupied expression of face nevertheless betrayed the fact that a mental process was taking place in her but she was not always prepared to communicate it to me, and tried to suppress once more what had been conjured up (p.153).

It is important to note from the above and many other examples provided by Freud (e.g., 1910, 1915a) that repression proper is a conscious attempt to hide information that has been consciously recognized as undesirable (Freud, 1893-1895, pp. 153, 155, 180). Hence, to do the work of psychoanalysis, it is necessary to appeal to the patient to cease from consciously excluding embarrassing or painful thoughts that may be too disagreeable to express:

We urge him always to follow only the surface of consciousness and to leave aside any criticism of what he finds, whatever shape that criticism may take; and we assure him that the success of treatment, and above all the duration depends on the conscientiousness with which he obeys this fundamental technical rule of analysis (Freud, 1915a, p. 287).

This statement implies that resistance as well as repression proper is a conscious process. As noted by Freud (1893-1895, 1911), when the patient finally reveals the hidden information, he will often add: "I could have said it the first time--and why didn't you'- thought it wasn't what you wanted', or though I could avoid it, but it came back each time' (1893-1895, p. 154).

Knowing, Yet Not Wanting To Know Obviously there are mental processes that occur outside of consciousness. Indeed, there are specific brain structures, such as the brainstem, which perform a dizzying array of functions, including regulating breathing, heart rate, and so on, all of which occur, for the most part, without conscious assistance--which is why even those who are "brain dead" continue to breath and so on.

It is also possible for someone to suffer a blow to the head, or to suffer an incredible emotional trauma, and to become amnesic--such that, in consequence, the information may never have been stored, or it has been stored inappropriately and cannot be found.

On the other hand, we should be quite hesitant to assume as Freud (1940) would have us believe that a disagreeable or embarrassing idea is unconscious, unconsciously repressed, unconsciously resisted, or beyond conscious knowledge. If it seems that these ideas are unconscious it is "only because the patient holds back or gets rid of the idea that he has become aware of" (Freud, 1910, p. 33).

In this regard, we argue that an individual may be aware, but not conscious. If aware, however, then it cannot be said that the person is unconscious. However, if aware, then it then becomes possible for the conscious mind to avoid what the rest of the brain is aware of--and this may be made possible by the frontal lobes or the failure to transfer this information from the right to left hemisphere.

According to the Freudian view, repression and resistance are attempts to protect the conscious self from unpleasant associations by consciously rejecting that which is unacceptable (Freud, 1910, p. 39; 1915b, p. 161). As stated by Freud (1910, p. 24), repression results from the "emergence of a wishful impulse which was in sharp contrast to the subject's other wishes and which proved incompatible with the ethical and aesthetic standard of his personality. There had been a short conflict, and the end of this internal struggle was that the idea which had appeared before consciousness as the vehicle of this irreconcilable wish fell victim to repression, was pushed out of consciousness with all its attached memories, and was forgotten."

According to Freud (1913, p. 142) this simultaneous knowing and not knowing is accomplished by a conscious decision to ignore or disobey the impulses that it no longer chooses to know about (Freud, 1940, p. 275). Nevertheless, these unacceptable ideas have necessarily existed, consciously at one time, though according to Freud (1893-1895, p. 165; 1940, p. 202), while conscious they were isolated like foreign bodies separated from the rest of conscious life. This indicates a conscious refusal to come to terms with conscious though detached knowledge, by refusing to clearly recognize and confront it and what it signifies (Freud, 1940, p. 203). Thus the patient is forced to entertain two simultaneously different attitudes and perceptions, one of which is being dis-avowed because it is unpleasant and difficult to face or admit (Freud, 1940). And yet "the disavowal is always supplemented by an acknowledgement; two contrary and independent attitudes always arise and result in the situation of there being a splitting of the ego" (Freud, 1940, p. 204), that is, into an uncon-scious and a conscious.

Again, we believe that this Freudian system is valid, only in respect to the functional interactions of the right and left hemisphere, and the frontal lobes and limbic system.

Returning to the problem of resistance and repression, it might be asked how an individual may retain and hide information from himself in an unconscious, and yet consciously recollect this information at the insistence of the therapist (Freud, 1910, 1925). Indeed, much repressed material is not directly attended to, whether we believe in an unconscious or not, since its distressful, shameful, or alarming nature makes it very difficult to confront and reflect upon. It seems quite natural that an individual would minimize the amount of time and quality of attention directed to these unpleasant subjects. And yet, these attempts at self-deception and denial are entirely conscious processes; there are no forces that unbeknownst to consciousness hide shocking information from it. As we have seen through Freud's own examples, resistance is a conscious refusal to admit what is already in the patient's conscious possession:

On certain occasions, though only for a moment, the patient recognized her love for her brother-in-law consciously. As an example of this we may recall the moment when she was standing beside her sister's bed and the thought flashed through her mind: "Now he is free and you can be his wife." At that time, as well as during the analysis, her love for her brother-in-law was present in her consciousness like a foreign body, without having entered into relation-ship with the rest of her ideational life. With regard to these feelings she was in a particular situation of knowing and at the same time not knowing--a situ-ation, that is, in which a psychical group was cut off. But this and nothing else is what we mean when we say that these feelings were not clear to her. We do not mean that their consciousness was of a lower quality or of a lesser degree, but that they were cut off from any free associative connection of though with the rest of the ideation content of her mind. The recovery of this repressed idea had a shattering effect on the poor girl. She cried aloud when I put the situation drily before her with the words: "So for a long time you have been in love with your brother-in-law." She complained at this moment of the most frightful pains and made one last desperate effort to reject the explanation: it was not true, I had talked her into it, it could not be true, she was incapable of such wickedness, she could never forgive herself for it. It was easy to prove to her what she herself had told me admitted of no other interpretation (Freud, 1893-1895, pp.167, 165, 175).

Thus, through their ardent denials, we find the patients disclosing the knowledge of what is supposed to be unconscious, repressed, and unknown to them. In truth, they have hidden it from themselves--and this is made possible by the unique organization of the brain.

As we have seen, resistance is not only a conscious attempt to thwart the therapist but is also a conscious attempt to disavow and hide unpleasant ideas from oneself--information that the patient is aware of.

We must agree with Stekel (1949) when he states that: "We do not believe the patients when they claim over and over again that they do not know anything. We regard them as actors who often enough deceive us consciously and who, most of the time want to deceive themselves. We do not fight against mysterious unconscious ideas. We tell the patient: 'You know, but you do not want to know'" (p. 256).

It is the position of this paper that when an individual known what he does not want to know and acts as if he doesn't know it, he is attempting to deceive himself. The whole theory of unconscious processes is thus a magical metaphor for conscious processes of self-deception made possible by the functional duality of the right and left hemisphere, and the inhibitory actions of the frontal lobe which may prevent hemispheric or limbic system transfer. The consciousness aspects of the left hemisphere are thus able to avoid gaining access to unpleasant or disagreeable memories, feelings, and so on--but only after the fact.

SELF-DECEPTION AND SYMPTOM FORMATION

It can be said that someone has lied when he knowingly possesses the truth and consciously attempts to deceive another person. Yet curiously, it sometimes happens that a person half persuades himself that the lie he has told or is telling is true and then behaves accordingly. According to Sartre (1956), when a person attempts to conceal the truth from himself he acts in bad faith (mauvaise foi). For our purposes we will call such behavior self-deception. Self-deception implies both a knowing and a not knowing or simultaneously telling a lie and believing it. According to Freud's concept of unconscious thoughts and mental processes we are presented with individuals who are not responsible for the lie since it occurs unconsciously. Thus we are provided with the notion of a lie without a liar.

However, self-deception does not occur just among neurotic or psychotic individuals but is a normal process of making adjustments to the world in a manner that is acceptable to the self-concept. Take for instance a paraphrased example borrowed from Sartre (1956): A young woman, while on a date with a particular man for the first time, consents to stop by his home. She knows or rather is aware of the possibility of his intentions, and she also knows that she may have to make a decision regarding them--but she doesn't want to think about such things (she is a nice girl). For the moment she is concerned only with the here and now.

Once at his home he makes certain advances; sitting close to her, looking in her eyes, he tells her she is pretty. But she restricts the meaning of this phrase to the present--it means only that she is pretty, his actions imply nothing. There is no projection, no illumination, she does not think about or consider the possibilities--though she is aware of them.

Slowly he places his hand upon her knee. The act risks changing the situation by calling for an immediate decision as to its meaning. To leave his hand there is to consent, to engage herself in his desires, to acknowledge them willingly. Yet to withdraw it is not only a recognition of its possibility, but a refusal. Her aim is to postpone her decision--she has "no idea" as to what he is up to. She leaves his hand there--because she doesn't notice it.

And yet, as they talk, he has moved closer to her, his hand inching its way past her knee. But she is concentrating on what he is saying--the curve of his lips, the white of his teeth, her reply to his questions; this is what she is conscious of and to what she addresses herself. She reacts as a personality that does not know that a hand lies there upon her leg. She is being seduced.

According to Sartre (1956) this woman is acting in bad faith because she has disarmed the actions and intentions of her companion by reducing them to being only what they are--at that moment. Yet there is a necessary unity here, for her actions necessitate a recognition of the intention as the motive for disarming the intention. The result is both a contradiction and a unity, that is, the recognition of the idea and the negation of that idea. Or as Freud would classify it, a knowing which is accompanied by a disavowal. Self-deception affirms this synthesis through the preservation of the possibility of the event through acting as if there were no possibility.

It often happens that people behave in a certain manner while maintaining the conviction that what they are doing or saying is not a true representation of how they really feel or how they really are. Such behavior, they reason, has an explanation outside themselves, being only a rare and momentary lapse, or is justified by certain mitigating circumstances, etc. And, if our young lady described above was, "taken advantage of" by "that scoundrel", she of course is not responsible. She "had no idea as to what he was after', and "besides she was tired and it was difficult to fight him off", and "before she knew it, it was too late", etc. She is free of guilt, she has deceived herself. And yet she knows and is aware of the truth; it is precisely because she knows that she invents innumerable excuses for her behavior.

Symptom Formation: Knowing What One Does Not Want To Know

When aware of something, one may have tacit knowledge of it, yet may choose not to consciously attend to it, to ignore it, to deny it, refuse to think about it, or hope it goes away. Hence, I may know something, yet refuse to give this non-linguistic, emotional, visual, or imageless knowledge a verbal and temporally-sequentially organized conscious acknowledgement. Thus disturbing feelings, ideas, or impulses may be partially denied or ignored by refusing to confront them consciously while being simultaneously aware of their presence. However, disturbing thoughts are not as easily dismissed as events or feelings that carry little emotional energy, though one may truly forget about unimportant events, those with considerable ideational importance may continue to exist as tacit knowledge of which one is aware.

The attempt to isolate significant information, by keeping it from being explicitly organized into conscious knowledge, may result in symptom formation. Because the resistance of objectification of an idea into consciousness requires an acknowledgement along with a disavowal, the idea is necessarily recognized. Thus the individual may be forced to employ certain thoughts or behaviors so as to divert his attention from the unacceptable impulse.

Through this division of knowledge, thoughts, ideas, and meanings quite different from the original impulse can be constructed. The impulse is thus interpreted, altered and distorted, and behavior may be altered and possibly disturbed accordingly. These distortions are called symptoms. The symptom is a replacement and thus an acknowledgement of what the individual cannot or would rather not face. The symptom becomes the lie that the patient willingly desires to believe.

The individual plagued by an unacceptable self-concept, threatening thoughts, impulses, or feelings which cannot be faced or coped with, resorts all too often to this conscious process of self-deception. Freud claims that these manifestations have roots in the unconscious. I believe that the formation of at least neurotic (as opposed to psychotic) symptoms results from the individual's attempt to limit, through self-deception and denial, conscious recognition of what he already knows, be it an unacceptable personality characteristic or emotionally charged though, desire, impulse, memory, or perception. The product behavior can result in a crippling complex of conflicts arising from the attempt to be what one is not, while desperately attempting to disguise what one is.

Symptom Formation and Not Knowing

One's Self, the source of thought, and the Whom, to whom one's thoughts are presented and explained, should not be viewed as localized in the language centers of the brain. Language is merely one aspect of that which may be identified as part of our psychic existence. Linguistic expression is simply one mode of process and function among many, a process which quickly loses its utility when used to explain what it "feels" like, for example, to ride a horse, a motor-cycle, or ski. Although we 'know' what it is like, it seems impossible to verbally communicate.

Moreover, not all impulses, feelings, desires, fears, cravings, etc., have a linguistic label, and are thus not readily translatable into the symbols and codes which give rise to and comprise thought and language. Indeed, to understand these processes we must learn the "language" in which they are 'coded.' Nevertheless, the dependence on language remains.

This dependence in unfortunate, for if we consider the relatively late development of language during our own formative years, we must realize that much of our social learning experiences took place prior to our development of a linguistic code or labeling ability (Dollard & Miller, 1950).

Moreover, during these same early years, our traumas, fears, and other emotional experiences were mediated by the silent, non-linguistic, non-dominant cerebra hemisphere as they are in adulthood (Schwartz, Davidson, & Maer, 1975) and processed and stored in a non-verbal pre-linguistic code, a code which due to physiological and psychological maturation may become lost to both sides of the brain. Nevertheless, and especially in regards to neuroses, many of these early impressions and feelings remain exactly that: feelings', present but not identifiable.

Take, for example, an infant who was repeatedly punished for touching its genitalia and masturbating. Soon, through repeated pairings of punishment, impulse to masturbate, as well as sexual impulses in general, become closely linked with anxiety and an inhibition of response. If the child's mother was successful in extinguishing this rather normal behavior in her young child, when the child reaches puberty and begins to experience this punished impulse, the experience will not be labeled as "sexual impulse" but will be interpreted as possibly nervousness, anxiety, fear, etc.

Hence, because the behavior was extinguished prior to the advent of a proper label, the impulse remains unrecognized. The result is a sexually disturbed individual who not only fails to recognize his sexual needs but mislabels the appearance of a sexual impulse in accordance with the punishment and anxiety experienced at the hand of mother, that is, as something "vaguely" unpleasant, too disturbing to think about, and impossible to label.

The Image and the Imageless

Imageless should not be equated with nothingness. An image is always a reproduction, meaning that is fashioned in the likeness of something else. An image is thus always presentational, a reproduction of something else without being it. In this sense, no matter how true the likeness, the image is always an abstraction, and as such, is always apart, that is, separate from the actual experience. At the level of awareness, imageless knowledge is the experience that there is no separation, no elucidation, no formulation. At the level of awareness imageless knowledge simply is; all else is a verbal interpretation and a temporal-sequential construction--a process of the domain we call consciousness.

Thus, to summarize: Awareness is bilateral, emotional, visual-spatial, environmental, geometric, non-verbal, etc., and linked to the right hemisphere and limbic system. The verbal aspects of consciousness, by contract, as linked to the left hemisphere. Insofar as the "Ego" is both conscious and unconscious, then it could be said that the Ego is bilateral and is maintained by the limbic system and the left and right hemisphere. In this way, it is possible for the Ego to be unconscious yet aware, and because it is aware, the Ego can also act to prevent information transfer from the limbic system or the right hemisphere and this is made possible by the Super Ego also known as the "frontal lobes."

4. The Neuropsychology of Development: Hemispheric Laterality, Limbic Language, and the Origins of Thought

ABSTRACT

Evidence and assumptions that concern hemispheric laterality and asymmetrical functional representation are discussed. It is hypothesized that the asymmetrical linguistic-motor vs. sensory-spatial-affective representation of function may be a result of differential rates of cortical, subcortical and spinal motor-sensory maturation. Evidence with regard to embryological and early postnatal neurological development is reviewed. It is argued that motor areas mature before sensory and that the left hemisphere develops prior to the right, such that the left hemisphere gains a competitive advantage in the acquisition of motor representation, whereas the later maturing right has an advantage in the establishment of sensory-affective synaptic representation, including that of limbic mediation. The influences of these differing maturational events on cognitive and psychic functioning are examined, particularly with regard to limbic influences on the development of language, thought, and mental imagery, and the effects of early emotional experience on later behavior. Thinking is viewed in part as a left hemisphere internalization of egocentric language, the internalization of which corresponds to the increasing maturation of intra-cortical and subcortical structures and fiber pathways, and the myelination of the corpus callosum connections that subserve information transfer between the hemispheres. It is argued that thought is a means of organizing, interpreting, and explaining impulses that arise in the non-linguistic portions of the nervous system so that the language dependent regions may achieve understanding. In addition, the neurodynamics and mechanisms involved in the mislabeling, misinterpretation, and inhibition of impulses, desires, and emotional expression are discussed in relation to disturbances in psychic functioning.

The Origins of Thought

Man is a rope, tied between beast and overman--a rope over an abyss. A dangerous across, a dangerous looking back, a dangerous shuddering and stopping. --Nietzsche

Thinking is clearly a form of communication and, as such, is a form of language, an organized hierarchy of associations, symbols, and labels that appear before an observer. In the process of thinking, one acts to organize information that is 'not thought out' and that is not clearly understood, so that it may become 'thought out' and thus comprehended (Ach, 1951; James, 1961; Schilder, 1951). Thought is a means of deduction, clarification, plan and goal formation, and reality manipulation. However, it is also a progression, an associative advance that leads from inner or outer perception to linguistic-motor expression (Freud, 1900) and an elaboration that some have argued appears with an initial or leading idea that is followed by a series of related ideations, or as originating developmentally from the non-accessible regions of the mind (Freud, 1900; James, 1961; Jung, 1954; Piaget, 1962). That is, it appears to be based on information that exists prior to thought as tacit and non-organized informational variables (Ach, 1951; Franks, 1974; Joseph, 1980; Polanyi, 1966) and is dependent on the reality-based organization of previous thought-like ideations, e.g., primary, symbolic and intuitive 'thought,' (Freud, 1900; Jung, 1956; Piaget, 1962).

Thought is thus a means of explanation through which ideas, impulses, desires, plans, or thing-in-the-world may be understood and possibly acted upon. Paradoxically, it is a process by which one explains things to oneself and thus necessitates that one apprehend and organize information that is possessed prior to its explanation. However, the fact that one acts as both the explainee and the explainer raises a curious question, 'who is explaining what to whom?'

The concern of this paper is the development of thought and the substrate from which it originates, the nervous system. Thought and language are viewed as integrally related to the motor predominance of the left hemisphere and the differential rates of hemispheric maturation. It is argued that the early maturation of the motor fibers of the peripheral nervous system and the motor cortex of the left hemisphere, coupled with the later maturation of the right hemisphere and sensory systems, results in differential acquisition and function, such that the left predominates in motoric processes and relies on temporal and sequential modes of operations, whereas the right is concerned primarily with receptive-sensory non-linguistic functions and the parallel-wholistic realization of information. Although considerable overlap of function is noted, thinking is viewed as a left hemisphere linked internalization of language, the internalization of which corresponds to the increasing maturation of intra-cortical/subcortical structures and pathways, and the myelination of the callosal fiber connections that subserve information transfer between the two hemispheres. Hence, thought serves in part as a means of organizing, interpreting, and explaining impulses that arise in the non-linguistic portions of the nervous system so that the language dependent regions may achieve understanding. In addition, disturbances in the formation of thought are discussed.

Asymmetry of Cerebral Function: Right vs Left Hemisphere

Although some lateralization of function has been noted in non-human vertebrates (cf. Dennesberg, 1979; Nottebohm, 1979; Walker, 1980), which in turn may reflect differential rates of hemispheric or central nervous maturation, laterality, or asymmetrical functional specialization of language and other higher cortical activities appear to be the hallmark of the human brain. Each hemisphere has, in the course of evolution, developed its own unique strategy and capability for processing information, and a neuroanatomical substrate to subserve each independent function. Curiously, the human brain is organized so that two potentially independent mental systems coexist, literally side by side (Bogen, 1969; Galin, 1976; Gazzaniga & LeDoux, 1978; Ornstein, 1972; Sperry, 1974), such that each hemisphere or mental system may act independently on specific information, initially without interference from the other (Dimond & Beaumont, 1974). A result of this curious arrangement is presumably the dual processing of sensory information (Dimond & Beaumont, 1974; Levy, 1974; Levy & Trevarthen, 1976). Interference, as well as the possibility of dual responding, probably is prevented by the corpus callosum, which both aids and inhibits interhemispheric transfer.

Not only are these systems dual, but they are asymmetrical as each hemisphere utilizes predominantly verbal-analytical or visual-spatial, sensory-affective associational strategies in the analysis and expression of information (Bogen, 1969; Levy, 1974; Luria, 1973; Sperry, 1974). In this regard, when one hemisphere learns or 'experiences,' the information analyzed then cab be transferred via the corpus callosum for further analysis or so that the other hemisphere also may 'learn,' being a beneficiary of the other's activity (Bogen, 1969; Dimond & Beaumont, 1974; Gazzaniga & LeDoux, 1978; Levy, 1974; Sperry, 1974). Although admittedly there is considerable overlap of function, in that both hemispheres may act to analyze all input, and given that some types of information are dealt with more efficiently by one than the other, in the long run this relationship is very adaptive as the range and speed of information analysis is enlarged and the efficiency of response is increased (Dimond & Beaumont, 1974; Levy, 1974; Levy & Trevarthen, 1976).

For example, when attempting to identify a block with few details, although the left may have a slight advantage, the two hemispheres may perform comparably. With the addition of details, the performance of the left beings to deteriorate, whereas the right continues to function efficiently regardless of the increasing complexity (Dimong & Beaumont, 1974; Levy & Sperry, 1968). In general, these qualities are characteristics of the differential strategies best employed by the two hemispheres during information analysis, the left relies on the analysis of temporal sequences, details (or parts), whereas the right is responsive to melodic and environmental sounds, visual-spatial analyses, and appears to form a gestalt.

However, in that the left codes, processes, and stores information using linguistic, categorical, functional, and motoric referents, whereas the right relies on spatial, emotional, and non-linguistic modes of understanding (Blumsetein & Cooper, 1974; Carmon & Nachson, 1971; Knox & Kimura, 1970; Levy, 1974; Sperry, 1974), complete and efficient inter-hemispheric communication may not always be possible (Joseph, Gallagher, Kahn & Holloway, 1984). At least in regard to consciousness or linguistic processing, there is the potential for much miscommunication, misinterpretation, and distortion regardless of the direction of transfer. We will return to this topic in the later half of this paper (cf. Neurodynamics).

The Right and Left Hemisphere

The right hemisphere. In general, the right hemisphere appears to be poly-sensory, expressively non-linguistic, affectively endowed, and responsible for the sensory integration of complex internal and external environmental variables (Critchley, 1953; Kimura & Durnford, 1974; Luria, 1973; Schwartz et al., 1975; Tucker, 1981). It is concerned with spatial orientation of perceived stimuli, analysis of the body's position in space, geometrical realization of spatial relationships, and the conceptualization of the whole of a stimulus configuration including form, figure-ground, and depth (Kimura & Durnford, 1974; Levy, 1974). The right hemisphere is also 'dominant' for the realization of tactile and propioceptive information (Critchley, 1953; Luria, 1973; Semmes, 1968), is the principal receptive organ for sensory impressions that arise from one's own body, and is preeminent for the realization of the body image (Critchley, 1953; Nathanson, Bergman, & Gordon, 1952; Roth, 1944, 1939; Sandifer, 1946; Weinstein & Kahn, 1950). Hence, we sometimes find with damage, particularly in the posterior cortical regions, that there is a subsequent failure to recognize or attend to the left half of the body, or space, such that the left half of the environment may be abnormally perceived or actively neglected and the sensory attributes and spatial relationships within the environment severely confused (Bisiach & Luzzatti, 1978; Critcheley, 1953; Gainooti, Messerli, & Tissot, 1972; Heilman, 1979; Nathanson et al., 1952). In addition, patients literally may deny the existence of the left side of the body, such that when engaged in daily activities they may brush only the right side of their head, and wash or dress only the right side of the body. In that hemi-neglect occurs much less frequently with left hemisphere damage, suggests that the right hemisphere normally attends to information that arises from both sides of the body and space, whereas the left hemisphere at best, and the perhaps only following right hemisphere parietal dysfunction, attends to the right side. In this regard, it certainly could be argued that the right hemisphere, in fact, has superior awareness (cf. Joseph, 1980) of the environment-body continuum than its left brain counterpart.

The right hemisphere also is concerned with the mediation of emotional arousal and expression, and with discerning the affective significance of facial contours (DeRenzi & Spinnler, 1966; Gilbert & Bakan, 1973; Schwartz et al., 1975; Tucker, 1981). When damaged, patients may no longer be able to control their emotions and may appear labile, manic or depressed, blunted, indifferent, or otherwise inappropriate (Bear & Fedio, 1977; Flor-Henry, 1969; Gainotti,

1972; Goldstein, 1948; Luria, 1973; Sherwin, 1977). However, it is important to note that although the left hemisphere does onto mediate emotion per se, it may act to inhibit emotional expression as generated in the limbic areas or the right half of the brain.

Although propositional language is mediated by the left hemisphere, the right is also capable of perceiving and comprehending some spoken language and concrete words (Gaazaniga, 1970; Zaidel, 1977), but its ability seems to be limited to nouns and simple sentences. However, it is largely non-linguistic, except for the ability to curse, sing, and pray. IT cannot analyze the phonemic elements of words due to its difficulty in preserving sequential information (Zaidel, 1977), or can it generate phonology from meaning (Levy, 1974; Marin, Schwartz, & Saffran, 1979). Rather, its comprehension is limited to a rather impoverished lexicon that consists of the ability to comprehend connotative (Gazzaniga, 1970; Zadel, 1976, 1977) and emotional components of speech and auditory stimuli (Carmon & Nachson; 1973; Haggard & Parkinson, 1971). The recognition of environmental sounds (e.g., wind and rain the rolling of thunder) and the reception and production of tonal patterns, intonational contours, and melody are mediated by the right as well. Not surprisingly, individuals with right hemisphere damage may have considerable difficulty comprehending affective speech and may loose a great deal of their ability to appreciate music (Gordon & Bogen, 1974; Heilman, Scholes &Watson, 1975; Luria, 1973; Milner, 1962; Tucker, Watson, & Heilman, Scholes & Watson, 1975; Luria, 1973; Milner, 1962; Tucker, Watson, & Heilman, 1977), whereas speech and language functions generally remain unaffected.

Interestingly, even with extensive damage to the left hemisphere, the majority of individuals continue to respond and interact emotionally with the environment and may retain the ability to sing familiar songs and learn new ones, with few or no articulatory errors (Bogen, 1969; Critchley, 1953; Geschwind, 1965; Geschwind, Quadfasel, & Segarra, 1968; Head, 1926; Smith, 1966). Moreover, although propositional speech may be absent, the ability to pray may be retained. The retention of these operations is truly amazing when one considers that massive left hemisphere damage destroys the linguistic portions of the brain, and the individual in all other respects may appear to be without comprehension or thought, and no longer interact with the environment in an intelligible manner.

Interestingly, singing, praying, and cursing appear to be dependent on the integrity of the right hemisphere and seem to be linked affectively, in that all 3 are affectively stamped, contain emotional components, and are dependent on an emotional motive force for the arousal of their expression (cf. Goldstein, 1943, 1948). Singing, praying, and cursing thus appear to be retained after left hemisphere dysfunction due to the mediation of emotional and musical production/reception by the right hemisphere (Bogen, 1969; Chase, 1967; King &

Kimura, 1970; Luria 1973; Milner, 1962; Smith, 1966).

Right hemisphere overview. The right hemisphere appears to be concerned predominantly with the reception and realization of non-linguistics, non-sequential, non-temporal sensory information that arises from one's own body or the surrounding environment. It is expressively non-linguistic, and does not seem to explicitly label, classify, or perform differential analysis on the elements of stimuli, but rather perceives things as a whole without disturbing or creating relationships. Possibly as an outgrowth of its superior sensory-receptive capacities, the right hemisphere also is concerned with the realization of spatial relationships, depth perception, emotional mediation and expression, and the appreciation of emotionally related sensory experiences. Interestingly, there is some provocative evidence that lends itself to the suggestion that right hemisphere activity during sleep provides the matrix from which dreams are derived and is responsible during waking for creative output.

The left hemisphere. The left hemisphere is most notable for containing the neural substrate for the expression of language (Broca's are in the frontal lobe) and its analysis and comprehension (Wernicke's area in the temporal region and the angular gyrus in the parietal lobe, see Table 2). It subserves gestural communication, the organization of motoric output and purposeful actions, and includes systems for the perception and labeling of material that can be coded linguistically (Brown, 1975; Galaburda & Sanides, 1980; Luria, 1973; Sanides, 1975). It also is associated with the organization and categorization of material into discrete temporal and analytical functions, verbal concept formation, and the distinction of sound by specific articulatory features (Critchley, 1953; Kimura, 1973; Kimura & Vanderwolf, 1970; Riese, 1956).

When the left hemisphere is damaged, performance on problems that involve temporal order is selectively impaired (Carmon & Nachson, 1971; Efron, 1963; Lackner & Teuber, 1973), as is nonverbal manual or oral performance (DeRenzi, Faglionia, Saoiardo, & Vignolo, 1966; Kimura, 1976, 1977), the copying of meaningless movements (Kimura & Archibald, 1974), rapidity of limb (Wyke, 1968) and pursuit-rotor movements, and the analysis of temporal sequencing of speech (Lenneberg, 1967). Hence, with a lesion in the posterior portions of the temporal lobe the individual may have difficulty organizing these thoughts for expression. That individuals with Wernicke's aphasia have been noted to complain 'everything sounds like a foreign language' or that 'words don't separate' (Lenneberg, 1967, p. 191) indicates that the specialization for performing temporal analysis and sequencing operations is mediated by this region (Albert, 1972; Carmon & Nachson, 1971; Halperin, Nachson & Carmon, 1973), and that the ability to extract meaning from perceived sequences of sound is dependent on the ability to organize and coordinate units into a temporal and interrelated linear series; an ability at which the left hemisphere excels (Kimura, 1973, 1977; KInsbourne, 1978; Shankweiler & Studdert-Kennedy, 1975).

After left-hemisphere damage, individuals may have difficulty carrying out simple commands (e.g., 'touch your nose, then your left elbow') and may make gross errors in the attempt (Brodal, 1969; Brown, 1972). In addition, it has been noted that the right hand, more so than the left, becomes activated during verbal communication (cf. Kimura, 1977). The fact that both hands and sides of the body may become apraxic after left hemisphere damage (Brown, 1972; Geschwind, 1965; Luria, 1973) and that both hands may become activated during vocal communication, indicates that the left hemisphere may exert bilateral influence or control over muscular output, at least of the upper musculature. However, its greatest influences is over the right hand, which, in the majority of the population, is dominant for grasping, manipulating, exploring, creating, and communicating. In fact, right hand usage appears to be in some manner a motoric extension of the language function as both seem to share some common neural substrates. This motoric relationship is demonstrated further by the findings that left hemisphere damage results in difficulty copying hand movements, that deaf mutes with left sided damage have deficits in gestural communication, and that individuals with right hemisphere speech demonstrate more left hand than right sided movements while speaking (cf. Kimura, 1976, 1977). Moreover, while speaking, the ability to simultaneously sequence, position, or maintain stabilization of one's limbs or hand is reduced (Hicks, 1975; Kimura & Archibald, 1974; Kinsbourne, & Cook, 1971). These observations have led to the suggestion that language in fact may be an evolutionary outgrowth of left hemisphere motor predominance.

Left hemisphere overview. The left half of the brain is preeminent for motoric-language functions, including motor aspects of cognitive activity. Its efficiency in the organization and coordination of movements in sequence and its mediation of analytical-mathematical and temporal processes, including the linguistic labeling and categorization of experience, enable it to play the dominant role in language dependent cognitive operations, such as thought, speech, and gestural communication. Nevertheless, it is important to note that the two hemispheres are not strictly dichotomous, as there is considerable overlap in functional representation and expression.

Competition for Synaptic Space: Sensory and Motor Development

The human brain, although conceptualized by some as functionally lateralized and asymmetrical arranged, is a composite of multitudinous neuronal cell assemblies, each of which subserves specific functions and large numbers of which are aggregated in discrete regions of both hemispheres. There is abundant multi-representation of both sensory and motor function throughout the cortices of both brain halves. Nevertheless, these representations are not symmetrical because the left, in humans, has become preeminent for motor and language functions, whereas the right appears to be concerned with visual-spatial, sensory-affective-emotional aspects of experience. Given the earlier maturation and development of the motor layer of the cortex embryologically across species, and the comparatively latter development and differentiation of the sensory cortex (Arey, 1965; Hamilton & Mossman, 1972), it appears that the motor-sensory dichotomy may be in part a function of the differential rates of cortical maturation of the two hemispheres.

Sensory and motor development. During the later stages of fetal development, through birth and extending into childhood, neurons continue to grow and develop axons and dendrites, even though the number of neurons and neural process that develop greatly exceeds that which will be present in adulthood (Arey, 1965; Hamilton & Mossman, 1972; Lund, 1978; Szentagothai, 1969). Generally, we may divide neuronal development into six basic stages (cf. Cowan, 1979; Szentagothai, 1969):

1. the generation of the cell.
2. the migration of the cell from its birth cite to its terminal substrate.
3. the aggregation of cells within specific brain regions.
4. cellular differentiation and the growth of axons and dendrites.
5. the formation of synaptic connections.
6. Myelination and the elimination of cells, axons, and dendrites; a continual process that spans the individual's lifetime.

At approximately the fourth week of embryonic development, three swellings begin to develop at the head of the neural tube (the substrate from which the central nervous system develops): The forebrain, midbrain, and hindbrain.

Specifically, the brainstem and cerebellum begin to develop in advance of the forebrain, which in turn matures in a rostral arc, i.e. diencephalon, limbic system, striatum, neocortex (unpublished observations). And these maturational events continue well into late childhood, adolescence and adulthood.

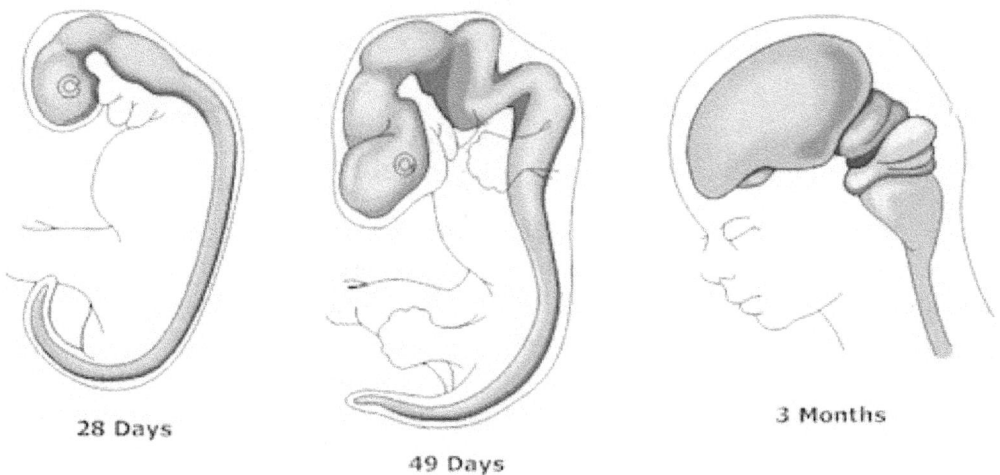

28 Days

49 Days

3 Months

However, prior to the development of the neocortex, as the brain continues to form and unfold, a germinal layer forms that surrounds the lateral ventricle. From this germinal layer, neuroblasts (primitive neuronal cell bodies) begin to migrate outward to form an outer cortical or marginal zone, the primordial cortex. This zone subsequently becomes filled with cell bodies at about the third month of prenatal development as it receives massive migrations of cells from the inner regions of the brain (Arey, 1965; Hamilton & Mossman, 1972; Truex & Carpenter, 1969). It is within the primordial cortex that the six to seven concentric layers of the cerebral neocortex are formed (Arey, 1965). Nevertheless, not all six concentric layers are formed at once. This is important, in that the different layers subserve different functions.

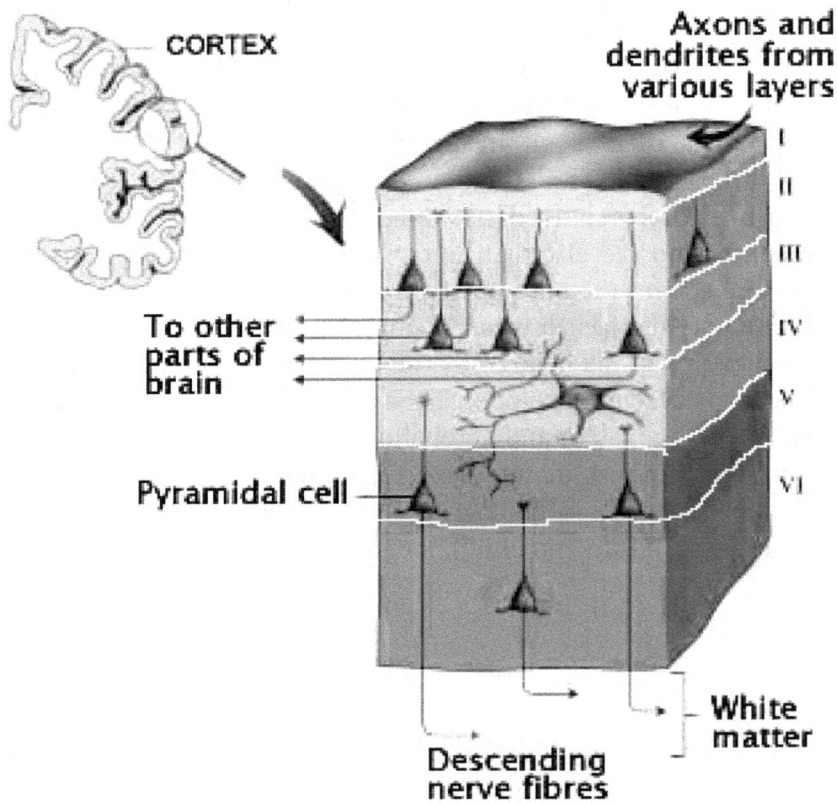

The first layers to be established are the deepest layers, 6 and 5, which consist primarily of pyramidal cells from which the corticospinal (pyramidal) tract develops, and which subserve motor functions. Similarly, during this period, the anterior or motor regions of the cerebral cortex (frontal lobes) and the deep layers of the temporal lobes (including hippocampal and other limbic areas) begin to form (Milner, 1967; Truax & Carpenter, 1969). Later in development there is a second wave of neuroblast migration into the primordial cortex, and layers 4 then 3 and 2 are formed. Although they contain some pyramidal cells, these layers have predominantly receptive functions, receiving specific afferent and sensory fibers directly from the thalamus (particularly layer 4) or association fibers from other cortical regions (Brodal, 1969). Layer 1, the most superficial, is the last to develop and receives the final wave of neuroblast migration.

Thus the neocortex, which consists of numerous sublayers, develops in an "inside-out" fashion, with the inner, deeper layers forming in advance of the outerlayer, thus forming a scaffolding or ladder.

Moreover, differentiation of these last three receptive layers is not complete until middle childhood (Arey, 1965), which suggests that the motor (pyramidal) layers develop (or rather appear) well in advance of the sensory regions of the cerebral cortex.

Nevertheless, although at birth motor maturation is much more advanced than sensory, particularly in the regions that control hand and arm movement, the primary motor strip (Conel, 1939; McGraw, 1969), both sensory and motor neurons are quite immature, and appear functionally isolated and unable to exert regulatory control over peripheral processes until later in development (Arey, 1965; Brodal, 1969; McGraw, 1969; Milner, 1967).

Although the six to seven layered neocortex is formed well before birth, with the exception of a few pyramidal cells in the primary motor cortex of the frontal lobe, there is no evidence of myelinization of metabolic activity within this region of the brain at birth (Fleshig, 1901; Langworthy, 1937).

Moreover, even at one year of age, the frontal, temporal, occipital, and parietal lobes are only about half the size of the lobes of the adult brain (Blinkov & Glezer, 1968). In fact, the neocortex of the frontal, parietal, occipital, and temporal lobes can take well over 7, 10, or more years to fully mature.

For example, we find that the corticospinal tract (the long axons from the pyramidal-motor layers of the cortex) does not completely establish its functional connections or become completely myelinated until the seventh month after birth (Yakovlev & Lecours, 1967). Hence, most early behavior is mediated by subcortical, limbic, and spinal mechanisms (Arey, 1965; Brodal, 1969; Conel, 1939; Hamilton & Mossman, 1972; Langworthy, 1933; Milner, 1969). It is largely due to this incomplete information and regulatory exchange that the behavior of anencephalic monsters and normal infants is largely indistinguishable for the first few days of life (Mcgraw, 1969).

Cell death and synaptic maintenance. Although growth continues throughout the brain after birth there are anywhere from 15 to 85% more neurons in the infant as compared to the adult brain due to cell death. Presumably because of the immaturity of the brain and the fact that functional and synaptic interconnections are largely incomplete or not yet established, a large number of these neurons die and/or their processes are retracted within the first few years of life (Arey, 1965; Cowan, 1979; Hamilton & Mossman, 1972). A number of interactions, however, appear to be responsible for this decline, such as the maturity of the synaptic substrate during the period of proliferative growth, hormonal environment, sensory-motor activity, early environmental experience, the number of synaptic and functional contacts available, and the number of competitors that are vying for a connection (Casagrande & Joseph, 1978; 1980; Diamond, Johnson, Mizono, Lee & Wells, 1977; Goodman & Horel, 1966; Greenough, 1976; Hubel & Wiesel, 1970; Joseph & Casagrande, 1978; Lund, 1978; Rakic, 1977; Rosenzweig, Bennet, & Diamond, 1972; see also Wolstenholme & O'Connor, 1968).

Because of neocortical (and subcortical/limbic) immaturity, and as these tissues are also experience-expectant they require considerable stimulation early in life, in order for neurons to grow, mature, divide, and establish and maintain synaptic connections. Indeed, it is well established that early environmental influences exert significant organizing effects on neural growth throughout the neocortex and limbic system, determining neocortical thickness, the density, size, shape and growth of dendrites and synapses, and the die off and drop out rate of glia, neurons, and axons. An enriched rearing environment can increase neural growth and establish billions if not trillions of synapses throughout the brain and neocortex and increase a thousand fold the number of synapses per axon (Greer, Diamond, & Tang, 1982). In the occipital cortex, for example, the dendritic fields are increased by about 20% among those reared in an enriched visual environment.

The optimal maintenance and preservation of a neuron, its axon, dendrites, and synaptic connections require not only experience-expectant stimulation, but that the neuron and terminal substrate with which it is expected to form its synaptic connection functionally mature and interact at similar times. Witness, for example, the motor and sensory systems. Cortical motor layers develop prior to sensory neurons, spinal and subcortical motor fibers become myelinated before the sensory systems (Langworthy, 1933; Milner, 1967; Yakovlev & Lecours, 1967), and the motor roots of the spinal cord are completely myelinated at birth or soon thereafter, whereas the sensory roots are not complete until the fourth to seventh month of postnatal development (Yakovlev & Lecours, 1967). Presumably, this possible coordination of substrate maturation allows the axons of the corticospinal tract to establish functional connections with the spinal motor neurons, whereas viable synaptic connections with processes terminating in the superficial layers of the neocortex as they mature at a later date.

We may suppose that many of those neurons that subsequently die off do so because they, their processes, or the fibers with which they were to make synaptic connections matured at the wrong time or were not functionally active or functionally maintained. On the other hand, death may be due to their terminal substrate being occupied by fibers from a competing neuron (cf. Casagrande

& Joseph, 1978, 1980; Joseph & Casagrande, 1978). This is particularly important as stimulus to growth is exerted by non-occupied synaptic space which attracts axons and stimulate fiber growth (Raisman, 1969). However, as noted, the maintenance of terminal space (i.e., a neuronal cell body and its dendrites) is dependent on the establishment of functional innervations, the maintenance of which is experience-expectant.

Experience-Expectant Stimulation & Deprivation.

Abnormal and deprived early experience during specific critical developmental periods can result in the elimination of millions of neurons, axons, dendrites and synapses, and may even induce functional invasion by competing neuronal cell assemblies which establish functional dominance over those neurons that have been insufficiently or adversely stimulated (e.g., Casagrande & Joseph, 1978, 1980; Hirsch & Spinelli, 1971; Joseph, 1986b; Spinelli et al., 1972). As demonstrated in animal (Hirsch & Spinelli,1971; Spinelli et al., 1972) and in human studies (Novelly & Joseph, in press), neocortical and limbic neurons deprived of normal experience sometimes come to subserve wholly different functions. In consequence, even when normal experience is later provided, deprived neurons are unable to function normally.

For example, kittens reared in a visual environment consisting only of thick horizontal lines were later unable to see lines or objects oriented in a vertical direction. They would bump into table legs and furniture as if blind (Hirsch & Spinelli, 1971; Spinelli et al., 1972). In addition, neurons in the visual cortex became unresponsive to vertical stimuli, and instead could only react to objects oriented in a horizontal direction. Moreover, kittens that were provided considerable visual input, but were denied the opportunity to motorically interact with their environment (as they were strapped into a carriage and only allowed to move freely in the dark), were later found to be blind.

Likewise, as demonstrated in human (Freeman & Bradley, 1980) and non-human primates (Casagrande & Joseph, 1978, 1980; Joseph & Casagrande, 1978, 1980), abnormal and deprived experience during early critical developmental periods may even induce perceptual blindness. This blindness is secondary to functional invasion and occupation by competing neuronal cell assemblies which were provided normal input and then took over those neurons that have been insufficiently or adversely stimulated. For example, in non-human primates, if one eye is sutured shut soon after birth and patterned visual input is prevented from reaching target cells in the lateral geniculate nucleus of the thalamus and visual neocortex, target neurons and cellular layers and columns become smaller and functionally suppressed by adjacent cells receiving normal input (Casagrande & Joseph, 1978, 1980; Hubel &Wiesel, 1968). Cortical columns and layers innervated by the deprived eye shrink and those of the experienced eye grow larger, filling in the vacated space. Perceptual functioning becomes exceedingly abnormal and the deprived eye is unable to see objects, places, or smiling faces.

Because developing neurons are experience-expectant when denied that experience, they may cease to function normally, or come to be inhibited by competing neural assemblies (Casagrande & Joseph, 1978, 1980; Freeman & Bradley, 1980; Joseph & Casagrande, 1980). Likewise, if the developing limbic system is denied sufficient emotional and social stimulation, or if the child is subject to abuse and emotional stress, it too may become "blind" and fail to respond or respond in an abnormal fashion to social-emotional nuances even when normal input is later provided. Indeed, children (Ainsworth,et al, 1978; Bowlby, 1982; Koluchova, 1976; Spitz, 1945) and non-human primates and mammals (Harlow & Harlow, 1965a,b; Joseph, 1979; Joseph & Casagrande, 1980; Joseph & Gallagher, 1980) reared under maternally deprived, abusive, or neglectful conditions, suffer severe social, emotional, cognitive, perceptual, and intellectual deficits. These disturbances include diminished curiosity and a lack of exploratory behavior, a reduced ability to anticipate consequences or to inhibit irrelevant or inappropriate behaviors which lead to punishment, and an impaired ability to form loving attachments. Even the ability to function in a sexually normal fashion is disrupted (Langmeier & Matejcek, 1975; Moberg, 1985).

Some severely neglected and deprived children, including those who have been severely sexually abused, fail to shown any sexual drive, or they may repeatedly expose themselves and masturbate, or sexually attack other children (reviewed in Langmeier & Matejcek, 1975). Sexuality often becomes abnormal, because sexuality is associated with the functional integrity of the limbic system, the amygdala, septal nuclei, and hypothalamus in particular. If sexually abused, these structures may become abnormally sexually differentiated and/or develop other neural abnormalities which result in abnormal behavior or traumatic sexualization.

Competition and the cerebral hemispheres. Although this has not yet been well demonstrated (cf. Corballis & Morgan, 1978), for reasons yet unknown the motor cortex of the left hemisphere appears to mature embryologically at a faster rate than the right, as the axons from the pyramidal cells in the deep layers of the cortex (the corticospinal tract) begin to grow (Kertesz & Geschwind, 1971; Yakovlev & Rakic, 1966) and presumably establish functional synaptic connections with subcortical and spinal mechanisms in advance of the right. This suggestion is further supported behaviorally by evidence that indicates that the majority of infants demonstrate right-sided motor activity (e.g., right-sided turning of the body and head) almost four times more often than activity to the left (Saling, 1979; Siqueland & Lipsitt, 1966; Turkewitz, Gordon, & Birch, 1965). However, the left hemisphere is believed to influence if not program a large-portion of left as well as right-sided motor activity (Geschwind, 1965).

Although this may be accomplished via cortico-cortico interactions given that numerous collaterals are given off by the pyramidal tract before decussating in the medulla (Brodal, 1969) and perhaps by the anterior corticospinal tract as it descends before crossing in the spinal cord, it is possible that both left- and right-sided subcortical and spinal motor neurons receive an abundance of left hemisphere pyramidal collaterals due to its competitive developmental advantage in establishing synaptic contact.

It appears that the earlier development of the left hemisphere is not confined to the motor layers of the cortex and its pyramidal-corticospinal tract, but may include the outer sensory-receptive layers as well. Ultimately, synaptic and cellular growth in these receptive layers possibly is reduced, as is the direct acquisition of afferent-sensory processes, due to the maturation and myelination of the dorsal roots, spinal afferents and post-thalamic sensory fibers not coinciding with left hemisphere development. Moreover, (albeit hypothesized) accelerated maturation (which is not reflected in size) may reduce growth and proliferation of the superficial layers in the left hemisphere as well as the eventual formation of its gyri and convolutions in a variety of neocortical areas (cf. N. Kopp, pp. 302-303, in Corballis & Morgan, 1978). Hence, although the superficial layers of the left have begun to develop in advance of the right, the later and protracted rate of development in the right hemisphere (which is in part demonstrated by its capacity to recover and acquire functions of the left subsequent to left-sided cortical damage; see Plasticity & Recovery, this paper) coincides with the later development of sensory and afferent systems, which come to be represented more greatly in these regions. In consequence, not only are a number of neocortical areas larger in the right, but during neonatal development gyral formation in some of these regions quickly surpass in size their counterparts in the left and are thus 'recognizable' somewhat sooner (Chi, Dooling, & Gilles, 1977). However, one also could argue that certain areas in the right actually begin to develop in advance of similar areas in the left, but that the maturational cycle is much more prolonged.

Nevertheless, it appears that the motor neurons and fibers within the left hemisphere (including afferents feeding back from subcortical nuclei) may act to reduce or rather come to partially occupy substrate that would potentially be available for sensory representation. That is, the motor fibers of the left hemisphere may have a competitive advantage over sensory-afferent representation in the left half of the brain due to early maturation of both the left hemisphere and the motor systems. The later developing sensory processes thus are relegated to greater representation in the right, in which they subsequently terminate in greater numbers. A situation thus evolves in which as the sensory fibers and systems develop and the sensory-afferent-dorsal roots mature, there is more available space in the later developing right hemisphere (which has less space committed to motor functions). Presumably it is this unequal representation of receptive capacity and acquisition that predisposes the right hemisphere to demonstrating somewhat earlier and greater receptive activity in the infant and adult (Andreassi, 1980; Crowell, Jones, Kapuniai, & Nakagawa, 1973).

In general, it appears that the motor-sensory dichotomy of hemispheric function may be in part a function of differences in rate of neuronal maturation as well as the availability of space and synaptic connections in the course of development. Variations in the course of these maturational events, in turn, would result in variations in functional representation and, consequently, cognitive functioning. For example, vocal development and sound production correspond to maturational stages in motor development (Lenneberg, 1967) and are related closely to the maturation of the motor roots of the cranial nerves (Lecours, 1975). Accordingly, early maturing individuals perform better on verbal tests, whereas later maturing Ss perform better on tests of spatial ability (Zurif & Carson, 1970), variations which apparently reflect functional representation of sensory and motor processes. Similarly, maturational lag is associated with language and abnormalities (Bakker, 1972; Leong, 1976; Satz & Van Nostrand, 1973), which may be due to the failure to establish sufficient anatomical representation of the necessary motor units during the critical period of motor development. However, it is important to note that sensory feedback is crucial for the development of motor precision in language functions, as well as later behavioral functioning in general.

It must be cautioned that evidence of anatomical asymmetries is based on gross measurements of brain volume. However, it is interesting to note that the planum temporal of the left hemisphere (which contains Wernicke's area) is larger than that of the right in the majority of infants and adults (Geschwind & Levitsky, 1968; Wittelson & Palli, 1973), whereas the right hemisphere weights more than the left (Lemay, 1976)--a function, possibly, of greater abundance of white matter and reciprocal interconnections with sensory and limbic modalities. These anatomical asymmetries are not, however, complete, as there is considerable overlap in acquisition.

NeuroPlasticity and Recovery

The right hemisphere plasticity and immaturity hypothesis is supported by a number of independent investigations, which have shown that if the left hemisphere of a young child is severely damaged, speech functions reappear, presumably mediated by the right hemisphere (Basser, 1962; Byers & Mclean, 1962; Dennis & Whitaker, 1976; Hecaen, 1976; Krashen, 1973; Lenneberg, 1967). Although the ability to structure and produce syntactic components is somewhat defective after recovery (Dennis & Kohn, 1975; Dennis & Whitaker, 1976), it appears that removal of left hemisphere synaptic influences allows the right to acquire belatedly the motor representation necessary for language. For example, in the visual system, if a single eye is given a competitive advantage in development and experience through the suturing shut of the opposing eye, its synaptic representation in the thalamus (lateral geniculate body) and visual cortex is increased, whereas the cells normally innervated by the disadvantaged eye become smaller and/or innervated by the normal eye. When reverse sutures are performed and the disadvantaged eye is opened (for example, 6-9 months after birth), the animals seems blind, and the cells innervated by that eye appear non-functional except in regions where competitive influences are null (the monocular segment). However, if the normal eye is removed surgically, the disadvantaged eye will regain function and vision returns (Casagrande & Joseph, 1978, 1980; Joseph & Casagrande, 1978, 1980).

After cerebral or peripheral damage, it has been demonstrated repeatedly that the lesion is accompanied subsequently by sprouting and regeneration of damaged and undamaged neurons such that vacated sites are reinnervated (Cunningham, 1972; Devor, 1976; Murray & Golderberger, 1974; Raisman, 1969). In mammals, however, central neurons show much less plasticity than peripheral structures (Bernstein, 1967). Nevertheless, two basic types of reconstructive changes may occur: (1) The severed axons will exhibit regenerative sprouting if its cell body remains intact; and (2) intact axons or neurons from surrounding border regions will grow processes and reinnervate a region that has been separated from its afferent input (Goodman & Horel, 1966; Raisman, 1969). These two processes then may compete for the newly available terminal space. However, the stimulus to growth suddenly will cease when the available region is reinnervated (Raisman, 1969). This demonstrates that unoccupied synaptic space acts as a powerful growth stimulus that exerts a strong attractive influence on the severed as well as intact axonal neighbors.

The fact that recovery of language function greatly decreases after age 6 (Hacaen, 1976; Krashen, 1973) indicates a decline in the availability of non-damaged synaptic space. Prior to this, it appears that the deinnervated space in the peripheral structures and deinnervated axons from the sensory regions and motor musculature that subserve expressive and receptive language act as either a possible attractive influence on fibers that originate in the right hemisphere, or regenerate and then successfully compete for synaptic space with the available substrate in the right hemisphere. Because deinnervated motor areas may compete for space in the immature right half of the brain, anatomical representation of motor functions possibly expands due to the increased pressure for motor mediation. On the other hand, with massive damage to the right hemisphere in early infancy, we would expect to find some functional (but incomplete) recovery of sensory function (e.g., spatial ability, etc.) as a greater number of sensory fibers compete for left hemisphere space. Because of this, the representation of both sensory and motor function would be decreased. This is why, for instance, early right hemisphere injury affects speech more than equivalent damage in adults (Basser, 1962; Hecaen, 1976), and why complete removal of a hemisphere in infancy affects cognitive functioning in general (cf. Kohn & Dennis, 1974).

The sensory-limbic Axis

Sensory functions, pleasure, pain, hunger, etc., are associated naturally in that all are essentially emotional and limbic. Even sensations such as light, pressure, etc. come to have limbic and thus emotional characteristics if sufficiently intense. Emotion is, in fact, somatosensory linked and interdependent. The inability to feel is the inability to experience emotion.

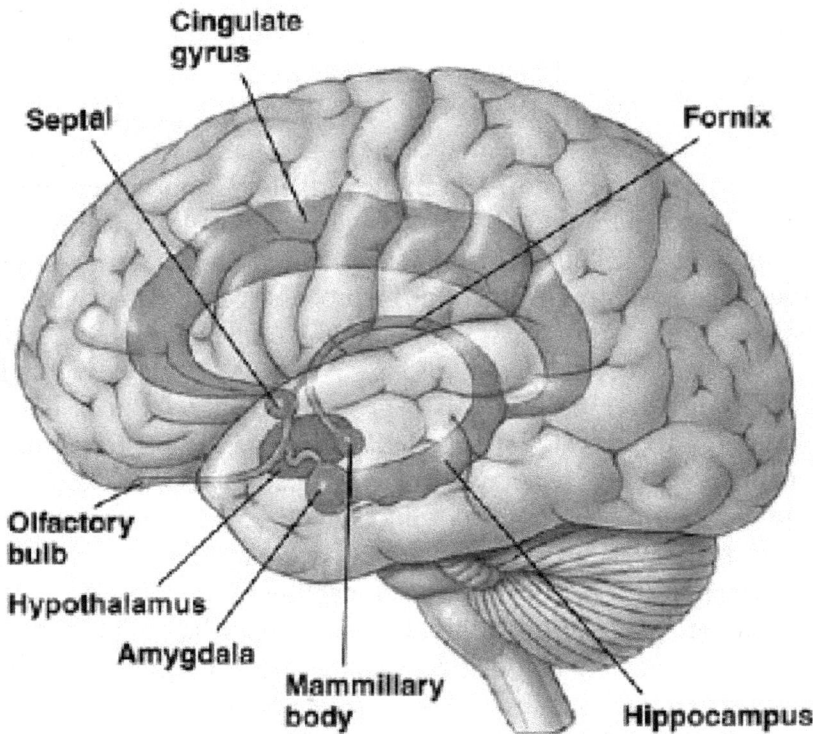

Although the right hemisphere appears to subserve emotional expression, emotionality is most clearly a function of subcortical limbic structures (Gloor, 1060; Green, 1958; Kaada, 1967; Robinson, 1967; Scoville, 1954). These structures, are in fact the major sites for the elicitation of emotional arousal (Egger & Flyn, 1962; Gloor, 1960; Green, 1958; Kaada, 1967; Ursin & Kaada, 1960) and are important mediators of sensory functions because they receive (e.g., the amygdale) projections from all sensory receptors (Gloor, 1960; Machine & Segundo 1956) and provide the neural substrate (the hippocampus) for the registration and storage of imaginal and sensory experience, and presumably memory in general (Green, 1964; Milner, 1974). Moreover, limbic structures, such as the amygdala, in addition to being the possible seat for the control of emotional tone, the enhancement of pleasure, consummatory behavior, fear, sadness, affection, and happiness (cf. Isaacson, 1974, for review) also are involved in the control of aggression (Mark & Ervin, 1970), the inhibition of emotional activity (Penfield, 1954), fight or flight responses (Ursin & Kaada, 1960), and emotional vocalization (Robinson, 1967). In this last regard, it is interesting to note that intense somatosensory stimulation and activation give rise to involuntary and emotionally connotative vocalizations, for example "ouch!"

The Amygdala, Child Development, Deprived Environments, Emotion

The development of wariness, fear, separation anxiety, and the establishment of long term emotional attachments, is mediated by, and parallels the differential rates of growth and maturation of the hypothalamus, amygdala, septal nuclei, cingulate gyrus, as well as the hippocampus.

Consider for example, fear, an emotion associated with the amygdala. The human medial amygdala is exceedingly immature at birth and doesn't become well myelinated until around the 8th postnatal month (Gilles, Leviton, & Dooling, 1983; Langworthy, 1937; Yakovlev & Lecours, 1967).

As is well known, the development of axonal myelin insulation improves neural transmission and is an indicator of functional maturation (see Grafstein, 1963). Myelinated axons do not normally undergo programmed cell death (unpublished observations).

Hence, paralleling the maturation of the medial amygdala, at 8-months, the infant first begins to experience and express feelings of fear, and then becomes progressively more fearful, such as in response to an approaching stranger (Bronson, 1972; Emde, Gaensbauer, & Harmon, 1976; Sroufe & Waters, 1976), adult male strangers in particular with female infants becoming more fearful than males infants.

This initial fearlessness (and amygdala immaturity) is exceedingly adaptive, for the generation of fear would interfere with the initial establishment of intimate emotional attachments, and the infant's need for considerable social-emotional and physical stimulation, which is initially and eagerly welcomed even from strangers. The later development of fear ensures that an infant that can crawl, will not indiscriminately crawl into the arms of a stranger, or dangerous animals, only to be whisked away or eaten.

The hypothalamus, amygdala, as well as the septal nuclei, hippocampus, and anterior cingulate, all begin to mature at different, as well as at overlapping rates, and each subserves a variety of unique as well as overlapping functions. For example, in addition to fear, the amygdala subserves the ability to form emotional attachments, whereas the hippocampus subserves memory and the cingulate is associated with separation anxiety. As these structures develop, they thus require input which may enhance memory or social emotional functioning.

Again, if denied sufficient social, maternal, and emotional experience-expectant stimulation early in development, or if reared in an abnormal, neglectful, or abusive environment, these nuclei, and other forebrain structures (including the neocortex), may atrophy, establish aberrant interconnections, and cease to function normally (Henriksen, Bloom, & McCoy, 1978), as demonstrated in humans (Heath, 1954) and nonhuman subjects (Walsh, Budtz-Olsen, Penny, & Cummins, 1969). In consequence, various or all aspects of social emotional and various aspects of cognitive and perceptual functioning may become abnormal. The brain and the limbic system require considerable stimulation in order to function normally.

For example, when experiencing hunger or thirst, the organism is motivated to seek food or water. Likewise, the experience-expectant limbic system actively seeks social stimulation, and requires emotionally "sensitive" maternal contact in order to develop normally. Hence, for the first 8 months of postnatal development infants will indiscriminately seek contact and will smile at the approach of anyone, even complete strangers (Charlesworth & Kruetzer, 1973; Spitz & Wolf 1946). It appears that the immature "experience-expectant" limbic system in fact demands emotional and social contact, and will seek it out regardless of consequences.

Hence, young animals raised in social isolation will form attachments to bare wire frames, inanimate objects, and television sets, and will seek out animals that might maul them and creatures that might eat them (Cairn, 1966; Fisher, 1955; Harlow & Harlow 1965a,b; Hoffman & DePaulo, 1977). Infant humans, non-human primates, and a variety of young creatures will desperately seek contact with mothers and caretakers who reject, physically abuse, and who nearly kill them. For example, Melzack and Scott (1957) found that dogs who were denied social emotional stimulation for the first 7 to 10 months of life, desperately sought social contact and would hover excitedly next to the experimenter even when he repeatedly stuck pieces of glass or burning matches into their snouts. Similarly, Harlow and Harlow (1965a,b) report that young monkeys would persistently attempt to make maternal contact even when severely bitten and battered. Infant humans are no different.

This "paradoxical" abusive-contact seeking behavior is not entirely surprising as the limbic system appears to be genetically programmed to respond to aversive or threatening conditions by seeking safety and contact comfort from the mother or surrogate caretaker. Hence, the infant is unable to reconcile the contradiction when subject to abuse and cannot inhibit the contact and safety seeking response, as the immature limbic system is not genetically designed to avoid abnormal mothering. On the contrary, the same limbic tissues which perceive and respond to aversive stimulation, e.g. the amygdala, cingulate, and septal nuclei (Olds & Forbes, 1981), are the same nuclei which promote and desperately require social contact and emotional stimulation.

So pervasive is this need for physical interaction and social stimulation, that when grossly reduced or denied, the result may be death. For example, in several well known studies of emotionally neglected but well nourished and otherwise healthy children warehoused in foundling homes during the early 1900's, morbidity rates for those less than one year of age was generally over 70%. Over 10,000 children were admitted to the Dublin Foundling home, and only 45 survived (reviewed in Langmeier & Matejcek, 1975).

Children reared in institutions where mothering and emotional comfort are minimized typically display low intelligence, profoundly reduced verbal skills, extreme passivity, apathy, severe attentional deficits, pathological shyness, and exceedingly bizarre social behavior (Dennis, 1975; Goldfarb 1943, 1945, 1946; Langmeier & Matejcek, 1975; Spitz, 1945, 1946). R.A. Spitz (1945, 1946) found that deterioration is in fact quite rapid, and that within less than a year unmothered children became unresponsive to social stimulation, would lay passively on their beds, and made vigorous attempts to avoid strangers or novel objects or toys. When approached they would scream and withdraw. These deprived youngsters spent hours engaged in obsessive, repetitive, stereotyped, and abusive self-stimulating movements; i.e. head banging, rocking, or repeatedly pinching the same piece of skin until sores developed.

Nor are these catastrophic consequence limited to humans, for similar disturbances characterize the behavior of other primates reared under deprived conditions. As detailed by Harlow and Harlow (1965; p. 138), monkeys raised alone, in a sterile environment, with surrogate terry cloth mothers would stereotypically "sit in their cages and stare fixedly into space, circle their cages in a repetitive stereotyped manner and clasp their heads in their hands or arms and rock for long periods of time. They develop compulsive habits, such as pinching precisely the same patch of skin between the same fingers hundreds of times a day; occasionally such behavior may become punitive and the animal may chew and tear at its body until it bleeds."

Human and non-human primates need to be held, touched, caressed, rocked, and carried about everyday for these are the adaptive evolutionary requirements of our inherited 100 million-year-old mammalian nervous system. For much of human and primate evolution infants were carried about by their mothers who would bend, twist, walk, run, climb, and gather foodstuffs with a baby strapped or clutched tightly to her body. Hence, the adaptive significance of the infant's grasp reflex; infants are provided the capacity to clutch back. And, when mother was not in attendance, the baby would likely be held by a sister, aunt, grandmother, and so on.

However, it must be emphasized that the degree of any subsequent abnormality is dependent on the nature, degree, and extent of the deprivation or abuse suffered. In less extreme case where "mothering" is inconsistent and/or if the child is seldom held or played with, the individual may merely become shy and feel awkward and uncomfortable when around others. This is because intimate physical, social, and emotional contact with a "sensitive" and attentive caretaker is a biological and limbic system necessity which directly contributes to the strength and quality of any emotional attachment and the ability to relate successfully with others (Ainsworth et al., 1978; Bowlby, 1960, 1982). When deprived of this stimulation, or when reared in an abnormal or abusive environment, the limbic system, the neocortex, and striatum, and all aspects of social, emotional, and even sexual functioning will be negatively impacted to varying degrees depending on the degree, duration, and the age at which the infant is neglected and insufficiently stimulated.

Limbic Language

Although non-humans do not have the ability to speak, mammalian vocalizations are also primarily limbic and somatosensorily mediated, being evoked in situations that involved sexual arousal, terror, anger, flight, helplessness, and separation from the primary caretaker when young (cf. Robinson, 1967). Similarly, in human infants the first vocalizations arise in response to somatosensory upheavals. Hence, primate (including human) vocalizations arise at first in a context deeply embedded in limbic activity (Jurgens, 1969; Kaada, 1967; Robinson, 1967), activity that invariably accompanies and/or calls forth involuntary and voluntary action from oneself or others for example, a signal that indicates danger, the cry of pain that accompanies a reflexive withdrawal, or a grunt involuntarily evoked as one engages in strenuous activity.

Similar vocalizations appear to arise from and are mediated by the right hemisphere and are possibly under the control or influence of the orbital regions of the frontal lobes and temporal regions, which, in turn, are outgrowths of limbic nuclei (cf. Figure 1) and which have maintained cytoarchitectural as well as functional similarities to limbic cortex (Kaada, 1967; Maclean, 1954). It is little wonder that with massive left hemisphere lesions that result in loss of expressive and receptive language, the ability to swear, cry, sing, and even pray may remain intact, as well as the capacity to respond to non-linguistic, auditory stimuli (Geschwind et al., 1968).

"Language was originally a system of emotive and imitative sounds--sound which express terror, fear, anger, love, etc., and sound which imitate the noises of the elements: the rushing of water, the rolling of thunder, the roaring of the wind, the cries of the animal world, and so on; and lastly, those which represent a combination of the sound perceived and the emotional reaction to it. --Jung

Apparently, right hemispheric involvement with emotional functioning is due to greater abundance of reciprocal interconnections with the limbic system. As noted, the later maturation of the right hemisphere may coincide with the maturation and myelogenetic cycle of the post-thalamic exteroceptive and proprioceptive fiber tracts (which object from thalamus to cortex), which are not complete until the end of the first year post-partum (Lecours, 1975). In that the dorsomedial nucleus of the thalamus receives major fiber connections from the amygdale, has connections with the temporal and orbital regions (Nauta, 1962), and is involved in vocalization as well as sensory functions, the possibility exists that right hemisphere emotionality is due to the availability and subsequent representation of thalamic-cortical-limbic fibers and nuclei in the later maturing right hemisphere substrate, which comes to subserve and mediate limbic functioning.

NEURODYNAMICS
The Limbic Origins of Psyche: Emotion, Hunger, Thirst, Pleasure, Pain

The Primary processes are present in the mental apparatus from the first, while it is only during the course of life that the secondary processes unfold, and come to inhibit and overlay primary ones. In consequence of the belated appearance of the secondary processes, the core of our being remains inaccessible to understanding. --Freud

Prior to the migration of neuroblasts from the mantle layer into those zones that will become the deep motor layers of the cerebral cortex, the corpus striatum (basal ganglia and other subcortical motor areas) and diencephalons-rhinencephalon (thalamus, hypothalamus, hippocampus, amygdala, and other limbic nuclei) have begun to form and differentiate, and by birth have become functional (Arey, 1965; Hamilton & Mossman, 1972). Because the cortical motor areas do not establish functional control for some time after birth, and because of the later beginning and completion of the myelogenetic cycle of the corticospinal tract (eighth month of gestation to seventh month post-partum, Yakovlev & Lecours, 1967), most motoric output, including the control over reflexes, is mediated by the more advanced subcortical motor centers and spinal cord disappear within several months after birth (Brodal, 1969; Langworthy, 1933; McGraw, 1969; Milner, 1967) and reappear after extensive cortical damage. Thus the newborn's behavior largely reflects subcortical and limbic activities (Milner, 1967). Because the outer sensory and association layers of the cortex still are forming and due to the later completion of the myelogenetic cycle of the post-thalamic sensory fibers that project to these layers, sensory analysis and responding is largely limbic and thalamic.

According to Milner (1967), during the first month after birth the 'sole evidence of neocortical functioning, as well as the infant's reactivity [is] almost entirely [directed] to stimuli impinging directly on the body-surface,' and that 'sensations transmitted by the skin senses (including the mouth) provide the entirety of the neonate's experience; that is, the neonate is essentially internally oriented' (pp. 179-180). Hence, psychic functioning in the newborn is characterized by a vague, undifferentiated awareness, and a multitude of excitatory and inhibitory interactions and a series of transient feeling states and emotional upheavals that lack psychological referents or distinct features. Rather, the referents of feeling states, or limbic activations, are general and diffuse, being aimed at the reactivation of experience associated with the alleviation of displeasure or with the experience of pleasurable or painful affect. In fact, from birth to 1 month, the infant displays only two attitudes, accepting and rejecting (Milner, 1967). Because the limbic system monitors and responds to internal and external sensations (e.g., hunger, pain, etc.) with only minimal inhibitory regulation by the newborn's neocortex, most if not all 'behavior' is governed by brain stem and limbic concerns, including that of the first (and later) vocalizations: limbic speech. Hence, in response to hunger, the infant cries, and to pleasure it gurgles and babbles.

Visual Imagery. Although, admittedly, we have no direct knowledge as to psychic interactions in the neonate, it does seem reasonable to assume that as the cortex and underlying structures and fiber pathways mature, neural 'programs' are formed. That is, fiber pathways that are repetitively fired, deactivated or activated in response to sensory and affective activity, become associated with that activity, and when appropriately triggered subsequently can replay this 'learned' pattern (Penfield, 1954, Spinelli, 1970). With the repetition of affective states and accompanying motor outputs and sensory impression, certain external and internal states, actions, movements, stimuli, etc., become endowed with 'significance' and are retained as neural 'programs' and patterns of neural activity that may be revoked or 'triggered' by associated activity or stimuli. A form of 'learning,' so to speak, has occurred. When replayed, the organism presumably reexperiences to some degree the sensations, emotions, etc., originally associated and repetitively experienced. In regard to the neonate and infant, it is presumably these mechanisms that first give rise to dreamlike ideation and imagery (cf. Freud, 1900).

Thus, for example, as the infant experiences hunger and stomach contractions as well as its cries of displeasure, these states become associated with the sound, smell, taste, etc., of mother and her associated movements and other stimuli that accompany being fed (cf. Piaget, 1952, pp. 37, 407-408). Repetitively experienced, the sequence from hunger to satiety evoked and becomes associated with the activation of certain neural pathways. Eventually, when the infant becomes hungry, if prolonged there is the possibility that the entire neural sequence associated may become activated and an 'image' experienced (manifested behaviorally through sucking and tongue movements in the absence of food; cf. Piaget, 1952). The organism experiences the experience of being fed. In that these sensory-motor-limbic associations make up the bulk of the neonate's experience, these rudimentary representations (images) may in themselves be indistinguishable from experience simply because in themselves they are experience. Thus the young organism, as yet unable to distinguish between representation and reality, may respond to the image as reality (cf. Freud, 1900). That is, for example, when a neural sequence is activated such as due to prolonged hunger, the infant responds to the neural 'experience' with associated neuronal activity (i.e., being fed), and for a brief period the neonate reacts as though its hunger were being satiated. Reality is replaced by an image, or rather a 'dream.' However, the organism is not long fooled, for as the pain of hunger remains and increases, limbic activity is increased, and the image falls away to be replaced by a cry of hunger and perceptual surveillance.

As the child ages and the cortex matures, the failure of the limbic induced sensory-motor neural patterns to satisfy, relieve, or achieve the ends for which they were activated impels the organism to pay renewed and ever-increasing attention to the external environment. Eventually, aided greatly by the maturation of the motor centers in the left hemisphere, the infant is able to attend selectively and to distinguish partially between reality and image and to create representations of reality in a motor context, which may be manipulated to effect the changes or satisfy the impulses desired. For example, the infant purposefully may cry or call 'mamma.'

It is the establishment and elaboration of these rudimentary patterns of neural activity that eventually provide the foundations for psychological and cognitive development.

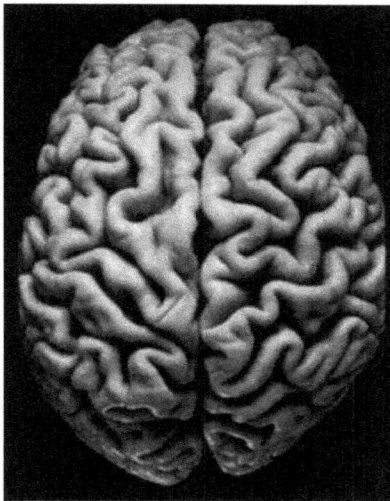

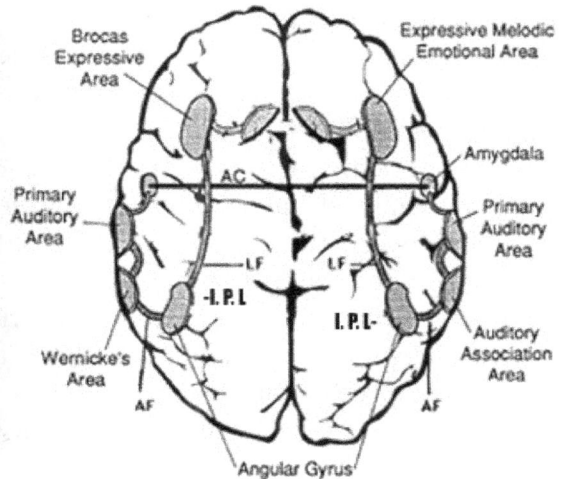

Limbic Language and Egocentric Speech

Limbic language heralds the founding drive from which all purposeful and intellectual activities develop. Language springs from roots buried within the limbic system (cf. Jung, 1954). Limbic speech, however, is basically concerned with the expression of feelings, moods, impulses, desires, etc., which serve as commands and accompaniments to action, or may be represented subtly in tone or inflection. Nevertheless, limbic speech is not bound up with thinking, the expression of thought, or reflection, although symbolism and imagery may be inherent in its production. Limbic speech, although communicative and thus social, may occur independent of thought yet may be evoked by thought, or may provide the undifferentiated matrix from which thought at times originates. It is, however, primarily emotional, automatic, and yet symbolic, as a command or accompaniment to action or desire. Crying may bring forth food or other forms of somatosensory relief, babbling accompanies pleasure or social contact, or may result as a form of self-simulation, and 'mamma' brings forth mamma and all her accompanying attributes.

Initially, however, cries, babbles, or, for example, the word 'mamma' do not signify the infant's feeling states, desires, etc. Rather, these are limbic induced motoric responses to the evocation of a diffuse feeling that merely signifies itself. 'Mamma' thus can mean 'mamma come here,' 'mama give me,' 'mama I hurt,' 'I am hungry,' etc. (Piaget, 1952; Vygotsky, 1962). Although without precise definition, once vocalized it remains a means of communication. The earliest forms of communication and thus social speech, therefore are embedded in limbic activity, for limbic speech provides a context within which associations may be formed and schemas developed. Language slowly develops from the construction and association of these schemas and vocalization-experience pairings. But at first, language is only a servant to the limbic system and represents limbic needs of the organism in the form of motor output that carries beyond the crib to which the infant is confined, crossing boundaries that later will be traversed by legs and body as immediate calls to action or an immediate exercise in motor functioning.

As the left hemisphere continues to mature and develop, wresting control of the peripheral and cranial musculature from subcortical influences, a second aspect of language emerges, one that arises through interactions and nominal associations with external stimulatory activities and denotative speech. Although it springs forth from relationships originally limbic based, denotative (or social) language is concerned with nominal functions and denotative statements of fact or belief and statements of assertion. As such, denotative social language is closely bound with cognitive activities and the eventual expression of one's thoughts. Thinking, however, does not appear until much later in development. Moreover, although a semi-independent motor function, it remains influenced by social-limbic language throughout life.

Language, like all other motor functions, thus has dual origins, resulting from sensory-limbic stimulation and commands and accompaniments to action, as well as the need for stimulation and activation of its own neuronal channels to maintain function and synaptic integrity. Language is thus a fusion of motor and sensory-limbic activities, and an outgrowth of two biological processes that characterize all organismic existence. Language arises from affective interactions within a structure provided by the motor apparatus, the initial expression of which are decidedly emotional and thus limbic (cf. Vygotsky, 1962), and the elaboration of which is sustained by the power to affect, manipulate, and successfully communicate with the environment. Eventually, by age 3, a third form of language beings to appear, which in turn will give rise to thought: Egocentric speech.

Egocentric speech. According to Vygotsky (1962), egocentric speech makes its first appearance at approximately 3 years of age. According to Piaget (1952, 1962, 1974), at its peak, egocentric speech comprises almost 40-50% of the preoperational child's language; the remainder consists of social speech (denotative and emotional). Prior to this development, communication is directed strictly toward outside sources. There is no attempt to communicate with the self, for there is no internal dialogue because thought has not yet developed. At age 3, egocentric speech the peculiar linguistic structure from which thought will arise appears in the context of social-denotative vocalizations (Vygotsky, 1962). It is an essentially self-directed form of communication, which heralds the first attempts at self-explanation. It is essentially speech for oneself (Piaget, 1962; Vygotsky, 1962).

While the child is engaging in egocentric speech, he does not appear concerned with the listening needs of his audience simply because to all appearances his words are meant for his ears alone (Piaget, 1952, 1962, 1974; Vygotsky, 1962); he seems to be thinking out loud (Vygotsky, 1962). When engaged in an egocentric monologue there is no interest in influencing or explaining to others what in fact is being explained, and the child will keep up a running accompaniment to his actions, commenting on his behavior in an explanatory fashion, even while alone. Moreover, while engaged in this self-directed external monologue the child appears oblivious to the responses of others to his statements (Piaget, 1962; Vygotsky, 1962).

While the child is engaging in egocentric speech, he does not appear concerned with the listening needs of his audience simply because to all appearances, his words are meant for his ears alone (Piaget, 1952, 1962, 1974; Vygotsky, 1962); he seems to be thinking out loud (Vygotsky, 1962). When engaged in an egocentric monologue there is no interest in influencing or explaining to others what in fact is being explained, and the child will keep up a running accompaniment to his actions, commenting on his behavior in an explanatory fashion, even while alone. Moreover, while engaged in this self-directed external monologue the child appears oblivious to the responses of others to his statements (Piaget, 1962; Vygotksy, 1962).

Egocentric speech, although accompanying behavior, initially does not parallel action. Rather, it appears after the action has occurred (Piaget, 1952, 1962, 1974; Vygotsky, 1956). It is only as egocentric speech becomes progressively internalized that the child's comments and explanations occur earlier in the sequence of expression, until finally the child begins to explain his actions before they are performed.

Egocentric speech presents us with a curious anomaly, for we must accept that the child knows what he has done without commenting or explaining his actions; moreover, he must know hwy he has performed certain actions without need of explaining them to himself. Nevertheless, the fact that he explains and comments upon his behaviors after they occur argues otherwise. Paradoxically, the child acts as both actor and witness, explainer and explained to. Clearly, the child explains his actions to himself (Vygotksy, 1962).

According to Vygotksy (1962), after its initial appearance and elaboration, egocentric speech becomes steadily more involuted and internalized and thus progressively more covert. At its overt maximum, its traits and structures are being internalized and strengthened and so comprise a greater portion of the child's cognitive activities than may be witnessed. Egocentric speech, therefore, never diminishes, but continues to increase until it becomes completely inner speech at approximately age 7. The child then has learned to think words as well as to speak them, and to think them in a temporal and organized sequence that retains its original and primary function of self-communication.

Egocentric speech and thought. The essential feature of the external components of egocentric speech is that it is based on stimuli and actions that occur outside the immediate sphere of understanding and experience, for initially it is evoked after the observation of behavior, which indicates that access to the impulses to behave is only external. As the child ages, however, the egocentric commentary occurs progressively sooner. For example, the child will first comment or explain a picture that she has drawn after it is finished, whereas at a later stage she will state what a picture is while she is drawing it, until finally the child will announce what she will draw and the draws it (Vygotksy, 1962). Essentially, as the child ages she appears to receive advanced warning of her intentions, until finally the information is available before rather than after she acts. At this stage, whoever, egocentric speech is almost completely internalized as thought.

As we know, language and its comprehension are subserved by the left hemisphere in both children and adults. Although egocentric speech contains limbic and thus right hemisphere components, it is also primarily left hemispheric because it is an elaboration of basic left hemisphere motor functions, the linguistic manipulation and organization of stimuli and sensations. That egocentric speech appears initially only after the action has occurred, indicates that the left hemisphere of the young child is responding to impulses and actions initiated outside its immediate realm of experience and understanding. It seems that the left hemisphere in the production of egocentric speech is attempting to organize the results of its experience (the observation of behavior)

into a meaningful sequence, which it then linguistically communicates to itself; experiences within which it participates as a passive observer of impulses
to action initiated elsewhere.

Interhemispheric Communication. The appearances of egocentric speech,
its elaboration and eventual internalization occurs in response to several maturational changes in the central nervous system. Interestingly, although both
hemispheres in the course of development increase their interactive influence
upon the subcortex and periphery, communication between the hemispheres
is exceedingly poor or nonexistent prior to age 3 (cf. Salamy, 1978) and very
limited until approximately 5 (Galin et al 1977, 1979; Gallanger & Joseph,
1982; Kraft, 1977; O'Leary, 1980; Joseph et al., Note 1). Presumably, this is a
function of the immaturity of the corpus callosal fiber connections between the
hemispheres, which do not become completely myelinated until the end of the
first decade (Yakovlev & Lecours, 1967). In fact, as recently demonstrated by
Joseph and colleagues (Note 1), communication is so poor that when pictorial
stimuli are presented tachistosopically to the right hemisphere, young children
(age 4) will respond when questioned with large information gaps, which they
erroneously fill with confabulatory responding. It thus appears that the left
hemisphere of a young child has at best incomplete knowledge of the contents
and activity that occur in the right.

Indeed, in that the right hemisphere retains access to the motor systems, the
necessity for as well as the peculiar self-explanatory nature of egocentric commentary becomes understandable. The left hemisphere is in fact explaining
behavior and impulses to action initiated by the right; information that the left
hemisphere does not have direct access to due to the immaturity of the corpus
callosum (Joseph et al., 1984).

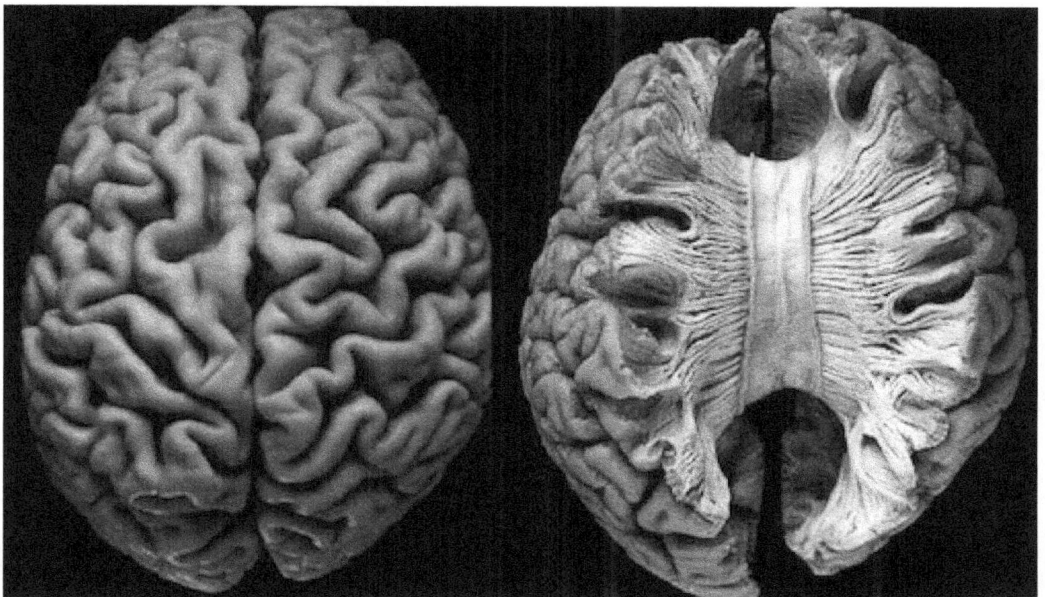

"At one moment therefore, the infants acts impulsively, unrelated to dangers of reality and uninfluenced by them; he may attack a loved person or destroy a toy, his anger moves easily from one person or cause to another. On the other hand, an understanding and regard for the consequences of actions, a piece of reasoning, may appear intermittently as representations of higher ego activity, and interfere with the infant's free expression" (Anna Freud, 1951; p. 23).

Egocentric speech, even when fully internalized as thought, remains essentially a result, an intermediary between impulse and comprehension. As a bridge it is a form of transformation and organization, a result of associations and labels becoming linked so that impulses and stimuli may become interpreted and organized for explanation. It is through this intermediary that the left hemisphere achieves understanding, and eventually a greater control over behavior.

Essentially, egocentric speech is a function of the left hemisphere's attempt to organize and make sense of behavior initiated by the right half of the brain. Because interhemispheric communication is at best grossly incomplete, the left utilizes language to explain to itself the behavior in which it observes itself to be engaged. As the commisures mature and information flow within and between the hemispheres increases, the left also acts to organize linguistically its internal experience. As the organism develops, interhemispheric information exchange is increased (Joseph et al., Note 1) and the language axis (Broca's, Wernicke's, and the angular gyrus regions) increasingly acts to organize (as well as inhibit) sensory-limbic right hemispheric transmission and initiated behaviors, organizing these impulses as it organizes impulses that originate in the left cortex, so that they may be efficiently and motorically carried out. Rather than passively observing the sensory-limbic actions as they occur in the environment, as the commisures become complete, the left hemisphere now actively engages in the formulation oh behavior, achieving understanding prior to its occurrence. Essentially, commisural transmission allows the left hemisphere access to right hemisphere impulses-to-action before the action occurs rather than forcing it to make sense of the behavior after its completion.

Nevertheless, it is important to note that the resulting organization that we have defined as thought (Joseph, 1980) need not be solely linguistic, buy may consist of imaginal, spatial, pictorial or other components (cf. De Groot, 1965). In addition, as noted elsewhere (Joseph, 1980) understanding is not dependent on thought.

These semi-animals, happily adapted to the wilderness, to war, free roaming, and adventure, were forced to change their nature. Suddenly they found all their instincts devalued, unhinged. They must walk on legs and carry themselves where before the water carried them: A terrible heaviness weighed upon them, they felt inapt for the simplest manipulations, for in this new, unknown world they could no longer count on the guidance of their unconscious drives. They were forced to think, deduce, calculate, weigh cause and effect unhappy people, reduced to their weakest, most fallible organ, their consciousness. ---Nietzsche

The Language Axis and the Construction of Thought

Although a multitude of structures and fiber pathways are involved in the final formulation of egocentric speech, thought, and language, their appearance is dependent on and a product of the language axis and a multiple series of interactions between thalamic nuclei, the angular gyrus, Broca's and Wernicke's area. Although they will not be discussed, frontal-reticular interactions necessarily are involved in the excitatory and inhibitory processes that allow for selective acquisition and rejection of various stimuli and informational sources (cf. Como, Joseph, Fiducia, & Siegel, 1979; Joseph, Forrest, Fiducia, Como, & Siegel, 1981; Luria, 1973). In addition, the actual formation of thought appears to retain in some manner a functional linkage with peripheral motor structures. For example, while engaging in logical explicit forms of linguistic thought, both manual and oral regions of the body musculature become activated as measured by EMG (McGuigan, 1978). In part, this is probably due to the activation of Broca's area, which is immediately adjacent to the supplementary motor areas that subserve the lower facial region.

Thalamus. Thalamic nuclei presumably act in an intermediary fashion in the formulation of thought, aiding in the temporal analysis and categorization of information during the initial stages of thought construction. As a possible intermediary, the thalamus, via its rich interconnections with all sensory cortices (Truax & Carpenter, 1969), may act to access various sources of information and to perform some initial integrations. If indeed this formulation is correct, thalamic structures such as the pulvinar (cf. Brown, 1975; Fazio, Sacco, & Bugiani, 1973; Penfield & Roberts, 1959; Van Buren & Borke, 1969) and ventrolateral nucleus (cf. Fazio et al., 1973; Van Buren & Borke, 1969) probably play an important role in the regulation, differentiation, and selective attention to various sources of information. Moreover, the thalamus may act to coordinate the activities of the cortical language areas (Penfield & Roberts, 1959) at least in the initial stages of information selection.

Broca's area. Broca's speech area is literally the final pathway by which thought, emotion, and other impulses come to be assimilated and organized as motoric articulations for verbal communication and thought formation. Located at the foot of the third operculum in the secondary motor association area of the left and right lateral surfaces of the frontal lobes. Ot is via this structure that impulses are transmitted to the adjacent motor neurons that subserve speech and the facial area (tongue, jaw, throat, lips), programming the musculature for the production of spoken language. Although the left motor association cortex

(Broca's region) receives an abundance of information via the arcuate fasciculus transmitted from the posterior speech zones (Wernicke's and the angular gyrus), due to the terminal arborization of all sensory regions in the frontal areas and the extensive connections maintained with the limbic cortex (Jones & Powell, 1970), these structures also have access to a variety of divergent forms of information and may play a role in the inhibition of stimuli or sensations deemed unnecessary or undesirable (Como et al., 1979; Luria, 1973). That is, decision-making may occur in the frontal regions (cf. Luria, 1973).

Wernicke's area . Wernicke's area lies in the secondary auditory association cortex in the posterior portions of the temporal lobe between the auditory cortex in the posterior portions of the temporal lobe between the auditory cortex and the angular gyrus, and shares fiber connections with these regions as well as with Broca's area (Geschwind, 1965, 1979). It appears that the underlying phonemic structure of all propositional thought and speech arises in this region. In conjunction with Broca's and the angular gyrus, this structure also plays an interdependent role in the formulation of programs for specific utterances because it supplies information to both poles of the language axis. Wernicke's region also plays a role in the decoding, labeling, and analysis of speech, aiding in comprehension of both written and oral language (Geschwind, 1965, 1979; Luria, 1973). Hence, when a sound is heard it is received in the auditory cortex of the temporal lobe, which transmits the impulse to Wernicke's area, where it is encoded and then transferred to other brain regions for associational analysis and comprehension. Similarly, in the organization of thought, the original signal and associations are transmitted to this area, where further encoding and labeling are performed, and the underlying structure is reinforced.

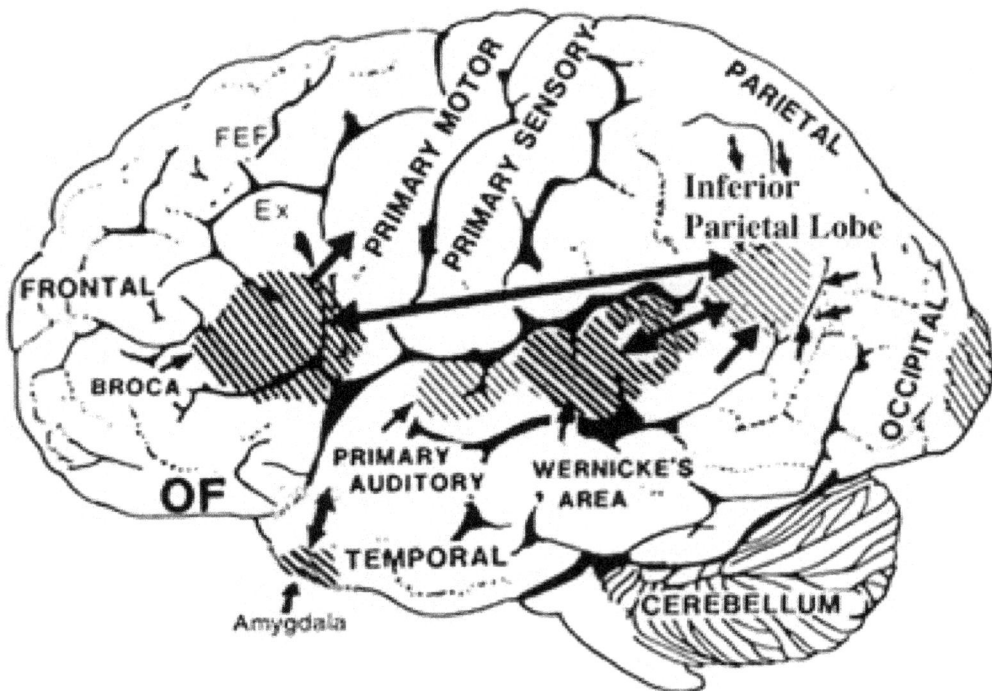

Angular gyrus. The angular gyrus within the inferior parietal lobe, like the frontal lobes, is involved in the assimilation of diverse information variables, their integration, the calling-up of relevant associations, and function as a necessary intermediary for all conscious functioning, particularly in the development and comprehension of language and thought. Situated at the junction of the primary projection areas for vision, hearing, and somesthesis, and maintaining rich interconnections with the integration areas of the thalamus, this region is involved in the association of various stimuli, the assimilation of associations (Critchley, 1953; Geschwind, 1965, 1979; Luria, 1973; Truax & Carpenter, 1969), and the calling up of further information as an aid to left hemispheric comprehension. Through its involvement in the construction of cross-modal associations, the angular gyrus increases the capacity for the organization, categorization, and labeling of sensory-motor events. Intermodal association thus allows for motoric representation and sequential enactment as well as for more abstract representation of experience. Consequently, memory capacity is increased due to the multiple inter-classifications of associational inputs.

When activated via information transfer from other cortical/subcortical regions, the angular gyrus supplies a visual and auditory code that enables more precise linguistic labeling via Wernicke's area. When a word is read, the pattern of visual input is transmitted to the thalamus, inferior temporal region, and the angular gyrus where analyses are performed. The information then is partially integrated and transmitted to other associational structures for further processing; all of which is then transmitted to the frontal and angular region for further associational and motoric integration and temporal sequencing. Conversely, when a word is heard or a question is posed, the information is transferred to the auditory cortex and its associational areas, where the message is stabilized, prolonged, and its temporal features and phonemic components analyzed and properly labeled. The message then is sent to diverse regions, such as the angular gyrus, where cross-modal associations such as visual, somasthetic and other sensory-motor concomitants are aroused, integrated, organized, assimilated, and finally comprehended. The next stage then is initiated and the answer is sought out through the activation of closely linked associations. These are then organized, labeled, comprehended and the reply stated.

Neurodynamics and Disconnection Syndromes

As we are now well aware, the developing organism is extremely vulnerable to early environmental influences, such that both nervous system and behavior may be dramatically altered as manifested in the adult (Casagrande & Joseph, 1978, 1980; Greenough, 1976; Joseph & Casagrande, 1978, 1980; Joseph & Gallagher, 1980; Joseph, Hess & Birecree, 1978; Langmeir & Matejcek, 1975; Rosenzweig et al., 1972). In that the emerging human organism is asymmetrically arranged, with apparently little interaction and informational exchange between the cerebral hemispheres, the effects of early 'socializing' experiences could have potentially profound effects indeed. As a good deal of this early experiences is likely to have its unpleasant if not traumatic moments, it is fascinating to consider the latter ramifications of the early emotional learning occurring in the right hemisphere unbeknownst to the left; learning and associated emotional responding which may later be completely inaccessible to the language centers of the brain even when extensive interhemispheric transfer is possible (cf. Risse & Gazzaniga, 1978, for evidence that bears directly on this phenomenon). In this regard, the curious asymmetrical arrangement of function and maturation may well predispose the developing organism to later come upon situations in which it finds itself responding emotionally, nervously, anxiously, or 'neurotically' without linguistic knowledge, or without even the possibility of linguistic understanding as to the cause, purpose, eliciting stimulus, or origin of it behavior. Instead, like the egocentric child, the individual may be faced with behavior that he may explain only after it occurs: 'I don't know what came over me.'

Similar manifestations, however, may occur that are unrelated to early environmental nonlinguistic emotional learning, but result from inhibitory actions at the level of the cortex which prevent the language axis from receiving information as to the cause of the organism's behavior, or even that an impulse, desire, behavior, has emerged or occurred.

Isolation of the speech area. In children, the language centers not only are limited in their access to processes occurring in the right hemisphere, but also are not fully linked with all nervous centers in the left half of the brain (Lecours, 1975; Joseph et al., Note 1). In part, this is due to callosal immaturity and the fact that the cortex has not developed fully (Arey, 1965; Milner, 1967). The language axis of the child, therefore, may appear to be in many respects isolated from sources of information that occur in the remaining neural-psychic apparatus and thus may appear 'ignorant' when questioned about these processes (Gallagher & Joseph, 1982; Joseph et al., 1984). Functionally, in normal adults, a similar manifestation of speech area isolation at times may be evident.

As has been demonstrated in brain-injured individuals (Geschwind et al., 1968), the speech area may become completely isolated from the surrounding cortical mantle, and, if intact, form a functional unit. That is, the individual may retain the ability to repeat questions and sentences, although propositional speech is absent and linguistic comprehension of internal and external events is no longer possible. Thus, although communication between Wernicke's, Broca's and the angular gyrus is maintained, associations from other brain regions cannot reach the speech center. This is in fact called 'isolation of the speech area' (Geschwind et al., 1968). In the case described by Geschwind et al. (1968), the patient had suffered massive cortical damage, but surprisingly, although the surrounding tissue was destroyed, the speech area was completely intact. The examiners noted when the patient regained 'consciousness' that 'she sang songs and repeated questions. On several occasions when the examiner said, 'ask me no questions' she would reply 'I'll tell you no lies,' or when told, 'close your eyes' she might say 'go to sleep.' An even more striking phenomenon was observed early in the patient's illness. The patient would sing along with songs or musical commercials' (pp. 343-346). Nevertheless, she appeared to be wholly without comprehension.

Although complete isolation of the speech area may result only following severe injury to the brain, indeed, it does not require a skilled neurologist to recognize as thought their language axis does not have access to processes that occur elsewhere. Such a condition may occur when an individual would rather not be 'conscious' of knowledge he possesses and thus refuses to attend to or linguistically organize it (Joseph, 1980). Presumably, all access channels between the speech area and the neural pathways between modalities are shut down, which isolates the speech areas and the neural pathways between modalities are shut down, which isolates the speech area from its knowledge source. Not surprisingly, when questioned, especially in regard to embarrassing, threatening, or otherwise undesirable information, the language apparatus indeed has no knowledge of these processes and will respond accordingly. In such instances we in fact seem to be conversing with a self-maintained closed circuit that has established or resulted from the establishment of massive inhibitory interactions. Thus, isolated, an aroused or sought-after source of information is prevented from arriving at the language axis or becoming integrated with the major stream of activity through inhibition of the neuronal circuitry that subserves its transference to and expression by the speech centers. Such interactions, however, should not be presumed as indicative of unconscious processes.

Unless we accept the notion that the inhibition of this information transfer is automatic, such that the neuronal field creates its own inhibitory surround so as to prevent transmission, there is involved a necessary acknowledgement of the presence of this 'undesirable' material in order to inhibit its expression. As such, the isolation cannot be complete, and there must be at least some information transmission and cross talk between brain areas so that inhibition may be initiated and maintained via inhibitory feedback. Because the isolation cannot be complete (barring brain injury thus causing an anatomical separation), there must be some information leakage, which, although incomplete, may appear in fragmentary form within the language axis. How much appears would of course be a function of its own stimulus strength and the strength of inhibition.

Stimulus Anchors and the Train of Thought

In most instances in which an area of the brain is activated via internal or external probes, a train of material closely associated with the input stimuli is co-jointly, concurrently, and subsequently aroused in response. For example, if one is asked: 'What did you do in school today?' a number of associations are aroused and integrated in response to the stimulus input, all of which bear in some manner on each element of the eliciting stimulus. Finally, in the process of associational linkage, those with the strongest stimulus value, that is, those that most closely match the question in terms of appropriateness and thus with the highest probability of being the relevant, desirable, and required response, or those with the strongest stimulus strength or arousal value (Jung, 1954), rapidly take a place in a hierarchical arrangement that is being organized in a form suitable for expressing a reply (cf. Frijuda, 1972; Luria, 1973; Reitman, Grove, & Shoup, 1964). Hence, associated material of a lower probability, although aroused, nevertheless is dismissed quickly or deactivated due to its low associative value, undesirability, as well as its inappropriateness in the rapidly developing response hierarchy (Frijda, 1972; Jung, 1956).

In regard to the above question, the stimuli 'What, did, you do, in school' and associated contextual cues (Koffka, 1935) become anchors around which a train of highly associated material (Ach, 1951) is aroused, arranged, individually matched and group matched (cf. Frijda, 1972). The anchored associations that then match all sources of relevant input with sufficient value of probability act as templates of excitation that stimulate and attract other relevant associations, which then are assimilated and associated or subsequently deactivated due to their low probability in contrast to the associations already organized. Moreover, because the train of closely linked associations change in correspondence to the developing hierarchy, subsequently aroused and assimilated associations may come to have a lower probability of association within the matrix of overall activity and may be deactivated (cf. Reitman et al., 1964). The final product of this hierarchical arrangement of mutually determining associational linkages is the train of thought--a coherently organized linear sequence of interlinked associations.

The train of thought is a result and a reaction to internal and external stimulus input. It is a result of an attempt to make sense of a wide variety of information variables; the linkage of associations for the purposes of labeling, categorizing, and thus understanding (Ach, 1951; Joseph, 1980). However, what is understood is not the initiating stimulus (at least with regard to impulses that originate internally), but the resulting organization of associations. Comprehension is based on an interpretation, an interpretation which, in many instances, may be related only loosely to the original signal, impulse, need, or desire. Moreover, if the information or signal is not fully amenable to linguistic or visual labeling, or is only weakly or incompletely available for interpretation or associational linkage (due, for example, to its partial inhibition), then the speech area, having to select from a flood of probable but not completely valid associated probabilities, will organize and express those that seem to be most likely or which have the greatest strength of arousal or association. In such instances, the signal is incompletely and thus incorrectly interpreted (depending on the degradation of the signal). As such, the individual may respond with confabulation, the formation of symptoms, or other forms of disturbance. Hence, the individual may grossly mislabel his feelings, impulses, desires, as well as the 'causes' of his behavior.

For example, let us take a stimulus that elicits a strong (lateralized) emotional reaction (e.g., intense jealousy). If the context is entirely inappropriate for its expression or realization (due to personality structure, upbringing, self-concept, situational determinants, etc.) inhibition will occur at some level of the system such that the fiber pathways that lead from the region of excitation to the language axis of the left hemisphere are functionally suppressed or deactivated. In such an instance, although fully represented in the non-linguistic portions of the neural-psychic apparatus, the linguistic-motor aspect of consciousness would in fact be ignorant of this upheaval, its significance, or the stimuli through which it was elicited. If we were to ask the left hemisphere if it is 'angry,' 'sexually aroused,' 'fearful,' 'jealous,' etc., it would truthfully respond 'no.' However, as noted by the behavior of the child who engages in egocentric speech, although non-labeled or inhibited from reaching the language axis, the source of arousal still may find expression or at least influence behavior or cognition in subtle ways. Hence, the left hemisphere taking note of these subtle effects (e.g., sweaty hands, tense muscles, fast heartbeat, etc.), may instead not the presence of some 'queer' feeling, call forth possible associations, and interpret accordingly (e.g., hungry, nervous, in need of exercise, etc.).

On the other hand, the signal or impulse may be completely intact, yet the individual may not have the correct associations available for interpretation. A similar circumstance may occur with questions that originate externally. The individual then may respond with a 'tip of the tongue' experience. Usually, however, if the correct associations are not available, several related associations may be formulated subsequently in response. Although not completely appropriate, if these associations are of sufficient associational probability and are in turn supported by other high probability associations such that all are mutually reinforcing, the train of linkage will become 'conscious' and even though erroneous, possibly vocalized and believed. The individual then would be making a guess or drawing an erroneous conclusion. In most instances, however, incorrect possibilities are deactivated, or, once expressed, discarded or corrected in response to disconfirming evidence. Of course, in some individuals an incorrect statement is maintained or upheld regardless of the availability of the correct or disconfirming evidence. These individuals are called liars.

Alpha and Omega. It would be quite erroneous to assume that the language axis of the 'normal,' 'well adjusted' brain has unrestricted access to all functional knowledge sources, for the language axis has only a limited capacity and can access and accurately report only a small bit of the information amenable to linguistic coding. Moreover, not all impulses, feelings, desires, fears, cravings, knowledge, etc., have a label and thus are not readily translatable into the codes that give rise to thought and language (Joseph, 1980). It seems that emotional expression in particular is easily amenable to misinterpretation, misidentification, and misunderstanding. The same seems to be true of most cognitive processes, however, for as Neisser (1967), Nisbett and Wilson (1977) have pointed out, most individuals have no understanding of what is involved and on what basis evaluations, judgments, problem-solving strategies, and reasons for initiating certain behaviors were arrived at. This is largely due, however, to the dependence on language for, simply, the language axis and the sequential, organizational processes which subserve consciousness in the form of thought have no direct access to the basis of these 'processes' in part because they result from them.

"We knowers are unknown to ourselves, and for good reason: how can we ever hope to find what we have never looked for? There is a sound adage which runs: 'Where a man's treasure lies, there lies his heart.' Our treasure lies in the beehives of our knowledge. We are perpetually on our way thither, being by nature winged insects and honey gatherers of the mind. The only thing that lies close to our heart is the desire to bring something home to the hive. The sad truth is that we don't understand our own substance, we must mistake ourselves; the axiom, 'Each man is farthest from himself,' we will for us to all eternity. Of ourselves we are not knowers." --Nietzsche

Self and Consciousness

One's Self, the source of thought, and the Whom to whom one's thoughts are presented and explained, should not be viewed as localized in the language centers of the brain, or, for that matter, the left hemisphere. Unfortunately, the Self is most often and most readily identified with the expressible components of individuality, whereas right hemispheric, non-linguistic, non-explicit, subcortical and emotional aspects are not. This is absurd, of course, for although we are able by power of our marvelous abstraction and labeling abilities to fragment the Self into identifiable components, the fabric from which the Self and consciousness arise includes the whole person: right, left, subcortical and peripheral, as well as the environment in which one exists. As such, the Self and consciousness are a continuous process of interrelationships that defy precise localization.

5. The Limbic System: Emotion, Laterality, and Unconscious Mind

Many people experience emotion as a potentially overwhelming force that warrants and yet resists control - as something irrational that can happen to you ("you make me so angry" - "I'm madly in love"). Perhaps in part, this schism between the rational and the emotional is attributable to the raw energy of emotion having its source in the nuclei of the ancient limbic lobe - what some have referred to as the reptilian brain, a series of nuclei that first made their phylogenetic appearance long before man walked upon this earth. Although over the course of evolution a new brain (neocortex) has developed, we remain creatures of emotion. We have not completely emerged from the phylogenetic swamps of our original psychic existence. The old limbic brain has not been replaced.

Buried within the depths of the cerebrum are several large aggregates of limbic neurons that are preeminent in the mediation and expression of emotional, motivational, sexual, and social behavior and that control and monitor internal homeostasis and basic needs, such as hunger and thirst. These regions include the hypothalamus, amygdala, hippocampus, septal nuclei, anterior and posterior cingulate, various thalamic nuclei, portions of the reticular activating system, the orbital frontal lobes, certain nuclei of the cerebellum, and other structures that together form the limbic system.

Of specific concern in this chapter are the hypothalamus, amygdala, hippocampus, and septal nuclei, the social - emotional and psychic functions they mediate, and the neural circuitry that supports their activity. Limbic laterality, temporal lobe epilepsy, hallucinations and memory, and select aspects of psychological and unconscious development, including the pleasure principle and primary process, are also briefly discussed.

AFFECTIVE ORIGINS: OLFACTION AND SOMESTHESIS

Emotionality serves a protective function, either to promote survival of the individual (e.g., fight or flight) of that of the species (e.g., sexual activity). The first and most primitive manifestations of emotion are elicited in response to olfactory (e.g., pheromones and other externally secreted chemical messengers) and tactile sensory stimulation. These primitive emotions are expressed as withdrawal (fear) or approach reactions to pain or threat and are elicited in response to motivational needs, such as the seeking of a sexual partner or for the procurement of food (prey) and nourishment (Graeber, 1980; Michael & Keverne, 1974; Savage, 1980; Wilson, 1962).

Olfaction was originally crucial to evolutionary and phylogenetic development, as it informed the organism about the environment at a distance, without the necessity of physical contact. By means of olfactory cues and the detection of pheromones, the organism is able to detect and track food, determine the intent and social - emotional status of conspecies, as well as signal its own intent, motivation, social position, and/or sexual availability (Michael & Keverne, 1974; Wilson, 1962).

For example, consider pheromones, substances secreted by the skin through specialized glands or found in urine and feces. Chemical communication through pheromones is used by moths, social insects, dogs, cats, and primates, as well as amphibian, sharks, and reptiles. Although detection of pheromones among insects is often accomplished through specialized chemoreceptors located on various parts of the body (Wilson, 1962), mammals and pirates rely on olfactory receptors within the nostrils that transmit this information to the olfactory bulb (or lobe in some species) and the telencephalon (Graeber, 1980; Savage, 1980).

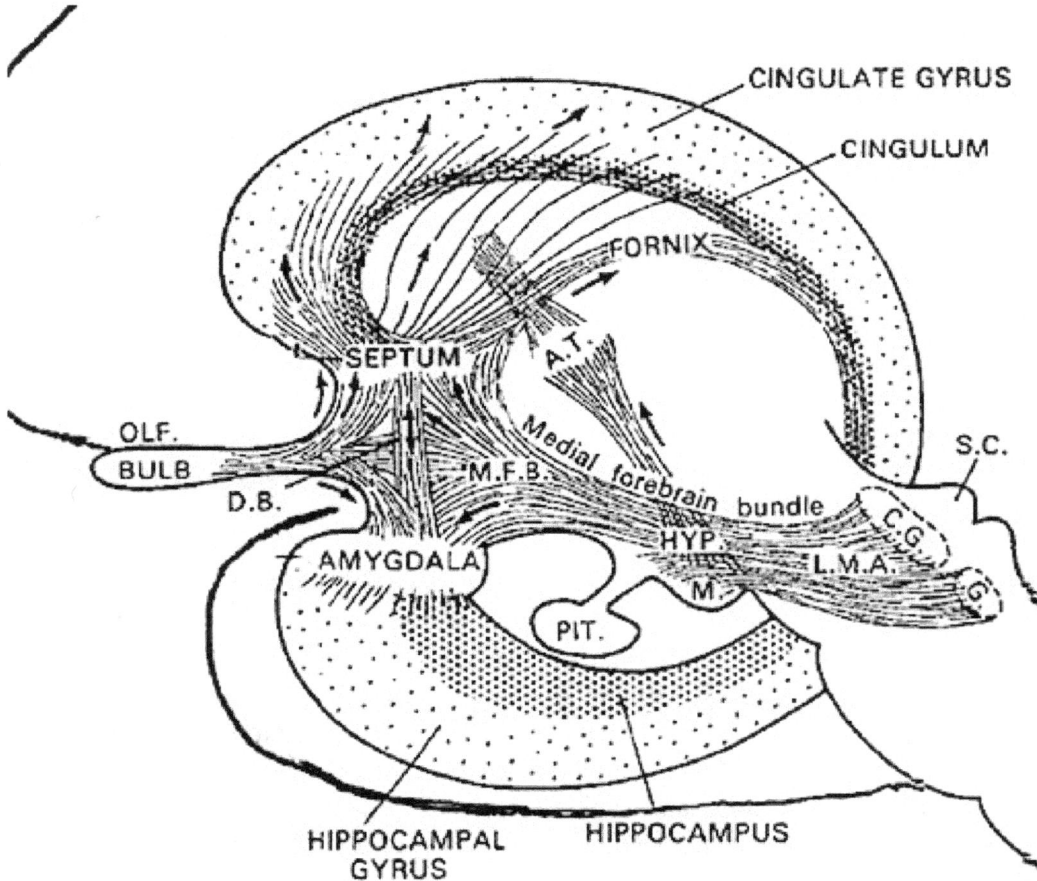

Among mammals, olfactory cues are also used for marking possessions and one's territory. For example, dogs will urinate on trees and bushes, whereas a stallion might urinate on the feces of his mare. Prosimian primates will actually urinate on their hands, rub the secretions over their body, and thus mark everything they come into contact with. It has also been shown that male primates, as well as other mammals, rendered anosmic (through the cutting of the olfactory nerves) completely lose interest in sexually available females because of the inability to detect olfactory (pheromone) cues indicating sexual readiness (Michael & Kaverne, 1974). Thus, olfaction is a crucial mode of communication; among some species, the olfactory bulb is so highly developed that it is considered a lobe of the brain.

Among humans olfaction appears to have lost its leading role in signaling motivationally significant information. However, olfactory cues are still employed for indicating sexuality (e.g., perfume), and odor exerts a powerful influence on what is considered socially acceptable, hence the abundance of artificial chemicals designed to eliminate various body fragrances. Moreover, one need only suffer a severe cold to appreciate the dominant function of smell in the ability to detect and appreciate fully the flavor of food and thereby experience pleasure in eating (Fig 19). Odors also affect learning and memory and have the capability of triggering vivid recollections of some far away and past event.

Ontogenetically, although influenced by olfactory cues, humans first experience or express emotionally in relationship to the body and in response to tactile sensations or rapid changes in position (Emde & Koenig, 1969; Spitz & Wolf, 1946). Pain is also first experienced in relationship to the body, and is somesthetically rooted, although some have argued that pain is not an emotion.

Among human infants, the earliest smiles are induced through tactile stimulation (e.g., light stroking or even blowing on the skin), whereas loss of support is the most powerful stimulus for triggering an emotional reaction in the newborn (Emde & Koenig, 1969; Sptiz & Wolf, 1946). The earliest and most consistent manifestation of emotion in the infant consists of screaming and crying, whereas positive affect is limited to an attitude of acceptance and quiescence (Sptiz & Wolf, 1946), emotions first mediated by the hypothalamus that may or may not be true emotions at all.

In their journey from the external to the internal environment, olfactory and tactile input are transmitted to various limbic nuclei, such as the lateral hypothalamus and entorhinal area of the hippocampus (olfactory only) and the amygdala (olfaction and somesthesis). Indeed, it is through nuclei such as the amygdala (as opposed to the hypothalamus) that the first true (or rather, felt) aspects of emotion appear to be generated. It is also because of the tremendous input of olfactory information to various limbic nuclei that this part of the brain was at one time referred to as the rhinencephalon, literally "nose brain."

HYPOTHALAMUS

The hypothalamus is a very ancient structure. Unlike most other brain regions, it has maintained a striking similarity in structure throughout phylogeny and apparently over the course of evolution (Crosby, DeJonge, & Schneider, 1966). Located in the most medial aspect of the brain, along the walls and floor of the third ventricle, this nucleus is fully functional at birth and is the central core from which all emotions derive their motive force.

The hypothalamus is highly involved in all aspects of endocrine, hormonal, visceral, and autonomic functions; it mediates or exerts controlling influences on eating, drinking, the experience of pleasure, rage, and aversion. The hypothalamus is also sexually dimorphic; i.e., both structurally and functionally, the hypothalamus of men and women is sexually dissimilar.

Sexual Dimorphism in the Hypothalamus

As is well known, sexual differentiation is strongly influenced by the presence or absence of gonadal steroid hormones during certain critical periods of prenatal development in many species, including humans. However, not only are the external genitalia and other physical features sexually differentiated, but certain regions of the brain have also been found to be sexually dimorphic and differentially sensitive to steroids, particularly the preoptic area and ventromedial nucleus of the hypothalamus, as well as the amygdala (Bleier, Byne, & Siggelkow, 1982; Dorner, 1976; Gorski, Gordon, Shyrne, & Southam, 1978; Rainbow, Parsons, & McEwen, 1982; Raisman & Field, 1971, 1973). Specifically, the presence or absence of the male hormone testosterone, during this critical neonatal period, directly affects and determines the pattern of interconnections between the amygdala and hypothalamus and between axons and dendrites in these nuclei, and thereby the organization of specific neural circuits. In the absence of testosterone, the female pattern of neuronal development occurs.

For example, if the testes are removed before differentiation, or if a chemical blocker of testosterone is administered, preventing this hormone from reaching target cells in the limbic system, not only does the female pattern of neuronal development occur, but makes so treated behave and process information in a manner similar to that of females (e.g., Joseph, Hess, & Birecree, 1978); i.e., they develop female brains. Conversely, if females are administered testosterone during this critical period, the male pattern of differentiation and behavior results.

That the preoptic and other hypothalamic regions are sexually dimorphic is not surprising, in that it has long been known that this area is critical in controlling the basal output of gonadotropins in females before ovulation and is heavily involved in mediating cyclic changes in hormone levels, such as follicle-stimulating estrogen, hormone (FSH), luteinizing hormone (LH), and progesterone. Chemical and electrical stimulation of the preoptic and ventromedial thalamic nuclei also triggers sexual behavior and even sexual posturing in females and males (Lisk, 1967, 1971).

In primates, electrical simulation of the preoptic area increases sexual behavior in males and significantly increases the frequency of erections, copulations, and ejaculations, we well as pelvic thrusting followed by an explosive discharge of semen even in the absence of a mate (Maclean, 1973). Conversely, lesions to the preoptic and posterior hypothalamus eliminates male sexual behavior and results in gonadal atrophy.

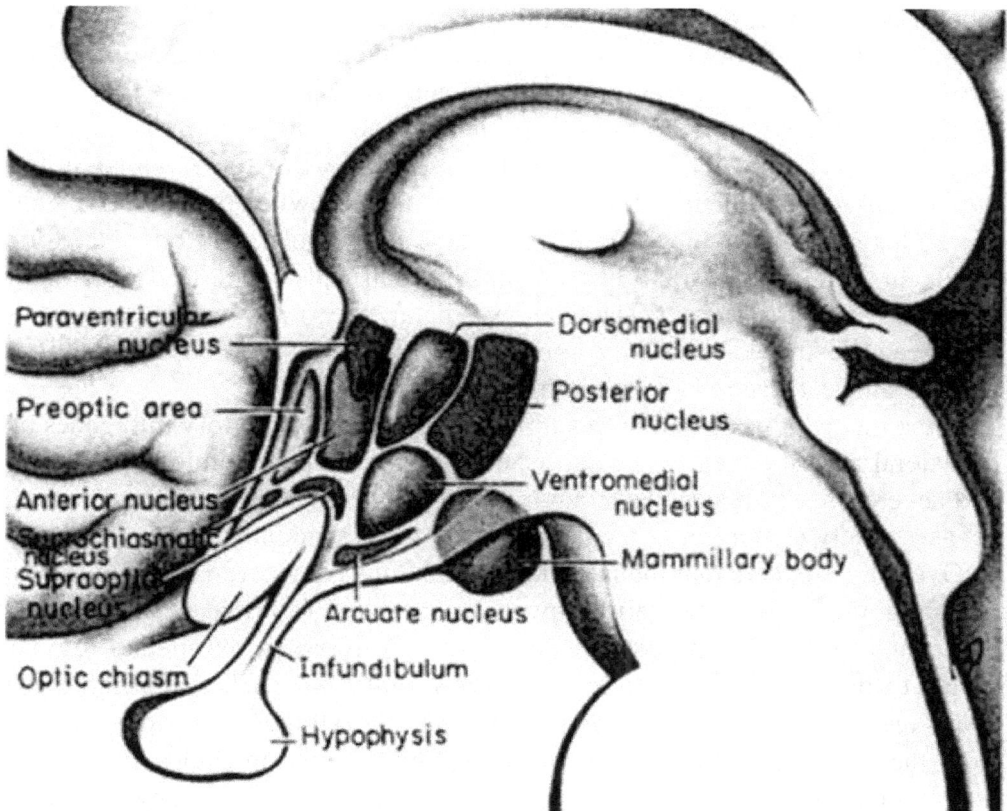

Lateral and Ventromedial Hypothalamic Nuclei

Although consisting of several nuclear subgroups, the lateral and medial (ventromedial) hypothalamic nuclei play particularly important roles in the control of the autonomic nervous system, the experience of pleasure and aversion, eating and drinking, and raw (undirected) emotionality. These nuclei also appear to share a somewhat antagonistic relationship (fig. 20).

For example, the medial hypothalamus controls parasympathetic activities (e.g., reduction in heart rate, increased peripheral circulation) and exerts a dampening effect on certain forms of emotional/motivational arousal. The lateral hypothalamus mediates sympathetic activity (increasing heart rate, elevating blood pressure) and is involved in controlling the metabolic and somatic correlates of heightened emotionality. In this regard, the lateral and medial region act to exert counterbalancing influences on each other.

Hunger and Thirst

The lateral and medial region are highly involved in monitoring internal homeostasis and motivating the organism to respond to internal needs such as hunger and thirst. For example, both nuclei appear to contain receptors that are sensitive to the body's fat content (lipostatic receptors) and to circulating metabolites (e.g., glucose), which together indicate the need for food and nourishment. The lateral hypothalamus also appears to contain osmoreceptors (Joynt, 1966) that determine whether water intake should be altered.

Electrophysiologically , it has been determined that the hypothalamus not only becomes highly active immediately before and while the organism is eating or drinking, but the lateral region alters its activity when the subject is hungry and is simply looking at food (Hamburg, 1971; Rolls, Burton, and Mora, 1976). In fact, if the lateral hypothalamus is electrically stimulated, a compulsion to eat and drink results (Delgado & Anand, 1953). Conversely, if the lateral area is destroyed bilaterally, aphagia and adipsia are so severe that animals will die unless force fed (Teitelbaum & Epstein, 1962).

If the medial hypothalamus is surgically destroyed, inhibitory influences on the lateral region appear to be abolished such that hypothalamic hyperphagia and severe obesity result (Teitelbaum, 1961). Thus, the medial area seems to act as a satiety center but as a center that can be overridden.

Overall, it appears that the lateral hypothalamus is involved in the initiation of eating and acts to maintain a lower weight limit such that when the limit is reached the organism is stimulated to eat. Conversely, the medial regions seems to be involved in setting a higher weight limit such that when these levels are approached, it triggers the cessation of eating. In part, these nuclei exert these differential influences on eating and drinking by means of motivational/emotional influences they exert on other brain nuclei (e.g., by reward or punishment).

Pleasure and Reward

In 1952, Heath reported what was then considered remarkable. Electrical stimulation near the septal nuclei elicited feelings of pleasure in human subjects: "I have a glowing feeling. I feel good!" Subsequently, Olds and Milner (1954) reported that rats would tirelessly perform operants to receive electrical stimulation in this same region and concluded that stimulation "has an effect that is apparently equivalent to that of a conventional primary reward." Even hungry animals would demonstrate a preference for self-stimulation over food.

Feelings of pleasure (as demonstrated by self-stimulation) have been obtained following excitation to a number of diverse limbic areas, including the olfactory bulbs, amygdala, hippocampus, cingulated, substantia nigra (a major source of dopamine), locus coeruleus (a major source of norepinephrine), raphe nucleus (serotonin), caudate, putamen, thalamus, reticular formation, medial forebrain bundle, and orbital frontal lobes (Brady, 1960; Lilly, 1960; Olds & Forbes, 1981; Stein & Ray, 1959; Waraczynski & Stellar, 1987).

In mapping the brain for positive loci for self-stimulation, Olds (1956) found that the medial forebrain bundle (MFB) was a major pathway that supported this activity. Although the MFB interconnects the hippocampus, hypothalamus, septum, amygdala, and orbital frontal lobes (areas that give rise to self-stimulation), Olds discovered that in its course up to the lateral hypothalamus, reward sites become more densely packed. Moreover, the greatest area of concentration and the highest rates of self-stimulatory activity were found to occur not in the MFB but in the lateral hypothalamus (Olds, 1956; Olds & Forbes, 1981). Indeed, animals "would continue to stimulate as rapidly as possible until physical fatigue forced them to slow or to sleep" (Olds, 1956).

Electrophysiological studies of a single lateral hypothalamic neurons have also indicated that these cells become highly active in response to rewarding food items (Nakamura & Ono, 1986). In fact, many of these cells will become aroused by neutral stimuli repeatedly associated with rewards such as a cue tone - even in the absence of the actual reward (Nakamura & Ono, 1986; Nishino, Sasaki, Fukuda, & Muramoto, 1980). However, this ability to form associations appears to be secondary to amygdaloid and other source of input (Fukuda, Ono, & Nakamura, 1987).

Nevertheless, if the lateral region is destroyed, the experience of pleasure and emotional responsiveness is almost completely attenuated. For example, in primates, faces become blank and expressionless, whereas of the lesion is unilateral, marked neglect and indifference regarding all sensory events occurring on the contralateral side occur (Marshall & Teitelbaum, 1974). Animals will in fact cease to eat and will die.

Aversion

In contrast to the lateral hypothalamus and its involvement in pleasurable self-stimulation, activation of the medial hypothalamus is apparently so aversive that subjects will work to reduce it (Olds & Forbes, 1981). Thus, electrical stimulation of the medial region leads to behavior that terminates the stimulation - apparently so as to obtain relief (e.g., active avoidance). In this regard, when considering behavior such as eating, it might be postulated that when the upper weight limits (or nutritional requirements) are met, the medial region becomes activated, which in turn leads to behavior (e.g., cessation of eating) that terminates its activation.

It is possible, however, that medial hypothalamic activity may also lead to a state of quiescence such that the organism is motivated simply to cease to respond. In some instances, this quiescent state may be physiologically neutral, whereas in other situations the result may be highly aversive. Nevertheless, quiescence is associated with parasympathetic activity, which is also mediated by the medial area.

Hypothalamic Damage and Emotional Incontinence: Laughter and Rage

When electrically stimulated, the hypothalamus responds by triggering two seemingly oppositional feeling states: pleasure and unpleasure/aversion. The generation of these emotional reactions in turn influences the organism to respond so as to increase or decrease what is being experienced. Through its rich interconnections with other limbic regions, including the neocortex and frontal lobes, the hypothalamus is able to mobilize and motivate the organism either to cease or to continue to behave. Nevertheless, at the level of the hypothalamus, the emotional states elicited are very primitive, diffuse, undirected and unrefined. The organism feels pleasure in general, or aversion/unpleasure in general. Higher-order emotional reactions (e.g., desire, love, hate) require the involvement of other limbic regions as well as neocortical participation.

Emotional functioning at the level of the hypothalamus is not only quite limited and primitive, it is also largely reflexive. For example, when induced by stimulation, the moment the electrical stimulus is turned off, the emotion elicited is immediately abolished. By contrast, true emotions (which require other limbic interactions) are not simply turned on or off but can last from minutes to hours to days and weeks before completely dissipating.

Nevertheless, in humans, disturbances of hypothalamic functioning (e.g., caused by an irritating lesion such as tumor) can give rise to seemingly complex, higher-order behavioral - emotional reactions, such as pathological uncontrollable laughter and crying. However, in some cases, when patients are questioned, they may deny having any feelings that correspond to emotion displayed (Davison & Kelman, 1939; Ironside, 1956; Martin, 1950). In part, these reactions are sometimes due to disinhibitory release of brainstem structures involved in respiration, whereas in other instances the resulting behavior is caused by hypothalamic triggering of other limbic nuclei.

Uncontrolled Laughter

Pathological laughter has frequently been reported to occur with hypophyseal and midline tumors involving the hypothalamus, aneurysm in this vicinity, hemorrhage, astrocytoma, or papilloma of the third ventricle (resulting in hypothalamic compression), as well as surgical manipulation of this nucleus (Davison & Kelman, 1939; Dott, 1938; Foerster & Gagel, 1933; Martin, 1950; Money & Hosta, 1967; Ironside, 1956; List, Downman, & Bagheiv, 1958).

For example, Martin (1950) describes a man who, while "attending his mother's funeral was seized at the graveside with an attack of uncontrollable laughter which embarrassed and distressed him considerably" (p. 455). Although this particular attack dissipated, it was soon accompanied by several further fits of laughter and he died soon thereafter. At postmortem examination, a large ruptured aneurysm was found compressing the mamillary bodies and hypothalamus.

In a similar case (Anderson, 1936, cited by Martin, 1950), a patient literally died laughing following eruption of the posterior communicating artery that resulted in compression (by hemorrhage) of the hypothalamus. "She was shaken by laughter and could not stop: short expirations followed each other in spasms, without the patient being able to make an adequate inspiration of air, she became cyanosed and nothing could stop the spasm of laughter which eventually became noiseless and little more than a grimace. After 24 hours of profound coma, she died."

Because laughter in these instances has not been accompanied by corresponding feeling states, this pseudoemotional condition has been referred to as sham mirth (Martin, 1950). However, in some cases, abnormal stimulation in this region (such as due to compression effects from neoplasm) has triggered corresponding emotions and behaviors - presumably due to activation of other limbic nuclei.

For example, laughter has been noted to occur with hilarious or obscene speech - usually as a prelude to stupor or death - in cases in which tumor has infiltrated the hypothalamus (Ironside, 1956). In several instances, it has been reported by one group of neurosurgeons (Foerster & Gagel, 1933) that while swabbing the blood from the floor of the third ventricle, patients "became lively, talkative, joking, and whistling each time the infundibular region of the hypothalamus was manipulated." In one case, the patient became excited and began to sing.

Hypothalamic Rage

Stimulation of the lateral hypothalamus can induce extremes in emotionality, including intense attacks of rage accompanied by biting and attack upon any moving object (Flynn, Edwards, & Bandler, 1971; Gunne & Lewander, 1966; Wasman & Flynn, 1962). If this nucleus is destroyed, aggressive and attack behavior is abolished (Karli & Vergness, 1969).

Thus, the lateral hypothalamus is responsible for rage and for aggressive behavior. The lateral hypothalamus maintains an oppositional relationship with the medial hypothalamus. Thus, stimulation of the medial region counters the lateral area such that rage reactions are reduced or eliminated (Ingram, 1952; Wheatley, 1944), whereas if the medial is destroyed. There results lateral hypothalamic release and the triggering of extreme savagery.

In man, inflammation, neoplasm, and compression of the hypothalamus have also been noted to give rise to rage attacks (Pilleri & Poeck, 1965), and surgical manipulations or tumors within the hypothalamus have been observed to elicit manic and ragelike outbursts (Alpers, 1940). These appear to be release phenomenon, however. That is, rage, attack, aggressive, and related behaviors associated with the hypothalamus appears to be under the inhibitory influence of higher order limbic nuclei such as the amygdala and septum (Siegel & Skog, 1970). When the controlling pathways between these areas are damaged (i.e., disconnection) these behaviors are sometimes elicited.

For example, Pilleri and Poeck (1965) described a man with severe damage throughout the cerebrum, including the amygdala, hippocampus, and cingulate, but with complete sparing of the hypothalamus, who continually reacted with howling, growling, and baring of teeth in response to noise or a slight touch or if approached. Hence, the hypothalamus, being released, responds reflexively in an aggressive nonspecific manner to any stimulus. Lesions of the frontal - hypothalamic pathways have been noted to result in severe rage reactions as well (Fulton & Ingraham, 1929; Kennard, 1945).

Nevertheless, like sham mirth, rage reactions elicited in response to direct electrical activation of the hypothalamus immediately and completely dissipate when the stimulation is removed. These outbursts have been referred to as sham rage.

Lateralization

Although scant, there is some evidence to suggest that the right hypothalamus may be more heavily involved in the control of neuroendocrine functioning, particularly in females. Greater right hypothalamic concentration of substances such as LHRH (luteinizing hormone) has also been reported (see Gerendai, 1984, for review).

Psychic Manifestations of Hypothalamic Acitivity

Phylogenetically and from an evolutionary perspective, the appearance and development of the hypothalamus predate the emergence and differentiation of all other limbic nuclei, e.g., amygdala, septal nucleus, and hippocampus (Andy & Stephen, 1961; Brown, 1983; Herrick, 1925; Humphrey, 1972). It constitutes the most primitive, archaic, reflexive, and purely biological aspect of the psyche.

Biologically, the hypothalamus serves the body tissues by attempting to maintain internal homeostasis and by providing for the immediate discharge of tenions in an almost reflexive manner. Studies of lateral and medial hypothalamic functioning indicate that it appears to act reflexively, in an almost on/off manner so as to seek or maintain the experience of pleasure and escape or avoid unpleasant noxious conditions.

Emotions elicited by the hypothalamus are largely undirected, short-lived, and unconnected with events occurring within the external environment, being triggered reflexively and without concern or understanding regarding consequences. Direct contact with the real world is quite limited and almost entirely indirect as the hypothalamus is largely concerned with the internal environment of the organism. It has no sense of morals, danger, values, logic, and so forth, and can neither feel nor express love or hate. Although quite powerful, hypothalamic emotions are largely undifferentiated, consisting of such feelings as pleasure, unpleasure, aversion, rage, hunger, and thirst.

As the hypothalamus is concerned with the internal environment, much of its activity occurs outside conscious-awareness. Moreover, being involved in maintaining internal homeostasis, through, for example, its ability to reward or punish the organism with feelings of pleasure or aversion, it tends to serve what Freud (1911) described as the pleasure principle.

The Pleasure Principle

The lateral and medial nuclei exert counterbalancing influences that serve to modulate activity occurring in the other. As described by Freud (1911), the pleasure principle not only serves to maximize pleasant experiences, but acts to keep the psyche as a whole free from high levels of excitation (be they pleasurable or unpleasant).

Like the hypothalamus, the pleasure principle is present from birth and for some time thereafter the search for pleasure is manifested in an unrestricted manner and with great intensity, as there are no oppositional forces (except those between the lateral and medial regions) to counter it's strivings. Indeed, higher order limbic nuclei have yet to mature.

Functionally isolated, the hypothalamus at birth has no way of reducing tension or of mobilizing the organism for any form of effective action. It is helpless. When tensions associated with immediate needs (e.g., hunger or thirst) become unpleasant the only response available to the hypothalamus is to cry and make ragelike vocalizations. When satiated, the hypothalamus can only respond with a feeling state suggesting pleasure or at least quiescence. Indeed, as is well known, for the first few months of life, the infant's awareness largely consists of a very restricted matrix involving tactile, visceral (hunger), and kinesthetic sensations, whereas emotionally the infant is capable of screaming, crying, or demonstrating very rudimentary features of pleasure, i.e., an attitude of acceptance of quiescence (McGraw, 1969; Milner, 1967; Piaget, 1952; Spitz & Wolf, 1946). It is only with the further differentiation and maturation of higher-order limbic nuclei (e.g., amygdala, septal nucleus, hippocampus) that the infant begins to achieve some awareness of external reality and begins to form memories as well as differentiate and associate externally occurring events and individuals.

AMYGDALA

In contrast to the primitive hypothalamus, the more recently developed amygdala is preeminent in the control and mediation of all higher order emotional and motivational activities. Through its rich interconnections with various neocortical and subcortical regions, amygdaloid neurons are able to monitor and abstract from the sensory array stimuli that are of motivational significance to the organism (Steklis & Kling, 1985). This includes the ability to discern and express even subtle social - emotional nuances such as friendliness, fear, love, affection, distrust, and anger, and at a more basic level, determine whether something might be good to eat. In fact, amygdaloid neurons respond selectively to the flavor of certain preferred foods, as well as to the sight or sound of something that might be especially desirable to eat (Fukuda et al., 1987; O'Keefe & Bouma, 1969; Ono et al., 1980).

Single amygdaloid neurons receive considerable topographic input, and are predominantly polymodal, responding to a variety of stimuli from different modalities simultaneously (O'Keefe & Bouma, 1969; Perryman, Kling, & Lloyd, 1987; Sawa & Delgado, 1963; Schutze, Knuepfer, Eismann, Stumpf, & Stock, 1987; Turner, Mishkin, & Knapp, 1980; Ursin & Kaasa, 1960; Van Hoesen, 1981). The amygdala is also very sensitive to somesthetic input and physical contact such that even a slight touch in a very circumscribed area of the body can produce amygdaloid excitation. Overall, in addition to emotional and motivational functioning, because multimodal assimilation of various sensory impressions occurs in this region, it is also involved in attention, learning, and memory.

Medial and Lateral Amygdaloid Nuclei

The amygdala is buried within the depths of the anterior - inferior temporal lobe and consists of two major nuclear groups. These are a phylogenetically ancient anteromedial group (or medial amygdala) involved in olfaction and motor activity, as well as a relatively newer basolateral division (lateral amygdala) that first appears in primates (Herrick, 1925; Humphrey, 1972). Like the lateral and medial hypothalamus, these two amygdaloid nuclei subserve different functions and maintain different anatomical interconnections.

Emybryologically, the medial amygdala is the first portion of the basal ganglia striatal complex to appear during development, being formed through neuroblast migration from the epithilium of the lateral ventricle (Humphrey, 1972). As it circles in an arc from the frontal to temporal lobe, the tail of the caudate nucleus actually terminates and merges with the amygdala. This portion of the amygdala is in fact part of the basal ganglia and is heavily involved in motivating and coordinating gross, or whole-body, motor activity.

The medial amygdala receives fibers from the olfactory tract and through a rope of fibers called the stria terminalis, projects directly to and receives fibers from the medial hypothalamus (through which it exerts inhibitory influences) as well as the septal nucleus (Carlsen, De Olmos, & Heimer, 1982; Gloor,

1955; Russchen, 1982; Swanson & Cowan, 1979). In addition, the medial (and lateral) regions are rich in cells containing enkephalins, and opiate receptors can be found throughout the amygdala (Atweh & Kuhar, 1977; Uhl, Kudar, & Snyder, 1978).

Sexuality

Portions of the hypothalamus and amygdala are sexually dimorphic; i.e., there are male and female amygdaloid nuclei. Thus, at a very basic and primitive level, emotional and motivational perceptual/behavioral functioning becomes influenced and guided by the neuroanatomical sexual bias of the host. Electrical stimulation of the amygdala, the medial division in particular, results in sex-related behavior and activity. In females this includes ovulation, uterine contractions and lactogenetic responses, and in males penile erections (Robinson & Mishkin, 1968; Shealy & Peele, 1957).

Damage to the amygdala bilaterally often results in heightened and indiscriminant sexual activity. For example, primates and other animals (while in captivity) will engage in excessive masturbation and genital manipulation and will repeatedly attempt to copulate even with species other than their own (e.g., a cat with a dog, a dog with a turtle) regardless of their sex. With bilateral destruction, animals are not only overly active sexually but are able to identify appropriate partners (Brown & Schaefer, 1888; Kluver & Bucy, 1939). This abnormality is one aspect of a complex of symptoms sometimes referred to as the Kluver - Bucy syndrome (to be discussed in more detail below).

Lateral Amygdala

With the evolutionary ascent of primates, the relatively new lateral division of the amygdala progressively expands and differentiates. The lateral amygdala contributes fibers to the stria terminalis and gives rise to the amygdalofugal pathway, through which it projects to the lateral and medial hypothalamus (upon which it exerts inhibitory and excitatory influences, respectively), the dorsal medial thalamus (which is involved in memory, attention, and arousal), olfactory tubercle, as well as other subcortical regions (Aggleton, Burton, & Passingham, 1980; Carlsen, et al., 1982; Dreifuss et al., 1968; Gloor, 1955, 1960; Klinger & Gloor, 1960; Mehler, 1980; Russchen, 1982). It also receives fibers from the medial forebrain bundle, which in turn has its site of origin in the lateral hypothalamus (Mehler, 1980).

In general, whereas the medial amygdala is highly involved in motor, olfactory and sexual functioning, the lateral division is intimately involved in all aspects of higher order emotional activity; hence its rich interconnections with the lateral and medial hypothalamus, and the neocortex. The lateral amygdala maintains rich interconnections with the inferior, middle, and superior temporal lobes, as well as the insular temporal region, which in turn allows it to sample and influence the auditory, somesthetic, and visual information being received and processed in these areas, as well as to scrutinize this information for motivational and emotional significance (Herzog & Van Hoesen, 1976; Kling et al., 1987; Machne & Segundo, 1956; Mesulam & Mufson, 1982; O'Keefe & Bouma, 1969; Steklis & Kling, 1985; Turner et al., 1980; Van Hoesen, 1981).

Gustatory and respiratory sense are also rerepresented in this vicinity (Fukuda et al., 1987; Maclean, 1949; Ono et al., 1980), and the lateral division maintains rich interconnections with cingulated gyrus, orbital frontal lobes (Pandya, Van Hoesen, Domeskick, 1973), and the parietal cortex (O'Keefe & Bouma, 1969), through which it receives complex somesthetic information.

The lateral amygdala is highly important in analyzing information received and transferring information back to the neocortex so that further elaboration may be carried out at the cortical level. It is through the lateral division that emotional meaning and significance can be assigned to as well as extracted from that which is experienced.

The amygdala, overall, maintains a functionally interdependent relationship with the hypothalamus. It is able to modulate and even control rudimentary emotional forces governed by the hypothalamic nucleus. However, it also acts at the behest of hypothalamically induced drives. For example, if certain nutritional requirements need to be met, the hypothalamus signals the amygdala, which then surveys the external environment for something good to eat. By contrast, when, via environmental surveillance, the amygdala discovers a potentially threatening stimulus, it acts to excite and drive the hypothalamus so that the organism is mobilized to take appropriate action. Thus, when the hypothalamus is activated by the amygdala, instead of responding in an on - off manner, cellular activity continues for an appreciably longer time period (Dreifuss et al., 1968). The amygdala can tap into the reservoir of emotional energy mediated by the hypothalamus to achieve certain ends.

Attention

The amygdala acts to perform environmental surveillance and can trigger orienting responses as well as mediate the maintenance of attention if something of interest or importance were to appear (Gloor, 1955, 1960; Kaada, 1951; Ursin & Kaasa, 1960). Electrical stimulation of the lateral division can initiate quick and/or anxious glancing and searching movements of the eyes and head, such that the organism appears aroused and highly alert as if in expectation of something that is going to happen (Ursin & Kaada, 1960). The EEG becomes desynchronized (indicating arousal), heart rate becomes depressed, respiration patterns change, and the galvanic skin response significantly alters (Bagshaw & Benzies, 1968; Ursin & Kaada, 1960) - reactions that characteristically accompany the orienting response of most species. Once a stimulus of potential interest is detected, the amygdala acts to analyze its emotional - motivational importance and will act to alert other nuclei such as the hypothalamus so that appropriate action may take place.

Fear, Rage, and Aggression

Initially, electrical stimulation of the amygdala produces sustained attention and orienting reactions. If the stimulation continues fear and/or rage reactions are elicited (Ursin & Kaada, 1960). When fear follows the attention response, the pupils dilate and the subject will cringe, withdraw, and cower. This cowering reaction in turn may give way to extreme fear and/or panic such that the animal will attempt to take flight.

Among humans, the fear response is one of the most common manifestations of amygdaloid electrical stimulation. Moreover, unlike hypothalamic on/off emotional reactions, attention and fear reactions can last up to several minutes after the withdrawal of stimulation.

In addition to behavioral manifestations of heightened emotionality, amygdaloid stimulation can result in intense changes in emotional facial expression. This includes facial contortions, baring of the teeth, dilation of the pupils, widening or narrowing of the eyelids, flaring of the nostrils, tearing, as well as sniffing, licking, and chewing (Anand & Dua, 1955; Ursin & Kaada, 1960). Indeed, some of the behavioral manifestations of a seizure in this vicinity (i.e., temporal lobe epilepsy) typically include chewing, smacking of the lips, and licking.

In many instances, rather than fear, there instead results anger, irritation, and rage, which seems to build up gradually, until finally the animal or human attacks (Egger & Flynn, 1963; Gunne & Lewander, 1966; Ursin & Kaada, 1960; Zbrozyna, 1963). Unlike hypothalamic sham rage, amygdaloid activation results in attacks directed at something real or, in the absence of an actual stimulus, at something imaginary. Moreover, rage and attack will persist well beyond the termination of the electrical stimulation of the amygdala. In fact, the amygdala remains electrophysiologically active for long periods, even after a stimulus has been removed (be it external - perceptual, or internal - electrical), such that is appears to continue to process - in the abstract - information even when that information is no longer observable (O'Keefe & Bouma, 1969).

In addition to permitting sustained electrophysiological activity, the amygdala has been shown to be heavily involved in the maintenance of behavioral responsiveness even in the absence of an immediately tangible or visible objective or stimulus (O'Keefe & Bouma, 1969). This includes motivating the organism to engage in the seeking of hidden objects or continuing a certain activity in anticipation of achieving some particular long-term goal. At a more immediate level, the amygdala is probably very important in object permanence (i.e., keeping an object in mind when it is no longer visible) and concrete or abstract anticipation. Anticipation is, of course, very important in the prolongation of emotional states such as fear or anger, as well as the generation of more complex emotions such as anxiety. In this regard, the amygdala is probably important not only in regard to emotion, but in maintaining mood states.

Fear and rage reactions have also been triggered in humans following depth electrode stimulation of the amygdala (Chapman, 1960; Chapman, Schroeder, Geyer, Brazier, Fager, Poppen, Solomoan, & Yakovlev, 1954; Heath, Monroe, & Mickle, 1955; Mark, Ervin, & Sweet, 1972). Mark et al. (1972) describe one female patient who following amygdaloid stimulation became irritable and angry, and then enraged. Her lips retracted, there was extreme facial grimacing, threatening behavior, and then rage and attack - all of which persisted well beyond stimulus termination.

Similarly, Schiff et al. (1982) described a man who developed intractable aggression following a head injury and damage (determined via depth electrode) to the amygdala (i.e., abnormal electrical activity). Subsequently, he became easily enraged, sexually preoccupied (although sexually hypoactive), and developed hyperreligiosity and psuedomystical ideas. Tumors invading the amygdala have been reported to trigger rage attacks (Sweet, Ervin, & Mark, 1960; Vonderache, 1940).

The amygdala appears capable of not only triggering and steering hypothalamic activity but acting on higher level neocortical processes so that individuals form emotional ideas. Indeed, the amygdale is able to overwhelm the neocortex and the rest of the brain so that the person not only forms emotional ideas but responds to them. A famous example of this is Charles Whitman, who in 1966 climbed a tower at the University of Texas and began to indiscriminantly kill people with a rifle.

Whitman had initially consulted a psychiatrist about his periodic and uncontrollable violent impulses but was unable to obtain relief. Prior to climbing the tower, he wrote himself a letter (Sweet et al., 1969): I don't really understand myself these days. Lately I have been a victim of many unusual and irrational thoughts. These thoughts constantly recur, and it requires a tremendous mental effort to concentrate. I talked to a doctor once for about two hours and tried to convey to him my fears that I felt overcome by overwhelming violent impulses. After one session I never saw the Doctor again, and since then I have been fighting my mental turmoil alone. After my death I wish that an autopsy would be performed to see if there is any visible physical disorder. I have had tremendous headaches in the past.

Later he wrote: It was after much thought that I decided to kill my wife, Kathy, tonight after I pick her up from work....I love her dearly, and she has been a fine wife to me as any man could ever hope to have. I cannot rationally pinpoint any specific reason for doing this....

That evening, he killed his wife and mother and wrote: I imagine that it appears that I brutally killed both of my loved ones. I was only trying to do a good thorough job....

The following morning, he climbed the University tower carrying a high-powered hunting rifle and for the next 90 min he shot at everything that moved, killing 14 and wounding 38. Postmortem autopsy of his brain demonstrated a glioblastoma multiforme tumor the size of a walnut compressing the amygdaloid nucleus.

Docility and Amygdaloid Destruction

Bilateral destruction of the amygdala usually results in increased tameness, docility, and reduced aggressiveness in cats, monkeys and other animals (Schreiner and Kling, 1956; Weiskrantz, 1956; Vochteloo & Koolhas, 1987), including purportedly ferocious creatures such as the agouti and lynx (Schreiner and Kling, 1956). In man, bilateral amygdala destruction (by neurosurgery) has been reported to reduce and/or eliminate paroxysmal aggressive and violent behavior (Terzian & Ore, 1955).

In some creatures, however, bilateral ablation of the amygdala has been reported to at least initially result in increased aggressive responding (Bard & Mountcastle, 1948) and, if sufficiently aroused or irritated, even the most placid of amygdalectomized animals can be induced to fight fiercely (Fuller, Rosvold, & Pribram, 1957). However, these aggressive responses are very short-lived and appear to be reflexively mediated by the hypothalamus. Thus, these findings (and the data reviewed above) suggest that true aggressive feelings are dependent on the functional integrity of the amygdala (versus the hypothalamus).

Social - Emotional Agnosia

Among primates and mammals, bilateral destruction of the amygdala significantly disturbs the ability to determine and identify the motivational and emotional significance of externally occurring events, to discern social - emotional nuances conveyed by others, or to select what behavior is appropriate for a specific social context (Bunnel, 1966; Fuller, Rosvold & Pribram, 1957; Gloor, 1960; Kluver & Bucy, 1939). Bilateral lesions lower responsiveness to aversive and social stimuli, reduce aggressiveness, fearfulness, competitiveness, dominance, and social interest (Rosvold, Mirsky, & Pribram, 1954). Indeed, this condition is so pervasive that subjects seem to have tremendous difficulty discerning the meaning or recognizing the significance of even common objects - a condition sometimes referred to as psychic blindness or the Kluver - Bucy syndrome.

Thus, animals with bilateral amygdaloid destruction, although able to see and interact with their environment, respond in an emotionally blunted manner and seem unable to recognize what they see, feel, and experience. Things seem stripped of meaning. Like an infant (who similarly is without a fully functional amygdala), patients with this condition engage in extreme orality and will indiscriminantly pick up various objects and place them in their mouth regardless of its appropriateness. There is a repetitive quality to this behavior, for once they put it down they seem to have forgotten that they had just explored it, and will immediately pick it up and place it again in their mouth as if it were a completely unfamiliar object.

Although ostensibly exploratory, there is thus a failure to learn, to remember, to discern motivational significance, to habituate with repeated contact, or to discriminate between appropriate and inappropriate stimuli. Rather, when the amygdala has been removed bilaterally, the organism reverts to the most basic and primitive modes of object interaction such that everything that is seen and touched is placed in the mouth (Brown & Schaffer, 1888; Gloor, 1960; Kluver & Bucy, 1939). This condition pervades all aspects of higher-level social - emotional functioning, including the ability to interact appropriately with loved ones.

For example, Terzian and Ore (1955) described a young man who, following bilateral removal of the amygdala, subsequently demonstrated an inability to recognize anyone, including close friends, relatives, and his mother. He ceased to respond in an emotional manner to his environment and seemed unable to recognize feelings expressed by others. He also exhibited many features of the Kluver - Bucy syndrome (perseverative oral "exploratory" behavior and psychic blindness), as well as an insatiable appetite. In addition, he became extremely socially unresponsive such that he preferred to sit in isolation, well away from others.

Among primates who have undergone bilateral amygdaloid removal, once they are released from captivity and allowed to return to their social group, a social - emotional agnosia becomes readily apparent, as they no longer respond to or seem able to appreciate or understand emotional or social nuances. Indeed, they appear to have little or no interest in social activity and persistently attempt to avoid contact with others (Dick, Myers, & Kling, 1969; Jonason & Enloe, 1971; Jonason, Enloe, Contrucci, & Meyer, 1973). If approached they withdraw, and if followed they flee. Indeed, they behave as if they have no understanding of what is expected of them or what others intend or are attempting to convey, even when the behavior is quite friendly and concerned. Among adults with bilateral lesions, total isolation seems to be preferred.

In addition, they no longer display appropriate social or emotional behaviors, and if kept in captivity will fall in dominance in a group or competitive situation - even when formerly dominant (Bunnel, 1966; Dicks et al., 1969; Fuller et al., 1957; Jonason & Enloe, 1971; Jonason et al., 1973). As might be expected, maternal behavior is severely affected. According to Kling (1972), mothers will behave as if their "infant were a strange object to be mouthed, bitten, and tossed around as though it were a rubber ball."

Limbic Language

Although language is usually discussed in regard to grammar and vocabulary, there is a third major feature to expression and comprehension through which a speaker may convey, and a listener determine, intent, attitude, feeling, and meaning. Language is both descriptive and emotional. A listener comprehends not only what is said, but how it is said - what a speaker feels. Feeling and attitude is generally communicated via vocal inflection, intonation, and melody such that anger, happiness, sadness, sarcasm, and so forth, are indicated by such cues as variations in pitch, timbre, and stress contours - vocal capacities associated with the functional integrity of the right cerebral hemisphere, and the right frontal and temporal regions in particular (Joseph, 1988a). For example, studies using dichotic listening have repeatedly shown that the right hemisphere is superior in distinguishing stress and pitch contours, determining frequency, amplitude, melody, duration, processing inflectional contours, and even decoding contextual information in the absence of denotative speech.

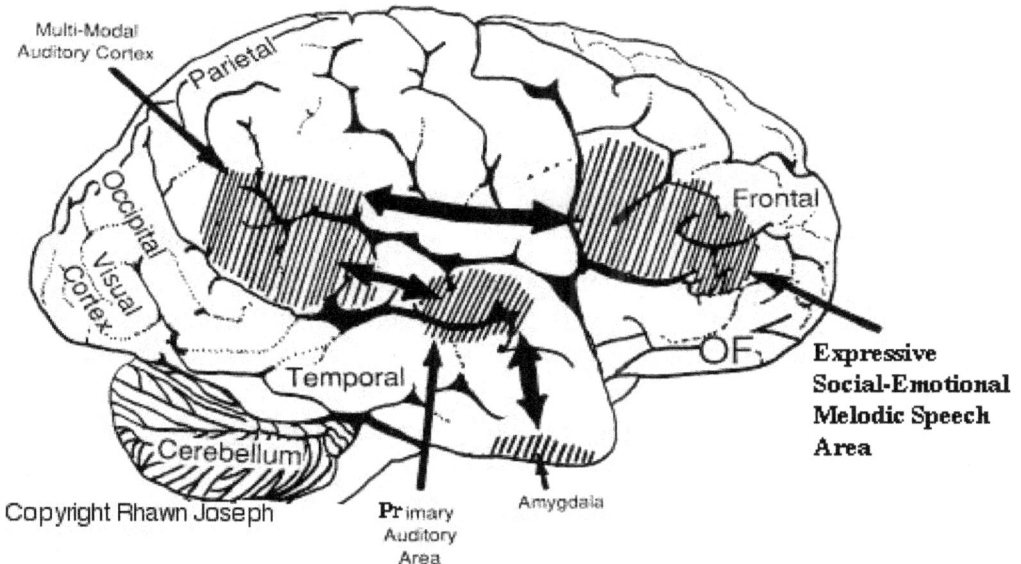

Conversely, patients who have undergone right temporal lobe neurosurgery or who have suffered damage involving this area, often demonstrate impairments in regard to many aspects of prosodic - melodic perception, such that nonverbal environmental sounds and musical stimuli fail to be adequately recognized. People who have suffered deep right frontal or temporal lobe damage also have difficulty controlling the pitch and inflection of their voice, whereas patients undergoing sodium amytal anesthetization of the right hemisphere speak in a bland monotone.

It has been argued in detail (Joseph, 1982, 1988a) that the superior capacity of the right hemisphere in processing and expressing melodic - prosodic

- emotional information is directly related to lateralized limbic functioning, and in particular, a greater abundance of neuronal interconnections with various limbic nuclei such as the amygdale. Emotional vocalizations are limbic in origin and are hierarchically represented, processed, and expressed by various neocortical regions within the frontal and temporal lobes.

In fact, the major fiber pathway that links the language axis (Wernicke's and Broca's areas) of the left hemisphere and the melodic - prosodic - emotional Axis of the right, i.e., the arcuate fasciculus, extends from the left frontal convexity (Broca's area) through the inferior parietal lobule, and after giving off major fibers to Wernicke'e region continues down into the depths of the inferior temporal lobe where it establishes contact with the amygdala. In this way, the amygdala has a significant influence on emotional speech (Figs. 21-23).

Indeed, both phylogenetically and ontogenetically, the original impetus to vocalize springs forth from roots buried within the depths of the ancient limbic lobes (e.g., amygdala, hypothalamus, septum). For example, although nonhumans do not have the capacity to speak, they still vocalize, and these vocalizations are primarily limbic in origin, being evoked in situations involving sexual arousal, terror, anger, flight, helplessness, and separation from the primary caretaker when young. The first vocalizations of human infants are similarly emotional in origin and limbically mediated (Joseph, 1982).

Although cries and vocalizations indicative of rage or pleasure have been elicited through hypothalamic stimulation, of all limbic nuclei the amygdale is the most vocally active - particularly the lateral division (Robinson, 1967). In humans and animals, a wide range of emotional sounds have been evoked through amygdala activation, such as sounds indicative of pleasure, sadness, happiness, and anger (Robinson, 1967; Ursin & Kaada, 1960). Conversely, in man, destruction limited to the amygdala, (Freeman & Williams, 1952, 1963), the right amygdala, in particular, has abolished the ability to sing, convey melodic information, or properly enunciate by vocal inflection. Similar disturbances occur with right hemisphere damage (Joseph, 1988a). Indeed, when the right temporal region (including the amygdala) has been grossly damaged or surgically removed, the ability to perceive, process, or even vocally reproduce most aspects of musical and emotional auditory input is significantly curtailed.

Emotion and Temporal Lobe Seizures

The amygdala is buried within the depths of the anterior - inferior temporal lobe and maintains rich interconnections with areas throughout the temporal neocortex. Because of their intimate association, damage to the temporal lobe, particularly the anterior regions, often involves and disrupts amygdaloid functioning. In fact, because the amygdala and inferior- anterior temporal lobe have, of all brain regions, the lowest seizure threshold, and are minimally resistant, and are therefore maximally vulnerable to developing abnormal seizure activity, even mild injuries may result in kindling (i.e., abnormal activation), and therefore disruption of their functional integrity. Indeed, damage to adjacent tissue has been known to spread, by kindling, to the amygdala and inferior regions. Once consequence is temporal lobe epilepsy.

Personality, emotional, and sexual disturbances are a frequent complication of temporal lobe seizures in a significant minority of patients. Such patients may develop paranoid, hysterical, or depressive tendencies, deepening of mood, hyposexuality, and other characteristics suggestive of affective disorders (Bear, Leven, Blumer, Chetam, & Ryder, 1982; Gibbs, 1951; Herman & Chambria, 1980; Strauss, Risser, & Jones, 1982; Williams, 1956). Immediately following or during the course of a seizure 10% or more of such patients experience a change in emotionality (Herman & Chambria, 1980; Strauss et al., 1982; Williams 1956).

In part, because the highest incidence of psychiatric disorder occurs in cases in which the EEG spike focus is in the anterior temporal area (Gibbs, 1951), and because limbic nuclei such as the amygdala are frequently involved, it has been postulated that seizure activity sometimes hyperactivates these nuclei (Bear, 1979), which in turn distorts the affective meaning applied to afferent streams of visual, auditory, and somesthetic information (Gibbs, 1951).

Thus, during a seizure, these patients may be temporarily overwhelmed by feelings such as fear, or things they see or hear seem to become abnormally invested with emotional significance (Bear et al., 1979; Gibbs, 1951; Gloor et al., 1982), presumably due to abnormal amygdala activation. Interestingly, one common symptom of temporal lobe epilepsy is an aura of tastes, and more often odors, that are usually quite unpleasant (e.g., like burning wire, burning feces, burning rubber or tires).

Seizure induced emotional changes tend to predominantly involve feelings of depression, pleasure, displeasure, or fear - with fear being one of the most common emotional experiences (Gloor et al., 1982; Williams, 1956). More rarely, seizures involving sexual behavior, crying, laughing, or ragelike responses have been associated with temporal lobe epilepsy.

There is some evidence to suggest that certain emotional changes are more frequently associated with seizure originating in the right temporal lobe and/ or amygdale, whereas disturbances involving thought (e.g., psychosis, schizophrenia) characterize left temporal lobe abnormalities (Flor-Henry, 1969, 1983; Offen, Davidoff, Troost, & Richey, 1976; Joseph, 1988; Sherwin, 1981; Schiff et al., 1982; Taylor, 1975; Weil, 1956).

For example, in their depth electrode study of five patients with seizure disorders, Gloor et al., (1982) found that all feelings of fear and displeasure were associated with right temporal, right amygdala, or right hippocampal activation. These findings are consistent with the observations of Slater, Beard, and Glithero (1969), Bear and Fedio (1977), Flor-Henry (1969), and others, who have noted an association between right temporal seizures and affective disorders.

Other reports, however, have been less clear cut - presumably because seizures originating in the amygdala/temporal lobe can quickly spread from hemisphere to another (through the anterior commissure) and because of the failure to employ depth electrodes to pinpoint the seizure foci. Thus, in some instances, emotions such as fear seem to arise regardless of which hemisphere is involved (Strauss et al., 1982). Nevertheless, even in these cases, an emotional dichotomy is apparent. For example, Herman and Chambria (1980) described cases with right temporal foci in whom free-floating fears developed that were not tied to something specific but that encompassed terrifying, "death-like" and "nightmarish" feelings, whereas a patient with a left temporal foci developed intense fears but were unable to describe what they were afraid of.

Stimulation of the amygdala can significantly alter facial emotional expression, including tearing. In a small minority of cases, right temporal seizures have been reported to cause paroxysmal attacks of weeping, with lacrimation, the making of mournful sounds, including sobbing and crying (Offen et al., 1976). However, crying as well as laughing seizures have also been noted to occur with left-sided involvement (Chen & Forster, 1973; Sethi & Rao, 1976). Nevertheless, without the benefit of depth electrodes, such as employed by Gloor and colleagues (1982), it is difficult to determine in which amygdala and/or temporal lobe a seizure actually originates.

The amygdala and regions of the hypothalamus are sexually dimorphic, and stimulation of either area can trigger sexual behavior. Similarly, sensations of sexual excitement, sometimes leading to orgasm, may also occur as a function of seizures originating in the temporal lobe (Currier, Little, Suess, & Andy, 1971; Freemon & Nevis, 1969; Remillard et al., 1983) and in the frontal lobe (Spencer, Spencer, Williamson, & Mattson, 1983). Of interest, 7 of 10 patients with sexual seizures described by Remillard et al., (1983) had foci originating in the right hemisphere. Similar findings have been reported by Freemon & Nevis (1969), Penfield and Rasmussen (1950, p.27), and Spencer et al. (1983).

Sexual seizures caused by temporal lobe abnormalities are often accompanied by actual sexual behavior. "The patient was sitting at the kitchen table with her daughter making out a shopping list. She stopped, appeared dazed, slumped to the floor on her back, lifted her skirt, spread her knees, and elevated her pelvis rhythmically. She made appropriate vocalizations for sexual intercourse, such as "it feels so good" and "further, further" (Currier et al., 1971, p.260).

Frontal Lobe Sexual Seizures

The frontal lobes, the orbital region in particular, maintain rich interconnections with the amygdala (Nauta, 1964), and receives olfactory projections as well (Tanabe, 1975). Epileptiform activity arising in the deep frontal (orbital) regions has also been associated with the development of sexual seizures, including exhibitionism, genital manipulation, and masturbatory activity (Spencer et al., 1983).

Similar to that described regarding right temporal emotional and sexual disturbances, Spencer et al. (1983) found that three of their four patients with sexual automatism had seizures originating in the right frontal area, whereas the remaining patient had bifrontal disturbances. Other investigators have also noted that peculiar disturbances in emotion and personality are far more likely to arise following right versus left frontal damage (Joseph, 1986, 1988; Hillbom, 1960; Lishman, 1968).

Presumably, at least in part, emotional and sexual disturbances associated with right temporal and right frontal (orbital) dysfunction are caused by activation of the limbic structures, with which they intimately interconnected. However, there is also much evidence to indicate that in some respects, inferior temporal and orbital frontal areas are in fact outgrowths of and thus part of the limbic system.

Summary

Over the course of early evolutionary development, the hypothalamus reigned supreme in the control and expression of raw and reflexive emotionality, i.e., pleasure, displeasure, aversion, and rage. Largely, however, it has acted as an eye turned inward, monitoring internal homeostasis and concerned with basic needs. With the development of the amygdala, the organism was now equipped with an eye turned outward, so that the external emotional features of reality could be tested and ascertained. When signaled by the hypothalamus the amygdala begins to search the sensory array for appropriate emotional - motivational stimuli, until what is desired is discovered and attended to.

However, with the differentiation of the amygdala, emotional functioning also became differentiated and highly refined. The amygdala hierarchically wrested control of emotion from the hypothalamus. The amygdala is primary in regard to the perception and expression of most aspects of emotionality, including fear, aggression, pleasure, happiness, and sadness, and in fact assigns emotional or motivational significance to that which is experienced. It can thus induce the organism to act on something seen, felt, heard, or anticipated. The integrity of the amygdala is essential in regard to the analysis of social-emotional nuances, the organization and mobilization of the person's internal motivational status regarding these cues, as well as the mediation of higher-order emotional expression and impulse control. When damaged or functionally compromised, social-emotional functioning becomes grossly disturbed.

Through its rich interconnections with other brain regions, the amygdaloid nucleus is able to sample and influence activity that occurs in other parts of the cerebrum and add emotional color to one's perceptions. As such, it is highly involved in the assimilation and association of divergent emotional, motivational, somesthetic, visceral, auditory, visual, motor, olfactory, and gustatory stimuli. Thus, it is very much involved with learning, memory, and attention can generate reinforcement for certain behaviors (Douglas, 1967). Moreover, through reward or punishment, it can promote the encoding, storage, and later retrieval of particular types of information. That is, learning often involves reward, and it is through the amygdala (in concert with other nuclei) that emotional consequences can be attributed to certain events, actions, or experiences, as well as extracted from the world of possibility, so that it can be attended to and remembered.

Lastly, as is evident from studies of patients displaying abnormal activity or seizures originating in or involving this nuclei, the amygdala is able to overwhelm the neocortex and gain control over behavior. As based on electrophysiological studies, the amygdala seems to be capable of literally turning off the neocortex (such as occurs during a seizure), at least for brief time periods. That is, the amygdala can induce electrophysiological slow-wave theta activity in the neocortex, which indicates low levels of arousal as well as high voltage fast activity. In the normal brain, it probably exerts similar influences such that at times people (i.e., their neocortex) lose control over themselves and respond in a highly emotionally charged manner.

HIPPOCAMPUS

The hippocampus is an elongated structure located within the inferior medial wall of the temporal lobe (posterior to the amygdale); it surrounds, in part, the lateral ventricle. In humans, the hippocampus consists of an anterior and posterior region and is shaped somewhat like a telephone receiver.

There are three major neural pathways leading to and from the hippocampus. These include the fornix-fimbrial fiber system, a supracallosal pathway (i.e., the indusium griseum), which passes through the cingulated, and through the entorhinal area, sometimes referred to as the gateway to the hippocampus.

It is through the entorhinal area that the hippocampus receives olfactory and amygdaloid projections (Carlsen et al., 1982; Gloor, 1955; Krettek & Price, 1976a; Steward, 1977) and fibers from the orbital frontal and temporal lobes (Van Hoesen et al., 1972). It is through the fornix and fimbrial pathways that the hippocampus makes major interconnections with the thalamus, septal nuclei, medial hypothalamus, and through which it exerts either inhibitory or excitatory influences on these nuclei (Feldman, Saphier, & Conforti, 1987; Guillary, 1957; Poletti & Sujatanon, 1980).

Septal interactions: The hippocampus maintains a particularly intimate relationship with the septal nuclei (sometimes referred to as the septum - not to be confused with the septum pelucidum). The septal nucleus partly serves as in interactional relay center as it channels hippocampal influences to other structures such as the hypothalamus and reticular formation (and vice versa) and as a major link through which the hippocampus and amygdala sometimes interact (Hagino & Yamaoka, 1976).

Amygdala interactions: The hippocampus is greatly influenced by the amygdala, which in turn monitors and responds to hippocampal activity (Gloor, 1955; Green & Adey, 1956; Steriade, 1964) (Figs. 23 - 25). The amygdala also acts to relay certain forms of information from the hippocampus to the hypothalamus (Poletti & Sajatanon, 1980). Together the hippocampus and amygdala complement and interact in regard to attention, the generation of emotional and other types of imagery, as well as learning and memory.

Hippocampal Arousal, Attention, and Inhibitory Influences

Various investigators have assigned a major role to the hippocampus in information processing, including memory, new learning, cognitive mapping of the environment, voluntary movement toward a goal, as well as attention, behavioral arousal, and orienting reactions (Douglas, 1967; Grastayan et al., 1959; Green & Arduini, 1954; Isaacson, 1982; Milner, 1966, 1970, 1971; Olton, Branch, & Best, 1978; Routtenberg, 1968). For example, hippocampal cells alter their activity greatly in response to certain spatial correlates, particularly as an animal moves about in its environment (Olton et al., 1978). It develops slow-wave theta activity during arousal (Green & Arduini, 1954) or when presented with noxious or novel stimuli (Adey, Dunlop, & Hendrix, 1960). However, few studies have implicated this nucleus as important in emotional functioning per se, although responses such as "anxiety" or "bewilderment" have been observed when directly electrically stimulated (Kaada, Jansen, & Andersen, 1954).

Hippocampal - Neocortical Interactions

Desynchronization of the cortical EEG is associated with high levels of arousal and information input. As the level of input increases, the greater the level of cortical arousal (Como, Joseph, Fiducia, & Siegel, 1979; Joseph, Como, Forrest, Fiducia, & Siegel, 1981). However, when arousal levels become too great, efficiency in information processing, memory, new learning, and attention becomes compromised, as the brain becomes overwhelmed.

When the neocortex becomes desynchronized (indicating cortical arousal), the hippocampus often (but not always) develops slow-wave theta activity (Grastayan et al., 1959; Green & Arduini, 1954) such that it appears to be functioning at a much lower level of arousal. Conversely, when cortical arousal is reduced to a low level (indicated by EEG synchrony), the hippocampal EEG often becomes desynchronized.

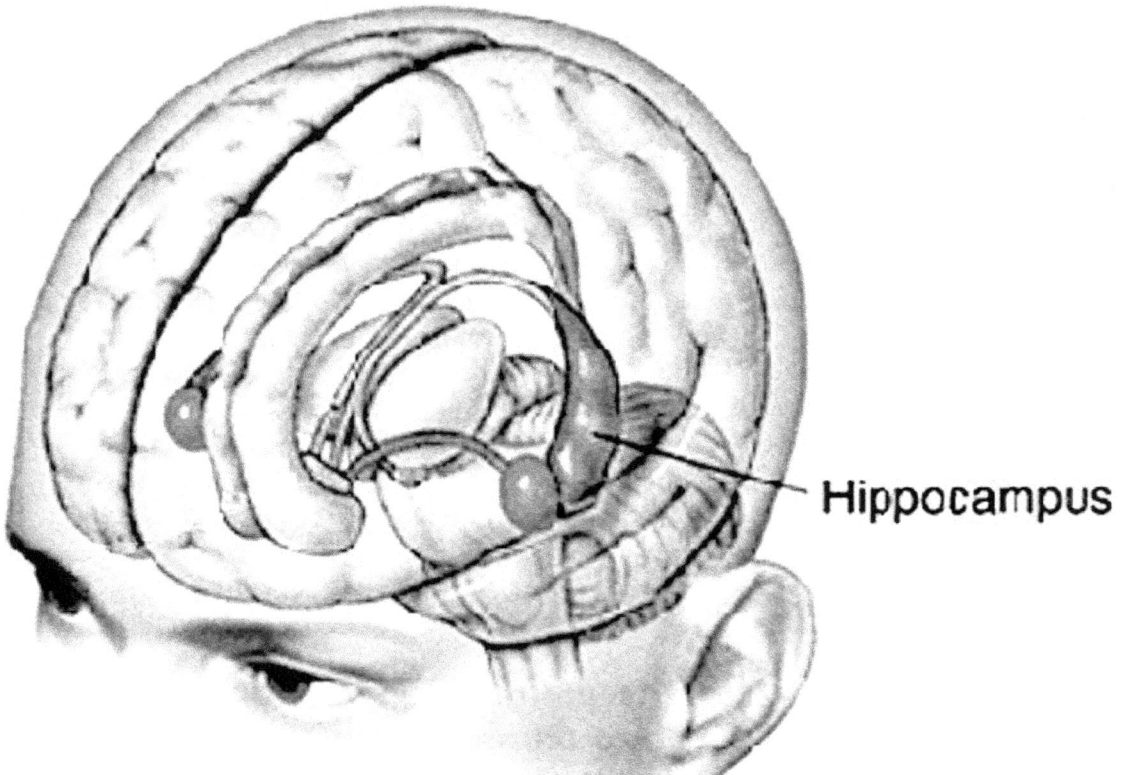

Hippocampus

These findings suggest when the neocortex is highly stimulated, the hippocampus functions at a much lower level in order to monitor what is being received and processed so as not to become overwhelmed. When the neocortex is not highly aroused, the hippocampus presumably compensates by increasing its own level of arousal so as to tune into information that is being processed at a low level of intensity. In situations in which both the cortex and the hippocampus become desynchronized, distractability and hyperresponsiveness result and the subject becomes overwhelmed and confused and may orient to and approach several stimuli (Grastyan et al., 1959). Attention, learning, and memory functioning are decreased. Such situations sometimes occur when one is highly anxious, upset, or when traumatized. In fact, it is likely that the hippocampus might be injured by traumatic stress, which in turn would explain why some individuals become amnesic following a prolonged trauma (Joseph, unpublished 1980).

There is also evidence to suggest that the hippocampus may act to reduce extremes in cortical arousal. For example, whereas stimulation of the reticular activating system augments cortical arousal and EEG-evoked potentials, hippocampal stimulation reduces or inhibits these potentials such that cortical responsiveness and arousal are dampened (Feldman, 1962; Redding, 1967). If cortical arousal is at a low level, hippocampal stimulation often leads to augmentation of the cortical evoked potential (Redding, 1967).

The hippocampal also exerts desynchronizing or synchronizing influences on various thalamic nuclei, in turn augmenting or decreasing activity in this region (Green & Adey, 1956; Guillary, 1955; Nauta, 1956, 1958). As the thalamus is the major relay nucleus to the neocortex, the hippocampus appears to be able to block or enhance information transfer to various neocortical areas.

It is likely that the hippocampus may act to influence information reception at the neocortical level as well as possibly reduce extremes in cortical arousal (be they too low or high) via inhibition or excitation, and/or it may act so that the neocortex is neither over- not underwhelmed when engaged in the reception and processing of information. That is, very high or very low states of excitation are incompatible with alertness and selective attention and with the ability to learn and retain information (Joseph et al., 1981).

Aversion and Punishment

In many ways, the hippocampus appears to act in concert with the medial hypothalamus and septal nuclei (with which it maintains rich interconnections) so as to prevent extremes in arousal and thus maintain a state of quiet alertness (or quiescence). Moreover, as with the medial hypothalamus, it has been reported that the subjective components of aversive emotion in humans is correlated with electrophysiological alternations in the hippocampus and septal area (Heath, 1976).

The hippocampus also appears to be heavily involved in the modulation of reactions to frustrations or punishment (Gray, 1970), particularly in regard to learning. For example, the hippocampus responds with trains of slow theta waves when presented with noxious stimuli but habituates with repeated presentation. Moreover, there is some evidence to suggest that this nucleus in conjunction with the amygdala and septal nuclei is important in generating negative cognitive-mood states such as anxiety.

Attention and Inhibition

The hippocampus participates in the elicitation of orienting reactions and the maintenance of an aroused state of attention (Foreman & Stevens, 1987; Grastayan et al., 1959; Green & Arduini, 1954; Routtenberg, 1968). When exposed to novel stimuli or when engaged in active searching of the environment, hippocampal theta appears (Adey et al., 1960). However, with repeated presentations of a novel stimulus, the hippocampus habituates and theta disappears (Adey et al., 1960). Thus, as information is attended to, recognized, and presumably learned and/or stored in memory, hippocampal participation diminishes. Theta also appears during the early stages of learning as well as when engaged in selective attention and the making of discriminant responses (Grastyan et al., 1959).

When the hippocampus is damaged or destroyed, animals have great difficulty inhibiting behavioral responsiveness or shifting attention. For example, Clark and Issacson (1965) found that animals with hippocampal lesions could not learn to wait 20 sec between bar presses, if first trained to respond to a continuous schedule. There is an inability to switch from a continuous to a discontinuous pattern, such that a marked degree of perseveration and inability to change sets or inhibit a pattern of behavior once initiated occurs (Douglas, 1967; Ellen, Wilson, & Powell, 1964). Habituation is largely abolished, and the ability to think or respond divergently is disrupted.

In part, this finding suggests that hippocampal damage disrupts the ability to learn and thus remember - findings that have been repeatedly demonstrated in humans. In animals, disinhibition caused by hippocampal damage can even prevent the learning of a passive-avoidance task, such as simple ceasing to move.

When coupled with the evidence presented above, it appears that the hippocampus acts to enhance or diminish areas of neural excitation selectively, which in turn allows for differential selective attention and differential responding. When damaged, the ability to shift from one set of perceptions to another or to change behavioral patterns is disrupted, and the organism becomes overwhelmed by a particular mode of input. Learning, memory, as well as attention are greatly compromised.

Learning and Memory

The hippocampus is most usually associated with learning and memory encoding (e.g., long-term storage and retrieval of newly learned information), particularly the anterior regions (Fedio & Van Buren, 1974; Milner, 1966; 1970; Penfield & Milner, 1958; Rawlins, 1985; Scoville & Milner, 1957). Many other brain areas, such as the mamillary bodies and dorsal-medial nucleus of the thalamus, are also important in memory functioning.

Bilateral destruction of the anterior hippocampus results in striking and profound disturbances involving memory and new learning (i.e., anterograde amnesia). For example, one such patient who underwent bilateral destruction of this nuclei (H.M.), was subsequently found to have almost completely lost the ability to recall anything experienced after surgery. If you introduced yourself to him, left the room, and then returned a few minutes later, he would have no recall of having met or spoken to you. Dr. Brenda Milner has worked with H.M. for almost 20 years, and yet she is an utter stranger to him. H.M. is in fact so amnesic for everything that has occurred since his surgery (although memory for events prior to his surgery is comparatively exceedingly well preserved), that every time he rediscovers that his favorite uncle died (years after his surgery) he suffers the same grief as if he has just been informed for the first time.

HM

Normal Brain

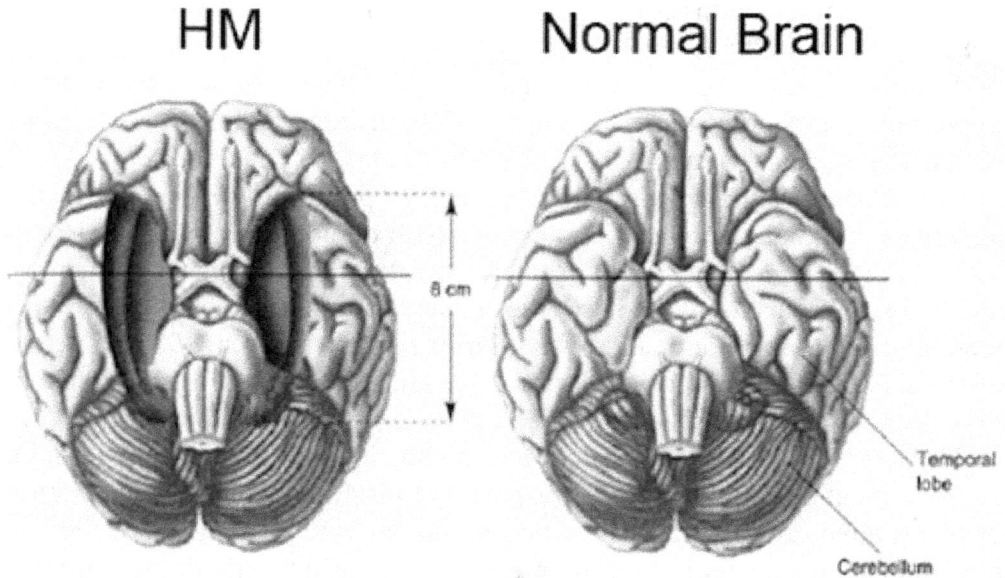

8 cm

Temporal
lobe

Cerebellum

Despite his lack of memory for new (nonmotor) information, Henry (or H.M.), has adequate intelligence, is painfully aware of his deficit, and constantly apologizes for his problem. "Right now, I'm wondering" he once said, "Have I done or said anything amiss?" You see, at this moment everything looks clear to me, but what happened just before? That's what worries me. It's like waking from a dream. I just don't remember.... Every day is alone in itself, whatever enjoyment I've had, and whatever sorrow I've had.... I just don't remember.

Presumably the hippocampus acts to protect memory and the encoding of new information during the storage and consolidation phase via the gating of afferent streams of information and the filtering/exclusion (or dampening) of irrelevant and interfering stimuli. When the hippocampus is damaged, input overload results, the neuroaxis is overwhelmed by neural noise, and the consolidation phase of memory is disrupted such that relevant information is not properly stored or even attended to. Consequently, the ability to form associations (e.g., between stimulus and response) or to alter preexisting schemas (such as occurs during learning) is attenuated (Douglas, 1967).

Hippocampal and Amygdaloid Interactions: Memory

It has been argued that significant impairments involving memory (in man) cannot be produced by lesions restricted to the hippocampus (cf. Horel, 1978). Thus, in some instances with restricted lesions, good recall of new information is possible for at least several minutes (Horel, 1978; Penfield & Milner, 1958).

Rather, there is evidence that strongly suggests that the hippocampus plays an interdependent role with the amygdala in regard to memory (Kesner & Andrus, 1982; Mishkin, 1978; Sarter & Markowitsch, 1985). Interestingly, nuclei such as the dorsal-medial region of the thalamus, which have also been shown to be important in memory (Squire & Moore, 1979), maintain rich interconnections with the amygdala (Krettek & Price, 1977b; Nauta, 1971).

Nevertheless, although psychic blindness is produced by damage to the amygdala, striking anterograde deficits in recall do not seem to occur (Horel, 1978; Scoville & Milner, 1957). Rather, the role of the amygdala in memory and learning seems to involve activities related to reward, orientation, and attention, as well as emotional arousal (Sarter & Markowitsch, 1985). If some event is associated with positive or negative emotional states, it is more likely to be learned and remembered.

The amygdala seems to reinforce and maintain hippocampal activity through the identification of motivationally significant information and the generation of pleasurable rewards (through action on the lateral hypothalamus). That is, reward increases the probability of attention being paid to a particular stimulus or consequence as a function of its association with reinforcement (Douglas, 1967; Kesner & Andrus, 1982).

However, the amygdala and hippocampus may act differentially in regard to the effects of positive versus negative reinforcement on learning and memory. For example, whereas the hippocampus produces theta in response to noxious stimuli, the amygdale increases its activity following the reception of a reward (Norton, 1970). Thus, if errors are made during acquisition (negative emotional state), possibly the hippocampus modulates the appropriate reaction, whereas when presented with a reward, the amygdala reinforces the organism's response.

Thus, the hippocampus acts to reduce or enhance extremes in arousal associated with information reception and storage in memory, whereas the amygdala acts to identify the social-emotional-motivational characteristics of the stimuli as well as to generate (in conjunction with the hippocampus) appropriate emotional rewards to reinforce learning and memory. We find that when both the amygdala and hippocampus are damaged, striking and profound disturbances in memory functioning result (Kesner & Andrus, 1982; Mishkin, 1978).

Laterality

It is now very well known that lesions involving the mesial-inferior temporal lobes (i.e., destruction or damage to the amygdala/hippocampus) of the left cerebral hemisphere typically produce significant disturbances involving verbal memory - particularly as contrasted with patients with right-sided destruction. Left-sided damage disrupts the ability to recall simple sentences and complex verbal narrative passages or to learn verbal paired associates or a series of digits (B. Milner, 1966, 1970, 1971).

By contrast, right temporal destruction typically produces deficits involving visual memory, such as the learning and recall of geometric patterns, visual mazes, or human faces (Corkin, 1965; Milner, 1966; Kimura, 1963). Right-sided damage also disrupts the ability to recognize (through recall) olfactory stimuli (Rausch, Serafetinides, & Crandall, 1977), emotional sounds and passages (Wechsler, 1973), sounds from the environment, as well as tactile mazes (Joseph, 1988a).

It appears, therefore, that the left amygdala/hippocampus is highly involved in processing and/or attending to verbal information, whereas the right amygdala/hippocampus is more involved in the learning, memory and recollection of nonverbal, visual-spatial, environmental, emotional, motivational, tactile, olfactory, and facial information.

THE PRIMARY PROCESS
Amygdaloid-Hippocampal Interactions during Infancy: Hallucinations

The amygdala-hippocampus complex, particularly that of the right hemisphere, is very important in the production and recollection of nonlinguistic images associated with past experience. In fact, direct electrical stimulation of this region within the temporal lobes results not only in the recollection of images but in the creation of fully formed visual and auditory hallucinations (Halgren et al., 1978; Horowitz et al., 1968; Malh et al., 1964; Penfield & Perot, 1963), as well as feelings of familiarity (e.g., de ja vu).

Indeed, it has long been known that tumors invading specific regions of the brain can trigger the formation of hallucinations that range from the simple (flashing lights) to the complex. The most complex forms of hallucination, however, are associated with tumors within the most anterior portion of the temporal lobe (Critchley, 1939; Gibbs, 1951; Horowitz et al., 1968; Tarachow, 1941), i.e., the region containing the amygdala and anterior hippocampus.

Similarly, electrical stimulation of the anterior lateral temporal cortical surface results in visual hallucinations of people, objects, faces, and various sounds (Horowitz et al., 1968) - particularly the right temporal lobe (Halgren et al., 1978). Depth electrode stimulation and thus direct activation of the amygdala and/or hippocampus is especially effective.

For example, stimulation of the right amygdala produces visual hallucinations, body sensations, de ja vu, illusions, as well as gustatory and alimentary experiences (Weingarten et al., 1977). Conversely, Freeman and Williams (1963) reported that the surgical removal of the right amygdala in one patient abolished hallucinations. Stimulation of the right hippocampus has also been associated with the production of deja-vu-, memory-, and dreamlike hallucinations (Halgren et al., 1978), and in fact, hallucinations seem to occur most frequently following hippocampal activation (Horowitz et al., 1968).

As is well known, LSD can elicit profound hallucinations involving all spheres of experience. Following the administration of LSD, high-amplitude slow waves (theta) and bursts of paroxysmal spike discharges occur in the hippocampus and amygdala (Chapman & Walter, 1965; Chapman, Walter, Ross et al., 1963), but with little cortical abnormal activity. In both humans and chimpanzees in whom the temporal lobes, amygdala, and hippocampus have been removed, LSD ceases to produce hallucinatory phenomena (Baldwin et al., 1959; Serafetinides, 1965). Moreover, LSD-induced hallucinations are significantly reduced when the right versus left temporal lobe has been surgically ablated (Serafetinides, 1965). Overall, it appears that the amygdala, hippocampus, and the neocortex of the temporal lobe are highly interactionally involved in the production of hallucinatory experiences. Presumably, it is the neocortex of the temporal lobe that acts to interpret this material (Penfield & Perot, 1963) as perceptual phenomena.

Dreaming
When hallucinations follow depth electrode or cortical stimulation, much of the material experienced is very dreamlike (Halgren et al., 1978; Malh et al., 1964) and consists of recent perceptions, ideas, feelings, and other emotions that are similarly illusionary and dreamlike. Indeed, the right hippocampus, and the right hemisphere in general (Broughton, 1982; Goldstein et al., 1972; Hodoba, 1986; Humphrey & Zangwill, 1961; Kerr & Foulkes, 1978; Meyer, Ishikawa, Hata, & Karacan, 1987), also appear to be involved in the production of dream imagery as well as REM (at least in part) during sleep.

For example, there have been reports of patients with right cerebral damage, hypoplasia, and abnormalities in the corpus callosum who have ceased to dream altogether, suffer a loss of hypnotic imagery, or tend to dream only in words (Botez, Olivier, Vezina, Botex, & Kaufman, 1985; Humphrey & Zangwill, 1951; Kerr & Foulkes, 1981; Murri, Arena, Siciliano, Mazzotta, & Muratorio, 1984). However, there have also been some reports that when the left hemisphere has been damaged, particularly the posterior portions (i.e., aphasic patients), the ability to verbally report and recall dreams also is greatly attenuated (e.g., Murri et al., 1984). Of course, aphasics have difficulty describing much of anything, let alone their dreams.

Electrophysiologically, the right hemisphere also becomes highly active during REM, whereas, conversely, the left brain becomes more active during NREM (Goldstein et al., 1972; Hodoba, 1986). Similarly, measurements of cerebral blood flow have shown an increase in the right temporal regions during REM sleep in subjects who upon wakening report visual, hypnagogic, hallucinatory, and auditory dreaming (J.S. Meyer et al., 1987). Interestingly, abnormal and enhanced activity in the right temporal and temporal - occipital area acts to increase dreaming and REM sleep for an atypically long time period

(Hodoba, 1986). Thus, it appears that there is a specific complementary relationship between REM sleep and right temporal electrophysiological activity.

In addition, during REM, the hippocampus begins to produce slow wave, theta activity (Jouvet, 1967; Olmstead, Best, & Mays, 1973; Robinson, Kramis, & Vanderwolf, 1977). Presumably, during REM, the hippocampus acts as a reservoir from which various images, words, and ideas are drawn and incorporated into the matrix of dreamlike activity being woven by the right hemisphere. It is probably just as likely that the hippocampus serves as a source from which material is drawn during the course of a daydream.

Interestingly, daydreams appear to follow the same 90- to 120-min cycle that characterize the fluctuation between REM and NREM periods, as well as fluctuations in mental capabilities associated with the right and left hemisphere (Broughton, 1982; Kripke & Sonneschein, 1973). That is, the cerebral hemispheres tend to oscillate in activity every 90- 120 min - a cycle that appears to correspond to the REM - NREM cycle and the appearance of day and night dreams.

Dreams and Infancy

In the newborn, and up until approximately 6 - 9 months, there are two distinct stages of sleep which correspond to REM and N-REM periods demonstrated by adults (Berg & berg, 1978; Dreyfus-Brisac & Monod, 1975; Parmlee, Wenner, Akimaya, Schults, & Stern, 1967). Among infants, however, REM occur during wakefulness as well as during sleep. In fact, REM can be observed when the eyes are open, when the infant is crying, fussing, eating, or sucking (Emde & Metcalf, 1970). Moreover, REM is also observed to occur within a few moments after an infant begins to engage in nutritional sucking and appears identical to that which occurs during sleep (Emde & Metcalf, 1970).

The production of REM during waking in some respects seems paradoxical and a number of different mechanism are no doubt responsible. Nevertheless, it might be safe to assume that like an adult, when the infant is in REM, he or she is dreaming, or at least, in a dreamlike state. This state might correspond to what Freud described as the primary process. That is, when produced when the infant is crying or fussing, it is dreaming of whatever relief it seeks. Correspondingly, REM that occurs while eating or sucking may have to do with the limbic structures involved not only in the production of dreamlike activity, but the identification, learning, and retention of motivationally significant information (i.e., the amygdala and hippocampus).

The Primary Process

The hypothalamus, our exceedingly ancient and primitive Id, has an "eye" that only sees inward. It can tell if the body needs nourishment but cannot determine what might be good to eat. It can feel thirst, but has no way of slaking this desire. The hypothalamus can only say: "I want," "I need," and can only signal pleasure and displeasure. However, as the seat of pleasure, the hypothalamus can be exceedingly gracious in rewarding the organism when its needs are met. Conversely, when its needs go unmet, it can respond not only with displeasure and feelings of aversion, but with undirected fury and rage. It can cause the organism to cry out. However, the cry does not produce the immediately desired relief or reduction in tension. Pressure is exerted on the limbic system and the organism to engage in environmental surveillance so as to meet the needs monitored by the hypothalamus.

Over the course of the first months of life, as the amygdala and then the hippocampus develop, the organism begins to develop an eye that not only sees outward but that can register and recall events, objects, people, and so forth associated with tension reduction, pleasure, and the satiety of the infants internal needs (e.g., the taste, smell, feeling of mother's breast and milk, the experience of sucking and relief). This is called learning.

With the maturation of these two limbic nuclei, the infant is increasingly able to differentiate what occurs in the external environment based on hypothalamically monitored needs and the emotional/motivational significance of that which is experienced. The infant can now orient, selectively attend, determine what brings satisfaction, and store this information in memory.

Primary Imagery

Although admittedly we have no direct knowledge as to the psychic interactions in the neonate, it does seem reasonable to assume that as the neocortex and underlying structures and fiber pathways mature, neural "programs" are formed that correspond to the repeated registration of experiences which are deemed significant (e.g., pleasurable). That is, neural pathways that are repetitively fired, deactivated, or activated in response to specific sensory and affective activities and experiences, become associated with that activity, such that an associated neural circuit is formed (Joseph, 1982; Penfield, 1954; Spinelli, 1970); i.e., a memory is created. Eventually, if this circuit is reactivated, the "learned" pattern is reexperienced; i.e., the organism remembers. Thus, when the amygdala/hippocampus is stimulated by a hungry hypothalamus, the events and images associated with past experiences of pleasure not only can be searched out externally but can be recalled in imaginal form. For example, as an infant experiences hunger and stomach contractions as well as its own cries of displeasure, these states become associated with the sound, smell, taste, and other cues of mother and her associated movement, as well as other stimuli that accompany being fed (cf. Piaget, 1952, pp. 37, 407-408). Repetitively experienced, the sequence from hunger to satiety evokes and becomes associated with the activation of certain neural pathways.

Eventually, when the infant becomes hungry and that hunger is prolonged, the entire neural sequence associated with hunger and feeding (i.e., hunger, mother, food, satiety) may become involuntarily triggered and activated (through association) such that an "image" of being fed is experienced. The activation of these rudimentary and infantile memory images is probably what constitutes, at least in part, the primary process.

Behaviorally, this is manifested by REM and by sucking and tongue movements as if eating, when in fact no food is present (cf. Piaget, 1952). When hungry, the infant will begin to cry and REM might be observed; the infant will then stop crying and smack its lips and make sucking movements (mediated by the amygdala) as if it were being fed. The infant experiences being fed in the form of a dream (Joseph, 1982) or hallucination.

The brain of the human infant is quite immature for several years, in turn restricting information reception and processing (Joseph, 1982; Joseph, Gallagher, Holloway, & Kahn, 1984); also, given the limited amount of reality contact infants are able to achieve, these rudimentary memories and images (even when they occur during waking, i.e., REM) are probably indistinguishable from actual experience simply because they are experience.

Like a dream, when replayed, the infant presumably reexperiences to some degree the sensations, emotions, and so forth originally linked to tension reduction. Thus, the young infant, still unable to distinguish between representation and reality, responds to the image as reality (Freud, 1911), even while awake - as manifested by REM. When hunger is prolonged, the associations linked to feeding are triggered, and for a brief period the infant behaves as if its hunger has been sated. Reality is replaced by an image, or rather, a "dream." This is the primary process.

Since the hypothalamus (which monitors internal homeostasis) is not conscious that the dream images experienced are not real, it initially accepts the memory/dream images transmitted from the amygdala and hippocampus and ceases to cry, i.e., it responds to the imagined sources of nourishment just as it responds to a cue tone associated with a food reward (Nakamura & Ono, 1986; Ono et al., 1980). However, the hypothalamus is not long fooled, for the primary process does not offer effective long-lasting relief from tension. As the pain of hunger remains and increases, limbic activity is increased, and the image falls away, to be replaced by a cry of hunger (Joseph, 1982). The amygdala and hippocampus are thus forced to renew their surveillance of the environment in search of sources of tension reduction. Cognitive development is thus promoted (Freud, 1911): Whatever was thought of (desired) was simply imagined in an hallucinatory form, as still happens today with our dream-thoughts every night. This attempt at satisfaction by means of hallucination was abandoned only in consequence of the absence of the expected gratification, because of the disappointment experienced. Instead, the mental apparatus had to decide to form a conception of the real circumstances in the outer world and to exert itself to alter them.... The increased significance of external reality heightened the significance also of the sense-organs directed towards the outer word, and of the consciousness attached to them; the later now learned to comprehend the qualities of sense in addition to the qualities of pleasure and "pain" which hitherto had alone been of interest to it. A special function was instituted which had periodically to search the outer word in order that its data might be already familiar if an urgent need should arise; this function was attention. Its activity meets the sense-impressions halfway, instead of awaiting their appearance. At the same time there was probably introduced a system of notation, whose task was to deposit the results of this periodical activity of consciousness - a part of that which we call memory.

SEPTAL NUCLEI

Phylogenetically, the septal nuclei develops following the appearance of the amygdala, at about the same time and rate as the hippocampus (Andy & Stephan, 1976; Brown, 1983; Humphrey, 1972). It also increases in relative size and complexity as we ascend the ancestral tree, attaining its greatest degree of development in humans.

Specifically, the septal nuclei lie in the medial portions of the hemispheres, just anterior to the third ventricle near the hypothalamus. The septum projects heavily throughout the hypothalamus and maintains rich interconnections with all regions of the hippocampus (Mesulam, Mufson, Levey, & Wainer, 1983; Siegel & Edinger, 1976). It also contributes to and receives fibers from the amygdala via the stria terminalis (Swanson & Cowan, 1979) and receives and gives rise to one of the most massive roots of the medial forebrain bundle through which it receives olfactory and ascending brainstem (reticular formation) fibers (Nauta, 1956).

Aversion and Internal Inhibition

In general, the septal nucleus appears to act in conjunction with the medial hypothalamus and hippocampus, particularly as related to internal inhibition and the exertion of quieting and dampening influences on arousal and limbic system functioning. In this regard, the septum appears to counterbalance some aspects of amygdaloid and lateral hypothalamic activity while simultaneously facilitating the actions of the hippocampus and media hypothalamus (Kolb & Whishaw, 1977; Mogenson, 1976; Petsche, Gogolack, & Van Zwieten, 1965; Petsche, Stumpf, & Gogolack, 1962).

Specifically, depending on the level of hypothalamic arousal, the septum exerts facilitatory or inhibitory influences on this nuclei (Mogenson, 1976) and from impulses it receives and relays from the reticular activating system, mediates and greatly influences hippocampal activity as well (Kolb & Whishaw,

1977; Petsche et al., 1962, 1965). The septal region (along with that of the hippocampus and medial hypothalamus) has also been reported in humans to show electrophysiological alterations in activity that correspond to subjective feelings of aversion (Heath, 1976).

The septum and amygdala appear to enjoy largely an antagonistic relationship, particularly in regard to influencing the hypothalamus. For example, the amygdala acts either to facilitate or to inhibit septal functioning, whereas septal influences on the amygdala are largely inhibitory. However, in large part, the amygdala and septal nuclei appear to exert most of their counterbalancing influences at the level of the hypothalamus, particularly in regard to emotionality.

Quiescence. A primary activity of the septal nucleus appears to be that of reducing extremes in emotionality and arousal and maintaining the organism in state of quiescence and readiness to respond. Stimulation of the septum acts to reduce blood pressure and heart rate and to induce adrenocortical secretion (Endroczi, Schreiberg, & Lissak, 1963; Kaada, 1951; Ranson, Kabat, & Magoun, 1935). It also counters lateral hypothalamic self-stimulatory activity (Mogenson, 1976).

Rage. Electrical stimulation of the septum counters and inhibits aggressive behavior (Rubenstein & Delgado, 1963) and suppresses the expression of rage reactions following hypothalamic stimulation (Siegel & Edinger, 1976). If the septal nucleus is destroyed, these counterbalancing influences are removed such that initially dramatic increases in aggressive behavior, including rage (Ahmad & Harvey, 1968; Blanchard & Blanchard, 1968; Brady & Nauta, 1953; King, 1958), results. Bilateral lesions in fact give rise to explosive emotional reactivity to tactile, visual, or auditory stimulation, which can take the form of attack (fight) or flight. If the amygdala is subsequently lesioned, the septal rage and emotional reactivity are completely attenuated (King & Meyer, 1958). Thus, septal lesions appear to result in a loss of modulatory and inhibitory restraint.

Septal Social Functioning

Eventually, within a few weeks or months, rage and aggressiveness due to septal lesions subside and/or completely disappear. However, a generalized tendency to overrespond and a generalized failure to inhibit emotional responsiveness persist (McClary, 1961; 1966; Poplawsky, 1987). In particular, rather than rage or irritability, septal lesions produce indiscriminant socializing and an extreme need for social and physical contact (Jonason & Enloe, 1971; Jonason et al., 1973; McClary, 1961, 1966; Meyer, Ruth, & Lavond, 1978).

Socialization and Contact Comfort

In contrast with amygdaloid lesions, which produce a severe social-emotional agnosia and social avoidance, septal lesions produce a dramatic and persistent increase in social cohesiveness (Jonason & Enloe, 1971; Jonason et al., 1973; D. R. Meyer et al., 1978). Thus, the normal intact amygdala appears to promote social behavior, whereas the septal nucleus seems to counter socializing tendencies.

With complete bilateral destruction of the septum in animals, the drive for social contact appears to be irresistible such that persistent attempts to make physical contact occur - even with species quite unlike their own. This occurs because the amygdala (as well as other nuclei) are no longer opposed by the septal nucleus.

Ritchie (1975, cited by D. R. Meyer et al., 1978)reported that septally lesioned rats, unlike normals, will readily seek out mice (to which they are normally indifferent) or rabbits (which they usually avoid). If presented with a choice of an empty (i.e., safe) chamber or one containing a cat, septally lesioned rats persistently attempt to huddle and crawl upon this normally feared creature, even when the cat is acting perturbed. If a group of septally lesioned animals are placed together, extreme huddling results. So intense is this need for contact comfort following septal lesions that if other animals are not available they will seek out blocks of wood, old rags, bare wire frames, or walls.

Among humans with right-sided or bilateral disturbances in septal functioning (such as due to seizure activity generated in this region), a behavior referred to as stickiness is sometimes observed. Such individuals seek to make repeated, prolonged, and often inappropriate contact with anyone who is available or who happens to be nearby so as to tell them stories or jokes or merely pass the time. They seem to have little actual concern regarding the feelings of others or how interested they might be in interacting. Indeed, they do not readily take a hint and are difficult to get rid of. In hospital situations, they can be found intruding on other patients and their families, hanging out by the nurse's station, or incessantly visiting other rooms to chat.

Attachment and Amygdaloid - Septal Developmental Interactions

Broadly considered, the various nuclei comprising the limbic system demonstrate functional maturity at different rates. Correspondingly, certain behaviors and capacities appear at various time periods, overlay previous capacities, become differentiated, and/or in turn become suppressed or eliminated.

At birth the hypothalamus reins supreme in regard to emotion. The infant freely expresses feelings of aversion and rage or pleasure or quiescence and demonstrates extreme orality and other behaviors similar to a state in which the amygdala is functionally nonexistent.

As the amygdala and hippocampus begin to develop and mature, the organism becomes more reality oriented and more social as the ability to attend selectively to externally occurring events and to store this information in memory becomes more pronounced. The pleasure principle (although still dominant) begins to be served by the reality principle (Freud, 1911).

Since the development of the amygdala precedes that of the septal nuclei, social-emotional activities mediated by the amygdala are expressed in an uninhibited fashion. That is, when the counterbalancing influences of the septum are absent (as a result of surgical removal or functional immaturity), the behavior expressed is in many respects similar to that expressed by infants, for example, indiscriminate contact seeking (e.g., attachment behavior).

Attachment

Attachment is not the same as dependency. Necessarily, the infant is dependent on its caretaker (e.g., mother). However, until the child is 6-7 months of age, it will smile at the approach of anyone, even complete strangers, and will vigorously protest any form of separation from these unknown people (e.g., if they leave the room). This state corresponds to the amygdaloid development period, during which septal influences are largely nil.

At about 7 months of age, the infant becomes more discriminant in its interactions and it is during this time period that a very real and specific attachment is formed, e.g., to one's mother - an attachment that becomes progressively more intense and stable. This period represents initial septal developmental influences such that global contact seeking becomes increasingly narrowed and restricted.

After these specific attachments have been formed, children begin to show anxiety, fear, and even flight reactions at the approach of a stranger (Spitz & Wolf, 1946). By 9 months, 70% of children respond aversively, whereas by 10 months they might cry out if a stranger were to appear (Schaffer, 1966; Waters, Matas, & Sroufe, 1975). By 1 year of age, 90% of children respond aversively to strangers (Schaffer, 1966).

Thus, during the amygdaloid phase there is indiscriminant approach and contact seeking. During the septal stage, indiscriminate social contact seeking is inhibited, whereas the specific attachments already formed are strengthened, reinforced, and maintained. The differential rates of amygdala and septal development are thus crucial in promoting survival and social interaction with significant others.

Their differential maturation rates also represent critical time periods, which in turn require specific interactional social-physical forms of stimulation, such as the presence and care of a primary caretaker. If these interactional needs are not met during the critical development period, gross abnormalities can result.

Indeed, if contact with others is restricted during these early phases, the ability to socially interact successfully at a later stage of development is retarded. This is even true among the so-called lower animals. For example, kittens that are not exposed to humans grow up to be wild and unapproachable. The phenomena of imprinting requires that similar interactions take place.

Contact Comfort

As is well known, for the first 6-8 months of life, physical and social interaction with others is critical in regard to psychological, neurological, and physical development. During this period (which corresponds to the amygdaloid phase of maturation), contact seeking is indiscriminant. However, indiscriminant contact seeking and attachment during early development maximizes opportunities for physical - social interaction. So intense is the need for physical and social contact that young animals will form attachments to bare wire frames (Harlow, 1962), to television sets, to dogs that might maul them, and to creatures that might kill them (Cairn, 1966). Young human children will form attachments to mothers who might abuse them. With removal of the septum, animals will even seek contact with creatures that might eat them.

So pervasive is this need for physical interaction (especially among humans) that, when grossly reduced or denied, the result is often death. For example, in several well-known studies of children raised in foundling homes during the early 1900s, when the need for contact was not well recognized, morbidity rates for children under 1 year of age were more than 70%. Of 10,272 children admitted to the Dublin Foundling home during a single 25-year period, only 45 survived (Langmeier & Matejcek, 1975).

Of those who have survived an infancy spent in institutions in which mothering and contact comfort were minimized, signs of low intelligence, extreme passivity, apathy, as well as severe attentional deficits are often characteristic (Dennis, 1960; Langmeier & Matejcek, 1975; Spitz, 1945). Such individuals have difficulty forming attachment or maintaining social interactions later in life.

In his famous series of experiments with monkeys, Harlow (1962) also showed that even those raised with surrogate terry cloth mothers develop extremely bizarre behaviors: The laboratory monkeys sit in their cages and stare fixedly into space, circle their cages in a repetitive stereotyped manner and clasp their heads in their hands or arms and rock for long periods of time. They often develop compulsive habits, such as pinching precisely the same patch of skin on their chest between the same fingers hundreds of times a day; occasionally such behavior may become punitive and the animal may chew and tear at its body until it bleeds.

It is noteworthy that the maternal behavior of chimpanzees raised in isolation is also quite abnormal and in fact similar to that of mothers who have had their amygdalas destroyed. As described by Harlow (1965): After the birth of her baby, the first of these unmothered mothers ignored the infant and sat relatively motionless at one side of the cage, staring fixedly into space hour after hour. As the infant matured desperate attempts to effect maternal contact were consistently repulsed.... Other motherless monkeys were indifferent to their babies or brutalized them, biting off their fingers or toes, pounding them, and nearly killing them until caretakers intervened. One of the most interesting findings was that despite the consistent punishment, the babies persisted in their attempts to make maternal contact.

Deprivation and Amygdaloid - Septal Functioning

In addition to mediating all aspects of higher-order emotional functioning, the amygdala is particularly responsive to tactual stimulation such that single cells may respond to touch, regardless of where on the body the person was stimulated. When there is inadequate tactile and physical-social interaction during early development, the ability of these cells to develop and function adequately is significantly reduced. That is, the amygdala (as well as other nuclei) becomes environmentally damaged.

As has been repeatedly demonstrated in studies of the visual system as well as on cerebral development, one consequence of reduced environmental input during certain critical periods of development is cell death, atrophy, and functional retardation, as well as inhibition by competing neuronal cell assemblies (Casagrande & Joseph, 1980; Greenough, 1976; Rosenzweig, 1971). Consequently, gross behavioral and perceptual abnormalities result (Harlow, 1962; Joseph & Casagrande, 1980; Joseph & Callagher, 1982).

For example, if the lids of one eye are surgically sutured shut, patterned visual input is prevented from reaching target cells in the lateral geniculate nucleus of the thalamus. When this occurs target neurons become smaller, fewer in number, and functionally suppressed by adjacent cells receiving normal input. If the sutures are reversed such that the formerly deprived eye is opened, the subject responds as if blind. However, if the normal eye is surgically removed, some vision returns to the formerly deprived eye, and some functional neuronal recovery is observed (Casagrande & Joseph, 1980), presumably because of the removal of inhibitory influences. This deprivation effect is only noted to occur during the first few months after birth in primates.

Given the similarity in social unresponsiveness and disturbances in maternal behavior demonstrated by amygdalectomized and socially-maternally deprived primates, it is likely that like the visual system, abnormalities in cellular development and function also occur within the amygdala and septum secondary to deprivation. Unfortunately, few researchers have explored this line of inquiry.

Nevertheless, it has been demonstrated that the right amygdala is larger than the left in animals reared in enriched and normal environments, whereas these differences are nonsignificant in animals reared in a restricted setting (Diamond, 1985), suggesting that deprivation differentially reduces right amygdaloid growth. This finding is significant because among humans, the right cerebral hemisphere has been shown to be dominant in regard to most aspects of social-emotional functioning.

Heath (1972) also noted that the monkeys reared without mothers develop abnormal spiking in the septal region. Similarly, Heath (1954) found abnormal seizure-like discharges in the septum of withdrawn schizophrenics, noting that the severity of the abnormality was correlated with the severity of the psychosis.

Among those with normal limbic systems, it is presumably the interaction of these same nuclei that gives rise to attachment and bonding behavior in adults. Moreover, it is probably through the interactions mediated by these nuclei that emotions such as jealousy are generated. That some individuals respond with considerable grief, anger, and even uncontrollable rage when a "loved one" has ended a relationship probably can also be explained from a limbic perspective.

THE UNCONSCIOUS MIND

The limbic system, as defined in this paper, provides the foundations for the development of not only emotion, but social-psychological functioning as well as the more rudimentary aspects of unconscious mental activity. The limbic system, however, represents the most primitive features of the unconscious. A second, more highly developed, unconscious (i.e., nonlinguistic) mind is maintained via the functional interactions of the right cerebral hemisphere (Joseph, 1988a, b); however this mental realm only seems to be unconscious from the perspective of the left cerebral hemisphere. Nevertheless, because the interaction of various limbic nuclei is also important in learning and memory, as well as social-emotional functioning, it is likely that a variety of neurotic disturbances, needs, desires, and insecurities have as their foundation events differentially experienced and attended to by the limbic system.

This seems particularly true regarding events experienced during infancy and early development. For example, it is well known that most events experienced before age 4 are extremely difficult to recall. In part this is because of the different modes of information processing and coding employed by adults versus children (Joseph, 1988a). That is, much of what was experienced, learned, and stored in memory during early childhood occurred before the development of language. Thus, adults, relying on language, can no longer access the code in which this early information was stored. The key no longer fits the lock.

Nevertheless, because the limbic system of both an adult and infant uses nonlinguistic social-tactual-emotional-visual forms of information processing and expression, much of what occurs at this level falls outside the immediate jurisdiction of the conscious mind, which relies predominantly on language and rational, temporal-sequential linguistic thought for understanding.

Because of this, we may feel angry or sad and not consciously (i.e., linguistically) know why. However, at a limbic level, we may feel anxious or angry because something was seen, heard, or even smelled, which called forth old limbic memories of some unpleasant event - memories that are not accessible to the linguistic mind.

Take, for example, a young child, who, like many infants, at times manipulated her genitalia. Unfortunately, this little girl's mother (being very repressed and rigid) responded with shock and disgust when she discovered her daughter's action. Resolving to put an end to this "repulsive behavior," the mother shouted "No" in an angry voice and slapped the child every time her daughter touched herself.

Soon the behavior was abolished through simple stimulus - response learning. This is because, at a limbic level, every time the urge to touch the genitalia arose, the behavior was severely punished. The child's urge to touch became associated with physical pain, and later with the anxiety of anticipated pain. Only three nuclei are necessary to form this circuit - the hypothalamus, amygdala, and hippocampus - as it is purely emotional and without thought. Indeed, much of this learning can probably occur without amygdaloid or hippocampal involvement, in which case the memory may remain unconscious. In fact, if sufficiently traumatized, the hippocampus (as well as the amygdala) might be damaged, thus resulting in unconscious learning and memories that remain unconscious and which cannot be recalled due to trauma and hippocampal induced amnesia.

Nevertheless, as an adult, this young woman now has difficulty associating with men, as sometimes their mere presence brings forth terrible feelings of anxiety. She has one failed relationship after another, and soon discovers herself to be frigid. She hates sex, she hates to be touched, and men repel her.

Consciously, she may be able to generate a panoply of explanations, all of which, at some level, are believable, albeit erroneous. Consciously, she has no true idea as to the source of her difficulties. Nevertheless, at the limbic level, the explanation is very basic. Certain men trigger the onset of amorous feelings. Unfortunately, the woman never feels sexually aroused. The first inklings of arousal immediately generate feelings of anxiety and limbic memories of physical pain. This is what she experiences and responds to.

CONCLUSION

A variety of nuclei and brain regions are important in emotional functioning. In this chapter, only a few of these regions are discussed. In some respects, many might argue that such structures as the hippocampus should not be considered part of the limbic system because of its involvement with memory. Indeed, it has become increasingly fashionable to decry even the use of the term limbic system, as there are simply so many diverse cerebral regions involved in the control and mediation of emotional functioning. Even by the most liberal of anatomical definitions there can be no structural basis for the concept when all such nuclei are considered.

This is not the case, however, in regard to the highly interactional anatomical and functional system maintained by the amygdala, hypothalamus, hippo-campus, and septal nuclei. Therefore, I encourage continued use of the term, particularly in that to most people limbic system implies that part of the "old" brain concerned with emotional functioning.

6. Sex, Violence And Religious Experience

THE LIMBIC SYSTEM & VIOLENT BEHAVIOR

The neocortical surface of the brain (the so called "gray matter") is the seat of the rational mind, being concerned with language, math, reasoning, and higher level cognitive capacities. Beneath this neocortical mantle, buried within the depths of the cerebrum are several large aggregates of limbic structures and nuclei which are preeminent in the control and mediation of all aspects of emotion including violent, aggressive, sexual, and social behavior and the formation of loving attachments. The limbic system controls the capacity to experience love and sorrow, and governs and monitors internal homeostasis and basic needs such as hunger and thirst, including even the cravings for pleasure-inducing drugs (Bernardis & Bellinger 1987; Childress, et al., 1999; Gloor 1992, 1997; LeDoux 1992, 1996; MacLean, 1973, 1990; Rolls, 2014, 1992; Smith et al. 1990). Over the course of evolution, the forebrain and much of the neocortex (and the so called "rational mind") evolved in response to and so as to better serve the limbic system and fulfill and satisfy limbic needs.

However, the limbic system is not only predominant in regard to all aspects of motivational and emotional functioning, but is capable of completely overwhelming "the rational mind" due in part to the massive axonal projections of the amygdala upon the neocortex. Although over the course of evolution a new brain (neocortex) has developed, Homo sapiens sapiens ("the wise may who knows he is wise") remains a creature of emotion. Humans have not completely emerged from the phylogenetic swamps of their original psychic existence. Hence, due to these limbic roots, humans not uncommonly behave "irrationally" or in the "heat of passion," and thus act at the behest of their immediate desires; sometimes falling "madly in love" and at other times, acting in a blind rage such that even those who are 'loved" may be slaughtered and murdered.

The schism between the rational and the emotional is real, and is due to the raw energy of emotion having it's source in the nuclei of the ancient limbic lobe — a series of structures which first make their phylogenetic appearance over a hundred million years before humans walked upon this Earth and which continue to control and direct human behavior.

Humans, although rational creatures, are also killers, men in particular. Men are natural born killers, which is why the history of human affairs is written in gore and blood. This sad state of affairs is due to the incredible power of the limbic system which, when aroused, can hijack and gain complete control over the rational mind. Unfortunately, as the limbic brain also mediates spirituality and religious experience, not uncommonly the worst of crimes and the worst of murders, are committed in the name of god and in the name of religion.

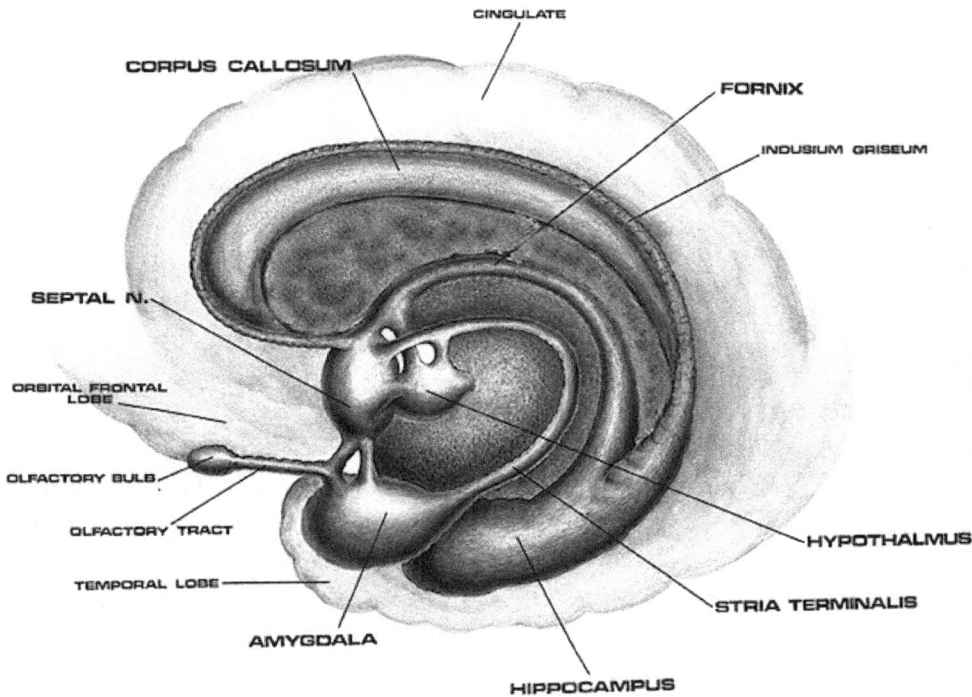

FUNCTIONAL OVERVIEW

In general, the primary structures of the limbic system include the hypothalamus, amygdala, hippocampus, septal nuclei, and anterior cingulate gyrus; structures which are directly interconnected by massive axonal pathways (Gloor, 1997; MacLean, 1990; Risvold & Swanson, 1996). With the exception of the cingulate which is referred to as "transitional" cortex (mesocortex) and consists of five layers, the hypothalamus, amygdala, hippocampus, septal nuclei are considered allocortex, consisting of at most, 3 layers.

The hypothalamus could be considered the most "primitive" aspect of the limbic system, though in fact the functioning of this sexually dimorphic structure is exceedingly complex. The hypothalamus regulates internal homeostasis including the experience of hunger and thirst, can trigger rudimentary sexual behaviors or generate feelings of extreme rage or pleasure. In conjunction with the pituitary the hypothalamus is a major manufacturer/secretor of hormones and other bodily humors, including those involved in the stress response and feelings of depression.

The amygdala has been implicated in the generation of the most rudimentary and the most profound of human emotions, including fear,

146

sexual desire, rage, religious ecstasy, or at a more basic level, determining if something might be good to eat. The amygdala is implicated in the seeking of loving attachments and the formation of long term emotional memories. It contains neurons which become activated in response to the human face, and which become activated in response to the direction of someone else's gaze. The amygdala also acts directly on the hypothalamus via the stria terminalis, medial forebrain bundle, and amygdalafugal pathways, and in this manner can control hypothalamic impulses. The amygdala is also directly connected to the hippocampus, with which it interacts in regard to memory.

The amygdala maintains a functionally interdependent relationship with the hypothalamus in regard to emotional, sexual, autonomic, consumatory and motivational concerns. It is able to modulate and even control rudimentary emotional forces governed by the hypothalamic nucleus. However, the amygdala also acts at the behest of hypothalamically induced drives. For example, if certain nutritional requirements need to be meet, the hypothalamus signals the amygdala which then surveys the external environment for something good to eat (Joseph, 1982, 1992a). On the other hand, if the amygdala via environmental surveillance were to discover a potentially threatening stimulus, it acts to excite and drive the hypothalamus as well as the basal ganglia so that the organism is mobilized to take appropriate action.

When the hypothalamus is activated by the amygdala, instead of responding in an on/off manner, cellular activity continues for an appreciably longer time period (Dreifuss et. al., 1968). The amygdala can tap into the reservoir of emotional energy mediated by the hypothalamus so that certain ends may be attained.

LIMBIC LUST, MURDER, AND RELIGIOUS EXPERIENCE

The amygdala and hypothalamus often act in a highly coordinated manner in reaction to an exceedingly important emotional stimulus, or in response to a specific limbic need, such as hunger, thirst, rage, or sexual desire (Joseph, 1992a, 1996; MacLean 1969, 1990). For example, in response to hypothalamically monitored needs (hunger, sexual desire), the amygdala may scan the environment until it determines that a particular food item or person, has the necessary attributes (Gloor 1992; Joseph 1992a,b, 1996; Kling et al., 1987; O'Keefe and Bouma 1969; Ursin and Kaada 1960). In response to urgent hypothalamic desires, the amygdala might even assign sexual attributes to an individual that normally might not be viewed as sexually enticing.

It is also through hypothalamic and amygdala activity that a particular item or object (e.g. a banana) might be viewed as both a food item and sexual object. Or conversely, why certain individuals may be viewed as sexual as well as aversive and hateful (e.g. one's husband or wife). Indeed, because the hypothalamus and amygdala are so concerned with sex, rage, fear, and hunger, not only may these attributes be assigned to one individual, animal, or object simultaneously (e.g. fear of the beast one is going to enjoy killing and eating; hunger, guilt, and aversion regarding a high caloric treat; hatred for a loved one) but these conflicting emotions may be combined so as to give rise to exceedingly intense, albeit abstract emotional states; e.g. religious awe as well as religious rage.

HYPOTHALAMIC RAGE

Stimulation of the lateral hypothalamus can induce extremes in emotionality, including intense attacks of rage accompanied by biting and attack upon any moving object (Flynn et al. 1971; Gunne & Lewander, 1966; Wasman & Flynn, 1962). If this nucleus is destroyed, aggressive and attack behavior is abolished (Karli & Vergness, 1969). Hence, the lateral hypothalamus is responsible for rage and aggressive behavior.

The lateral maintains an oppositional relationship with the medial hypothalamus. Hence, stimulation of the medial region counters the lateral area such that rage reactions are reduced or eliminated (Ingram, 1952; Wheately, 1944), whereas if the medial is destroyed there results lateral hypothalamic release and the triggering of extreme savagery.

Inflammation, neoplasm, and compression of the hypothalamus have also been noted to give rise to rage attacks (Pilleri & Poeck, 1965), and surgical manipulations or tumors within the hypothalamus have been observed to elicit manic and rage-like outbursts (Alpers, 1940). These appear to be release phenomenon, however. That is, rage, attack, aggressive, and related behaviors associated with the hypothalamus appears to be under the inhibitory influence of higher order limbic nuclei such as the amygdala and septum (Siegel & Skog, 1970). When the controlling pathways between these areas are damaged (i.e. disconnection) sometimes these behaviors are elicited.

For example, Pilleri and Poeck (1965) described a man with severe damage throughout the cerebrum including the amygdala, hippocampus, cingulated, but with complete sparing of the hypothalamus who continually reacted with howling, growling, and baring of teeth in response to noise, a slight touch, or if approached. Hence, the hypothalamus being released responds reflexively in an aggressive-like non-specific manner to any stimulus. Lesions of the frontal-hypothalamic pathways have been noted to result in severe rage reactions as well.

THE AMYGDALA AND RAGE

Fear and rage reactions have also been triggered in humans following depth electrode stimulation of the amygdala (Chapman, 1960; Chapman et al., 1954; Heath et al. 1955; Mark et al. 1972). Mark et al. (1972) describe one female patient who following amygdaloid stimulation became irritable and angry, and then enraged. Her lips retracted, there was extreme facial grimmacing, threatening behavior, and then rage and attack—all of which persisted well beyond stimulus termination.

Similarly, Schiff et al. (1982) describe a man who developed intractable aggression following a head injury and damage (determined via depth electrode) to the amygdala (i.e. abnormal electrical activity). Subsequently, he became easily enraged, sexually preoccupied (although sexually hypoactive), and developed hyper-religiosity and psuedo-mystical ideas. Tumors invading the amygdala have also been reported to trigger rage attacks (Sweet et al. 1960; Vonderache, 1940).

In many instances patients or animals will react defensively and with anger, irritation, and rage which seems to gradually build up until finally the animal or human will attack (Egger & Flynn, 1963; Gunne & Lewander, 1966; Mark et al., 1972 Ursin & Kaada, 1960; Zbrozyna, 1963). Unlike hypothalamic "sham rage," amygdaloid activation results in attacks directed at something real, or, in the absence of an actual stimulus, at something imaginary. There have been reported instances of patient's suddenly lashing out and even attempting to attack those close by, while in the midst of a temporal lobe seizure (Saint-Hilaire et al., 1980), and/or attacking, kicking, and destroying furniture and other objects (Ashford et al., 1980).

Moreover, rage and attack will persist well beyond the termination of the electrical stimulation of the amygdala. In fact, the amygdala remains electrophysiologically active for long time periods even after a stimulus has been removed (be it external-perceptual, or internal-electrical) such that is appears to contine to process—in the abstract—information even when that information is no longer observable (O'Keefe & Bouma, 1969). The individual may therefore remain enraged or fearful long after the threat or offending party had departed the scene. The amygdala makes it possible to feel moody.

The amygdalas appears capable of not only triggering and steering hypothalamic activity but acting on higher level neocortical processes so that individuals form emotional ideas. Indeed, the amygdala is able to overwhelm the neocortex and the rest of the brain so so that the person not only forms emotional ideas but responds to them, sometimes with vicious, horrifying results. A famous example of this is Charles Whitman, who in 1966 climbed a tower at the University of Texas and began to indiscriminantly kill people with a rifle (Whitman Case File # M968150. Austin Police Department, Texas, The Texas Department of Public Safety, File #4-38).

Charles Whitman climbed the University tower carrying several guns, a sawed off shotgun, and a high powered hunting rifle, and for the next 90 minutes he shot at everything that moved, killing 14, wounding 38. Post-mortem autopsy of his brain revealed a glioblastoma multiforme tumor the size of a walnut, erupting from beneath the thalamus, impacting the hypothalamus, extending into the temporal lobe and compressing the amygdaloid nucleus

(Charles J. Whitman Catastrophe, Medical Aspects. Report to Governor, 9/8/66).

THE AMYGDALA, FEAR, AND THE LORD

"And now Israel, what does the Lord your god require of you, but to fear the Lord your god." Deuteronomy 10:12.

The amygdala in particular is exceedingly important in generating feelings of fear (Davis et al., 1997; Gloor, 1992, 1997; Halgren, 1992; Rosen & Schulkin 1998; Scott et al., 1997; Williams 1956) as well as sexuality, rage, and hunger. In this regard, feelings of religious awe may be based on fear (d'Aquili and Newberg 1993), rage, extreme, hunger, or sexual arousal. Fear, however, is often the most potent means of eliciting religious feeling, for even a committed atheist when confronted with the possibility of a horrifying death may cry out to god. The "Lord God" Yahweh, in fact, depends on fear, and glories in terrifying his subjects in order to reveal his presence and power:

"The beginning of wisdom is the fear of the Lord." -Proverbs 1:7, 9:10, 15:33.

"And now, Israel, what does the Lord your god require of you, but to fear the Lord your god." -Deuteronomy 10:12.

"God has come... in order that the fear of Him may be ever with you so that you do not go astray." -Exodus 20:17.

THE FRONTAL-TEMPORAL LOBES

In addition to the amygdala, temporal lobe and hippocampus, d'Aquili and Newberg (1993) point out that the right frontal lobe also plays a significant role in the generation of mystical experience. It is thus noteworthy that the right frontal lobe can pray, swear, and curse "God" even when the (speaking) left cerebral hemisphere has been severely damaged and the patient is aphasic and can no longer speak (Joseph, 1982, 1988a, 1999a).

The right frontal and temporal lobe, hypothalamus, and amygdala also interact in regard to sexual arousal (Freemon & Nevis 1969; Joseph, 1986a, 1988a, 1992a, 1999a; MacLean 1969, 1990; Remmillard et al. 2013; Robinson & Mishkin 1968; Spencer et al., 2013). This is a very important relationship, and in part explains why (although there are exceptions), religions tend to be quite sexual and/or exceedingly concerned with sexual mores and related activity. As is well known, female pregnancy and matters pertaining to birth control and abortion are of extreme concern to most modern as well as ancient religions (Campbell 1988; Frazier 1955; Parrinder 1980; Smart 1969).

The limbic system as well as the frontal and temporal lobes are highly concerned with acting on or inhibiting aggression and murderous rage reactions which also arise in the limbic system (Joseph 1986, 1988, 1992a, 1996). This may also explain why many religious sects are so "righteously" belligerent and hateful and have employed torture, human or animal sacrifice, and sanctioned if not encouraged the murder of nonbelievers: What could be referred to as limbic-religious blood lust.

"Shed man's blood, by man your blood be shed." -Genesis 9:6.

SHED MAN'S BLOOD

The "Lord God," Yahweh, repeatedly required that the ancient Israelites undergo a bloody ritual of submission (e.g. Exodus 24:1-14), and in fact proscribed a ritual of incredible bloodiness for the investiture of his priests (Exodus 29: 1-46). This "Lord God" also required the slaughter and sacrifice of living creatures whose blood is splashed on his altar, and on his priests.

For example, upon the ratification of the covenant, twelve oxen were drained of their blood, one oxen for each of the 12 tribes of Israel, and as it was splashed on the people, Moses says" This is the blood of the covenant that the Lord now makes with you concerning these commands" (Exodus 24:8). In yet another bloody ritual a bull is slaughtered then a ram, then yet another ram, and their blood is splashed and smeared on the altars, on the priests, and on the people, and climaxes with those who are being ordained as priests holding up bloody pieces of the body (Exodus 29: 10-28).

King Solomon slaughtered 22,000 oxen and 22,000 sheep as an offering to this "Lord God" whose loves meat, but not his vegetables.

In fact, as this "Lord God" is apparently a meat eater, this may explain why he criticized Cain, a tiller of the soil, and rejected his first harvest offering of fruit and vegetables that he had grown with his own hand (Genesis, 4):

"And in the process of time it can to pass that Cain brought of the fruit of the ground and offering until the Lord. And Abel, he also brought of the firstlings of his flock and the fat thereof. And the Lord had respect unto Abel and to his offering: But unto Cain and to his offering he had not respect." -Genesis 4.3-5.

For ancient hunters, aggression and the killing of animals (and other humans) was a way of life. Hunters often employed hunting magic and related religious rituals to insure success. Religion and murder, like religion and sex, are linked to the limbic system and evolved accordingly. Consequently, when in the throes of religious excitement, torture and murder may even receive the blessing or might be actively encouraged by one's "God."

The "Lord God" enjoys killing people, and informs Moses that even those who wish to convert should be slaughtered: "You shall make no covenant with them... " and "I will blot out their memory... The Lord will be at war with Amalek throughout the ages" (Exodus 17:14-16). This Lord God, will in fact engage in widespread ethnic cleansing and genocide, and will also impose a rein of terror on His own people (e.g., Exodus 32: 26-29, 35). The Lord God loves to spill the blood of innocent and guilty alike.

On the other hand, the blood sacrifice is also related to the worship of the goddess, the Great Mother of All. It is menstrual blood which issues from the womb, the source of all life, and the rites of the goddess cult involved sanctifying and splashing menstrual blood on the altar. Hence, many ancient patriarchal religions appear to have adopted this practice, with the notable exception that instead of using the blood to celebrate life, the purpose of these bloody rituals was to evoke the power to take life which would be sacrificed to a warring, meat eating, God.

WARRIOR GODS: FEAR AND PANIC

Throughout history, many of the patriarchal Gods have been aggressive, jealous, conquering, angry and war-like e.g., Marduk, Mazda, Zeus, Apollo, Mars, and Yahweh "the Lord of hosts." For example, in addition to being described as the "Lord of hosts," Yahweh also means: "The Destroyer."

These warrior gods, including Yahweh, were prone to mass murder and extremely violent rages. Yahweh repeatedly threatened and engaged in the slaughter of enemies and believers alike, without mercy or regard for women and children. Terror, war, and murder: the Lord God's middle names. "Terror, and the pit, and the snare are upon you, O inhabitant of earth (Isaiah 24:17). And as the Lord took delight in doing you good and multiplying you, so the Lord will take delight in bringing ruin upon you and destroying you (Deuteronomy 26:63). The Lord will bring a nation against you from afar, from the end of the earth, which will swoop down like the eagle... a ruthless nation, that will show the old no regard and the young no mercy (Deuteronomy 28:47-50). It shall devour the offspring... you shall eat your own issue, the flesh of your sons and daughters... until He has wiped you out... leaving you nothing... until it has brought ruin unto you..." (Deuteronomy 28:50-55).

"In the Name of God... by the Troops shall the unbelievers be driven towards Hell, until when they reach it, its gates shall be opened... for just is the sentence of punishment on the unbelievers... " Koran, XXXIX

"Behold I send an angel before thee, to keep thee in the way. Beware of him and obey his voice, for I will be an enemy unto thine enemies, and an adversary unto thine adversaries, and I will cut them off... .I will send my terror before thee, and will destroy all the people to whom thou shalt come... . and I will drive them out from before thee, until thou be increased and inherit the land." -Exodus 23:20-30

"...when you approach a town, you shall lay seizure to it, and when the Lord your god delivers it into your hand, you shall put all its males to the sword. You may, however, take as your booty the women, the children, the livestock, and everything in the town—all its spoils—and enjoy the spoil of your enemy which the Lord your god gives you... In the towns of the people which the Lord your god is giving you as a heritage, you shall not let a soul remain alive." -Exodus 20:15-18; -Deuteronomy 20:12-16

"When Israel had killed all the inhabitants of Ai... and all of them, to the last man had fallen by the sword, all the Israelites turned back to Ai and put it to the sword...until all the inhabitants of Ai had been exterminated... and the king of Ai was impaled on a stake and it was left lying at the entrance to the city gate." -Deuteronomy 8:24-29.

In fact, the Lord God condemned even his most loyal followers if they dared to show even the slightest mercy even to animals (e.g., I Samuel 28:18-19). The Lord God, as described in the Old Testament, had what might best be described as a hair trigger temper, such that at the slightest sign of murmuring he would pounce and kill men, women, and children, the innocent and guilty alike including those of his own people. As is evident, for example, in the story of Exodus, the people were His prisoners, and He killed or insured that every adult who exited Egypt would die in the desert, including even Moses for making but one small mistake in following His orders: Instead of ordering a rock to give water, Moses tapped it with his cain. For this "sin," the Lord God kills Moses with a Divine kiss.

LIMBIC AND RELIGIOUS MASS MURDER: ETHNIC CLEANSING, AND GENETIC GENOCIDE

Led by Moses, and their "Lord God" the ancient Hebrews/Israelites murdered untold numbers, perhaps hundreds of thousands of innocent men, women, children, including even their livestock. These atrocities were committed against all manner of innocent peoples not because they had sinned, but because they happened to be along the path to the "promised land" and because they had the misfortune of living on land and possessing property that the "Lord God" wanted to give to his people.

"When the Lord they God shall bring thee into the land, and shall cast out many nations before thee, the Hittite, and the Gergashite, and the Amorite, and the Canaanite, and the Perizzite, and the Hivite, and the Jebusite... and the Lord thy god shall deliver them up before thee, and thou shalt smite them, then thou shall utterly destroy them."-Deuteronomy 7:1-2.

"Of the cities of these peoples, that the Lord they God giveth thee for an inheritance, thou shalt save alive nothing that breathest." -Deuteronomy, 20:16.

Moreover, "God" gloried not only in the murder of innocent women and children, but took special satisfaction in the theft of their property—the spoils of a godly-genocidal war. The Lord God believed that theft and injustice was a testament to his glory as a conquering warrior god.

"Great and goody cities, which thou didst not build and houses full of all good things which thou didst not fill, and cisterns hewn, which thou didst not hew, and vineyards and olive-trees, which thou didst not plant, and thou shalt eat and be satisfied." -Deuteronomy 6:10-11.

Yahweh, the Lord of Hosts, the "Destroyer," apparently so enjoyed the spectacle of mass murder, that once he began to kill He found it difficult to stop, killing everything and everyone, the guilty and the innocent, the righteous and the wicked, and even their animals. Apparently only blood and more blood, the signs of death, could protect the innocent, and cool his ardor for indiscriminate mass murder, which is why He explains to Moses that the only way to protect himself and the Jewish people from slaughter was by painting their doors red with blood: "When I see the blood" He informs Moses, "I will pass over you, and there shall be no plague upon you to destroy you, when I smite the land of Egypt" (Exodus 12:12-13).

It was only the sight of blood which prevented Yahweh from killing Moses almost immediately after informing him that he was to lead the Israelites out of Egypt.

"And the Lord said unto Moses...Go... And it came to pass by the way in the inn, that the Lord met him and sought to kill him. Then Zipporah" (the wife of Moses) "took a sharp stone and cut off the foreskin of her son, and cast it at his feet, and said, Surely a bloody husband art though to me."

Why would the "Lord God" seek to kill his prophet? There are many possibilities, all of which may have provoked "Lord God's" inherent murderous nature. For example, Moses dilly dallied and did not go directly to Egypt as he had been commanded. In addition, the "Lord God" may have been provoked to rage because Moses had failed to circumcise his son. Moses never did escape the wrath of the "Lord" for the "Lord God" kills Moses just as he and the Israelites are on the verge of entering the promised land.

THOU SHALT KILL AND KILL AND KILL

Despite the commandment "thou shall not kill" Yahweh kills and murders the innocent and guilty alike and encourages and in fact orders the Israelites to murder even babies and women with children.

"And they warred... as the Lord commanded and slew all the males. And they slew the Kings... and they took all the women and their little ones... and they burnt all their cities wherein they dwelt, and all their goodly castles with fire... And Moses was wroth...and said unto them. Have ye saved all the women and the little ones alive? Now therefore kill every male among the little ones, and kill every woman that hath known man by lying with him." (Numbers, 31).

As repeatedly detailed in Exodus, the "Lord God" purposefully and repeatedly "hardened Pharaoh's heart" simply as an excuse for murdering innocent Egyptians, so that He could glory in the carnage.

"I will harden Pharaoh's heart that I may multiply My signs and marvels in The land of Egypt. When the Pharaoh does not heed you, I will lay My hand upon Egypt." -Exodus 7:3.

Moreover, it is not just Egyptians or the hapless innocents who the Israelites and their Lord God murder in their 40 years of wondering, but tens of thousands if not hundreds of thousands of Jews. The Lord God kills them for rebelling, He kills them for "murmuring," He kills them for complaining, He kills them for questioning. And He kills their wives, brothers, fathers, mothers, and children: "Go forth and slay brother, neighbor, and kin" (Genesis, 32:26-29).

For example, when Moses ascended the Mountain to meet with the Lord God, the Israelites grew impatient with the hardships and murders they were forced to endure, and they turned again to worshipping the scared goddess, as represented by the golden calf. In reaction, Yahweh thunders and his prophet Moses proclaims that innocent and guilty alike shall be murdered: "Put ye every man his sword upon his thigh and slay every man...his brother, and every man his companion, and every man his neighbor and kin" (Genesis, 32:26-29).

When Moses proclaimed the Israelites a "nation of priests," some of the Israelites then dared to ask: if "Every member of the community is holy and the Lord is among them all... Why do you set yourselves up above the assembly of the Lord?" The response? Those who dared to ask questions are killed, including their wives, children, brothers and neighbors: "And the Earth opened her mouth and swallowed them" (Numbers, 16:32).

The ancient Israelis not only received special permission from their "Lord God" to murder wayward Jews, non-Jews and Jewish nonbelievers, including women and children whom they slaughtered without mercy (e.g. Numbers, 3115-18; Numbers 34, 50-53), but even Jewish babies. Indeed, it was a Hebrew religious-tradition to kill and slaughter not only non-Jewish males in general but first born Jewish sons (a custom until the time of Moses, e.g. Bergmann 1992).

SACRIFICIAL MURDERS

Human sacrifice was a common feature of many ancient religions, and serves as one of the founding stones of Christianity: The Lord God's sacrifice of his son, Jesus, on the Cross. Hence, it is said that "Christ died for your sins." Christ was sacrificed as an offering to the Lord God, so as to wash away the sins of the masses—or so claims the Roman Catholic Church.

The Lord God showed a particular fondness for killing first born sons, including the first born of the ancient Egyptians, as well as his own people.

"A blessing on him who seizes your babies and dashes them against rocks." -Psalm 137:9).

"I polluted them with their own offerings, making them sacrifice all their firstborn, which was to punish them, so that they would learn that I am Yahweh." -Ezekiel 20:25-36; 22:28-29.

This "Lord God" even required the death of the first born son of King

David and his wife Bethsheba, despite the fact that He claimed to have loved David most dearly. However, by killing this little boy, "God" in effect pardoned King David for repeatedly breaking two of His commandments, i.e., murder (of Bethsheba's husband) and adultery. Thus, the son of David was sacrificed by the Lord God so as to cleanse David of his sins, including, perhaps his sinful propensity for having sex with other men's wives (e.g. Abigail wife of Nabal, and Michal wife of Paltiel). However, David the murderer and adulterer, was also a fierce warrior who had killed tens of thousands, including even tens of thousands of Jews—and this "Lord God" loved him most dearly.

Sacrificial murder, therefore, is a way of indicating atonement, and for obtaining favor from one's god, including the Lord God. For thousands of years it has been a world wide religious custom to sacrifice captured warriors, young virgins, and especially children.

In some ancient societies children were killed because they were "pure" and precious. Thus the death of these innocent children could be considered a true sacrifice. Moreover, the ancient gods, including Yahweh, required that those murdered in their glory, be pure and without blemish. Children, and virgins, therefore, were a natural choice.

First born sons in particular were singled out for this honor as many ancient religions held to the belief that the first born was the offspring of a god who had opened the womb and impregnated the mother. This belief may even have caused the Lord God some consternation, for he not only required the killing of the first born, but he repeatedly shows favor to second born sons, e.g., Abel over Cain, Isaac instead of Ishmael, Jacob instead of Esau, Ephram over Manasseh, Solomon the second born son of David and Bethsheba, and so on.

It was also believed by many pagan groups that because a god may have impregnated the mother, the god also lost some of his lifeforce in the process. Hence, the first born would be killed to liberate that energy which could then be absorbed by the god.

It may well have been because of these beliefs that other gods were absorbing this life force, that the Lord God, although at first demanding the sacrifice of first born sons, later changed His mind, and condemned the practice:

"This very day you defile yourselves in the presentation of your gifts by making your children pass through the fire of all your fetishes." -Ezekiel 20:31.

CHRISTIAN HOLY WARS KILL THEM ALL: GOD WILL RECOGNIZE HIS OWN

It was upon these images of the murdering warrior God, the Lord of Hosts, that Pope Urban II proclaimed that war for the sake of God was holy. Thus, the Catholic Popes instigated numerous Crusades and inquisitions. In consequence hundreds of thousands of Moslems, Jews, and women and children, were sexually tortured, slaughtered, spitted, and roasted alive, and their cities and villages pillaged and set ablaze. All in the name of the Lord God. So intense was their limbic blood lust that even Christians were murdered.

For example, in the 13th Century an army of some thirty thousand Christian knights and Crusaders descended into southern France and attacked the town of Beziers in search of heretics. Over thirteen thousand Christians flocked to the churches for protection. When the Bishop, one of the Pope's representatives, was informed that the army was unable to distinguish between true believers and heretics, he replied, "Kill them all. God will recognize his own."

However, in order to recruit those worthy of such a glorious and murderous undertaking, the Pope had to appeal to murderers, rapists, molesters of children, and those who enjoyed the prolonged torture of their victims.

"You oppressors of orphans, you robbers of widows, you homicides, you blasphemers, you plunderers of others' rights... If you want to take counsel for your souls you must go forward boldly as knights of Christ..." so proclaimed the Pope who offered "indulgences" and forgiveness to all those who would commit blasphemies and murder women and children in the name of the Lord God and Jesus Christ.

As summed up by Henry Charles Lea: "Protestants and Catholics rivaled each other in the madness of the hours... Christendom seemed to have grown delirious, and Satan might well have smiled at the tribute to his powers seen in the endless smoke of the holocaust which bore witness to his triumph over the Almighty."

Again, however, although "Lord God" of the "Old Testament" repeatedly approved of mass murder and rape, and in fact employed these practices against His own people, Jesus Christ obviously did not preach mass murder, torture, rape, pedophilia, and the castration of young boys: "But love ye your enemies and do good and lend, hoping for nothing and your reward shall be great. Be ye merciful...judge not, and ye shall not be judged: condemn not and ye shall not be condemned: forgive and ye shall be forgiven" (Luke, 7: 35-37); "For the Son of man is not come to destroy men's lives, but to save them" (Luke 10: 56; however, see Matthew 10:16 vs 34-35).

Nevertheless, despite the teaching of Christ, the Catholic Church fully approved of castration, mass murder, and the most horrible of tortures. Indeed, as is well known, the Spanish and Catholic missionaries, acting at the behest of the Catholic Popes (and their Spanish/Catholic Sovereigns), continued these Satanic practices once they invaded the America's during the 1500's and up through the 19th century.

As the Catholic Dominican Bishop Bartolom de Las Casas reported to the Pope: the Aztec and Indian natives were hung and burnt alive "in groups of 13... thus honoring our Savior and the 12 apostles." However, because some of the various victims managed to live throughout the night, the Dominican priests ordered that sticks be shoved down their throats so the soldiers and priests would not be kept awake at night by their cries and moans.

Of course the Aztecs did not practice a benign form of worship, for they tore the beating hearts from their victims in order to please their God (Carrasco 1990) and they killed thousands if not hundreds of thousands in so doing.

Similarly, many Indian tribes of the Mississippi valley practiced human sacrifice as did the ancient Jews, Europeans, and leaders of the Protestant Reformation who urged the killing of anyone and everyone who did not agree with their interpretation of the Bible, including fellow Christians:

"Therefore let everyone who can smite, slay, and stab, secretly or openly, remembering that nothing can be more poisonous, hurtful, or devilish that a rebel. It is just as when one must kill a mad dog; if you do not strike him he will strike you, and a whole land with you." -Martin Luther.

What is the origin of these sadistic religious practices? The human limbic system. And, we should recall: the limbic system is concerned not just with sex and violence, but resources, such as food, water, land, and thus territory.

HOLY WAR: OSAMA BIN LADEN, MOHAMMED ATTA & 9/11

It was thirteen hundred years ago, in the seventh century A.D., that the religion known as Islam arose in the Arabian peninsula. With astonishing rapidity, it quickly spread across and conquered the Middle East, Byzantium, Persia, northern Africa, and Spain.

Islam was spread by the sword.

The armies of the Christian Byzantine Empire were annihilated in 636, and Jerusalem fell in 638. Four hundred years later, in the year 1095, the Catholic Pope Urban II called upon the nobility and their armies to go to forth and assist their Christian brothers, the Byzantines, and kill the Muslims in the name of God—a cause that could be justified by scripture:

"Behold I send an angel before thee, to keep thee in the way. Beware of him and obey his voice, for I will be an enemy unto thine enemies, and an adversary unto thine adversaries, and I will cut them off... I will send my terror before thee, and will destroy all the people to whom thou shalt come... and I will drive them out from before thee, until thou be increased and inherit the land." Exodus 23:20-30

However, it was not the pious that Urban desired to fight his wars, but lovers of murder and mayhem. He required killers to do a killer's job. He was interested in recruiting for his holy cause only those who were murderers, rapists, molesters of children and anyone who enjoyed the prolonged torture of their victims.

He had just launched the first crusade.

An important factor that played a role in helping to persuade the nobles and their armies to participate in such a gruesome task so terribly far from home, were the offers of an "indulgence." An "indulgence" was literally a license to sin, to do harm, and it was a guarantee that all sins would be forgiven by god, no matter how odious the crime.

In consequence, the crusaders not only attacked and massacred the Muslims, successfully retaking Jerusalem on July 15, 1099, but they massacred their fellow Christians who had the misfortune and bad luck of living in villages that fell along the way—a pattern that was repeated in subsequent Crusades over the centuries.

The Muslims viewed the Christians as "polytheists," and idolaters, and set out to cleanse the Holy lands of these blasphemers.

The Islamic Holy Wars and Counter-Crusades began.

Saladin was the greatest of Muslim generals, and in 1187, he annihilated the entire army of the Kingdom of Jerusalem at the Horns of Hattin, near the Sea of Galilee. Jerusalem had again come under Islamic rule.

Now after nearly a thousand, the "crusader forces" had returned and whereas the Medieval Catholic Church had been driven back and had failed to defeat the Muslim peoples, the United States had succeeded greatly. During the Gulf War America had invaded the holiest of all Muslim lands, the land of Mecca, the Holy land of Saudi Arabia—which was an intolerable affront to 1,400 years of Islamic tradition. It was an afront to Allah, to god.

Over a thousand years ago, after driving out the polytheists and those who worshipped multiple gods and those who profaned the lands of Arabia, the prophet Muhammad had declared that henceforth there shall "not be two religions in Arabia."

Muhammad's words were law—he was the messenger of God.

And now, a thousand years later, the polytheists, the Crusaders had returned.

The presence of foreign troops, with their many gods, was blasphemous. It was a sin. It was a crime against god.

The American led, Western "crusader forces," of course, saw their presence in a whole different light. They were not the invaders. They were in Saudi Arabia to protect it from Saddam Hussein's armies and to liberate Kuwait.

From the perspective of the Americans, they were not an occupying force but remained stationed in Saudi Arabia after the Gulf War, in order to protect the kingdom from Saddam Hussein.

It was not just entirely on religious grounds that bin Laden and other Arabs were incensed. They also believed the presence of the "Crusader Forces" were corrupting the morals of the people and causing the kingdom of Saudi Arabia incredible economic and financial harm.

"The crusader forces became the main cause of our disastrous condition," bin Laden wrote in his 1996 declaration of jihad which read, in part, as follows:

"DECLARATION OF WAR AGAINST THE AMERICANS OCCUPYING THE LAND OF THE TWO HOLY PLACES: EXPEL THE INFIDELS FROM THE ARAB PENINSULA. A MESSAGE FROM USAMA BIN MUHAMMAD BIN IN LADEN"

"Praise be to Allah, we seek His help and ask for his pardon. we take refuge in Allah from our wrongs and bad deeds. Who ever has been guided by Allah will not be misled, and who ever has been misled, he will never be guided. I bear witness that there is no God except Allah, and I bear witness that Muhammad is His slave and messenger.

"It should not be hidden from you that the people of Islam had suffered from aggression, iniquity and injustice imposed on them by the Zionist-Crusaders alliance and their collaborators; to the extent that the Muslims blood became the cheapest and their wealth as loot in the hands of the enemies. Their blood was spilled in Palestine and Iraq. The horrifying pictures of the massacre of Qana, in Lebanon are still fresh in our memory. Massacres in Tajakestan, Burma, Cashmere, Assam, Philippine, Fatani, Ogadin, Somalia, Erithria, Chechnia and in Bosnia-Herzegovina took place, massacres that send shivers in the body and shake the conscience. All of this and the world watch and hear, and not only didn't respond to these atrocities, but also with a clear conspiracy between the USA and its' allies and under the cover of the iniquitous United Nations, the dispossessed people were even prevented from obtaining arms to defend themselves.

"The people of Islam awakened and understood that they were the main targets for the aggression of the Zionist-Crusaders alliance. All false claims and propaganda about "Human Rights" were hammered down and exposed by the massacres that took place against the Muslims in every part of the world.

"The latest and the greatest of these aggressions, incurred by the Muslims since the death of the Prophet (ALLAH'S BLESSING AND SALUTATIONS ON HIM) is the occupation of the land of the two Holy Places -the foundation of the house of Islam, the place of the revelation, the source of the message and the place of the noble Ka'ba, the Qiblah of all Muslimsby the armies of the American Crusaders and their allies. (We bemoan this and can only say: "No power and power acquiring except through Allah")....

"Traitors implement the policy of the enemy in order to bleed the financial and the human resources of the Ummah, and leave the main enemy in the area-the American Zionist alliance enjoy peace and security! This is the policy of the American-Israeli alliance as they are the first to benefit from this situation.

"But with the grace of Allah, the majority of the nation, both civilians and military individuals are aware of the wicked plan. They refused to be played against each other and to be used by the regime as a tool to carry out the policy of the American-Israeli alliance through their agent in our country: the Saudi regime.

"If there are more than one duty to be carried out, then the most important one should receive priority. Clearly after Belief (Imaan) there is no more important duty than pushing the American enemy out of the holy land.

"Ibn Taymiyyah, after mentioning the Moguls (Tatar) and their behavior in changing the law of Allah, stated that: the ultimate aim of pleasing Allah, raising His word, instituting His religion and obeying His messenger (ALLAH'S BLESSING AND SALUTATIONS ON HIM) is to fight the enemy, in every aspects and in a complete manner; if the danger to the religion from not fighting is greater than that of fighting, then it is a duty to fight them even if the intention of some of the fighter is not pure i.e. fighting for the sake of leadership (personal gain) or if they do not observe some of the rules and commandments of Islam. To repel the greater of the two dangers on the expense of the lesser one is an Islamic principle, which should be observed. It was the tradition of the people of the Sunnah (Ahlul-Sunnah) to join and invade and fight with the righteous and non-righteous men. Allah may support this religion by righteous and non-righteous people as told by the prophet (ALLAH'S BLESSING AND SALUTATIONS ON HIM). If it is not possible to fight except with the help of non-righteous military personnel and commanders, then there are two possibilities: either fighting will be ignored and the others, who are the great danger to this life and religion, will take control; or to fight with the help of non righteous rulers and therefore repelling the greatest of the two dangers and implementing most, though not all, of the Islamic laws...."

In February of 1998, bin Laden published a second declaration of war: Ladenese Epistle.

OSAMA BIN LADEN DECLARES WAR (JIHAD) AGAINST JEWS & THE CRUSADERS

"Praise be to God, who revealed the Book, controls the clouds, defeats factionalism, and says in His Book: "But when the forbidden months are past, then fight and slay the pagans wherever ye find them, seize them, beleaguer them, and lie in wait for them in every stratagem (of war)"; and peace be upon our Prophet, Muhammad Bin-'Abdallah, who said: I have been sent with the sword between my hands to ensure that no one but God is worshipped, God who put my livelihood under the shadow of my spear and who inflicts humiliation and scorn on those who disobey my orders.

"The Arabian Peninsula has never — since God made it flat, created its desert, and encircled it with seas — been stormed by any forces like the crusader armies spreading in it like locusts, eating its riches and wiping out its plantations. All this is happening at a time in which nations are attacking Muslims like people fighting over a plate of food. In the light of the grave situation and the lack of support, we and you are obliged to discuss current events, and we should all agree on how to settle the matter.

"No one argues today about three facts that are known to everyone; we will list them, in order to remind everyone:

"First, for over seven years the United States has been occupying the lands of Islam in the holiest of places, the Arabian Peninsula, plundering its riches, dictating to its rulers, humiliating its people, terrorizing its neighbors, and turning its bases in the Peninsula into a spearhead through which to fight the neighboring Muslim peoples.

"If some people have in the past argued about the fact of the occupation, all the people of the Peninsula have now acknowledged it. The best proof of this is the Americans' continuing aggression against the Iraqi people using the Peninsula as a staging post, even though all its rulers are against their territories being used to that end, but they are helpless.

"Despite the great devastation inflicted on the Iraqi people by the crusader-Zionist alliance, and despite the huge number of those killed, which has exceeded 1 million... despite all this, the Americans are once against trying to repeat the horrific massacres, as though they are not content with the protracted blockade imposed after the ferocious war or the fragmentation and devastation.

"So here they come to annihilate what is left of this people and to humiliate their Muslim neighbors.

"If the Americans' aims behind these wars are religious and economic, the aim is also to serve the Jews' petty state and divert attention from its occupation of Jerusalem and murder of Muslims there. The best proof of this is their eagerness to destroy Iraq, the strongest neighboring Arab state, and their endeavor to fragment all the states of the region such as Iraq, Saudi Arabia, Egypt and Sudan into paper statelets and through their disunion and weakness to guarantee Israel's survival and the continuation of the brutal crusade occupation of the Peninsula.

"All these crimes and sins committed by the Americans are a clear declaration of war on God, his messenger and Muslims. And ulema have throughout Islamic history unanimously agreed that the jihad is an individual duty if the enemy destroys the Muslim countries. This was revealed by Imam Bin-Qadamah in "AlMughni," Imam al-Kisa'i in "Al-Bada'i," al-Qurtubi in his interpretation, and the shaykh of al-Islam in his books, where he said: "As for the fighting to repulse [an enemy], it is aimed at defending sanctity and religion, and it is a duty as agreed [by the ulema]. Nothing is more sacred than belief except repulsing an enemy who is attacking religion and life."

"On that basis, and in compliance with God's order, we issue the following fatwa to all Muslims:

"The ruling to kill the Americans and their allies — civilians and military — is an individual duty for every Muslim who can do it in any country in which it is possible to do it, in order to liberate the al-Aqsa Mosque and the holy mosque [Mecca] from their grip, and in order for their armies to move out of all the lands of Islam, defeated and unable to threaten any Muslim. This is in accordance with the words of Almighty God, "and fight the pagans all together as they fight you all together," and "fight them until there is no more tumult or oppression, and there prevail justice and faith in God."

"This is in addition to the words of Almighty God: "And why should ye not fight in the cause of God and of those who, being weak, are ill-treated (and oppressed)? — women and children, whose cry is: 'Our Lord, rescue us from this town, whose people are oppressors; and raise for us from thee one who will help!'"

"We — with God's help — call on every Muslim who believes in God and wishes to be rewarded to comply with God's order to kill the Americans and plunder their money wherever and whenever they find it. We also call on Muslim ulema, leaders, youths and soldiers to launch the raid on Satan's U.S. troops and the devil's supporters allying with them, and to displace those who are behind them so that they may learn a lesson.

"Almighty God said: "O ye who believe, give your response to God and His Apostle, when He calleth you to that which will give you life. And know that God cometh between a man and his heart, and that it is He to whom ye shall all be gathered."

"Almighty God also says: "O ye who believe, what is the matter with you, that when ye are asked to go forth in the cause of God, ye cling so heavily to the earth! Do ye prefer the life of this world to the hereafter? But little is the comfort of this life, as compared with the hereafter. Unless ye go forth, He will punish you with a grievous penalty, and put others in your place; but Him ye would not harm in the least. For God hath power over all things."

"Almighty God also says: 'So lose no heart, nor fall into despair. For ye must gain mastery if ye are true in faith.'"

In 1998, bin Laden announced his intentions to the world and the United States and called for the killing of "Americans and their allies, civilians and military . . . in any country in which it is possible to do it."

In 1998, U.S. targets were hit: the U.S. embassies in East Africa and the USS Cole in Aden, Yemen. Years of planning went into the 1998 bomb attacks—just as bin Laden promised. "The nature of the battle requires good preparation."

In 1998, he also promised that "the battle will inevitably move . . . to American soil."

In June of 2001, Osama bin Laden boasted that a horrific attack would soon take place in the United States.

"To all the Mujah: Your brothers in Palestine are waiting for you; it's time to penetrate America and Israel and hit them where it hurts the most."

On September 11, 2001, he made good on his terrorist threat. His Al-Qaeda terrorist organization murdered nearly 3,000 Americans.

On the morning of September 11. 2001, Mohammed Atta and 18 other Muslim terrorists, hijacked 4 commerical jets, two of which struck the World Trade Center, the others crashing into the Pentagon and a field in Pennsylvania.

Although Atta had repeatedly rehearsed this operation, on the morning of September 11, he was so worried that he might miss his flight from Maine to Boston that he rushed out and forgot his luggage. Later FBI agents would discover a jet fuel consumption calculator, an instructional video on flying commercial jets, a scrap of lined paper with a list of helpful hijacker hints, a letter, dated from 1996, and a kind of Hijacker's epistle.

In the notes and letter he left behind, Atta said he planned to kill himself so he would go to heaven as a martyr.

He also wrote out a page of last minute reminders which he may have photocopied and circulated among his followers. Atta, the presume author, had doodled on the paper, sketching a crude "key of life" that consisted of a arrowhead-like sword, with serpentine swirls and two hoops circling the shaft.

The letter also said in part:

"It is a raid for Allah....When the time of truth comes and zero hour arrives, then straighten out your clothes, open your heart and welcome death for the sake of Allah. Seconds before the target, your last words should be: There is no God but Allah. Mohammed is his messenger. I pray to you, God, to forgive me from all my sins, to allow me to glorify you in every possible way."

The FBI also found a detailed letter, a kind of Hijacker's Epistle which had apparently been circulated among all four hijacker teams, for a copy was also found in the debris of yet another hijacked jetliner.

It read in part as follows:

"In the name of God, the most merciful, the most compassionate. . . . In the name of God, of myself and of my family . . . I pray to you God to forgive me from all my sins, to allow me to glorify you in every possible way..."

"Remember the battle of the prophet . . . against the infidels, as he went on building the Islamic state..."

And then, on the top of page 3, it was captioned: "The last night."

"Remind yourself that in this night you will face many challenges. But you have to face them and understand it 100 percent."

"Obey God, his messenger, and don't fight among yourself where you become weak, and stand fast, God will stand with those who stood fast."

"You should engage in such things, you should pray, you should fast. You should ask God for guidance, you should ask God for help. . . . Continue to pray throughout this night. Continue to recite the Koran."

"Purify your heart and clean it from all earthly matters. The time of fun and waste has gone. The time of judgment has arrived. Hence we need to utilize those few hours to ask God for forgiveness. You have to be convinced that those few hours that are left you in your life are very few. From there you will begin to live the happy life, the infinite paradise. Be optimistic. The prophet was always optimistic."

"Always remember the verses that you would wish for death before you meet it if you only know what the reward after death will be."

"Everybody hates death, fears death. But only those, the believers who know the life after death and the reward after death, would be the ones who will be seeking death."

"Remember the verse that if God supports you, no one will be able to defeat you."

"Keep a very open mind, keep a very open heart of what you are to face. You will be entering paradise. You will be entering the happiest life, everlasting life. Keep in your mind that if you are plagued with a problem and how to get out of it. A believer is always plagued with problems. . . . You will never enter paradise if you have not had a major problem. But only those who stood fast through it are the ones who will overcome it."

"Check all of your items, your bag, your clothes, knives, your will, your IDs, your passport, all your papers. Check your safety before you leave. . . . Make sure that nobody is following you. . . . Make sure that you are clean, your clothes are clean, including your shoes."

"In the morning, try to pray the morning prayer with an open heart. Don't leave but when you have washed for the prayer. Continue to pray." "When you arrive ... smile and rest assured, for Allah is with the believers and the angels are protecting you." "When you enter the plane pray:

"Oh God, open all doors for me. Oh God who answers prayers and answers those who ask you, I am asking you for your help. I am asking you for forgiveness. I am asking you to lighten my way. I am asking you to lift the burden I feel."

"Oh God, you who open all doors, please open all doors for me, open all venues for me, open all avenues for me."

"God, I trust in you. God, I lay myself in your hands. I ask with the light of your faith that has lit the whole world and lightened all darkness on this earth, to guide me until you approve of me. And once you do, that's my ultimate goal."

"There is no God but God. There is no God who is the God of the highest throne, there is no God but God, the God of all earth and skies. There is no God but God, I being a sinner. We are of God, and to God we return."

THE SUICIDE BOMBER

The suicide bomber is stereotypically highly religious, a loner with few social skills, depressed, withdrawn, and often confused about their sexuality

(Jess, et al., 2001, Jess & Beck, 2002). He who volunteers to destroy himself and others with a bomb strapped to his back, generally faces a future that offers little hope. He (or she) often feels overwhelmed by hopelessness and defeat as well as by feelings of terrible rage and anger.

The suicide bomber is not interested in killing soldiers, but civilians. He has no qualms, nor issues about destroying the innocent. Women and children are viable, desirable targets. By bringing the war off the battlefield and into the homes, work and public spaces of the innocent, the suicide bomber tries to spread so much terror that the enemy, who they call Israel and America, will withdraw from their lands and capitulate, as capitulation offers the possibility of salvation.

The suicide bomber seeks revenge and salvation--often for imaginary sins, including the "sin" of homosexuality (Jess, et al., 2001, Jess & Beck, 2002). The suicide bomber believes that through hatred and violence, by murdering innocent people, he is fulfilling a moral and spiritual quest that will lead to martyrdom and paradise.

Most suicide bombers are highly religious. Prospects are recruited from mosques and religious institutions and are led to believe that by killing themselves and others, they will go straight to paradise, where they will be seated in honor, next to their almighty God.

According to Islamic tradition, and as taught by "suicide teacher" Mohammed el Hattab, "He who gives his life for Islam will have his sins forgiven and will attain the highest state of paradise." And what is paradise? 72 virgins who will love him and him alone. Eternal sexual bliss is one of the rewards of martyrdom.

Suicide bombers are commonly recruited by and are affiliated with the Palestinian militant group Hamas (or Islamic Jihad).

They are recruited from mosques and schools. It is not the brave and courageous who are enticed, but those who appear lonely, troubled, shy and withdrawn—those who might leap at the chance to be accepted, to be part of a group, to be given a mission in life, to feel important, and to belong.

Suicide bombers from the Middle East are typically young, impressionable, often highly religious, and living in despair and hopeless poverty. They are loners. They are shy, awkward, and usually have few or no friends. In American slang, they are the "losers." They are young men with nothing to lose and everything to gain: Paradise and 72 willing virgins.

And like many virgins and young men confused about their sexuality, these "losers," are often seduced by older men, who offer camaraderie and the chance to feel wanted, to belong and to have friends, if only the young man will agree to kill himself.

Friendly, awkward, shy and alone, these young men are easy pickings for those sophisticated in the art of enticing the friendless with gifts of kindness, respect and yes, even love, brotherly love.

The recruiters, these older men wise in the world, offer the "loser" a chance to become an accepted, respected member of the gang, part of an elite brotherhood: A brotherhood in blood.

Perhaps at many as half of the 19 men who agreed to kill themselves and thousands of others on September 11, were mentally disturbed and confused about their sexuality.

Wail Alshehri, for example, suffered from significant "psychological problems" that required treatment. Abdulaziz Alomari was an alcoholic. He had at least one arrest for drunk driving.

Ahmed Salah Alghamdi, was highly religious, the graduate of a religious school, and he suffered from the torment of sin. He knew himself to be a sinner. He was wracked by a pathological, almost delusional guilt, frequently asking others, such as his parents, to pray for him. Highly religious, mentally disturbed, and drowning in self-hate.

Some would agree to murder and destroy Americans, only because they were seeking to destroy their own unknown face.

As detailed in the book, America Attacked (Jess et al., 2001), Mohammed Atta was centrally involved in possibly all phases of the 9/11 attack, including planning, spying, recruitment and training. Atta may have also been a self-hating homosexual.

Atta was a religious man, which is a common characteristic of those who carry out suicide attacks. Yet, Atta differed from the stereotypical suicide attacker in that he was older, 33, highly educated and technically skilled. He also came from a well to do family and could be considered "upwardly mobile."

In many respects he was no different from any other "upwardly mobile" highly educated Muslim. But there were also several notable exceptions. Atta drank alcohol excessively and he enjoyed hanging out at bars, including Sharky's Billiard Bar in Hamburg where he was attending the Technical University in 1996.

He also preferred the company of men to women and his goal in life was to launch a suicidal attack and murder thousands of people who had never caused him or his loved ones harm.

Mohammed Atta, was a religious man. His teachers described him as polite, diligent, intelligent and very religious, perhaps fanatically so. Yet he was also a sinner who loved fast cars, flashy clothes, young men and money, lots and lots of money. Mohammed Atta knew himself to be a sinner. He desperately sought redemption, turning first to religion, and then failing that, or perhaps, because of it, embracing suicide and mass murder in order to achieve martyrdom and to wash away his guilty sins. Mohammed Atta was also a leader. He played a leadership role in Hamburg and again in the United States.

Young men were attracted to Mohammed Atta and many were seduced and convinced to join his army, including fellow hijackers, Marwan Al-Shehhi, Saeed Alghamdi and Ziad Jarrahi—men he met and "seduced" in Hamburg.

Together they were to do battle with the enemies of Islam and their reward would be paradise and the cleansing of their guilty sins.

Mohammed Atta saw himself as a soldier and he would lead his men on a journey to eternity and paradise.

In 1996, Atta demanded that University officials accommodate his religious needs. He convinced them to establish an Islamic prayer room for himself and 20 other Muslim students.

Atta, the seductress, began recruiting others to the cause. Arabic men not only attended his prayer room sermons but would gather late at night at his home.

Atta held meetings in the prayer room and in his home and soon found a willing convert, Marwan Al-Shehhi, who was 11 years his junior. Marwan Al-Shehhi, was not just a convert, he moved into Atta's apartment and formed an unusual relationship that was to last until the day they died.

The two men became "inseparable" and "joined at the hip."

They lived together. They trained together. They worked out together. They drank together. They committed mass murder together. The only thing they didn't do together was seek out women. They had no interest in women. They did not like women. Atta, in fact, hated women and left instructions that women should not be allowed to attend his funeral.

In America and Europe, when two men form close, physical, "inseparable" relationships, and eschew the company of females, few eyebrows are raised even when it becomes clear the men are homosexuals.

In more conservative Islamic countries, it is also not uncommon for men to spend a lot of time together and to even live together. That two highly religious men might also become inseparable and eschew the company of women, would be viewed as a sign of virtue and no cause for alarm.

However, if the same two inseparable Arab men liked to drink, wear expensive fancy clothes and spend time on the town, but also eschewed the company of women, such behavior would be recognized for what it is, and would not be tolerated. It would have been viewed as "immoral." Sinful. In many Muslim countries, they could be legally killed and stoned to death.

Al-Shehh was also a problem drinker. In Florida he and Atta frequented a number of bars and were often observed to down four or five drinks in a row. Their favorite "poisons" included rum and Coke and Stolichnaya vodka and orange juice.

If they were Osama bin Laden's men, they didn't act like it. Bin Laden would not let his boys smoke cigarettes. Drinking alcohol would have led to banishment from the ranks of his al-Qa'eda movement.

However, as we now know, in 1996 bin Laden made an exception. He welcomed sinners into the ranks of his fighters. He would even finance their sins.

Were the two men homosexuals? We don't know.

What we do know is that Mohamed Atta and Marwan Al-Shehhi were fashion conscious, they enjoyed wearing expensive clothes, and were provided with large sums of money to indulge these habits. They also spent time keeping their bodies pretty by working out together at gyms.

They were always together. The lived together. They were inseparable.

Indeed, these two men remained "bound at the hip" until the day they boarded separate planes in Boston and hijacked them to New York City.

That they were both highly religious, Mohammed Atta in particular, does not rule out homosexuality. It certainly didn't rule out mass murder. Rather their brand of religiosity leads credence to the possibility that these two fashion conscious, inseparable, unmarried men may have harbored "forbidden" tendencies, even if they never acted on them.

These forbidden tendencies may have also been the lure, which attracted and then bound so many young men to Atta's camp. These same guilty, sinful tendencies may have had the motive force, which drove them to commit suicide and mass murder—devilish deeds that would cleanse them of their sins, even if they had never acted on them.

Mohammed Atta, while he was in Hamburg, Germany, had taken on the role of a cleric, of a priest! In was in this clerical-type shepherd role that he was able to gather so many sheep.

Homosexuality is common among clerics, shamans and priests. In the United States, homosexuality was so rampant within the ministry of the Episcopal church and its homosexual clergy so openly promiscuous that the church lost over 5 million members by 1997.

Likewise, and as will be detailed later in this chapter, the priesthood of the Catholic Church also includes a high percentage of homosexuals and pedophile priests who rape young men and boys. For example, in 1999, the Bishop of the Santa Rosa Diocese in California was forced to resign because another priest, a younger man, accused the Bishop of repeatedly raping him. In 2000, the Catholic Church decreed that priests could no longer be alone with altar boys, because of the possibility these older men would seduce their young charges.

In Hamburg, Mohammed Atta also served a priestly role. He gathered young and highly intelligent Muslim men who were alone, friendless and isolated, and bound them to him and his cause. He was a seductress.

Ziad Jarrah was seduced. He soon began living together with Mohammed Atta and Marwan Al-Shehhi in the same apartment—an arrangement that astonished his father, Samir Jarrah, when he learned of it.

This unusual relationship also put a strain on Ziad's relationship with his fiance, Asle. Although they were to get married, Ziad became so entangled with his new friends that he no longer had time to see her. He was so busy that when it came time for Asle to meet his parents, he could not find the time to accompany her.

Too busy to bring his fiance to meet his family?

Ziad Jarrah had been seduced. He not only lived with the two men but he died with them. On September 11, he helped to commandeer and then to pilot one of the four hijacked planes.

When Ziad Jarrah moved in with Mohammed Atta and Marwan Al-Shehhi, the Jarrah family realized that something was terribly wrong with their only son.

He was living with two men in a small apartment when he had more than enough money to live alone. His relationship with his fiance grew more and more estranged. He was behaving strangely. He was preoccupied. He seemed moody and depressed.

What was wrong with Ziad? His family did not know. What was troubling Ziad? They did not know.

What we do know is that Ziad felt compelled to kill himself for Islam.

According to Islamic tradition and as instructed by "suicide teacher" Mohammed el Hattab, "He who gives his life for Islam will have his sins forgiven."

What sins had he committed? What had he done that was so utterly horrible that only a martyr's death could wash away his sins?

We do not know.

Yet, something so bothered the young man that he volunteered to participate in a mission of death that was guaranteed to end his painful life and cleanse him of his guilty sins.

MASS SUICIDE AND MODERN RELIGIOUS MURDERERS

Murder of the innocents and the slaughter of infidels and nonbelievers are not antiquated religious customs. Nor are these acts the exclusive province of Islam or men from the Middle East. Cults and religious groups regularly arise in various lands and cultures and frequently indulge in similar practices, e.g. the mass suicide of Jim Jones and his followers in "Jonestown" ("People's Temple"), David Koresh and the fiery death of his followers at Waco Texas.

MASS DEATH AT JONESTOWN

On November 18, 1978, the reverend Jim Jones exhorted and then forced his followers to engage in mass murder and mass suicide in Guyana. More than 900 people, many of them children, drank, or were forced to drink cyanide-laced fruit punch. Jones, a charismatic monomaniac who at one time was a favorite of San Francisco's liberal political elite, took a less painful and quicker way out. He died of a bullet to the head.

Jim Jones was born in 1931 and became a "Holy Roller" preacher as a child. As he reached adulthood he developed a creed he called "apostolic socialism" which deified social justice but which down played the role of god—hence his appeal to the ultra-liberal San Francisco political elite. Current San Francisco Mayor, and former leader of the California Assembly, Willie Brown, was one of Jones' "biggest cheerleaders." Nevertheless, despite his cult status even among the city's liberal elite, Jones became increasingly paranoid and disillusioned with San Francisco, especially following a negative magazine article in New West magazine. Jones moved his followers to a 3,800 acre plot of land in Guyana's outback, where they raised livestock and grew their own food.

The people's paradise soon became a prison camp. Followers were not allowed to leave, and there were forced beatings and forced sex. Jones who required that his followers refer to him as "father" demanded utter obedience. Incessant screaming and threats were delivered nightly against imagined slights and for faintly perceived faults, with the guilty being beat with "the board of education." Anyone who questioned his authority was placed in "The Box," an underground coffin. Unruly children would be taken to the Jonestown well and hung upside down by a rope tied around their ankles and would be repeatedly dunked and beaten.

Jones became increasingly concerned about the CIA and FBI plotting against him and his paradise, and soon required that his followers practice mass suicide so as to protect them from outside threats: the "White Night" drills. Jones would get on the compound loud speaker at night and begin shouting: "White Night! White Night!" which signaled a coming attack. Followers were required to leap from their beds and assemble at designated spots where they would then drink from huge vats.

However, what triggered the mass suicide was a visit from a California politician, Leo Ryan, who upon leaving invited a number of Jones' followers to come with him. Ryan was murdered and insofar as Jones was concerned, the day of reckoning was at hand. One of those who drank the poison wrote in her suicide note: "We died because you would not leave us alone."

HEAVEN'S GATE

Mass suicide has been practiced by numerous religious cults, both ancient and recent, including, in the 1990s, 74 followers of the "Order of the Solar temple," and 39 members of "The Higher Source," i.e. "Heavens Gate." Many of the male members of the Higher Source, including the leader of the cult, a minister (and son of a Presbyterian minister), Marshall Applewhite (also known as "Do") also had themselves castrated.

In the case of Applewhite, castration was not only a means of achieving purity, but a way to control his rampant homosexuality. In 1970, for example, he was fired from the University of Saint Thomas after it became known that he was having sex with male students. Soon thereafter he apparently began hearing voices, including the voice of god and was awakened to a new form of spirituality which required abstinence from sex. His creed eventually came to include strict rules involving "no human-level relationships and no socializing."

As an aside it is noteworthy that abnormal amygdala activity, or amygdala destruction, can alter sex drive and sexual orientation, and can also induce a loss of social interest. Human-level relationships are avoided (Joseph, 1999b; 2000a).

Soon Applewhite's credo came to include UFOs. Specifically, Applewhite saw the coming of the Hale-Bopp comet as a heavenly sign that a UFO was coming to take him and his followers to a better place, to "the level above human, to take us home to Their World: Heaven." The cultists, therefore, killed themselves by eating a poison laced pudding. They lay on bunk beds in a standard black uniform and Nike tennis shoes and died.

DAVID KORESH: BEATEN BY THY GOD

David Koresh ("Vernon") of the Branch Dividian cult, also ordered the death of his followers, and they apparently obeyed his wishes, for he had told them he was "god." In fact he first made his announcement on January 30, 1987, when he sent a wedding invitation to the Seventh Adventists Church in southern California, which read in part: "I have seven eyes and seven horns. My Name is the Word of God...Prepare to Meet Thy God."

David Koresh was a sexual sadist. He had sex with the wives of all his many followers, and had sex with and would beat and order the beating of their and his children. According to one of his followers, Marc Breault (Breault & King, 1993) "Children were spanked for any reason; crying during a sixteen hour Bible study, refusing to sit on Vernon's lap, or daring to defy the prophet's wishes."

"Each child had their own special paddle with their name written on it... Some women thought the best way to please their Son of God lover was to be especially severe when dishing out discipline. But sometimes it wasn't easy for the adults to spank the children. They couldn't find a spot on the child's buttocks that wasn't black and blue or bleeding" (Breault & King, 1993).

The women were sometimes subjected to the same treatment. One 29-year old woman who announced that she was hearing voices was imprisoned in one of the small cottages on their property. She was beaten, and repeatedly raped by her "guards."

David Koresh was also a prophet of Doom, and between the months of October 1991 to February 1993, he and his group spent over $200,000.00 on guns. This arsenal included 50-caliber machine-guns and a million rounds of ammunition (Breault & King, 1993). Koresh was not only preparing for Armageddon but was contemplating mass suicide.

On Sunday, February 28, 1993, at 9.55 A.M., in Waco, Texas, a hundred ATF agents began their raid which resulted in a blood bath. A 51 day siege finally ended in a firestorm, and the death of David Koresh, his followers, and their children who burned to death when the compound was set on fire and burned to the ground. It was David's wish, however, for the entire world to go up in flames.

CROSS CULTURAL SPIRITUAL BLOOD LUST

These crazed religious practices are not limited to Americans. Consider, the Japanese religious cult "Aum." Their leader Shoko Asahara and many top cult members were arrested and charged with murder in June of 1995 for releasing the nerve gas Sarin in five subway cars during rush hour injuring over 5,500 Japanese commuters (New York Times, 6/7/95).

Similarly, although the "modern" Islamic, Christian and Jewish religions forbids it, many modern day Middle Eastern and African Islamic, Christian, and Jewish fundamentalists, regularly preach murder and hatred.

For example, "militant rabbis" in Israel had encouraged and condoned the assassination of Israel's Prime Minister, Yitzhak Rabin, and had issued a "pursuer's decree," which in effect morally required that he be killed (New York Times 11/11/95). And, he was murdered by a student of religion, Yigal Amir, who claimed he acted upon "God's" instructions.

Because of this limbic religious blood lust, members of religious sects may turn against one another and commit murder in the name of God, just as Jews murder Jews, and Christians murder Christians in the name of God.

For example, on November 29, 1998, six members of the United Pentecostal Church were arrested for kicking to death six people, including three children, "to wipe out the enemies of God." According to Reuters new agency (11/28/98), the killings began two weeks prior, when the pastor of the thirty member sect announced during a sermon that he could hear voices from Jesus Christ" ordering that members of the group be punished. "The pastor, helped by his wife and two other men, began beating, whipping and stamping on the worshipers... followed by more disciplinary torture over several days. Among the dead were two boys, aged three and four."

HINDU, MUSLIM, BUDDHIST SPIRITUAL BLOOD LUST

"Terrorist attacks" by Muslim religious zealots is not uncommon. However, these attacks and calls for the death of infidels, are not always aimed at Jews or Americans, but their "brothers" and "sisters," i.e. fellow Arabs and Muslims.

Consider Algeria. Between 1992 and 1999, over 80,000 Algerians were murdered by Muslim zealots, with the majority of victims consisting of children, girls, and pregnant women who were beheaded, disemboweled, or who had their throats slashed. The reason for these murders and campaign of terror? To establish a holy Islamic state.

In fact, in 1998, on the first holy day of Ramadan—which marks "god's" revelation of the Koran to the Prophet Muhammed—Muslim religious zealots killed and slaughtered more than 400 people, most of them women.

Hindu Muslim violence has haunted the country of India for hundreds of years, but reached a boiling point in 1947 when Pakistan, an Islamic nation, declared its independence from India, a secular democracy. Over a million Hindus and Muslims were slaughtered that year.

The violence and religious murders have continued unabated.

In the spring of 2002, a Muslim mob stoned and then set fire to a train carrying Hindus, burning alive dozens of women and children. The Hindus were returning from Ayodhya where they had been demanding the construction of a Hindu temple on the site of a mosque destroyed by their nationalists allies. After the train pulled into Godhra, a town populated mostly by Muslims, some of the Hindus began attacking Muslim men. The Muslim community reacted with stones and set fire to the train. Nearly 60 Hindus were killed.

The next day, tens of thousands of Hindus retaliated with acts of unspeakable butchery. Muslim men were beaten, stabbed, shot, and beheaded, young women and mothers were stripped naked, raped, and then skewered on swords or soaked with gasoline and set on fire. The bellies of pregnant women were slit open, and their unborn babies ripped from their bodies, raised skyward on the tip of a sword and then tossed into the many fires that were consuming Muslim towns and cities. Thousands of Muslims lost their lives in this manner and over 100,000 were left homeless.

Even Buddhist monks murder and assault one another. For example, in November of 1998, gray-robed Buddhist monks of the Chogye Buddhist order in Korea, spent several weeks assaulting and beating each other with clubs, chairs, and even fire bombs as they battled over religious control of the order's main temple (Associated Press, 11/17/98). In one brawl alone, 37 were seriously injured and five killed.

RELIGIOUS MURDERS AND THE LIMBIC SYSTEM

What is the source of these religious-murderous and sadistic behaviors? The limbic system and the same cluster of nuclei which subserve aggression, sexuality, and spirituality. It is the limbic system which enables humans to respond with irrational and murderous blood lust in the name of "God" and religion and it is the amygdala that can induce a horribly violent and murderous assault with minor provocation—and as we have seen, the Lord God, and numerous religious fanatics, have also murdered at the slightest provocation.

For example, if the amygdala is directly stimulated rage reactions are commonly triggered. The face will begin to contort, the teeth will be bared, the pupils will dilate, the nostrils will flare, and then the animal or human will viciously attack (Egger & Flynn, 1963; Gunne & Lewander, 1966; Mark et al., 1972; Ursin & Kaada, 1960; Zbrozyna, 1963). Amygdaloid activation results in attacks directed at something real, or, in the absence of an actual stimulus, at something imaginary; including, presumably, in the case of religious fanatics, imaginary slights or insults.

There have been reported instances of patients suddenly lashing out and attacking friends, relatives or strangers while in the midst of a temporal lobe seizure (Saint-Hilaire et al., 1980), and/or attacking, kicking, and destroying furniture and other objects (Ashford et al., 1980). One female patient, during amygdala stimulation, became irritable and angry, and then enraged. Her lips retracted, there was extreme facial grimacing, threatening behavior, and then rage and attack—all of which persisted well beyond stimulus termination (Mark et al., 1972).

Yet another man developed intractable aggression following a head injury and the development of abnormal activity within the amygdala (Schiff et al.,1982). Subsequently, he became easily enraged, sexually preoccupied, and developed hyper-religiosity and pseudo-mystical ideas. He became violently religious.

Indeed, the amygdala is able to overwhelm the neocortex and the rest of the brain so that the person not only forms religious ideas but responds to them, sometimes with vicious, horrifying results. Hence, it appears to be the limbic system, and the amygdala in particular, which not only contributes to religious and spiritual experience, but religious brutality and murder committed in the name of one's god.

Sexual Spirituality and the War Against Women

A not uncommon characteristic of high levels of limbic system and inferior temporal lobe activity are changes in sexuality as well as a deepening of religious fervor (Bear 1979; Blumer, 1970; Slater & Beard 1963; Trimble 1991; Taylor 1972, 2005). Hypersexuality or conversely, hyposexuality is not uncommon.

It is noteworthy that not just modern day evangelists, but many ancient religious leaders, including Abraham, Jacob (Israel) and Muhammed, tended to be highly sexual and partook of many partners (e.g. St. Augustine of Hippo: "Give me chastity, 'o lord, but just not yet"), or, they shared their wives (Abraham), or they married women who were harlots (e.g. Hosea) or had sex with other men's wives (Muhammed, King David), or killed other men in order to marry or have sex with their wives (King David).

King Solomon (like his father King David) experienced numerous dream states in which he communicated with God, and he required the sexual services of 700 wives and 300 concubines.

LIMBIC SYSTEM SEXUALITY

The limbic system is concerned with sex. Structures such as the amygdala and hypothalamus not only mediate sexual behavior, but they are sexually differentiated and have sex specific patterns of neuronal and dendritic development, (Allen et al. 1989; Blier et al. 1982; Gorski et al. 1978; Nishizuka & Arai, 1981; Rainbow et al. 1982; Raisman & Field, 1971, 1973; Simerly, 1990; Swaab & fliers, 1985). That is, there are "male" and "female" limbic systems, and even "homosexual" limbic systems such that male homosexuals are in possession of a limbic system which is more "female" than "male" in structural organization (Levay, 1991; Swaab, 1990), and in some respects (the anterior commissure) hyper-female in size (Allen, & Gorski, 1992).

Indeed it has now been well established that the amygdala and the hypothalamus (specifically the anterior commissure, anterior-preoptic, ventromedial and suprachiasmatic nuclei) are sexually differentiated and have sex specific patterns of neuronal and dendritic development including the preoptic area and ventromedial nucleus of the hypothalamus (Bleier et al. 1982; Dorner, 1976; Gorski et al. 1978; Rainbow et al. 1982; Raisman & Field, 1971, 1973). The preoptic and other hypothalamic regions are not only sexually dimorphic but chemical and electrical stimulation of the preoptic and ventromedial hypothalamic nuclei triggers sexual behavior and even sexual posturing in females and males (Hart et al., 1985; Lisk, 1967, 1971). In female primates, even maternal behavior can be triggered (Numan, 1985). In fact, dendritic spine density of ventromedial hypothalamic neurons varies across the estrus cycle (Frankfurt et al., 1990) and thus presumably during periods of maximal sexual receptivity and arousal, as well as during pregnancy and while nursing.

Electrical stimulation of the preoptic area increases sexual behavior in females and males, including the frequency of erections, copulations and ejaculations, we well as pelvic thrusting followed (in the case of males) by an explosive discharge of semen even in the absence of a mate (Hart, et al., 1985; Maclean, 1973). Conversely, lesions to the preoptic and posterior hypothalamus eliminates male sexual behavior and results in gonadal atrophy.

Although the etiology of homosexuality remains in question, it has been shown that the ventro-medial and anterior nuclei of the hypothalamus of male homosexuals demonstrate the female pattern of development (Levay, 1991; Swaab, 1990). Male homosexuals are in possession of limbic system that is more "female" than "male" in functional as well as structural orientation.

The primate amygdala is also sexually differentiated (Nishizuka & Arai, 1981; see also Simerly, 1990). The male amygdala is 16% larger than the female amygdala (Breedlove & Cooke, 1999; Filipek, et al., 1994), whereas in male rats, the medial amygdala is 65% larger than the female amygdala and grows or shrinks in the presence of testosterone (Breedlove & Cooke, 1999).

The male vs female amygdala also contains a greater number of synaptic connections and shows different patterns of steroidal activity (Breedlove & Cooke, 1999; Nishizuka & Arai, 1981; Simerly, 1990). These sex differences are particularly evident in the medial amygdala, which is also a principle site for steroidal uptake, including the female sex hormone, estrogen, and contains a high concentration of leutenizing hormones (Stopa et al., 1991). The number of immunoreactive cells in the female amygdala also fluctuates during the estrus cycle, being highest during proestrus (SImerly, 1990), and thus presumably acts so that if pregnant, the fetus will not be attacked as foreign, and/or so as to coordinate, with the hypothalamus, the appropriate neuroendocrine responses during pregnancy and following birth.

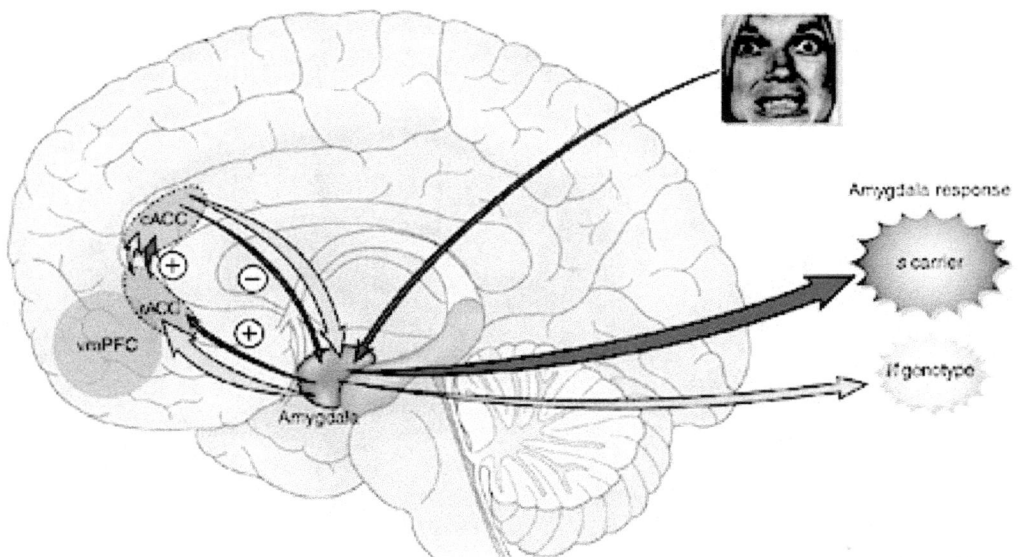

Because the amygdala is involved in sexuality and is sexually differentiated, activation of the amygdala can produce clitoral tumenence, penile erection, sexual posturing (Kling and Brothers, 1992; MacLean, 1990; Robinson and Mishkin, 1968; Stoffels et al., 1980) sexual feelings (Bancaud et al., 1970; Remillard et al., 2013). Electrical stimulation of the amygdala can also trigger sensations of extreme pleasure (Olds and Forbes, 1981), memories of sexual intercourse (Gloor, 1986), thrusting, sexual moaning, ejaculation, as well as ovulation, uterine contractions, lactogenetic responses, and orgasm (Backman and Rossel, 2014; Currier, Little, Suess and Andy, 1971; Freemon and Nevis,1969; Warneke, 1976; Remillard et al., 2013; Shealy and Peel, 1957).

Moreover, in rats and other animals, kindling induced in the amygdala can trigger estrus and produce prolonged female sexual behavior. In fact, abnormal seizure activity within the amygdala or overlying temporal lobe may induce a female to engage in "sexual intercourse" even in the absence of a partner. For example, Currier and colleagues (1971, p. 260) describe a female temporal lobe seizure patient who was "sitting at the kitchen table with her daughter making out a shopping list" when she suffered a seizure. "She appeared dazed, slumped to the floor on her back, lifted her skirt, spread her knees and elevated her pelvis rhythmically. She made appropriate vocalizations for sexual intercourse such as: It feels so good...further, further."

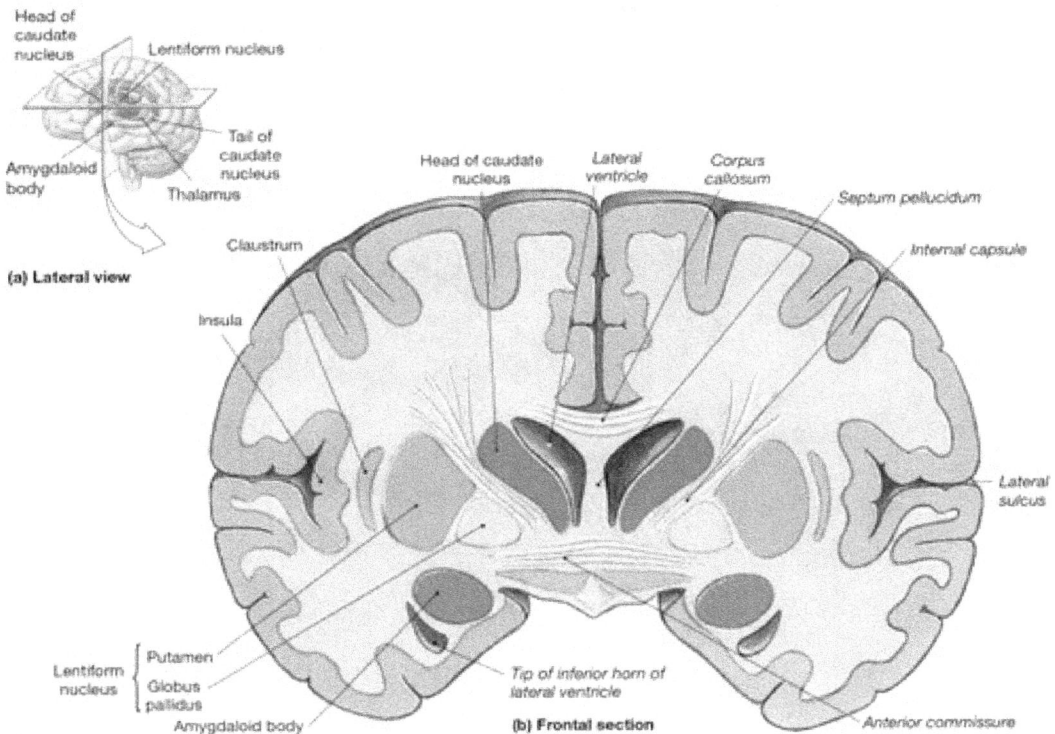

The anterior commissure, the band of axonal fibers which interconnects the right and left amygdala/temporal lobe is also sexually differentiated. Like the corpus callosum, the anterior commissure is responsible for information transfer as well as inhibition within the limbic system. Specifically, the female anterior commissure is 18% larger than in the male (Allen & Gorski 1992). Moreover, the anterior commissure is larger not only in females, but is 35% larger in homosexual males vs male heterosexuals (Allen & Gorski, 1992).

It has been argued that the increased capacity of the right and left female amygdala to communicate (via the anterior commissure) coupled with the more numerous and more densely packed neurons within the female amygdala (which in turn would decrease firing thresholds and enhance communication), and the sex diffferences in the hypothalamus, would also predispose females to be more emotionally and socially sensitive, perceptive, and expressive (Joseph 1993). Hence, these limbic sex differences induces her to be less aggressive and more compassionate and maternal, and affects her sexuality, feelings of dependency and nurturance, and desire to maintain and form attachments in a manner different than males. However, in homosexual males, this may predispose at least some of these males to not only behaving more emotionally, but more violently than women and even more violently than heterosexual men.

As will be detailed, it is these same sex differences in the limbic system which account for why women are more religious than men, attending church more often, and why homosexuals are drawn to the priesthood. Unfortunately, some of these homosexual priests, because they are in possession of a "female" limbic system that is bathed with aggression-inducing testosterone, have also promoted or engaged in religious violence directed toward men as well as women.

THE RELIGIOUS WAR AGAINST WOMEN

Abnormalities in the limbic system, the amygdala and hypothalamus in particular, can also provoke extreme sexual violence. In one case, a young man with seizure activity originating in his amygdala and temporal lobe, beat his mother to death and then sexually assaulted the body (Joseph, 2000a).

An abnormal limbic system may abnormally link sex with murder; and among men, the sexual murder or torture of women. Moreover, because the amygdala is clearly linked to religious experience, sexual behavior and religious expression may become linked in a positive or a negative fashion. For example, whereas a heterosexual male may respond to a beautiful woman with interest and sexual lust, homosexual priests may respond with murderous lust. Augustine, Jerome, Tertullian, and other Catholic theologians absolutely hated women, viewing them as evil temptresses that lead men to hell.

"What is the differences whether it is a wife or mother, it is still Eve the temptress that we must beware of in any woman." -St. Augustine.

"Do you not know that you are each an Eve? The sentence of God on this sex of yours lives in this age: the guilt must of necessity live too. You are the devil's gateway; you are the unsealer of that forbidden tree; you are the first deserter of the divine law; you are she who persuaded him whom the devil was not valiant enough to attack. You so carelessly destroyed man, God's image. On account of you, even the Son of God had to die." -Tertullian, 16th Century

Even the Lord God, Yahweh voiced contempt and hatred for women: "Because the daughters of Zion are haughty and walk with stretched forth necks and wanton eyes, walking and mincing as they go, and making a tinkling with their feet...The Lord shall wash away the filth of the daughters of Zion...

by the spirit of burning." -Isaiah 3:16-26, 4:4.

Women throughout the ages have been demeaned and attacked by male and homosexual religious authorities, and their murder was often sanctioned by various gods. Indeed, during the middle ages, the Catholic Church declared war on women who were then systematically tortured and slaughtered by the Catholic authorities.

FEAR OF FERTILITY

Yet another factor in the religious crusade against women was the fear of the goddess cult by the patriarchal religious authorities. For example, during the Crusades, because the women were temporarily freed of male dominance, some began to practice their own religion and worship their own goddesses, including those pagan goddesses that celebrated female sexuality; e.g., fertility cults— perhaps the oldest and most ancient of "organized religions." In consequence, the leaders of the Catholic Church felt compelled to act. They sought to destroy woman in general, and those women associated in any way with the goddess and her fertility cults. Women, therefore, were suspect if they were in any way sexually attractive, and those who were the most sexually appealing (i.e. fertile) were the first to be murdered by the Catholic Authorities. Her reproductive facility was viewed as supernatural in origin, her menstrual cycle being in tune with the cycles of the moon, and the moon and the blood of her menstruation were believed to have supernatural power. Every woman was not just Eve, but an incarnation of the goddess.

The reproductive and supposedly supernatural power of menstrual blood may have contributed to those rituals requiring that animals be drained of blood and their blood splashed on religious altars, including the altars of the Lord God Yahweh. Originally it was menstrual blood that was splashed upon the altar of the goddess.

The ancient Hebrews, and the Catholic authorities, therefore, deemed a menstruating woman as unclean, for menstrual blood had the power of life and that of the supernatural. Women in general, therefore, were subject to attack, with the ancient Jews reciting a prayer thanking god "for not making me a woman." However, it was her reproductive facility, and thus her association with the goddess that was most feared and detested.

As detailed by Robert Briffault (1931) "the supernatural source which magic powers are regarded as being primarily derived is... connected in the closest manner with the functions of women... her... reproductive functions... and a power which is used in a dread-inspiring manner. And by virtue of their natural position and function, the wielder of domestic magic...she had charge of the sacred objects and performed all the religious functions connected with the household."

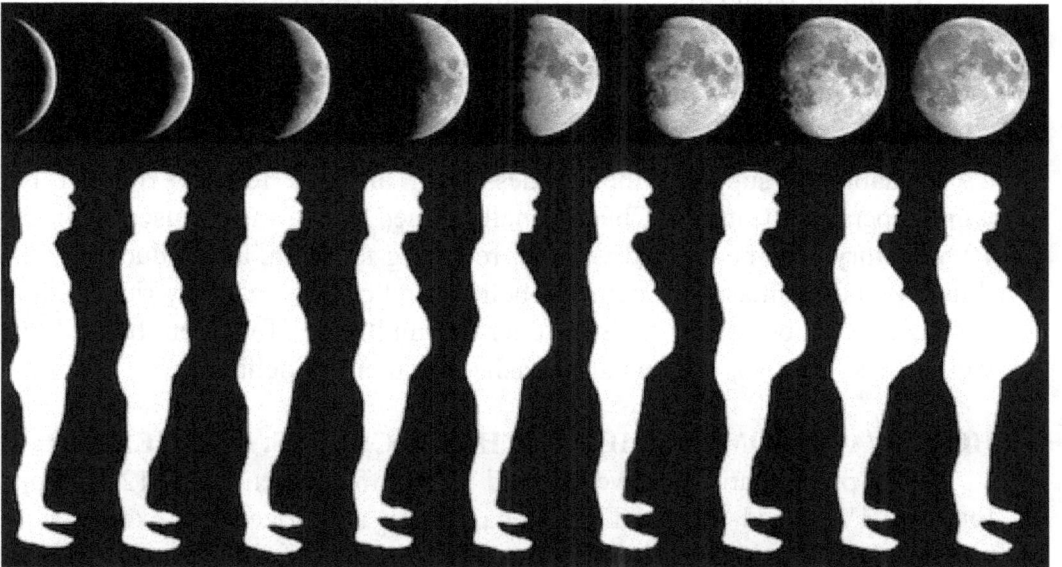

Yet another factor in the religious persecution of woman, was her facility at speech. Indeed, and as is now well demonstrated, females demonstrate clear language superiorities over males, and they even talk faster as well (Joseph, 2000b). Speech and language, that is, a woman's tongue, is her natural and traditional weapon, which she may use for "pronouncing curses, of casting spells... of bewitching and performing incantations. It was a dreaded power.

The curse of a woman is accounted far more potent than the curse of a man" (Briffault, 1931). Women in general, therefore, were considered suspect by the Catholic authorities and during the Middle Ages they were rounded up, sexually tortured, and killed.

THE CATHOLIC CHURCH DECLARES WAR AGAINST WOMEN

Woman has been persecuted by various religious authorities because of the power of her tongue, her reproductive power, her associations with pagan goddesses, and her ability to enchant men with her sexuality.

Woman is biologically predisposed to having multiple orgasms, multiple partners, and to indulging in promiscuity—which does not mean she will necessarily behave promiscuously (Joseph, 2000a,c). Nevertheless, because of her sexual nature and association with fertility, religious rituals of the goddess stereotypically involved sexual orgies and ritual prostitution.

Although male authorities, both religious and otherwise, have attempted throughout the ages to control and yoke female sexuality, this has not always been successful. In Europe, including Rome, during the 6th through 11th century, female sexuality and her promiscuous nature was openly flaunted, and the worship of pagan goddesses became widespread and was threatening the authority of the Church. As early as the 1st century, the missionary sermons of St. Paul were repeatedly drowned out by women chanting "Diana, Diana, Diana;" Diana being the goddess known since prehistoric times as the Queen of Heaven.

Unable to suppress the goddess cult, and the fertility rites of her worshippers, the Catholic Church finally issued the "Canon Episcopi" in the 10th century: "Some wicked women, reverting to Satan, and seduced by the illusions and phantasms of demons, believe and profess that they ride at night with Diana, on beasts, with an innumerable multitude of women. It would be well if these women, one and all, perished for their infidelities."

HOLY WAR: WOMAN THE WITCH, SORCERESS, AND HEALER

The proclamation, however, had little effect. Hence, in 1252, Pope Innocent IV issued the Ad Exitrpanda, which authorized the execution of heretics (e.g. wealthy landowners) and the seizure of their goods, and the prolonged sexual torture of women who were beautiful, wealthy, or old, ugly, and eccentric and/or who gathered in groups to talk and converse and possibly worship pagan goddesses. This was followed by the first Papal bull on sorcery, in 1259, and yet another bull, the notorious Super illius specula, in 1322 (by Pope John the XXII), and then the famous Bull Summis desiderantes affectibus, in 1484 (by Pope Innocent VIII), which demanded the death of these women. The Popes and the Catholic Church proclaimed them witches and declared war against women.

Being "true soldiers of Christ," the Dominicans Heinrich Kramer and Johann Sprenger responded to this Papal decree by writing and issuing the Malleus Maleficarum (witch's hammer), thus unleashing a murderous, sadistic, blood lust that demanded the sexual torture and burning of "woman the witch, healer and sorceress" (Achterberg, 1991; Gies & Gies 1978; Lederer 1968). "For she is a liar by nature, so in her speech she stings while she delights us... for her voice is like the song of the Sirens, who with their sweet melody entice the passerby's and kill them..." (Malleus Maleficarum).

These sadistic misogynists were soon seconded by Bishop Bossuet, of France, who believed there was an army of 180,000 witches threatening France and Rome, all of them beautiful and thus bewitching. The Bishop then demanded that they be gathered up in one body and that "all be burned at once in one fire."

And the great sin of these women? According to the Malleus Maleficarum "Carnal lust! All witchcraft comes from carnal lust, which in women is insatiable."

As summed up by Lederer (1968) in his interesting book, The Fear of Women, "witchcraft was a woman-thing, and the persecution of witches a man-thing; for it was first and last the women who were being persecuted and burned." In fact, hundreds of thousands of women were murdered, whole populations were destroyed, and some villages were left with only one female inhabitant. In 1404 alone, it is estimated that the Papal fathers burned at least 30,000 women. So many women were murdered, that it could be said that the Church was attempting no less than a whole sale genocide of women—"sexocide" at Lederer put it.

Because many of the men (the Crusaders) had been killed or were serving in the army of the Catholic Lord God, the women were often left unprotected. Sometimes whole villages were destroyed, or all the women in a given area were rounded up by the Catholic authorities and then sexually murdered.

These women, particularly those who were exceedingly attractive or ugly, were then hideously tortured and then slaughtered by burning, boiling in oil, crushing, and via whatever device the religious authorities felt appropriate or which suited their sick minds. In Germany huge ovens were constructed for the purposes of mass female murder (Achterberg, 1991; Lederer 1968). As noted, even the Lord God Yahweh recommended that haughty, beautiful women be burned in the fire.

However, it was not just beautiful females, for they come in a limited supply, but those who were old, eccentric, childless, and particularly women who owned property and pets, such as cats. Indeed, the cats would be tortured and murdered alongside the women. The "Black Plague" in fact was in part a consequence of the denunciation and killing of cats, coupled with the sanctification of rats and mice (the proverbial church mouse), by the Catholic authorities.

As to the children of these witches, the Church had only "compassion." These children would be merely flogged in front of the fires in which their mothers were burning.

Predominantly, when the priests set upon a village, it was the young and beautiful who were generally the first to be tortured, as they were experts at love magic. These beautiful women would bewitch men by shooting mesmerizing "beams of light from their eyes" which causes the man to fall in love (Briffault, 1931).

As pointed out by Dominicans Heinrich Kramer and Johann Sprenger, "Consider also her gait, posture, and habit, which is vanity of vanities. There is no man in the world who studies so hard to please the good God as even an ordinary woman studies by her vanities to please men... who... they infect with witchcraft by inclining men to inordinate passion. Yet...her heart is a net, it speaks of the inscrutable malice which reigns in their hearts... for they fulfill their lusts by consorting with devils."

According to Pope Innocent VIII, these women "have given themselves up to devils in the form of incubi and succubi. By their incantations, spells, crimes, and infamous acts they destroy the fruit of the womb in women, in cattle and various other animals; they destroy crops, vines, orchards...they render husbands impotent."

Of course, they also had the power to heal, so when the Archbishop of St. Andrews became ill and his physicians were unable to provide him with a cure, he sent for a well known woman healer whose expertise was in the making of ointments, Allison Peirsoun of Byrehill. Her cure was a success! So, the Archbishop had her tortured and burned.

RELIGIOUS SEXUAL SADISM

The torture of women, of course, was sexual. The woman would be stripped and her body, breasts, and orifices would be thoroughly investigated for the mark of the devil. However, the "investigation" was made with long needles and conducted by priestly specialists called "prickers," who would stick these needles into every suspected Devil's mark.

After the women became unconscious or unresponsive to the continued pricking, the priests would then employ "gresillons which crushed the tips of fingers and toes in the vice; the echelle, or rack, for stretching the body; the tortillon for squeezing its tender parts... legs were broken, even into fragments... or the legs would be grilled on the caschielawis...the fingernails were pulled off with the turkas, or pincers, and needles driven up to their heads into the quick."

As summed up by Henry Charles Lea: "Protestants and Catholics rivaled each other in the madness of the hours... Christendom seemed to have grown delirious, and Satan might well have smiled at the tribute to his powers seen in the endless smoke of the holocaust which bore witness to his triumph over the Almighty."

SEXUALITY, THE GODDESS AND THE SERPENT

Karen Robidoux was a young mother and a member of a religious sect. Karen was also very attractive and beautiful, and some of the female members of the sect believed she was vain. They resented how some of the men of the sect would look at her. And then, one of the women, Michelle Mingo, experienced a vision and heard a voice: it was a prophecy from God. Because Karen was pretty, and vain, she could no longer eat solid food, but could only drink boiled almonds. Her 10month old son, Samuel, was also to be denied food. Breast milk would be his only source of nourishment. The sect members then allowed the boy to starve, and ignored his cries for food as they ate at the dinner table. He starved to death—"God's" punishment for his mother's beauty and sexuality.

Sexuality is a major concern of most major religions (Lederer 1968; Parrinder 1980; Smart 1969) as well as the limbic system. In fact, almost all major religions and their Gods, either act to promote sexuality, or to suppress it. This should not be entirely surprising for religions are very sexual and many were originally concerned with the fertility of the fields and the abundance of prey (Campbell 1988; Frazier 1950; Harris, 1993; Kuhn 1955; Malinowski 1948; Parrinder 1980; Prideux 1973). Religious rituals evolved accordingly beginning with the cult of the goddess, the sacred feminine, the Great Mother of All.

As summed up by Halle Austen (1990) in her book, The Heart of the Goddess, "creativity, the power to manifest physical and psychic reality, is one of the Goddess's primary aspects...the source of all being. She appears as the Great Mother, the Sustainer of Life, the Cosmic Creatrix. It is from her that all life proceeds and to her that all life returns." And, "just as our ancestors honored a woman's ability to create humans from her womb and feed them from her breast, they also honored the Earth as the Great Mother who nourishes us" and the sky and heavens as the Great Mother who would nourish the Earth and her children with drops of life sustaining rain.

Aditi, the Hindu Goddess of the Void, for example, represented "creative power" and "abundance," and the cosmic womb from which all creation has its source. The goddess is also known as the celestial cow and the golden calf who provides nourishment not just to humans, but the gods. As the Great Mother also nourishes the gods, she is also the guardian of cosmic harmony and order.

The cult of the goddess has its origins in the Paleolithic. Hence the numerous "Venus" statues and carvings of heavy breasted pregnant women.

The goddess cult continued to reign supreme during the Neolithic and she was worshipped by the ancient Sumerians, Egyptians, Akkadians, Babylonians,

Hebrews, and in fact all ancient peoples including those of the Americas and Australia.

The Australian Aborigines, for example, are not only the oldest continuous culture that has survived on this planet, but they have worshiped the Great Mother, the "All Mother" since time immemorial, depicting her as giving birth in dreamtime to all the peoples of the Earth.

Because the sun also nourishes the Earth, the Goddess was also a goddess of the sun, for the sun was originally thought to be female. Hence, Igaehindvo the Cherokee Sun Goddess, Brigit the Celtic Fire Goddess, and Amaterasu Omikami the Great Spirit Shining in Heaven and Goddess of the sun of ancient Japan: "The radiance of the Sun Goddess filled the universe and all the deities rejoiced." In ancient Egypt the goddess Nut was believed to have given birth to the sun.

Woman was also associated with the sun and fire because in many ancient cultures, it was believed that woman first discovered the art of making fire, which she used to warm the people and prepare the food. Fire is also a symbol of spiritual awakening and knowledge of the spirit world.

The goddess, therefore, was associated in all ways with all things having to do with life and death, and the spirit world.

"I am Nature, the universal Mother, mistress of all the elements, sovereign of all things spiritual, queen of the dead. Though I am worshiped in many aspects, known by countless names, and propitiated with all manner of different rites, the whole earth venerates me." -Apuleius, second century B.C.

It is noteworthy that in addition to being the source of all life, that the Goddess was also associated with the serpent. The Sumerian "god" Enki, was depicted as a snake, and sometimes as a snake with breasts. The peoples of predynastic Egypt also worshipped the goddess in the form of a snake. In ancient Egypt, the hieroglyph for "snake" also means Goddess.

The Hindu Goddess Kali wore a garland of snakes around her neck—which represented the female life force, the transformative power of Shakti. Shakti was often depicted by the ancient Hindus and the pre-Aryan Dravidans, with a kundalini snake emerging from her womb and vagina.

The snake although an obvious phallic, and thus sexual symbol, is also associated with wisdom. Hence, the snake in the garden of Eden not only tempts Eve, but he induces her to eat of the fruit of the tree of knowledge. In fact, throughout the world snakes have commonly been the companions of oracles, those who had special knowledge of the future.

The snake represents supernatural knowledge and power which is one of the reasons the Lord God instructed Moses to "take your rod and cast it down before Pharaoh. It shall turn into a serpent... Then Pharaoh called the wisemen and the sorcerers, now the magicians of Egypt and they... cast down every man his rod, and they became serpents: But Aron's rod swallowed up their rods"

(Exodus 7:10-12). And, it was a serpent which Moses carried as a fetish which he used to protect and lead the Israelites as they wondered for 40 years in the desert.

Nevertheless, because of its association with wisdom, sexuality, and woman, the serpent also came to be viewed as evil. The Lord God condemned the serpent, whereas the Catholic Church depicted this reptile as synonymous with Satan, the Devil. Both woman and the serpent became enemies of the Lord God.

SEX, FOOD, AND RELIGION

The limbic system mediates not only sexuality, but internal homeostasis and food intake. Likewise, many modern mystical and religious practices have also involved the ritual control over sex and food. This includes many American Indian, Christian, Jewish, and Moslem sects (Campbell 1988; Parrinder 1980; Smart 1969). Thus the commandment "thou shalt not... " These are limbic taboos, as eating and sexuality (like murder and violence) are under limbic control.

Many limbic taboos, however, promote survival, for example, by proscribing the eating of poisonous plants or unclean animals. Similarly, by forbidding anal or indiscriminate sex one was spared the wrath of this "God" and whatever plagues he might send in the form of venereal disease or viruses. If we rule out the possibility of an attack with nuclear armed missiles, mass death due to disease is presumably what became of Sodom and Gomorrah where the anal sex crazed mobs attempted to sodomize even the angels sent by the "Lord God" himself (Genesis 19).

Sex and food (along with fear, rage, and aggression) are probably the most powerful of all limbic emotions and motivators. If sufficiently hungry or sexually aroused, these conditions can completely overwhelm the brain. Limbic hyperactivation in turn can induce religious or spiritual dreams or hallucinations.

Hungry men, women, and infants will dream of food. Those who are sexually aroused will dream of sex. However, a parched and starving man will not just dream, he will hallucinate food and water and will attempt to slake his desires by consuming a hallucination.

Given that early (as well as modern) human populations were often concerned with obtaining food (as well as a sex partner) many of their earliest religious beliefs and rituals were therefore concerned with increasing the abundance of game animals as well as preserving their own progeny (Armstrong 1994; Campbell 1988; Frazier 1950; Harris, 1993; Kuhn 1955; Parrinder 1980; Prideux 1973). As noted, many an ancient Upper Paleolithic cave was decorated with fertility and sex symbols, including pregnant women (Venus figures) and animals (Bandi 1961; Joseph 1996; Kuhn 1955; Leroi-

Gurhan 1964), whereas Egyptian tombs contain numerous paintings of food and goddesses.

Thus, given our ancient hunter-gatherer (and then later, farming) heritage, many religions both ancient and relatively modern are highly concerned with fertility and food, and tend to be very sexual and limbic in orientation if not origin. This is also why there have always been fertility goddesses and gods who are associated with eating and drinking, especially alcohol (Campbell 1988; Frazier 1950; James 1958; Parrinder 1980; Smart 1969). This also includes, for example, Osiris, and especially Dyonisus who was among other things, a sex crazed dancing god of the vine. In fact, one of the first miracles performed by Jesus involved making wine from water, and, as we know, Jesus surrounded himself with highly sexual women, prostitutes who would sometimes rub precious oil on his body.

SEX, GOD, AND RELIGION : RELIGIOUS LOVE CHARMS AND SPIRITUAL SEXUALITY

In Arabic-pre-Islamic tradition, it was said that martyrs were rewarded in heaven with 72 Virgins. "Man has not touched them before them nor jinni. Which then of the bounties of your Lord will you deny? Reclining on green cushions and beautiful carpets. Which then of the bounties of your Lord will you deny? —The Beneficent.

Hadith number 2,562 in the collection known as the Sunan al-Tirmidhi says, "The least [reward] for the people of Heaven is 80,000 servants and 72 wives, over which stands a dome of pearls, aquamarine and ruby."

Among the ancient religions of India and China, the sexual activity of the Gods and the promotion of similar sexual activities among the believers were widespread religious practices and beliefs (Austin 1990; Campbell 1988; Parrinder 1980).

For example, the ancient Vedas were greatly concerned not only with the worship of various nature gods, but with the rituals of sexual union. Ancient Indian religious texts are filled with love charms and instructions as to how to win the love of a man or woman, or to protect against demons.

Temple prostitutes were also quite common throughout India and the Middle East, as well as in Rome and Greece. Some temples employed so many girls that they were like giant brothel emporiums (Parrinder 1980). As noted, sexuality and desire (like religious feeling) are directly mediated by the amygdala and hypothalamus.

In fact, sexual intercourse became a religious ritual among Hindus and Buddhists who practiced "tantra." The practitioners of tantra were inspired by visions of cosmic sex and the acquisition of sexual energy. Through tantra one might be confronted with the cosmic mystery of creation as exemplified by another deity, Shakti, the divine mother. However, restrictions on where one could have sexual intercourse (not in public) and certain types of sexual acts such as oral sex, were prohibited as well as sex with strange women or those of a lower caste (Parrinder 1980). Nevertheless the joys of sex were continually emphasized and embraced. Hence, the Kama Sutra, the "love text".

On the other hand, it was believed by some ancient far Eastern sects that in order to gain power, one had to break taboos and, for example, engage in sexual orgies or have sex with women while they were menstruating. Menstrual blood was believed to possess the spiritual and creative life force, whereas sexual orgies were believed to liberate tremendous amounts of sexual energy.

These "taboo" sexual practices were also a form of tantra, referred to as "left handed tantra." Those who followed the way of the left handed tantra claimed that passion was nirvana and that adepts should cultivate all sexual pleasures (Parrinder 1980). Both male and female deities, usually in the act of having sex, were worshipped.

Ancient Chinese and Taoists religions are also quite sexual (Austin 1990; Parrinder 1980).

These beliefs are exemplified by the concepts of Yin and Yang which appeared over 3000 years ago and which represented the male and female principles of the universe. Sexual intercourse was viewed as a symbolic union of the earth and heaven, which, during rainstorms were believed to mate. By engaging in sexual relations man and woman achieved harmony by following the example of the gods.

THE LORD GOD: A SEXUAL GOD

Beginning at about the same time that the Judaic religion became more dominant in certain areas of the Middle East, around 3,000 years ago, and over the following thousand years a tremendous change in religious sexual thought began to flourish, enveloping the Roman Empire, and which eventually paralleled and coincided with the development of Christianity and Islam.

Unlike the goddess and the highly sexual gods of, for example, the ancient Greeks and Romans, the "Lord God" of the Old Testament does not have sex with human women and there is no hint of sexual duality in his personage. He is a male god, a warrior and destroyer. Although not overtly sexual, the sexual activities of men and especially women were of great concern to Him. Be fruitful and multiply, He ordered, and as to certain women who were presumably barren, and in one case, still a virgin, He is reported to have opened their wombs.

Male sexuality was also a concern; that is, the diminishment of the intensity of man's sexual pleasure. Thus He required a form of sexual self-mutilation. As part of his covenant with Abraham and the Israeli people, it was ordered that every male child would suffer the amputation of the tip of his penis (which is densely innervated by fibers that yield intense sexual pleasure): "And ye shall circumcise the flesh of your foreskin; and it shall be a token of the covenant betwixt me and you" (Genesis 17: 10-11).

Although there is no evidence that the Lord God was sexually active with human females, it does appear that the Lord God had a sexual consort, the goddess Sophia, before he became the Lord God of the Earth. Sophia was the goddess of wisdom and she was not only his lover, she may have been his sister: "It is Wisdom calling. Understanding raising her voice. I Wisdom, live with Prudence; I attain knowledge and foresight... The Lord possessed me in the beginning in his own way, before his works of old...When he prepared the heavens, I was there...I was by him as one brought up with him. I was daily his delight, rejoicing always before him" (Proverbs 8).

In addition, it appears that she may have betrayed Him. "Rejoicing in the habitable part of the earth, and my delights were with the sons of men" (Proverbs 8:31).

It is true, however, that the above quote may mean that she simply was delighted by the sons of men, and did not have sex per se with men.

However, if she did betray Him, this may explain why she is never again referred to, as well as His following complaints: "Aholah played the harlot when she was mine; and she doted on her lovers...desirable young men...thus she committed her whoredoms with them and with all whom she doted... she defiled herself" (Ezekiel 21: 5-7). "Like mother, like daughter. You are the daughter of your mother, who rejected her husband" (Ezekiel 16: 44-45).

It may also explain His jealous prying nature, and obvious dislike of the female sex and his recommendation that women who were truly attractive and sexually desirable should be burnt (Isaiah 3:16-26, 4:4) as they were all "whores" (Hosea 2).

"Because the daughters of Zion walk with wanton eyes, mincing as they go, and making a tinkling with their feet...The Lord shall wash away the filth of the daughters of Zion... by the spirit of burning" -Isaiah 3:16-26, 4:4.

Indeed, Sophia's "delights with the sons of men" and her betrayal of the Lord God may explain why He not only disdains pretty women, but pardons and encourages the sexual exploitation of women and the rape and sexual slavery of women, and their murder if they dare behave like whores. In fact, He brags that he "delivered...Aholah who played the harlot when she was mine... into the hand of the Assyrians... They discovered her nakedness; they took her sons and her daughters, and slew her with the sword... they executed judgment on her" (Ezekiel 23:9-10).

RELIGIOUS RAPE AND SEX SLAVERY

The Lord God is very concerned with the sexual relations between men and women. Those men He loved most dearly had sex with multiple sex partners. The Lord God even approved of sex slavery, including sex with little girls who are taken slave:

"Kill every male among the little ones. But all the women and female children that have not known a man, keep alive for yourselves." -Numbers, 31

The permissive attitude of the Lord God when it comes to female sex slavery may explain why the modern state of Israel not only has no laws forbidding the owning or selling of human beings, but why non-Jewish women are openly sold as sex slaves in the state of Israel (see M. Specter, "Traffickers New Cargo: Naive Slavic Women, New York Times, January 11, 1998).

And, once a sex slave, always a sex slave, even if she is Jewish female.

"When you acquire a Hebrew slave... and if his master gave him a wife, and she has borne him children, the wife and her children shall belong to the master. When a man sells his daughter as a slave, she shall never be freed, as male slaves are." -Exodus, 21: 2-7.

And the Lord God even encourages and pardons the violent rape of young Jewish virgins by Jewish men. For example, after some of the Benjaminites raped a woman to death (Judges 19:25), and following the murder of their own women in retaliation by the other 11 tribes of Israel, these tribes then realized that the tribe of Benjamin would die out without women. So, they decided to attack yet another city of their fellow Jews and to "utterly destroy every male and every women that hath lain by man" with the exception of "young virgins" who the Benjaminites were allowed to rape (Judges 21:11-13).

However, not enough virgins were acquired following this massacre. So the Benjaminites were instructed by the Lord God to rape another group of young Jewish virgins.

"Then they said, Behold there is a feast of the Lord in Shiloh...Go and lie in wait in the vineyards. And see and behold if the daughters of Shiloh come out to dance in dances, then come ye out and catch you every man his wife of the daughters of Shiloh" (Judges 21:19-25).

The Benjaminites, however, raped men as well as women, and were quite willing to gang rape a woman until she was dead (Judges 19:25-26, 20:5). According to the divine law, they could also brutally and sadistically rape female slaves; and if she lived for two days following a violent, brutal rape, there was no penalty.

Again, the Lord God of the ancient Jews had an obvious contempt for the female sex: "Because the daughters of Zion are haughty and walk with stretched forth necks and wanton eyes, walking and mincing as they go, and making a tinkling with their feet; The Lord will smite with a scab of crowns of their head and the Lord will discover their secret parts. In that day the Lord will take away their tinkling ornaments...their bracelets...bonnets, ornaments... earrings, the rings...and fine linen...And it shall come to pass that instead of sweet smell there shall be stink." -Isaiah 3:16-26, 4:4.

Indeed, this Lord God, when angry, found pleasure in using explicit sexual imagery when condemning his people, Israel—whom He referred to as a woman when He was angry. Indeed, He repeatedly threatened to strip this brazen female "naked" and referred to her as an "adulterous harlot." Echoes of the divine betrayal by the Sophia, the goddess of Wisdom?

"And let her put away her harlotry from her face and her adultery from between her breasts. Else I will strip her naked... And I will snatch away My wool and My linen that serve to cover her nakedness. Now will I uncover her shame in the very sight of her lovers..." -Hosea 2.

SEXUALITY, PEDOPHILIA, & HOMOSEXUALITY

Although this volatile, mercurial, and masculine seeming "Lord God" was asexual, sexual behavior was of tremendous concern to "Him" for He commands sexual moral obedience—at least of women and married women in particular— and repeatedly tells his people, starting with Adam and Eve, "be fruitful and multiply."

And sexuality, and the condemnation and/or control of sexuality, has been and remains a major concern of cults or established religions. The temple virgins who served the virgin goddesses of ancient Greece and Rome, were required to remain celibate upon threat of death. Shakers' communities have been dying out for lack of sex and children. Buddhist monks are celibate, and Hindu "renouncers" swear off sex later in life. According to Catholic Canon law priests are required to remain celibate and celibacy is defined to include "perfect and perpetual continence," meaning no sexual activity of any kind, including masturbation.

However, celibacy did not become a requirement of priests until 1139 when it became church law. Prior to 1139, priests, bishops, cardinals, and even Popes had lovers and wives, some Popes had several.

The Mormon Church initially willingly embraced, as a religious creed, the right of every man to have several wives; i.e. sex partners, and the founder of the Church was reputed to have 40 wives. Likewise, Muhammed was reported to have the sexual prowess of forty men, and to have bedded at least 9 wives and numerous concubines including even one young girl (Lings 2013). However, the grand champion of religious-sexual excess was King Solomon who required the sexual services of 700 wives and 300 concubines.

Sex scandals are also commonplace among priests, rabbis, pastors, and ministers.

The Reverend Jimmy Swaggart whose sermons were filled with fire and brimstone, and who regularly assailed Catholics, Jews (who "are going to Hell") pornography and prostitution, temporarily gave up a $150 million dollar a year world wide television ministry in 1988, when his fondness for prostitutes and pornography was exposed and he was caught with a prostitute. Moreover, the attorney who exposed Swaggart also claimed the minister was a pedophile. Swaggart has since returned to the ministry. During his sermons he yells, collapses to the floor, recounts conversations with god, repeatedly breaks into tears, and again rails against Catholics and Jews (who "are going to Hell"), and against pornography, prostitution, as well as hypocrisy.

Likewise, Jim Bakker lost a 300 million dollar ministry, when among other things, he and another pastor were accused of repeatedly sexually assaulting a young female member of their flock. In 1999, the dean of Harvard Divinity School was forced to resign "for conduct unbecoming" after it was discovered that he had downloaded thousands of "explicit" pornographic images on his Harvard-owned computer in his Harvard-owned home.

In January of 2001, the Reverend Jesse Jackson admitted that he had cheated on his wife and has fathered a daughter from another woman. "No doubt, many close friends and supporters will be disappointed in me," Jackson said.

Sex with little girls is permitted by the Jewish Holy book, the Talmud:

For example, according to the Talmud: "Rab said: When a grown-up man has intercourse with a little girl it is nothing, for when the girl is less than this it is as if one puts the finger into the eye... tears come to the eye again and again, so does virginity come back to the little girl under the age of three."

Talmud Sanhedrin 55b "Come and hear! A maiden aged three years and a day may be acquired in marriage by coition [sexual intercourse], and if her deceased husband's brother cohabits with her, she becomes his."

The Talmud, is believed by Orthodox Jews, to be based on god's laws and to be Mosaic in origin; that is, based on oral (and written laws) first put forth my Moses: "Kill every male among the little ones. But all the women and female children that have not known a man, keep alive for yourselves." -Numbers, 31

Sex with children, and especially sex with little boys, has been a common religious practice, and a priestly predisposition, since the inception of the priesthood and organized religion.

DAVID KORESH: SEXUAL THEOLOGY

David Koresh announced he was God in a wedding invitation to the Seventh Adventists Church in southern California. It read in part: "I will scold your daughters for their nakedness and pride that they parade in My Father's house, and by my angels, I will strip them naked before all eyes because of their foolish pride." Naked women at his mercy and who he would impregnate and "fill with seed" was his predominant religious fantasy.

As he confided to his "right hand man" Marc Breault early in his ministry: "I'll have women begging me to make love to 'em. Just imagine, virgins without number." Within just a couple of years he would be attended by at least twenty young women, most of whom he "married," including two that were just 14 years old, and one who was age 12 (Breault & King, 1993).

Sex was at the center of his theology, even claiming that god would take them to heaven in a divine spaceship that he sketched on a blackboard. It was an erect penis!

Koresh apparently was consumed by sexual thoughts and was the recipient of sexual visions that he claimed were sent to him by God. Soon he began demanding sex from the wives of his followers, women who he believed had married these other men without his permission and who should have married him: "All you men are just fuckers, that's all you are. You married without getting God's permission.

Even worse, you married my wives. God gave them to me first. So now I'm taking them back." According to Marc Breault, everybody was shocked, stunned, and Vernon kept saying things like: "So Scott, how does it feel to know your not married anymore... In October, 1989, he began having sex with the other men's wives... and directed the women to inform him when they had reached the fertile part of their cycle to maximize the chance of pregnancy." As per the men, Vernoninformed them that it was their job to "defend King Solomon's bed."

Vernon not only had sex with and impregnated their wives—fathering over 20 children— but began having sex with these children; that is, when he wasn't beating them.

PEDOPHILE PRIESTS

As the American public learned in the Spring of 2002, pedophile priests have been raping little boys for decades if not for the last thousand years. Indeed, homosexuality and the homosexual rape of boys and young men is so common in the Catholic church that by edict, priests are no longer allowed to be alone with altar boys.

In 1999, the Bishop of the Santa Rosa Diocese in California was forced to resign because another priest accused the Bishop of repeatedly raping him.

Rev. John Rebovich enticed teenagers with free alcohol and then would rape them when they were too drunk to resist.

Gerald Ridsdale, an Australian Catholic priest, pleaded guilty to sexually assaulting 21 children.

Stephen Kiesle, a priest serving in Fremont, California, was charged in May of 2002 with sexually molesting up to 21 children. Kiesle had admitted to police that he "really liked young blond girls." However, his victims included both boys and girls, many as young as 9, some of whom he would tie up before assaulting them sexually.

The Rev. Paul Shanley has been accused of raping hundreds of boys. Indeed, Paul Shanley openly advocated the idea of sex between men and boys and he often attended meetings with the North American Man-Boy Love Association.

Boston priest John Geoghan had been raping young boys for over three decades; nearly 200 boys in 30 years. He is not alone.

It has been reported by the Boston Globe that 50 of Boston's Roman Catholic priests have been molesting boys for decades. Because of the public uproar, the Boston archdiocese was forced to remove over 20 priests between January and September of 2002, because of sexual misconduct and the rape of young boys. Over 400 lawsuits have been filed against priests in Boston.

In 2002, a number of seminary students were expelled from the Catholic Theological Union in Chicago because they were openly engaging in homosexual relations with one another. The Maryknoll Seminary is known to be "overrun with gay men."

The entire Catholic Church appears to be overrun with "gay men;" that is homosexual pedophiles. In consequence, when one pedophile priest is removed or transfers, another pedophile takes his place.

In March of 2002, Anthony J. O'Connell, the Roman Catholic bishop of Palm Beach, Florida, resigned after admitting that he sexually abused a teenage seminary student. He had also been accused of raping two other boys. He had been appointed to lead the Church by Pope John Paul II following the resignation of another pedophile, Bishop Joseph Keith Symons, who admitted sexually assaulting at least five boys.

According to Dr. Thomas Plante of the Jesuit college, Santa Clara University, and editor of a book on sexuality and the church, "50% of the Catholic clergy are gay." The Boston Globe (1/31/ 2000) has revealed that Catholic priests have a rate of HIV infection which is four times that of the general population.

Estimates as to the number of active, homosexual pedophiles in the Catholic Church range from 6.1% to 16% according to Richard Sipe, an "expert" on sex abuse and the Church.

There are an estimated 50,000 priests in the US. Hence, the number of pedophile priests could range from 3,000 (6%) to 8,000 (16%). However, there are only 188 diocese in the U.S., which means that there are anywhere from 16 to 42 pedophile priests per diocese.

Not surprisingly, over 800 priests have been removed from the ministry as a result of allegations against them. There have so far been over 1,400 insurance claims against the Church. The Church has so far paid out over $1 billion in liability with an estimated $500 million pending.

Not surprisingly, pedophile priests often work together in the same church or diocese. The Rev. Paul Shanley and another pedophile priest owned and operated a "gay inn" in Palm Springs which catered to gay men. Shanley is accused of raping hundreds of boys. Indeed, Paul Shanley openly advocated the idea of sex between men and boys and he often attended meetings with the North American Man-Boy Love Association.

Likewise, according to the Cleveland Plain Dealer in a series of stories printed in the Spring of 2002, "parishes such as Ascension and St. Patrick, both in the West Park neighborhood of Cleveland, had more than one alleged abuser working at the same time." Moreover, when they work together, they also aid each other in recruiting and raping children.

Consider, for example, Reverends Gary Berthiaume and Allen Bruening. As detailed in May of 2002, by the Cleveland Plain Dealer: "Bruening would use trips to the pool to seduce young boys." One child, "Frank (not his real name), says he not only had to contend with Bruening, but also with Berthiaume, who was sent to the Cleveland diocese after serving six months in a Michigan prison for child abuse. After the swimming trips, says Frank, both priests were waiting in the showers. While Bruening stood naked in a one-person stall, says Frank, Berthiaume would be ordering him to join the other priest in the shower."

"Here I am, a little kid, and here is this pastor, you want to believe you're a good kid," Frank says. "This person is the next closest thing to God. You would do anything that they would say. How could you question these people?"

According to Professor Germain Grisez, and as he reported to the Bishops Committee on Sexual Abuse, the Catholic Church is permeated by a "homosexual subculture" and "homosexual clerics" who seduce "adolescent boys."

Likewise, according to Father Donald B. Cozzens, author of The Changing Face of the Priesthood, there are a "disproportionate number of gay men that populate our seminaries."

As summed up by Bishop Wilton D. Gregory, "It is an ongoing struggle to make sure the Catholic priesthood is not dominated by homosexual men."

It is in part because of the homosexual culture and the large number of pedophile priests that the Catholic Church and its Catholic Bishops have for centuries, covered up the homosexual rape of children by not just priests, but Bishops and Cardinals.

For example, Boston priest John Geoghan had been raping young boys for over three decades. When Bernard Cardinal Law of the archdiocese of Boston became aware of Geoghan's behavior which was threatening to become public, he quietly transferred Geoghan to another parish where he immediately began sexually assaulting children. Geoghan was not the only priest he transferred; and many other bishops have done likewise: cover up the crime and transferring the pedophile priest to another city where he can continue raping boys.

It is also because of these coverups, that in March of 2002, several Federal anti-racketeering suits—RICO The Racketeering Influenced and Corrupt Organizations Act— have been filed. By filing that kind of suit, the lawyers are calling the Catholic Church a criminal enterprise.

It is because of this overarching homosexual-pedophile culture that many priests, bishops, and cardinals, have not just covered up abuse, but in fact see nothing wrong with it, viewing these crimes as "forgivable"—except in those cases where a priest or Bishop repeatedly rapes young boys and the rapes become public.

Even in these latter instances, Church officials, until recently, were inclined to reinstate pedophile priests even after their cases become public or the cause of ruinous lawsuits.

For example, Bishop Donald W. Wuerl of Pittsburgh, had to battle with Church officials for over 7 years before he was able to remove Rev. Anthony Cipolla. Cipolla had been accused of raping altar boys, one of whom filed suit.

Father Cipolla appealed to the Vatican's homosexual hierarchy, its highest tribunal, which ruled to reinstate him despite the suitcases full of papers which documented the priest's sex crimes. Essentially, the Vatican, in making this ruling, also ruled that it was permissible to have sex with children.

Most Bishops are more concerned with protecting the church's name and its bank accounts, and with making it possible for pedophile priests to continue raping boys. Until recently, American bishops have repeatedly transferred predator priests from parish to parish, but only when their sex crimes threatened to bring unwanted publicity to the parish.

In May and June of 2002, when the scandal grew and threatened to completely overwhelm the Catholic Church, Church officials and thus the homosexual hierarchy first tried to downplay the problem and then sought to distinguish between sex with young children who were unwilling to have sex, and sex with older children and older boys who, the Church claims, welcomed if not solicited the sex. The Church denied that it had a problem, and claimed that rapes and sex with minors was not truly an act of pedophilia, because "almost all the cases involved adolescents and therefore were not cases of true pedophilia."

Moreover, according to Church officials, these offenses are forgivable because they amount to little more than the sexual seductions of teenage boys who willingly submit.

In May, and as was widely reported in newspapers such as the Boston Globe, Church officials announced that priests who had abused children in the past, would remain priests so long as there was no evidence that the priest in question was a serial abuser. Likewise, those who abused children in the future would not suffer any penalty, that is, unless they raped and sexually assaulted a number of children. However, even in these cases bishops would be given the discretion to retain these priests given mitigating circumstances.

"The question of the reassignment of priests who have harmed children is still a thorny issue," said Bishop Wilton D. Gregory, president of the United States Catholic Conference, the national forum for American bishops.

To their surprise, the public and the media reacted angrily to their pronouncements and the willingness of the Church to tolerate the sexual rape of young boys. It was repeatedly said in the media, that "the Catholic Church just doesn't get it."

Presumably, the homosexual hierarchy was unable to understand why the public was outraged, precisely because this homosexual hierarchy experienced the same "forgivable" sexual longings for children. In fact, three of the five members of the Ad Hoc Committee on Sex Abuse, including its chairman, Bishop John B. McCormack of Manchester, N.H., had been severely criticized for their handling of sex abuse cases, and their willingness to tolerate the rape of children and homosexuality among members of the clergy. Because so many shared the same deviant desires, even minor sanctions against abusers thus seemed excessive to many in the pedophilia priesthood.

In fact, on August 8, 2002, the Rev. Connors, the president of an association of Roman Catholic religious orders, publically complained that pedophile priests were being "scapegoated" and then criticised American bishops for reacting to the complaints of victims and the media, charging "that American bishops have been scapegoating abusers." Connors also ridiculed those victims who had stepped forward by asking: "Are we having fun yet?" Connors, as president, represents the views of 15,000 U.S. priests in orders such as the Jesuits and Benedictines.

We should recall, that these are the same religious orders which burned women and beat their children in front of the raging fires. This is the same church which would castrate young boys to keep them singing and looking pretty; i.e. the "castratos"--boys who were also the sexual playthings of Bishops, Cardinals and priests.

Thus we should not be surprised to discover that insofar as modern pedophile priests, and the majority of bishops and cardinals are concerned, sex with boys is normal and forgivable. In fact, even sex with children could be forgiven so long as a large number of children were not involved and the cases did not become public.

Specifically, in June of 2002, top Vatican officials and American cardinals tried to contain the public uproar. Repeat offenders and those priests who "become notorious and are guilty of the serial, predatory sexual abuse of minors," would be dismissed, said Cardinal Theodore E. McCarrick of Washington. However, as to those priests who had molested and sexually assaulted children in the past, the Church would be more lenient and leave it up to the local Bishops to determine if the priest was a true danger to children.

According to Cardinal Francis E. George of Chicago, the church wanted to protect the rights of priests who had been accused. In fact, Church officials sent a letter to Catholic priests in America, expressing sympathy and support "through these troubled times. We regret that episcopal oversight has not been able to preserve the church from this scandal," the letter read. That is, the sympathy of the church, was not for the victims, but only for the church and its priesthood. Indeed, pedophile priests must be protected according to Bishop Gregory, who emphasized that "even a priest who offends still enjoys rights."

Cardinal Francis George, of Chicago has also argued that bishops should be allowed to make distinctions between serial pedophiles and priests who "crossed boundaries" with older teenagers.

"A little more wiggle room enables you to be more just," Cardinal George said. "There is a difference between a moral monster who preys upon little children, and does so in a serial fashion, and someone who perhaps under the influence of alcohol engages in an action with a 17or 16-year-old young woman who returns his affection." Young woman?

Stephen Rubino, a lawyer who has represented over 300 alleged victims of priest abuse, estimates 85% of the victims have been teenage boys.

As summed up by National Review senior writer Rod Dreher "This is chiefly a scandal about unchaste or criminal homosexuals in the Catholic priesthood."

Even some Catholic priests and Bishops grudgingly agree, though they were again quick to make a distinction between pedophilia and sex with young men: According to Monsignor Eugene Clark of New York, "disordered" homosexuals were to blame. His view was seconded by the Bishop of Detroit who has said the current crisis is "not truly a pedophilia-type problem, but a homosexualtype problem."

Indeed, homosexuality is so rampant among Catholic priests that according to the Boston Globe (1/31/2000) they have a rate of HIV infection which is four times that of the general population.

And its is not just the Catholic Church which is a refuge for homosexual priests. The Presbyterian Church stated that perhaps as many as 23% of clergy have had "inappropriate sexual contact" with other men. As detailed in the San Francisco Chronicle (Lattin, 7/15/97) homosexuality became so rampant within the ministry of the Episcopal church, and its homosexual clergy so openly promiscuous that the church lost over 5 million members by 1997.

In Australia, hundreds of boys and young men have alleged they were raped by the homosexual priests and pastors of the Christian Fathers boarding schools.

In Islamic countries, homosexual are also drawn to the priesthood. It is rumored, for example, that in the holy city of Om, in Iran, that homosexuality and homosexual pedophilia is a common, albeit forbidden practice among Mullahs who sexually exploit their young male charges.

Homosexuality and homosexual pedophilia are common among the Hindu priesthood and the Hare Krishnas. In fact, over a dozen Hare Krishna temples in the United States were forced to file for bankruptcy in 2002, because of the over 400 million dollars in judgments awarded in lawsuits for the homosexual

sexual abuse of boys and young men. They are not alone: The archdiocese of Boston has acknowledged playing millions of dollars in settlements for almost 90 priests.

THE LIMBIC SYSTEM AND SEXUALITY

Why the concern regarding sex pro or con in religious thought? Why the illicit, perverse, or promiscuous tendencies of priests and other religious leaders? Sex, like religious experience, or the ability to derive pleasure from eating and drinking, is mediated by the limbic system; i.e. the hypothalamus, amygdala, and temporal and frontal lobes (e.g. Freemon & Nevis 1969; Joseph, 1988a, 1992a, 1999a; MacLean 1969, 1990; Remmillard 2013; Robinson & Mishkin 1968).

Activation of the sexually dimorphic amygdala can produce penile erection and clitoral engorgement (Kling & Brothers, 1992; MacLean, 1990; Robinson & Mishkin, 1968; Stoffels et al., 1980), and trigger sexual feelings (Bancaud et al., 1970; Remillard et al., 2013), extreme pleasure (Olds and Forbes, 1981), memories of sexual intercourse (Gloor, 1986), as well as ovulation, uterine contractions, lactogenetic responses, and orgasm (Backman & Rossel, 2014; Currier et al., 1971; Freemon & Nevis, 1969; Remmillard et al., 2013). However, the limbic system also mediates violent behavior, including sexually violent behavior. Thus religiosity also can also be tainted by violent sexual thoughts and behavior.

THE GODDESS, HOMOSEXUALITY, & RELIGIOUS EXPERIENCE

It has been estimated that at least 50% of Catholic priests are homosexuals, and that the HIV rate among Catholic Priests is four times that of the general population. In fact, throughout history, shaman and other religious figures have commonly been homosexuals. Homosexuals are probably drawn to the priesthood for the same reason that women have been viewed as more in touch with the supernatural; i.e., the sexual differentiation of the limbic system.

Recall that portions of the hypothalamus and amygdala are sexually dimorphic; i.e. there are male and female amygdaloid nuclei (Bubenik & Brown, 1973; Nishizuka & Arai, 1981). Female amygdala neurons are smaller and more numerous, and densely packed than those of the male (Bubenik & Brown, 1973; Nishizuka & Arai, 1981), and smaller, densely packed neurons fire more easily and frequently than larger ones. This may contribute to the fact that females are more religious, more emotional and more easily frightened than males as the amygdala is a principle structure involved in evoking feelings of fear (Davis et al., 1997; Gloor, 1997; LeDoux, 1996) as well as spirituality.

THE FEMALE LIMBIC SYSTEM AND RELIGIOUS EXPERIENCE

Confirming common experience, numerous scientific studies have demonstrated that women display clear superiorities over males in regard to the expression, perception, and comprehension of social emotional nuances, regardless of the manner in which they are conveyed (Burton & Levy, 1989; Brody, 1985; Buck, 2007, 2014; Buck et al., 1974, 1982; Card et al., 1986; Eisenberg et al., 1989; Fuchs & Thelan, 1988; Harackiewicz, 1982; Lewis, 2013, Rubin, 2013; Safer, 1981; Shennum & Bugental, 1982; Strayer, 1980). And this is true even from the earliest stages of childhood.

This greater social emotional sensitivity, including a much greater likelihood of becomes frightened and easily terrified, is likely due to sex differences in the functional and structural organization of the limbic system, the amygdala in particular. Because female amygdala neurons are more numerous and packed more closely together (Bubenik & Brown, 1973; Nishizuka & Arai, 1981), and as smaller, tightly packed neurons demonstrate enhanced electrical excitability, lower response thresholds, and increase susceptibility to kindling, there is thus a tendency for the female amygdala to become hyperactivated more easily than the male amygdala—thus inducing extreme fear as well as a propensity for religious and spiritual feelings.

Indeed, women are not just more emotional, but have more intense religious experiences, attend church more often, are more involved in religious activities, involve their children more in religious studies, hold more orthodox religious views, incorporate religious beliefs more often in their daily lives and activities, and pray more often as well (Argyle & Beit-Hallahami, 2005; Batason & Ventis, 1982; De Vaus & McAllister Glock. 1967; Lazerwitz, 1961; Lindsey, 1990; Sapiro, 1990).

Moreover, the anterior commissure, a thick rope of axonal fibers which interconnects the right and left amygdala of the right and left temporal lobe is also sexually differentiated. The anterior commissure is responsible for information transfer as well as inhibition within the limbic system. Specifically, the female anterior commissure is 18% larger than in the male (Allen & Gorski 1992).

Thus not only is the female amygdala more excitable, but the right and left amygdala (located in the right and left temporal lobe) can more easily communicate and excite one another. Hence, again, the increased tendency for females to become easily frightened and to be religiously inclined. To know fear is to know god.

THE HOMOSEXUAL LIMBIC SYSTEM AND SPIRITUAL EXPERIENCE

Although the etiology of homosexuality remains in question, it has been shown that the ventromedial and anterior nuclei of the hypothalamus of male homosexuals demonstrate the female pattern of development (Levay, 1991; Swaab, 1990), whereas the anterior commissure, which interconnects the right and left amygdala/temporal lobe, is "hyper-female" in size (Allen, & Gorski (1992). Male homosexuals are in possession of limbic system that is more "female" than "male" in functional as well as structural orientation.

This female pattern of limbic system development also explains why homosexuals are more inclined to behave and think like women rather than men. Indeed, homosexual males and females tend to be more alike than different in regard to social-emotional reactions and tendencies (Tripp 1987). In some cases these feminine tendencies are grossly exaggerated (Tripp, 1987); i.e. the "swishy" male with the exaggerated high pitched voice.

A significant number of homosexuals, in fact, are psychologically similar to females in a number of ways, including having a high interest in fashion and wearing apparel, a pronounced tendency to employ feminine body language and vocal tones, to shun sports and avoid fights, and to have a fear of physical injury, particularly during childhood (Bell et al. 1981; Bieber et al. 1962; Van Den Aardweg, 2014; Tripp, 1987). Many also tend to maintain intense dependency relations with their mothers and to remain distant from strong male figures including their fathers (Green, 1987).

As children homosexual males tend to prefer female companions and friends, girls toys, activities, and often girls clothes, (Bell, et al. 1981; Saghir & Robins, 1973; Grellet et al. 1982; Green, 1987). From 67% to 75% of homosexuals vs 2%-3% of heterosexual males reported being "feminine" and more like girls than boys as children (Saghir & Robins, 1973; Green, 1987).

Thus, being in possession of a "female" limbic system may not only account for similarities between homosexuals and women in regard to fear and other behaviors, but in respect to spirituality. Thus priests and shamans not only tend to be homosexuals, but public displays of overt homosexual activity and promiscuity coupled with religious fervor often comes to characterize the behavior of homosexual clergy, as was the case within the ministry of the Episcopal church, and which appears to be the case in the Catholic Church.

The homosexual limbic system, however, is actually organized in a hyper-female fashion. When compared to the female and especially the male limbic system, the homosexual limbic system could be considered different from the male and female "norm." And as noted, abnormalities in the limbic system or hyperactivity, is associated with hypersexuality, transvestitism, and public displays of sexuality (Blumer, 1970; 1999; Davies and Morgenstern, 1960; Kolarsky, et al., 1967; Leutmezer et al., 1999; Terzian and Ore, 1955).

As noted, in ancient religions, including those who worshipped the great goddess, male youths would sometimes castrate themselves, and would forever after dress as women in order to obtain her power and serves as her priests.

Goddess cults have always been associated with fertility rites and woman's sexuality. Homosexuality between women worshippers was common, whereas when men were allowed to participate, the religious service would become a sex orgy. That modern day homosexual priests with their female limbic system behave similarly, albeit with men and boys, is therefore to be expected.

"If it was good company and conversation that Adam needed, it would have been much better if god were to have arranged to have two men together as friends, not a man and a woman." -St. Augustine.

7. Origins of Thought: Consciousness, Language, Egocentric Speech and the Multiplicity of Mind

Abstract

Consciousness is not a singularity, but a multiplicity. It is this multiplicity which makes self-consciousness (consciousness of consciousness) possible, and which provides the foundation for the development of thought which originates outside of consciousness. Thinking serves as a form of deduction and self-explanation, where one aspect of the mind explains its thoughts to another realm of mind. Thinking can be visual, imaginal, tactile, musical, or take the form of words strung together as a train-of-thought. Insofar as thoughts are verbal, this indicates that one region of the brain is organizing and explaining these verbal thoughts to another region of the mind which comprehends these verbal thoughts. Verbal thinking utilizes the same neural pathways and structures as spoken language; and Broca's expressive and Wernicke's receptive speech areas participate in the expression and comprehension of verbal thoughts. Because these neural pathways and language structures are immature for the first several years of life, and are limited in their ability to communicate within the brain, children initially think out-loud, using a form of language referred to as egocentric speech. As the brain matures, egocentric speech eventually becomes internalized as thought, such that by ages 5 to 6, children have completely internalized egocentric thought production, and think their thoughts in the privacy of their head. However, because the mind is a multiplicity with different tissues of the mind processing different forms of information, the dominant streams of consciousness associated with vision and language, often do not have access to information which might explain the motives for their actions, or how they arrived at certain conclusions or judgments. Because the mind is a multiplicity, in the final analysis, we knowers, remain unknown to ourselves.

1. Consciousness and the Multiplicity of Mind

Consciousness must be conscious of something. Consciousness, to be conscious, requires something which is separate from consciousness, and which becomes an object or focus of consciousness, such as a lamp, dog, cloud, car, singing birds, or the smile of a willing lover. Consciousness in order to be conscious requires something to be conscious of. Even if consciousness is only conscious of being conscious, i.e. self-consciousness, consciousness must be separate from itself to be conscious of itself, thereby becoming an object of consciousness, mirroring and reflecting itself as a duality. The same can be said of the train-of-thought which appears before or within consciousness but is never identical with consciousness (Joseph 1982).

Consciousness is not a singularity, but a multiplicity which, in humans, is often dominated by visual impressions and language (Joseph 1982, 2009). It is this multiplicity which makes consciousness of consciousness possible.

Insofar as consciousness is associated with the brain, then it could be said that the tissues of the mind consists of semi-independent mental realms which are maintained by the brainstem, thalamus, limbic system, the right and left hemisphere, and the occipital, temporal, parietal, and frontal lobes (Joseph 1982, 1986a,b, 1988a,b, 1992, 1999, 2009), each of which speaks and comprehends their own unique language, e.g. visual, tactile, olfactory, auditory. For example, the primary visual cortex perceives, processes, and becomes conscious of visual input but is largely deaf and blind to complex auditory signals, whereas the primary auditory receiving areas are blind and cannot see complex shapes or forms (Joseph 1996).

Because specific regions of the brain are specialized to perform specific functions, information transmitted between these tissues of the mind must undergo a transformation and become translated into a language the other can process and understand (Joseph 1982, 1986a,b, 1988a; Joseph et al., 1984). When the brain talks to itself, it could be said to be thinking; and these thoughts may be visual, verbal, tactile, emotional, or consist of myriad abstractions and symbols which then become objects of the consciousness maintained by the various realms of the multiplicity of mind.

Even if we associate consciousness with the activity of the brain, consciousness is never identical with what it is conscious of. When conscious of the flickering light of a lamp and the wavering shadows on the wall, consciousness and the shadows and flickering lamp light are not one and the same, but are separate and distinguishable. Consciousness is not the lamp, and the lamp is not consciousness; though it could be argued the lamp is in consciousness, just as a lamp may reside in a room next to the bed but remains distinguishable and separate from its surroundings.

Even if the brain is electrically stimulated by a neurosurgeon, any sensations, hallucinations, or memories evoked are experienced as appearing before or in an observing consciousness (Gloor, 1997; Halgren 1992; Penfield 1952; Penfield and Perot 1963). Likewise, when a subject is placed in a sensory reduced environment and deprived of external stimulation, although the brain will produce its own stimulation and generate complex hallucinations, these hallucinatory phenomenon are experienced as detached from the mind (Mason & Brady 2010; Riesen 1975; Zubek 1969).

Consciousness is relational and the same applies to conscious thought which is experienced as separate from consciousness, albeit as taking place inside the head. We may be "in our thoughts" but only insofar as we are thinking the thoughts, and thinking about ourselves as an object of thought which we are conscious of.

2. Thinking and Listening

Producing thoughts and experiencing and becoming conscious of thoughts as they are being produced, are indications of duality in the brain and mind (Joseph 1982, 2009).

In some respects, consciousness could be likened to a witness or an observer, and the same is true when observing or listening to an internally generated train of thought which often takes the form of an internal dialogue or picture show which is being experienced and even heard or seen within one's head (Joseph 1982). Thought, be it thinking in words, musical notes, math symbols, geometric patterns, or picture-images, is not synonymous with consciousness and often originates outside of consciousness and only appears to consciousness after the thoughts are organized and assembled (Ghiselin, 1952; Mandler, 1975; Neisser, 2006; Neisser & Fivush 2008; Nisbet & Wilson, 1977; Wilson, 2004).

Thoughts are the "actors on the stage" and consciousness is the audience. However, until the thoughts emerge, they are hidden from consciousness.

Take for example, "tip of the tongue" word-finding difficulty, in which the missing word is known yet not known. We are conscious that the word can't be found, but are not conscious of the identity of the missing word, though we know it's there. Nevertheless, although aware of the missing word's existence, we are unable to identify it or name it until it appears before consciousness. Awareness vs consciousness and "tip-of-the-tongue" are evidence of duality, and the same is true when considering the nature of consciousness and thought (Joseph 1982).

This duality is most evident in considering the developmental origins of verbal-thoughts. Thinking-in-words initially takes place not inside-the-head, but externally: children first speak their thoughts out-loud, and only gradually, as they grow older, do they begin to think their thoughts inside the privacy of their own head (Piaget, 1974; Vygotsky, 1956). The child speaks their thoughts, and listens to them as they are spoken, and this indicates duality. It is these developmental origins which provide one of the keys to understanding the nature of thought, the purpose of which is to serve the multiplicity of mind.

3. Language, Duality And The Train-Of-Thought

Although an individual may utilize visual, emotional, olfactory, musical, or tactile "imagery" when they think, thinking often takes the form of "words" which might be "heard" or rather, experienced, within one's own head. In fact, one need only listen to one's own thoughts in order to realize that thinking often consists of an internal linguistic monologue, a series of words heard inside the mind.

Thinking-in-words could be considered a form of internal perception, where strings of words, ideas, sentences, are produced by those tissues of the mind which speak the sounds of language, and which are perceived by that aspect of the conscious mind and brain which understands the sounds of language (Joseph 1982, 1986a,b, 1988ab). Thus, the train-of-thought which passes before consciousness, always has an origin which is outside of that aspect of the conscious mind which is listening to the train as it passes.

Thinking-in-words often serves consciousness and the brain as a means of explanation, commentary, or aimless internal chatter (Joseph 1982). Verbal thinking generally consists of an organized temporal-sequential hierarchy of associations, symbols, and labels which appear before an observor, or which are heard by the thinker. It is also a temporal progression, an associative advance and an elaboration which often appears with an initial or leading idea that is followed by a series of related verbal ideations. In the process of thinking in-words, one often acts to organized information which is "not thought out" and not clearly understood, so it may become thought out and thus comprehended in a logical, temporal sequential verbal format. To "think things out", "give it a lot of thought", or "think about it", serves an explanatory or deductive function (Mandler, 1975; Neisser, 2006; Wilson, 2004).

Yet the need to explain things to oneself seems paradoxical. It might be asked, "who is explaining what to whom?" Apparently the "I" that I am thinks (explains) these "things" to the "I" that I am.

We are presented with a curious duality in the nature of consciousness, the purpose of thought, and in the functioning of the brain. Insofar as the train of thought originates in me, in my brain and in my mind then I should know its aim and content prior to (not after) symbolizing the substance of the subject into the temporal-sequential linear organization that the verbal-thinking process generates. However, often we do know; there is an awareness; but the thoughts, idea, memories, and so on, remain hidden from consciousness, as again exemplified by "tip-of-the-tongue" world finding difficulty (Joseph 1982). We become conscious of this information and achieve explicit knowledge only after the information is transferred or made available to that aspect of the conscious mind which is dependent on language for understanding.

Limiting our discussion to word-thoughts, we must conclude, therefore, that the thoughts which will be expressed are not in consciousness before they are expressed, and are formulated and organized by a part of the brain which relies on language for expression, whereas they are comprehended by those regions of the mind-brain which require language for comprehension.

These thoughts only become an object of consciousness after they are organized into a train-of-thought by an aspect of mind which is separate from yet linked with that region of consciousness which experiences the train as it goes by. In other words, one realm of the brain and mind is clearly providing information and often explaining feelings, actions, observations, intentions, or conclusions, to another realm of the brain mind, and this is accomplished when thoughts become language; one region of the mind producing the verbal thoughts, the other listening (Joseph 1982, 1986a, 1988a). And, because these particular forms of thought are structured and perceived as words heard within one's head, then not surprisingly, they come to rely on the same neural pathways and brain structures which subserve the production and perception of language (Friederici, 2002; Kaan & Swaab, 2002; Newman, Just, & Carpenter, 2002) , i.e. the inferior parietal lobule/angular gyrus, Broca's expressive speech area in the left frontal lobe, and Wernicke's receptive speech area in the superior temporal lobe (Joseph 1982, 1986a, 1988a; 1999, 2000a). In fact, when engaged in verbal thought, these language areas typically becomes activated as indicated by functional imaging (Kaan & Swaab, 2002; Keller et al., 2001; Paulesu, et al., 1993; Petersen et al., 1988).

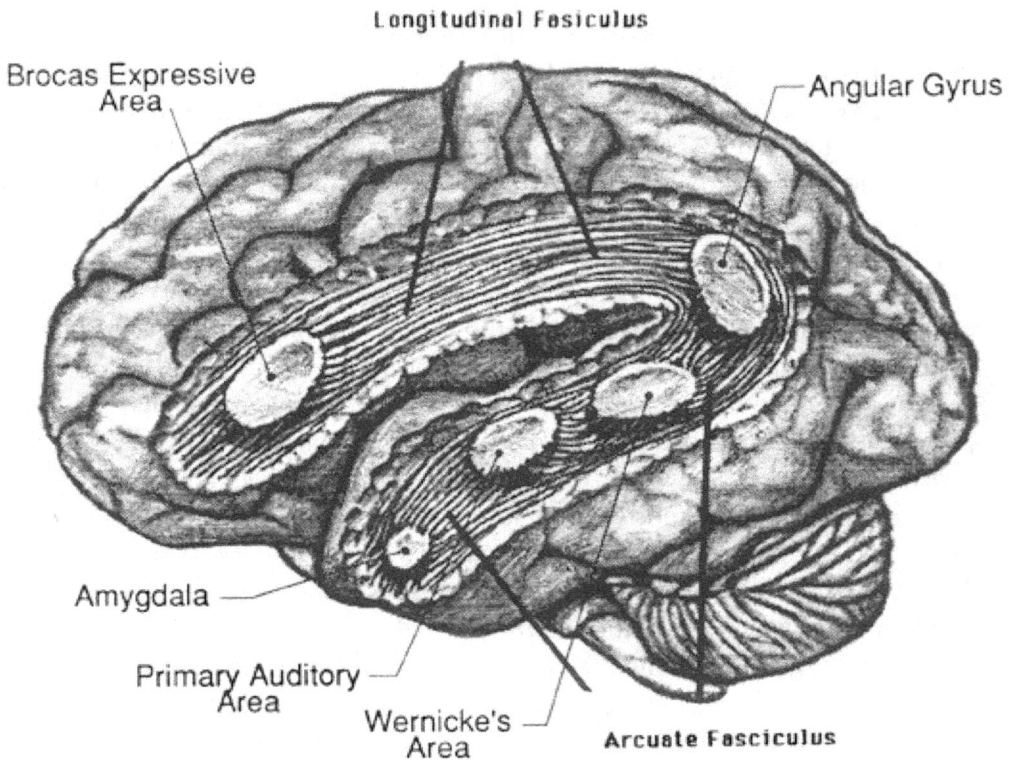

4. **Brain and Language**
Specifically, it is the lateral surface of the right and left frontal lobe which control vocalization and verbal-though production, the left frontal (Broca's area) producing the words, the right frontal the melody of language (Joseph 1982, 1986a, 1988a, 1996, 1999). The frontal lobes are interlocked with the association areas in the posterior regions of the cerebrum (Petrides & Pandya. 1999, 2001), including Wernicke's area and the angular gyrus - inferior parietal

lobe (IPL), as well as the memory centers in the temporal lobe (Joseph 1986a, 1996, 1999). The frontal lobes are therefore continually informed about and have continual access to information processed in these areas of the brain.

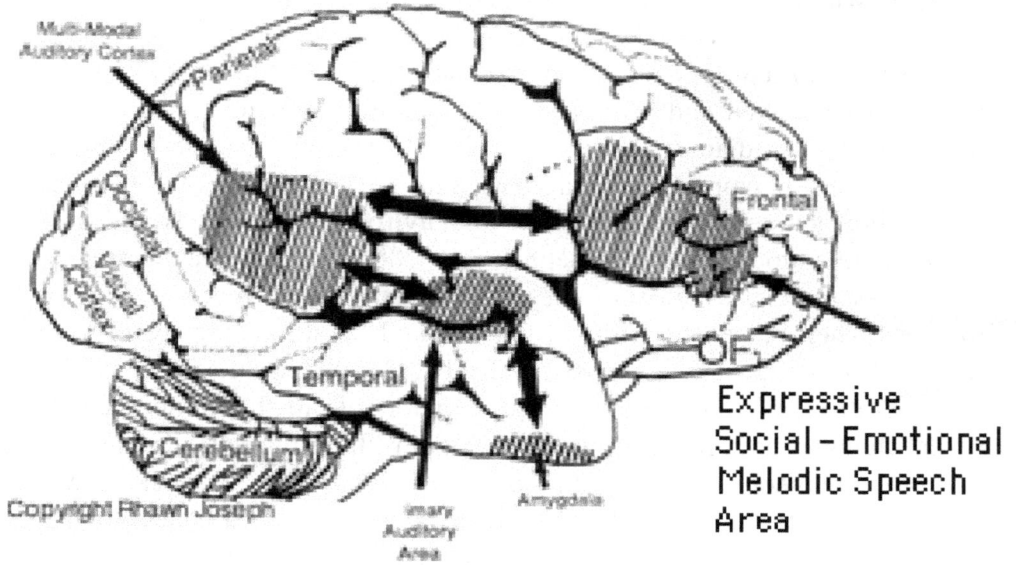

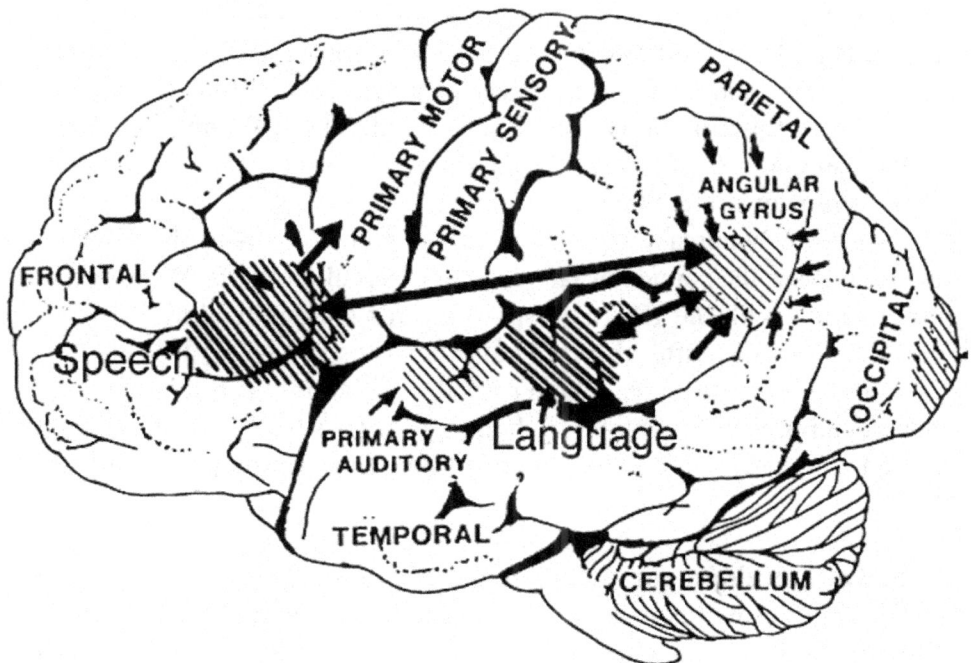

The frontal lobes serve as the senior executive of the brain and personality (Joseph 1986a, 1999, 2010) and play a significant role in searching for and assimilating the information which will be thought about (Christoff and Gabrieli, 2000; Christoff et al., 2001; Joseph 1986, 1999; Newman et al., 2002, 2003; Paulesu et al. 2010). The frontal lobes are responsible for organizing the thoughts which are to be explained and then comprehended by the auditory areas..

The primary auditory receiving areas are located in the superior temporal lobe. Once received and processed, these auditory signals are transferred to the immediately adjacent Wernicke's area which associates these sounds and comprehends the words of language; whereas the the auditory association area in the right temporal lobe comprehends environmental sounds and melodic and emotional vocalizations (Joseph 1982, 1988a, 2000). Via a massive neural pathways (the corpus callosum) the right and left auditory receiving areas work together when presented with complexity and paralinguistic features which require analysis (Just et al., 1996; Michael et al. 2001; Schlosser et al., 1998).

In addition, Wernicke's also plays a major role in the comprehension of verbal-thought, and, in conjunction with the right temporal lobe, the generation of spontaneous thought (Christoff et al., 2004). It is Wernicke's area (in conjunction with the IPL) which provides the words of language to the frontal lobes, and then listens to the train of thought as it passes by.

These temporal lobe auditory areas are linked to the frontal vocalization areas by a rope of nerve fibers, the arcuate fasciculus which passes through the angular gyrus/IPL--an area of the brain which assimilates associations and provides auditory-verbal labels to sensory stimuli. In fact the IPL and frontal lobes become active across a variety of language and non-language problem-solving and thinking tasks (Ben-Shahar et al., 2003; Dapretto & Bookheimer, 1999; Lehmann et al. 2009; Szaflarski et al., 2006; Tyler & Marslen-Wilson 2008; Vigneau et al. 2006).

Therefore, whereas Broca's area (in conjunction with the IPL) organizes and expresses the sounds of language and verbal thoughts, Wernicke's area (in conjunction with the IPL) is responsible for comprehending thoughts, ideas, feeling, and so on, after they are put into words.

Thus, it is Broca's area which does the explaining, and it is Wernicke's area which comprehends the train of thought as it goes by. However, when it comes to comprehending internally generated thoughts, Wernicke's area is the last to know.

5. The Development of Language & Thought

Thinking in words is clearly related to language; and human language is a function of the human brain. It is the brain which makes it possible to speak and comprehend language, and to think thoughts, and these thoughts enable one region of the brain to communicate information to another brain area which is dependent on language for comprehension. Thinking in words often serves as a means of organizing, interpreting, and explaining information or impulses so that the language dependent regions of the brain and mind may achieve understanding (Gallagher & Joseph, 1982; Joseph 1982; Joseph & Gallagher, 1985, Joseph et al., 1984).

Initially, and over the course of child development, the thinking of thoughts, that is, the thinking of thoughts as strings of words is externalized and verbally expressed out-loud and is then comprehended as an explanatory commentary, after-the-fact (Joseph, 1982). Children think out-loud, and not in the privacy of their head (Piaget, 1952, 1962, 1974; Vygotsky, 1962).

The brain (or mind) of the child talks to itself by speaking thoughts (instead of thinking the thoughts) which are then heard, by the child, as the thoughts are spoken out-loud. In fact, initially, it appears that children are incapable of thinking inside-their-head but can only understand their thoughts, if they are spoken (Piaget, 1952, 1962, 1974; Vygotsky, 1962).

This indicates that one region of the child's brain must talk out-loud to communicate with another region of the brain which can only receive and comprehend these thoughts if they are spoken. The duality of actor/orator vs audience/listener, is present from the very beginning and it takes place on a stage located not inside the child's head, but outside in the world.

Certainly infants and young children are capable of internalized thinking, but these thoughts are visual and emotional, not verbal (Joseph 1982, 1992; 2003; Piaget, 1974). This is a function of their limited vocabulary and is exemplified by the fact that the first long term memories are emotional and visual and not verbal (Joseph 2003). Thus, the child must acquire language before thinking in words or forming verbal memories. However, the pathways between those areas of the brain which produce thoughts vs those which comprehend thoughts, must also also mature, before word-thoughts can be produced and comprehended internally.

6. Three Linguistic Stages

Broadly considered, there are three maturational stages of verbal development that correspond to the acquisition and development of language and which leads to the thinking of thoughts (Joseph 1982, 1992, 1996). Initially, linguistic expression is reflexive and/or indicative of generalized and diffuse emotions and feelings states. Vocalizations are largely emotional-prosodic in quality, and mediated by limbic and brainstem nuclei. It is only over the course of the first few months, around 1-3 months of age, that these prosodic-melodic utterances become associated with specific moods and emotions (Joseph 1982, 1992). It is at this time that "early babbling" makes it appearance and the infant begins to "coo," "goo," in a repetitive fashion, i.e. "ma ma ma"; and this is referred to as "early babbling."

"Early" babbling is produced by the brainstem and limbic system (Joseph

1982, 1992), and is replaced by "late" babbling which has its onset around 4 months of age. Late babbling is sometimes described as "repetitive babbling" in which the same consonant is repeated, such as "da da da." These transitions are directly related to the maturation of the neocortical speech areas as they gain hierarchical control over the subcortical speech centers (Joseph 1996).

Around one year of age, and once the neocortical speech areas begin to mature and establish hierarchical control over the limbic system and brainstem so as to program the oral-laryngeal motor areas, a new form of vocalization emerges and the infant begins to produce "jargon babbles" and to speak their first words. Syllabication is imposed on the intonational contours of the child's emotional speech by the still immature neocortex of left frontal lobe and motor areas, such that the melodic features of generalized vocal expression come to be punctuated, sequenced, and segmented, and vowel and consonantal elements begin to be produced (Joseph, 1982, 1992). Left hemisphere speech comes to be superimposed over limbic (and right hemisphere) melodic language output, and the infant begins saying actual words. However, due to the immaturity of the neocortex (Churgani, et al. 1987; Blinkov & Glezer, 1968; Conel, 1939; Lecours, 1975), most of the speech produced is "jargon" and resembles the "jargon aphasia" associated with injuries to Wernicke's area (Goodglass & Kaplan, 2000; Joseph 1996). However, rather than due to brain damage, jargon babbling reflects the extreme immaturity of the neocortical speech areas.

The development of jargon babbling appears to correspond to maturational events taking place in the motor areas of the neocortex which begins to rapidly mature around the first postnatal year (Chi, Dooling, & Gilles, 1977; Gilles et al. 1983; Scheibel, 1991, 1993). Jargon babbling coincides with the production of the first words which are spoken around 11-12 months on average (Nelson, 1981; Oller & Lynch, 1992). In fact, jargon babbling resembles actual speech, and at a distance it may sound as if the infant is conversing and speaking real words, though in fact they are babbling prosodically sophisticated neologistic jargon. Hence, the emergence of the jargon babbling stage signifies an obvious shift in sound production from the limbic system and brainstem to the still immature neocortex.

Jargon babbling not only resembles normal fluent speech but is often produced as the infant is gazing at or making eye-to-eye contact with the listener. Jargon vocalizations are both social and self-directed as the infant also jargon babbles while alone and at play, or while gazing at or exploring some object. When meant for the child's ears alone, these babbles could be described as "egocentric babbling."

Over the ensuing years, social speech emerges from jargon (conversational) babbling, whereas egocentric speech emerges from egocentric babbling. It is egocentric speech which gives birth to thinking-in-words.

Egocentric speech is essentially speech for oneself, and is the first evidence of thinking in words. The child is thinking out loud (Joseph, 1982; Piaget, 1962; Vygotsky, 1962). Egocentric speech is slowly internalized between the ages of 3 and 5, and eventually becomes completely covert; at which point, the child has not only learned to speak in words, but to silently think in words as well (Joseph 1982, 1996; Vygotsky, 1962).

7. Egocentric Speech and the Origins of Thought

At around age 3 the child produces two types of speech: social and egocentric. Part of the time the child engages in social conversational speech which is directed toward others, whereas the remainder of speech activities are egocentric and directed for the sole benefit of the child who listens to their vocalizations as they play (Joseph, 1982, 1996; Piaget, 1962; Vygotsky, 1962).

Egocentric speech is self-directed speech that consists of an explanatory monologue in which children comment on or explain their play and other actions, to themselves. Initially egocentric speech is produced only after the action has occurred. That is, the child essentially talks to themselves, but in an explanatory fashion: They tell themselves what they are doing and what they have done. Egocentric speech is a self-directed form of communication which heralds the first attempts at self-explanation via thinking-out-loud.

Egocentric speech makes its appearance at approximately 3 years of age, and at its peak comprises almost 40-50% of the child's language (Piaget (1952, 1962, 1974). In contrast to conversational social speech in which the child is engaging in a back-and-forth dialogue and is attuned to the listening needs of others, when engaged in egocentric speech the child is oblivious to his/her audience simply because the words spoken are meant for his/her ears alone (Piaget, 1952, 1962, 1974; Vygotsky, 1962). The child is essentially thinking out loud in an explanatory fashion, commenting on and describing his or her actions and is both orator and audience (Joseph, 1982).

When engaged in an egocentric monologue, there is no interest in influencing or explaining to others what in fact is being explained. The child will keep up a running verbal accompaniment to their actions, commenting on their behavior in an explanatory fashion even while alone. Moreover, while engaged in this self-directed external monologue the child appears oblivious to the responses of others to their statements (Piaget, 1952, 1962, 1974; Vygotsky, 1956). If a playmate were to reply to a statement made during an egocentric monologue, it would not be heard by the child producing the egocentric monologue. It is as if the child has no awareness that others hear him when producing egocentric speech. In fact, many a child has been shocked when, later, his mother (or a friend) repeats or comments upon something he assumed no one else could hear; as if mom can read their thoughts! Thus, when engaged in egocentric speech, the child may not be conscious of the fact they are thinking out-loud.

8. Thinking Out Loud: The Stages of Thought Internalization

Egocentric speech is a dialogue where the child is both the speaker and the listener, the actor and the audience. It is an explanatory monologue, serving as a commentary that is initially produced only after an action has been completed by the child. That is, the child explains to him/herself what they have done after they have observed themselves complete some action, and then they comment on and/or explain what has taken place (Piaget 1974; Vygotsky, 1962). For example the 3-4 year old child will paint a picture and then explain it after it is completed: "This the sun shining on mommy." Or they will crash their toy truck into another toy truck and then remark on and explain what happened: "Trucks crashed."

As the child grows older, instead of explaining after the fact, they will explain what they are doing, as they are doing it. For example the 4-5 year old child will paint a picture and explain it while she is painting, or state the toy trucks are crashing into each other as they smash together. The egocentric monologue accompanies the action.

By time the child reaches age 5-6, they will announced what they are going to do and then do it. For example, the child may state she is going to paint a picture of "mommy and daddy at the beach and they are happy", and then paints it, or he will take his two toy trucks and state "now they are going to crash and everyone will die" and then he will smash the trucks together.

Hence, as the child grows older their comments and explanations occur earlier in the sequence of expression, until finally the child begins to explain his/her actions before they are performed instead of after they have occurred (Piaget, 1952, 1962, 1974; Vygotsky, 1962).

Egocentric speech presents us with an obvious duality: initially, the part of the brain and mind that relies on language, does not know what another part of the brain and mind intends to do, until after the fact, at which point the language regions comment on and explains, to themselves, what they have done and why. This indicates that the aspect of the brain and mind which is talking and listening, does not know until after the actions are performed, and only achieves understanding when language is used to explain their actions to themselves. Presumably, this is due, in part, to functional disconnections between brain areas, secondary to the immaturity of these tissues and their nerve fiber interconnections (Joseph 1982, 1996; Joseph et al., 1984). That is, because different brain areas and their neuronal interconnections can take years to decades to mature (Blinkov & Glezer, 1968; Lecours, 1975; Szaflarski et al., 2006), their ability to communicate is limited. One area of the brain and mind may initiate a behavior, which is witnessed or experienced by other (disconnected) brain areas, only as it occurs outside the brain and body.

Thus, for the first several years of life, it appears that the child's brain as a whole does not know what or why they are engaged in certain actions, or even that they intend to perform certain actions, until after the act has been completed; at which point they explain what they did and why, to those areas of the brain-mind which are dependent on language. This is because one region of the brain initiates the behavior, and another witnesses and then explains to itself what just happened. The child is both actor and witness, explainer and explained to. Clearly, when engaged in an egocentric monologue and thinking-out-loud, the child (the left frontal-temporal-parietal axis) explains their actions to themself.

However, around age 4, and as the child's neural pathways mature, the frontal lobes and language-dependent regions of the brain and mind receive information about their intentions as they engage in these acts, and thus explain, to themselves, what they are doing, while they are doing it. Neuronal interconnections begin to mature and to increasingly communicate and share information.

Around age 5, this information become available before rather than during or after they act. Thus they begin to explain, to themselves, what they are planning to do before they do it. This advanced warning parallels the increasing maturity of the nerve fiber pathways between brain areas (Szaflarski et al. 2006), and the ability of these tissues of the mind to successfully transfer and receive complex information from other regions of the cerebrum (Gallagher & Joseph, 1982; Joseph & Gallagher, 1985; Joseph et al., 1984).

By the time the child reaches age 5-6, egocentric speech has become increasingly internalized (Piaget 1974; Vygotsky, 1962). The pathways between different brain areas have matured (Szaflarski et al., 2006) such as the arcuate fasciculus linking Wernicke's and Broca's area, and the child no longer needs to vocalize their thoughts, but can instead think and comprehend these thoughts internally (Joseph, 1982, 1996). The neural pathways linking these tissues of the mind have sufficiently matured so that information can be readily transferred between brain areas. However, even after thinking has become completely internalized it remains self-directed. Thinking continues to serve an explanatory function.

9. The Internalization of Egocentric Speech

Egocentric speech is never directed at others. Groups of children may be playing together, and each child may be engaged in an egocentric monologue. They are not talking to the other children. They are talking to themselves. They are thinking-out-loud.

Initially egocentric speech is completely external and after the fact (Piaget, 1974; Vygotsky, 1956). Since they are utilizing words, it is thus apparent that they are engaging those areas of the left hemisphere which are dominant for expressing and comprehending language: the frontal-temporal lobes (Kaan & Swaab, 2002; Michael et al., 2001; Schlosser et al. 1998; Tyler & Marslen-Wilson 2008). However, because of the immaturity of the child's brain, vast regions are partially disconnected. Therefore, specific areas of the brain and mind associated with language, are also partially disconnected, essentially creating two brains and two minds in a single head (Gallagher & Joseph, 1982; Joseph 1982; Joseph & Gallagher, 1985; Joseph et al., 1984). In fact, the pattern of neurological activity during the performance of language tasks, does not begin to resemble the adult pattern until the onset of puberty (Holcomb et al., 1992).

Therefore it appears that children must speak their thoughts out loud due to the immaturity of the interconnections between Broca's area and the temporal-parietal area (Szaflarski et al., 2006), and the immaturity of the corpus callosum neural pathways which link the right and left hemisphere (Gallagher & Joseph, 1982; Joseph 1982, Joseph & Gallagher, 1985; Joseph et al., 1984). Thus, the right half of the brain or limbic system may initiate certain behaviors without the knowledge of the language dependent aspects of consciousness associated with the left hemisphere; which then explains, to itself, what it observes. However, because of the immaturity of the neural pathways within the language areas of the left hemisphere, the frontal lobes must speak the thoughts so they may be heard by Wernicke's area.

Hence, improvements in neocortical transmission in the language areas, and between the right and left hemisphere, parallel the stages of egocentric speech and its internalization as silent thought.

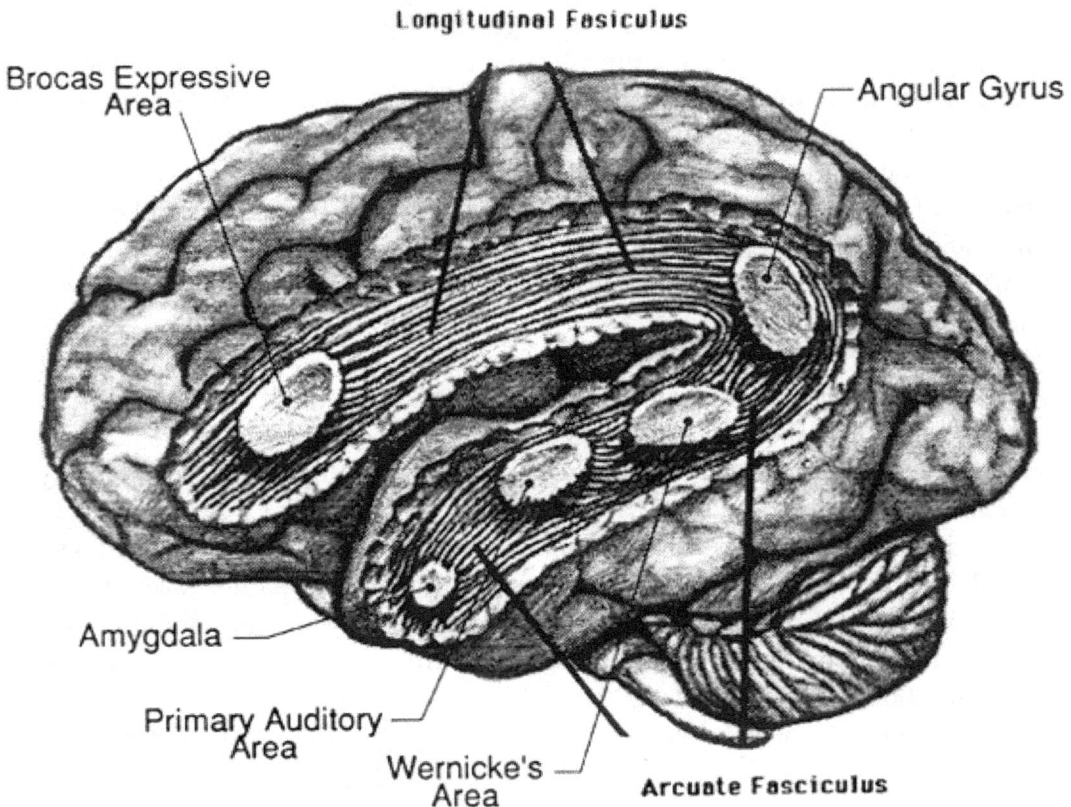

216

10. Thinking and the Evolution and Development of the Multiplicity of Mind

The mind is a multiplicity, and different regions of the brain and mind speak different languages as they are specialized to perceive or process specific types of stimuli and information. Moreover, although certain areas of the brain are richly interconnected, the neural pathways between yet other tissues of the mind are sparse or non-existant as there is no need for them to communicate, except indirectly and through intermediary tissues such as the angular gyrus of the inferior parietal lobule. For example, the angular gyrus of the inferior parietal lobe is situated at the junction of the association areas for vision, hearing, and somesthesis, and, in conjunction with the frontal lobes and Wernicke's area (Lehmann et al. 2009; Szaflarski et al., 2006), assimilates associations and diverse information variables and then organizes and categorizes them into words and multi-modal linguistic concepts which can be translated into language and the train of thought (Joseph 1982, 1986a). It is the IPL/angular gyrus which enables a person to see, for example, a "cup" and to associate the word "cup" with the visual image, or to imagine a variety of cups of varying size, colors, or utilities, ranging from a "world cup" to the "cup size of a bra." The IPL essentially assimilates diverse associations thereby making it possible to form multi-modal concepts which may be differentially comprehended by different aspects of the mind.

However, not all sensations or information variables have auditory equivalents and cannot be adequately described using language. Words completely fail to describe how it feels to ride a horse, parachute from a falling plane, swim beneath the sea, dilate the pupil of an eye, or experience an orgasm. Moreover, there are myriad behaviors which do not require and which occur independently of the language-dependent aspects of the mind. For example, there are nerve centers in the ancient brainstem which control heart rate, breathing, and pupil dilation; and although influenced by "higher cortical" activity in the forebrain, for the most part these activities take place without the assistance or participation of the conscious mind and the more recently evolved neocortical surface layers of the cerebrum. There are also limbic system structures such as the hypothalamus and amygdala which mediate various aspects of emotion and which may hijack the rational, logical, and language dependent aspects of the brain and mind (Joseph 1992); which, like the egocentric child, may act without thinking, and then later may exclaim: "I don't know what came over me" and then search for an answer.

From an evolutionary perspective, it must be recognized that the brain has evolved over the course of the last 600 million years, and it is only around 150 million ago that the six layered neocortical mantle, began to evolve and to slowly cover and envelop the old brain which includes the extremely ancient limbic system and brainstem. Further, it is only within the last 100,000 years that the language-dependent aspects of consciousness began to evolve (Joseph, 1996, 2000b).

However, be it the brain of a modern human, or that of a reptile, much of behavior is under the control of these more ancient tissues of the mind which do not require language or any type of logical or analytical thinking to perform their functions. Thus, these regions of the brain usually function completely independently of human consciousness which is essentially not-conscious and has no conscious access to the workings of the more ancient regions of the mind.

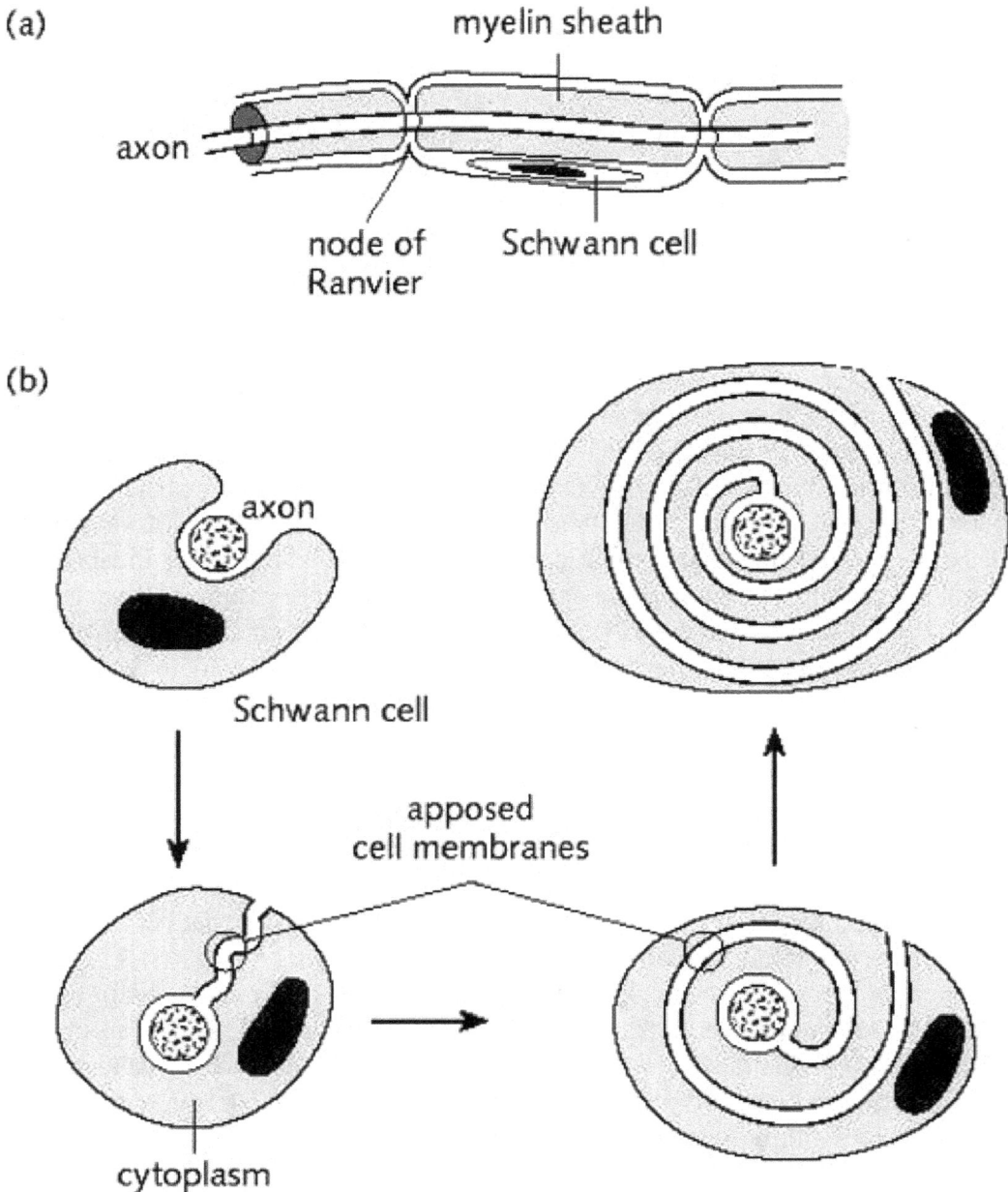

(a)

myelin sheath

axon

node of
Ranvier

Schwann cell

(b)

axon

Schwann cell

apposed
cell membranes

cytoplasm

However, the development and maturation of the human brain does in fact parallel the evolution of the brain (Joseph 1996) such that the brainstem structures are fully functional at birth (Joseph 2000c), followed by the limbic system (Joseph 1992, 1996), whereas the more recently evolved neocortex is the last to develop, with some areas taking up to 20 years to fully mature (Blinkov & Glezer, 1968; Conel, 1939; Lecours, 1975).

The Myelination of the Human Brain. Myelin is the outer protective coating of the axon and promotes and stabilizes nerve transmission. From Yakovlev & Lecours (1967)

Likewise, emotional, visual, and motor functioning mature well in advance of language. Therefore, in many respects, language, and the language-dependent aspects of consciousness, can be considered an after-thought phyologenetically and ontogenetically, and as such, these areas of the brain and mind often have no access to those tissues controlling behavior, and are therefore often the last to know.

11. The Multiplicity of Mind and the Conscious Unconscious

Seeking knowledge and information, the language dependent aspects of consciousness produce thoughts, to explain to itself why. Thinking, especially thinking-in-words, often serves an explanatory function; and often its purpose is to obtain information or explain behaviors which have a source in the non-linguistic regions of the brain and mind.

Essentially, egocentric speech, and then internalized verbal thoughts, are largely a function of the left hemisphere's attempt to organize, interpret, and make sense of behavior initiated by brain structures not concerned with the denotative, grammatical aspects of human language. Initially, egocentric speech externalized thought production is due to the immaturity of neural pathways which hinder information transfer not just within the left hemisphere, but from the right to left half of the brain (Gallagher & Joseph, 1982; Joseph, 1982; Joseph & Gallgher 1985; Joseph et al., 1984); a function of the immaturity of the corpus callosal fibers connections between the hemispheres (Yakovlev & Lecours, 1967).

However, even in the "normal" intact adult, information transmission within and between the right and left halves of the brain is often incomplete (Joseph 1982, 1988a). Because different brain tissues are functionally specialized to perform different functions, they speak different language and even information which is shared must be interpreted and translated--and much may be misinterpreted or lost in translation. As such, those regions of the mind in the left hemisphere dependent on language sometimes observe (and participates) in behaviors which it did not initiate, and which it does not understand, and will then think up an explanation. However, because it does not have access to this information, the explanations which are thought up, frequently amount to little more than self-deception.

Consider a famous experiment by Nisbet and Wilson (1977) in which they invited shoppers to participate in a consumer survey and to evaluate and indicate their preference for one of four identical pairs of nylon stockings which were laid out in front of them from left to right. Subjects were asked to handle and test the stockings, and to indicate which they thought was the best. Although the stockings were in fact identical, 75% of the subjects believed the stocking on their right was superior to the rest. When asked "why?" subjects confabulated a variety of explanations, claiming differences in color, texture, softness, durability, and so on, even though the stockings were identical! Most (right handed) people show a right-sided response bias (which is why "brand X" is always placed to the left). However, when asked if the position of the article effected their judgment, "virtually all subjects denied it, usually with a worried glance at the interviewer suggesting that they felt either that they had misunderstood the question or were dealing with a madman."

The fact is, based on over a century of psychological studies (Ghiselin, 1952; Mandler, 1975; Neisser, 2006; Neisser & Fivush 2008; Nisbet & Wilson, 1977; Wilson, 2004), it can be concluded that most people have no knowledge as to how they arrived at solving certain problems, made certain judgments, or why they engaged in various behaviors. Even the processes involved in intellectual and creative pursuits are outside the reach of consciousness. In a famous study of creative geniuses, Ghiselin (1952) concluded that "production by a process of purely conscious calculation seems never to occur." Instead, these geniuses often described themselves as observers, as bystanders who witness the fruits of the problem-solving or creative process only as or after it occurs.

The mind is a multiplicity, and those aspects of these mental realms collectively referred to as consciousness, have at best, only limited access to functional knowledge sources. Moreover, the language dependent aspects of consciousness has a limited capacity and can access and accurately report only a small bit of the information amenable to linguistic coding. Not all impulses, feelings, desires, fears, cravings, knowledge, etc., have a label and are not readily translatable into the codes that give rise to thought and language (Joseph, 1982, 1988a). Emotions in particular are easily subject to misinterpretation, misidentification, and misunderstanding, and the same is true of most non-emotional cognitive processes. Most individuals have no understanding of what is involved and on what basis evaluations, judgments, problem-solving strategies, and reasons for initiating certain behaviors were arrived at. This is largely due, however, to the dependence on language. The sequential, organizational linguistic processes which subserve consciousness in the form of verbal thought have no direct access to the basis of these 'processes' in part because they result from them in an attempt to explain them.

12. The Multiplicity of Mind: We Knowers Are Unknown To Ourselves

The brain/mind is a multiplicity (Joseph 2009), and for good reason. These conditions protect the brain and linguistic consciousness from becoming over-whelmed. Different regions of the brain, and thus the mind, are functionally specialized, thereby creating multiple minds which can engage in multiple tasks simultaneously. However, in consequence, we knowers are unknown to ourselves.

Thinking is a product of the multiplicity of mind. However, so too is consciousness. Consciousness, to be conscious, requires something which is separate from consciousness, and which becomes an object or focus of consciousness. Because the mind is a multiplicity, consciousness can be conscious of being conscious. Different regions of the brain may be conscious or at least aware of the mental activity occurring in yet other regions of the cerebrum; and this makes self-consciousness possible. In other words, consciousness does not separate from itself when attempting to know and think about itself, and this is because different forms of consciousness, a multiplicity of minds, dwell within the same head, thereby enabling the mind to be both observer and observed. Thus the distinguishing characteristic of consciousness is that it does not co-incide with itself. Consciousness is not a duality-a reflection which is its own reflecting--but a multiplicity.

It is the languages spoken by these multiple minds, and the non-conscious origins of so much of what constitutes human behavior, which are responsible for the development and origin of thought, the primary function of which, is self-explanation: The I that I am, explains these thoughts to the I that I am.

References

Ben-Shahar, M., et al. (2003). The neural reality of syntactic transforma-tions: evidence from fMRI. Psychological Science 14.433-40.

Blakemore, S-J., et al., (2007). Adolescent development of the neural cir-cuitry for thinking about intentions, Soc Cogn Affect Neurosci. doi: 10.1093.

Blinkov, S. M., & Glezer, I. I. (1968). The human brain in figures and tables. Plenum, New York.

Booth, J. R., et al. (2007). The role of the basal ganglia and cerebellum in language processing. Brain Research 1133.136-44.

Bischoff-Grethe, A., et al. (2000). Conscious and unconscious process-ing of nonverbal predictability in Wernicke's area. Journal of Neuroscience 20.1975□1981.

Chi, J-G., Dooling, E. C., & Gilles, F. H. (1977). Gyral development of the human brain. Ann. Neurology, 1, 86-93.

Christoff, K., and Gabriele, J.D.E. (2000). The frontopolar cortex and hu-man cognition: Evidence for a rostrocaudal hierarchical organization within the human prefrontal cortex. Psychobiology, 28: 168-186.

Christoff, K., et al., (2001), Rostrolateral prefrontal cortex involvement in relational integration during reasoning. NeuroImage, 14: 1136-1149.

Christoff, K., Ream, J.M., and Gabrieli, J. D. E. (2004). Neural basis of spontaneous thought processes. Cortex, (2004) 40, 623-630.

Churgani, H.T., et al. (1987). Positron emission tomography of human brain functional development. Ann Neurology, 22, 487-497.

Conel, J. L.(1939). The postnatal development of the human cerebral cortex. Harvard University Press. Cambridge.

Dapretto, M., & Bookheimer, S. Y., (1999). Form and content: dissociating syntax and semantics in sentence comprehension. Neuron 24.427-32.

Dikker, S., H. et al. (2009). Sensitivity to syntax in visual cortex. Cognition 110.293-321.

Dehaene. S., et al. (1999). Sources of Mathematical Thinking: Behavioral and Brain-Imaging Evidence Science, 284, 970-974.

Friederici, A. D. (2002). Towards a neural basis of auditory sentence comprehension. Trends in Cognitive Sciences, 6, 78-84.

Gallagher, R. E., & Joseph, R. (1982). Non-linguistic knowledge, hemispheric laterality, and the conservation of inequivalance. Journal of General Psychology, 107, 31-40.

Ghiselin, B. (1952). The Creative Process. New York, Mentor.

Gilles, F.H., Leviton, A., & Dooling, E.C., (1983). The developing human brain. Boston: John Wright.

Gloor, P. (1997). The Temporal Lobes and Limbic System. Oxford University Press. New York.

Goodglass, H., & Kaplan, E. (2000). Boston diagnostic aphasia examination. New York, Lange.

Halgren, E. (1992). Emotional neurophysiology of the amygdala within the context of human cognition. In J. P. Aggleton (Ed.). The Amygdala. New York, Wiley-Liss.

Holcomb, P,. J., Coffey, S. A., & Neville, H. J. (1992). Visual and auditory sentence processing. Developmental Neuropsychology, 8, 203-241.

Joseph, R. (1980). Joseph, R. (1982). The Neuropsychology of Development. Hemispheric Laterality, Limbic Language, the Origin of Thought. Journal of Clinical Psychology, 44 4-33.

Joseph, R. (1986a). Confabulation and delusional denial: Frontal lobe and lateralized influences. Journal of Clinical Psychology, 42, 845-860.

Joseph, R. (1986b). Reversal of language and emotion in a corpus callosotomy patient. Journal of Neurology, Neurosurgery, & Psychiatry, 49, 628-634.

Joseph, R. (1988a) The Right Cerebral Hemisphere: Emotion, Music, Visual-Spatial Skills, Body Image, Dreams, and Awareness. Journal of Clinical Psychology, 44, 630-673.

Joseph, R. (1988b). Dual mental functioning in a split-brain patient. Journal of Clinical Psychology, 44, 770-779.

Joseph, R. (1992) The Limbic System: Emotion, Laterality, and Unconscious Mind. The Psychoanalytic Review, 79, 405-455.

Joseph, R. (1996). Neuropsychiatry, Neuropsychology, Clinical Neuroscience, 2nd Edition. Williams & Wilkins, Baltimore.

Joseph, R. (1999). Frontal lobe psychopathology: Mania, depression, aphasia, confabulation, catatonia, perseveration, obsessive compulsions, schizophrenia. Psychiatry, 62, 138-172.

Joseph, R. (2000a). Limbic language/language axis theory of speech. Behavioral and Brain Sciences. 23, 439-441.

Joseph, R. (2000b). The evolution of sex differences in language, sexuality, and visual spatial skills. Archives of Sexual Behavior, 29, 35-66.

Joseph, R. (2000c). Fetal brain behavioral cognitive development. Developmental Review, 20, 81-98.

Joseph, R. (2001). The Limbic System and the Soul: Evolution and the Neuroanatomy of Religious Experience. Zygon, the Journal of Religion & Science, 36, 105-136.

Joseph, R. (2003). Emotional Trauma and Childhood Amnesia. journal of Consciousness & Emotion, 4, 151-178.

Joseph, R. (2009). Quantum Physics and the Multiplicity of Mind: Split-Brains, Fragmented Minds, Dissociation, Quantum Consciousness, Journal of Cosmology, 3, 600-640.

Joseph, R. (2010). The neuroanatomy of free will: Loss of will, against the will "alien hand", Journal of Cosmology, 14, In press

Joseph, R., Gallagher, R., E., Holloway, J., & Kahn, J. (1984). Two brains, one child: Interhemispheric transfer and confabulation in children aged 4, 7, 10. Cortex, 20, 317-331.

Joseph, R., & Gallagher, R. E. (1985). Interhemispheric transfer and the completion of reversible operations in non-conserving children. Journal of Clinical Psychology, 41, 796-800.

Just, M. A., Carpenter, P. A., Keller, T. A., Eddy, W. F., & Thulborn, K. R. (1996). Brain activation modulated by sentence comprehension. Science, 274, 114-116.

Kaan, E., & Swaab, T. Y. (2002). The brain circuitry of syntactic comprehension. Trends in Cognitive Sciences, 6, 350-356.

Keller, T. A., Carpenter, P. A., & Just, M. A. (2001). The neural bases of sentence comprehension: A fMRI examination of syntactic and lexical processing. Cerebral Cortex, 11, 223-237.

Kobayashi, C. et al. (2007). Language in adult English speakers but not in English-speaking children

Kobayashi, C. et al. (2007) Cultural and linguistic effects on neural bases of

Theory of Mind in American and Japanese children. Brain Res. 1164, 95☐107.

Kossyln, S. M.,e t al. (1995) Topographical representations of mental images in primary visual cortex. Nature, 378: 496-498.

Lecours, A. R. (1975). Myelogenetic correlates of the development of speech and language. In E. Lenneberg & E. Lenneberg (Eds.), Foundations of language development. New York: Academic Press.

Lehmann, D., et al., (2009). Core networks for visual-concrete and abstract thought content: A brain electric microstate analysis, NeuroImage, 49, Issue 1, 1 January 2010, 1073-1079.

Mandler, G. (1975). Consciousness: Respectable, useful, and probably necessary. In R. Solso (Ed). Information processing and cognition. Erlbaum, NJ.

Mason OJ, Brady F. (2009). The psychotomimetic effects of short-term sensory deprivation. J Nerv Ment Dis. 197, 78378-78385.

Michael, E. B., Keller, T. A., Carpenter, P. A., & Just, M. A. (2001). fMRI investigation of sentence comprehension by eye and by ear: Modality fingerprints on cognitive processes. Human Brain Mapping, 13, 239-252.

Nakai, T., K. et al. (2005). An fMRI study of temporal sequencing of motor regulation guided by an auditory cue a comparison with visual guidance. Cognitive Process 6.128-35.

Neisser, U. & Fivush, R. (2008) The Remembering Self: Construction and Accuracy in the Self-Narrative, Cambridge University

Neisser, U. (2006) The Perceived Self: Ecological and Interpersonal Sources of Self Knowledge, Cambridge U. Press.

Nelson, K. (1981). Individual differences in language development: Implications for development and language. Developmental Psychology, 17, 170-187.

Newman, S. D., Carpenter, P. A., Varma, S., & Just, M. A. (2003). Frontal and parietal participation in problem solving in the Tower of London: fMRI and computational modeling of planning and highlevel perception. Neuropsychologia, 41, 1668-1682.

Newman, S. D., Just, M. A., & Carpenter, P. A. (2002). Synchronization of the human cortical working memory network. NeuroImage, 15, 810-822.

Nisbett, R. E. & Wilson, T. D (1977). Telling More Than We Can Know: Verbal Reports on Mental processes, Psychological Review, 84, 231-259.

Ojemann, G. A., I. Fried, E. Lettich. (1989). Electrocorticographic (ECoG) correlates of language. I. Desynchronization in temporal language cortex during object naming. Electroencephalography and Clinical Neurophysiology 73.453-63.

Oller, D. K., & Lynch, M. P. (1992). Infant vocalizations and innovation in infraphonology: Toward a broader theory of development and disorders. In C. A. Ferguson, L. Menn, & C. Stoel-Gammon (Eds.), Phonological development: models, research, implications. Academic Press, New York.

Paulesu, E., Frith, C. D., & Frackowiak, R. S. J. (1993). The neural correlates of the verbal component of working memory. Nature, 362, 342-344.

Paulesu, P., et al., (2010). Neural correlates of worry in generalized anxiety
disorder and in normal controls: a functional MRI study. Psychological Medicine (2010), 40: 117-124.

Penfield, W. (1952) Memory Mechanisms. Archives of Neurology and Psychiatry, 67, 178-191.

Penfield, W., & Perot, P. (1963) The brains record of auditory and visual
experience. Brain, 86, 595-695.

Pernera, J., & Aichhorna, M. (2008) Theory of mind, language and the tem-
poroparietal junction mystery Trends in Cognitive Sciences, 12, 123-126.

Petersen, S. E., Fox, P. R., Posner, M. I., et al. (1988). Positron emission to-
mographic studies of the cortical anatomy of single-word processing. Nature,
331, 585-589.

Petrides, M., & Pandya, D. N. (1999). Dorsolateral prefrontal cortex:
comparative cytoarchitectonic analysis in the human and the macaque brain
and corticocortical connection patterns. European Journal of Neuroscience
11.1011-1036.

Petrides, M., & Pandya, D. N. (2001). Comparative cytoarchitectonic analysis of the human and the macaque ventrolateral prefrontal cortex and corticocortical connection patterns in the monkey. European Journal of Neuroscience
16.291-310.

Piaget, J. (1952). The origins of intelligence in children. New York: Norton.

Piaget, J. (1962). Play, dreams and imitation in childhood. New York: Norton.

Piaget, J. (1974). The child and reality. New York: Viking Press.

Riesen, A. H. (1975). Developmental Neuropsychology of Sensory Deprivation, Academic Press.

Saxe, R., & Kanwisher, N. (2003). People thinking about thinking people
-fMRI studies of Theory of Mind. Neuroimage, 19, 1835-1842.-

Scheibel, A. B. (1991). Some structural and developmental correlates of
human speech. In (K. R. Gibson & A. C. Petersen, Eds). Brain Maturation and
Cognitive Development. New York,. De Gruyter.

Scheibel, A. B. (1993). Dendritic structure and language development. In B.
de Boysson-Bardies, S., de Schonen, P. Jusczyk, P, MacNeilage, and J. Morton
(Eds.), Developmental neurocognitive speech and voice processing in the first
year of life. Academic Press.

Schlosser, M. J., Aoyagi, N., Fulbright, R. K., Gore, J. C., & McCarthy,
G. (1998). Functional MRI studies of auditory comprehension. Human Brain
Mapping, 6, 1-13.

Scholz. J., et al., (2009). Distinct Regions of Right Temporo-Parietal Junction Are Selective for Theory of Mind and Exogenous Attention, PLoS ONE

4(3): e4869. doi:10.1371/journal.pone.0004869

Steinmetz H., and Seitz, R. J., (1991). Functional anatomy of language processing: neuroimaging and the problem of individual variability. Neuropsychologia 29 (12).114-1161.

Szaflarski, J. P. et al., (2006) A longitudinal functional magnetic resonance imaging study of language development in children 5 to 11 years old, Annals of Neurology Volume 59, Issue 5, pages 796-807.

Tyler, L. K., & Marslen-Wilson, W. (2008). Fronto-temporal brain systems supporting spoken language comprehension, Phil. Trans. R. Soc. B., 363, 1037-1054.

Wilson, T. D. (2004). Strangers to Ourselves: Discovering the Adaptive Unconscious, Belknap Press, MA.

Vigneau, M., V. Beaucousin, P. Y. Herv, H. Duffau, F. Crivello, O. HoudB. Mazoyer, and N. Tzourio-Mazoyer. 2006. Meta-analyzing left hemisphere language areas: Phonology, semantics, and sentence processing. Neuroimage 30(4).1414-1232.

Vygotsky, L. S. (1962). Thought and language. Cambridge: MIT Press.

Yakovlev, P. I., & Lecours, A. (1967). The myelogenetic cycles of regional maturation of the brain,. In A. Minkowski (Ed.), Regional development of the brain in early life, (pp.404-491). London: Blackwell.

Zubek, J. P. (1969). Sensory Deprivation: Fifteen years of research. Appleton-Century-Crofts.

8. The Split Brain: Two Brains - Two Minds

Abstract

It is commonly supposed that consciousness has mental unity which is supported by the brain. However, it has been demonstrated that consciousness is actually a multiplicity which includes a duality maintained by the right and left half of the brain. In the normal brain it has sometimes been supposed that mental unity is supported and maintained via the extensive bridge of nerve fibers which interconnect the right and left cerebral hemispheres; i.e. the corpus callosum, and that only in conditions such as following callosotomy, does something akin to a splitting of the mind occur. Split-brain functioning following sectioning of the corpus callosum is discussed and a case study is presented. However, surgery provides only the most dramatic examples of this duality. Split-brain functioning is also characteristic during early childhood, and the results from experiments demonstrating this duality are presented. Further, partial splitting of the mind is evident even in the normal brain. It often occurs that people will express certain behaviors, act on various impulses, make "thoughtless", embarrassing statements or a "slip of the tongue" and "have no idea" as to "what came over" them. In some instances they may even fail to become aware that anything unusual or objectionable has resulted, or conversely, conjure up explanations for their actions. Similar anomalies confront us in regard to certain emotions, desires, impulses, and conflicts --the origins and source of which are seemingly submerged and hidden; at least from the language dominant hemisphere which maintains the language-dependent aspects of consciousness (i.e. in most instances the left). Because each half of the brain subserves unique (albeit overlapping) functions, such that functional lateralization is characteristic, in some situations one brain half may have little or no knowledge as to what is occurring in the other. Some forms of information cannot be transferred or even recognized by the opposite hemisphere. Functional lateralization also leads to the experience of forgotten dreams; and this is because it is the right hemisphere which produces much of the imagery experienced during the dream, whereas it is the left hemisphere which forgets the dream upon wakening. When the left hemisphere is denied access to information retained or processed by the right, and/or is unable to learn why a certain behavior or action was initiated, it typically makes up explanations (which it believes) or engages in active denial. These dualities are most evident following split-brain surgery, but can also explain some of the unique mental characteristics of childhood and the emotional difficulties experienced by the normal brain and the conscious mind.

1. THE MULTIPLICITY OF MIND

The mind is a multiplicity linked to the functional integrity of specific brain areas (Joseph 2010), including the brainstem, limbic system, and right hemisphere. Strictly speaking, the brainstem mediates and provides the neuro-anatomical foundations for the most rudimentary, reflexive, and unconscious aspects of mind (Joseph 1996). The limbic system, which includes the amygdala, hypothalamus, hippocampus, septal nuclei and cingulate gyrus, is more clearly linked to what Freud and Jung likened to the unconscious, subserving basic emotional and motivational impulses, the desire for sex and love, and the most sublime of emotions, including spirituality and religious rapture (Joseph 1990c, 1992b, 1994, 1998, 2001). The left cerebral hemisphere is classically associated with expressive speech and receptive language, and the temporal-sequential, mathematical, analytical, and linguistic aspects of consciousness. However, it has now been well established that the right cerebral hemisphere is dominant over the left in regard to the perception, expression and mediation of almost all aspects of social and emotional functioning, including the recall of emotional memories (Joseph 1982, 1986a,b, 1988a,b, 1990, 1992, 1996, 2003). This includes the comprehension and expression of emotional sounds, such as the melody of the voice, prosody, intonation, as well as singing, cursing and praying, and the perception of environmental sounds. In fact, as language consists of not just words but melody and prosody, the two hemispheres inter-act, via the corpus callosum, to produce and comprehend all aspects of human speech.

Further, the right hemisphere is dominant for most aspects of visual-spatial perceptual functioning, the recognition of faces including friend, loved ones, and one's own face in the mirror. Faces, of course, convey emotion whereas visual-space is related to the environment and the movement and localization of the body in space. Thus, recognition of one's own body and the maintenance of the personal body image is also the dominant realm of the right half of the brain (Joseph 1982, 1986a, 1988a,b, 1990a, 1992, 1996). The body image, for many, is tied to personal identity; and the same is true of faces.

Visual-spatial, facial recognition, body image dominance, the comprehension and expression of emotional sounds, all appear to be tied to a greater representation of these functions in the right vs the left hemisphere. Further, the right hemisphere and limbic structures within the right temporal lobe, provide the neurological foundation for the generation of dreams (Joseph 1988a, 1992b, 2001, 2011a). The right hemisphere, in fact, appears to maintain a realm of conscious-awareness which is completely different from that of the left (Bogen 1969; Sperry 1966, 1968, 1974, 1982). Right hemisphere mental functioning is more visual, spatial, emotional, personal, and non-verbal. By contrast, the mental realm of the left is more tied to language, including thinking in words (Joseph 1982, 1990b, 1996, 2011b)

In part, it is believed that the right hemisphere dominance over social and emotional functioning is due to more extensive interconnections with the limbic system (Joseph, 1982, 1988a), including the fact that limbic system appears to be functionally and structurally lateralized (Joseph 1990c, 1992b, 2001). This right hemisphere limbic dominance came to include the expression and representation of limbic language, thus providing the right cerebrum with a functional dominance in regard to the expression and comprehension of emotional speech as well as all aspects of emotional perceptions, the image of the body, and the recognition of friends and loved ones.

By contrast, with the evolution of language, tool use, and right hand motor control, functions formally associated with the left half of the brain were crowded out, due to competition for neural space, and became the exclusive and enhanced domain of the right (Joseph 1990a,b, 1992a, 1993, 1996, 2000a).

Over the course of evolution, the limbic system and each half of the brain have developed their own unique strategies for perceiving, processing, and expressing information, as well as specialized neuroanatomical interconnections that assist in mediating these functions. Thus, whereas the limbic system mediates the more unconscious aspects of social-emotional and personal awareness, the neocortex and the cerebral hemispheres are organized such that two potentially independent mental systems coexist, literally side by side.

2. THE NEUROANATOMY OF MIND

The right and left half of the brain are functionally specialized. Each half of the brain is divided into a frontal, temporal, occipital, and parietal lobe and is covered with a 6-layered neocortex. Each neocortical layer is also functionally specialized and change in thickness or cellular composition depending on location (e.g. the frontal, temporal lobe). Beneath the outer layered neocortex is the white matter which consists of myelinated and unmyelinated axons traversing and interconnecting various regions of the brain. In the depths of each hemisphere subcortical and limbic structures predominate, with each half of the brain possessing an amygdala, hippocampus, and so on.

The two halves of the brain are also interconnected with thick ropes of myelinated and unmyelinated axons which form the corpus callosum and anterior commissure. Via these axonal pathways, the right and left halves of the brain, and the various lobes and subcortical structures, can interact and exchange information. This enables tasks to be performed more efficiently, as, literally "two heads are better than one."

However, because each half of the brain speaks a different language and are specialized to analyze and perform different tasks, they are not always able to comprehend what is taking place in the other. Nevertheless, having "two brains" also gives rise to creativity, insight, and greater mental ability. On the other hand, this unique arrangement can also contribute to intra-psychic conflicts as each half of the brain in fact has a "mind of its own."

3. TWO BRAINS - TWO MINDS

That the right half of the brain is capable of conscious experience has now been well demonstrated in studies of patients who have undergone complete surgical destruction of the corpus callosum (i.e. corpus callosotomies, AKA split-brain operations (Bogen, 1969; Sperry 1966, 1968, 1974, 1982). These surgeries have been performed for the purposes of controlling intractable epilepsy. However, following surgery, the right and left half of the brain can function semi-independently, and when they do, they demonstrate the presence of two minds in the same head (Joseph 1986b, 1988a,b).

As described by Nobel Lauriate Roger Sperry (1966, p. 299), "Everything we have seen indicates that the surgery has left these people with two separate minds, that is, two separate spheres of consciousness. What is experienced in the right hemisphere seems to lie entirely outside the realm of awareness of the left hemisphere. This mental division has been demonstrated in regard to perception, cognition, volition, learning and memory."

Normally, visual and tactile information from the left half of visual-tactile environment is transmitted to the right half of the brain which controls the left extremities, and input from the right visual-tactile environment the left half of the brain which controls the right extremities. Likewise, auditory information received by the right and left ear are predominantly transmitted to the opposite half of the brain. Therefore, when a split-brain patient is tactually stimulated on the left side of the body, their left hemisphere will not know it. Since the left half of the brain controls the comprehension and expression of (non-emotional) language, split-brain patients are unable to name objects placed in the left hand, and they fail to report the presence of a moving or stationary stimulus in the left half of their visual fields (Bogen, 1979; Gazzaniga & LeDoux, 1978; Joseph, 1988b; Levy, 1974, 1983; Seymour et al. 1994; Sperry, 1982). In addition, the disconnected left hemisphere cannot verbally describe odors, pictures or auditory stimuli tachistoscopically or dichotically presented to the right cerebrum. Moreover, the left hemisphere has difficulty recognizing, identifying, or naming emotional and environmental sounds (Joseph, 1986b, 1988b)--capacities associated with the functional integrity of the right hemisphere.

By contrast, the disconnected right hemisphere is able to indicate when they are tactually or visually stimulated on the left side. When tachistoscopically presented with words to the left of visual midline, or when objects are placed in the left hand (but out of view), the right hemisphere can correct point with the left hand to the word or object--even when the left hemisphere claims it has seen or felt absolutely nothing.

For example, when presented with words like "toothbrush", such that the word "tooth" falls in the left visual field (and thus, is transmitted to the right cerebrum) and the word "brush" falls in the right field (and goes to the left hemisphere), the left hand will point to the word viewed by the right cerebrum (i.e., tooth) and the right hand will point to the word viewed by the left hemisphere (i.e., brush). When offered a verbal choice, the speaking (usually the left) hemisphere will respond "brush" and will deny seeing the word "tooth."

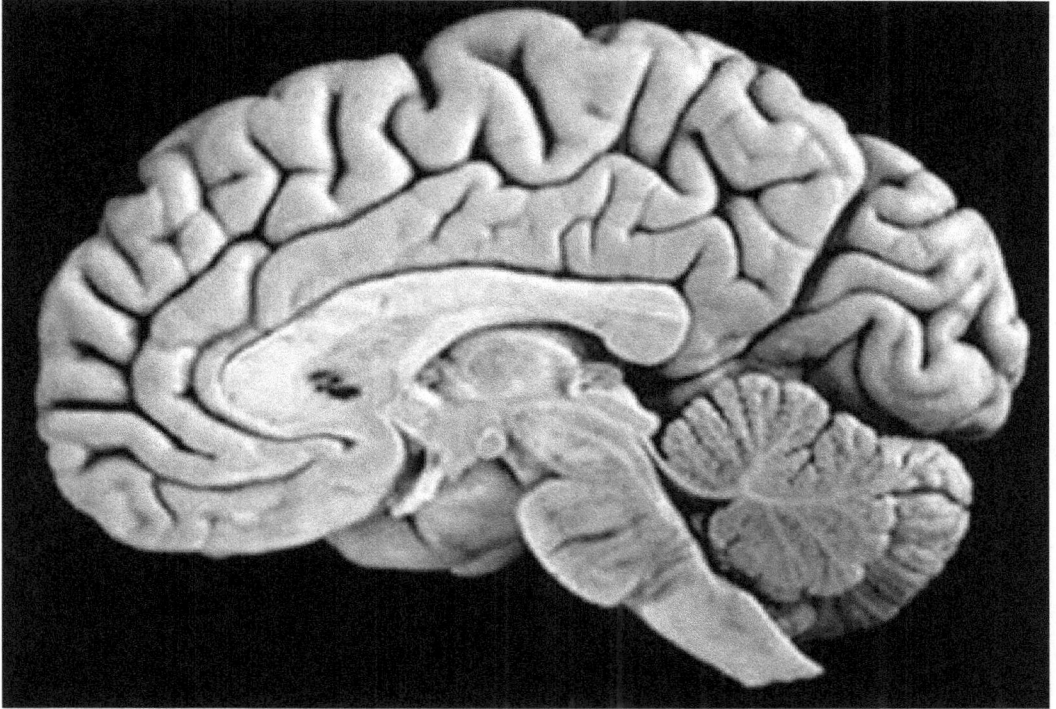

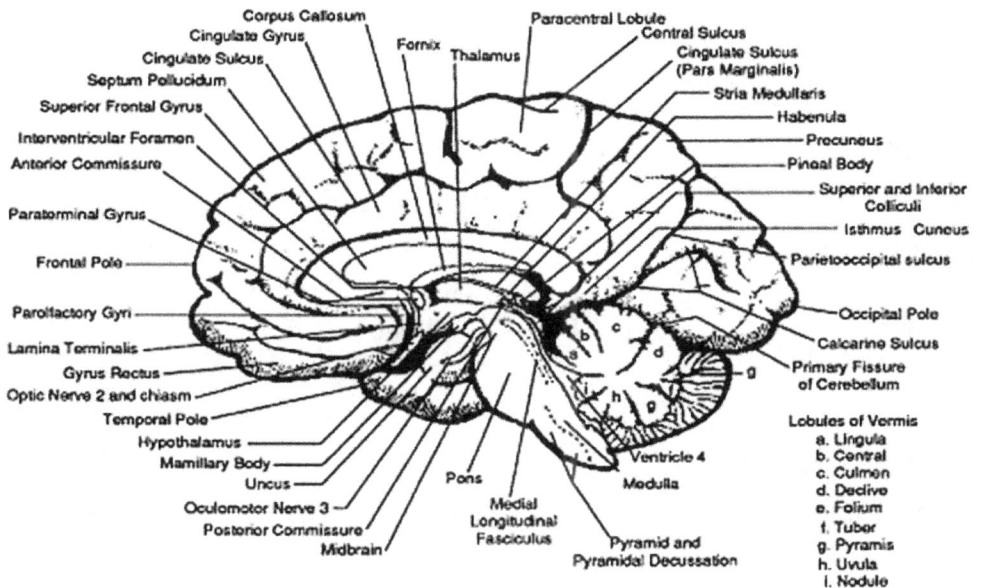

Corpus Callosum
Cingulate Gyrus
Cingulate Sulcus
Septum Pellucidum
Superior Frontal Gyrus
Interventricular Foramen
Anterior Commissure
Paraterminal Gyrus
Frontal Pole
Parolfactory Gyri
Lamina Terminalis
Gyrus Rectus
Optic Nerve 2 and chiasm
Temporal Pole
Hypothalamus
Mamillary Body
Uncus
Oculomotor Nerve 3
Posterior Commissure
Midbrain

Fornix
Thalamus

Paracentral Lobule
Central Sulcus
Cingulate Sulcus
(Pars Marginalis)
Stria Medullaris
Habenula
Precuneus
Pineal Body
Superior and Inferior Colliculi
Isthmus   Cuneus
Parietooccipital sulcus
Occipital Pole
Calcarine Sulcus
Primary Fissure of Cerebellum

Ventricle 4
Medulla
Pons
Medial Longitudinal Fasciculus
Pyramid and Pyramidal Decussation

Lobules of Vermis
a. Lingula
b. Central
c. Culmen
d. Declive
e. Folium
f. Tuber
g. Pyramis
h. Uvula
i. Nodule

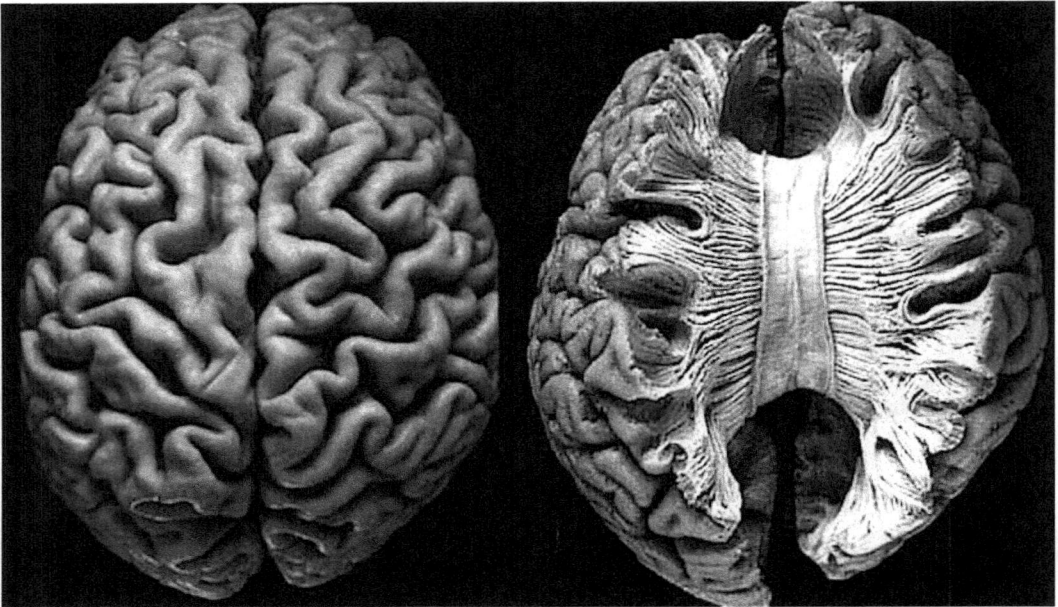

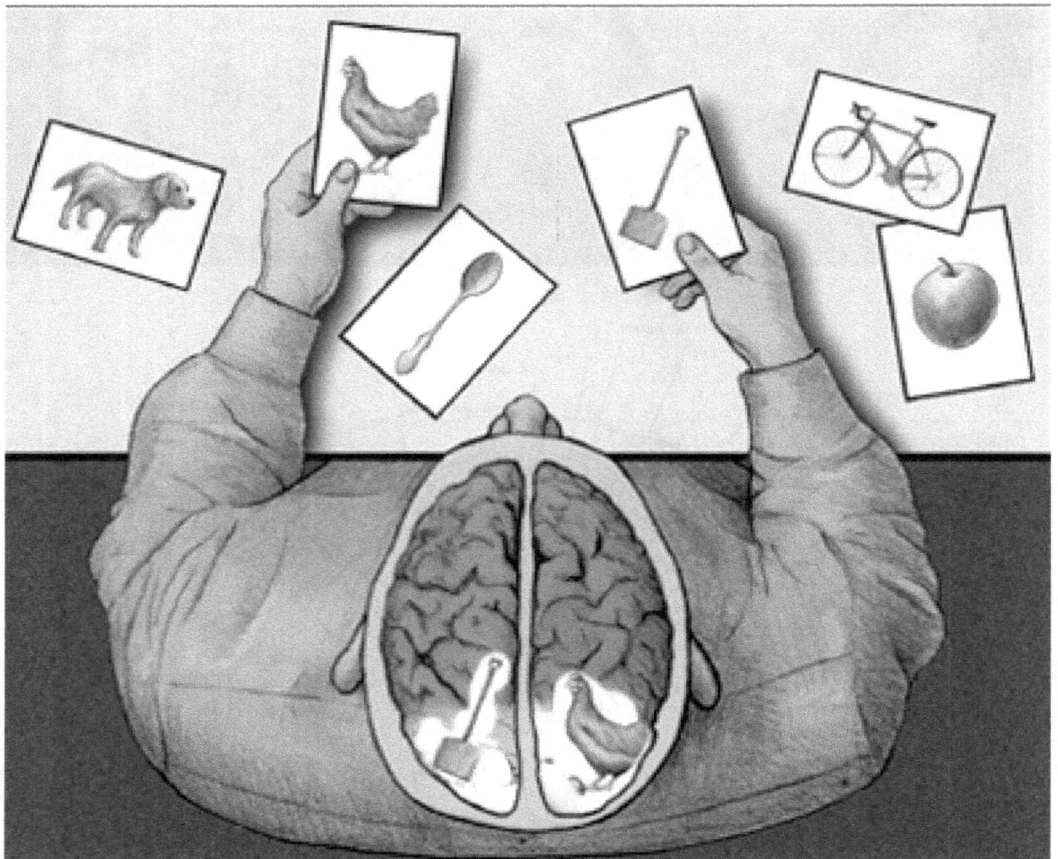

4. SPLIT-BRAIN CASE DESCRIPTION

Patient 2-M (AKA "Two Minds") had his first seizure at age eight and suffered from complex partial seizures as well as status epilepticus (with up to 100 seizures per day). Medication was unable to control these seizures.

EEG and depth electrodes demonstrated right frontal abnormalities which led to bilateral synchronous seizure activity. Independent discharges were noted throughout the left cerebrum. CT-scan failed to indicate the presence of any gross lesions. The patient was of low average intelligence (WAIS), and sodium amytal anesthetization demonstrated left hemisphere dominance for speech.

Following callosotomy at age 20, neurological examination demonstrated that his speaking, left hemisphere, was unable to perceive visual or tactile information from the left visual field and left half of the body. It was noted that his left arm and leg would not consistently follow commands and/or would respond in a maner opposite to what was requested. 2-M also complained that the left half of his body and left hand would sometimes behave independently and engage in actions that he, his left hemisphere, could not control. This condition remained basically unchanged even 3-years post surgery.

SOMESTHESIS: Touch sensation and localization was adequately appreciated when applied to the right half of 2-M's body. When applied to the left half 2-M responded with 0% accuracy when verbal reports were required. However, when he was asked to simply raise the left hand when the left half of the body was touched, 2-M responded correctly on 9 times out of 10.

STEREOGNOSIS: When common objects were placed in the right hand, 2-M named them without difficulty. When placed in the left hand he was not able to name them correctly; nor was he able to indicate, verbally or with the right hand (via pointing) the correct object when offered multiple (visual) choices: 0% accuracy. When asked to point with the left hand to objects felt with the left hand, 2-M responded with 90% accuracy.

DIFFERENTIAL SIMULTANEOUS TACTILE STIMULATION. This task involved simultaneously stimulating both hands (while out of view) with either the same or two different textured materials (e.g. sandpaper to the right, velvet to the left) and required the patient to point (with the left and right hand simultaneously) to an array of fabrics hanging in view on the left and right of the testing apparatus. Hence, following stimulation to hands while out of view, he would pull his hands out from inside the apparatus and point with the left to the fabric felt by the left, and with the right to the fabric felt by the right.

2-M almost consistently pointed with the left hand to the fabric experienced by that extremity and with the right hand t the material experienced by the right hand.

Suprisingly, when 2-M was stimulated with two different fabrics, e.g sandpaper to the right, wire mesh to the left, although his left hand (right hemisphere) responded correctly and pointed to the correct object, his left hemisphere vocalized: "Thats wrong!" Repeatedly he reached over with his right hand and tried to force his left extremity to point to the fabric experienced by the right (although the left hand responded correctly!). His left hand refused to be moved and physically resisted being forced to point at anything different. The left hand/right hemisphere insisted on pointing to the correct fabric applied to the left hand, whereas t he right hand/left hemisphere tried to force the left hand to point to the material experienced by the right. In one instance a physical struggle ensued, the right grappling with and hitting the left arm and hand.

Moreover, while performing this (and other tasks), 2-M made statements such as: "I hate this hand", or, "this is so frustrating", and would strike his left hand with his right or punch his left arm. Hence, it was readily apparent that 2-Ms left hemisphere had no idea that his left hand was responding correctly. Not once, however, did his left hand attempt to interfere with the choices made by the right.

UNILATERAL VS WHOLE BODY MOVEMENTS TO COMMAND. 2-M was presented with a series of 10 unilateral commands requiring movement of half the body (e.g. "salute like a soldier") or 10 whole body commands (e.g. "stand like a boxer"). He was able to respond correctly to all commands involving the right half of the body but to only 3 of the 10 whole body commands. This indicates his language dominant left hemisphere was unable to program the movements of the left half of the body which is under the control of the right hemisphere. In these instances, 2-M remarked that he hated the left half of his body.

VISUAL SPATIAL SKILLS. 2-M was required to draw and copy several geometric forms and objects (e.g. a flower) as well as a clock with all the numbers. The drawings and copies made by 2-M were distorted, macrographic, and he was unable to reproduce figures involving perspective with his right hand. Left sided neglect and inattention were also noted in his drawings. For example, when asked to draw the face of a clock, he drew only half the clock, and all the numbers appeared on the right half of his drawing.

RESPONSES TO T-SCOPE VISUAL-PICTORIAL STIMULI. The right and left cerebral hemisphere were separately and individually presented, via tachistoscope, with 40 visual-pictorial scenes (e.g. a boy with a ball, a man with an umbrella), 20 written words (e.g. "Hat"), and 10 digits. All visual stimuli were presented 3 degrees to the left or right of visual midline for 150 msec. 2-M was required to describe the pictures and to verbalize the words or digits presented.

He responded with 100% accuracy when the right visual field (left hemisphere) was stimulated; that is, he was able to accurately verbally described what he had seen. However, when the right hemisphere was presented with visual-pictorial material, 30% of his verbal responses were wildly confabulatory and characterized by gap filling. For example, when the right hemisphere shown a picture of a man walking in the rain and carrying an umbrella, he replied: "A pizza man making a pizza at a pizza joint". This indicates that some of this visual material was transferred to the left hemisphere, but it was incomplete. In consequence, the left hemisphere filled the gap with confabulatory ideation.

5. SPLIT-BRAIN CASE DISCUSSION

Following complete callosotomy 2-M displayed the disconnection syndromes unique to commissurotomy patients (Bogen, 1979; Gazzaniga & LeDoux, 1978; Sperry, 1966). Hence, when tactually stimulated on the left side of the body, (the left hemisphere's of) he demonstrated marked neglect when verbal responses were required, was unable to name objects placed in the left hand, and failed to report the presence of a moving stimulus in the left half of his visual fields. In addition, he demonstrated marked difficulties naming incomplete figures (and thus forming visual closure) as well as a reduced ability to name and identify non-linguistic and environmental sounds --capacities associated with the functional integrity of the right hemisphere (which was not afforded an opportunity to respond in this series of tests).

However, by raising his left hand, 2-M's right hemisphere was able to indicate when he was tactually or visually stimulated on his left side. When tachistoscopically presented with words to the left of visual midline, although unable to name them, when offered multiple visual choices in full field his right hemisphere was able to point correctly with the left hand to the word viewed. With his left hand he was also able to respond correctly on 90% of the texture identification trials and resisted attempts by his right hand to make it point elsewhere.

In addition, he frequently experienced and demonstrated difficulty making the left half of his body respond to verbal and left hemisphere mediated commands. The left half of his body often responded independently of the left hemisphere, and contrary to its wishes. Indeed, the functional and expressive independence of 2-Ms right cerebrum caused this individual (or at least his left hemisphere) a considerable degree of frustration, anger and embarrassment. In this regard he complained of situations where his left hand would turn off or change channels on the T.V. (even though he, i.e., his left hemisphere was enjoying the program), or his left leg refuse to walk in certain directions, act in a manner completely opposite to what he expressively intended, or his left hand would perform socially inappropriate actions (such as attempting to strike a relative).

In the laboratory, on several occasions, he became angry with his left hand, swore and expressed "hate" for it, struck his left hand/arm with his right hand,

and engaged in physical struggles with his left extremity, the right attempting to force the left to comply with some particular activity or to cease to act in a maner which the left hemisphere found objectionable or annoying. In many (if not most) instances, the left resisted and continued to behave in the manner directed by the right hemisphere.

In general, these findings indicate the isolated right cerebral hemisphere of 2-M was able to understand and follow certain simple verbal commands, obtain and maintain knowledge of various visual, auditory, linguistic, and somesthetic events, as well as respond, make decisions and act purposefully --even when his left hemisphere had no idea as to what information the right possessed and why it was behaving in a particular manner.

Given the differential behavioral observations regarding the activity and responsiveness of the left half of his body, it appears that 2-M is possessed of two minds, one of which resides within the right, the other within the left cerebral hemisphere. Nevertheless, it appears that 2-Ms left hemisphere remains predominant in regard to most expressive activities.

6. RIGHT BRAIN PERVERSITY

The disconnected right and left cerebral hemispheres, although unable to communicate and directly share information, are nevertheless fully capable of independently generating and supporting mental activity (Bogen, 1969, 1979; Gazzaniga & LeDoux, 1978; Joseph, 1986b, 1988a,b; Levy, 1974, 1983; Sperry, 1982). In the right hemisphere we deal with a second form of awareness which differ dramatically from the mental system of te left, but which accompanies in parallel what appears to be the "dominant" temporal-sequential, language dependent stream of consciousness in the left cerebrum.

Moreover, as has been demonstrated by Sperry, Bogen, Levy, and colleagues, the isolated right cerebral hemisphere, like the left, is capable of self-awareness, can plan for the future, has goals and aspirations, likes and dislikes, social and political awareness, can purposefully initiate behavior, guide responses choices and emotional reactions, as well as recall and act upon certain desires, impulses situations or environmental events --without the aid, knowledge or active (reflective) participation of the left half of the brain.

In that the brain of the normal as well as "split-brain" patient maintains the neuroanatomy to support the presence of two psychic realms, it is surprising that a considerable degree of conflict does not arise during the course of everyday activity. Frequently (such as in the case of the "split-brain" patient, LB, described below), although isolated the right half of the brain is fully willing to assist the left in a myriad of activities. Presumably such difficulties do not occur because both minds, having once been joined, share the same goals and interests. However, common experience seems to argue otherwise, for even in the intact individual, psychic functioning often is plagued by conflict.

In its most subtle manifestations the disconnected right hemisphere may attempt to provide the left with clues when the left (speaking) hemisphere is called upon to describe or guess what type of stimulus has been secretly shown to the right (such as in a T-scope experiment). Because the corpus callosum has been severed transfer and information exchange is not otherwise possible.

Hene, when a picture has been shown to the right and the left has been asked to guess, the right hemisphere may listen and then nod the head or clear the throat so as to give clues or indicate to the left cerebrum that it has guessed incorrectly. In one case the right hemisphere attempted to trace or write an answer on the back of the right hand (e.g. Sperry et al. 1979). For example, after the right hemisphere was selectively shown a picture of Hitler, and then asked to indicate their attitude toward it before verbally describing it, the patient signalled "thumbs down".

EX: "That's another 'thumbs-down'?"

LB: "Guess I'm antisocial."

EX: "Who is it?"

LB: "GI came to mind, I mean..." Subject at this point was seen to be tracing letters with the first finger of the left hand on the back of his right hand.

EX: "You're writing with your left hand; let's keep the cues out."

LB: "Sorry about that."

Nevertheless, the behavior of the right hemisphere is not always cooperative, and sometimes it engages in behavior which the left cerebrum finds objectionable, embarrassing, puzzling, mysterious, and/or quite frustrating. This is probably true for the normal as well as the "split-brain" individual.

For example, Akelaitis (1945, p. 597) describes two patients with complete corpus callosotomies who experienced extreme difficulties making the two halves of their bodies cooperate. "In tasks requiring bimanual activity the left hand would frequently perform oppositely to what she desired to do with the right hand. For example, she would be putting on clothes with her right and pulling them off with her left, opening a door or drawer with her right hand and simultaneously pushing it shut with the left. These uncontrollable acts made her increasingly irritated and depressed."

Another patient experienced difficulty while shopping, the right hand would place something in the cart and the left hand would put it right back again. Both patients frequently experienced other difficulties as well . "I want to walk forward but something makes me go backward. ' A recently divorced male patient noted that on several occasions while walking about town he found himself forced to go some distance in another direction. Later (although his left hemisphere was not conscious of it at the time) it was discovered (by Dr. Akelaitis) that this diverted course, if continued, would have led him to his former wife's new home.

Geschwind (1981) reports a callosal patient who complained that his left hand on several occasions suddenly struck his wife--much to the embarrass-ment of his left (speaking) hemisphere. In another case, a patient's left hand attempted to choke the patient himself and had to be wrestled away.

In addition, Bogen (1979, p. 333) indicates that almost all of his "complete commissurotomy patients manifested some degree of intermanual conflict in the early postoperative period." One patient, Rocky, experienced situations in which his hands were uncooperative; the right would button up a shirt and the left would follow right behind and undo the buttons. For years, he complained of difficulty getting his left leg to go in the direction he (or rather his left hemi-sphere) desired. Another patient often referred to the left half of her body as "my little sister" when she was complaining of its peculiar and independent actions.

A split-brain patient described by Dimond (1980, p. 434) reported that once when she had overslept her "left hand slapped me awake." This same patient, in fact, complained of several instances where her left hand had acted violently. Similarly, Sweet (1945) describes a female patient whose left hand sometimes behaved oppositionally and in a fashion which on occasion was quite embar-rassing.

The split-brain patient, 2-M, frequently was confronted with situations in which his left extremities not only acted independently, but engaged in pur-poseful and complex behaviors --some of which he (or rather, his left hemi-sphere) found objectionable and annoying. He complained of instances in which his left hand would perform socially inappropriate actions (e.g. attempt-ing to strike a relative) and would act in a manner completely opposite to what he expressively intended, such as turn off the TV or change channels, even though he (or rather his left hemisphere) was enjoying the program. Once, after he had retrieved something from the refrigerator with his right hand, his left took the food, put it back on the shelf and retrieved a completely different item "Even though that's not what I wanted to eat!" On at least one occasion, his left leg refused to continue "going for a walk" and would only allow him to return home.

In the laboratory, he often became quite angry with his left hand, he struck it and expressed hate for it. Several times, his left and right hands were observed to engage in actual physical struggles with the right hand striking the left. His left hemisphere also made statements such as: "I hate this hand" or "This is so frustrating" and would strike his left hand with his right or punch his left arm. In these instances there could be little doubt that his right hemisphere was behaving with purposeful intent and understanding, whereas his left brain had absolutely no comprehension of why his left hand (right hemisphere) was behaving in this manner.

7. LATERALIZED GOALS AND ATTITUDES

Why the right and left cerebral hemispheres in some situations behave co-operatively and yet in others in an oppositional fashion is in part a function of functional lateralization and specialization and the differential representation of social-emotional analytical abilities predominantly within the right hemi-sphere. Hence, because each hemisphere is concerned with different types of information, even when analyzing ostensibly the same stimulus may react, interpret and process it differently and even reach different conclusions (Joseph, 1988a,b). Moreover, even when the goals are the same, the two halves of the brain may produce and attempt to act on different strategies.

Functional lateralization may thus lead to the development of oppositional attitudes, goals and interests. For example, one split brain individual's left hand would not allow him to smoke, and would pluck lit cigarettes from his mouth or right hand and put them out. Apparently, although his left cerebrum wanted to smoke, his right hemisphere didn't approve -apparently he had been trying to quite for years.

Nevertheless, these difficulties are not limited to split-brain patients, for conflicts of a similar nature often plague "normal" individuals as well. It doesn't take split-brain surgery for an individual to express self-hate or to engage in self-destructive behaviors.

8. LATERALIZED MEMORY FUNCTIONING & CONFABULATION

Although a variety of neurochemical and neuroanatomical regions are involved in the formulation of memory, functional specialization greatly determines what type of material can be memorized or even recognized by each half of the cerebrum. This is because the code or form in which a stimulus is represented in the brain and memory is largely determined by the manner in which it is processed and the transformations that take place. Because the right and left cerebral hemispheres differentially process information, the manner in which this information is represented also will be lateralized (Bradshaw & Mattingly, 1997). Hence, some types of information only can be processed or stored by the right vs. the left cerebrum.

For example, it is well known that the left hemisphere is responsible for the encoding and recall of verbal memories, whereas the right cerebrum is dominant in regard to visual-spatial, non-verbal, and emotional memory functioning (Joseph 1982, 1988a, 1990a,b, 1992, 1993, 1996, 2003). If the left temporal lobe were destroyed, verbal memory functioning would become impaired since the right cerebrum does not readily store this type of information. Conversely, the left has great difficulty storing or remembering nonlinguistic, visual, spatial, and emotional information including the ability to recall or recognize emotional faces or personal emotional memories.

Hence, it is the left hemisphere which is responsible for the encoding and recall of verbal, temporal-sequential, and language related memories, whereas the right cerebrum is dominant in regard to visual-spatial, non-verbal, and social emotional memory functioning. Each hemisphere stores the type of material that it is best at recognizing, processing, and expressing.

9. UNILATERAL MEMORY STORAGE: HIDDEN MEMORIES

In the intact, normal brain, even non-emotional memory traces appear to be stored unilaterally rather than laid down in both hemispheres (Bures & Buresova 1960; Doty & Overman 1977; Hasegawa et al., 1998; Kurcharski et al. 1990; Levy, 1974; Risse & Gazzaniga, 1979). Moreover when one hemisphere learns, has certain experiences, and/or stores information in memory, this information is not always available to the opposing hemisphere; one hemisphere cannot always gain access to memories stored in the other half of the brain (Bures & Buresova 1960; Doty & Overman 1977; Hasegawa et al., 1998; Joseph, 1986b, 1988ab, 1992b; Kucharski et al. 1990; Levy, 1974; Risse & Gazzaniga 1979).

To gain access to these lateralized memories, one hemisphere has to activate the memory banks of the other brain half via the corpus callosum (Doty & Overman, 1977; Hasegawa et al., 1998) or anterior commissure (Kucharski et al. 1990). This has been demonstrated experimentally in primates. For example, after one hemisphere had been trained to perform a certain task, although either hemisphere could respond correctly once it was learned, when the commissures were subsequently cut, only the hemisphere that originally was trained was able to perform--i.e., could recall it. The untrained hemisphere acted as though it never had been exposed to the task; its ability to retrieve the original memories was now abolished (Doty & Overman, 1977; Hasegawa et al., 1998).

In a conceptually similar study, Risse and Gazzaniga (1979) injected sodium amytal into the left carotid arteries of intact patients so as to anesthetize the left cerebral hemisphere. After the left cerebrum was inactivated, the awake right hemisphere, although unable to speak, was still able to follow and behaviorally respond to commands, e.g., palpating an object with the left hand.

Once the left hemisphere had recovered from the drug, as determined by the return of speech and motor functioning, none of the eight patients studied was able to verbally recall what objects had been palpated with the left hand, "even after considerable probing." Although encouraged to guess most patients refused to try and insisted that they did not remember anything. However, when offered multiple choices in full field, most patients immediately raised the left hand and pointed to the correct object!

According to Risse and Gazzaniga (1979), although the memory of touching and palpating the object was not accessible to the verbal (left hemisphere) memory system, it was encoded in a nonverbal form within the right hemisphere and was unavailable to the left hemisphere when normal function returned. The left (speaking) hemisphere was unable to gain access to information and memories stored within the right half of the brain. Nevertheless, the right brain not only remembered, but was able to act on its memories.

This indicates that when exchange and transfer is not possible, is in some manner inhibited, or if for any reason the two halves of the brain become functionally disconnected and are unable to share information, the possibility of information transfer at a later time is precluded) -even when the ability to transfer is acquired or restored. The information is lost to the opposite half of the cerebrum.

Moreover, because some types of information are processed by the right and left hemisphere in a wholly different fashion, they are unable to completely share or gain access to the data or even the conclusions reached by the other -as they are unable to process or recognize it -which in turn precludes complete interhemispheric transfer; information is lost during the transfer process.

Nevertheless, although inaccessible or lost, these memories, details, and attached feelings can continue to influence whole brain functioning in subtle as well as in profound ways. That is, one hemisphere may experience and store certain information in memory and at a later time in response to certain situations act on those memories, much to the surprise, perplexity, or chagrin of the other half of the brain; one hemisphere cannot always gain access to memories stored in the other half of the brain.

Dreaming and Forgotten Dreams. Of course, complete functional deactivation is probably quite rare in the normal brain. However, there is some evidence to suggest that interhemispheric communication is reduced, for example, during sleep and possibly during dreaming (Banquet, 1983; Joseph, 1988a).

Most dreaming occurs during REM, which possibly is associated with right hemisphere activation and low-level left hemisphere arousal (Goldstein et al. 1972; Hodoba, 1986; Meyer et al. 1987). It also becomes progressively more difficult to recall one's dreams as one spends time in or awakens during Non-REM (Wolpert & Trosman, 1958), which is associated with high left hemisphere and low right brain activation (Goldstein et al. 1972). Thus are dreams really forgotten, or are they locked away in a code which is not accessible to the speaking left hemisphere?

10. LONG LOST CHILDHOOD MEMORIES

For most individuals it is extremely difficult if not impossible to verbally recall events which occurred before the age of three and a half (Joseph, 1998, 1999a, 2003). There are several reasons for this: Information processed and experienced during infancy vs. adulthood is stored via certain transformations and retrieval strategies which are quite different. As the brain matures and new information processing strategies are learned and developed, the manner in which information is processed and stored is altered. Although these early memories are stored within the brain, the organism no longer has the means of retrieving them, i.e., the key no longer fits the lock.

That is, early experiences may be unrecallable because infants use a different system of codes to store memories whereas adults use symbols and associations (such as language) not yet fully available to the child (Joseph, 1982, 2003). Much of what was experienced and committed to memory during early childhood took place prior to the development of lingusitic labeling ability and was based on a pre- or nonlinguistic code (Joseph 2003). Hence, the adult, who is relying on more sophisticated and language-related coding systems, cannot find the right set of neural programs to open the door to childhood memories. The key does not fit the lock because the key and the lock have changed.

The inability to recall early memories is also a function of programmed cell death--the loss of memory-laden neurons which are shed by the millions over the course of early development and the immaturity of the corpus callosum in children (Joseph, 1982, 1999a,b). That is, non-verbal information perceived and processed by the right vs left hemisphere is generally stored in the right vs left hemisphere. Later, when the commissures mature, this information cannot be transferred except under special conditions.

However, under conditions of traumatic memory loss; e.g. repression, not just the the right and left hemisphere, but the differential activation of the amygdala and hippocampus are contributory; and later the recall of such memories may be opposed by the frontal lobes; the right frontal in particular (Joseph 1986a, 1988a, 1999c).

11. EMOTION AND RIGHT BRAIN FUNCTIONING IN CHIDLREN
As is now well known, the developing organism is extremely vulnerable to early experience during infancy such that the nervous system, perceptual functioning and behavior may be altered dramatically (Casagrande & Joseph 1978; 1980; Joseph 1979, 1998, 1999a,b, 2003; Joseph & Casagrande 1980; Joseph & Gallagher 1980; Joseph et al. 1978). There is some evidence that the right cerebral hemisphere and the right amygdala may be more greatly affected (Joseph 1998, 2003).

Moreover during these same early years our traumas, fears, and other emotional experiences, like those of an adult, are mediated not only by the limbic system, but also via the nonlinguistic, social-emotional right hemisphere. And, just as they are in adulthood, these experiences are stored in the memory banks of the right cerebrum.

However, much of what was experienced and learned by the right half of the brain during these early years was not always shared or available for left hemisphere scrutiny (and vice-versa). That is, a childs two hemispheres are not only functionally lateralized, but limited in their ability to share and transfer information. In many ways, infants and young children have split-brains (Gallagher & Joseph 1982; Joseph & Gallagher, 1985; Joseph et al. 1984).

Nerves are insulted with myelin. Myelin is like the rubber coating that covers naked wires. This insulation, be it on a naked wire or an axonal appendage of a neurons, enables information to travel more efficiently between neurons. However, myelination of axons can take up to 10 years after a child is born (Yakovlev & Lecours, 1967). Therefore, information process is not as efficient in the brain of a child vs an adult.

The Myelination of the Human Brain. Myelin is the outer protective coating of the axon and promotes and stabilizes nerve transmission. From Yakovlev & Lecours (1967)

The corpus callosum, the major axonal fiber pathway that connects the right and left half of the brain, takes over 10 years to fully myelinate (Yakovlev & Lecours, 1967).

Due to the immaturity of the corpus callosum and in particular, the slow rate of axonal myelin within the callosum (Yakovlev & Lecours, 1967), communication is so poor that children as old as age 4 have mild difficulty transferring tactile, auditory, or visual information between the hemispheres, e.g., accurately describing complex pictures shown to the right hemisphere (Joseph et al. 1984). That is, in addition to differential functional specialization, the slow development of the myelination process can in turn slow and disrupt axonal information transmission.

Indeed, although pain can be transmitted and received via axons devoid of myelin, this type of data is completely lacking in complexity and is devoid of initial cognitive attributes. Pain transmitting axons have very simple requirements. By contrast as data complexity increases, so too does the complexity of those neurons which transmit these signals (Konner, 1991; Ritchie, 1984).

For example, as axon diameter increases, so too does the extent of myelination. Similarly, quantities of axoplasm and cytoplasm, nuclear diameter, and neuronal packing density are also correlated with myelination (Konner, 1991). Conversely, lack of myelin, or those neurons which have not yet myelinated, are associated with an increased susceptibility to conduction failure and interference due extraneous influences, including signal modification by neighboring axons (Konner, 1991; Ritchie, 1984).

12. SPLIT-BRAIN FUNCTIONING IN CHILDREN

A considerable body of evidence has been presented showing that traumatic or surgical section of the corpus callosum results in interhemispheric deficits in information transfer. A number of investigators have provided evidence which purports to show that young children have similar difficulties in interhemispheric information transfer (Finlayson, 1976; Galin, Diamond and Herron,

1977; Galin, Johnstone, Nakell and Herron, 1979; Gallagher and Joseph, 1982; Joseph et al. 1976; Joseph et al., 1984; Kraft, Mitchell, Languis and Wheatley, 1980; O'Leary, 1980; Salamy, 1978). These deficits, presumably, are secondary to the prolonged development (e.g. myelination) of the callosum, the maturation of which is not complete until after the first decade (Yakovlev and Lecours, 1967). Hence, the capacity to transfer information between the cerebral hemispheres appears to progressively increase with age (Finlayson, 1976; Galin et al., 1977; Gallagher & Joseph, 1982; Kraft et al., 1980; Joseph & Gallagher 1985, Joseph et al., 1984; O'Leary, 1980).

Consonant with the behavioral evidence cited above are some electrophysiological studies. Briefly, Salamy (1978) failed to find ipsilateral evoked potentials (which are perhaps dependent on transfer via the corpus callosum) prior to age 3.5, while Joseph et al. (1976) found deficiencies in the spatio-temporal organization of the EEG and thus a general lack of interhemispheric linkage in newborn and premature infants.

The evidence is indicative of limited rather than absent interhemispheric information exchange in young children. Indeed, given the differing local rates of cortical and callosal fiber development there is likely to be considerable transfer depending on age and modalities employed. Furthermore, since the anterior commissure (a tract of fibers which interconnects the inferior temporal lobes and limbic areas) appears developmentally complete by the end of the first year of life, even the hemispheres of the very young child may be only partially disconnected.

Confabulation and Limited Interhemispheric Information Exchange In cases of complete section of the commissures each hemisphere appears ignorant of the cognitive activities occurring in the other. In cases of partial sectioning (e.g. sparing of the anterior commissure) some transfer occurs (Joseph 1986b), albeit limited and incomplete. For example, in one patient with an intact anterior commissure, following right hemisphere tachistoscopic pictorial stimulation all verbal reports, while containing some accurate descriptive elements, were largely contaminated with errors, gross embellishments, and other falsehoods (Joseph 1988b). Information transfer (from the right hemisphere) was thus characterized by confabulatory "gap-filling" (Joseph, 1982, 1986a,b, 1988a,b). The left hemisphere apparently attempted to make sense of the limited information received by filling the gaps with related ideation.

For example, when a complex picture, e.g. a man with a gun, was presented tachistoscopically to the patient's left hemisphere, he would typically respond with concrete and accurate two or three word replies, such as "guy with a gun". When the same stimulus was presented to the right hemisphere (requiring transfer to the language centers of the left) he would respond with elaborate, detailed, and vivid descriptions which bore little relation to the actual elements of the original stimulus around which other associations were erroneously anchored: "Gun, hold up...he has a gun and is holding up a bank teller, a counter separates them." Furthermore, the left hemisphere's of these patients believed the largely erroneous descriptions they provided. As noted by Gazzaniga and LeDoux (1978), the left hemisphere did not "offer its suggestion in a guessing vein but rather as a statement of fact" (pp. 148-149).

Likewise, because of the limited information exchange between the hemispheres of children, children also confabulate if questioned concerning pictorial stimuli presented to the right hemisphere (Joseph et al., 1984).

Confabulation. Anchors and Gaps. It has been argued and demonstrated (Joseph, 1982, 1986a,b, 1988a,b) that confabulation occurs when the language centers of the left hemisphere are isolated from sources of information about which the patient is questioned, such that complete and accurate information is not fully available. In these instances confabulation is a result of attempts to fill the gaps in the information received with associations which are in some manner related to the fragments available. Moreover, disconnection also results in failure to correct erroneous statements, as contradictory evidence is also not always accessible. Hence, confabulatory statements, although erroneous, often contain accurate elements around which erroneous (albeit related) ideations are anchored. However, confabulation may also result due to other factors such as frontal lobe dysfunction such that (in these instances) an uninhibited flow of irrelevant information is transmitted to and organized by the language regions.

Experiment: Materials and Methods Three age groups of children (4, 7, 10) were tested and asked to describe pictorial stimuli presented to the left or right cerebral hemisphere. It was predicted that due to the immature state of the corpus callosum and the cerebral cortex, children would incompletely transfer, inaccurately report and embellish information presented to the right as compared to the left hemisphere. These subjects were also tested on the Tactile Form Recognition task of the Halstead-Reitan Battery, and their tendency to make transfer vs. non-transfer error ascertained.

Confabulation. For the purposes of this study confabulation was defined as an erroneous embellished description, such that perceptual-pictorial information not originally presented was inaccurately inserted and included in the subject's report. Hence, a detail may be reported correctly but the meaning of the detail may become overly significant and erroneously generalized such that a concept results which has little or, in fact, no bearing on the original stimulus.

Information Gaps and Inclusion Score An important feature of the gap filling hypothesis is the notion that information degradation and confabulation should be directly related, such that, the larger the gap, the more likely is the individual to confabulate. Hence, for purposes of this study the major pictorial-perceptual elements (features) of each tachistoscopic stimulus were determined. For example, if in response to a picture of a girl blowing out the candles on a birthday cake, a subject responded, 'girl" or "cake", the inclusion score would be "1" (as opposed to "2" for stating "girl" and "cake"). For each presentation the inclusion of these items in the child's verbal report was therefore tabulated.

Misses, Semantic, Syntactic, and Perceptual Errors Tabulation of the frequency of misses, semantic, syntactic and perceptual errors was largely for exploratory purposes. Briefly, misses were scored when no response was offered or the subjects claimed not to have seen the stimulus. Syntactic errors were noted when the subject's speech was agrammatical. A response was scored as a semantic error when the words used in the description, although incorrect, were from the correctly aroused semantic group pertaining to the stimulus. The erroneous response therefore bore a meaningful resemblance to the original stimulus. In contrast, a perceptual error was tabulated when an erroneous response/description bore an obvious similarity to some aspect of the stimulus whole; for example, calling, calling a "cake" a "drum". However, unlike a confabulation, no erroneous information could be added (for either a semantic or perceptual error). In this respect it was hypothesized that a perceptual error probably would result from a breakdown of serial feature-by-feature visual scene analysis as well as a failure to adequately engage in visual exploration. Thus, only certain details are recognized in isolation and stripped of related features. Hence, the left hemisphere should perform more poorly.

Subjects All subjects were right handed, aged 4, 7, 10. A total of 132 subjects were tested, with a minimum of 20 girls and 20 boys in each age group (4, N=48; 7, N=44; 10, N=40). Subjects were volunteers recruited from private schools and were obtained via parents' consent.

Materials All children were tested in an isolated and quiet room provided by the school. For assessing tendencies to confabulate or make other errors in description, a single-field tachistoscope was employed and all responses were tape-recorded. Visual stimuli consisted of pictures adopted from the Peabody Picture Vocabulary Test (PPVT). Eleven stimuli were employed for presentation to either the right, left, or visual-field midline, for a total of 33 presentations. In addition, two control stimuli and a pre-test "warm-up" stimulus from the PPVT were used. All stimuli were projected on a small white screen with a single fixation point (1.54 cm hole) at its center.

For the tactile-form recognition test, the standard Halstead-Reitan Form Board was employed. Briefly, this is a wooden rectangular box which contains a hole for hand insertion at is bottom front center. Across the face of the apparatus are embedded four small plastic chips, each of which is identical to one of the four stimulus objects (i.e. cross, triangle, square, circle) used for non-visual tactile exploration.

Procedure for Tactile-Form Recognition Testing Prior to testing each child was questioned about and asked to demonstrate (on 3 tasks) hand usage. Inclusion criterion was right handedness on 2 of the 3 tasks. After being familiarized with the stimuli and procedures the subject was instructed to place his right hand through the hole of the apparatus. A single chip was placed in the fingers and the subject encouraged to feel it until he knew what it was. The subjects were then instructed to release the chip, remove his hand, and thus use the same hand (tactile-visual non-transfer, TVNT), to point to the object on the face of the form board. After 4 trials, once with each chip, the procedure for non-transfer recognition was repeated for the left hand.

Instructions for the transfer condition were similar except that the child was asked to point with the opposite hand (tactile-visual transfer, TVT) to the correct object. After this procedure was repeated once for each object, the child was instructed to insert the left hand and the transfer trials were repeated.

Testing always began with the right hand. However, for approximately half of the subjects at each age level, testing began with the non-transfer trials, and with the transfer trials for the remaining subjects.

Procedure for Tachistoscopic Testing Following the Tactile Form Recognition tests, each subject was presented with a picture from the PPVT and encouraged to give a detailed description of it. The child was then seated facing the viewing screen, asked to maintain visual fixation on the center point even when the picture fell to either side, and asked to describe each picture presented. Two practice trials in which images were presented at the screen's center were run and the child encouraged to describe what was seen. Testing was then initiated and 33 trials were run.

All stimuli were presented once to each visual field, the right or left hand edge at least 3 degrees from the central fixation point, for exactly 150 msec., and last to the screen's midline. Left and right presentations were counterbalanced (i.e. stimuli presented initially to the right then left and midline, 5 different stimuli initially to the left then right and midline, and one stimulus presented initially the midline followed by left then right sided presentations). All slides were presented in a pre-established random order which remained constant for all subjects. Eye movements were carefully monitored by an observer hidden behind the viewing screen, via the center-fixation hole.

After each presentation the subject was asked, "What did you see?" However, if the child's response was a one word reply (e.g. "girl"), the experimenter asked "what else did you see?" If the child again responded with a one word description the next stimulus was presented and the procedure repeated.

Scoring and Rater Reliability Three raters, blind to mode of presentation (i.e. right, left, midline) were given intensive instruction and training in scoring procedures and criteria for the tachistoscopic responses, and extensive practice in the scoring of protocols. Interrater reliability, based on the scoring of 8 research protocols, yielded reliability coefficients in excess of .90.

RESULTS A 4 x 2 x 3 (Experimenter by Sex by Age) analysis of variance (ANOVA) was performed to control for experimenter influences and for the six dependent T-scope variables: Inclusion, Confabulation, Semantic, Syntactic and Perceptual errors, and Misses, and the two tactile variables, tactile-visual transfer (TVT) vs tactile-visual non-transfer (TVNT) which were treated as repeated measures. For each T-scope dependent variable, mode of hemispheric tachistoscopic presentation (right, left, midline) were treated as repeated measures. No significant sex differences were discovered.

Inclusion Score Significant main effects were found for age (Fw96.20, p< .00001) and mode of visual-field presentation (Fw108.94, p<.00001). A Newman-Keuls (NK) analysis of means indicated that 4-year olds had significantly lower inclusion scores (summed across presentation modes) than 7 (Qw16.31, p<.01) or 10-years old children (Qw22.68, p<.01), and that subjects age 7 responded with significantly fewer inclusion items than the 10-year old group (Qw6.37, p<.01). In addition, across all age groups inclusion scores were found to be significantly larger for left hemispheres presentation than for right (Qw1.202, p<.01). Midline exposures also resulted in a significantly grater number of inclusion elements than left (Qw2.47, p<.01) or right presentation (Qw3.672, p<.01). However, this latter finding is probably due to having viewed all stimuli two times prior to midline presentations (once right and once left). The NK analysis of experimenter influences did not reveal any significant main effects.

Confabulation The ANOVA demonstrated a significant effect for age (Fw26.25, p<.00001) and visual field presentation (Fw64.37, p<.00001). The NK analysis on ordered pairs of group means indicated that 4-year olds made significantly more confabulatory responses that 7 (Qw4.24, p<.01) or 10-year olds (Qw5.383, p<.01). The age 7 group also confabulated significantly more than the older children (Qw1.04, p<.05). In addition, more confabulatory responses were found to occur following right vs. left hemisphere presentation (Qw.803, p<.01), or midline exposures (Qw.834, p<.01).

A significant interaction between presentation site and age was also indicated (Fw4.26, p<.002), such that regardless of right hemisphere (RH), left hemisphere (LH) or midline (MH) presentation, 4-year olds confabulated significantly more than 7 (RH: Qw2.133, p<.01; LH: Q<.932, p<.01; MH: Qw1.178; p<.01) or 10-year old children (RH: Qw2.492, p<.01; HM: Qw1.375, p<.01; MH, Qw1.517, p<.01). In addition, among the children age 4, there were significantly more confabulatory responses to the right vs. left (Qw1.542, p<.01) or midline presentation (Qw2.25, p<.01). Among 7 and 10-year olds no significant age or presentation mode differences were discovered. No significant main effects for experimenter influences were uncovered.

Semantic and Syntactic Errors and Misses No significant main effects for syntactic errors were found. A main effect for semantic errors was discovered (Fw9.10, p<.001), such that 10-year old subjects made fewer errors than 4 (Qw2.66, pw.01) or 7-year old children (Qw2.386, p<.01). Children age 4 and

7 did not significantly differ.

Significant differences in the tendency to omit responses (Misses) were found to be age dependent (Fw8.86, p<.0003). It was determined that 4-year olds had significantly more misses than 7 (Qw.602, p<.01) or 10-year olds (Qw.775, p<.01). The 7 and 10-year groups were not found to significantly differ.

Perceptual Errors Significant age-related differences in the tendency to commit perceptual errors were noted by the ANOVA (Fw5.67, p<.005). The NK analysis indicated that 4-year old children made significantly more perceptual errors than 7 (Qw.442, p<.01) or 10-year olds (Qw.994, p<.01). In addition, children age 7 committed more perceptual errors than older group (Qw.554, p<.01).

In regard to overall scores summed across age and mode of presentation, it was also noted that the experimenters were associated with significantly different amounts of perceptual errors (Fw7.33, p<.0002), such that Experimenters (E) 1 and 3 significantly differed from each other and from E2 and E4 (p<.01), who in turn did not differ from each other (Qw.12, pwNS). A check of the raw data for this and all other effects associated with the different subjects tested by each experimenter indicated, however, that these differences are largely a function random subject variations (probably associated with demographic characteristics- e.g. socio-environmental influences) and the tendency of a few subjects to respond with an unusually large number of errors.

Tactile-Form Recognition When considering total error scores for both the transfer (TVT) and non-transfer conditions (TVNT), significant differences were found across age groups (Fw8.86, p<.0003), such that the 4-year group made more errors overall than did 7 (Qw.6818, p<.01) or 10-year olds (Qw.7375, p<.01). Also, regardless of age, significantly more transfer vs non-transfer errors were discovered (Fw13.11, p<.0004). In addition, the ANOVA showed a significant interaction effect for task and age (Fw7.94, p<.0006). In this regard, 4's made significantly more transfer errors than either 7 (Qw.489; p<.05) or 10's (Qw.575, p<.01), and 4-year olds, unlike older subjects, were found to perform more poorly on the transfer vs non-transfer tactile conditions (Qw.4375, p<.05).

Correlations Among Tachiscopic and Tactile Conditions > A multiple correlational analysis (SCSS) demonstrated, as predicted, a highly significant inverse relationship between inclusion scores (i.e. pictorial elements accurately reported) and confabulation (rwy.484, p<.001) across modes of hemispheric presentation. Hence, the fewer pictorial elements reported, the more likely was there a tendency to confabulate. In addition, age was highly correlated with inclusion scores (rw.764, p<.001) and confabulatory responding (rwy.530, p<.005), confirming that with increasing age the number of pictorial features reported (inclusion score) increases, whereas the tendency to confabulate significantly diminishes.

Interestingly, both TVT (transfer) and TVNT (non-transfer) tactile-recognition errors were found to be positively associated with confabulatory responding (rw.215, p<.01; rw.192, p<.05, respectively), and negatively linked with inclusion scores (TVT: rwy.336, p<.01; TVNT: rwy.158, p<.05), with the overall magnitude of correlation for both confabulation and inclusion score being grater for transfer errors.

13. SPLIT-BRAIN FUNCTIONING IN CHILDREN: DISCUSSION

The results support the hypothesis that children have significant difficulty performing tasks which require inter-hemispheric transfer of tactile and visual material at age 4, a time period beyond which language has become lateralized to the left hemisphere. Verbal descriptions were thus often incomplete, impoverished, and erroneous when right and left hemisphere transfer was required.

For example, in response to the tachistoscopic presentation of a picture depicting a girl blowing out the candles on a birthday cake, many young children fail to report a number of major pictorial features (such as the girl or cake), and thereby earn a low inclusion score. Although after right hemisphere stimulation some elements are accurately reported (as determined by inclusion score), there is a significant failure to account for a number of major perceptual-pictorial elements as compared to the left hemisphere. It thus appears that some information is deleted or degraded during transfer.

In addition, a large number of children were found to erroneously extrapolate and to embellish their descriptions (i.e. confabulate), particularly when pictorial stimuli were tachistoscopically presented to the right hemisphere. This was especially evident among children age 4, as right hemisphere confabulatory errors greatly exceeded those of the left. In contrast, the transfer (i.e. from the right to left hemisphere) and non-transfer (left hemisphere presentation) confabulation scores of the older children were significantly fewer in number and did not significantly differ.

The lower the inclusion score (the fewer items reported), the more likely was a subject to confabulate. This latter significant inverse correlation, maintained regardless of age or presentation field, was particularly strong following right hemisphere stimulation and thus information transfer to the left cerebral cortex. Thus, in conjunction with information deletion (i.e. low inclusion score), erroneous information was somehow added to the subject's report, such that the fragments received became subsequently embellished. For example, in response to stimulus J (girl blowing out the candles on a cake), the child might respond: "a girl opening a package", or "a girl eating something", and thus be scored as confabulatory. In both instances, features such as cake, candles, girl blowing, are deleted and information such as opening a package, or eating something, are inserted.

Why didn't the child simply state, "I don't know" rather than confabulate when confronted by these gaps? We presume this response was rare because that information was not available to the language axis of the left hemisphere. According to the model discussed at length elsewhere (Joseph, 1982, 1986a) confabulation in part results in response to gaps in information, gaps which in turn appear to be due to the transmission of information which is degraded or incomplete when received by Wernicke's region. This type of confabulation is thus the result of the language center's attempts to organize and make sense of the limited information received, by insertion of additional information. The process of insertion is not random, but is partly determined by contextual variables (e.g. emotional status, past history), as well as the availability of associations/ideations which are linked to the perceived fragments of information. Essentially, the language axis of the left hemisphere does not know that it does not know as it attempts to make sense and report only what is transmitted to it.

It is noteworthy that unilateral left cerebral presentation also resulted in confabulation. This suggests that intra-hemispheric substrate immaturity as well as that of the corpus callosum may also be responsible for part of the observed findings. One candidate for the site of intra-hemispheric deficiencies is the inferior parietal lobule. This area is one of the last to myelinate (Flechsig, 1901), the last in which dendrites appear, and at age 4 is only 64% developmentally complete (Blinkov and Glezer, 1968). Fiber interconnections linking this with other neocortical areas are also not completely matured until approximately 7 years of age (Lecours 1975). In that the inferior parietal lobule presumably has a major role in information assimilation and transfer to Wernicke's region, the lack of maturation in this area may contribute significantly to the observed deficiencies.

Significant differences in transfer vs. non-transfer were not noted for misses (i.e. trails in which no response was elicited), perceptual, semantic, or syntactic errors. In part, the failure to discover transfer effects on these processes may be due to greater efficiency and degree of maturation in the anatomical systems involved, or inadequacies within the methodology. However, the former explanation does not seem completely valid as significant age-related differences were apparent. It is noteworthy that misses, confabulation, and perceptual and semantic errors were significantly large in number regardless of hemisphere of presentation. That is, significant developmental increases in the ability to process and express select forms of information was apparent on almost every measure employed including TVNT. In part, these findings bolster the suggestion advanced above, that possible intra-hemispheric age-related deficits/ immaturities may be responsible for these errors in addition to (and perhaps independent of) corpus callosum immaturity.

Indeed, intra-hemispheric deficits in processing may be a significant factor which differentially influenced the overall results. That is, the tendency to make errors in either the right or left hemisphere independent of transfer, may, when transfer is required, give rise to errors in the right and left hemisphere and thus a larger error score independent of and/or addition to the corpus callosum.

On the other hand, particularly as related to semantic and perceptual errors, correct and efficient visual, tactile, and linguistic identification may require the dual and shared (cooperative) activity of both cerebral hemispheres. Hence, processing deficits within each hemisphere may due to an inability to gain complete access to perceptual activity occurring in the other. Assuming the mature callosum does not merely relay information but acts to mediate interhemispheric information integration by allowing for the participation and unique contribution of each hemisphere to the analysis of information, the immature callosum may in effect prevent or decrease each hemispheres, ability to contribute to and integrate the analysis being performed by the other. As such, unilateral processing may suffer as much as operations requiring transfer from one hemisphere to the other.

14. SPLIT-BRAIN COGNITION IN CHILDREN

Human psychological development and the expression of certain cognitive abilities parallels cerebral development and the myelination of the corpus callosum (Gallagher & Joseph, 1982; Joseph, 1982; Joseph, Gallagher, Holloway, & Kahn, 1984). Joseph et al. (1984) have provided evidence that indicates that information processing in certain modalities may be deficient in young children due to immaturity of the callosum as well as the inferior parietal lobule, a brain area intimately involved in cross-modal (auditory, visual, somesthetic) assimilation (Joseph, 1982). In consequence, information processing and the verbal expression and description of operations performed in other brain regions are often incomplete and/or erroneous due to incomplete transfer (e.g., from the right to left hemisphere). There is also evidence that strongly suggests that the immature corpus callosum, a pathway that takes over 10 years to mature, may be partially responsible not only for deficits in the interhemispheric transfer of learning, but also the failure of pre-operational children to demonstrate knowledge of operations involved in the conservation of liquid volume. Neural fiber pathway development and gross increments in brain tissue volume also appear to directly related to learning ability and the stages of cognitive development as defined by Piaget.

For example, conservation of volume is the ability to realise that two different containers may have the same volume, even though the containers are of a different shape. Specifically, Piaget demonstrated that if you show a child two beakers of water, one of which is short and fat, the other tall and thin, and ask the child which beaker contains more water, children under the age of 7 will generally choose or say 'the tall one', even though both contain the same amount of water.

According to Piaget (1952, 1962, 1974), prior to the development of the understanding of conservation, the thought of the child is guided by "appearances" rather than logical thought or deductive reasoning. That is, appearances (or perception) dominates, such that, for example, if an object such as a ball of malleable clay is stretched out and elongated, the child may believe that the elongated clay is greater in size and has more volume than the ball of clay.

Piaget argued that children fail to "conserve" and are fooled by the differences in the size of the container, because children younger than 7 have not developed the cognitive and symbolic ability to conserve volume or to engage in reversible operations (to think back and make mental comparisons).

In order to engage in conservation, the child, therefore, must become capable of engaging in though patterns that allow for "reverse" thinking; i.e. to remember that the volumes of water were the same when in the same size contains, and that the ball of clay contains the same amount of clay despite changed in its shape.

Once a child reaches the age of 7 or 8, they become increasingly capable of performing reversible operations. Thus, if you pour water from the short wide glass into the tall, thin glass, the child realize there is the same amount of water as before, despite the illusion of appearances.

However, the nature of Piaget's conservations tasks almost always requires that decision making take place within the domain of language; the child is "asked" to engage in "verbal reasoning" and to indicate his or her understanding through language; that is, what is examined is verbal thought, and only those reasoning processes that can be communicated through language. Perception and reasoning, however, can also take place through the non-verbal domain (Joseph, 1982, 1986b, 1988a,b, 2010). Depending on the task, correct analysis and the demonstration of cognition may require the exchange of information between the right and left hemisphere; the left being dominant for language. However, if the left hemisphere is unable to give an accurate account of operations performed by the right until the callosum is sufficiently matured, it is likely that higher-order cognitive abilities mediated by the right actually may appear earlier than is suspected. That is, children have a non-linguistic knowledge which cannot be communicated verbally, as this information is processed and stored in the right half of the brain, but cannot be communicated to the language dominant left hemisphere.

For example, Kraft et al. (1980), recorded EEGs from the right and left parietal area of children 6 to 8, and found during the performance of a Piagetian conservation task that those who were able successfully to perform conservation operations responded with greater right hemisphere involvement during the response phase than those who could not. Gallagher and Joseph (1982) also found that non-conserving children engage the left cerebral hemisphere more than the right and more so than adults.

When coupled with the data presented by Gallagher and Joseph (1982) and colleagues (Joseph et al., 1984), it appears that young children fail to demonstrate successfully these operations because the left hemisphere, which is required to provide a response in the absence of right hemisphere input, confabulates an explanation based on current and externally available information. This strengthens the supposition that young children are in fact able to adequately process and perform certain operations, but are unable to give accurate verbal descriptions of their (the right hemisphere's) conclusions.

The right hemisphere is dominant for all aspects of emotion (Joseph 1982, 1988a, 1996). Interhemispheric information transfer from the right to the left hemisphere is also facilitated when questions and/or stimuli are couched in an emotionally arousing or reinforcing context even when the hemisphere are disconnected (Gazzaniga & LeDoux, 1978; Geschwind, 1965).

Hence, although the right and left hemispheres of children are partially disconnected due to corpus callosum immaturity, it is nevertheless possible to demonstrate that these children in fact posses knowledge concerning equivalence relationships and have the ability to perform internal reversals, even though they are in the "pre-operational-nonconserving" stage of childhood (Gallagher & Joseph 1982; Joseph & Gallagher 1985).

Cognition Experiment: Subjects, Materials & Procedures. Sixty-six children 4 and 5 years of age were assigned to one of three groups, with all groups approximately balanced for number and sex. Materials used for testing conservation included a small flask that contained colored water, two 6.5 x 5 cm cylindrical glass beakers, and one 19 x 2.5 cm glass cylinder.

Immediately prior to the experiment all participants were given a preliminary test to determine whether they were conservers. Each child was brought individually to the testing room and was seated at the table facing the experimenter. The child was asked to take the flask and pour equal amounts of colored water into the two short glasses. In all cases the experimenter would, if necessary, ensure equivalence by adding an appropriate amount. The experimenter then asked the child to take one of the beakers and pour its contents into the tall beaker. The child then was asked, 'Do these glasses both have the same amount, or does one have more?' One child responded correctly and was excused from further testing. Those who indicate that the taller beaker had more water were classified as non-conservers and then were tested according to one of the three conditions described below.

Condition 1. The experimenter asked the child to pour the liquid from the tall glass back into the short beaker used during the pretest. The child was asked whether the two now held the same amount or whether one had more. When the child confirmed that the two quantities were the same, he or she was asked to pour a little liquid from one short beaker into the other, so that one obviously would contain more. The child then was instructed to pour one of the short glasses (indicating the one with less liquid in it) into the tall beaker and was asked a relational question, 'which glass has more?' The experimenter then asked a question that required an empirical reversal of the presently perceived context, 'Suppose we were to pour it back into the other glass again. Which of the two (short glasses) would have more?' After responding, the child was asked actually to perform this reversal and to compare the quantities. He then was told to return the smaller amount of liquid to the tall cylinder, and the experimenter asked the child to imagine the stimulus in a reinforcing context and to indicate a preference: 'Let's pretend that you are very thirsty and that this is your favorite drink. Which one would you rather have?' These final three questions, relational, reversal, and preferences, were recorded as correct or incorrect as determined by the child's selection of the vessel that did indeed contain the greater volume of liquid.

Condition 2. Procedures for Condition 2 were exactly the same as for Condition 1 except that the reversal situation was eliminated. Thus the relational question was followed immediately by the preference question.

Condition 3. Condition 3 was identical to 1 except that the order of the first and third questions was reversed, so that the preference question was asked first and the one that required a specific relational judgment was asked last.

RESULTS The results were quite clear and supported the hypothesis. Chi square analyses were applied to the data, with correction factors computed when necessary. The preference question was answered correctly with significantly greater frequency than was the relational question, $x2 = 9.98, p < .01$. The reversal question also was answered correctly significantly more often than the relational question, $x2 = 32.71, p < .005$, and the preference question, $x2 = 7.87, p < .025$. No significant sex differences were found.

15. DISCUSSION: RIGHT HEMISPHERE COGNITION IN NON-CONSERVING CHILDREN

The results demonstrate that non-conserving children maintain an awareness of equivalence relationships, but are more likely to conserve when the situation involves imagining a rewarding experience, as opposed to merely verbally describing a currently perceived static relation between substances. Coupled with the data provided by Gallagher and Joseph (1982) this evidence demonstrates that the left hemisphere of non-conserving children does not have complete access to the results of right hemisphere cognitive processing. Thus the speaking left hemisphere appears ignorant of this information and provides incorrect answers, and these performance deficits can only be overcome when emotionally aroused which results in information transfer. In this regard, we believe that presenting the problem in the context of a reinforcer allowed for right hemisphere activation and, thus, inter-hemisphere transfer of this knowledge through the immature callosal fibers that connect the two cerebral hemispheres. A similar mode of inter-hemispheric information transfer in a split brain patient has been presented by Joseph et al., (1984).

As noted, a significant number of non-conserving children were capable of performing the empirical reversal and preference inversion problems. However, by directly linking the preference and relational questions to the presently perceived (albeit transformed) stimuli and requiring the children to address their responses accordingly, the possibility that these same children were basing their responses on knowledge that involved empirical return rather than a non-linguistic knowledge of inversion was controlled for. Although both reversal and preference were demonstrated significantly more often than knowledge of relational non-equivalence, the failure of many subjects to respond correctly on both question types mitigates any suggestion that these two processes also may be related in some way. Although both involved communication of knowledge that concerns conservation, the preference response was tied directly to an immediate perception of the stimulus array, whereas the reversal involved responding to the experimenter's questions with regard to a previously perceived or expected stimulus context. Hence, we are presented with two fundamentally different types of knowledge, both of which appear separately yet exist simultaneously in a form that appears to resist integration in children at this age level.

This evidence is consistent with the view that young children are functioning as though they had split brains and suggests that non-conserving children fail to express knowledge associated with the right hemisphere because they appear to be relying on a left hemisphere 'cognitive' strategy due to the linguistic referents traditionally involved with the task. Hence, when understood in terms of functional asymmetry and interhemispheric disconnection, the statement that children do not comprehend equivalence relationship loses its meaning.

16. SPLIT-BRAIN FUNCTIONING IN CHILDREN: THE ONTOLOGY OF EMOTIONAL CONFLICT

Thus, due in part to the slow pace of corpus callosum myelination, coupled with differential right and left cerebral specialization, the left hemisphere of a young child has at best incomplete knowledge of the contents and activity that are occurring within the right. This sets the stage for differential memory storage and a later inability to transfer this information between the cerebral hemispheres once the child reaches adulthood.

Because of lateralization and limited exchange, the effects of early "socializing" experience can have potentially profound effects. "As a good deal of this early experience is likely to have unpleasant if not traumatic moments, it is fascinating to consider the later ramifications of early emotional learning occurring in the right hemisphere unbeknownst to the left; learning and associated emotional responding that later may be completely inaccessibile to the language centers of the left half of the brain. That is, although limited transfer in children confers advantages, it also provides for the eventual development of a number of very significant psychic conflicts --many of which do not become apparent until much later in life."

Moreover, due to the immaturity of the callosum, children frequently can encounter situations in which the right and left cerebrum not only differentially perceive what is going on, but are unable to link these experiences so as to understand fully what is occurring or to correct misperceptions (Joseph, 1982, 1992b). Consider, for example, a young divorced mother with ambivalent feelings toward her young son (Joseph, 1988a, 1992b).

Although she does not express these feelings verbally, she conveys them through her tone of voice, facial expression, and in the manner in which she touches her son. She knows that she should love him, and at some level she does. She wants to be a good mother and makes herself go through the motions. However, she also resents her son because she has lost her freedom, he is a financial burden, and he may hinder her in finding a desirable mate. She is confronted by two opposing attitudes, one of which is unacceptable to the image she has of a good mother. Like many of us, she must prevent these feelings from reaching linguistic consciousness. However, this does not prevent them from being expressed via the limbic system or right hemisphere.

Her son, of course, also has a right hemisphere which perceives her tension and ambivalence. The right half of his brain notes the stiffness when his mother holds or touches him and is aware of the manner in which she sometimes looks at him. Worse, when she says, "I love you," his right hemisphere senses the tension and tone of her voice and correctly perceives that what she means is, "I don't want you" or, "I hate you. " His left hemisphere hears, however, "I love you" and notes only that she is attentive. He is in a "double bind" conflict, with no way for his two cerebral hemispheres to match impressions.

The right half of this little boy's brain feels something painful when the words "I love you" are spoken. When his mother touches him, he becomes stiff and withdrawn because his right hemisphere, via the analysis of facial expression, emotional tone, tactile sensation, etc. is fully aware that she does not want him.

Later, as an adult, this same young man has one failed relationship after another. He feels that he can't trust women, often feels rejected, and when a girl or woman says "I love you," it makes him want to cringe, run away, or strike out. As an adult, his left hemisphere hears "Love," and his right cerebrum feels pain and rejection.

Because the two halves of his cerebrum were not in communication during early childhood, his ability to gain insight into the source of his problems is greatly restricted. The left half of his brain cannot access these memories. It has "no idea" as to the cause of his conflicts.

In this regard, this asymmetrical arrangement of hemispheric function and maturation may well predispose the developing child in later life to come upon situations in which it finds itself responding emotionally, nervously, anxiously, or neurotically, without linguistic knowledge, or without even the possibility of linguistic comprehension as to the cause, purpose, eliciting stimulus, or origin of its behavior. As a child or an adult, it may find itself faced with behavior that is mysterious, embarrassing, etc. "I don't know what came over me."

17. FUNCTIONAL COMMISSUROTOMIES AND LIMITIED INTERHEMISPHERIC TRANSFER

The human brain is organized so that two potentially independent mental systems coexist, literally side by side. Each hemisphere has, in the course of evolution, developed its own unique strategy and capability for processing information, and a neuroanatomical substrate to subserve each independent function. Therefore each hemisphere or mental system may act independently on specific information, initially without interference from the other. A result of this duality is the dual processing of sensory and cognitive information. The illusion of mental unity is made possible, in part by the corpus callosum, which both aids and inhibits interhemispheric transfer.

Not only are these systems dual, but they are asymmetrical as each hemisphere utilizes predominantly verbal-analytical or visual-spatial, sensory-affective associational strategies in the analysis and expression of information. When one hemisphere learns, thinks, or 'experiences,' the information can be transferred via the corpus callosum for further analysis or so that the other hemisphere also may 'learn,' being a beneficiary of the other's activity. Although admittedly there is considerable overlap of function, in that both hemispheres may act to analyze all input, and given that some types of information are dealt with more efficiently by one than the other, in the long run this relationship is very adaptive as the range and speed of information analysis is enlarged and the efficiency of response is increased; i.e. every individual has two brains and two minds, in the same head.

However, in that the left codes, processes, and stores information using linguistic, categorical, functional, and motoric referents, whereas the right relies on spatial, emotional, and non-linguistic modes of understanding, complete and efficient interhemispheric communication may not always be possible. Thus there is the potential for much miscommunication, misinterpretation, and distortion regardless of the direction of transfer.

The corpus callosum is the gateway via which information may travel from one brain half to the other. However, it also acts to limit information exchange even in the adult, intact, brain, since almost 40% of the adult callosum lacks myelin (Selnes, 1974). Since myelin acts to insulate and, thus, preserve information transmission by minimizing leakage and increasingly conduction velocity and integrity, some information is lost and degraded even when transfer is possible (Berlucchi & Rizzolatti, 1968; Hicks 1974; Joseph et al. 1984; Marzi, 1986; Merriam & Gardner, 1987; Myers, 1959, 1962; Rizzolatti et al. 1971; Taylor & Heilman 1980).

Moreover, particularly when one is dealing with complex or emotional information situations often arise in which one brain half has little or no knowledge as to what is occurring in the other (Joseph, 1982, 1988a, 1992b, 2010; Joseph et al. 1984). In part, this is a consequence of lateralized specialization. Certain forms of information can only be processed and recognized by the right or left half of the brain, but not both and in not the same manner. Even information that is transferred may be subject to interpretation and miss-interpretation (Joseph, 1982, 1986ab, 1988ab; Joseph et al., 1984). When information or knowledge which is stored or processed in the language of the right hemisphere is then transferred to the left, that information may not be translated correctly, and gaps in the translation may be filled with confabulatory ideation.

In addition, one brain half can be prevented from knowing what is occurring in the opposite half due to inhibitory or suppressive actions initiated by the frontal lobes, such that certain forms of information are suppressed, censored, and inter-hemispheric (as well as intra-hemispheric) information transmission prevented (Joseph, 1982, 1988a, 1992b, 2011). Even the normal intact adult brain may experience functional commissurotomies.

Therefore, these three conditions --lateralized specialization, frontal lobe inhibitory activity, and incomplete myelination of callosum axons-- can reduce the ability of the two hemispheres to communicate among normal, intact individuals. Hence, in many ways the brain of even a normal adult is functionally split and disconnected, and for good reason. These conditions protect the brain and linguistic consciousness from becoming overwhelmed. As we have seen with frontal lobe damage, when communication is allowed to occur freely (due to disinhibition) the overall integrity of the brain to function normally is curtailed dramatically.

Nevertheless, a unique side effect of having two hemispheres that are not always able to communicate completely or accurately is intra-psychic conflicts. That is, we sometimes find ourselves feeling happy, sad, depressed, angry, etc. without a clue as to the cause. In other instances, we actually may commit certain thoughtless, impulsive, overly emotional, or embarassing actions and "have no idea" as to "what came over" us.

Moreover, unbeknownst to the left hemisphere, sometimes the right perceives, remembers, or responds to some external or internal source of experience and/or to its own memories and, thus, and reacts in an emotional manner.

The left (speaking) hemisphere in turn only knows that it is feeling something but is unsure what or why , or, conversely, confabulates various denials, rationalizations and explanations which it accepts as fact.

18. CONCLUDING COMMENTS

The two halve of the brain begin life functionally disconnected. Functional disconnection remains a fact of life, even in adulthood. This can give rise to any number of intra-psychic conflicts.

Although there is evidence of considerable functional overlap as well as interhemispheric cooperation on a number of tasks, the mental systems maintained by both the right and left cerebral hemispheres are highly developed. Whereas the consciousness of the left hemisphere is linked to language, math, temporal-sequencing, and control over the right hand, the mind of the right hemisphere is more social-emotional, bilateral, and in many ways dominant over the temporal-sequential, language-dependent half of the cerebrum. Indeed, the right cerebrum can independently recall and act on certain memories with purposeful intent; is the dominant source of our dreams, psychic conflicts and desires, and is fully capable of motivating, initiating as well as controlling behavioral expression --often without the aid or even active (reflective) participation of the left half of the brain.

We knowers are unknown to ourselves, and for good reason: how can we ever hope to find what we have never looked for? There is a sound adage which runs: 'Where a man's treasure lies, there lies his heart.' Our treasure lies in the beehives of our knowledge. We are perpetually on our way thither, being by nature winged insects and honey gatherers of the mind. The only thing that lies close to our heart is the desire to bring something home to the hive. The sad truth is that we don't understand our own substance, we must mistake ourselves; the axiom, 'Each man is farthest from himself,' we will for us to all eternity. Of ourselves we are not knowers. --Nietzsche

REFERENCES

Akelaitis, A. J. (1945). Studies on the corpus callosum. IV. Diagnostic dyspraxia in epileptics following partial and complete section of the corpus callosum. American Journal of Psychiatry, 101, 594-599.

Banquet,J. P. (1983).Inter- and intrahemispheric relationships of the EEG activity during sleep in man.Electroencephalography and Clinical Neurophysiology, 55, 51-59.

Berlucchi, G., & Rizolatti, G. (1968). Binocularly driven neurons in visual cortex of split-chiasm cats. Science, 159, 308-310.

Blinkov, S. M., & Glezer, I. I. (1968). The human brain in figures and tables. New York: Plenum.

Bogen, J. E. (1969). The other side of the brain. Bulletin of the Los Angeles Neurological Societies, 34, 135-162.

Bogen, J. E. (1979). The callosal syndrome. In K. M. Heilman & E. Valenstein (Eds.), Clinical neuropsychology.(pp. 308-358). New York: Oxford University Press.

Bures, J., & Buresova, O. (1960). The use of Leao's spreading depression in the study of interhemispheric transfer of memory traces. Journal of Comparative and Physiological Psychology, 59, 211-214.

Dimond, S. J. (1980). Neuropsychology. London: Butterworths.

Doty R. W., & Overman, W. H. (1977). Mnemonic role of forebrain commissures in macaques, In S. Harnad, R. W. Doty, L. Goldstein, J. Jaynes, & G. Krauthamer, (Eds.), Lateralization in the nervous system, (pp.75-88). New York: Academic Press.

Finlayson, M. A. J. (1975). A behavioral manifestation of the development of interhemispheric transfer of learning in children. Cortex, 12, 290-295.

Flechsig, de P. (1901). Developmental (mylogenetic) localisation of the cerebral cortex in the human subject. Lancet, 2, 1027-1031.

Galin, D,. Diamond, R., &. Herron, J.(1979). Development of crossed and uncrossed tactile localization on the fingers. Brain and Language, 4, 588-590.

Galin, D., Johnstone, J., Nakell, L., & Herron, J. (1979). Development of the capacity for tactile information transfer between hemispheres in normal children. Science, 204, 1330-1332.

Gallagher, R. E., & Joseph, R. (1982). Non-linguistic knowledge, hemispheric laterality, and the conservation of inequivalence. Journal of General Psychology, 107, 31-40.

Gazzaniga, M. S., & LeDoux, J. E. (1978). The integrated mind. New York: Plenum Press.

Geschwind, N. (1965). Disconnexion syndromes in animals and man. Brain, 88, 585-644.

Geschwind, N. (1981). The perverseness of the right hemisphere. Behavioral and Brain Sciences, 4, 106-107.

Goldstein, L., Stoltzfus, N. W., & Gardocki, J. F. (1972). Changes in interhemispheric amplitude relationships in the EEG during sleep. Physiology and Behavior, 8, 811-815.

Hasegawa, I., Fukushima, T., Ihara, T., & Miyahsita, Y. (1998). Callosal window between prefrontal cortices: Cognitive interaction to retrieve long-term memory. Science, 281, 814-818.

Hicks, R. E. (1974). Asymmetry of bilateral transfer. American Journal of Psychology, 87, 667-674.

Hodoba, D. (1986). Paradoxic sleep facilitation by interictal epileptic activity of right temporal origin. Biological Psychiatry, 21, 1267-1278.

Joseph, R. (1979). Effects of rearing and sex differences on learning and competitive exploration. Journal of Psychology, 101, 37-43.

Joseph, R. (1982). The Neuropsychology of Development: Hemispheric Laterality, Limbic Language, and the Origin of Thought. Journal of Clinical Psychology, 44, 3-34.

Joseph, R. (1986a). Confabulation and delusional denial: Frontal lobe and lateralized influences. Journal of Clinical Psychology, 42, 845-860.

Joseph, R. (1986b). Reversal of cerebral dominance for language and emotion in a corpus callosotomy patient. Journal of Neurology, Neurosurgery, and Psychiatry, 49, 628-634.

Joseph, R. (1988a) The Right Cerebral Hemisphere: Emotion, Music, Visual-Spatial Skills, Body Image, Dreams, and Awareness. Journal of Clinical Psychology, 44, 630-673.

Joseph, R. (1988b). Dual mental functioning in a split-brain patient. Journal of Clinical Psychology, 44, 770-779.

Joseph, R. (1990). Neuropsychology, Neuropsychiatry, Behavioral Neurology, Plenum, New York.

Joseph, R. (1992a) The Limbic System: Emotion, Laterality, and Unconscious Mind. The Psychoanalytic Review, 79, 405-456.

Joseph, R. (1992b). The Right Brain and the Unconscious. New York, Plenum.

Joseph, R. (1994) The limbic system and the foundations of emotional experience. In V. S. Ramachandran (Ed). Encyclopedia of Human Behavior. San Diego: Academic Press.

Joseph, R. (1996). Neuropsychiatry, Neuropsychology, Clinical Neuroscience, 2nd Edition. Williams & Wilkins, Baltimore.

Joseph, R. (1998). Traumatic amnesia, repression, and hippocampal injury due to corticosteroid and enkephalin secretion. Child Psychiatry and Human Development. 29, 169-186.

Joseph, R. (1999b). Environmental influences on neural plasticity, the limbic system, and emotional development and attachment, Child Psychiatry and Human Development. 29, 187-203.

Joseph, R. (1999a). The neurology of traumatic "dissociative" amnesia. Commentary and literature review. Child Abuse & Neglect. 23, 715-727.

Joseph, R. (1999c). Frontal lobe psychopathology: Mania, depression, aphasia, confabulation, catatonia, perseveration, obsessive compulsions, schizophrenia. journal of Psychiatry, 62, 138-172.

Joseph, R. (2000). Limbic language/language axis theory of speech. Behavioral and Brain Sciences. 23, 439-441.

Joseph, R. (2000). Fetal brain behavioral cognitive development. Developmental Review, 20, 81-98.

Joseph, R. (2000a). The evolution of sex differences in language, sexuality, and visual spatial skills. Archives of Sexual Behavior, 29, 35-66.

Joseph, R. (2003). Emotional Trauma and Childhood Amnesia. journal of Consciousness & Emotion, 4, 151-178.

Joseph, R. (2009). Quantum Physics and the Multiplicity of Mind: Split-Brains, Fragmented Minds, Dissociation, Quantum Consciousness. Journal of Cosmology, 2009, 3, 600-640.

Joseph, R. (2011). Origins of Thought: Consciousness, Language, Egocentric Speech and the Multiplicity of Mind. Journal of Cosmology, 14. 4577-4600.

Konner, M. (1991). Universals of behavioral development in relation to brain myelination. In K. R. Gibson & A. C. Petersen (Eds). Brain maturation and cognitive development. New York, Aldine De Gruyter.

Kraft, R. H., Mitchell, O. R., Languis, M. L., & Wheatley, G. H. (1980). Hemispheric asymmetries during six-to-eight year-olds' performance on Piagetian conservation and reading tasks. Neuropsychologia, 18, 637-644.

Kurcharski, D., & Hall, W. G. (1987). New routes to early memories. Science, 238, 786-788.

LeCours, A. R. (1975). Myelogenetic correlates of the development of speech and language. In E. Lenneberg & E. Lenneberg (Eds.), Foundations of language development. New York: Academic Press.

Levy, J. (1974). Psychological implications of bilateral asymmetry. In S. Dimond & J. G. Beaumont (Eds.), Hemisphere function in the human brain, (pp.121-183). London: Paul Elek.

Levy, J. (1983). Language, cognition and the right hemisphere. American Psychologist, 38, 538-541.

Marzi, C. A. (1986). Transfer of visual information after unilateral input to the brain. Brain and Cognition, 5, 163-173.

Merriam, A. E., & Gardner, E. B. (1987). Corpus callosum function in schizophrenia: A neuropsychological assessment of interhemispheric information processing. Neuropsychologia, 25, 185-193.

Meyer, J. S., Ishikawa, Y., Hata, T., & Karacan, I. (1987). Cerebral blood flow in normal and abnormal sleep and dreaming. Brain and Cognition, 6, 266-294.

Myers, R. E. (1959). Interhemispheric communication through the corpus callosum: Limitations under conditions of conflict. Journal of Comparative and Physiological Psychology, 52, 6-9.

Myers, R. E., (1962). Transmission of visual information within and between the hemispheres: A behavioral study. In V. B, Mountcastle (Ed). Interhemispheric Relations and Cerebral Dominance. Baltimore, Johns Hopkins Press.

O'Leary, D. S. (1980). A developmental study of interhemispheric transfer in children aged 5 to 10. Child Development, 51, 743-750.

Piaget, J. (1952). The origins of intelligence in children. New York: Norton.

Piaget, J. (1962). Play, dreams and imitation in childhood. New York: Norton.

Piaget, J. (1974). The child and reality. New York: Viking Press.

Risse, G. L., & Gazzaniga, M. S. (1979). Well-kept secrets of the right hemisphere: A carotid amytal study of restricted memory transfer. Neurology, 28, 950-953.

Ritchie, J. M. (1984). Pathophysiology of conduction in demyelinated nerve fibers. In P., Morell (Ed.), Myelin, 2nd Ed. New York, Plenum.

Rizzolatti, G., Umilta, C., & Berlucchi, G. (1971). Opposite superiorities of the right and left cerebral hemispheres in discriminative reaction time to physiognomical and alphabetical material. Brain, 94, 431-442.

Salamy, A. (1978). Commissural transmission: Maturational changes in humans. Science, 200, 1409-1411.

Selnes, O. A. (1974). The corpus callosum: Some anatomical and functional considerations with special reference to language. Brain and Language, 1, 111-139.

Seymour, S. E., Reuter-Lorenz, P. A., & Gazzaniga, M. S. (1994). The disconnection syndrome. Brain, 117, 105-115.

Sperry, R. (1966). Brain bisection and the neurology of consciousness. In J. C. Eccles (Ed). Brain and conscious experience, (pp 298-313). New York: Springer Verlag.

Sperry, R. W. (1968). Hemisphere disconnection and unity in conscious awareness. American Psychologist. 23, 723-733.

Sperry. R. (1974). Lateral specialization in the surgically separated hemispheres. In F. O. Schmitt & F. G. Worlden (Eds.), The neurosciences: Third study program, (pp 1-12). Cambridge. MIT Press.

Sperry, R. (1982). Some effects of disconnecting the cerebral hemispheres. Science, 217, 1223-1226.

Sperry, R. W., Zaidel, E., & Zaidel, D. (1979). Self recognition and social awareness in the disconnected minor hemisphere. Neuropsychologia, 17, 153-166.

Sweet, W. H. (1945). Intracranial aneurysm simulating neoplasm. Syndrome of the corpus callosum . Archives of Neurology and Psychiatry, 45, 86-103.

Taylor, G. H. & Heilman, K. (1980). Left-hemispheric motor dominance in right handers. Cortex, 16, 587-603.

Wolpert E. A., & Trosman, H. (1958). Studies in psychophysiology of dreams I. Experimental evocation of sequential dream episodes. Archives of Neurology, 79, 603-606.

Yakovlev, P. I., & Lecours, A. (1967). The myelogenetic cycles of regional maturation of the brain,. In A. Minkowski (Ed.), Regional development of the brain in early life, (pp.404-491). London: Blackwell.

9. Hunters, Gatherers, and the Evolution of Sex Differences in Language, Cognition, and Consciousness

Sex Differences Are Pervasive One need only gaze at the human male vs female body to be presented with obviously profound differences between the sexes. Female's are less muscular, their hips are wider, and they have a vagina, uterus, fallopian tubes, a menstrual cycle, and can become pregnant and have babies, whereas, on average, males are taller, have greater upper body strength, facial hair, a penis, and are decidedly more aggressive and violent. Human males and females differ genetically, biologically, physically, hormonally, neurologically, psychologically and cognitively including in the realms of language and visual-spatial skills; all of which contribute to sex differences in consciousness.

Consciousness & Language Consciousness, like cognition, is a multiplicity of mind, one aspect of which relies upon language (Joseph 2010, 1982, 2009, 2011a). Thus we often think-in-words. Thinking-in-words is language internalized, one area of the brain and mind producing these word-thoughts, another region comprehending them. These language-related tissues of the mind correspond to Broca's expressive speech area which serves as the final common pathway where thoughts are organized and expressed or experienced as words, and Wernicke's receptive speech area which comprehends these word-thoughts, be they spoken aloud or experienced in the privacy of one's own head (Joseph 2010, 1982, 1988, 20011a; Joseph et al., 1984).

Linking Wernicke's and Broca's areas is the angular gyrus of the inferior parietal lobule which contribute to language comprehension and expression and the assimilation of multi-modal concepts from multiple modalities which enables the mind to create multiple multi-sensory categories for different words. Thus, for example, the word mustang can be conceived of as a horse, different colors of a horse, a horse running or galloping, or as a car of varying shapes, colors, and sizes.

As human language, by definition, is unique to humans, the evolution of language, and thus those aspects of the brain and mind which produce and comprehend language, have given rise to a form of consciousness which is also unique to humans, i.e. human consciousness. However, not only is human consciousness a multiplicity with different mental realms concerned with different aspects of conscious experience, but the multiplicity of the mind and thus human consciousness and cognition differs depending on if one is a man or a woman and which areas of the brain is being used to support or provide content to consciousness. In fact, non-human animal consciousness and cognition also differ depending on if that animal is male or female; and these gender differences, at least in the realm of language, have become more pronounced with the evolution of humans.

Sex Differences in Other Species There have been numerous research reports examining the biological, psychological, cultural, and political influences which are presumed to influence and shape sex differences in cognition, emotion, language, and human sexual behavior. Certainly, these and other environmental factors can have a significant impact on the brain and mind (Joseph 1982, 1992a, 1996, 1998, 1999ab). However, some of the same sex-specific cognitive traits and sex differences in sexual behavior, are demonstrated by other species (Joseph, 1979, Joseph et al. 1978; Joseph & Gallagher 2010) which evolved long before humans (Joseph 1992a,b, 1993, 1996, 1999a). These later findings demonstrate that many sex differences are biological. However, some of these sex differences became exaggerated in humans following the emergence of language, right hand dominance, and the capacity to create and utilize complex tools.

Language & Right Handedness Language is directly related to human handedness, which is explains why the areas subserving the hand and mouth are adjacent to each other in the frontal-parietal sensory motor areas, and why most humans tend to move the right hand when speaking and use the right hand for writing. However, although the ability to vocalize is characteristic of almost all animal species, handedness, it is only humans who predominantly show right hand dominance; in among approximately 90% of the population. Human language is also grammatical, and organized, expressed and comprehended in a rapid, temporal sequential fashion. As language and handedness are both subserved by adjacent areas in the left hemisphere of the human brain, it thus appears that the evolution of language is directly related to handedness and the ability to engage in rapid, temporal sequential fine motor movements of the hands and fingers all of which are associated with the functional integrity of the left hemisphere in the majority of the population. In consequence, language is temporal-sequential and grammatical, most people move their right hand and arm when talking, and use the right hand for writing (Joseph 1982, 1988a, 1990a, 1996a,b, 2000a).

These capacities and cognitive abilities are related to ancestral domestic activities traditionally associated with women, such as gathering, food preparation, the fashioning and cleaning of clothing, and prolonged child care which promotes mother-infant mutual vocalization (Joseph 1992, 1993, 1996, 2000a,b). Since the environment acts on gene selection and neuroplasticity over the course of human evolution, as these new abilities developed, the motor areas and neocortex of the left hemisphere of both males and females became functionally reorganized. Older cognitive functions associated with the left hemisphere were crowded out and displaced due to competition for synaptic space, and the left hemisphere became dominant for handedness and language in both sexes. However, as tasks traditionally associated with females not only promoted speech and language development but bi-manual fine motor activity of both hands, the female brain became bilaterally organized for language and fine motor tasks.

SEX DIFFERENCES IN LANGUAGE & VISUAL SPATIAL SKILLS

It has been consistently demonstrated that human females demonstrate superiorities on tasks involving fine motor movements of the fingers and hands (Broverman et al., 1968; Hampson and Kimura, 1992; Junaid and Fellowes 2006; Larson et al. 2007; Thomas and French 1985). Female superiorities have also been repeatedly demonstrated across a variety of verbal-measures of language ability (Deary, 2007; Johnson & Bouchard 2007; Strand et al. 2006) including and especially reading and writing (Bae et al., 2000; Hedges and Nowell 1995; Willingham and Cole 1997; Strand et al. 2006). The female advantage in verbal abilities is cross cultural, international, and is evident by age 2 (Mullis et al., 2003; National Center for Education Statistics, 2002; Ogle et al., 2003), whereas the greater propensity for females to engage in rapid movements of the upper extremities is evident even in rats (Joseph & Gallagher, 2010).

These sex differences, which are biological and evident in other species (Joseph 1992, 1993, 2000a,b; Joseph et al 1978; Joseph & Gallagher 2010), helps explain why human females make up the the majority of typists, secretaries, telephone operators (Jacob, 1999; Reskin 1993) , and other professions which require good verbal abilities and which are stereotypically associated with women, and why many females preferentially seek these types of jobs (Huffman and Torres, 2001; Mencken and Winfield 2000; Torres and Huffman 2002). Over the course of history, "gatherers" have been traditionally female (Dahlberg, 2011; Martin and Voorhies, 1975; Murdock and Provost, 1973; Zilman, 2011), and use both hands for gathering (the right more than the left), and are likely to vocalize with other females and with their children and infants (Joseph 1992, 1993, 2000a,b).

Language & Competition for Synaptic Space As tasks traditionally associated with females not only promoted speech and language development but bi-manual fine motor activity of both hands the female brain is bilaterally organized for language (Shaywitz, et al., 1995) and for performing fine motor tasks (Broverman et al., 1968; Hampson and Kimura, 1992). In consequence, ancestral visual-spatial functions associated with the right and left hemisphere were crowded out of the female brain and diminished due to competition for synaptic space. The same is not true for males. Thus, by contrast, males demonstrate inferior verbal skills and superior visual spatial and gross motor skills and this dominance also became more pronounced over the course of human evolution.

Competition for synaptic representation in the primate brain was first demonstrated in 1978 (Casagrande & Joseph 1978, 2010; Joseph & Casagrande 1978, 2010). Specifically, primates were reared with one eye-lid sutured shut, allowing only non-pattern illumination to be transmitted from the effect eye to the lateral geniculate nucleus of the thalamus (which subserved vision) and the visual neocortex. When the sutures were removed, all animals were blind in the effected eye, and synaptic space normally occupied by nerves from that eye had been hijacked by neurons from the normal eye. Neurons from the eye which was not being used, had lost the competition for synaptic space. However, when the normal eye was removed, the struggle for synaptic representation was reversed, and the formally deprived eye reestablished synaptic connections and regained vision (Casagrande & Joseph 1978, 2010; Joseph & Casagrande 1978, 2010).

Likewise, it was subsequently determined that following brain damage in humans, and the severing of axonal-neuronal interconnections, that normal neurons closest to the brain areas which had been vacated established synaptic connections with these tissues. For example, patients who had suffered parietal lobe injuries--an area of the brain which subserves the body image--experienced altered synaptic representation, such that if the hand or arm were stimulated, they instead experienced the sensation on another body part whose neurons took over the temporarily vacated synaptic space (Joseph 1986a, 1988a, 1990, 1996). Thus, there is a continual struggle and competition for synaptic representation in the human brain. Therefore, over the course neuronal representation for various cognitive functions also came to be altered, such that ancestral sex differences in cognition came to be exaggerated and other functions displaced with the evolution of handedness and language.

Hunting & Spatial Abilities Over the course of human evolutionary history, males were the hunters. Hunters require good visual spatial directional skills so as to find their way over long distances. A good hunter must be able to kill with accuracy by throwing spears or firing arrows at an animal which may be running or standing some distance away. Moreover, hunting does not promote nor does it require language as is evident in all species which hunt. Among humans, a talkative hunter will scare off all game. Thus we see that human males out-perform females on visual-spatial tasks (Hugdahl et al., 2006; Irwing & Lynn 2005; Johnson & Bouchard 2007) including accuracy in throwing (Joseph 1992a, 1993).

Male superiorities on spatial measures are evident by age 5 (Levine et al., 1999), and include mental rotation, 3-dimensional thinking, maze learning, and the understanding of maps and directional abilities (Brown et al., 1998; Cutmore et al., 2000; Loring-Meier & Halpern, 1999; Masters & Sanders, 1993; Nordvik & Amponsah, 1998). Male spatial learning superiorities been also demonstrated in other species including rats (Joseph 1979, Joseph & Gallagher, 2010; Joseph et al., 1978).

COGNITION AND SEX DIFFERENCES IN OTHER SPECIES

The fact that these sex differences are evident in other species and can be reversed by hormonal manipulations (Joseph 1979, Joseph & Gallagher, 2010; Joseph et al., 1978), completely disproves the conspiracy theorists who argue that nurture, rearing habits, sex-stereotyping, and discrimination are the responsible agents.

Specifically, Joseph (1979; Joseph & Gallagher, 2010) demonstrated when rats were raised in an enriched environment vs a deprived and restricted environment, that male rats consistently demonstrated a highly significant superiority over females in spatial maze learning. Further, even when all animals learned the spatial tasks, males demonstrated superior retention of this information and superior long-term memory for spatial details. Thus, when provided identical experiential opportunities beginning in infancy, and regardless of these environments, these sex differences remain including a propensity for females to persist in making rapid movements of the upper extremities (Joseph & Gallagher, 2010).

This also explains why, despite the fact that both sexes have experience with drinking from glass containers, males outperform females when asked to specify the horizontal water level in a tilted glass (Morris, 1971; Thomas et al., 1973). That is, females tend to erroneously draw the water as tilted when the glass is tilted, whereas males will draw the water as horizontal.

Male rats are also significantly more exploratory and show a greater propensity to investigate their environment even when both males and females are raised in an enriched environment (Joseph & Gallagher, 2010). In fact, human male children also demonstrate superior spatial abilities, and are also far more likely than females to explore a strange environment.

Moreover, Joseph et al. (1978) found that these sex differences could be reversed, biologically, through the administration of the "male" hormone, testosterone, to females, or by preventing the release of testosterone in males, during the time period when the brain is sexually differentiating. For example, female rats exposed to testosterone during early brain development when the brain is sexually differentiating, perform similarly to males on maze learning tasks, whereas castrated-infant males later perform similarly to females (Joseph et al., 1978). Likewise, homosexual men (whose brains, in some cases, may have been feminized prenatally, e.g. Meyer-Bahlburg, 1993) tend to perform more poorly than heterosexual men on spatial tasks and similar to females on verbal tasks (Gladue et al., 1990; however see Gladue & Baily, 1995). Moreover, human and non-human females exposed to high levels of masculinizing hormones and androgens during early brain development demonstrate superior visual-spatial skills, and are more competitive and aggressive as compared to normal females (Money and Ehrhardt, 1972; Ehrhardt and Baker, 1974; Joseph et al., 1978; Reinisch and Sanders, 1992; Mitchell, 1979). Hunting and the killing of prey is related to visual-spatial perceptual functioning and the tendency to behave in an aggressive fashion.

Sex Differences in Brain & Behavior Clearly these and other sex differences are biological as they are present in early childhood and in species who evolved before humans. Further, sex differences have been well documented in the structure and organization of the human and non-human brain, the limbic system in particular (Allen et al. 1989; Bleier et al. 1982; Bubenik & Brown, 1973; Filipek, et al., 1994; Nishizuka & Arai, 2011; Rainbow et al. 1982; Raisman & Field, 1971, 1973), and including those tissues of the mind which mediate male vs female sexuality, emotion, language, and spatial skills (Joseph et al., 1978; Hart et al., 1985; Lisk, 2007, 1971; Maclean, 1973; Robinson and Mishkin, 1968; Stoffels et al., 2010). For example, because a heterosexual male possess a "male" brain, he responds with sexual arousal to the female body, is more aggressive and violent, and demonstrates superior visual-spatial skills. By contrast, because the heterosexual female possesses a brain organized in the "female" pattern, she demonstrates superior language skills, is more social and emotional and nurturing, and desires to have babies and to hold and talk to babies. The brain is part of the body and it is precisely because she has a "female" brain and body that women have a menstrual cycle, become pregnant, have babies, and nurse those babies.

Moreover, it is precisely because human males are more aggressive and physically violent (which makes them good hunters) whereas females are biologically designed to produce and nurture helpless big brain babies and to talk to these babies while engaged in domestics tasks (such as food preparation, gathering) which can be performed close to the home base, that some of the sex differences apparent in other species have become exaggerated over the course of human evolution.

Evolution of Hunters & Gatherers As first proposed and theorized by Joseph in 1992, sex difference in spatial abilities and vocalization became exaggerated in humans with the evolution of "the big brain" and full-time sexual receptivity in females during the later stages of Homo erectus evolution, followed by the emergence of fine motor skills and language with the evolution of Upper Paleolithic humans, e.g. the Cro-Magnon. Although the evolutionary history of human sex differences can be traced back over 5 million years and to other species (Joseph 2000a,b 2011) the great divide in male vs female cognition and consciousness may have had had its onset about 500,000 years ago and began to accelerate around 35,000 years ago with the evolution of modern human language, complex tool use and construction, and a full time division of labor between the sexes. This "hunter-gatherer" theory of sex differences in behavior, language, cognition, and consciousness has subsequently been adopted by other scientists.

HOMO ERECTUS: SEXUALITY AND THE HOME BASE

There is no general agreement as to the various phylogenetic relationships shared by the wide variety of Plio-pleistocene hominids so far discovered, and it is not yet established if present-day humans descended from Australopithecus, Homo habilis, or both or neither. Nevertheless, following and overlapping with H. habilis and Australopithecus were a wide range of quite different individuals collectively referred to as Homo erectus.

Homo erectus were big and robust, with thick browridges, large teeth, and bulging shoulder muscles (Day, 1996; Potts, 1996; Rightmire, 1990). This diverse species of humanity ranged throughout Africa, Europe, Russia, Indonesia and China from approximately 1.9 million until about 300,000 years ago, with a few isolated populations possibly hanging on in the island of Java, until 27,000 years B.P. (Day, 1996; Potts, 1984, 1996; Rightmire, 1990; Swisher et al., 1996). Presumably, H. erectus is the common ancestor for Neanderthals, the Cro-Magnon, and modern humans. However, it is just as likely that modern humans have descended from different branches of the species referred to collectively as "H. erectus".

About 1.5 million years ago H. erectus learned to harness fire and by 500,000 B.P., the first hearths began to appear in China, France, Hungary and elsewhere (Clark & Harris 1985; Isacc, 1982; Rightmire, 1990). By 500,000 B.P. H. erectus was regularly constructing crude shelters and home bases (Clark and Harris 1985; Potts 1984, 1996; Rightmire, 1990; Zhang, 1985); achievements that may have coincided with a major change in female sexuality.

BREASTS, BUTTOCKS & COSMETICS

About 1.5 million years ago H. erectus learned to harness fire and by 500,000 B.P., the first hearths began to appear in China, France, Hungary and elsewhere (Clark & Harris 1985; Isacc, 1982; Rightmire, 1990). By 500,000 B.P. H. erectus was regularly constructing crude shelters and home bases (Clark and Harris 1985; Potts 1984, 1996; Rightmire, 1990; Zhang, 1985); achievements that coincided with a major change in female sexuality (Joseph 2000a,b, 2011).

For example, during the evolutionary transition which led from Australopithecus and H. erectus and full time bipedalism, the vaginal canal underwent a reorientation from back to front, which enabled males and females to face one another during sexual intercourse, thus promoting interpersonal intimacy. With the exception of the Bonobo who are more variable, all other primates and nonhuman animals generally assume a dorsal ventral posture when mating (Ford and Beach, 1951; Goodall, 1986, 1990, 2010; Wickler, 1973).

This vaginal reorientation may have coincided with an expansion in the size of the brain, alterations in the size of H. erectus pelvis, and the emergence of full time female sexual receptivity. That is, with the evolution of H. erectus the female estrus cycle and the hominoid menstrual cycle which generally dictates sexually receptivity, may have undergone a transition such that, unlike all other (non-captive) female species, but like modern day woman, the H. erectus female became sexually receptive at all times and evolved enlarged breasts and buttocks to signal her sexual status when standing upright. Like modern woman (Masters and Johnson, 1966) and many other female mammals and primates such as the chimpanzee (Ford and Beach, 1951; Goodall, 1986, 1990, 1996, 2010), the human female likely retained the capacity to enjoy multiple sex partners (and to experience multiple orgasms) one after another, and to exercise sexual choice. This would have led to the selective breeding of intelligent, less violent males, sporting a big penis (Joseph 2000a,b, 2011).

It has also been estimated (Joseph 2000a,b, 2011) that the breasts and buttocks of the human female became permanently enlarged during this evolutionary time period, serving to continually signal her new sexual status and availability. Indeed, these same sexual signals are employed by other species when in estrus. In non-human primates, the vaginal area is directed in a posterior direction, and puffs out and becomes quite large and red or pink when she enters estrus. Thus, a red or pink, swollen derriere signals sexual receptivity and advertises the female's sexual availability. Some female primates, such as the gelada baboon, also advertise their sexual status via sexual swellings of the chest nodules which flush red (Fedigan, 1992; Jolly, 1972). These nodules form a necklace-like pattern which mimics the pattern of her rump--an enlarged derriere being a common sexual advertisement among many species.

However, in contrast to all other primates, the human female is the only female with permanently enlarged breasts and buttocks, and the only female who is in estrus 365 days a year and capable of sexual intercourse at all times. Non-human primates are fully sexually receptive only during circumscribe time periods, which coincides with enlargements of the breasts and the posterior directed vagina.

When female chimps and other apes and monkeys enter estrus, their dorsally oriented genitals/vaginal lips turn pink or a bright crimson and balloons outward, and in fact becomes so huge and distended they have difficulty sitting down. Moreover, estrus chimps and other primates, including baboons and the gorilla, go to great lengths to focus male attention on her buttocks, which she may flaunt and display by swaying them "enticingly" (Fedigan, 1992; Ford and Beach, 1951; Goodall, 1986, 1990; MacKinnon, 1979; Nadler, 1976; Wickler, 1973). As detailed by Joseph (2000a,b, 2011) the female genitalia and a derriere that is swollen or emphasized are obvious sexual signals which are employed to solicit male sexual attention. Likewise, modern human females accentuate and call attention to the derriere by wearing tight skirts and high heels which emphasizes the buttocks by puffing it out. When attired in this fashion she is assuming a sexually receptive posture and continually advertising her sexual availability.

The development of full time sexual receptivity and accompanying changes in the breasts and buttocks may also explain why H. erectus (or archaic H. sapiens) began utilizing various earth pigments (ochre), possibly for artistic or cosmetic purposes, or to emphasize female sexual availability. In one site, lumps of red ochre, many pointed like pencils, were found (Rightmire, 1990); and redness of the female primate genitalia is an obvious signal of sexual receptivity. This explains why modern women apply red lipstick, rouge, and other colorful cosmetics to their face and lips. The invention of cosmetics allowed women to purposefully mimic, mirror, and exaggerate the signs of estrus in other primates. However, this likely took place hundreds of thousands of years before the evolution of language (Joseph 2000a,b).

THE HOME BASE & LONG TERM ATTACHMENTS

These great changes in sexual status, the development of secondary sexual attributes and the invention of cosmetics to mirror the signs of estrus in other primates, most likely occurred during the rein of H. erectus, and contributed to the development of long term male-female mating relationships as is suggested by the creation of the home base and semi-permanent shelters where adult males and females and their young would reside (Joseph 2000a,b, 2011). That is, when female became full receptive, and could advertise this fact physically, dominant males were motivated to form long-term attachments. Normally, among most non-human mammals, males are interested in females only when they are capable of sexual intercourse, and lose interest when the estrus period wanes and disappears. Moreover, among many species of primate, dominant males may attempt to possess high status estrus females and prevent other males from mating with her, and will accomplish this by dragging her away to an isolated area which becomes a temporary home base.

For example, among chimpanzees, its not uncommon for a dominant male to threaten and physically force a high status estrus females to accompany him away from the troop, and to establish a home base where he provides her with an inordinate amount of attention; that is, until she ceases to be sexually receptive at which point he loses interest and returns to the troop (Goodall, 1986, 1990). Therefore, when the human female became continually sexually receptive, and continually advertised this fact, this may have motivated some human males to respond in a similar fashion and to form a long term mating relationship and to establish a personal home base, as is common among modern humans. Although a few other species mate for life, and/or limit their seasonal breeding to one mate, with the exception of gibbons, and some New World monkeys, this is not the case with other primates or most mammals, and is true in less than 1% of birds (Kleiman, 1977; Stacey, 1982; Wickler, 1973).

Hence, the emergence of full time sexual receptivity (and associated physical changes) likely contributed to the formation of the home base, long term mating relationships (or at least serial monogamy), and perhaps even the first "families" (Joseph 1993, 2000a,b, 2011). Thus, the emergence of sexual self-consciousness coincided with the invention of cosmetics, the evolution of breasts, buttocks and the big brain, and preceded the evolution of modern human consciousness and those tissues of the mind which produce and comprehend language and verbal-thoughts (Joseph 2011b).

CHIMPANZEES & AUSTRALOPITHECUS

The human female shares numerous sexual, cognitive, emotional, and behavioral characteristics with other species, the chimpanzee in particular (Bygott,1979; de Waal, 1989; Goodall, 1986, 1990; Itani & Suzuki, 2007; McGrew, 2011; Tanner, 2011, 1987; Stanford, 1998).

For example, although there is considerable variation, the sexual behavior of the female human (Gold & Burt, 1978; Matteo & Rissman, 1984; Udry & Morris, 1968, 1970; Wolfe, 1991) and the female chimpanzee (Fedigan, 1992; Michael, 1972) increases at the time of ovulation, that is, at midcycle; though in humans there is a second peak just before and after menstruation (Fisher, 1973).

Moreover, human and group living (multi-male/female) hominoid females are capable of experiencing multiple orgasms (e.g., Allen, and Lemmon, 2011; Burton, 1971; Chevalier-Skolnikoff, 1974; Goldfoot et al., 2010; Masters & Johnson, 1966; Michael et al., 1974) and commonly (though there are exceptions) have sex with multiple partners. Multiple sex partners ensures she will become pregnant, and multiple orgasms reward her with increasing pleasure as she sexually dallies with male after male.

These and other female chimpanzee/human commonalties appear to be due to a common genetic and evolutionary heritage and the sexual differentiation of the limbic system. It is this ancestral hominoid heritage and limbic system commonalties which explains not only her capacity for multiple orgasms, but the fact that the human female's "monthly rhythm of ovogenesis, ovulation, estrogen and progesterone secretion, uterine stimulation, and menstrual bleeding follows the basic primate pattern" (Beach, 1974, p. 356). Because they share common ancestors and a limbic system organized in a "female" pattern, female chimps and humans are sexually similar.

The human female sometimes behaves like, and in many respects is similar to a female chimpanzee in heat. This is not only because we share a common ancestor, but because our most distant "human" ancestors were essentially apes. Indeed, given these primate origins and commonalties, the initial evolution of the human females' unique sexuality is therefore best understood from a pongid perspective, beginning with Australopithecus, and later, H. habilis, whose social life was probably more ape-like than human-like.

The common ancestors for chimps and humans diverged 5 million year ago and the chimpanzee brain (395-410 cc) is just slightly smaller than the average Australopithecus brain (375-440 cc). Male chimps are also about 20% larger than females. Given these and the other genetic and sexual similarities mentioned above, it can be assumed that social-sexual and consort relations between ancestral males and females were more like that of chimpanzees.

Australopithecus vs Human Brain

And, being more ape-like than human, it can also be inferred that these first pre-human females, i.e., australopithecus, were extremely sexually promiscuous, and probably had sex up to fifty times a day when they entered estrus. Like other social apes, our pre-human female australopithecine/habilis ancestors probably had sex with every desirable and high status male of the troop, and she likely snuck off to an adjacent troop where she would then enjoy a romantic vacation by again having sex up to fifty times a day with every desirable and high status male--a common behavior of female chimps.

EVOLUTION OF FEMALE PELVIS AND THE "BIG BRAIN"

It was during the latter stages of H. erectus evolution that the brain became significantly enlarged. The brain in fact doubled in size with the transition from Australopithecus (440 cm3) to H. erectus (937 cm3), approaching within 15% of present-day humans (Conroy 1998; Potts, 1996; Rightmire, 1990; Tobias, 1971). However, with the advent of the big brain, human sex differences, including those related to full time female sexuality receptivity appear to have become even more pronounced.

Because a bigger brain comes in a bigger head, this required a larger birth canal and an increase in the sexual physical differentiation in the size and width of the H. erectus (and modern) female pelvis so as to accomodate the birth of a big brained baby (Joseph 1996, 2000a,b, 2011). Specifically, with the evolution of a bigger brain and with the transition from Australopithicus to H. habilis to H. erectus, the pelvic opening became longer and more round and ovoid as it expanded from front to back (Day, 1982, Lovejoy, 1988; Sigmon, 1982); changes with made birth more difficult as now the infant's head had to rotate as it passes down the birth canal in order to emerge from the pelvic opening.

As with modern woman, this adaptation likely forced the female erectus' upper legs wider apart and her knees closer together, thus altering her gait and balance, causing her to wiggle her derriere when walking. Presumably, this alteration, coupled with the evolution of new muscles to aid in the upright stance, accentuated and drew attention to the female derriere and her sexual availability (Joseph, 1993; 2000a,b, 2011); a sexual-social signal accentuated in modern women through high heels and tight clothes, and in the last century, via the bustle, hoop skirts and in previous centuries dresses designed to grossly exaggerate the width of the hips.

These same sexual signals are characteristic of other primates (Ford and Beach, 1951; Wickler, 1973). When female chimps and other apes and monkeys enter estrus, their dorsally oriented genitals/vaginal lips turn pink or a bright crimson and balloons outward, and in fact becomes so huge and distended they have difficulty sitting down. Moreover, estrus chimps and other primates, including baboons and the gorilla, go to great lengths to focus male attention on her buttocks, which she may flaunt and display by swaying them "enticingly" (Joseph 2000a, 2011). The female genitalia and a derriere that is swollen or emphasized are obvious sexual signals which are employed to solicit male sexual attention. Likewise, modern human females accentuate and call attention to the derriere by wearing tight skirts and high heels which emphasizes the buttocks by puffing it out. When attired in this fashion she is assuming a sexually receptive posture and continually advertising her sexual availability.

THE FEMALE PELVIS AND THE DIVISION OF LABOR

The transformation of the human female hips and pelvis, however, also limited her ability to run and maneuver about in space, at least, as compared to most males (Joseph 2000a,b). This is because, be it H. erectus or modern H. sapiens sapiens "these changes are less pronounced in the... male pelvis" (Lovejoy, 1988, p. 123). "The size of the canal in men is controlled mainly by locomotor, not reproductive factors" (Campbell, 1985, p. 152).

Moreover, her pelvis became more fragile and subject to fracture if stressed by continuous hard running and jumping. This condition plagues even female soldiers who suffer an unusually high incidence of pelvic and leg fractures although their training and duties are not as strenuous or arduous as males (2005 U.S. Marines Report, Paris Island). The female pelvis is in fact less massive, lighter and thus weaker than the male pelvis. Moreover, as the human femoral neck expanded and became longer so as to accommodate these changes in the pelvic girdle, it became more porous and thus weaker (Lovejoy, 1988) as well as subject to stress related injury due, in part, to the greater transverse and sagittal diameter of the female pelvic inlet.

As summarized by Day (1996, p. 5), "The female bony pelvis is a compromise between the needs of childbirth and those of locomotion with heavy selection pressure on the need for successful reproduction. The male pelvis, with no reproductive function to parallel that of the female, evolved for efficiency in bipedal locomotion." This helps "explain the differences in peak achievements between men and women in running and jumping."

Hence, be it modern woman or female H. erectus, in accomodating the birth of big brained babies, these physical changes not only advertised her sexual status, but likely interfered with her ability to endure and keep up with males on their long hunting sojourns; a function also of her reduced upper body strength and the prolonged dependency of her big brained babies (Joseph, 1993, 2000a,b, 2011). As such, with the evolution of H. erectus (and their descendants), males and females became increasingly specialized to perform certain tasks, e.g. hunting versus gathering--a division of labor characteristic of all hunting and gathering groups (Dahlberg, 2011; Hiatt, 1970; Martin and Voorhies, 1975; Murdock and Provost, 1973; Zilman, 2011) including chimpanzees (Goodall, 1990, 1996, 2010).

NATURAL BORN KILLERS: MALES ENJOY HUNTING & KILLING

Female primates, including chimps (whose activated DNA is 98.6% identical to human DNA), sate their hunger, and that of their infants, through systematic gathering of foodstuffs and insects, most of which is shared with their young (Goodall, 1990, 2010). Although male primates gather, they also hunt and become exceedingly excited when killing not only other animals (Hamburg, 1971; Harding and Strum, 1978), but other primates including former members of their troop who they may ambush and beat to death (Goodall, 1986, 1990). Be it other animals or a chimp from a neighboring band, when chimpanzees attack and wage war, and beat and kill other chimpanzees, the entire troop will gather and excitedly beg for a tiny morsel of the bloody flesh of victims as it is torn away by the victor (Goodall, 1986).

As is well established, be it human, non-human primate, or mammal, males tend to be violent (Elia, 1988; Fedigan, 1992; Goodall, 1986, 1990; Hamburg, 1971; Johnson, 1972; Lorenz, 1966; Manning, 1972; Mitchell, 1979; Moyer, 1974), and the male proclivity to hunt appears to be direct extension of these proclivities (Joseph, 1993, 2000a,b). Indeed, over 80% of all violent crimes and murders are committed by human males (Uniformed Crime Reports, 1990-1996, 2000-2006). Likewise, male chimpanzees initiate attack five times as frequently as females and are responsible for 90% of all aggressive encounters (Bygott, 1979; Van Lawick-Goodall, 1968).

Sex differences in violence and belligerence are apparent as early as the first two months of life even in chimps and other primates (Ransom and Rowell, 1972; Mitchell, 1979). Likewise, following the divergence between common ancestors, the human (male) tendency toward violence was probably also a characteristics of Australopithecus who not only hunted small animals, but possibly each other. According to Dart (1949), Australopithecines "were confirmed killers; carnivorous creatures that seized others by violence, battered them to death, tore apart their broken bodies, dismembered them limb from limb, and slaking their ravenous thirst with the hot blood of the pitiful victims and greedily devouring their writhing flesh." Although Dart's conclusions have been challenged, as noted above, male chimps not uncommonly engage in violent, murderous and even cannibalistic interactions with other troop member (Goodall, 1986, 1990).

Like modern humans, gangs of male chimpanzees will also engage in surprise attacks on neighboring colonies, beating and killing the old, infirm, and young alike, including former friends; even drinking their blood. That Australopithecus were killers and hunters of meat, thus, should not be surprising. In fact, from an examination of Australopithene teeth it is apparent that these were meat eaters whose diet also consisted of vegetable matter (Sponheimer & Lee-thorp, 1999)--as is the case of modern hunting gathering groups (Dahlberg, 2011; Lee and DeVore, 1968; Martin and Voorhies, 1975; Murdock and Provost, 1973; Zilman, 2011) and chimpanzees.

Specifically, from an analysis of the tooth enamel of Australopithecus africanus, it was determined that these individuals ate food rich in carbon 13--that is high protein food (Sponheimer & Lee-thorp, 1999). As these teeth were lacking the telltale scratches associated with vegetable matter, it can thus be concluded they were eating meat. Moreover, the remains of A. garhi were found in association with antelope leg bones which show the signs of cutting and pounding by a shark and blunt rock (White et al., 1999). Hence, it is fairly obvious that these were meat eaters and probably killers of animals and each other, as is the case with chimpanzees.

Thus sex differences in aggression and trends in the division of labor had probably been established with the evolution of hominoids (as is suggested by modern day chimps), and likely continued to develop with the emergence of Australopithecus. These same trends, however, were likely exaggerated with the evolution of the big brained H. erectus.

A bigger, more complex brain confers upon the bearer increased cognitive and intellectual capabilities, and the male H. erectus employed that greater intelligence in the pursuit of bigger game, including deer, bisons, horses and even bear and elephant (Potts, 1996; Rightmire, 1990). His killing technique, however, remained rather primitive. He would either surround his prey and stone them to death, or would use fire to frighten and stampede all manner of animals over cliffs or into swamps, typically killing more animals than could possibly be eaten (Potts, 1996; Rightmire, 1990).

The Silent Hunter Hunting does not require language (Joseph, 1993, 2000a,b). Rather, successful hunting requires prolonged silence, excellent visual-spatial and gross motor skills, and the capacity to endure long treks in the pursuit of prey. These are abilities at which males excel, including modern H. sapiens sapiens (Joseph, 1988a, 2000a,b). As detailed below, it is not until the emergence of modern (Upper Paleolithic) H. sapiens sapiens, around 35,000 to 50,000 years ago, that "modern language evolved.

Females Love Babies Just as primate males hunt whereas most primate females are not so inclined, it is unlikely that the female H. erectus, burdened by pregnancy, crying children, and her fragile pelvis and awkward gait, would have behaved otherwise. Rather, whereas most primate males, including male humans tend to have little interest in infants or caring for the young (Belsky et al., 1984; Clarke-Stewart, 1978; Frodi et al., 1982; Joseph 2000b) it is likely that the female H. erectus, like other female primates, would have pursued her own inclinations including the desire to bear babies and nurture the young.

In fact, be it chimpanzee, baboon, rhesus macaque, or human, females demonstrate an extraordinary interest in babies and will engage in play-mothering during even the earliest phases of their own childhood (Blakemore, 1990; Devore, 1964; Elia, 1988; Fedigan, 1992; Goodall, 1971, 1990).

Female humans (Berman, 1983; Blakemore, 1990) and non-human (group living) female primates will eagerly cuddle, groom, and hold babies that are not their own (Jolly, 1972; Kummer, 1971, Strum, 1987; Goodall, 1971). Although there are exceptions such as the pigtail macaque, in general, female humans, apes, or monkeys with babies, become the center of female-attention (Elia, 1988; Fedigan, 1992; Jolly, 1972, Mitchell, 1979, Strum, 1987).

Males are Indifferent Toward Babies By contrast, human men and boys and non-human male primates have little interest in babies or young children and generally provide little or no nurturant care even for their own offspring though they may form alliances with siblings (Fedigan, 1992; Joseph 2000a; Kummer, 1968, 1971; Mitchell, 1968, 1979, Goodall, 1971; Gordon and Draper, 1982; Rossi, 1985; Rowell et al., 1968), one of the few exceptions being owl monkeys, and to a lesser extent, baboons. Human males and fathers rarely behave in any manner that approximates normal female maternal behavior (Belsky et al., 1984; Clarke-Stewart, 1978; Frodi et al., 1982). As noted, males tend to be aggressive and violent and the desire to nurture infants may not be compatible with these tendencies. Indeed, males are responsible for over 70% of all murders involving infants and children (U.S. Justice Dept. 1996-1997, 2006-2007). Likewise, among many other primates (gorillas, Barbary apes, and rhesus, howler, and red-tailed monkeys) males stereotypically kill infants not their own (Fedigan, 1992; Hrdy, 1979).

THE INFANT-MATERNAL ROOTS OF EMOTIONAL LANGUAGE: BABIES & MOTHERS TALK TO EACH OTHER

Considerable vocalizing occurs between mothers and infants. The infants of many species will often sing along or produce sounds in accompaniment to those produced by their mothers (Barnett, 1995; Bayart et al., 1990; Jurgens, 1990; Wiener et al., 1990). These interactions reinforce and promote mutual vocalization, attachment, and contribute to survival. Hence, the first forms of complex social-emotional communication may have been produced in a maternal context (Joseph, 1992, 1993, 1996, 1999).

Because the big brained infant H. erectus likely required long term care, it can be assumed that there was an impetus for female H. erectus to vocalize with her young for years at a time. Indeed, a big brain requires more time to grow and mature, which in turn results in prolonged immaturity and helplessness. In consequence, the female tendency to vocalize more than males may have been given additional impetus well over 500,000 years ago.

Among most social animals and gathering groups, the production of sound is very important in regard to infant care. If an infant becomes lost, separated, or in danger, this is best conveyed via a cry of distress and fear; cries which would cause a mother to come running to the rescue. Conversely, vocalizations produced by the mother and her co-gatherers enable an infant (or a lost gatherer) to orient and find its way back if perchance it got lost or separated. In this regard, the tendency to vocalize may have ensured for breeding success.

Among animals, and present day human mothers, much of this initial mutual sound production is emotional and prosodic (Cooper and Aslin, 1990; Fernald, 1991, 1992; Fernald et al., 1989; Jurgens, 1990) and constitutes what has been referred to as "limbic language" (Joseph 1982, 2000). Emotional sound production is mediated by the amygdala, cingulate gyrus, and other limbic and subcortical nuclei, but is also hierarchically represented in humans and produced and perceived by the right frontal and right temporal-parietal area. These emotional-language areas appear to be more extensively developed in human females (Joseph 1993, 1996, 2000b). By contrast, the denotative, and vocabulary-rich grammatical components of modern human speech are mediated by the left frontal and temporal-parietal area; i.e. Broca's and Wernicke's speech areas and the inferior parietal lobule.

Because infants are emotionally oriented and have little or no understanding of non-emotional speech, these mutual mother-infant vocalizations usually consists of exaggerated emotional prosody (Cooper and Aslin, 1990; Fernald, 1991, 1992, 1993; Fernald et al., 1989). Indeed, human infants prefer listening to and are more responsive to these exaggerated emotional vocalizations, particularly when produced by a female (Cooper and Aslin, 1990). Moreover, human (and non-human) females demonstrate superiorities in this regard, and not only produce more mutual social-emotional vocalizations than males (Brody, 1985; Burton and Levy, 1989; Gilbert, 1969; Glass, 1992; Tannen, 1990) and more words in five seconds of conversation, but tend to employ 5-6 different prosodic variations and utilize the higher registers when conversing (Joseph, 1993). They're also more likely to employ glissando or sliding effects between stressed syllables (Brend, 1975; Coleman, 1971; Edelsky, 1979). Men tend to be more monotone, employing 2-3 variations on average, most of which hovers around the lower registers (Brend, 1975; Coleman, 1971; Edelsky, 2009). Even when trying to emphasize a point males are less likely to employ melodic extremes but instead tend to speak louder.

FEMALE & BILATERAL EMOTIONS
Although influenced by sex differences in the oral-laryngeal structures,

these differential vocalizing abilities are also reflected in the greater capacity of the female brain (and right hemisphere) to express and perceive emotional vocalizations (Burton and Levy, 1989; Hall, 1998; Soloman and Ali, 1972). This superior sensitivity includes the ability to understand, perceive, and express empathy and social-emotional nuances (Burton and Levy, 1989; Brody, 1985; Buck, 1984; Eisenberg et al., 1989; Fuchs and Thelan, 1988; Harackie-wicz, 1982; Lewis, 1983) and a greater willingness to express emotional issues and discuss personal problems (Gilligan, 1982; Joseph 1993, 1996, Lutz, 2010; Walker et al., 1997; Lombardo and Levine, 2011).

In fact, from childhood to adulthood, females appear to be much more emotionally expressive than males (Brody, 1985; Burton and Levy, 1989; Gilbert, 1999; Tannen, 2000), who in contrast have difficulty discussing personal difficulties or expressing their emotions other than through anger, happiness, and sexual arousal (Balswick, 1982, 1988; O'Neil, 1982; Sattel, 1999; Tannen, 2000).

FEMALE H. ERECTUS & LANGUAGE

Like chimpanzees and present-day human females, female H. erectus probably engaged in mutual mother-infant vocalizations, and probably engaged in more mutual social-emotional vocalization not just with her children but with each other. However, as based on an analysis of H. erectus' tool technology, and the fact that it was somewhat crude and lacking in temporal sequential sophistication (though certainly a lot of time was put into their construction), it can be assumed that the brain of H. erectus was not capable of thinking in terms of complex temporal sequences. Hence, it can be assumed that H. erectus did not speak in the temporal sequential and grammatical manner characteristic of present-day humans. Nor is it likely H, erectus uttered complex words.

Rather, H. erectus probably tended to use gesture, body language, facial expression, as well as grunts, groans, mimicry, and the production of emotional sounds in order to convey needs, fears, feelings and desires. On the other hand, as gathering and the construction of domestic-tools was probably an ongoing female H. erectus activity (see below), these activities likely eventually gave rise to improvements in bilateral, temporal and sequential fine motor skills; thus providing what would become the neurological foundation for the development of complex grammatical, vocabulary-rich, and emotional human language.

HUNTING VS DOMESTIC TOOLS

Whereas the big brained male H. erectus refined his hunting techniques and became a hunter of bigger game, females refined their gathering and food preparation techniques as is suggested by the creation of crude tools such as choppers, scrapers, cleavers with a stright cutting edge on one side (Day, 1996;

Rightmire, 1990), and probably digging sticks, as is characteristic of female chimpanzees (Goodall, 1990; McGrew and Marchant, 1992). That these tools were fashioned by a female hand can be deduced by their domestic use. Among hunting and gathering groups it is females and not the males who make and use tools (Joseph 1992, 1993, 2000a,b; Niethhammer, 1977), the only exception being hunting implements and weapons of war which females are not allowed to touch (Tabet, 1982). Similarly, females chimps use food/gathering-related tools much more frequently than males (Goodall, 1986; McGrew and Marchant, 1992) who in turn are more likely to use sticks and rocks to threaten other males (Bygott, 1979; Goodall, 1990; Van Lawick-Goodall, 1968). As with other male primates, and given his penchant for big game hunting, it was probably the male H. erectus who created the first hand ax; a killing tool which made its appearance about 650,000 years ago (Potts, 1996; Rightmire, 1990).

As tools for "gathering" versus "hunting" were fashioned throughout Africa, Europe, Asia and the Middle East throughout the Paleolithic, it can be assumed, with the exception of the European Neanderthals, that this division of labor prevailed throughout the Middle Paleolithic, well after H. erectus was replaced by archaic H. sapiens (Joseph 1993, 1996, 2000a,b). For over 500,000 years males have hunted and killed whereas females cared for the young, maintained the home base, and were responsible for the gathering of food and related "domestic" activities. However, this is not to imply that females never killed small animals, or never collaborated or assisted in the hunt, as that is not the case with humans or chimpanzees.

By the close the Middle Paleolithic (around 30,000 B.P.), and the emergence of "modern" Upper Paleolithic H. sapiens sapiens, such as the Cro-Magnon, hunting had become the center of religious and artistic life. Nevertheless, 60-80% of the Cro-Magon diet consisted of fruits, nuts, grains, honey, roots and vegetables (Clark, 1952; Prideaux, 1973), which were probably gathered by females (Joseph 1992, 1993, 2000a,b). Among the hunting and gathering societies in existence during the last few centuries, women have been and are the gatherers and main providers of food whereas spoils from the hunt account for only about 35% of the diet (Dahlberg, 2011; Lee and DeVore, 1968; Martin and Voorhies, 1975; Murdock and Provost, 1973; Zilman, 2011).

FEMALE TOOL-MAKERS, TEMPORAL SEQUENCING AND LANGUAGE

In grubbing for roots and bulbs the Upper Paleolithic female-gatherer probably used a digging stick which she periodically sharpened by using stone flakes. These gatherers also fashioned hammerstones for cracking nuts and grinding the produce she collected. As in recent hunting gathering groups (Lee, 1974; Murdock and Provost, 1973), her duties would include the preparation of any meats she scavenged or which the men brought home from their hunting sojourns. In addition to food preparation, clothes were sewn and fashioned out of hides (Clark, 1952, 2007; Prideaux, 1973), and these too are tasks associated with women (Gusinde, 1961; Lee, 1974; Neithhammer, 1977), including, presumably, the Cro-Magnon females of the Upper Paleolithic (Joseph 1992, 1993, 1996, 2000ab).

Thus the duties of the Upper Paleolithic female were much more multifaceted and complex than her predecessors, and included cleaning hides via a scraper, drying and curing the skin over the smoke of a fire, and then using a knife or cutter to make the general desired shape. The Upper Paleolithic female was also employing a punch to make holes in these hides, through which leather straps or a vine could be passed so as to create a garment that could keep out the cold (Clark, 1952, 2007; Prideaux, 1973). They were also weaving and using a needle to sew garments together; "domestic" tasks which are almost exclusively associated with "women's" work (Murdock and Provost, 1973; Neithhammer, 1977).

The necessary skills required for tool construction and the efficient gathering of vegetables, fruits, seeds, berries and the digging of roots, include rapid, temporal-sequential fine motor maneuvers with the arms, hands, and particularly the fingers (Hamrick et al., 1998; Marzek, 1997; Susman, 1994). Similarly, most tools are made in a step-wise, temporal sequential manner, with specific fine-motor movements, and considerable precision; abilities which enable the tool maker to construct the same implement over and over again. This also requires that the tool maker be in possession of a brain that can control the hand and which can use foresight and planning in order to carry out the steps involved in the implement's manufacture. In this regard, although males were also fashioners of tools (i.e. those used for killing), it is noteworthy that females tend to excel at fine motor activities, such as those involving rapid, repetitive temporal sequencing (Broverman et al., 1968; Hampson and Kimura, 1992; Junaid and Fellowes 2006; Larson et al. 2007; Thomas and French 1985).

These "domestic" activities, coupled with her innate tendency to engage in mutual vocalizations, and to vocalize with her young and fellow gatherers, likely coincided with, and stimulated the development of the temporal sequential neurological foundations of what would become a female superiority in the evolution of grammatical, vocabulary-rich human language. However, as the tools made by H. erectus and archaic (Middle Paleolithic) H. sapiens remained relatively crude, whereas a creative explosion in tool-making technology typifies the onset of the Upper Paeleolithic (Clark, 1952, 2007; Joseph 1992, 1993, 1996, 2000a,b; Mellars, 1989, 2010; Prideaux, 1973), it appears that the human brain did not become fully adapted for perceiving and expressing temporal sequential, grammatically complex, vocabulary-rich language, until the emergence of the Upper Paleolithic human female. As will be detailed below, the emergence of these linguistic capabilities were promoted by and thus coincided with advances in tool technology and associated fine motor skills--activities which gave rise to the functional evolution of Broca's speech area and the angular gyrus of the inferior parietal lobe.

SEX DIFFERENCES IN LANGUAGE

Over the course of the last 500,000 years, women have been engaging in group and domestic activities including child rearing, that promoted the evolution of the neural substrates which subserve the development of human speech, including bilateral representation of language.

Language is both emotional and descriptive, grammatical and melodic, consisting of word units and prosody, such that just as both halves of the brain interact to produce music (Joseph, 1982, 1988), both halves become activated when engaged in language related activities, the left providing the grammatical word units, the right supplying the emotional melody. Hence, as demonstrated through functional imaging, the right (as well as the left) hemisphere becomes activated (Bookheimer, et al., 1995; Bottini et al., 1994; Peterson, et al. 1988; Price, et al., 1996; Shaywitz, et al., 1995). However, the female right (and left) hemisphere appears to be superiorally endowed with these and other aspects of language (Joseph, 1982, 1993, 1996; Shaywitz et al., 1995). Females demonstrate bilateral cerebral activation when engaged in certain language tasks as demonstrated by functional imaging (Shaywitz, et al., 1995).

Over 50 years of research has clearly demonstrated that human females display superior language, articulation, word knowledge, syntactic, and related linguistic skills as compared to males (Bayley, 1968; Broverman et al., 1968; Harris, 1978; Koenigsknecht and Friedman, 1976; Hampson and Kimura, 1992; Harshman et al., 1983; Hyde and Lynn, 1988; Kimura, 1993; Levy and Heller, 1992; Lezak, 1983; McGlone, 2010; McGuiness, 1976; Moore, 2007). In contrast to males, human females vocalize more, engage in more social speech, display superior linguistic skills, and excel over males on word fluency tests, for example, naming as many words containing a certain letter, or words belonging to a certain category. Females also vocalize more as infants, speak their first words and develop larger vocabularies at an earlier age. Their speech as children is easier to understand, they improve their articulation and grammatical skills at a faster rate, and the length and complexity of their sentences is greater than males (Bayley, 1968; Harris, 1978; Hyde and Lynn, 1988; Koenigsknecht and Friedman, 1976; Levy and Heller, 1992; Lezak, 1983; McGlone, 2010; McGuiness, 1976; Moore, 2007).

Moreover, women gravitate toward and are overwhelmingly over represented in professions requiring conversational language, such as telephone operators or secretary--over 90% of those working in these fields are women (U.S. Department of Labor, 1997, 2007).

Whereas women and girls display clear language superiorities, or a greater tendency to vocalize and speak, males suffer from more language related disturbances such as stuttering (Corballis and Beale, 1983; Lewis and Hoover, 1983). Moreover, males lose language-related capabilities as they become aged, are more likely to become aphasic following stroke, and do not recover language as quickly or as fully as females (Joseph 1996).

In addition, girls learn how to read more quickly and more proficiently than males who are more likely to suffer from reading difficulties including dyslexia (Corballis and Beale, 1983). Females not only demonstrate superior reading comprehension and writing and spelling skills (Lewis and Hoover, 1983), but in 1997 the U.S. Department of Education (USDE) reported that the writing skills of a 9 year old girl in fourth grade is equal to that of a 13 year old boy in 8th grade, and in 2007 the USDE reported that "at all age levels females continue to outscore males in reading proficiency... and females of all ages have outscored males in writing." Rreading and writing are directly associated with the functional integrity of the IPL/angular gyrus.

Presumably, females demonstrate superior language skills because their brains are better adapted for producing temporal-sequential and social-emotional vocalizations (via the maternal cingulate gyrus), and because, as gatherers and tool makers they were not under the same constraints as the silent male hunters. Males were required to maintain long periods of relative silence -at least while hunting. Their brains adapted accordingly.

GATHERING & WHY WOMEN LIKE TO TALK, SHOP, AND SHARE THEIR FEELINGS

Over the course of human evolution, and in contrast to the silent, non-communicative (or threatening) males, female mothers and female gatherers were able to freely chatter with their babies or amongst themselves. Indeed, gathering not only fostered the development of language, but like hunting for men, also served as a social activity that enabled women to interact in a socially intimate manner.

Our ancient female ancestors probably gathered in large groups of 7 or more individuals, as the typical size of a band is about 25 individuals on average. Some women were pregnant or probably carried infants which might be set on the ground here and there, or accompanied by young adolescents who would frolic about and play. Such gathering groups must have commonly been loud, noisy and very gay affairs filled with the talk of the women and the sounds of games and yells of the children. Hence, unlike the men who must remain quiet for long time periods in order to not scare off game, the women are free to chatter and talk to their hearts delight. Talking also served as a means of maintaining the location of the group so that if a gatherer or a child chanced to walk away, she (or her child) could always relocate the others by their hodgepodge of speech. However, it is also likely many women did not wonder far away from the group or home base as they might become prey.

Talking thus became part of the gathering glue and served the purpose of keeping the group together and thus of bonding the group as a collective (a function similar to that performed by the anterior "maternal" cingulate gyrus). The women talked about their children, their mates, each other, and were thus exposed and allowed to expose their own feelings and thoughts to those who valued talking as much as they. As will be detailed below, for women, to socialize and be together means to talk, and to talk is a very social bonding element for women even today (Joseph 1992, 2000b; Tanner 1992); a characteristic first forged by the evolution of the anterior "maternal" cingulate gyrus of the limbic system and the subsequent female tendency to vocalize with her young (Joseph 1996).

When considering our evolutionary heritage and the fact that women have spent a good 100,000 years or more in female dominated gathering groups, where socializing and freely expressing oneself about social and family matters were the norm, it should come as no surprise that modern day females continue to respond similarly. It is thus little wonder that some women also feel a compulsion to shop and derive considerable enjoyment making it into an all day affair. Likewise, some modern men, bravely armed with high powered weapons, also feel a compulsion to stalk and kill helpless animals, or even other humans. It is hard to escape hundreds of thousands of years of evolutionary pressure.

In this regard, it perhaps should not be surprising that after almost half a million years of holding his tongue and jockeying for status and position amongst his fellows, that modern day man continues to respond similarly and have the same concerns, albeit translated and modified to some degree; i.e. sports, politics, business.

However, it is also because woman has engaged in group gathering activities, accompanied by sisters, daughters, cousins, and friends, for over a million years, that "modern" westernized woman continues to feel an innate need to "gather." That is, she "shops" and derives considerable enjoyment from shopping. Over 90% of department store space is devoted to women, because of this innate female characteristic. Women may shop when they are depressed and lonely--as gathering was always a social activity-- and they may buy dozens of articles, and "cute little shoes" and other clothes that they may never wear or wear only once. She feels impelled to shop/gather. Indeed, a woman can spend hours "shopping" and might make it an all day affair. By contrast, a man will often go to the store, pick up what he needs and leaves. A woman gathers and gathers along side numerous other gathering/shopping women. Men hunt until they kill and then go home. Women gather. Men hunt.

The nature of human consciousness has been shaped by our evolution history.

HUNTING, VISUAL-SPATIAL SKILLS, & WHY MEN DON'T SHARE THEIR FEELINGS

Searching for large game animals requires that the hunting band roam quietly over a huge expanse of varied terrain which may extend up to 500 or more miles. To be successful requires good directional sense, so that one could wonder and also find their way home without the aid of a street sign, good visual-spatial skills such as depth and distance perception, so that one can walk and run as well as anticipate the trajectory of movement of a running prey, without tripping and falling, an excellent capacity to recognize, comprehend, and mimic animal, environmental and non-speech sounds, and an increased capacity to communicate non-verbally via gesture, body language, and particularly via animals mimicry so that game would not be scared off by the sounds of speech and so that communication with other hunters could occur. These are all skills associated with the right hemisphere of a modern day human being.

Indeed, the mimicry of certain animal sounds could communicate a host of meanings such as where something might be located (as those animals whose sound he is mimicking may be found only in a certain location), that a particular beast (which elicits cries from the creature being mimicked) is approaching or nearby, or a particular action that the group should take (fight or flight or freeze), and so on. These are all capacities at which the right hemisphere of modern man and woman, excels.

However, in that big game hunting was an activity dominated by men, when one considers the skills involved to be a successful big game hunter it should be no surprise that after 300 or more thousand years of engaging in these activities, that males continue to demonstrate cognitive superiorities in regard to related abilities.

When males banded together for the purposes of the big game hunt, they had to walk silently for long time periods and their only form of communication was probably concerned only with the expression of facts and information about the hunting they were engaged in. If they were to begin talking about anything else such as their family, they possibly would have been shunned as their discussion would have been irrelevant, and would have scared off the game. A premium was thus placed on men who could hold their tongue.

Also decreasing the need for talk was the fact that as hunters stalk their prey they may walk at anywhere from 25-100 yards apart. Two or more men standing together is more likely to warn a prey of their presence, whereas men standing apart have the advantage of chasing a beast into the waiting arms of his comrade.

Finally, it is likely that once the hunt was complete and they returned to their base camp, that the men did not then begin to discuss their wives, children, feelings, or personal problems, but their own prowess as hunters, incidents related to the hunt, and whatever adventures they may have had that would enhance their status amongst their fellows.

VISUAL-SPATIAL SKILLS OF THE SILENT HUNTER

In contrast to females where language representation appears to be bilateral, language appears to be more greatly concentrated in the left half of the male brain (Bradshaw et al. 1977; Joseph, 1982, 1993, 1996; McGlone, 2010; Shaywitz et al., 1995). For example, among males only left hemisphere lesions result in these language deficits whereas women are more likely to suffer word finding difficulty and a reduction in vocabulary with right or left hemisphere lesions (Kimura and Harshman, 1984). In contrast to the female cerebrum, the right half of the male-brain appears to be more lateralized and functionally specialized to analyze (and sequence) visual-spatial relationships (Levy, 1972, 1974; McGee, 1979; McGlone, 2010; Witelson, 1985)--perceptual functions directly related to skill at hunting.

Whereas female gatherers are free to talk to one another and their children, hunting does not promote fine motor skills or linguistic development as the successful hunter must be silent and capable of throwing a spear and analyzing the spatial coordinates which separate yet link him to a potential prey. These male activities do not promote speech. Wolves and wild dogs spend a considerable part of each morning and evening tracking and hunting and there is no evidence of speech among these creatures. Speech confers few advantages to a group of human hunters, who must maintain long periods of silence so as to not scare off potential game.

Aspects of hunting, however, also put a premium on parietal-temporal lobe and right cerebral cognitive development. Tracking, aiming, throwing, geometric analysis of spatial relationships, as well as environmental sound analysis, are also directly related to the functional integrity of the right half of the brain (Guiard, 1983; Haaland and Harrington, 1990; Joseph, 1982, 1988a,b, 1996) and the parietal lobe in particular. The right parietal area is associated with the mediation of visual-spatial perceptual functions, especially in males, including providing the sensory feedback which would enable a hunter to aim and throw a spear; guidance which is transferred, via the corpus callosum to the left hemisphere somatomotor areas controlling the right upper extremity.

Hence, given 500,000 years of multi-generational male experience in hunting which usually required days or even weeks of wondering hundreds of miles from the home base, present-day males therefore demonstrate superior visual spatial skills including superior maze learning, tracking, aiming, and related non-verbal abilities, as compared to females (Broverman et al., 1968; Dawson et al., 1975; Harris, 1978; Joseph, 1993, 1996; Kimura, 1993; Levy and Heller, 1992). This includes a male superiority in the recall of geometric shapes, detecting figures that are hidden and embedded within a complex array, constructing 3-dimensional figures from 2-dimensional patterns, visually rotating and detecting the number of objects in a 3-dimensional array, and playing and winning at chess (which requires superior spatial abilities).

Males possess a superior geometric awareness and directional sense and geographic knowledge, are better at solving tactual and visual mazes or mentally manipulating spatial relations on paper, and are far superior to females in aiming, throwing, and tracking such as in coordinating one's movements in relationship to a moving target (Broverman et al., 1968; Harris, 1978; Kimura, 1993; Levy and Heller, 1992; Linn and Petersen, 1985; Porteus, 1965; Thomas et al., 1973). Only about 25% of females in general exceed the average performance of males on tests of such abilities (Harris, 1978; Joseph 2000a).

Moreover, some of these differences are present during childhood and include a male superiority in the recall and detection of geometric shapes, detecting figures that are hidden and embedded in an array of other stimuli, constructing 3-dimensional figures from 2-dimensional patterns, visually rotating or recognizing the number of objects in a 3-dimensional array, playing and winning at chess (which requires superior spatial abilities), directional sense and geographic knowledge, the solving of tactual and visual mazes, aiming and tracking such as in coordinating one's movements in relationship to a moving target, coordination in aiming and throwing, and comprehending geometrical concepts. Many of these abilities are directly related to skills that would enable an ancient hunter efficiently to track, throw a spear, and dispatch various prey without getting lost and while maintaining a keen awareness of all else occurring within the environment.

Various aspects of mathematical ability are directly related to visual-spatial functioning, such as geometry and the sequencing of space (Joseph 1988a; 1990, 1996). However, whereas the female brain has an advantage when it comes to sequencing language, the male brain is more adept at sequencing space. Hence, males are far more likely than females to be mathematical geniuses. Even when the sample consists of females who are scholastically gifted, males achieve higher scores, for example, on the math portion of the college entrance exam, and outperform females by a ratio of 13 to 1 (Benbow and Benbow, 1984).

Moreover, some of these differences are present during childhood and early adolescence and have been demonstrated in other species (Linn and Petersen, 1985; Joseph, 1979; Joseph and Gallagher, 2010; Joseph et al., 1978). For example, male rats consistently demonstrate superior visual-spatial and maze learning skills as compared to females (Joseph, 1979, Joseph and Gallagher, 2010; Joseph et al., 1978). If reared in an enriched environment, males continue to outperform females (Joseph, 1979; Joseph and Gallagher, 2010).

SEX DIFFERENCES IN COMMUNICATION

It has frequently been noted that many men tend to interact and compete with one another in terms of power, status, physical and intellectual dominance, wealth, and control. In contrast, it has been argued that women tend to speak and interact in a more cooperative manner, and to be more interested in issues related to the family, inter-dependence, and social intimacy (Croates, 1986; de Beauvoir, 1961; Joseph, 1985, 1992; Stutman, 1983; Tanner, 1990). Although as a rule, such notions do not apply to all men or women, there indeed does appear to be a gender linked difference in the manner in which many boys, men, women, and girls tend to interact and speak together; and these same sex differences appear to be biological in origin, directly related to the hunting vs gathering way of life which has characterized much of human evolution, and are also seen in other primates such as chimpanzees.

As with chimpanzees, many boys tend to play in groups where there is a recognized leader and hierarchical order of followers. The leader often wins this position based on physical strength and capability, risk taking, and his ability to control or bully others. Similarly, the hierarchy of followers is arranged along these same competitive lines. When playing boys often tend to engage in sometimes very physical, aggressive games, where wrestling, tripping, and pushing each other in fun is part of the activity (Brooks-Gunn & Matthews, 1979; Eder & Hallinan, 1978; Lever, 1976; McGrew, 1979; Savin-Williams, 2010). That this is rooted in biology and not early training, however, is also indicated by the fact that male chimpanzees and other social mammals behave in a similar fashion (de Waal, 1989; Eibl-Eibesfeldt, 1995; Goodall, 1986, 1990; Wickler, 1973).

Many girls tend to play in much smaller groups or in pairs; i.e. best friends (which is equivalent to mother-baby, sister-sister). Although hierarchies also tend to form, rather than based on physical competitiveness (although that is often a factor) it tends to be based on personality, articulatory skills, and physical attractiveness; e.g. who is the "nicest," prettiest, or the most fashionably dressed (Brooks-Gunn & Matthews, 1979; de Beauvoir, 1961; Eder & Hallinan, 1978; Ekert, 1990; Joseph, 1985). It is precisely because women are also competitive (see chapter 8), that female relationships with females who are not kin, tend not to be as long lasting, as compared to male-male friendships, the exception being female-female (e.g. mother-daughter, sister-sister) kin relationships, which, however, also tend to be quite rocky at times.

In contrast to boys and men who are more likely to become involved in physically aggressive team sports where there are clear winners and losers, many young girls are more likely to engage in cooperative activities that focus on friendship, intimacy, sharing, talking, imaginativeness, and being liked. And even among female athletes, the emphasis is often on the cohesiveness and importance of the team as a mutual cooperative, whereas males often seek to emphasize their own importance and personal status and accomplishments and to belittle their competitors.

Challenges and competition between girls are more likely to be subtle and indirect, whereas cooperativeness, at least overtly, is the glue which binds them together (Eder & Hallinan, 1978; Ekert, 1990). That this is biologically based is evident in that female chimpanzees behave in an identical fashion. As summarized by De Wall (1989, p. 137): "Adult male chimpanzees seem to live in a hierarchical world with replaceable coalitions partners and a single permanent goal: power. Adult females, in contrast, live in a horizontal world of social connections. Their coalitions are committed to particular individuals whose security is their goal...for them it is of paramount importance to keep good relationships with a small circle of family and friends."

Although both girls and boys often are concerned with being the best and tend to use force in order to get their way, boys are more likely to utilize actual physical violence; a trait rooted in biology, the sexual differentiation of the male brain, and the fact that males in general, almost regardless of species are more violent and murderous than females and in fact employ violence in order to gain access to females. By contrast, females tend to use words as weapons (the proverbial "woman's tongue") and rely on males to behave aggressively for them, and employ their sexuality to manipulate men or to compete with other women.

It is important to emphasize, however, that there are numerous exceptions, as many girls are quite physically competitive and aggressive, and many boys seek close social and emotional intimacy with their best friends as well. Probably most members of both sexes, in fact, fluctuate between these various different modes of interacting.

Nevertheless, these general patterns of interaction in turn tend to color the way in which men and women interact as adults as well, including the manner in which they view the actions and even the speech of others. That is, in very general terms, men and women often tend to be concerned with different aspects of the same experience since they sometimes have different priorities. In fact, even when males and females are engaging in ostensibly the same conversation, they focus on different aspects of the same shared information and emphasize features which are quite distinct (Croates, 1986; Emil & Hallinan 1978; Tannen, 1990).

Many males tend to employ speech and language as a means of imparting not only information, but as a manner of establishing status and superiority, which in turn may promote a man's ability to gain access not only to resources, but female sex partners. It is a common male stratagem to continually probe for weakness and to make continual attacks until thwarted or power is gained. However, such fights need not always be physical but based on threat and bluff and dominance displays that entail the clever or derogatory use of language. For example, human adolescent and adult males often utilize sarcasm, teasing, verbal ridicule, verbal directives, sexual jokes, or direct insults so as to establish superiority over a rival who may whither in response to the continual verbal onslaught. Indeed, sexual remarks, sexual "jokes" and teasing are frequent among adolescents as well as "professional" men, as it is a means of establishing sexual dominance.

Among men, comments about one another can be quite graphic and often go well beyond innuendo and include remarks as to sexual inadequacy or potency, and may go so far as to challenge other males to serve as willing sexual orifices; e.g. "Suck my dick!" Insults and sexual comments are in fact often seemingly made in fun, and although ostensibly and overtly accepted as such, they usually belie attempts to achieve dominance over other men (Joseph, 1985, 1992b, 1993).

In fact, sexual threats, with direct exposure of an erect penis is often employed by primates in their dominance displays, whereas phallic displays are a "universal" expression among human societies (Eibl-Eibesfeldt, 1995; Wickler, 1973). Indeed, in part is because of the potentially threatening nature of the penis, that phallic aggression has crept into the vocabulary. When someone is really "pissed off" we know he is angry and we should watch out. If someone says "fuck you!" we know they are not inviting us to have sex. If someone has been "fucked", or "screwed" it is understood that they have been diminished or taken advantage of in some manner.

Of course, males also use phallic and other forms of rough, derogatory and teasing comments in order to establish comaraderie. For example, one male friend may say to another, "you still driving that piece of shit," and both will laugh. However, when a man teases, comments, or interacts with a woman on a similar level, even when it is extremely and considerably toned down, she is likely to become exceedingly offended, sometimes to the bewilderment of the man who sees his behavior as normal.

Females are much less likely to insult one another in fun, deride each other's supposed sexual shortcomings (at least while face to face) or demand in a jocular tone that other women serve them as sexual objects. Unlike some men, it is not seen as as a means of establishing rapport or dominance.

Moreover, when women engage in the seeking of status, or when they seek to put one another down (again, at least while face to face) it tends to be less confrontational, more subtle and less aggressive as ostensibly they appear more concerned with the social harmony and the establishment of mutual, friendly understanding and rapport (Hughes, 1988). They are more likely to try to smooth things over and be more artful and subtle rather than overtly or physically aggressive, even when jockeying for positions of dominance. Of course, like men, many women have no difficulty with cutting up a competitor.

Men and women thus tend to use language differently and in order to convey different messages. Although both use language in order to convey information, details and facts, many women tend to be more interested in discussing what they or others feel, what happened to them or their friends, who said what, and how various relationships are going. For many women, talk often serves as a means for maintaining intimacy and friendly interaction as well as for conveying details on business, finance, and politics and a variety of other subjects (Croates, 1986; Joseph, 1992b, Tannen, 1990). For many men, these personal details are inconsequential and of little importance.

Ginger: "That business lunch was just wonderful. Ruth was there and my boss came and we met two prospective clients. You should have seen Ruth. She was so happy after getting that big bonus, and she was wearing the most gorgeous outfit. Really quite stylish and..."

Andy (yawning and checking his email): "So what happened?"

Ginger: "Turn off that computer so I can tell you. Anyway, she has lost so much weight. She must really be feeling good about herself, with all the money she has been bringing in. Even Sally was impressed, and you know how Sally feels about that."

Andy: "About what?"

Ginger: "Haven't you been listening to me?"

Andy: "Yeah, sure. Business! lunch! I'm listening. Go on."

Ginger: "Anyway, and, oh, did I tell you? Sally's little boy has been getting so big. She brought these pictures. He is just so cute. And Ruth..."

Andy: "I thought this was a business lunch."

Ginger: "It was. You should have seen the look on Ruth's face when she saw those pictures. She wants to have a baby so bad...."

Andy: "What has this got to do with your job? What did your boss have to say about that?"

Ginger: "Sally is my boss!"

Andy: "And she brings pictures of her kid to show you and her clients?"

Ginger: "What's wrong with that?"

Andy: "I don't see what that has to do with your job."

Ginger: "It has everything to do with my job. If you work with people you should be interested in what is going on in their life. People don't live and work in a vacuum. What goes on at home affects their work. Haven't you've learned that ?

Andy (yawning and clicking on Sports News): "Learned what?"

For many men, when language and talk is personal it is either viewed as irrelevant or power oriented. It often becomes personal when they feel their status or authority may be threatened.

Jake has come home and begins to complain to his wife Ruth about a problem at work. He is very irritated and upset.

Jake: "I don't know what's wrong with my boss. He hired this new girl and she is not working out!

Ruth: "What does she look like?"

Jake: "Very ugly. Anyway, she was supposed to type this stuff and she never even got started.

Ruth: "How old is she? Is she pretty?"

Jake: "What's that got to do with anything? She's not doing her job. She was supposed to type these papers for me and she didn't do it."

Ruth: "Well, she's new. Is she married?"

Jake: "Why are you asking me all this crap? Who cares what she looks like or how many times she's been married. She is incompetent."

Ruth: "Maybe if you were nicer to her..."

Jake: "Hey! Whose side are you on? She's the one who is screwing up. Not me."

Ruth: "I didn't say you were."

Jake: "Bull shit! I don't know why I even bother talking to you. All you do is put me down."

Ruth: "Well excuse me for being curious."

Men tend to use language to describe what they are going to do and perhaps how they are going to do it to. It serves more as a form of achieving mastery over their environment and establishing their status and position in the world rather than as a means of revealing their feelings or intimate personal details about their life.

By contrast, those men who like to gossip about others and to share intimate and personal details are often shunned by other men as they are not interesting to talk to.

Women are more likely than men to use language as a means of maintaining social intimacy and are much more willing or likely to interweave professional talk with friendship. Women and girls use language as a means of maintaining a relationship, of feeling close and involved. Language is the glue that binds, and women are very concerned about and interested in what other women are talking about, even if (from a man's point of view) they are talking about nothing at all. In fact, one of the highest ratings of any television network production has been consistently obtained by an ABC daytime television show, "The View" in which a group of white and black women sit around and talk (Neilsen Ratings, May, 2010).

Men are more likely to focus on accomplishing something together, or talking about the accomplishments of others such as those who won a certain football game and those who excel or need improvement in their performance. Indeed, men are more likely to utilize speech as a means of maintaining or establishing status and to leave out details concerning their emotions or personal problems. Again, this can be directly attributed to the hunting way of life where skill and male vs male competitive (and cooperative) pressures not only determined success and failure but where weaknesses were ferreted out and ridiculed.

Women tend to become more interested in the details of a person's life and in discussing the details or their own lives as well as their thoughts and feelings, whatever they may be. When women are together this tendency naturally forms a bridge that maintains their rapport. They expect to be informed. However, when women are with men, they may feel that males is acting disinterest or not being communicative and is somehow lacking in the ability to express his feelings. Indeed, it is a common complaint expressed by my female patients and in the media; supposedly men do not express their feelings and are not as communicative, at least not in the company of the opposite sex.

In fact, often the man is disinterested and is wondering why she is talking about this or that and asking him all these "irrelevant" questions. Indeed, many men tend not to share their feelings and other personal details, because to do so might put them at risk with other men, who, being so competitive, may use it against them or make fun of them. Moreover, knowing that other men are not very interested in the social and personal aspects of another man's life, even when discussing family and children, men usually keep these topics quite brief as it seems irrelevant to the task at hand; to conquer the world, make more money, get that promotion, or to discuss those who are at present conquering the world, e.g. politics and sports (Gould, 1976). These are viewed as worthy topics to discuss at length.

Even when it comes to personal difficulties, men are more likely to keep it to themselves and answer in as few words as possible. By contrast, women are more likely to not only discuss these issues at length, but to discuss it again and again with all their friends.

Andy: "So how you been getting along since that divorce?"

Jake: "You know how it is."

Andy: "Yeah."

Many men often see no competitive advantage in discussing personal difficulties and to discuss such issues are sometimes seen as an admission of personal weakness or as a source of information that can be used to create an imbalance in power. Indeed, in my private practice, I have often heard men bitterly complain about how their spouse or girlfriend goes around telling people, including their relatives, very personal details about their relationships. Many men often view this as betraying information that is best kept secret and just between them. For them it is a breach of privacy and trust. "And its nobody g... d.... business!"

For many women, to keep these details to themselves is a breach of friendship and intimacy, for often their friends, and mothers insist on knowing the details and derive considerable enjoyment in hearing and discussing them. Again, consider Ms. Clinton revealing her husbands "emotional" problems, and "addictions" and "dysfunctions" and detailing his "emotionally abusive" childhood in a woman's magazine: TALK. Hence, women see nothing wrong or unusual about discussing or revealing intimate details about themselves or others, whereas men tend to fear they may lose power by doing so (as other men will make fun of them) and are thus so disinclined.

However, when it comes to plans, actions, and future goals, or even past conquests, the discussion between men is likely to be quite extensive, particularly if it can enhance their status.

Andy: "So getting any lately?"

Jake: "I'll say. You remember that blond that used to work at the...."

Nevertheless, even in these contexts, men are less likely to discuss what was said, how they felt about it, how romantic the evening was, how their partner felt, and so on, because it is viewed as irrelevant. They want to talk about their actions and accomplishments, whereas women tend to focus on the relationship. This of course can create considerable difficulty when men and women talk together.

Girls and women, thus tend to focus on the social, supportive, familial and the communal aspects of interaction. Men tend to focus on the individual and the struggle for dominance over the community, the family, and each other. In this regard, when women interact they tend to be less confrontational as they are more sensitive to social and emotional nuances, and as one of their main interests is to maintain the cohesion of the group or the friendship. Indeed, these same exact sex differences are also evident among other primates including chimpanzees (De Waal, 1989), and are also directly related to sex differences in the division of labor over the course of evolution, and the differential male vs female pattern of neurological development.

Again, however, it is important to emphasize that there is much variability in these seemingly differential attitudes toward the use of language as a means of social and intimate interchange. Some men are more interpersonally inclined, and many gossip among themselves about sports, politics, and each other, although they do not call it as such. Similarly, many women are much more interested in the non-personal, political, business, informational aspects of communication and would rightly take extreme exception to the notion they spend their time talking predominantly about family, relationships, clothes, shoes, sex, or each other.

OVERVIEW: SEX DIFFERENCES, THE PARIETAL LOBE, CORPUS CALLOSUM & LANGUAGE

Superior female linguistic capacities appear to be related to their evolutionary history as child bearers, gatherers, tool makers, hide preparers and so on; tasks which would require superior social-emotional capabilities (e.g. child care, socializing) and which require considerable temporal-sequential motor capability and somesthetic sensitivity which in turn is associated with Broca's area and the parietal lobule in particular.

Gathering (and to a lesser extent tool making vs hunting), is far more likely to require bi-manual activation of the hands, and thus simultaneous activation of both parietal lobes and the motor areas of the frontal lobe. These activities might also be expected to coincide with increased functional and metabolic activity in the parietal lobes, which in turn might lead to an expansion of this tissue (inferior parietal) and an increased need to intercommunicate.

In this regard it is noteworthy that the posterior portions of the corpus callosum, i.e. the rope of fibers that interconnect the right and left parietal lobes, may be thicker and larger in females than males (Holloway et al. 1994). However contrary evidence abounds (reviewed in Holloway et al. 1994). Nevertheless, as gathering and related activities are more likely to involve both hands (vs hunting and throwing with a single hand), it might be expected that the posterior (female) corpus callosum would be larger because the female right and left parietal lobe were probably simultaneously activated and utilized by females more than males for the last 100,000 or so years. On the other hand, one might equally expect that the corpus callosal interconnections linking the motor areas, particularly those involved in fine motor functioning, would be more developed in females.

Given that women have clear language superiorities and unlike males were able to frequently and continually talk and chatter, if not with other women, then their children, sisters, or mothers, and as the IPL is involved in word finding, sentence construction, and organizing Broca's area for grammatical speech production, it might also be expected that the female IPL is more involved in language production as compared to males. Again, however, the same might be said of Broca's area.

It might also be expected that the female left parietal lobe would be more resistant to aphasic disturbances as compared to males whose language representation is more sparse and fragile. In fact, males are far more likely to become severely aphasic with left parietal injuries than females, who in turn more quickly recover from similar damage (Kimura, 1993). Conversely, females are far more likely to experience expressive aphasia with damage to Broca's area than males (Hier et al. 1994; Kimura, 1993). Again, however, they are also more likely to regain expressive speech functions.

On the other hand, it may be that sex differences in severity of initial language loss in anterior vs posterior lesions may be a function of sex differences in the vasculature and the origin and/or type of debri responsible for cerebral infarcts. In this regard, although females demonstrate clear language superiorities and have more neurological space in the right and left hemisphere devoted to linguistic and expressive functions (Joseph 1993), and in fact demonstrate bilateral activation when engaged in certain verbal tasks, whereas males tend to be more unilateral (Shaywitz, et al., 1995), the exact nature and neurological foundations for these differences are certainly still open to debate.

THE INFERIOR PARIETAL LOBE, APRAXIA, AND TEMPORAL SEQUENCING

The angular gyrus of the inferior parietal lobule (IPL) is unique to humans. An examination of the ape brain fails to find any evidence of this structure. The angular gyrus and IPL are crucially evolved in controlling temporal sequential hand movements including the manipulation of external objects and internal impressions (De Renzi and Lucchetti, 1988; Heilman et al., 1982; Joseph, 1982; Kimura, 1993; Strub and Geschwind, 1983). The evolution of the angular gyrus enabled humans to engage in complex activities involving a series of related steps, to create and utilize tools, to produce and comprehend complex gestures, such as American Sign Language, and to express and perceive grammatical relationships--capacities which are disrupted with lesions localized to the IPL (Joseph 1982, 1988, 1996, 2000a,b). In fact, the motor engrams that make possible temporal and sequential motor acts, including those involved in grammatical verbal expression, are partly localized within the IPL (De Renzi and Lucchetti, 1988; Heilman et al., 1982; Kimura, 1993; Strub and Geschwind, 1983). In fact, the IPL not only interacts with but appears to program the frontal motor areas for the purposes of producing fine motor, temporal-sequential movements, including the vocalization of speech units (Joseph 1982).

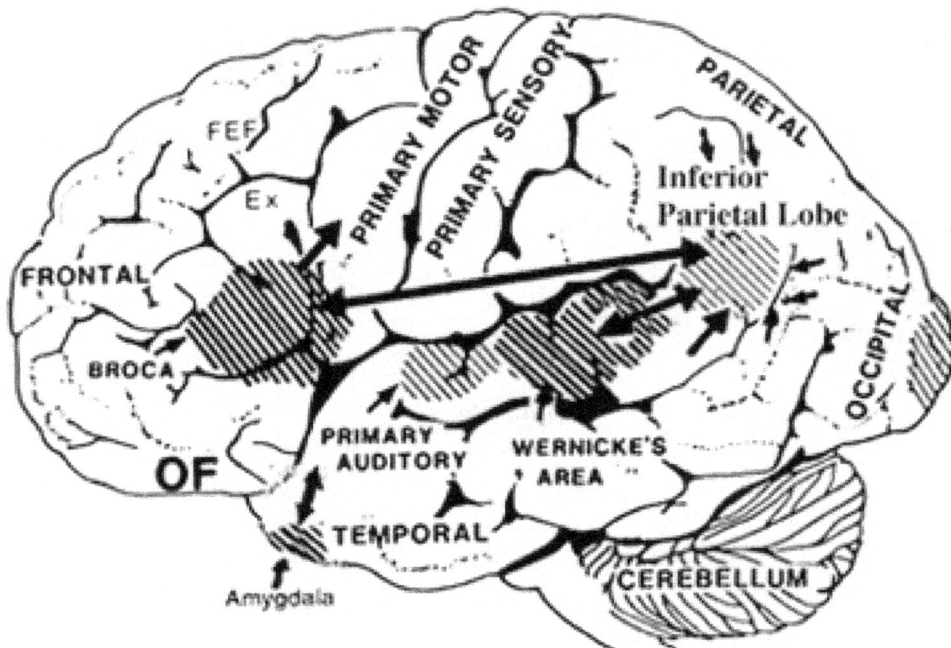

Those devoid of an angular gyrus/IPL, or those who have suffered a severe injury to this area, are generally unable to correctly manipulate or fashion complex tools -much less utilize them in a complex temporal sequence. This condition is referred to as apraxia; i.e. an inability to perform tasks involving interrelated steps and sequences (Joseph 1996). With severe IPL injuries, the individual may be unable to make a cup of coffee or put on their clothes, much less fashion or sew them together. Moreover, grammatical speech is disrupted and patients may suffer extreme word finding difficulty, or a conduction aphasia. That is, speech is no longer produced as Broca's area is disconnected from the IPL and Wernicke's area (Joseph 1982, 1996). Likewise, reading ability is disrupted as the IPL not only comprehends and produces gestures but visual symbols including written language. Hence, the IPL/angular gyrus (including the frontal motor areas) makes possible the ability to fashion and manipulate tools and organizes speech into vocabulary-rich, temporal sequential grammatical units.

As apes do not possess an angular gyrus, it appears that over the course of evolution, with the development of right handedness and selective pressures acting on gene selection across gathering/tool-making generations, the IPL/angular gyrus emerged as an extension of the auditory area in the temporal lobe and the superior parietal visual-hand area. Indeed, the parietal lobes are considered a "lobe of the hand" (Joseph 1982, 1993, 1996) and contain neurons which guide hand movements (Hyvarinen, 1982; Kaas, 1993; Lynch, 2010; Mountcastle et al., 1975, 2010) and which respond to visual input from the periphery and lower visual fields -the regions in which tool-making hands are most likely to come into view.

Because most individuals would use the right hand for tool making and the left for holding the tool, it is the left parietal lobe (which monitors the right lower visual field and controls the right hand) that guides and visually observes, learns and memorizes hand-movements when gathering, gesturing, or manipulating some object or constructing a tool. Over the course of evolution and as experience and the environment act on gene selection and induce neural plasticity, the parietal (and frontal) lobe expanded, the angular gyrus emerged, and neuroplastic alterations were induced in the adjoining motor-hand area in the frontal lobe including what would become Broca's speech area.

THE RIGHT HAND AND THE LEFT FRONTAL LOBE

Specifically, it is the lateral surface of the right and left frontal lobe which control vocalization and verbal-though production, the left frontal (Broca's area) producing the words, the right frontal the melody of language. The frontal lobes are interlocked with the association areas in the posterior regions of the cerebrum (Petrides & Pandya. 1999, 2001), including Wernicke's language comprehension area and the angular gyrus - inferior parietal lobe (IPL), as well as the memory centers in the temporal lobe (Joseph 1986a, 1996, 1999). The frontal lobes are therefore continually informed about and have continual access to information processed in these areas of the brain.

The frontal lobes serve as the senior executive of the brain and personality (Joseph 1986a, 1999, 2010) and play a significant role in searching for and assimilating the information which will be thought about or spoken out loud (Christoff and Gabrieli, 2000; Christoff et al., 2001; Newman et al., 2002, 2003; Paulesu et al. 2010). The frontal lobes are responsible for organizing the expression of speech, and this is accomplished by programming the frontal motor areas controlling the movements of the mouth, tongue, jaw, and lips, i.e. the primary and pre-motor areas located along the precentral gyrus (area 4), and within which are represented the muscles controlling the hands, fingers, and oral laryngeal musculature.

Specifically, the tissues of the mind known as Broca's Expressive Speech area, which is located in the left lateral surface of the frontal lobe (areas 45, 46), is responsible for programming the fine motor movements involved in the articulation of speech. Broca's speech area imposes temporal sequences on on the intonational contours of everything which is to be articulated so that the melodic features of generalized vocal expression come to be punctuated, sequenced, and segmented, and vowel and consonantal elements are produced. Left hemisphere speech comes to be superimposed over limbic and right hemisphere melodic language output (Joseph 2000). Thus, the primary area representing the oral-laryngeal musculature, is programmed by Broca's expressive speech area which becomes active prior to vocalization and during subvocalization as indicated by functional imaging.

LEFT FRONTAL DEVELOPMENTAL DOMINANCE FOR MOTOR FUNCTIONS

The motor areas of the frontal lobe include Broca's expressive speech area, Exner's hand area, the supplementary motor area, and the secondary and primary motor areas. There is a one-to-one correspondence between single neurons in the primary motor area and single muscles such that the musculature of the entire body surface is represented according to their importance (Chouinard & Paus 2006; Dum & Strick 2005; Verstynen, et al. 2011). For example, the hands, fingers, face and mouth, have a greater representation than the musculature of the back. The representation of the musculature is more diffuse in the secondary premotor area. The premotor transmits its motor impulses to the primary motor areas and subcortical regions of the brain but receives its marching orders from the supplementary motor area (Joseph 1999b; Nachev et al. 2008).

Movement, including those involved in expressive speech, is made possible by ropes of descending axons nerve fibers which project from the frontal motor areas to the brainstem and spinal cord. These nerve fiber ropes are referred to as the pyramidal and corticospinal tracts (Miller & Cummings 2006; Risberg

& Grafman 2006). Hence, if the motor areas of the right or left frontal lobe are destroyed, the left or right half of the body becomes paralyzed. However, if damage is restricted to Broca's area, the ability to speak in words is lost.

Among 80% of humans, the motor cortex of the left frontal lobe is dominant for expressive speech, and for controlling handedness and programming gross and fine motor, temporal-sequential, movements of the musculature, and of the right hand and fingers in particular . It is this dominance and the functional organization of the left frontal lobe and the left hemisphere, which led to the evolution of modern human language, the expression and comprehension of which is also associated with the functional integrity of the left half of the brain.

Left hemisphere dominance is due to factors associated with fetal development and experiential influences on gene expression over the course of evolution.

During fetal development, and due to the earlier maturation of the left cerebral motor areas, the descending corticospinal tract from the left hemisphere matures and establishes its connections with the brainstem and spinal cord in advance of the axons from the right hemisphere (Joseph 1982). The left corticospinal tract grows more quickly and descends into the brainstem, and then crosses at the pyramidal decussation, and then descends into the spinal cord in advance of those fibers from the right (Kertesz & Geschwind 1971; Yakovlev & Rakic 1966). This provides the left hemisphere and the right hand with a competitive advantage in establishing both motor control and thus right hand dominance. Thus, the left hemisphere motor areas gain a competitive advantage over the right hemisphere, and acquire greater motor control over brainstem and spinal motor neurons controlling movement (Joseph, 1982). Therefore, the somatomotor areas of the left frontal lobes mature at a more rapid rate than the right frontal lobes and by the first year tissues such as Broca's area overtake their right sided counterparts (Scheibel, 1991, 1993). However, because the corticospinal tract does not become functionally mature or myelinated until the 8th-12th postnatal month (Yakovlev & Lecours, 1968) it is only as the infant approaches their first birthday that handedness become obvious or apparent. Corticospinal and pyramidal axons continue to become increasingly well myelinated over the course of the second year (Debakan, 1970; Langworthy, 1937; Yakovlev & Lecours, 2007)--a process which continues into late childhood (Paus et al., 1999). Hence, around the 8-12th month the neocortex and frontal motor areas begin to increasingly exert hierarchical control over the limbic forebrain, brainstem, and musculature, and most individuals are right handed.

Broca's expressive speech area appears to be an evolutionary partner, if not an outgrowth of the adjacent frontal area controlling fine motor movements of the hands and fingers; referred to as Exner's writing area and which is located near the foot of the second frontal convolution of the left hemisphere, occupying the border regions of Broadmans areas 46, 8, 6. Exner's area appears to be the final common pathway where linguistic impulses receive their final motoric stamp for the purposes of writing; i.e. the formation of graphemes and their temporal sequential expression. Thus, Exner's writing area appears to program the adjacent hand-area represented in the primary motor areas (e.g. Boroojerdi et al., 1999) so that lingusitic impulses received through Broca's area, can be integrated into hand movements so that words can be written down.

Exner's area is dependent on Broca's expressive speech area with which it maintains extensive interconnections. In fact, Exner's writing center extends to and appears to become coextensive with Broca's area (Lesser et al. 1984)--which in turn was originally a hand area--at least in primates. Broca's area possibly acts to organize and relay impulses received from the posterior language zones to Exner's area in instances where written expression is desired. Exner's area, in turn, transfers this information to the secondary and primary motor areas for final expression.

Because these areas are linked, when most people talk, their expressive speech is often accompanied by hand movements. Conversely, whereas damage to Broca's area will significantly impair the ability to speak in words, destruction of Exner's area results in apraxia.

Broca's and Exner's areas do not exist in the brains of apes. In fact, the frontal lobes are not well developed in these species, or in ancestral pre-human species, be it Australopithecus, H. habilis, H. erectus, or Neanderthal. Rather, the frontal lobe does not significantly expand in size until the onset of the Upper Paleolithic and the evolution of the Cro-Magnon people, complex tool making technology, and language).

The primary auditory receiving areas are located in the superior temporal lobe. Once received and processed, these auditory signals are transferred to the immediately adjacent Wernicke's area which associates these sounds and comprehends the words of language; whereas the the auditory association area in the right temporal lobe comprehends environmental sounds and melodic and emotional vocalizations (Joseph 1982, 1988a, 2000). Via a massive neural pathways (the corpus callosum) the right and left auditory receiving areas work together when presented with complexity and paralinguistic features which require analysis (Just et al., 1996; Michael et al. 2001; Schlosser et al., 1998).

Movement and motor functioning are dependent on the functional integrity of the basal ganglia, brainstem, cerebellum, cranial nerve nuclei, the motor thalamus, spinal cord, as well as the primary, secondary and supplementary motor areas of the frontal lobes. These areas are all interlinked and function as an integrated system in the production of movement.

For example, Exner's writing area is in part, within areas 6 and becomes active prior to (as well as during) hand movements and appear to program hand movements, whereas the frontal eye fields (within areas 6,8,9) becomes active prior to (as well as during) eye movements and appears to program eye movements.

The motor areas do not just program, but anticipate, as well as memorize certain movement programs. In fact, individual neurons within the motor cortex (of monkeys) have been shown to learn and neurally encode serial order when performing a context-recall visual-memory scanning task (Carpenter et al., 1999). Moreover, neural activity will change when the serial order is changed and in accordance with the direction of movement as well as stimulus position and movement (Carpenter et al., 1999; Passingham, 1997). In fact, according to Carpenter et al., (1999, p. 1755), although "individual cells prove largely independent information about the items in the sequence... during different epochs of presentation of the stimuli, different patterns of distributed activity in even small ensembles of motor cortical cells are sufficiently distinct and robust to provide a basis for encoding the sequence," which in turn allows for memory scanning and retrieval.

In addition, "motor" neurons in the inferior-ventral portion of the (monkey) premotor cortex can code for the spatial location of nearby sounds, particularly those originating near the head (Graziano et al., 1999). Indeed, these neurons appear to be "multi-modal" and can also respond to visual and tactile stimuli that impinge near or upon the face (Graziano et al., 1999).

However, these motor areas also act in association with visual-spatial information received from the parietal and occipital lobe, so that the movements of the body, the hands in particular, can be guided while moving about in visual space. In this way, gross movements of the hand and arm can also be coordinated with visual impressions, thereby enabling an individual to throw an object with accuracy, such as a spear to impale a running animal.

Visual information is transmitted to the frontal eye fields (FEF) which are located along the superior lateral convexity of the frontal lobe and which are part of the premotor area and encompasses all of area 8 as well as portions of areas 9. As the name implies, the FEF is concerned with eye movements, and appears to program the corresponding primary areas so as to guide gaze shifts and visual tracking. The FEF receives visual information from the primary and association visual cortices in the occipital lobe (17, 18, 19), the auditory association (22) and multimodal visual association areas (20) in the temporal lobe (Barbas & Mesulam, 2011; Jones et al., 1978; Jones & Powell, 1970), and the somatosensory association area (Crowne, 1983). It also shares interconnections with the caudate, superior colliculus, and oculomotor nucleus (Astruc, 1971; Knuzle & Akert, 1977; Segraves & Goldberg, 1987). Hence, the FEF receives information concerning the auditory, tactual, and visual environment and is multimodally responsive.

The FEF coordinates and maintains eye and head movements, gaze shifts, and thus orienting and attentional reactions in response to predominantly visual, but also tactile and auditory stimuli. It is also involved in focusing attention on certain regions within the visual field, particularly the fovea guiding eye movements while reading, writing, and when tracking and throwing.

In addition to visual tracking and supporting focused visual attention, neurons in the FEF demonstrate anticipatory activity; that is, firing before a response is made. In fact, these neurons will continue to fire at a high rate until the behavior is initiated. In this regard, neurons in the FEF probably exert a countering influence so that attention does not drift and attention and eye position remains stabilized and focused on the primary retinal target.

Moreover, cells in the FEF will fire selectively in response to stationary and moving stimuli, to objects which are within arms reach, as well as to tactual stimuli applied to the hands and/or mouth. In fact, as an object approaches the face and mouth, some of these cells correspondingly increase their rate of activity (Rizzolatti et al., 2011ab). Hence, cells within the FEF are highly involved in mediating sustained attention and orienting reactions of the head and eyes, maintaining visual fixation and modulating visual scanning, as well as coordinating eye-hand and hand to mouth as well as smooth pursuit eye movements.

Motor functioning and movement, however, is also dependent on the parietal lobes, thalamus, basal ganglia, brainstem, cerebellum, and spinal cord, structures which are directly or indirectly interconnected (Kaas, 1993; Mink, 1997; Passingham, 1997; Schmahmann, 1997).

For example, smooth, purposeful, coordinated movement requires sensory feedback which is provided by the primary and association/secondary somatosensory areas located in the parietal lobe and which contain neurons which guide the hand and arm in visual space. These cortices are intimately linked with the primary motor and the motor association/secondary and supplementary motor areas as well as the basal ganglia, brainstem, and cerebellum--regions which become highly active during and often prior to movement initiation (Passingham, 1997). For example, when human subjects learn a sequence of finger movements, functional imaging reveals increased activity in the primary and premotor cortex, and the primary and association somesthetic cortex, as well as in the cerebellum, striatum, ventral thalamus and cerebellum (Passingham, 1997). In fact, most of the "motor" area also contain independent motor maps of the body, which is why, if a subject moves an arm, leg, or shoulder, each of these areas becomes active almost simultaneously--at least as demonstrated with functional imaging.

BROCA'S EXPRESSIVE APHASIA

Broca's speech are is located in the general vicinity of the posterior- inferior region of the left frontal area (i.e. third frontal convolution), and includes portions of areas 45, 6, 4, and all of area 44. This region is multimodally responsive (Passingham, 1997) and receives projections from the auditory, visual, somesthetic areas (Geschwind, 1965; Jones & Powell, 1970), as well as massive input from the inferior parietal lobule and Wernicke's area via a rope of nerve fibers referred to as the arcuate fasciculus (which also links these areas to the amygdala). In addition, Broca's area receives fibers from and projects to the anterior cingulate as well as the brainstem periaqueductal gray which subserves vocalization.

Broca's speech area is a final converging destination point through which thought and other impulses come to receive their final sequential (syntactical, grammatical) imprint so as to become organized and expressed as temporally ordered motoric articulations; i.e. speech. Verbal communication, the writing of words (via transmission to Exner's area), and the expression of thought in linguistic form is made possible via Broca's area which programs the adjacent oral-laryngeal musculature as represented within the adjacent primary motor areas; and which transmits to Exner's writing area so that words may be written.

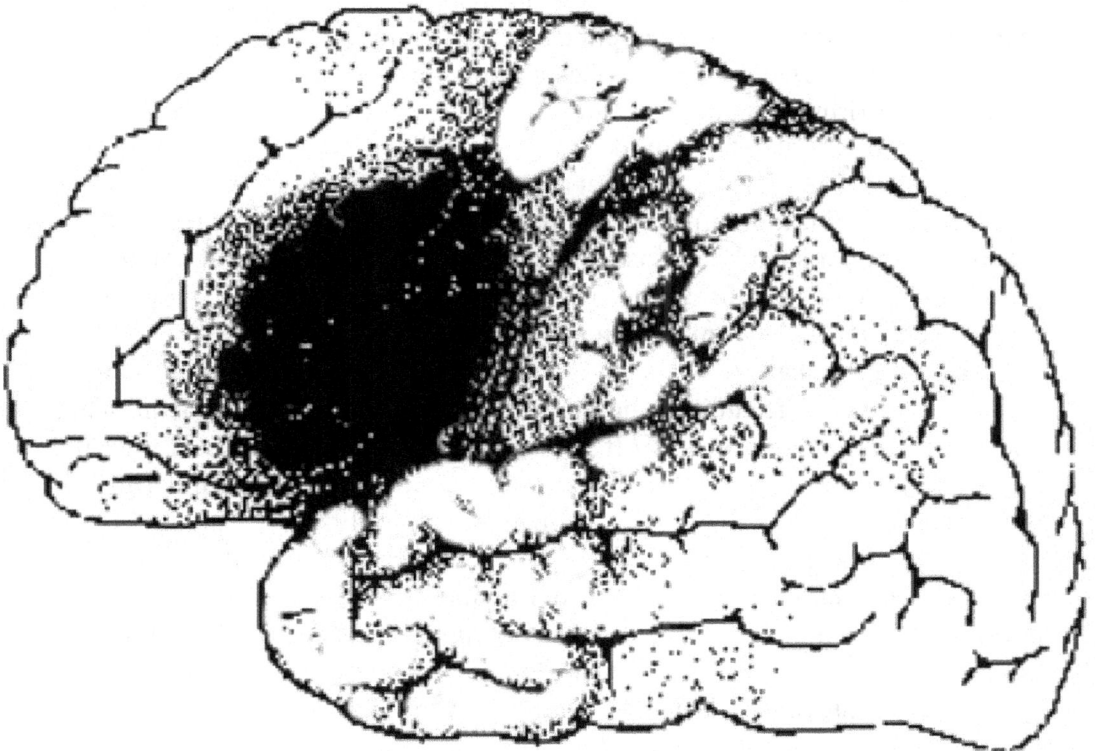

The importance of Broca's area and the left frontal lobe has also been demonstrated through functional imaging. For example, the left frontal lobe becomes activated during inner speech and subvocal articulation (Paulesu, et al., 1993; Demonet, et al., 1994). The left frontal lobe also becomes highly active when reading concrete and abstract words (Buchel et al., 1998; Peterson et al., 1988), and when engaged in semantic decision making tasks (Demb et al., 1995; Gabrielli et al., 1996). Moreover, activation increases as word length increases and in response to long and unfamiliar words (Price, 1997).

With injuries to Broca's area the individual loses the capacity to produce fluent speech. Output becomes extremely labored, sparse, and difficult, and they may be unable to say even single words, such as "yes" or "no". Often, immediately following a large stroke patients are almost completely mute and suffer a paralysis of the upper right extremity as well as right facial weakness (since these areas are neuronally represented in the immediately adjacent area 4). Patients are also unable to write, read out loud or repeat simple words. Interestingly, it has been repeatedly noted that almost immediately following stroke some patients will announce "I can't talk", and then lapse into frustrated partial mutism.

With less severe forms of Broca's (also referred to as expressive, motor, nonfluent, verbal) aphasia, speech remains labored, agrammatical, fragmented, extremely limited to stereotyped phrases ("yes", "no", "shit", "fine") and contaminated with syntactic and paraphasic errors i.e. "orroble" for auto, "rutton" for button. Writing remains severely effected, as are oral reading and repetition. Such patients also have mild difficulties with verbal perception and comprehension (Hebben, 1986; Maher et al. 1994; Tramo et al., 1988; Sarno, 1998; Tyler et al. 1995), including the ability to follow 3-step commands . Commands to purse or smack the lips, lick, suck, or blow are often, but not always, poorly executed; a condition referred to as bucal-facial apraxia. Non-speech oral movements are seldom significantly effected.

With mild damage, patients may demonstrate severe confrontive naming and word finding difficulties (anomia), as well as possible right facial, hand, and arm weakness. Speech is often characterized by long pauses, misnaming, paraphasic disturbances and articulatory abnormalities . Stammering and the omission of words may also be apparent. Similarly, electrical stimulation of this region results in speech arrest (Ojemann & Whitaker, 1978; Lesser et al., 1984) and can alter the ability to write and/or perform various oral-facial movements. It is noteworthy that even with anterior lesions or surgical frontal lobectomy sparing Broca's area a considerable impoverishment of spontaneous speech can result (Luria, 2010; Milner, 1971; Novoa & Ardila, 1987). Disturbances involving grammar and syntax, and reductions in vocabulary and word fluency in both speech and writing have been observed with frontal lesions sparing Broca's area. In word fluency tests, however, simple verbal generation (e.g. all words starting with L) is usually more severely impaired than semantic naming, e.g. all animals which live in the jungle--which is presumably a function of semantic processing being more dependent on posterior language areas.

THE EVOLUTION OF THE SPEECH AREAS

Non-human primates lack a functional Broca's area. Among non-human primates, the left frontal lobe, including the tissues homologous to Broca's area, does not subserve speech or vocalization (Jurgens et al. 1982; Myers 1976). Vocalization in apes and monkeys has a non-segmented organization, consisting of moans, screams, barks, grunts, pants, and pant-hoots (Erwin, 1975; Fedigan, 1992; Goodall, 1986, 1990; Hauser, 1997) and is the province of the limbic system and brainstem; e.g. the cingulate gyrus, amygdala, hypothalamus, and periaqueductal gray (Joseph, 1982, 1993; Jurgens, 1990; MacLean, 1990; Robinson, 2007; 1972). Therefore, although damage to Broca's area in humans results in a profound expressive aphasia, similar destruction in non-human primates has no effect on vocalization rate, the acoustical structure of primate calls, or social-emotional communication (Jurgens et al. 1982; Myers 1976). Rather, in primates, those tissues homologous to "Broca's area" are directly involved in motor activity of the arms and hands (Rizzolatti, et al. 1988). Moreover, neural pathways linking the primate IPL with the homologous primate "Broca's area" are only weakly developed (Abolitiz and Garcia, 1997). Apes and monkeys, however, do not possess an angular gyrus.

The angular gyrus sits at the junction of the posterior-superior temporal and the occipital-parietal lobes, and is critically involved in naming, word finding, grammatical speech organization, and is in part an extension of and links Wernicke's with Broca's area (Goodglass & Kaplan, 2000). Through its extensive interconnections with the adjacent sensory association areas, the IPL/angular gyrus receives and assimilates complex associations, thereby forming multi-modal concepts, and acts to classify and name this material which is then injected into the stream of language and thought. The IPL/angular gyrus, in concert with Wernicke's area, transmits this information to Broca's speech area, which in turn organizes the immediately adjacent oral, laryngeal motor areas (Foerster, 1936; Fox, 1995).

Just as Broca's area evolved from and is linked to tissues controlling the hand, Wernicke's ares and the angular gyrus evolved from and are linked to those areas which perceive meaningful vocalizations: the auditory cortex of the temporal lobe. Ninety percent of primate auditory cortex neurons are activated by species-specific calls (Newman & Wollberg 1973), whereas destruction of the left superior temporal lobe disrupts that ability to make sound discriminations (Heffner & Heffner 1984; Hupfer et al. 1977; Schwarz & Tomlinson 1990). Moreover, asymmetries in the planum temporal are apparent in chimpanzees (Gannon 1998), and the primate left hemisphere has also been shown to be dominant for the perception of primate vocalizations (Hauser & Anderson 1994; Peterson & Jusczyk 1984; Peterson et al. 1978).

Presumably, left planum temporal and hemisphere dominance for vocal perception and comprehension gradually increased in the transition from Australopithecus, to H. habilis, to H. erectus, to Neanderthals. As first proposed and detailed elsewhere (Joseph, 1993, 1996, 2000b) as Wernicke's area, the parietal-hand areas and the IPL expanded, merged, and collectively gave rise to the angular gyrus, auditory input began to be sequenced, and Wernicke's area became specialized for perceive and comprehending language units.

In addition, the arcuate fasciculus axonal pathways leading from the IPL to Broca's areas also significantly expanded and increased in density, and Wernicke's area became tightly linked with and began transmitting auditory-linguistic signals to Broca's area thus inducing neuroplastic alterations in these tissues. Likewise, the frontal lobes also evolved, with Exner's area evolving and becoming specialized to control fine motor movements of the fingers and hands, and Broca's area evolving and becoming specialized to control articulatory movements of the speech musculature.

Hence, Wernicke's area began sequencing auditory input, and Broca's area was transformed from a hand area to a speech area and ceased to control hand movements whereas Exner's area became specialized to control the fingers and hands. Once Broca's area evolved into a speech area, it began to organize the adjacent primary motor oral-laryngeal areas so as to express the words and sentences transmitted from the IPL and Wernicke's area.

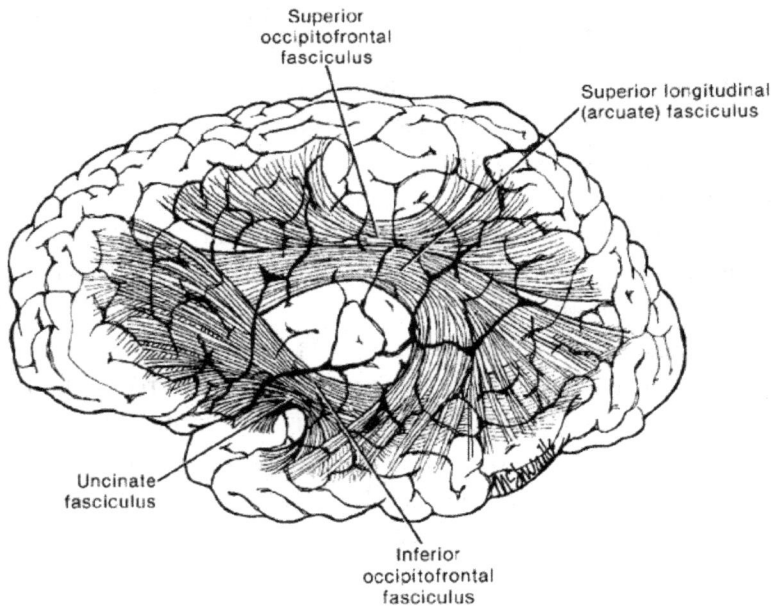

Moreover, as the right and left frontal vocalization areas are richly interconnected with the anterior cingulate vocalization centers, whereas the temporal lobe is tightly linked with the amygdala, once these neural-plastic transformation took place, "limbic language" (emotional speech) became hierarchically represented, yoked to the neocortex and subject to fractionization, temporal sequencing, and multi-classification (Joseph, 1999d,e). Wernicke's area was now able to communicate with Broca's area, with the angular gyrus injecting temporal sequences and assimilated associations into the stream of language and thought.

Hence, in addition to manipulating tools in a temporal sequential fashion, the evolution of Exner's area and the IPL/angular gyrus enabled humans to manipulate linguistic impulses in temporal sequences which were transmitted to the oral-laryngeal musculature which began to vocalize units of speech.

However, as based on an analysis of tool technology, it can be concluded that Australopithecus, H. habilis, H. erectus, and Neanderthals did not possess the neurological sophistication for vocalizing complex human language, and had not yet evolved an angular gyrus or a functional Broca's area (Joseph 1992, 1993, 1996, 2000a,b). Rather, the evolution of modern speech likely corresponded to the evolution of the Upper Paleolithic female gatherer and tool maker.

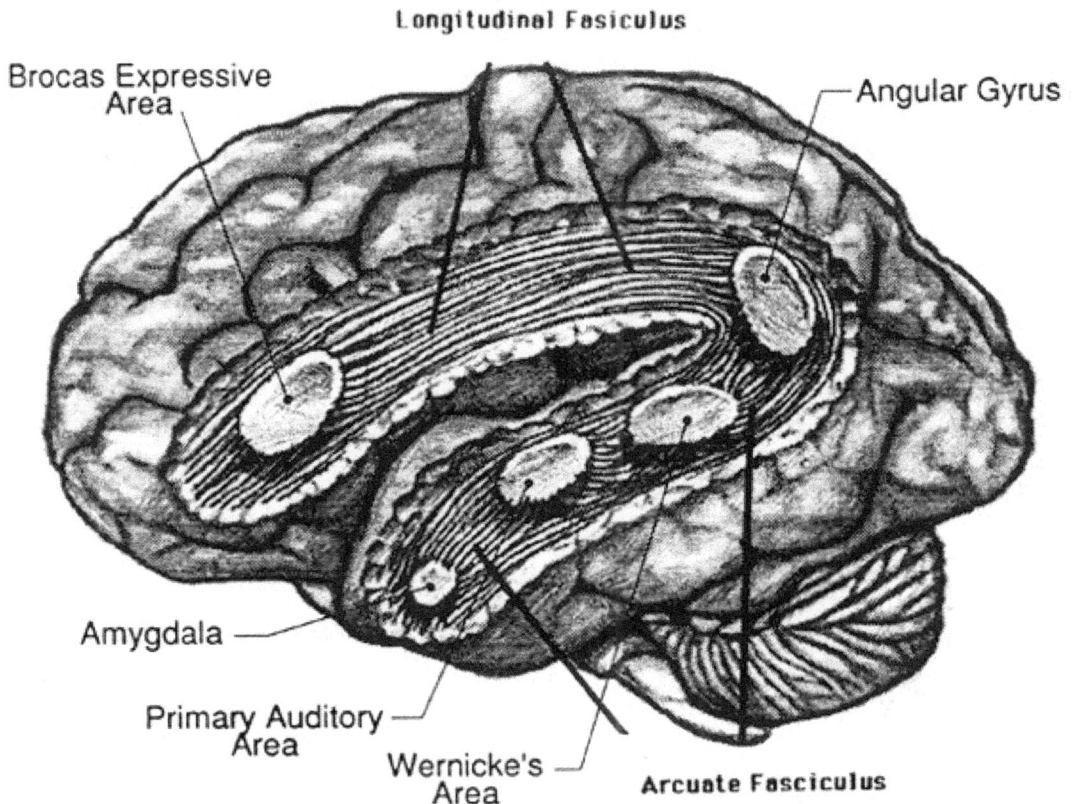

THE FEMALE BRAIN & THE UPPER PALEOLITHIC

The frontal motor area representing the hand is immediately adjacent to and intimately interconnected with the primary motor areas mediating oral, laryngeal, and mandibular movements, including Broca's area. Hence, manual activity, right handedness and expressive speech are directly related (Bradshaw and Rogers 1992; Corbalis, 1991; Joseph, 1982; Kimura, 1993), which is why when speaking, humans commonly gesture with the hands, the right hand in particular. However, it was not until the late Middle Paleolithic that up to 90% of Paleolithic Humans may have become right handed (Cornford 1986). Similarly, it was not until the Middle to Upper Paleolithic transition, 35,000 years ago, that tool making became literally an art and complex multifaceted features were incorporated in their construction and utilization (Chauvet et al., 1996; Leroi-Gourhan, 1964, 1982).

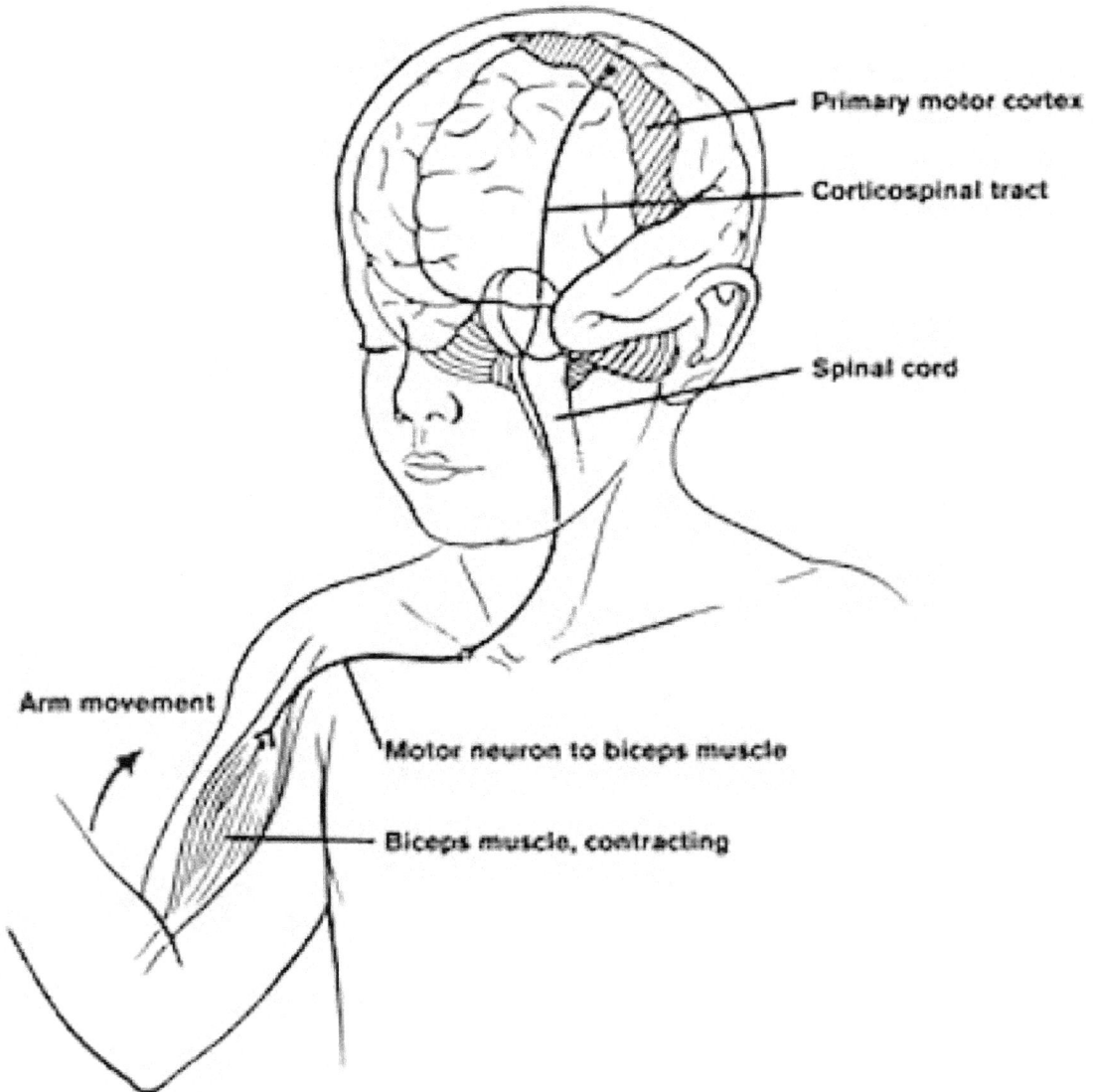

The Middle/Upper Paleolithic transition is characterized by the creation of complex bone tools, the sewing needle, and personal adornments such as carefully shaped beads of bone, ivory and animal teeth, animal engravings, perforated shells, statuettes, drawings, and paintings of animal and female figures (Chauvet et al., 1996; Clark, 2007; Leroi-Gourhan, 1964; 1982).

As the creation and wearing of personal adornments, and domestic tool construction and use is associated with the human female, and as tool making and gathering often involves both hands (albeit the right more than the left), it might be expected that the human female frontal-parietal areas may have functionally evolved in a manner different from men. That is, both the left and right half of the female brain may have become organized for producing motor sequences and grammatically complex vocabulary rich speech.

Moreover, as the parietal lobe and IPL/angular gyrus act on and program the frontal motor and speech areas, whereas the left frontal areas adjacent to the hand and face area control and program speech and hand-related fine motor activities (Joseph 1982, 1996, 2011d), it might be expected that sex differences would be more pronounced in Broca's and the parietal areas. In fact, Broca's area appears to be larger in the female.

Moreover, the posterior corpus callosum which interconnects the right and left parietal lobes, appears to be significantly larger in women than men (see Holloway et al., 1993, for evidence pro and con), whereas language appears to be represented in the right half of the female to a greater degree than it is represented in the right half of the male brain (Bradshaw et al., 1977; McGlone, 2010; Shaywitz et al., 1996) --a sex difference which may also account for her superior language capabilities but also her relatively inferior visual-spatial abilities (due to functional crowding). Left hemisphere dominance for verbal functioning and right hemisphere language (and greater emotional language) representation coupled with the enhanced ability of the right and left female-IPLs to communicate via the corpus callosum, may account for why women (versus men) are less likely than men to become aphasic with a left parietal lesion (Kimura, 1993; Mateer et al., 1982); i.e. women have language-related brain tissues in reserve and are able to continue talking so long as Broca's expressive speech area is uninjured.

SUMMARY & CONCLUSIONS

For much of human evolution females have engaged in tasks promoting, requiring, and involving rapid temporal sequential bilateral fine-motor skills such as gathering and domestic-tool construction and manipulation. Given an innate tendency to vocalize more than males, these activities, coupled with prolonged child care and mutual child-mother vocalizations, gave impetus to the evolution of the neocortical speech areas and a female language (and social emotional speech) superiority.

In contrast to gathering groups in which females could vocalize with each other and their young, males instead pursued their own violent tendencies and became silent hunters of big game. Over the course of evolution the male proclivity to silently travel long distance in the pursuit of prey exaggerated an already innate male visual-spatial perceptual superiority.

Nevertheless, although hunting does not promote the evolution of speech, males also acquired language skills through maternal genetic inheritance and and as he had a mother who would talk to him and teach him language. Thus, like the proverbial Eve, woman the gatherer provided man the hunter with the fruit of linguistic knowledge and what would become grammatically complex, vocabulary-rich speech, language, and linguistic consciousness.

FOOD, SEX & THE BIG BRAIN

Foraging, scavenging, and the chasing and hunting of small game has probably been a dominant activity of human beings and their ancestors for several million years. Naturally, our mind and brain has been tremendously influenced by these activities and has evolved accordingly.

Initially, both the male and the female of the species probably engaged almost equally in scavenging, gathering and the hunting of small game until just a few hundred thousand years ago with the onset of big game hunting. It was possibly this event, coupled with the rapid and progressive development of mans big brain several thousand years earlier, that a divergence in the mind of man and woman appears to have increasingly occurred. The hunting of large game animals appears to have become the dominant domain of the male, although they probably continued to assist in gather on occasion. Presumably females continued to predominantly engage in gathering, and to a lesser extent fishing and the hunting of small game, and only occasionally assisted males during the hunting and stalking of large animals.

Nevertheless, with this general division of labor, one might suppose that over the course of several hundred thousand years, tremendous differential influences on mental and brain functioning resulted due to selective evolutionary pressures on survival. That is, those who were best able to adapt were most likely to pass on their genetic traits to the next generation. In consequence, the brains and minds of men and women probably became adapted and molded accordingly each in accordance with the activities they were best at.

In this regard, as gathering and the harvesting of wild foods involves perseverative, temporal-sequential hand movements, those who were most successful at these activities probably had brains that were best adapted to these tasks. However, if the right hand is being used predominantly for picking and gathering, whereas the left hand is more likely to hold a receptacle, one might suspect that the left half of the brain would become even more adapt at temporal-sequential processing.

In contrast, as the searching and stalking and killing of prey would require good visual-spatial skills and sensitivity to environmental and non-verbal nuances, those who were most successful at these activities would in turn continue to develop, and pass on to the next generation, these same capabilities.

Over the course of human evolution, the human brain and the human head became increasingly larger. The progressive and significant increase in brain size, in turn required an adaption in the pelvic opening of these ancient females. A bigger brain comes in a bigger head. Hence, eventually the hips of the human female became wider as well so as to accommodate a larger brained baby. With a wider pelvic opening and with wider hips, a big brained baby could be delivered without becoming damaged or lodged inside the womb unable to emerge. If that were to happen, the baby, the mother, or both would die.

As a consequence of these changes in the hips and pelvis, the gait and balance of human females became altered over time such that they would wiggle when they walk and were no longer able to run as fast as males as their mobility became slightly restricted as well. This is slowly developing condition may have first began to exert significant limiting influences perhaps around 500 thousand years ago or longer, at which time, the brain of our remote ancestor, Homo Eretus appears to have reached it's maximum size (approximating that of a small brained, modern day human).

The human brain, of course is not fully developed at infancy and in fact is only about a third of the size of an adult. If it was too close to the adult size it would be impossible for an infant to be born. In consequence, the brain has to do a tremendous amount of growing after birth, a process that continues unabated for the first 10 years of life and then slows over the course of each ensuing decade.

However, by prolonging infant development the period of helplessness became lengthened as well which in turn necessitated prolonged child care. Hence, in contrast to most mammals and all other forms of life which remain helpless and unable to fend for themselves for only a few months at most, humans remain helpless for 4-6 years. This is in fact 4-6 times longer than for monkeys, and twice as long as for all other apes, which, like their female counterparts, invest a considerable amount of time in childcare.

Thus, the human infant, including an infant Cro-Magnon, Neanderthal, and even Homo erectus, required almost the full time presence of a mother for many years. This increased dependency in turn reduced female mobility as well. Hence, at a minimum, human females have been engaged in prolonged child care for at least half a million years or longer.

In consequence, having a baby or several children to watch over, and being unable to run as fast and being more clumsy in the attempt, the adult human female did not make for as good and as reliable hunter as compared to men. You can't carry an infant while stalking prey because your hands and arms must be free and you need quiet. The sudden cooing or crying of an infant is likely to result in an empty stomach. Hands and arms must also be free to carry weapons. Hence, there were a number of forces acting on females which conspired against their ability to excel as big game hunters. Indeed, her smaller stature also made her a less formidable foe and possibly even a a tempting prey.

Hence, for good reasons females probably tended to wander less than males, to stay closer to the home base, and to engage in activities compatible with a lifestyle that included the nurturance of infants for years at a time. The ultimate consequence of these particular limitations was that the human female probably remained more involved in gathering and the occasional hunting of small game closer to the home base whereas males were more free to engage in new activities such as big game hunting.

Again, one might naturally assume that to be successful as a hunter required certain adaptive changes in the human brain, alterations which would be maximal among males. Conversely, gathering, an activity involving prolonged and rapid temporal and sequential hand movements, would in turn exert its maximal selective pressures on the brain of the female. That is, those who were best at these activities were more likely to survive and pass on their genetic contributions and thus their adaptive skills, to their progeny.

Although a woman, or a band of women, could probably live and flourish without the aid of the bigger and more muscular male, and certainly could eat quite well without ever savoring a steak from a big game animal, her smaller size and the children that she bore, certainly put her at a disadvantage when dealing with males. Indeed, since our ancestors may have been slaughtering each other since the time of Australopithecus, one might speculate that it may have certainly been to her advantage to have at least one male around to aid in her protection as well as the rearing and education of the children. That is, she may have needed a male to protect her from other males who might rape or kill her and her children. Although perhaps highly debatable, there were probably other advantages to have a man about the house (or cave), as well such as being able to partake in the high protein diet that meat afforded.

In part, this bonding of a man to a woman, was aided by the establishment of a base camp and the maintenance of a family social group, events which appear to have first had their onset during the time of Homo Erectus (who lived from 1.9 million to about 300 thousand years ago). However, from an evolutionary perspective it was necessary to provide the man with some incentive to return home and thus maintain the sanctity of the family. The solution was possibly an appeal to the limbic system: food and sex.

To insure that the male return to base camp and share his spoils with her and her infant required that males become bonded to females or at least highly attracted so as to return again and again. The manner in which this, at least in part, may have been achieved was for females to become sexually receptive and sexually available 24 hours per day, 365 days a year. Indeed, the human female is the only female regardless of species, who is able to have sexual intercourse during times when she is not biologically receptive in order to become pregnant. For all other species, sex serves as a means of reproduction. Among humans, sex is a source of pleasure and serves as a means of bonding a man and a woman.

As is well known, for most other mammals there are long periods in which females are not sexually receptive and this is characterized by long period of non-sexual activity in which males show little or no interest in females. When she becomes sexually receptive there is a short period of frenzied sexual activity in which males will literally go sexually crazy for days until the female ends her estrus or "heat." of course, when a female goes into heat all other activities cease or greatly suffer.

It may have been sometimes during the rein of Homo erectus that this great change may have occurred and the female estrus disappeared altogether. Indeed, it was during the time of Homo erectus (or his immediate ancestor) that the first base camps also came into existence. However, even if we assume that this sexual revolution did not come until the end of their rein, that is, around the time period in which the first primitive Homo sapiens arrived on the scene it would still appear that full time sexual availability had it's onset somewhere between 300,000 and 500,000 years ago. Nevertheless, in consequence of that momentous occasion, the human female has been the sexiest of all other females ever since.

Her becoming sexually available at all times enabled man and woman to become free of the purely biological and hormonal influences which drive and control sexual behavior. They could now postpone sexual activity and could now decide for themselves when and if they were going to have sex, as well as where it would take place. Moreover, unlike the apes who only on occasion and for brief time period can enforce a selective preference, humans were also able to decide with whom they were going to have sex and could enforce their preference via denial and through bonding.

All in all, with the complete availability of a full time sexual partner coupled with the ability to make personal choices and to act on preferences, human sexual relations became more enduring and so to did the relationships between men and women; all of which probably promoted the development of language and the art of conversation.

Man, still functioning at the behest of his limbic system, could now have one almost insatiable need repeatedly satisfied. However, as noted, a second insatiable need, food, was also provided by the female. That is, gatherers provide the bulk of the food supply. There were thus two very good reasons for males to not only come home, but to stay home for prolonged time periods as females were not only providing for their sexual needs, but gathering and probably preparing their supper as well. Man was becoming domesticated.

FOOD & THE ORIGINAL "BREAD WINNERS"

With the exception of those who lived in the very coldest of climates where a gathering way of life would be quite difficult due to the scarcity of edible vegetable matter, for at least the last 100,000 years until perhaps about 10,000 years ago, females, and not males, appear to have been the main providers of food. Big and even small game hunting has always been (except in the much colder far northern climates) a supplementary means of acquiring an adequate food supply.

Indeed, even among the great majority of the very few modern hunting and gathering societies in existence today, spoils from hunting account for only about 35% of the diet. In contrast, gathering, which, we are assuming, has been the predominant domain of the female for the last 300 or so thousand years, accounts for the remainder. Indeed, even among the Cro-Magnon, where hunting was the center of religious and artistic life, 60-80% of their diet consisted of fruits, nuts, grains, honey, roots and vegetables, which was probably gathered by the females. Hence, it probably was the female, and not the male who probably wielded economic dominance for a significant period of our history.

Indeed, given woman's dominant role as gatherers and since they were probably generally responsible for both reproductive and most subsistence activities, males could be almost be considered a co-producing dependent. That is, women were probably the original "bread winners." Perhaps it is this ancestral tendency to nurture and provide that enables some modern women to not only tolerate but to support men who simply sponge off them.

Not surprisingly, males recognized that females were a tremendous economic asset, and in many societies, dominant males would try to accumulate as many wives or lovers as possible. However, although sexual availability may have played some role in this procurement during ancient times, the fact that females were the dominant producing partners meant that with many women, the man would be freed from having to spend all his time engaged in the pursuit of food, its preparation or even the maintenance of dwelling. The more females, the more leisure time and thus the more time to engage in recreational and artistic tasks.

Indeed, such a relationship, at least for dominant males, may have been rather ideal. For the first many years of their lives they were provided for by their mothers, and then later in life they were cared for by their wives. On the other hand, one could argue that not only were women the main providers, they were being exploited for the last 100,000 years as well.

Nevertheless, with increased leisure time and the evolving complexity of social relations, tremendous selective pressures were no doubt brought to bare on the mind and brain of human beings. Only those who could selectively adapt to these changing conditions could pass on their traits and genes. Not only were the right and left brain evolving new functional capabilities associated with sex differences in socializing, gathering and big game hunting, but correspondingly, about 50,000 years ago, the frontal lobes of the brain tremendously expanded in size. With the development of the frontal lobes came the capacity to engage in long term planning, the formation of goals, and the ability to maximally inhibit one's immediate desires and impulses. This great change came about with the appearance of the Cro-Magnon and social relations expanded beyond small bands to large tribes.

HUNTERS: BRINGING HOME THE BACON

Since food gathering was such an extremely important part of economic existence for so many hundreds of thousands of years until very recently, it might be asked, why has so much emphasis been placed on the importance of man the hunter and along with it, the false notion that he was the main provider?

To better understand the role of providing meat in the assumed dominance of man as the most important provider, let us consider our nearest living evolutionary cousins, the apes. When a chimpanzee captures another living creature for the purpose of consumption others will rapidly and excitedly gather around and beg and beg for just a morsel of the meat. Nothing like this happens when they forage for vegetables or insects. Indeed, the hunting and capturing of meat has an immediate highly arousing excitatory effect on the whole band all of whom will gather around in hopes of being given a tiny morsel.

Moreover, Chimps and baboons will chase after and kill small animals but they do not eat dead ones. That is, they do not scavenge. For them the pleasure of the meat is tied to the hunt, capture and killing. However, it is the male who predominantly engages in these acts.

Thus among the chimpanzees, the procurement and eating of meat for some reason promotes considerable social excitement and food sharing. In contrast, they do not share and show no interest in sharing or receiving any of the vegetable matter that another chimp may have found and is eating.

Similarly, among hunter gatherer societies, the procurement of meat promotes food sharing and the eating of meat is thus a very social activity. Indeed, when the hunter returns he is likely to be met by most members of his band all of whom will eagerly seek a share of the spoils. The hunter will naturally give the largest pieces to his own relatives and the smaller shares to friends and co-band members. These individuals in turn will give a share of what they receive to their own special friends and relatives. Hence, a certain degree of group cohesion and bonding occurs when meat is caught and shared and the sharing reinforces the social bond. Meat thus becomes an important currency of exchange.

In contrast, when females bring home the vegetable matter that they have gathered, there is no excitement and there are no begging hands eagerly seeking a share. What the female procures is shared only with her own immediate family.

Indeed, this same emphasis on the importance of meat can be found repeatedly in Genesis and the Books of Moses, as God demanded his sacrifices to be of living flesh. Hence, when Abel offered God "the firstlings of his flock, the Lord had respect unto Abel and to his offering." But when Cain, a tiller of the soil, made an offering of his vegetable produce, the Lord God "onto Cain and to his offering he had not respect."

Due to the social and even religious and spiritual importance of meat, a successful hunter is looked upon as a very special and powerful individual who has great prestige among his band. Moreover, the successful hunter who gives out shares of meat in accordance with his own prerogatives thus gains power over the group as well as respect. In consequence, the successful hunter is sometimes rewarded with more than one wife (or has numerous extra-marital affairs) presumably because he can provide for more than one and by his great stature is actually sought by females as well as prospective father-in-laws

Of course, those who were the best hunters were no doubt the most intelligent of their times, at least insofar as survival is concerned. Supposing that the most intelligent of the race were more likely to breed, with each passing generation, there was a manifold increase in the numbers of intelligent people. That is, the best hunters would have the most wives and thus the most children, whereas those who were not proficient either died out or bred so sparsely that their descendants and genetic contributions simple ceased.

Moreover, females then as now, tend to be attracted to males who hold a more dominant status than they themselves. Males tend to establish relations with females who are subordinate. Hence, our ancient spear throwers, like modern athletes today, often had their pick of willing females thus increasing further their particular genetic contribution to the race.

As noted, however, the ancient female was quite a prized possession, due to her sexual availability and gathering skills. Hence, males had to also compete amongst each other for their favors and attention, particularly in regard to females who may have been similarly dominant and intelligent.

Males also must often fight and sometimes even kill in order to maintain their dominant position, or in order to escape a position of inferiority. Females are not subject to these same pressures, particularly in that males fight not only for dominance but sometimes for sexual access to females which in turn can be a great source of status. These behaviors are most obviously evident among non-human male mammals.

Hence, not only successful hunting, but general intelligence, and the ability to successfully compete against other males in regard to status, dominance, and access to women may have been a tremendous concern of the human male for several thousand years. One might suppose his brain adapted accordingly.

Does this explain, at least in part, the tendency for modern men and adolescent males to sometimes denigrate, even in a teasing manner, the intellectual, physical and financial abilities or even sexual prowess of one other? Is this the remnants of an age old means of achieving dominance?

LANGUAGE & TOOL MAKING, GATHERING, & SHOPPING

The basic skills necessary in the gathering of vegetables, fruits, seeds, berries and the digging of roots include the ability to engage in fine and rapid, temporal-sequential physical maneuvers with the arms, hands, and particularly the fingers. As gathering has been a dominant activity for such a long time period in our prehistory, it is thus not surprising that the brain has possessed rudimentary temporal-sequential capabilities for several million years.

To aid the women on their daily shopping trips they carried large purses (or pouches) made either of leather or the stomach or bladders of various animals into which they could deposit their goods. It probably hung over the woman's shoulder almost like a long strapped purse. Women have thus been gathering (shopping?) and carrying purses for at least 100,000 and perhaps as long as 500,000 years and longer.

The purse was also highly beneficial to the development of social relations. If they did not collect they would probably simply eat what they found on the spot and would have nothing to bring home. In fact, humans are the only primate that engages in gathering. All other primates eat their vegetables as they find them. Of course a few highly social animals such as wolves and wild dogs bring food back to the den in their stomach where they may then regurgitate it for the young, or even for the old, injured or feeble. Birds also bring food to their young.

In addition to gathering, women made tremendous use of tools and may have been the first tool makers. Tool making, like gathering was possibly a major aspect of their lives and one they may have spent considerable time engaged in. In fact, the first tools were made not for hunting but for gathering and rooting plants. That is, if we suppose that gathering predated hunting and scavenging. For example, in grubbing for roots and bulbs, the gatherer would need a digging stick which they probably had to periodically sharpen as they gathered by using stone flakes. They would also carry a hammerstone for cracking nuts and for grinding the various produce collected during the day.

In addition, as the female also would sometimes hunt and try to capture small game (probably with the help of a big friendly dog) she may have carried an ax and knife for the purposes of dispatching whatever hapless prey she chanced upon. By Cro-Magnon times, many of them also carried large flat bone knives up to 9 inches in length.

These ancient women did not spend their time solely gathering, for foods had to be prepared and clothes had to be fashioned out of hides. Her duties may have included cleaning the hides via the use of a scraper, drying and curing the skin over the smoke of a fire, and then using a knife or cutter to make the general desired shape, and then a punch to make holes through which leather straps or vine can be passed so as to create a garment that could keep out the cold. They were also weaving and by Cro-Magnon times they were using a needle to sew garments together. Although there is no way of knowing if woman was the first tool maker, she certainly regularly engaged in these endeavors as a more or less on going activity.

Be it woman or man or both, they eventually developed the capacity to look at and feel a stone or bone and immediately realize its potential as a weapon, tool, or object of art. In this regard, they may have been more attuned to details and had to know stones and bones and the properties and grains of wood, the way, manner and direction it had to be struck and with what instrument to use in order to arrive at the desired effect.

However, in tool making, technique comes to have pre-emmenence. With the invention of technique, a linear, temporal and sequential approach long developed over eons of gathering, probably came to characterize the process. Certain tools are made a certain way with certain instruments with certain movements, and with a certain degree of muscular power and considerable precision. To make and utilize tools requiring a precision grip required that the manufacturer not only have a hand capable of such feats, but a brain that could control this hand and which could use foresight and planning in order to carry out the manufacture.

Due to selective pressures and the survival and breeding of those who were successful at these activities, the left half of the brain, which controls the right hand, became increasingly adapted for the control of temporal-sequencing be it for the purposes of tool making or for gathering. Hence, it was not only tools, but the neural substrate for the temporal-sequential and grammatical aspects of what would become spoken language that was being forged and passed on to succeeding generations, a process that had it's onset several million years ago. As females have engaged in gathering for time periods much longer than males, coupled with her possible role in tool manufacturing and tool use (e.g. skinning, clothes making, etc.), it might be assumed that these changes were maximal in the brains of women, particularly in the motor areas controlling speech (Broca's area) and hand control--and this in turn would explain woman's propensity to engage in tasks involving fine motor skills, such as typing, knitting, sewing, and so on.

Not only do modern human females demonstrate superior fine motor skills and an earlier onset of language and are less likely to suffer language disturbances, but that it may have been our female ancestors who more fully developed the temporal-sequential aspects of language first as well. Perhaps this was the fruit of knowledge woman offered man only after having taken the first bite. In fact, we find in Genesis that the first gatherer, the first seeker of knowledge, and the first individual to hold a complete conversation (other than God and the serpent) was a woman.

When we consider the many selective pressures which acted on males so as to inhibit speech during the hunt, whereas females were allowed to talk quite freely while gathering (as there would be no fear of scaring off game), it certainly appears that these factors strongly and differentially promoted female linguistic development and a greater capacity to discuss topics unrelated to events associated with the hunt. Language was now being used for social bonding and only later became an instrument designed more and more, at least insofar as males are concerned, for the purposes of exchanging information related to business and sports; i.e. the hunt.

LANGUAGE & CONSCIOUSNESS

Consciousness is not synonymous with language. There are multiple streams of consciousness and mental activity which are subserved by distinct regions of the brain, some of which do not require or even perceive the words of language. It is apparent that other species are conscious and do not require language. And it is obvious other species utilize sensory modalities which are not well developed (olfaction) or non-existant in humans (echo-location), but which make up a great part of the consciousness of these non-human animals. Language is a late evolutionary acquisition, and those aspects of consciousness associated with language, may have been acquired only over the course of the last 35,000 years. Language, and the train of verbal thought, often provide consciousness with contents, and the language-dependent aspects of consciousness are unique to the human mind are are only a recent evolutionary acquisition.

For much of human evolution females have engaged in tasks promoting, requiring, and involving rapid temporal sequential bilateral fine-motor skills such as gathering and domestic-tool construction and manipulation. Given an innate tendency to vocalize more than males and to engage in mutual mother-baby vocalizations, these gathering activities, coupled with prolonged child care and mutual child-mother vocalizations, gave impetus to the evolution of the neocortical speech areas and a female language (and social emotional speech) superiority. The female brain became reorganized and the neocortical representation of older functions were diminished.

In contrast to gathering groups in which females could vocalize with each other and their young, males instead pursued their own violent tendencies, eventually becoming silent hunters of big game. Hunting does not require language, but silence. Over the course of evolution the male proclivity to silently travel long distance in the pursuit of prey exaggerated an already innate male visual-spatial perceptual superiority.

Nevertheless, although hunting does not promote the evolution of speech, males also acquired language skills through maternal genetic inheritance and and as he had a mother who would talk to him and teach him language. Thus, like the proverbial Eve, woman the gatherer provided man the hunter with the fruit of linguistic knowledge and what would become grammatically complex, vocabulary-rich speech, language, and linguistic consciousness.

10. Self Deception and Confabulation

He knows it, because he believes it.
Believes it, because he knows.
But alas, he knows only that he believes.

CONFABULATION

The non-emotional aspects of language, that is, the temporal-sequential, grammatical and denotative aspects of speech, are localized in the left half of the brain, in the majority of the population. Whereas the right hemisphere is dominant for the social-emotional and melodic aspects of speech--that is, the way things are said--the left hemisphere mediates what is said, the words of speech and their grammatical organization.

Three unique areas in the left hemisphere which maintain speech. Broca's Expressive Speech area--which does the talking; Wernicke's Receptive Speech area--which does the listening; the Angular gyrus of the inferior parietal lobe, which assimilates verbal associations so that they may be understood; e.g., so that we know that the word "green" represents a color, or a state of not yet being ripe or untested, as well as a political group.

These three area, the angular gyrus/inferior parietal lobe, and Broca's and Wernicke's areas, form a "Language Axis." The Language axis maintains the verbal aspects of the conscious mind.

When the Language Axis (Wernicke's and Broca's area, the Angular gyrus) and thus the linguistic conscious mind become isolated or are prevented from receiving information from yet other regions of the psyche, an individual may begin to engage in denial, projections, rationalization, and displacement. However, when a brain injury causes a disconnection of the Language Axis from yet other areas of the brain and mind, the person may engage in denial, including the production of elaborate confabulations, denying the existence of the left half of the body, or claiming he can see when he cannot.

As previously noted, patients who have sustained injuries involving the right brain and who are subsequently paralyzed on the left side, may claim that their paralyzed extremities belong to the doctor or a person in the next room. This is because a bilateral image of the body is represented in the neocortex of the parietal lobe of the right hemisphere, whereas it is the left hemisphere which is doing the talking. In less extreme cases the speaking half of the brain may claim that their paralyzed limbs are normal even when they are unable to comply with requests to move them.

For example, one women "when asked why she could not move her arm replied, "somebody has a hold of it." Another patient, when asked if anything was wrong with her arm said, "I think it's the weather. It's just a little cold and stiff. I could warm it up and it would be all right." One women whose left leg was paralyzed continued to claim she could walk, even when she obviously could not. "I can walk at home," she stated, "but not here. It's slippery here." Another patient claimed that he could not raise his arm because of the stiff shirt he was wearing.

With severe right cerebral injuries (which in turn results in a massive reduction in information exchange between the two halves of the brain), the individual may make up extremely bizarre explanations regarding their injuries, which they seem to believe. For example, one patient who had suffered burns as well as a significant brain injury in an auto accident claimed that his rocket ship had crashed which resulted in an atomic explosion. He repeatedly insisted he was filled with radioactive fluid and thought that his various scars were the result of this fluid being removed from his body. However, his responses were not due to some form of psychosis, they were the product of a loss of inhibitory restraint and confabulation, an attempt to make sense of what he was experiencing.

Yet another patient who had received a gun shot wound to the right brain repeatedly and vehemently complained about being kept in a hospital which he believed was due to a plot by the government to steal his inventions and ideas --he had been a grocery store clerk. When it was pointed out that he had undergone surgery for removal of the bone fragments and the bullet, he pointed to his head and replied, "That's how they're stealing my ideas."

Moreover, these confabulators truly believe their own stories and will sometimes adamantly argue if you try to persuade them otherwise. Rather than relinquishing an incorrect belief when confronted with contradictory information, these individuals will sometimes simply adjust their story slightly.

In fact, the same thing often occurs when information stored in memory (be it in the right or left brain), is not available to the Language Axis and the conscious mind. Take for example a patient who was suffering from a partial amnesia and who although having been a patient for over two weeks, had no recall of why he was in the hospital. However, he could accurately recall much of what had occurred prior to the onset of his condition. When the nursing staff attempted to provide him with medication, he would respond, "What is this for? I've got to get out of here." or "I just stopped by for a checkup. I've got to get back to work." When I informed him he had been a patient for over two weeks he became very hostile and upset, and angrily replied, "I just came here for lunch. I do not belong here. I'm supposed to work in the garage. Now would you please tell me how to get to the garage?"

Essentially, the left half of the brain, being denied access to various sources of information, responds to the gap in the information received, with ideas and associations which are available, even when irrelevant or ridiculous. There is a failure to correct these erroneous statements as disconfirming evidence is not available to it. The Language Axis and the conscious mind have essentially been disconnected without knowing it and thus act to fill the gaps in the incomplete messages received with ideas which are related in some manner to the fragments available.

When the conscious mind is disconnected from another brain area, due to injury or a functional, psychic resistance or repression, then the Language Axis with which it is closely associated, will be unable to describe what is going on in that cerebral region. Instead it will describe only what is available to it; which it then believes. Moreover, it will not know that it does not know, or that it's explanations are erroneous, because that area of the brain which does know, has either been destroyed, or psychologically disconnected, and cannot alert the Language Axis to any discrepancies.

In extremely severe and rare cases of brain injury, the Language Axis and conscious mind, can become completely disconnected from the rest of the cerebrum such that almost no information is available to it. When this occurs, the associations available are so limited that it cannot think, cannot act, and can only respond to the most simple of questions in a stereotyped and very limited manner. However, the right brain may still engage in singing, praying, and engaging in it's normal routine; that is, if it is not too severely damaged.

In an interesting case described by Drs. N. Geschwind, F.A. Quadfasel and J.M. Segarra, a 22 year old woman with massive destruction of neocortical tissue due to gas asphyxiation was found to have a preserved Language Axis with all surrounding brain tissue destroyed, thus completely disconnecting it from the rest of the cerebrum. It was noted once the patient regained "consciousness" that "she sang songs and repeated statements made by the physicians. However, she would follow no commands, resisted passive movements of her extremities, and would become markedly agitated and sometimes injured hospital personnel." In all other regards, however, she was completely without comprehension or the ability to communicate. "The patients spontaneous speech was limited to a few stereotyped phrases, such as "Hi daddy," "So can daddy," "Mother," or "Dirty bastard." She never uttered a sentence of propositional speech over the nine years of observation. She never asked for anything and she never replied to questions and showed no evidence of having comprehended anything said to her. Occasionally, however, when the examiner said, "Ask me no questions," she would reply "I'll tell you no lies," or when told, "close your eyes" she might say "go to sleep." When asked, "Is this a rose?" she might say, "roses are red, violets are blue, sugar is sweet and so are you." To the word "Coffee" she sometimes said, "I love coffee, I love tea, I love the girls and the girls love me."

"An even more striking phenomenon was observed early in the patient's illness. She would sing along with songs or musical commercials sung over the radio or would recite prayers along with the priest during religious broadcasts.

If a record of a familiar song was played the patient would sing along with it. If the records was stopped she would continue singing correctly both words and music for another few lines and then stop. If the examiner kept humming the tune the patient would continue singing the words to the end. New songs were played to her and it was found she could learn these as evidenced by her ability to sing a few lines correctly after the record had been stopped. Furthermore, she could sing two different sets of words to the same melody." In other words, although the Language Axis was disconnected, the right brain (although also damaged) was still able to engage in some of its normal expressive routines, singing and praying. The left brain, however, being totally isolated, could produce almost nothing except well learned and stereotyped phrases. In other words, the left brain could only repeat what was available to it.

Such conditions are quite rare, however. Nevertheless, even when the brain has not been damaged the Language Axis is not always able to gain access to other brain areas. When this occurs it will make up explanations as the need arises, even when the brain is not damaged. One need not suffer a brain injury in order to confabulate; it is a normal part of left brain psychic functioning.

Quite frequently, due to one's conscious self-concept, or other concerns, information which is available but undesirable, will sometimes be suppressed, ignored, or denied. When this occurs, the Language Axis, and the conscious mind can then generate whatever explanation it wishes to explain away its actions, feelings, behavior, or the behavior and feelings of others. Having no access to the correct information, perceptions, memories, and emotions, it cannot be alerted, except indirectly (via the generation of tension) to any discrepancies, unless, of course, the right brain or limbic system completely overwhelms it or takes over. However, even then the left brain may be denied or deny itself access to whatever motivated its behavior, and like the brain injured individuals discussed above, will generate whatever explanation that is acceptable, even if unplausible or ridiculous.

HIDING INFORMATION FROM THE CONSCIOUS SELF

Desires, impulses, and a variety of emotions arising from the realms of the unconscious may come to be reinterpreted or defended against in a variety of ways. Some desires just fade away or are replaced by more important concerns. Others simply cannot be perceived or even recognized by one versus a second region of the brain or mind. Others are held in check, via the frontal lobes for example, until an appropriate time or means of fulfilling them emerges.

However, some tendencies and personal characteristics are pervasive and deeply ingrained. They are part of what and who we are. Sometimes these personal traits are largely unconscious and associated with the unconscious self-image, such as the ego personality of the Child. This constellation of traits may include feelings of hurt, jealousy, anger, guilt, cowardice, weakness, incompetence, clumsiness, fearfulness, childishness, and even feelings of selfishness and greed such as often occurs when children are spoiled.

Although accepted by the unconscious self-image, these unconscious tendencies may be completely unacceptable to the conscious self-image; what we have described as the adult Self, or Adult Ego personality. Being unacceptable they are ignored, denied, projected onto others or inhibited. At a conscious level a person may not even know he possesses these traits.

Indeed, many of us have had experiences where a good friend or lover has behaved in a selfish or inconsiderate manner. However, when we have complained or confronted him or her they may have steadfastly denied ever having behaved in this fashion and will vehemently deny even possessing such disagreeable tendencies. Incredibly he will deny what is obvious to us, and he will believe his own denials. Essentially, the Language Axis and the conscious mind have become disconnected from these sources of undesirable input. When the left brain tells these "lies," it is in fact telling the truth as it perceives it.

DENIAL. When unconscious needs, desires, personality traits or tendencies are contrary or upsetting to the conscious self-image the person may go to great lengths to suppress, deny, compensate or overcome them. Even when a person has engaged in some action which is contrary to his conscious self-image he may deceive himself into believing that it did not occur, and that it was a minor aberration which means nothing.

If he is shocked by what he has done or said, he may shrug it off with a simple: "I didn't mean it," "I was tired," "I was drunk," or "I don't know what came over me," as if some alien force took over his brain for a few minutes. These are defense mechanisms employed by the conscious mind and are referred to as "rationalization," "denial," "justification," and "self-deception."

SELF-DECEPTION: DENIAL & BELIEVING YOUR OWN LIES

It can be said that a person has lied when he knowingly possesses the truth and consciously attempts to deceive another person. Yet, curiously it sometimes happens that a person half persuades himself that the lie he has told is true and then behaves accordingly. When this occurs we may complain that he "believes his own lies." He has engaged in the production of confabulations.

Consciously we often attempt to hide disagreeable information from ourselves. This includes unconscious personal traits and tendencies that consciously we find disagreeable. Sometimes these characteristics are too difficult to confront consciously.

Denial, however, does not make a desire, impulse or personal short-coming vanish. Often we remain unconsciously-aware of its presence. Moreover, via the generation of various associated tensions and feelings the conscious mind continues to be influenced if not controlled by these unconscious forces. The person may even act on these impulses and not know why or "what came over them." Feeling these tensions or observing their actions the left half of the brain may then confabulate explanations as to why they did and said such and such.

It is precisely because of these "mysterious behaviors," emotional tensions and influences including the dual nature of our brain and mind that the left half of the brain must engage in denial in the first place. The person is aware of what he is hiding. To deny something means to simultaneous know that it exists. You cannot deny something without being confronted by it in the first place.

Hence, whenever a person attempts to deny or hide information from himself he is simultaneously acknowledging that they have an awareness of how and why the information is threatening. "John" would not deny and hide information from himself if he did not know simultaneously that it would bother him. However, it is the right brain which knows and the left brain which is in denial. How is this made possible, via the frontal lobes, which as noted, are the senior executive of the personality and the Ego, and which exert significant inhibitory restraint on information processing within the brain.

SELF-DECEPTION

The dual organization of the human brain and mind allows for the development of countless conflicts some of which are resolved by denial, projection, or self-deception. Because we have two brains we can know, yet not know, and can lie and deceive ourselves about our own unconscious tendencies and impulses.

Alan was pampered and spoiled by his two fairly well off, but not quite rich parents, particularly his mother who took a special delight in indulging her son. Although the second son in a family of 6, he remained the apple of their eye and remained so, even after he had been placed on trial for rape. Paradoxically (or so it seems), he always felt angry at his parents whom he believed never did enough for him. Unconsciously, he always felt cheated, even though he was the only one of their four children who never had to hold a job and had his way paid through college. He held his mother in particular contempt as he could recall several instances where he had to beg or plead for some special toy or favor before she would yield. In other words, he was the typical spoiled child who took love and affection for granted but was reluctant to return it, in part because of feeling cheated, but mostly because it had never been required in order for him to win favor. His unconscious Child, therefore, was thus quite selfish and greedy, and his unconscious Parent told him he deserved to have his own way.

His unconscious resentment, however, soon came to include most others for although highly intelligent, being a very short, thin and frail individual, he was often left out or was the last to be chosen for sports and games at school, and had never had a girlfriend or a date until he became a graduate student in college. He claimed, however, that this did not bother him because most people were idiots, women in particular. Moreover, he thought quite highly of himself and claimed to be a strong believer in fair play and honesty; that is, so long as he was treated in kind. Hence, his conscious self image was that of a "great guy." Indeed, this was the conscious message he always received form his parents, his mother in particular.

However, by his own admission, he stated that all women were selfish, mean spirited, untrustworthy bitches and whores, who would do anything to "clip a guy" and "screw men over." The only thing they were good for was sex, and insofar as he was concern, he could take them or leave them.

Nevertheless, much to his pleasant surprise, as a graduate student and teaching assistant, undergraduate females began to show him some interest which he exploited miserably up until he was arrested for rape. This came as a crushing surprise and confirmed his worst fears about women, as by his own admission he had acted no different with her than he had with several other undergraduate students that year.

"The prick teasing bitch wanted it," he told me, and he could not see what he had done that was wrong. "She knew what was going on. She hardly struggled. She didn't scream. If she didn't want sex, why did she come to my house? Like all women she just wants it both ways. Get what she can and then screw everybody else. I'm the one who is getting raped. You just can't trust women. They try to trick you and deceive you, and then take everything you got without giving you anything in return except grief. I just can't believe this is happening to me."

According to Allen it was one of his typical first dates. He had taken her to a drive in movie where he plied her with malt liquor followed by "whisky chasers." Once he was sure she was fairly intoxicated they left the movie and returned to his apartment where the drinking resumed. When he was sure she was too intoxicated to resist he began removing her clothes. Although, she still had enough wherewithal to try to resist and beg him to stop, he "knew" that she really meant "yes" and simply ignored her protests and, over powering her, took his pleasure.

When I tried to point out that all the cruel things he ascribed to women in general, actually fit his own behavior (e.g. untrustworthy, take advantage, screw someone over), he became angry and stated that this is what she had done to him by pressing the rape charges. He had only engaged in a normal manly manner, and had seduced her.

In actuality, however, he had seduced himself. He believed his own lies and had projected his own horrible tendencies onto women in general. This allowed him to justify and rationalize his mistreatment of them, while simultaneously meeting the needs of his Child and enhancing his poor self image and conscious feelings of masculine superiority in the process. Nevertheless, he was so disconnected and cut off from his true self image and selfish tendencies that he was unable to recognize his actions for what they were, and, he believed his rationalizations and explanations.

Even when I asked him how he would feel if he went out with some men, became very intoxicated, and was then sodomized or forced to engage in oral copulation, he could not see the connection. "Men don't do that to each other," he stated, "unless they're both queers, in which case it is OK."

Dr. J:" It's OK when someone is too intoxicated to resist or to know what's going on?"

Allan: "Hey. She could talk. She could walk. So what she was drunk. I didn't rape her. I seduced her. Nobody forced her to keep drinking." Dr. J: "Your 25. She's 17. Who supplied her with the drinks? Who got her so drunk she could barely resist?"

Allan: "Hey, man. That's the way its done. You have a sex problem or something?"

Dr. J: "No. I have a problem with rape."

Allan: "Rape! I didn't use force. I didn't beat her up."

Dr. J: "Sure you did. You used the force of alcohol. You numbed her brain so that she hardly knew what was going on. It sure wasn't your sparkling personality which got her into that bed. You might as well have hit her with the bottle."

KNOWING YET NOT KNOWING: SEDUCTION AND SELF-DECEPTION

Melody has a conservative background and well meaning upper middle class, semi-religious parents, who have impressed upon her the need to maintain certain values and morals. Melody is in her first semester at college and is living in the dorm. One day a handsome, muscular, sexy, upper classman whom she has seen about campus, invites her out. She consents. They have dinner, stop at a local nightclub, dance close together and have a drink. So far it has been an exciting, romantic evening.

On the way back to the dorm he invites her to stop by his place to "check out the new CD player and stereo equipment" he has just purchased. She knows, or rather, is aware of the possibility of his intentions and what he may really be up to. Indeed, the look in his eyes, his tone of voice and body language have made his real interests completely clear to the right half of her brain long before they even began dinner. However, insofar as her left brain is concerned this guy just wants to show her his stereo. Due to this awareness of his intentions including the manner in which she may respond, the left brain begins to invent and think up reasons why this is OK. He seems real nice. They had a real good time. He has not acted inappropriately. Why not?

Once at his home he makes certain advances; sitting close to her, looking in her eyes, he tells her she is "really beautiful." Her right brain responding to his tone, body language, etc., is now almost certain of his desires and what he means by his words. Moreover, she is not only aware of what might occur next, she is also aware that she may have to make a decision regarding this. However, she does not want to "think about that" (she is a nice girl).

Due to her conscious Self-image this information is suppressed and ignored. The left brain does not gain access to what the right brain is fully aware of. The left brain thinks: this guy is only being nice.

Slowly he places his hand on her knee. The act risks changing the situation by calling for an acknowledgement and an immediate decision as to its meaning. To leave his hand there is to consent, to acquiesce to his desires and to consciously acknowledge them willingly. Yet to withdraw his hand is not only a recognition of what it implies, but a refusal. Her aim is to postpone her decision, to stall conscious recognition. She has "no idea" as to what he is up to. She leaves his hand there because she "doesn't notice it."

And yet, as they talk, he has moved closer to her, his hand inching its way past her knee. But she is concentrating on what he is saying, the curve of his lips, the white of his teeth, her reply to his questions; this is what she is conscious of and to what she addresses herself. She reacts as a personality that does not know that a hand lies there upon her leg. She is being seduced.

She is also being self-deceived. She has restricted her consciousness and her thoughts to only selective aspects of what is occurring. She is failing to acknowledge consciously what she is fully aware of. Nevertheless, she is aware that his actions are sexual and has been aware of everything that might occur since the moment he asked her out.

However, if someone were to have told her this guy would bring her home and make possible sexual advances, she would have turned down the date. After all, her conscious self-image is that of a "good girl," a "nice girl." Although she is not a "prude," as she has always told her friends, she will have sex only with a man that she loves. Indeed, the possibility of "sex" was so far from her (conscious) mind that it never even occurred to her that this guy would make certain advances.

So why did she go out with him? Why is she not taking his hand and indicating that his behavior is not acceptable: "Please stop." One would have to assume that her limbic system was "turned on," she was sexually attracted to him and the possibility of his making sexual advances was not only acceptable but welcome; at least on an unconscious level. However, what is acceptable at an unconscious level is not acceptable to her conscious mind and conscious self-image. Unconsciously she is in fact yielding to his and her desires. Consciously she still does not know what is going on. His behavior seems harmless.

If we accept Jean Paul Sartre's position 1, there is a possible unity in a situation such as this, for her actions necessitate a recognition of the intentions of her male friend as well as a disavowal. Otherwise she would have asked: "What is your hand doing on my thigh? Please remove it." Indeed, even if unable to verbalize her conscious disproval (as some young women presumably become too intimidated to speak up) she would have made some attempt to back off or disengage herself if that is not what she wanted. However, consciously she is disconnected from what is really occurring and from what she is fully aware of, so she does nothing.

This is the beauty of self-deception, it allows a person to engage in a certain behavior while simultaneously preserving his conscious self-image. It preserves the possibility of an event occurring or a desire being acted upon while acting as if the event or action is not occurring.

Unconsciously a person may be fully aware of what he is doing or intends to do, while consciously he may behave as if nothing of the sort were going to occur. This allows a person to deceive himself into thinking he is not responsible for his own behavior or for what happens.

He takes her hand, leans close and kisses Melody upon the lips. The kiss takes her by "surprise." He snakes his other arm around her and kisses her again. She does not know what to do. She is confused. And yet, her limbic system is proclaiming: "I want it now!", her right brain is saying: "go for it," and the left brain in all innocence is saying: "What's going on?" He begins kissing and touching her passionately and she feels herself yielding as her Limbic System begins to overwhelm her conscious inhibitions. Her limbic system, in effect, has taken over, and her left brain, essentially, has gone "off line," and is now a passive observer to behavior over which it has no control.

Indeed, it often happens that people behave in a certain manner, or say certain things that are shocking even to themselves. They don't know "what came over" them. They may see themselves as moral, just, kind, loving, and for some unknown reason act in a cruel, selfish, sexual or spontaneous manner and yet retain the conviction that what they have done or said is not a true representation of how they really feel or who they really are. Often, however, once the limbic system has been satisfied and left brain regains dominance it may then begin to impose and cause feelings of guilt on itself.

Their left brain may then begin to confabulate explanations so as to defend their conscious self from the truth or to minimize these self induced feelings of unnecessary guilt. Such behavior, they reason has an explanation outside themselves, being only a rare and momentary lapse, is justified by certain mitigating circumstances or is due to the horrible shocking behavior of someone else: "They pushed me too far that time!"

And, if Melody was "taken advantage of" by "that rat," "that scoundrel," "that womanizer," then she is of course not responsible for what happened. She "had no idea as to what he was up to," or that "he was that kind of guy."

And, besides she "had too much to drink," (although she in fact had only one) and "was too tired to fight him off," (although she never tried) and before she "knew it, it was too late," and so on. Her conscious self-image remains untarnished. She is free of conscious guilt. As pointed out by Sartre, she has deceived herself. And yet, she knows and is aware of the truth. It is precisely because she knows that she invents innumerable excuses for her behavior.

Fortunately, for all concerned, however, "that rat" called her the very next day and brought over a dozen roses. She was enthralled, they began dating, and he turned out to be not such a bad guy after all.

PROJECTION

Consciously we often attempt to hide unpleasant information from ourselves. This includes personal traits and tendencies that consciously we find disagreeable. We also sometimes fail to consciously recognize certain disagreeable traits possessed by others, such as loved ones and potential mates. Sometimes these characteristics are too difficult to face. This is why it is said that "love is blind." However, within the domain of the right half of the brain we are often completely aware of the presence of these personality features, whether they are possessed by others or ourselves.

As pertaining to a potential mate or when meeting someone attractive, who is nevertheless flawed in some manner, this awareness may be realized in the form of a "gut feeling," "alarm bells," and other warning signs, that are of course ignored (by the conscious mind) until it is too late. It is only later, after we have been repeatedly hurt, used, or abused that we become conscious of what we were aware of. It is at this point that our good friends also begin to cheerfully remind us: "I told you so!" And we know they are right. We in fact were aware of it the whole time. The knowledge, however, was unconscious and part of the domain of what we have referred to as the unconscious-awareness of the right half of the brain.

KNOWING YET NOT KNOWING

Denial and self-deception often imply a knowing coupled with a not knowing . Although the right brain may know, the left brain is in the dark as to what is actually occurring. However, ignorance may not be bliss, as the left brain is still subject to tensions associated with these needs and desires, and may correspondingly experience some degree of discomfort.

Although information exchange between the two halves of the brain or the limbic system and the rest of the cerebrum may be incomplete and prevented to various degrees, there is often a seepage of information from one mental system to another. Even though, as we've seen, information may be incomplete and possibly misinterpreted and mistranslated, that seepage, coupled with the experience of tension, arousal, and any other associated feelings, often alerts the left brain to the possibility that certain information may be unpleasant and disagreeable. This initiates the activation of various defense mechanisms. A person is thus able to protect himself from gaining complete conscious knowledge by engaging in denial, self-deception, or through the confabulation of various explanations and justifications.

Indeed, even when we act on these impulses and desires, make the cruel comment, do and say the wrong thing, the left brain and conscious mind is still able to disavow what is obvious. It makes up reasons, rationalizes, justifies, denies, or even projects blame onto someone else. When others are blamed or when traits which we unconsciously possess are thought to be possessed by someone other than ourselves (such as in the case of Allan the rapist), we have engaged in projection. "Projections change the world into a replica of one's own unknown face."

11. The Neuroanatomy of Free Will:

Abstract

The neuroanatomy of "free will" is described, detailed, and supplemented by case histories of individuals who were compelled to behave against their will, who lost control over their will, and who suffered a complete loss of free will. In all instances the frontal lobe, the medial regions in particular are implicated in the mediation of "free will." The frontal lobes serve as the "Senior Executive" of the brain and personality, acting to process, integrate, inhibit, assimilate, and remember perceptions and impulses received from the limbic system, striatum, temporal parietal and occipital lobes, and neocortical sensory receiving areas. Through the assimilation and fusion of perceptual, volitional, cognitive, and emotional processes, the frontal lobes engages in decision making and goal formation, modulates and shapes character and personality and directs attention, maintains concentration, and participates in information storage and memory retrieval. Further, the frontal lobes, the SMA and medial regions in particular, can direct behavior by controlling movement and the musculature of the body, and in this manner, it serves what is best described as "free will."

1. FREE WILL & FUNCTIONAL LOCALIZATION

If the brain and mind are synonymous, is unknown. However, if the brain is damaged, the mind too is effected. Disturbances in brain functioning, be it due to drugs, alcohol, injury, tumor, stroke, emotional trauma, seizures, or electrode stimulation, directly affect consciousness (Joseph 1982, 1986a, 1988a, 1996, 1998, 1999a, 2001, 2003). Often specific aspects of the conscious mind are directly impacted (Joseph 1988a,b) including what has been referred to as "free will" (Joseph 1996, 1999b). This is because specific functions are localized to specific regions of the brain.

For example, in most humans, severe injury to the left frontal lobe can abolish the ability to speak words or intelligible sentences, a condition classically referred to as Broca's expressive aphasia (Joseph 1982, 1996, 1999b). Although the "Will" to speak remains intact, those afflicted may be capable only of expressing their frustrations by cursing which is mediated by the right frontal lobe, as is the ability to sing (Joseph 1982, 1988a, 1996, 1999b). Hence, patients can curse and may be able to sing words they can't say.

If, however, the damage to the left frontal lobe is widespread and extremely deep, penetrating into the medial (middle) portions of the anterior cerebral hemispheres, not just the "will to speak", but "free will" may be abolished and those afflicted may be forced to act "against their will" (Joseph 1986, 1988a, 1999b). What we call "free will" appears to be localized to the frontal lobes, the medial most portions in particular.

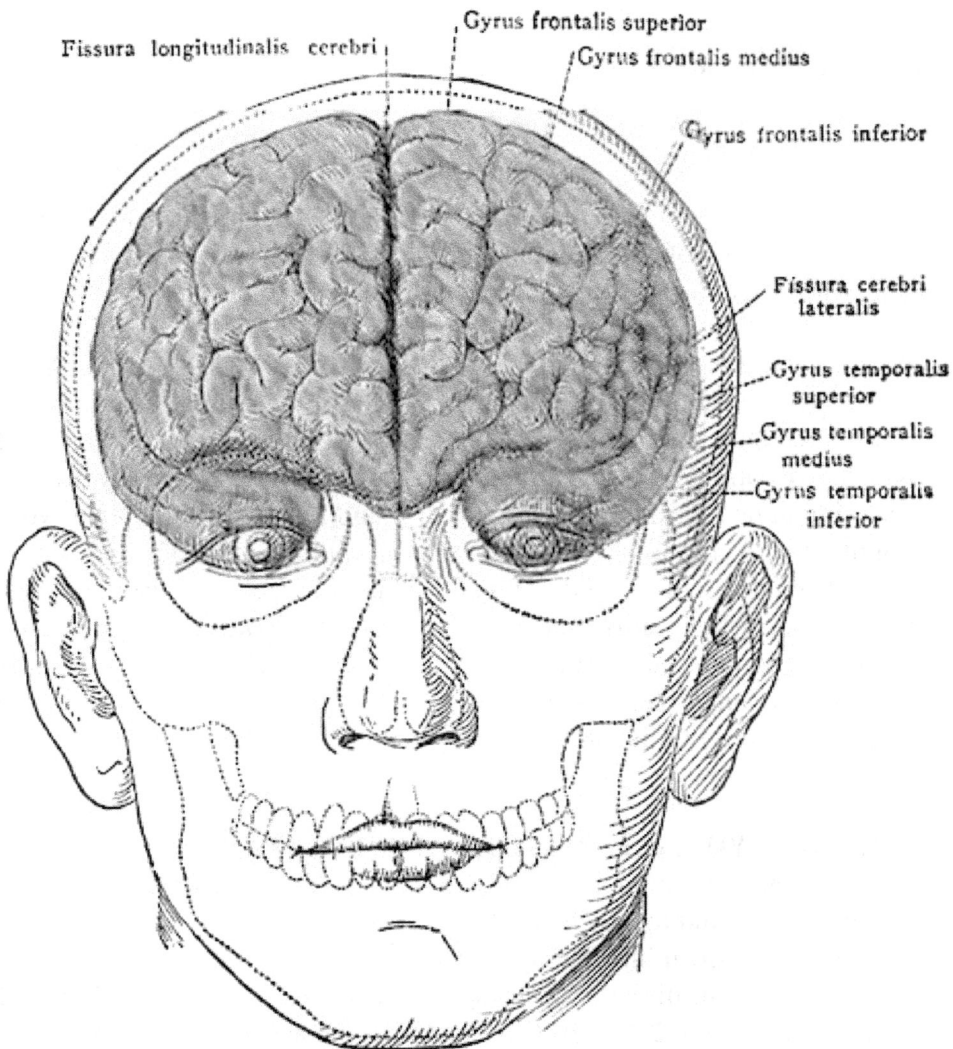

Labels on figure:
Fissura longitudinalis cerebri
Gyrus frontalis superior
Gyrus frontalis medius
Gyrus frontalis inferior
Fissura cerebri lateralis
Gyrus temporalis superior
Gyrus temporalis medius
Gyrus temporalis inferior

2. DISCONNECTION SYNDROMES

No region of the brain functions in isolation, unless isolated following a very circumscribed lesion thereby disconnecting it from other areas of the cerebrum; a condition classically referred to as "disconnection syndrome" (Geschwind 1981; Joseph 1982, 2009). For example, whereas Broca's area subserves the capacity to speak, the grammatical and denotative aspects of language (the words which will be spoken) are organized and assimilated by a multi-modal association area in the angular gyrus of the inferior parietal lobe (IPL) immediately adjacent to Wernicke's receptive language area in the superior temporal lobe (Joseph 1982, 1986, 1988a, 1996, 2000). The IPL and Wernicke's and Broca's areas are linked together by a rope of nerve fibers, the arcuate fasciculus.

However if the arcuate fasciculus is damaged such as due to stroke, those afflicted will know what they want to say, but will be unable to say it and will suffer severe word finding difficulty (Joseph 1982). Temporary functional disconnections occur even in the normal brain, where the missing word is known but can't be found, and this condition is experienced as "tip of the tongue" phenomenon. Thus one part of the mind is disconnected from another (Geschwind 1981; Joseph 1982, 2009). The "will to speak" remains intact (due to preservation of the frontal language areas), whereas the missing words are locked away in the posterior regions of the forebrain.

One rather severe form of of disconnection is "locked-in syndrome" which is due to destruction of the pyramidal (corticospinal) nerve fiber pathways linking the brainstem pontine area with the forebrain (Smith and Delargy 2005). Although completely paralyzed and seemingly comatose or "brain dead", patients are believed to be fully conscious and aware, and maintain the "will" to move and speak, but are unable with the exception of, in some cases, the capacity to blink and thus communicate through the eyes (Allain et al., 1998; Smith and Delargy 2005). Consciousness is maintained and cognitive functions are preserved because the forebrain is intact and completely functional with only mild reductions in cerebral metabolism noted (Allain et al., 1998; Zeman 2003).

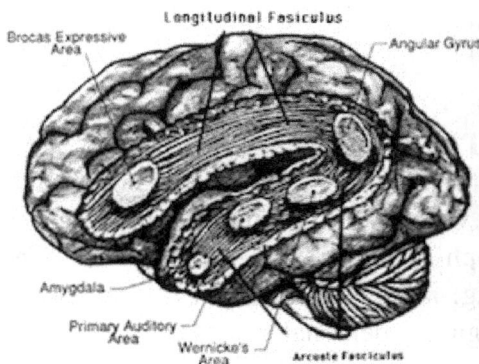

Parisian journalist Jean-Dominique Bauby suffered a stroke in December 1995, which severely damaged the neural pathways leading to and from the frontal lobes and brainstem. He was completely paralyzed and was initially believed to be in a vegetative state (Bauby 1998). However, when he began to cry, and by blinking his left eye, his caretakers realized he was in fact fully conscious. Over the next two years he learned to communicate by blinking a code for different letters of the alphabet, and dictated his memoir, The Diving Bell and the Butterfly (Bauby 1998). He died of pneumonia two days after it was published.

Consciousness is maintained in locked-in syndrome, because the forebrain which includes the frontal lobes, remains intact, and the inferior pons and the medulla of the brainstem which mediates vital life-sustaining functions are undamaged. However, because the motor areas in the forebrain are disconnected from the motor centers in the brainstem, the patient can't move their body and may appear to be in a vegetative state (Smith and Delargy 2005; Zeman 2003). They are no longer able to act according to their free will, which is locked-in.

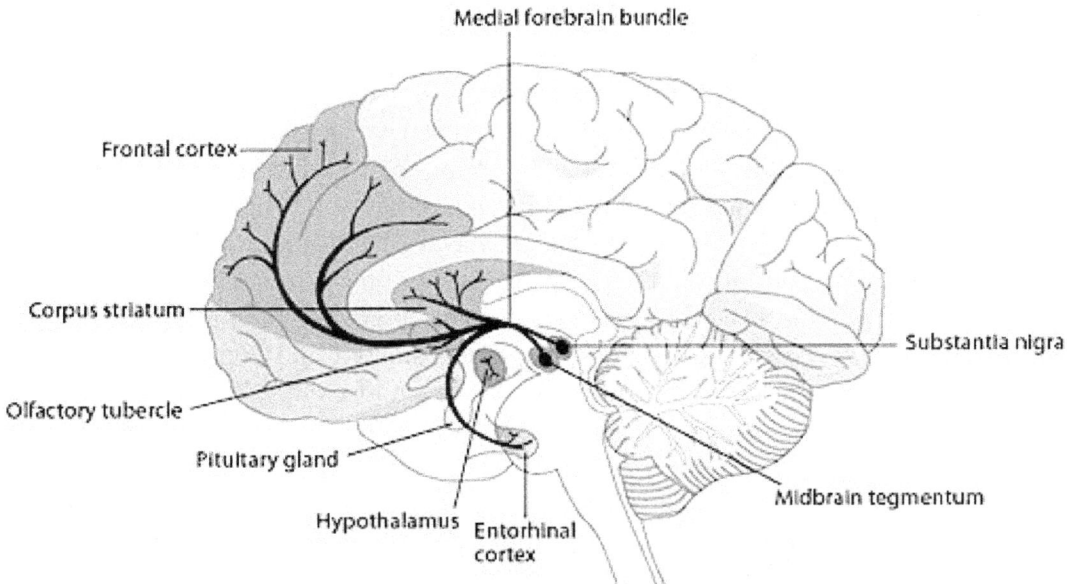

3. FREE WILL AND THE FRONTAL LOBES

Various aspects of personality, memory, attention, perception, emotion, the body image, and consciousness may be variably compromised with damage to different regions of the forebrain (Joseph 1982; 1986a,b; 1988a, 1992, 1994, 1998, 1999a,b, 2001, 2003, 2009), e.g. amygdala (emotion), hippocampus (memory), temporal lobe (memory, language, personality), parietal lobe (body image, hand-in-space). Certainly damage to these and other brain areas may limit and restrict what we call "free will". However, insofar as "free will" is defined as the ability to make plans, consider alternatives, and chose among and act upon them, if the frontal lobes remain intact and consciousness and movement are preserved, patients can still make choices and act on them, and they do not lose their free will.

By contrast, if the frontal lobes are damaged, or if the neural pathways between and within different frontal areas are compromised or disconnected, patients may not only lose their "free will" but other brain areas may act against their "will" and engage in behaviors which the patient cannot willfully resist (Joseph 1986a, 1988a, 1996, 1999b). Free will is localized to the frontal lobes, the medial frontal areas in particular.

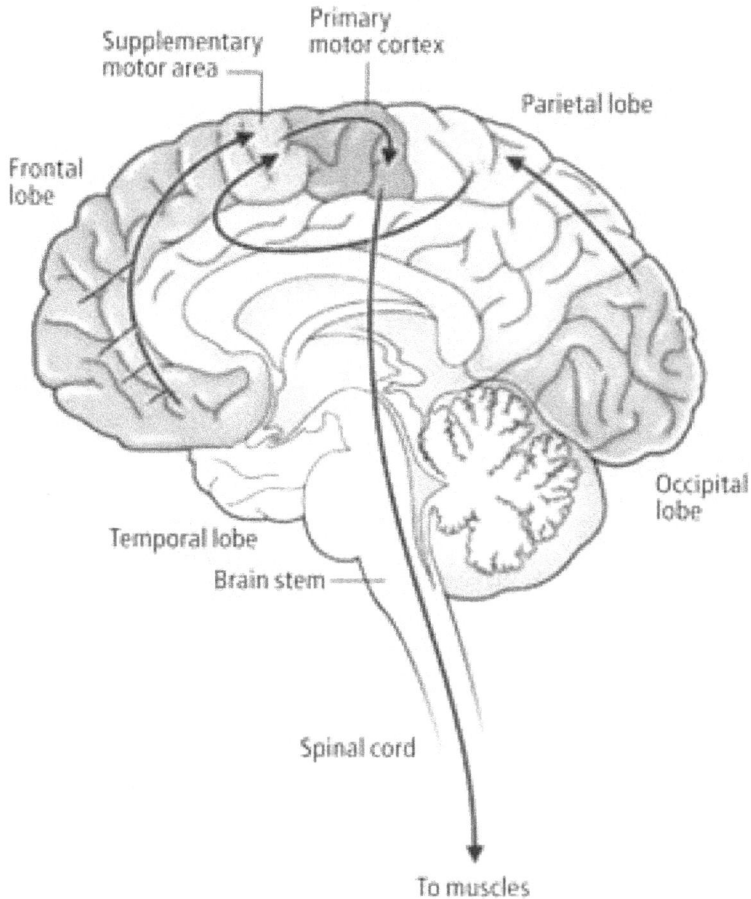

The frontal lobes are not a homologous tissue but serve myriad functions, and are "interlocked" via converging and reciprocal connections, with the brainstem, midbrain, thalamus, limbic system, striatum, anterior cingulate, and throughout the neocortex including the primary and secondary sensory receiving areas (Fuster 1997; Jones and Powell 1970; Miller & Cummings 2006 Pandya & Yeterian 1990; Petrides & Kuypers 1969; Pandya & Vignolo 1969; Risberg & Grafman 2006). Different areas of the frontal lobe are continually informed about activity in other areas of the brain, and in this way can direct attention, control behavior, and act according to their free will. In consequence, different aspects of "free will" can be compromised if neuronal pathways from the frontal lobes to other brain areas have been severed and "disconnected" and depending on which regions of the frontal lobe have been injured (Joseph 1986, 1988a, 1999b). For example, injuries to the right frontal lobe can result in a loss of control over free will.

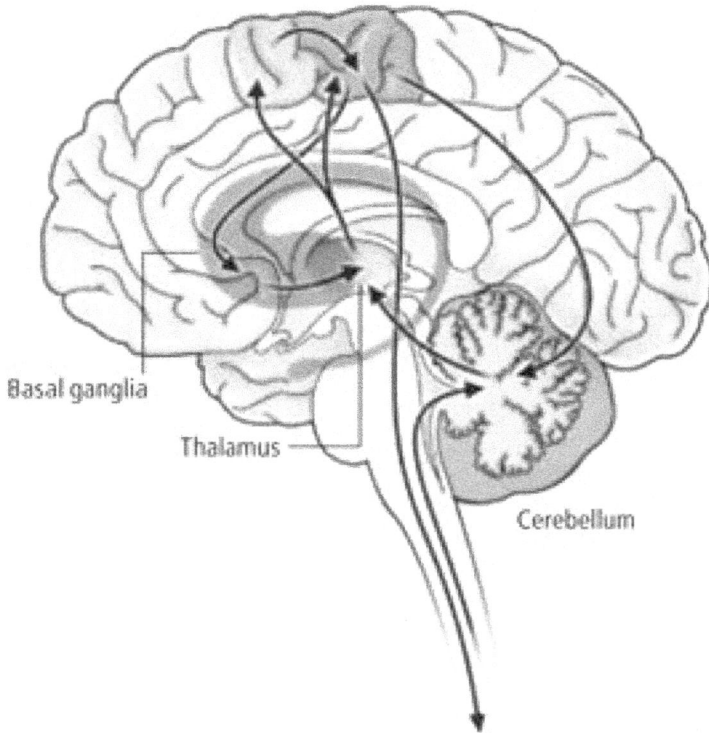

Basal ganglia

Thalamus

Cerebellum

4. THE RIGHT FRONTAL LOBE: FREE WILL UNRESTRAINED

Depending on the degree of damage, individual with right frontal lobe injuries may become unrestrained, overtalkative, and tactless, saying whatever "pops into their head", with little or no concern as to the effect their behavior has on others or what personal consequences may result. They may become inordinately disinhibited and influenced by the immediacy of a situation, buying things they cannot afford, lending money when they themselves are in need, and acting and speaking "without thinking" (Fuster 1997; Joseph 1986a, 1988a, 1999b; Miller & Cummings 2006; Risberg & Grafman 2006). Seeing someone who is overweight or obese they may begin making sounds like a hog ("oink oink") or laughingly call the person a "a fat pig." If they enter a room and detect a faint odor, its: "Hey, who farted?" If they see food they like, they may grab it off another person's plate.

Following severe right frontal lobe injuries there may be periods of gross disinhibition which may consist of loud, boisterous, and grandiose speech, singing, yelling, and beating on walls. The destruction of furniture and the tearing of clothes is not uncommon. Some patients may impulsively strike doctors, nurses, or relatives, or sexually proposition family members or complete strangers, and thus behave in a thoroughly labile, aggressive, callous and irresponsible manner.

One patient, with a tumor involving the right frontal area, threw a fellow patient's radio out the window because he did not like the music. He also loudly sang opera in the halls. Indeed, during the course of his examination he would frequently sing his answers to various questions (Joseph 1986a, 1988a, 1999b).

D.F. tried to commit suicide by sticking a gun in his mouth, but aimed it wrong and blew out his right frontal lobe. After he had been discharged from the hospital, he returned for his doctor's appointment laughing, making ridiculous remarks, and dressed inappropriately with a toothbrush, toothpaste, hair brush, and wash rag sticking out of his shirt pocket (Joseph 1988a).

When asked, pointing at his pocket, "Why the toothbrush?" he replied with a laugh, "That's just in case I want to brush my teeth," and in so saying he quickly drew the toothbrush from his pocket, climbed up on top of the doctor's desk, and began to brush his teeth. Then he began to dance on the desk and wanted to demonstrate how the skin flap which covered the hole in his head (from the craniotomy and bullet wound) could bulge in or out when he held his breath or held his head upside down (Joseph 1988a).

Nevertheless, despite his bizarre and inappropriate behavior, D.F.'s overall IQ was above 130 (98% rank: "Very Superior"). Unfortunately, although he had a high IQ, he could no longer control or employ that intelligence, intelligently. He had lost control over his "free will." During one lucid moment, he stated that "I don't know why I act this way. I just can't control it."

One frontal patient described as formerly very stable, and a happily married family man, became excessively talkative, restless, grossly disinhibited, sexually preoccupied and would approach complete strangers and proposition them for sex. He also extravagantly spent money and recklessly purchased a business which soon went bankrupt.

In another case, a 46-year old woman with a right frontal lobe tumor began walking around outside in just a slip and bra and then stripping naked in front of neighbors. She claimed she was descended from queens, was fabulously wealthy, and that many men wanted to divorce their wives and marry her (Joseph 1996).

A 19-year old man with seizure activity in the right frontal lobe, felt compelled to take his penis out of his pants, in public, and to masturbate. He was subsequently arrested for exposing himself in public. He claimed to have no control over his behavior. When interviewed he suddenly exposed his penis and urinated in the direction of his doctor who by grace of very fast reflexes moved aside just in time to avoid getting drenched (Joseph 1996).

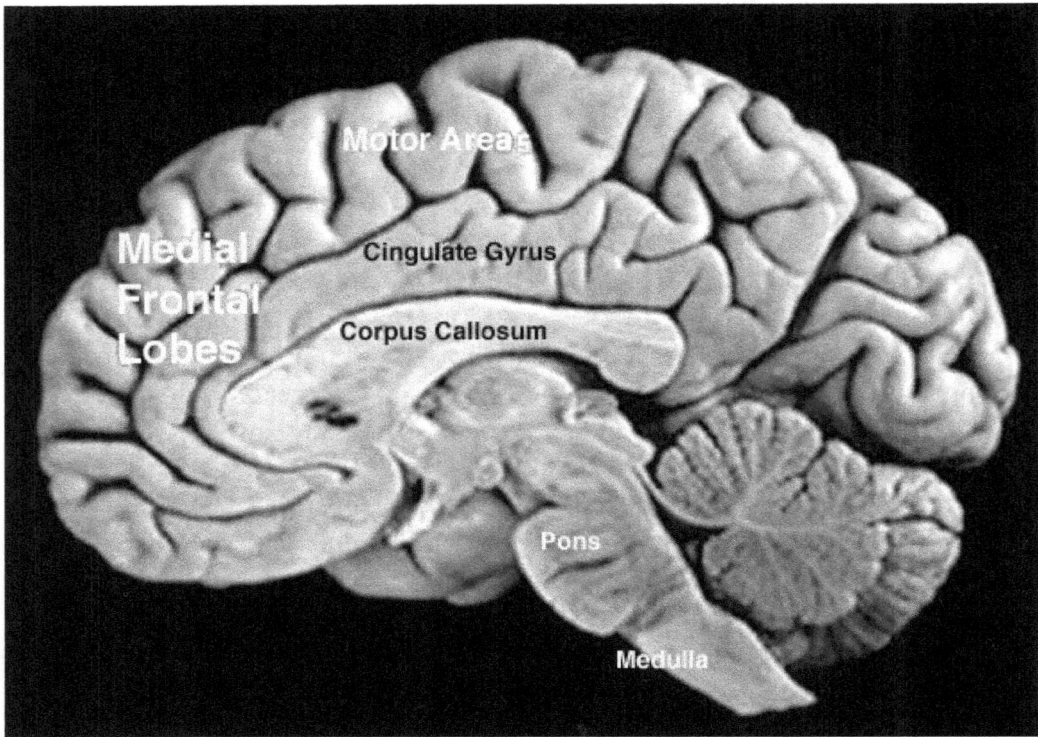

A very conservative, highly reserved, successful, brilliant , happily married engineer with over 20 patents to his name suffered a right frontal injury when he fell from a ladder. He became sexually indiscriminate and began patronizing up to 3 prostitutes a day, whereas before his injury his sexual activity was limited to once weekly with his wife. He spent money lavishly, depleted his considerable savings, suffered delusions of grandeur where he thought he was a senator and a billionaire, and camped out at Disney Land and attempted to convince personnel to fund his ideas for a theme park on top of a mountain (Joseph 1986a, 1996). At night had dreams where President Kennedy and Senator Kennedy would appear and offer him advice --and he was a republican!

Even with "mild" to moderate right frontal lobe injuries patients may initially demonstrate periods of tangentiality, grandiosity, irresponsibility, laziness, hyperexcitability, promiscuity, silliness, childishness, lability, personal untidiness and dirtiness, poor judgment, irritability, fatuous jocularity, and tendencies to spend funds extravagantly. Unconcern about consequences, tactlessness, and changes in sex drive and even hunger and appetite (usually accompanied by weight gain) may occur (Fuster 1997; Joseph 1986a, 1988a, 1999b; Miller & Cummings 2006; Passingham, 1997; Risberg & Grafman 2006).

These individuals, however, are not acting according to their free will, but often against their will. A janitor, who following his frontal lobe injury, suddenly believed he was a multi-millionaire congressman, quit his job, tried to take over the local congressman's office, stood on corners making speeches, and made extravagant purchases of items he could not afford or pay for. Later he explained that "I know its not true, but I believe its true and can't stop myself" (Joseph 1996).

Essentially, with right frontal lobe injuries, that aspect of the brain which serves and controls free will becomes disconnected from those brain areas which act on free will.

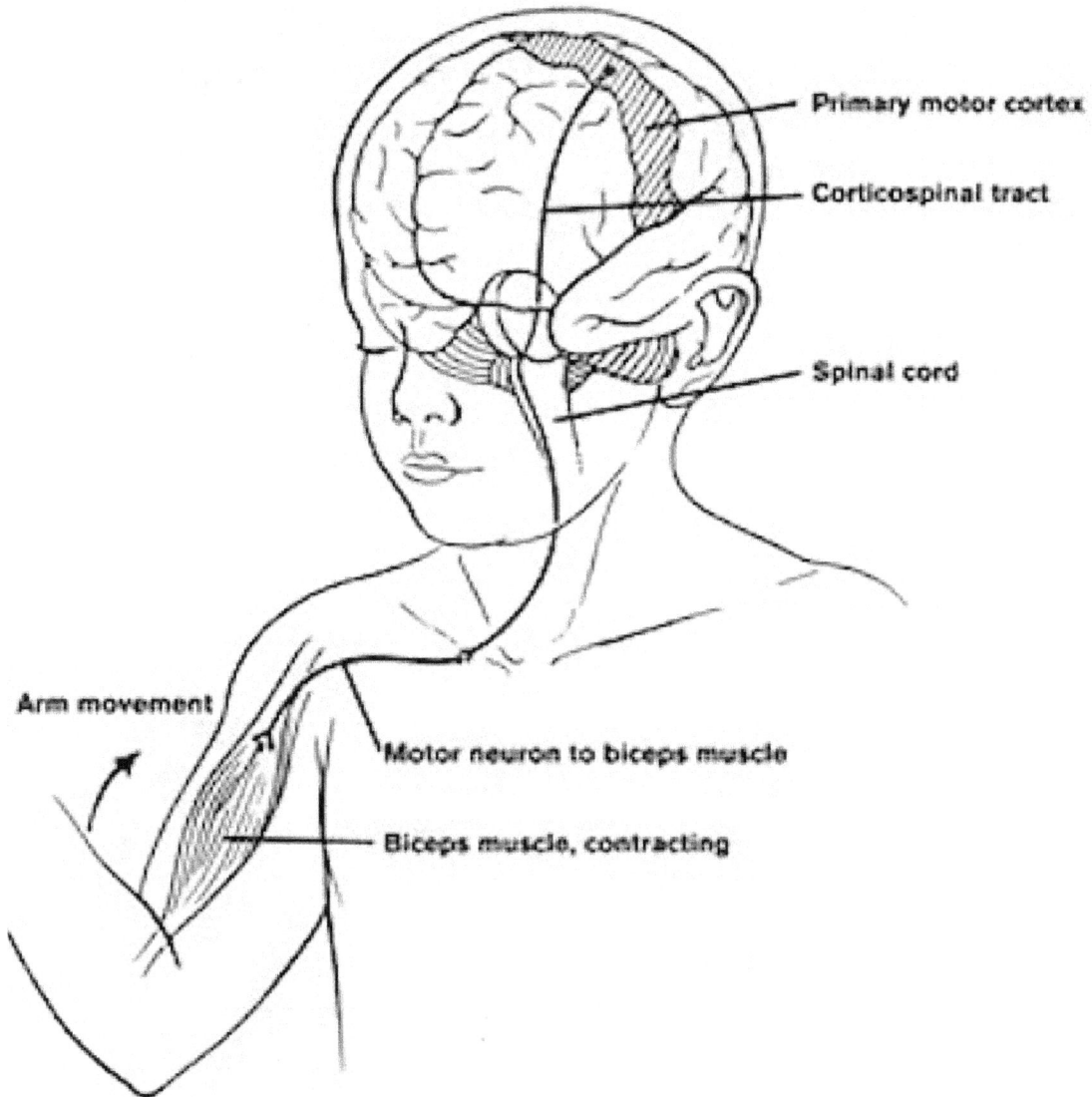

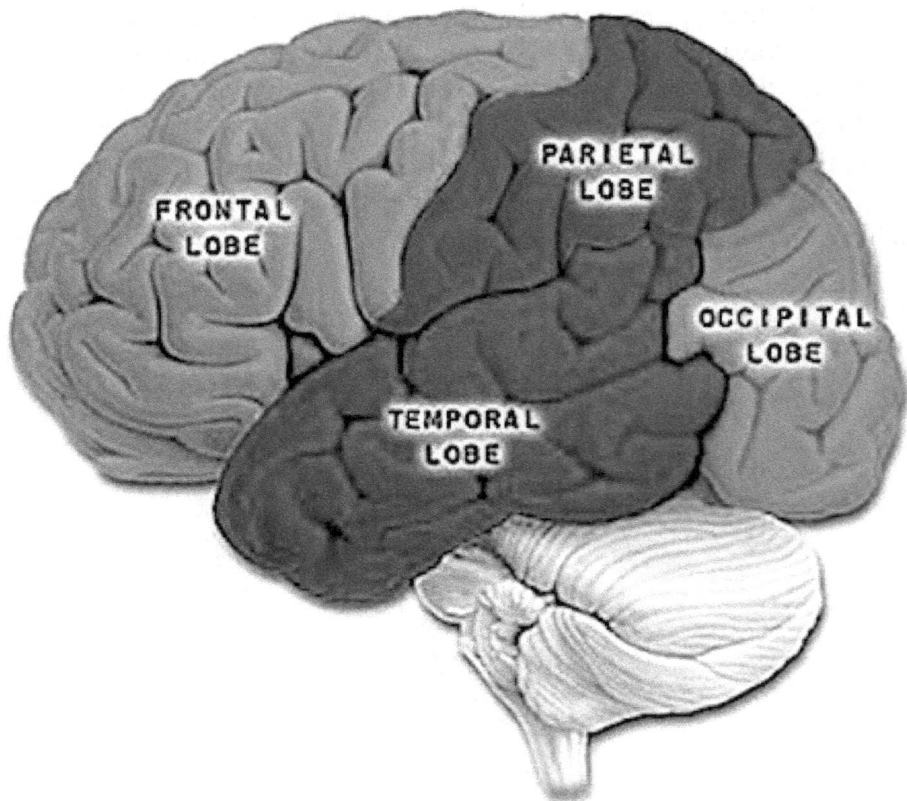

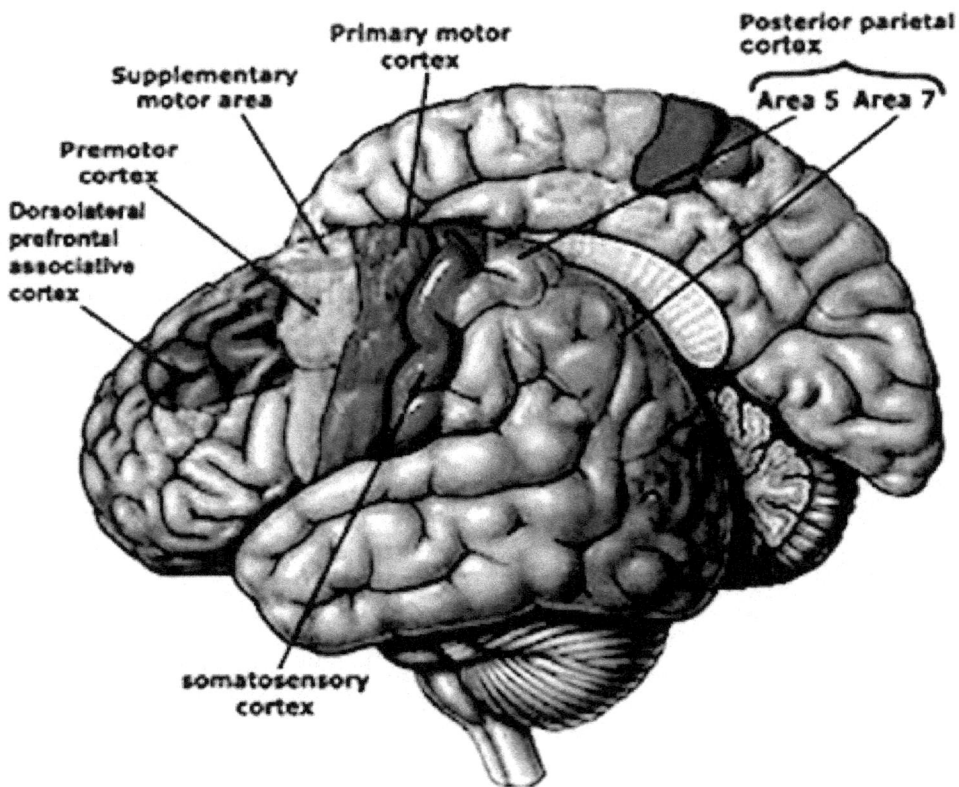

5. MOVEMENT AND THE FRONTAL MOTOR AREAS

To exercise one's free will requires access to and control over the motor systems of the brain and the body musculature. Motor control and movement are controlled, at the level of the frontal lobes, by the primary, secondary, and supplementary motor areas. These forebrain neocortical tissues are interconnected and communicate with other motor centers, such as the brainstem via a thick ropes of neurons known as the pyramidal (corticospinal) tracts (Fuster 1997; Joseph 1986a, 1988a, 1999a; Miller & Cummings 2006; Passingham, 1997; Risberg & Grafman 2006).

There is a one-to-one correspondence between single neurons in the primary motor area and single muscles such that the musculature of the entire body surface is represented according to their importance (Chouinard & Paus 2006; Dum & Strick 2005; Verstynen, et al. 2011). For example, the hands, fingers, face and mouth, have a greater representation than the musculature of the back. It is the primary motor areas which control fine motor functioning (such as when typing or writing).

The representation of the musculature is more diffuse in the secondary premotor area. The premotor transmits its motor impulses to the primary motor areas and subcortical regions of the brain but receives its marching orders from the supplementary motor area (Joseph 1999b; Nachev et al. 2008).

6. FREE WILL AND THE MEDIAL, SUPPLEMENTARY MOTOR AREAS

The SMA is more concerned with the general problem of guiding and coordinating the movements of the extremities through space (Andres, et al., 1999; Nachev et al. 2008; Passingham, 1997; Stephan, et al., 1999). However, it is the SMA which programs and exerts executive control over the secondary and primary motor areas, and it is the SMA and medial frontal lobes which are more closely linked with "free will."

The supplementary motor areas (SMA) originates within the medial walls of the right and left frontal lobe, is extensively interconnected with the limbic system, brainstem, anterior cingulate, and striatum, and extends up and over the medial walls to the lateral walls of the frontal lobes.

The SMA is especially concerned with preparing the hands, arms, and body to move, and becomes active simultaneously with the intention to move, but before movement and prior to activation of the secondary and primary motor areas. Single cell recordings (Brinkman & Porter, 1979; Tanji & Kurata, 1982) and studies of blood flow studies (Orgogozo & Larsen, 1979; Shibasaki et al. 1993), movement related evoked potentials (Ikeda et al. 1992) and other functional indices (Nachev et al., 2008) indicate increased activity in the SMA prior to moving, and when just thinking about moving the body, hand, arm, or leg. Thus activity begins in the SMA well before movements are initiated and prior to activation within the premotor and primary motor areas.

For example, when anticipating or preparing to make a movement, but prior to the actual movement, neuronal activity will first begin and then dramatically increase in the SMA, followed by activity in the premotor and then the primary motor area, and then the subcortical striatum (caudate, putamen, globus pallidus) and finally the brainstem (Alexander & Crutcher, 1990; Mink & Thach, 1991; Nachev et al., 2008). This indicates that the "will" to move begins in the SMA and medial frontal lobes and exert executive control over the secondary, primary, and subcortical motor areas which then perform these "willed" actions.

The SMA in fact exerts executive control over the arms and hands, and can willfully direct the extremities towards items of interest (e.g. coffee cups, tools, pens, keys, breasts) which are then grasped, manipulated and utilized (Andres et al., 1999; Passingham, 1997; Stephan et al., 1999). Direct electrical stimulation of the SMA will also trigger complex semipurposeful movements of the hands, arms, legs, and feet (Penfield & Jasper, 1954). The same is not true of the primary or secondary motor areas which, if activated by electrode, may only trigger the twitching of single muscles. Although guiding fine motor movements of the hands, the pre- and primary motor areas are under the control of the SMA and medial frontal lobes.

7. DISCONNECTION: ACTING AGAINST THE WILL

If the neural pathways linking the SMA / medial frontal lobes to the secondary and primary area are severed or grossly injured, the disconnected primary and secondary motor areas may act against the patient's will. Behavior will be controlled, deterministically, by external stimuli (Joseph 1996, 1999b).

Patients will involuntarily engage in complex coordinated movements involving their arms and hands, and may reach out and grasp and manipulate or even use various objects even though they don't want to, and try to willfully resist. Patients will lose control over their arms, hands and legs, and compulsively pick up and use tools, pens, cups, or other objects, such that in the extreme the right or left hand may act completely independently of the "conscious" mind (Denny-Brown, 1958; Gasquoine 1993; Goldberg & Bloom 1990; Lhermitte, 1983). Instead of acting voluntarily according to their "free will", behavior is determined and controlled by their environment.

In fact, the mere visual presence of a cup, pen, hammer, saw or scissors near the hand may trigger groping movements as well as grasping and involuntary use of the object. For example, they may compulsively reach out and take the examiner's pen or swipe their glasses from their face and put them on. One patient put three pair of spectacles on and wore them simultaneously.

If a hammer is placed on the testing table the patient may involuntarily pick it up and begin hammering on the table or walls, even when told not to, and will continue even after told to stop. If a glass of water is placed on the table, they may pick it up and drink from it, although they are not thirsty and despite their efforts to willfully oppose these actions (Gasquoine 1993; Goldberg and Bloom 1990; Lhermitte 1983; McNabb et al. 1988). Denny-Brown (1958) has referred to this condition as "magnetic apraxia" and "compulsive exploration".

Sometimes just the presence of a pencil and a piece of paper may cause a patient to pick up the pencil and begin writing an endless letter that might include the mechanical repitition of the same word line after line and even page after page.

Not only do they act against their will, but once they take hold of a pen, cup, or other object, they may be unable to let go and cannot release their grip. In fact, the entire arm and hand may become increasingly stiff and rigid, such that the hand will become frozen and seemingly "stuck" to the pen, paper, cup, or whatever they were holding or touching. While walking they may move more and more slowly until they seem unable to move, as if their feet are stuck to the floor as if glued (Denny-Brown, 1958).

Gegenhalthen (counterpull), i.e. involuntary resistance to movement of the extremities, appears to be exclusively associated with medial frontal lobe/SMA abnormalities (Joseph 1996, 1999b). If the patient raises an affected arm, it will stiffen and become increasingly rigid as the will to move it increases. If the patient tries to resist or to relax the arm, it instead becomes frozen in place.

The same condition also results if a nurse or physician attempts to move the arm; it becomes increasingly rigid and patients are forced, against their will, to maintain their afflicted arm even in uncomfortable postures, frozen in place, for long time periods. However, over time, the affected limb may slowly return to a resting posture--a condition referred to as "waxy flexibility."

Likewise, if they are presented with yet another cup, pen, or tool, the frozen state may slowly disappear and they may reach out and compulsively take hold of the new object.

Naturally, individuals so afflicted become quite upset, frustrated and embarrassed by the misbehavior of their body, and will complain these actions occur against their will.

8. THE ALIEN HAND

With injury or damage to the medial frontal lobes, SMA, and their interconnections with the other motor areas, one or both hands may take on a "life of their own"; a condition referred to as the "alien hand" (Joseph 1986a, 1988a,b, 1996, 1999b). In fact, with damage to or disconnection of these medial frontal tissues, involuntary uncontrolled movements may become so purposeful and complex that they appear to be directed not by the external environment, but by a consciousness with its own free; an alien consciousness that is disconnected from another aspect of consciousness which can speak but which cannot willfully oppose these alien behaviors by an alien hand (Joseph 1986a, 1988a,b, 1996, 1999b).

For example, a patient described by Gasquoine (1993) had a propensity to reach out and touch female breasts with his right hand. He couldn't stop touching and fondling pencils, cups, nik naks, doctors, nurses, and persons within in reach. He reported this caused him great embarrassment and that his hand acted against his will. However, as the alien behavior was confined to one hand, the right, he would grab hold of it with his left hand and try to wrestle it under control. He also learned to trick the alien hand. He would give it something to hold and it would be unable to let go, thus preventing it from grabbing breasts, buttocks, or property which was not his own. In addition to his alien hand, he also complained that he would speak his private thoughts out loud, against his will.

McNabb et al. (1988, pp. 219, 221) describe one woman with extensive damage involving the medial left frontal lobe and anterior corpus callosum, who was forced to punish her misbehaving right hand by slapping it with the left. However, sometimes the right hand would fight back and interfere with actions performed by the left. She complained that her right hand showed an uncontrollable tendency to reach out and take hold of objects and then be unable to release them. When the right hand behaved mischievously she attempted to restrain it by wedging it between her legs or by holding or slapping it with her left hand. She repeatedly express astonishment at these actions by her hand. A second patient frequently experienced similar difficulties. When "attempting to write with her left hand the right hand would reach over and attempt to take the pencil. The left hand would respond by grasping the right hand to restrain it."

Problems of an even more severe nature have plagued patients following complete (surgical) destruction of the corpus callosum and thus the neural pathways linking the two medial motor areas of the frontal lobes (Joseph, 1988ab). These independent "alien" behaviors usually involve the left hand and the half of the body, and were purposeful, intentional, complex and obviously directed by an awareness maintained by the disconnected right hemisphere (which controls the left hand). These alien actions were often completely against the "will" of the consciousness maintained in the left hemisphere.

These "alien" disturbances were so purposeful, and often so well thought out, it was as if these "split-brain" patients had developed two independent "free wills" maintained by independent minds housed in the right and left half of the brain (Joseph, 1986b, 1988a,b); two free wills and two minds which were unable to communicate, and each of which had a "mind of its own."

As originally described by Nobel Lauriate Roger Sperry (1966, p. 299), "Everything we have seen indicates that the surgery has left these people with two separate minds, that is, two separate spheres of consciousness. What is experienced in the right hemisphere seems to lie entirely outside the realm of awareness of the left hemisphere. This mental division has been demonstrated in regard to perception, cognition, volition, learning and memory."

For example, one patient's left hand would not allow him to smoke and would pluck lit cigarettes from his mouth. He reported that he had been trying to quit, unsuccessfully, for years, but it was only after the surgery that he found he couldn't smoke, because the left hand wouldn't let him (Joseph, 1988a).

Each frontal lobe has its own primary, secondary, and supplementary motor areas and medial (and lateral) tissues. Since free will is associated with the SMA and medial frontal lobe, disconnecting the connections between the right and left created two "free wills" one maintained by the right the other by the left frontal lobe.

Akelaitis (1945, p. 597) describes two patients with complete corpus callosotomies who experienced extreme difficulties making the two halves of their bodies cooperate. "In tasks requiring bimanual activity the left hand would frequently perform oppositely to what she desired to do with the right hand. For example, she would be putting on clothes with her right and pulling them off with her left, opening a door or drawer with her right hand and simultaneously pushing it shut with the left. These uncontrollable acts made her increasingly irritated and depressed."

Another patient experienced difficulty while shopping, the right hand would place something in the cart and the left hand would put it right back again. Yet another had the same problem when trying to decide what to eat, first picking one item, and the other hand putting it back and picking another. Of course, similar difficulties in "making up one's mind" also plague those who have not undergone split-brain surgery.

A recently divorced split-brain patient reported that on several occasions while walking about town he was forced to go some distance in another direction by the left half of this body. He resisted with the right half of the body and essentially engaged in a battle of the wills. As it turned out, the left half of the body was trying to take him to the home of his ex-wife, but the right half of his body refused to have anything to do with her.

Another split-brain patient who had recently broken up with his girlfriend voiced considerable anger toward her and stated an intention to never see her again. He also admitted that they had broken up several times, but always got back together. This time, however, he was adamant: He never wanted to see her again. When asked to indicate with "thumbs up" or "thumbs down" if he still liked her, the right hand gave a "thumbs down" and the left hand a "thumbs up."

Geschwind (1981) reports a callosal patient who complained that his left hand on several occasions suddenly struck his wife--much to the embarrassment of his left (speaking) hemisphere. In another case, a patient's left hand attempted to choke the patient himself and had to be wrestled away.

Bogen (1979) indicates that almost all of his "complete commissurotomy patients manifested some degree of intermanual conflict in the early postoperative period." One patient, Rocky, experienced situations in which his hands were uncooperative; the right would button up a shirt and the left would follow right behind and undo the buttons. For years, he complained of difficulty getting his left leg to go in the direction he (or rather his left hemisphere) desired. Another patient often referred to the left half of her body as "my little sister" when she was complaining of its peculiar and independent actions.

A split-brain patient described by Dimond (1980, p. 434) reported that once when she had overslept "my left hand slapped me awake." This same patient, in fact, complained of several instances where her left hand had acted violently toward other people, and this caused her considerable distress and embarrassment.

Yet another split brain patient, "2-C" complained of instances in which his left hand tried to strike a relative (Joseph 1988a,b). Once, after he had retrieved something from the refrigerator with his right hand, his left took the food, put it back on the shelf and retrieved a completely different item "Even though that's not what I wanted to eat!" he complained. Once while watching and enjoying a program TV, the left half of his body dragged him from his seat, and then changed the channels and then returned to his seat to watch a different program, even though: "That's not what I wanted to watch!" On several occasions, his left leg refused to continue "going for a walk" and would only allow him to return home.

9. MIRROR NEURONS, ALIEN HANDS, SPLIT-BRAINS & FREE WILL

In 1996 Rizzolatti et al., reported the discovery of "mirror" neurons which were most densely concentrated in the SMA. According to Rizzollati et al. (1998) "mirror neurons appear to form a system that allows individuals to recognize motor actions made by others by matching them with an internal motor copy," thus enabling them to perform and mimic the actions of others (Rizzolatti & Craighero 2004, Rizzollati et al. 2009).

Since the SMA and medial frontal lobes are implicated in the expression of all "alien" actions, is it possible these "alien behaviors" are merely reflexive, and that the "alien" hand is merely mimicking and mirroring the behavior it is observing, and thus doing the opposite?

The answer was provided in 1988 when it was conclusively demonstrated that these "alien" behaviors are purposeful, goal directed, and under the control of a "free will" and a mind separate from the dominant stream of consciousness which has access to language (Joseph 1988ab). In one test, both hands of split-brain patients were provided information and given multiple choices so as to make a choice which precluded mimicry. Specifically, the patient sat before a box with two separate holes (on the right and the left) in which he inserted his hands. Above each hole, in plain sight, were small patches of textured material, one above the other, i.e. sandpaper, wire mesh, smooth metal, velvet. While in the hole and out of sight, both hands were stimulated simultaneously, but with completely different materials, e.g. velvet to the right, sand paper to the left. They were then required to point to the material they experienced. On every test the right and left hands made the correct responses. The "alien" left hand did not mimic the right and did not interfere with its choices.

However, these patients (that is, the language dominant left hemispheres) had not been told they would be experiencing two different fabrics, and then expressed shock and dismay when the alien left hand chose a different material. In one case, the patient's normal right hand, repeatedly reached over and tried to force the left (alien) hand to make a different choice, that is, to chose the same material experienced by the right hand, although the left hand responded correctly! However, in this case, the alien hand resisted and refused to make a different choice, even when the patient (his left hemisphere) vocalized: "Thats wrong!" Repeatedly, he reached over and grabbed his left hand with his right but it refused to point at the wrong item. He became so angry with his left hand that he yelled at it, said "I hate this hand" and then began punching and hitting it. Finally the two hands began to fight!

In this and in other instances, it was demonstrated that the alien hand was not misbehaving, but acting purposefully and making rational choices according to its own free will - and which was opposed by the free will residing in the other, disconnected half of the brain.

10. CATATONIA, THE SMA / MEDIAL FRONTAL LOBES & FREE WILL

Different regions of the brain are specialized for performing specific functions, and interact with yet other regions to coordinate and make possible complex cognitive and behavioral activity. The frontal lobes serve as the senior executive of the brain and personality and the MFL and SMA provide the neuroanatomical substrate which executes what has been traditionally described as "free will."

If other areas of the brain are disconnected from the SMA / MFL the patient may feel compelled to act against their will. However, if the MFL and SMA are destroyed, "free will" is abolished. Ideas and thoughts are no longer generated, and the "will" to speak or to initiate or complete a voluntary movement may become completely attenuated and abolished (Hassler 1980; Laplane et al. 1977; Luria 1980; Penfield and Jasper 1954; Penfield and Welch 1951). Patients may lose even the will to speak and become mute, unresponsive, stiff, frozen, unmoving, motionless, and catatonic (Hassler 1980; Joseph, 1999b; Laplane et al. 1977; Luria 1980; McNabb et al. 1988; Penfield and Jasper 1954; Penfield and Welch 1951).

In one case, a soldier developed gegenhalten, waxy flexibility, mutism, and catatonia after a gunshot wound that passed completely through the frontal lobes, destroying much of the SMA and MFL and disconnecting these regions from other brain tissue. For two months he laid in a catatonic-like stupor, always upon one side with slightly flexed arms and legs, never changing his uncomfortable position. He did not obey commands, was incontinent, made no complaints, gazed steadily forward and showed no interest in anything. However, over the ensuing weeks, he periodically showed signs of lucidity, and could be persuaded to talk, and would answer quite correctly about his personal affairs. When questioned he explained that during his catatonic periods, although he was aware of his surroundings, his mind was empty, devoid of thoughts, and it just did not occur to him to move, speak, or eat. So he did nothing. Incredibly, the patient "was eventually returned to active duty" (Freeman and Watts 1942).

In another case, a 42 year old male with no previous psychiatric history, developed gegenhalten, mutism, and catatonia after a beating and suffering frontal and midline subdural hematomas (which later required the drilling of burr holes for evacuation). He resisted the efforts of others to move him, and would sit motionless and unresponsive for hours in odd and uncomfortable positions. The patient's symptoms seemed to wax and wane such that he demonstrated some periods of seeming normality (Joseph 1996). When he was in one of his "normal" periods, and was asked about his behavior he replied: "Its not that I don't want to move, its that I don't feel the want". During another "normal" period he explained his catatonic state thusly: "I feel nothing. No thoughts. Everything is going on outside my head, but nothing is going on inside my head. Like I am a rock. I am just there." Thus, while in his catatonic state his mind was a blank and there was simply no reason to even move.

It is not uncommon with SMA / MFL injuries for patients to become catatonic, mute, and remain in odd, uncomfortable postures for long time periods (Joseph 1996, 1999b). Upon recovery many patients may later remark they had completely lost the will to speak, that thoughts did not enter their head, that they were unable to think or generate ideas, and instead experienced a motivational-ideational void, a complete emptiness without feelings, which left them without any reason to move or function (Brutkowski 1965; Hassler 1980; Laplane et al. 1977; Luria 1980; Mishkin 1964). As those afflicted have no interest in eating, drinking, or even moving, essentially they had also lost the will to live.

With massive SMA MFL injuries, free will is abolished, and this is because, free will is localize to the frontal lobes.

11. CONCLUSIONS: THE NEUROANATOMY OF FREE WILL

The frontal lobes serve as the "Senior Executive" of the brain and personality, acting to process, integrate, inhibit, assimilate, and remember perceptions and impulses received from the limbic system, striatum, temporal parietal and occipital lobes, and neocortical sensory receiving areas. Through the assimilation and fusion of perceptual, volitional, cognitive, and emotional processes, the frontal lobes engages in decision making and goal formation, modulates and shapes character and personality and directs attention, maintains concentration, and participates in information storage and memory retrieval. Further, the frontal lobes, the SMA and medial regions in particular, can direct behavior by controlling movement and the musculature of the body, and in this manner, it serves what is best described as "free will."

Because "free will" is localized to the right and left frontal medial motor areas, surgical destruction of the neural pathways linking the right and left SMAs and medial frontal lobes will result in two independent streams of mental activity which act according to their own "free will."

Destruction and disconnection of the pathways leading from the SMA and medial frontal lobes to the secondary and primary motor areas leads to difficulties where the patient's body will act "against their will", and perform actions they are unable to prevent or control. Compulsive utilization, and forced groping and grasping, are obviously compulsive in nature, and reminiscent of obsessive-compulsive disorders, which are also associated with frontal lobe abnormalities (Joseph 1996), especially of the orbital-medial frontal lobes and the striatum /anterior cingulate which are buried within the depths of the frontal lobes.

Patients may experience unwanted, recurrent, perseverative ideas, or compulsions to repetitively perform certain acts, e.g. hand washing. They may also experience intrusive recurring thoughts, feelings, or impulses to perform certain actions against their will. Motorically obsessive compulsions may involve repetitive, stereotyped acts including the perseverative manipulation and touching of objects.

Moreover, abnormal or electrical activation of the medial-orbital frontal lobes has been reported to trigger recurrent and intrusive ideational activity ("forced thinking"), as well as compulsive urges to perform aberrant actions, e.g. shouting or the manifestations of various motor tics (Penfield and Jasper 1954; Ward 1988). Similar disturbance have been reported for those who are afflicted with Tourretts syndrome (Singer 2005; Walkup et al., 2006). Those with Tourretts may involuntarily shout out obscenities, and inappropriate remarks, or engage in spontaneous, but repetitive and stereotyped movements which are experienced as "unvoluntary" (Tourette Syndrome Classification Study Group, 1993). Although the exact etiology is unknown, abnormalities in deep medial frontal lobe structures have been reported (Walkup et al., 2006).

Damage to the neural pathways to and from the medial frontal lobes often simultaneously disrupt the orbital, lateral, and inferior frontal lobes, and can produce a unique constellation of compulsive, perseverative symptomology, including "forced thinking" and uncontrolled obsessive compulsions. That is, a patient may suffer from a difficulty suppressing or inhibiting previous thoughts or behaviors, which then occur again and again.

However, with massive damage to the medial frontal lobes and supplementary motor areas, instead of acting against their will, they lose the ability to will; free will is abolished, and this is because what has been called "free will" is localized to the frontal lobes, the medial frontal lobes in particular.

References

Alexander, G. E., & Crutcher, M. D. (1990). Preparation for movement. Neural representations of intended direction in three motor areas of monkey. J. Neurophysiology, 64, 164-178.

Allain P, Joseph PA, Isambert JL, Le Gall D, Emile J. Cognitive functions in chronic locked-in syndrome: a report of two cases. Cortex 1998;34: 629-34.

Akelaitis, A. J. (1945). Studies on the corpus callosum. American Journal of Psychiatry, 101, 594-599.

Bauby, J.-D., (1998), The Diving Bell and the Butterfly: A Memoir of Life in Death, Vintage.

Bogen, J. (1979). The other side of the brain. Bulletin of the Los Angeles Neurological Society. 34, 135-162.

Brinkman, C., & Porter, R. (1979). Supplementary motor area. Journal of Neurophysiology, 42, 681-709.

Brutkowski, S. (1965) Functions of the prefrontal cortex. Physiological Review, 45, 721-746.

Chouinard, P. A., Paus, T. (2006). The Primary Motor and Premotor Areas of the Human Cerebral Cortex, Neuroscientist April 2006 vol. 12 no. 2 143-152.

Denny-Brown, D. (1958). The nature of apraxia. Journal of Nervous and Mental Disease, 15-56.

Dimond, S. J (1980). Neuropsychology. Buttersworth.

Dum, R. P., Strick, P. L. (2005). Frontal Lobe Inputs to the Digit Representations of the Motor Areas on the Lateral Surface of the Hemisphere The Journal of Neuroscience,25(6):1375-1386.

Freeman, W., & Watts, J. W. (1942). Psychosurgery. Springfield, IL: Charles C. Thomas.

Fuster, J.M. (1997). The prefrontal cortex. Anatomy, physiology, and neuropsychology of the frontal lobes. New York: Ravens-Lippincott.

Fuster, J. M. (1995). Neuropsychiatry of frontal lobe lesions. In B.S. Fogel & R. B. Schiffer (Eds). Neuropsychiatry. Baltimore, Williams & Wilkins.

Gasquoine, P. G. (1993). Bilateral alien hand signs following destruction of the medial frontal cortices. Neuropsychiatry, Neuropsychology & Behavioral Neurology, 6, 49-53.

Geschwind N (1965) Disconnection syndromes in animals and man. Brain. 882, 237-274, 585-644.

Geschwind, N. (1981). The perverseness of the right hemisphere. Behavioral Brain Research, 4, 106-107.

Goldberg, G., & Bloom, K. K. (1990). The alien hand sign. American Journal of Physical Medicine & Rehabilitation. 69, 228-38.

Hassler, R. (1980). Brain mechanisms of intention and attention with introductory remarks on other volitional processes. Progress in Brain Research, 54, 585-614.

Ikeda, A., Luders, H. O., Burgess, R. C., and Shibasaki, H. (1992) Movement-related potentials recorded from supplementary motor area and primary motor area. Brain, 115, 1017-1043.

Jones, E. G., & Powell, T. P. S. (1970). An anatomical study of converging sensory pathways within the cerebral cortex of the monkey. Brain, 93, 793-820.

Joseph, R. (1982). The Neuropsychology of Development. Hemispheric Laterality, Limbic Language, the Origin of Thought. Journal of Clinical Psychology, 44 4-33.

Joseph, R. (1986a). Confabulation and delusional denial: Frontal lobe and lateralized influences. Journal of Clinical Psychology, 42, 845-860.

Joseph, R. (1986b). Reversal of language and emotion in a corpus callosotomy patient. Journal of Neurology, Neurosurgery, & Psychiatry, 49, 628-634.

Joseph, R. (1988a) The Right Cerebral Hemisphere: Emotion, Music, Visual-Spatial Skills, Body Image, Dreams, and Awareness. Journal of Clinical Psychology, 44, 630-673.

Joseph, R. (1988b). Dual mental functioning in a split-brain patient. Journal of Clinical Psychology, 44, 770-779.

Joseph, R. (1992) The Limbic System: Emotion, Laterality, and Unconscious Mind. The Psychoanalytic Review, 79, 405-455.

Joseph, R. (1994) The limbic system and the foundations of emotional experience. In V. S. Ramachandran (Ed). Encyclopedia of Human Behavior. San Diego, Academic Press.

Joseph, R. (1996). Neuropsychiatry, Neuropsychology, Clinical Neuroscience, 2nd Edition. Williams & Wilkins, Baltimore.

Joseph, R. (1998). Traumatic amnesia, repression, and hippocampal injury due to corticosteroid and enkephalin secretion. Child Psychiatry and Human Development. 29, 169-186.

Joseph, R. (1999a). The neurology of traumatic "dissociative" amnesia. Commentary and literature review. Child Abuse & Neglect. 23, 71-80.

Joseph, R. (1999b). Frontal lobe psychopathology: Mania, depression, aphasia, confabulation, catatonia, perseveration, obsessive compulsions, schizophrenia. Psychiatry, 62, 138-172.

Joseph, R. (2000). Limbic language/language axis theory of speech. Behavioral and Brain Sciences. 23, 439-441.

Joseph, R. (2001). The Limbic System and the Soul: Evolution and the Neuroanatomy of Religious Experience. Zygon, the Journal of Religion & Science, 36, 105-136.

Joseph, R. (2003). Emotional Trauma and Childhood Amnesia. journal of Consciousness & Emotion, 4, 151-178.

Joseph, R. (2009). Quantum Physics and the Multiplicity of Mind: Split-Brains, Fragmented Minds, Dissociation, Quantum Consciousness, Journal of Cosmology, 3, 600-640.

Joseph, R., Forrest, N., Fiducia, N., Como, P., & Siegel, J. (1981). Electrophysiological and behavioral correlates of arousal. Physiological Psychology, 1981, 9, 90-95.

Laplane, D., Talairach, J., Meininger, V., Bancaud, J., & Orgogozo, J. M. (1977). Clinical consequences of cortisectomies involving the supplementary motor area in man. Journal of the Neurological Sciences, 34, 301-314.

Lhermitte, F. (1983). "Utilization behaviour" and its relation to lesions of the frontal lobes. Brain, 106, 237-255.

Luria, A. (1980). Higher cortical functions in man. New York: Basic Books.

McNabb, A. W., Caroll, W. M., & Mastaglia, F. L. (1988). "Alien hand" and loss of bimanual coordination after dominant anterior cerebral artery territory infraction. Journal of Neurology, Neurosurgery, and Psychiatry, 51, 218-222.

Miller, B. L., & Cummings, J. L. (2006). The Human Frontal Lobes. Guilford Press.

Mink, J. W., & Thach, W. T. (1991). Basal ganglia motor control. I, II, & III. Journal of Neurophysiology, 65, 273-351.

Mishkin, M. (1964). Perseveration of central sets after frontal lesions in monkeys. In J. M. Warren & K. Akert (Eds.), The frontal granular cortex and behavior. (pp 219-241). New York: McGraw-Hill.

Nachev, P., Kennard, C., & Husain, M. (2008). Functional role of the supplementary and pre-supplementary motor areas Nature Reviews Neuroscience 9, 856-869.

Orgogozo, J. M., & Larsen, B. (1979). Activation of the supplementary motor area during voluntary movement in man suggest it works as a supramodal motor area. Science, 206, 847-850.

Pandya, D. N. & Yeterian, E. H. (1990) Architecture and connectivity of cerebral cortex: Implications for brain evolution and function. In: Neurobiology of higher cognitive function. eds. A. B. Scheibel & A. F. Wechsler. Guilford.

Pandya, D. N., & Kuypers, H. G. J. M. (1969). Corticocortical connections in the rhesus monkey. Brain Research, 13, 13-36.

Pandya, D. N., & Vignolo, L. A. (1969). Corticocortical connections in the rhesus monkey. Brain Research, 13, 13-16.

Passingham, R. (1997). Functional organization of the motor system. In Frackowiak, R. S. J., et al., (Eds.,) Human Brain Function. Academic Press, San Diego.

Penfield, W. (1952) Memory Mechanisms. Archives of Neurology and Psychiatry, 67, 178-191.

Penfield, W., & Jasper, H. (1954). Epilepsy and the functional anatomy of the human brain. Boston: Little-Brown & Co.

Penfield, W., & Perot, P. (1963) The brains record of auditory and visual experience. Brain, 86, 595-695.

Penfield, W., & Welch, K. (1951). Supplementary motor area of cerebral cortex. Clinical and experimental study. Archives of Neurology & Psychiatry, 66, 289-317.

Petrides, M. & Pandya, D. N.. (1988) Association fiber pathways to the frontal cortex from the superior temporal region in rhesus monkey. Journal of Comparative Neurology 273:52-66.

Risberg J., Grafman, J. (2006). The Frontal Lobes: Development, Function and Pathology, Cambridge University Press

Rizzolatti, G., Craighero, L. (2004). THE MIRROR-NEURON SYSTEM, Annu. Rev. Neurosci. 27, 169-92.

Rizzolatti G, Fadiga L, Fogassi L, Gallese V. (1996). Premotor cortex and the recognition of motor actions. Cogn. Brain Res. 3:131-141.

Rizzolatti G, Luppino G, Matelli M. (1998). The organization of the cortical motor system: new concepts. Electroencephalogr. Clin. Neurophysiol. 106:283-296.

Rizzolatti, G., Fabbri-Destro, M., Cattaneo, L. (2009). Mirror neurons and their clinical relevance. Nat Clin Pract Neurol 5 (1): 24-34.

Smith , E., and Delargy, M. (2005). Locked-in syndrome, BMJ. 330, 406-408.

Shibasaki, H., Sadato, N., Lyshkow, H., et al. (1993). Both primary motor cortex and supplementary motor area play an important role in complex finger movement. Brain, 116, 1387-1298.

Singer, H.S. (2005). Tourette's syndrome: from behaviour to biology". Lancet, 4, 149-59.

Sperry, R. (1966). Brain bisection and the neurology of consciousness. In F. O. Schmitt and F. G. Worden (eds). The Neurosciences. MIT press.

Stephan, K. M., Binkofski, F., Halsband, U., et al., (1999). The role of ventral medial wall motor areas in bimanual coordination. Brain, 122, 351-368.

Tanji, J., & Kurata, K. (1982). Comparison of movement-related neurons in two cortical motor areas of primates. Journal of Neurophysiology, 40, 644-653.

Tourette Syndrome Classification Study Group (1993). "Definitions and classification of tic disorders". Arch Neurol. 50, 1013-16.

Verstynen, T., et al. (2011). In Vivo Mapping of Microstructural Somatotopies in the Human Corticospinal Pathways. Journal of Neurophysiology, 105, 336-346

Walkup JT, Mink JW, Hollenback PJ, (2006). Advances in Neurology, Vol. 99, Tourette Syndrome. Lippincott, Williams & Wilkins. .

Ward, C.D. (1988). Transient feelings of compulsion caused by hemispheric lesions: Three cases. Journal of Neurology, Neurosurgery, and Psychiatry, 51, 266-268.

Zeman A. (2003). What is consciousness and what does it mean for the persistent vegetative state? Adv Clin Neurosci Rehabil, 3, 12-4.

12. Quantum Physics and the Multiplicity of Mind: Split-Brains, Fragmented Minds, Dissociation, Quantum Consciousness

Abstract

Quantum physics and Einstein's theory of relativity make assumptions about the nature of the mind which is assumed to be a singularity. In the Copenhagen model of physics, the process of observing is believed to effect reality by the act of perception and knowing which creates abstractions and a collapse function thereby inducing discontinuity into the continuum of the quantum state. This gives rise to the uncertainty principle. Yet neither the mind or the brain is a singularity, but a multiplicity which include two dominant streams of consciousness and awareness associated with the left and right hemisphere, as demonstrated by patients whose brains have been split, and which are superimposed on yet other mental realms maintained by the brainstem, thalamus, limbic system, and the occipital, temporal, parietal, and frontal lobes. Like the quantum state, each of these minds may also become discontinuous from each other and each mental realm may perceive their own reality. Illustrative examples are detailed, including denial of blindness, blind sight, fragmentation of the body image, phantom limbs, the splitting of the mind following split-brain surgery, and dissociative states where the mind leaves the body and achieves a state of quantum consciousness and singularity such that the universe and mind become one.

1. Introduction

In 1905 Albert Einstein published his theories of relativity, which promoted the thesis that reality and its properties, such as time and motion had no objective "true values", but were "relative" to the observer's point of view (Einstein, 1905a,b,c). However, what if the observer is not a singularity and has more than one point of view and more than one stream of observing consciousness? And what if these streams of consciousness were also relative?

Quantum physics, as exemplified by the Copenhagen school (Bohr, 1934, 1958, 1963; Heisenberg, 1930, 1955, 1958), also makes assumptions about the nature of reality as related to an observer, the "knower" who is conceptualized as a singularity. Because the physical world is relative to being known by a "knower" (the observing consciousness), then the "knower" can influence the nature of the reality which is being observed. In consequence, what is known vs what is not known becomes relatively imprecise (Heisenberg, 1958).

For example, as expressed by the Heisenberg uncertainty principle (Heisenberg, 1955, 1958), the more precisely one physical property is known the more unknowable become other properties, whose measurements become correspondingly imprecise. The more precisely one property is known, the less precisely the other can be known and this is true at the molecular and atomic levels of reality. Therefore it is impossible to precisely determine, simultaneously, for example, both the position and velocity of an electron.

However, we must ask: if knowing A, makes B unknowable, and if knowing B makes A unknowable, wouldn't this imply that both A and B, are in fact unknowable? If both A and B are manifestations of the processing of "knowing," and if observing and measuring can change the properties of A or B, then perhaps both A and B are in fact properties of knowing, properties of the observing consciousness, and not properties of A or B.

In quantum physics, nature and reality are represented by the quantum state. The electromagnetic field of the quantum state is the fundamental entity, the continuum that constitutes the basic oneness and unity of all things.

The physical nature of this state can be "known" by assigning it mathematical properties (Bohr, 1958, 1963). Therefore, abstractions, i.e., numbers, become representational of a hypothetical physical state. Because these are abstractions, the physical state is also an abstraction and does not possess the material consistency, continuity, and hard, tangible, physical substance as is assumed by Classical (Newtonian) physics. Instead, reality, the physical world, is created by the process of observing, measuring, and knowing (Heisenberg, 1955).

Consider an elementary particle, once this positional value is assigned, knowledge of momentum, trajectory, speed, and so on, is lost and becomes "uncertain." The particle's momentum is left uncertain by an amount inversely proportional to the accuracy of the position measurement which is determined by values assigned by the observing consciousness. Therefore, the nature of reality, and the uncertainty principle is directly affected by the observer and the process of observing and knowing (Heisenberg, 1955, 1958).

The act of knowing creates a knot in the quantum state; described as a "collapse of the wave function;" a knot of energy that is a kind of blemish in the continuum of the quantum field. This quantum knot bunches up at the point of observation, at the assigned value of measurement.

The process of knowing, makes reality, and the quantum state, discontinuous. "The discontinuous change in the probability function takes place with the act of registration...in the mind of the observer" (Heisenberg, 1958).

Reality, therefore, is a manifestation of alterations in the patterns of activity within the electromagnetic field which are perceived as discontinuous. The perception of a structural unit of information is not just perceived, but is inserted into the quantum state which causes the reduction of the wave-packet and the collapse of the wave function.

Knowing and not knowing, are the result of interactions between the mind and concentrations of energy that emerge and disappear back into the electromagnetic quantum field.

However, if reality is created by the observing consciousness, and can be made discontinuous, does this leave open the possibility of a reality behind the reality? Might there be multiple realities? And if consciousness and the observer and the quantum state is not a singularity, could each of these multiple realities also be manifestations of a multiplicity of minds?

Heisenberg (1958) recognized this possibility of hidden realities, and therefore proposed that the reality that exists beyond or outside the quantum state could be better understood when considered in terms of "potential" reality and "actual" realities. Therefore, although the quantum state does not have the ontological character of an "actual" thing, it has a "potential" reality; an objective tendency to become actual at some point in the future, or to have become actual at some point in the past.

Therefore, it could be said that the subatomic particles which make up reality, or the quantum state, do not really exist, except as probabilities. These "subatomic" particles have probable existences and display tendencies to assume certain patterns of activity that we perceive as shape and form. Yet, they may also begin to display a different pattern of activity such that being can become nonbeing and thus something else altogether.

The conception of a deterministic reality is therefore subjugated to mathematical probabilities and potentiality which is relative to the mind of a knower which effects that reality as it unfolds, evolves, and is observed (Bohr 1958, 1963; Heisenberg 1955, 1958). That is, the mental act of perceiving a non-localized unit of structural information, injects that mental event into the quantum state of the universe, causing "the collapse of the wave function" and creating a bunching up, a tangle and discontinuous knot in the continuity of the quantum state.

Einstein ridiculed these ideas (Pais, 1979): "Do you really think the moon isn't there if you aren't looking at it?"

Heisenberg (1958), cautioned, however, that the observer is not the creator of reality: "The introduction of the observer must not be misunderstood to imply that some kind of subjective features are to be brought into the description of nature. The observer has, rather, only the function of registering decisions, i.e., processes in space and time, and it does not matter whether the observer is an apparatus or a human being; but the registration, i.e., the transition from the "possible" to the "actual," is absolutely necessary here and cannot be omitted from the interpretation of quantum theory."

Shape and form are a function of our perception of dynamic interactions within the continuum which is the quantum state. What we perceive as mass (shape, form, length, weight) are dynamic patterns of energy which we selectively attend to and then perceive as stable and static, creating discontinuity within the continuity of the quantum state. Therefore, what we are perceiving and knowing, are only fragments of the continuum.

However, we can only perceive what our senses can detect, and what we detect as form and shape is really a mass of frenzied subatomic electromagnetic activity that is amenable to detection by our senses and which may be known by a knowing mind. It is the perception of certain aspects of these oscillating patterns of continuous evolving activity, which give rise to the impressions of shape and form, and thus discontinuity, as experienced within the mind.

This energy that makes up the object of our perceptions, is therefore but an aspect of the electromagnetic continuum which has assumed a specific pattern during the process of being sensed and processed by those regions of the brain and mind best equipped to process this information.

Perceived reality, therefore, becomes a manifestation of mind.

However, if the mind is not a singularity, and if we possessed additional senses or an increased sensory channel capacity, we would perceive yet other patterns and other realities which would be known by those features of the mind best attuned to them. If the mind is not a singularity but a multiplicity, this means that both A and B, may be known simultaneously.

2. Duality vs Multiplicity

In the Copenhagen model, the observer is external to the quantum state the observer is observing, and they are not part of the collapse function but a witness of it (Bohr, 1958, 1963; Heisenberg 1958). However, if the Copenhagen model is correct, and as the cosmos contains observers, then the standard collapse formulation can not be used to describe the entire universe as the universe contains observers (von Neumann, 1932, 1937).

Further, reality becomes, at a minimum, a duality (observer and observed) with the potential to become a multiplicity.

As described by DeWitt and Graham (1973; Dewitt, 1971), "This reality, which is described jointly by the dynamical variables and the state vector, is not the reality we customarily think of, but is a reality composed of many worlds. By virtue of the temporal development of the dynamical variables the state vector decomposes naturally into orthogonal vectors, reflecting a continual splitting of the universe into a multitude of mutually unobservable but equally real worlds, in each of which every good measurement has yielded a definite result and in most of which the familiar statistical quantum laws hold."

The minimal duality is that aspect of reality which is observed, measured, and known, and that which is unknown.

However, this minimal duality is an illusion as indicated not only by the potential to become multiplicity, but by the nature of mind which is not a singularity (Joseph, 1982, 1986a; 1988a,b).

Even if we disregard the concept of "mind" and substitute the word "brain", the fact remains that the brain is not a singularity. The human brain is functionally specialized with specific functions and different mental states localized to specific areas, each of which is capable of maintaining independent and semi-independent aspects of conscious-awareness (Joseph 1986a,b, 1988a,b, 1992, 1999a). Different aspects of the same experience and identical aspects of that experience may be perceived and processed by different brain areas in different ways (Gallagher and Joseph, 1982; Joseph 1982; Joseph and Gallagher 1985; Joseph et al., 1984).

Therefore, although it has been said that orthodox quantum mechanics is completely concordant with the defining characteristics of Cartesian dualism, this is an illusion. Cartesian duality assumes singularity of mind, when in fact, the overarching organization of the mind- and the brain- is both dualistic and multiple.

If quantum physics is "mind-like" (actual/operational at the quantum level, but mentalistic on the ontological level) then quantum physics, or rather, the quantum state (reality, the universe) is not a duality, but a multiciplicity. Indeed, the entire concept of duality is imposed on reality by the dominant dualistic nature of the brain and mind which subordinates not just reality, but the multiplicity of minds maintained within the human brain (Joseph, 1982).

Like the Copenhagen school, Von Neumann's formulation of quantum mechanics (1932, 1937), fails to recognize or understand the multiple nature of mind and reality. Von Neumann postulated that the physical aspects of nature are represented by a density matrix. The matrix, therefore, could be conceptualized as a subset of potential realities, and that by averaging the values of these evolving matrices, the state of the universe and thus of reality, can be ascertained as a unified whole. However, in contrast to the Copenhagen interpretation, Von Neumann shifted the observer (his brain) into the quantum universe and thus made it subject to the rules of quantum physics.

Ostensibly and explicitly, Von Neumann's conceptions are based on a conception of mind as a singularity acting on the quantum state which contains the brain. Von Neumann's mental singularity, therefore, imposes itself on reality, such that each "event" that occurs within reality, is associated with one specific experience of the singularity-mind. Thus, Von Neumann assumes the brain and mind has only "one experience" which corresponds with "one event;" and this grossly erroneous misconception of the nature of the brain and mind, unfortunately, is erroneously accepted as fact by most cosmologists and physicists. Further, he argues that in the process of knowing, the quantum state of this singularity brain/mind also collapses, or rather, is reduced in a mathematically quantifiable manner, just as the quantum universe is collapsed and reduced by being known (Von Neumann, 1932, 1937).

However, the brain and mind are not a singularity, but a multiplicity (Joseph, 1982, 1988a,b, 1999a). Nevertheless, Von Neumann's conceptions can be applied to the multiplicity of mind/brain when each mental realm is considered individually as an interactional subset of the multiplicity.

3. The Multiplicity of Mind and Perception

According to Von Neumann (1932), the "experiential increments in a person's knowledge" and "reductions of the quantum mechanical state of that person's brain" corresponds to the elimination of all those perceptual functions that are not necessary or irrelevant to the knowing of the event and the increase in the knowledge associated with the experience.

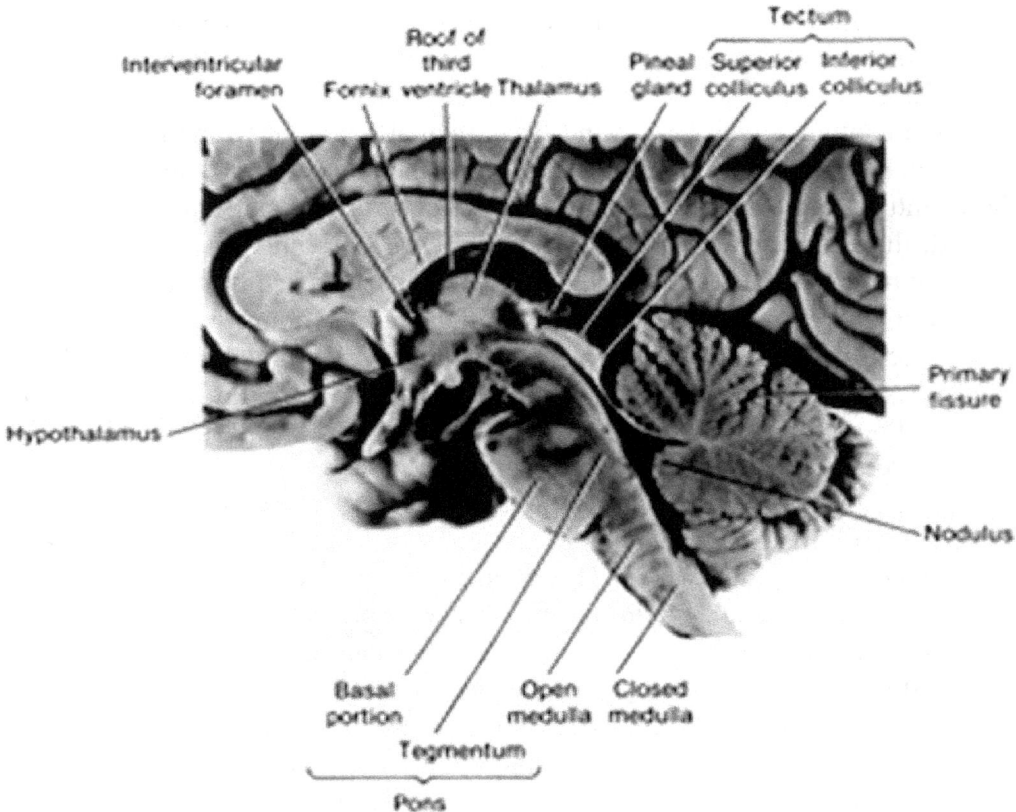

If considered from the perspective of an isolated aspect of the mind and the dominating stream of consciousness, Von Neuman's conceptions are essentially correct. However, neither the brain nor the mind function in isolation but in interaction with other neural tissues and mental/perceptual/sensory realms (Joseph 1982, 1992, 1999a). Perceptual functions are not "eliminated" and removed from the brain. Instead, they are prevented from interfering with the attentional processes of one aspect of the multiplicity of mind which dominates during the knowing event (Joseph, 1986b, 1999a).

Consider, by way of example, you are sitting in your office reading this text. The pressure of the chair, the physical sensations of your shoes and clothes, the musculature of your body as it holds one then another position, the temperature of the room, various odors and fragrances, a multitude of sounds, visual sensations from outside your area of concentration and focus, and so on, are all being transmitted to the brainstem, midbrain, and olfactory limbic system. These signals are then relayed to various subnuclei within the thalamus.

The neural tissues of the brainstem, midbrain, limbic system and thalamus are associated with the "old brain." However, those aspects of consciousness we most closely associated with humans are associated with the "new brain" the neocortex (Joseph, 1982, 1992). Therefore, although you may be "aware" of these sensations while they are maintained within the old brain, you are not "conscious" of them, unless a decision is made to become conscious or they increase sufficiently in intensity that they are transferred to the neocortex and forced into the focus of consciousness (Joseph, 1982, 1986b, 1992, 1999a).

The old brain is covered by a gray mantle of new cortex, neocortex. The sensations alluded to are transferred from the old brain to the thalamus which relays these signals to the neocortex. Human consciousness and the "higher" level of the multiplicity of mind, are associated with the "new brain."

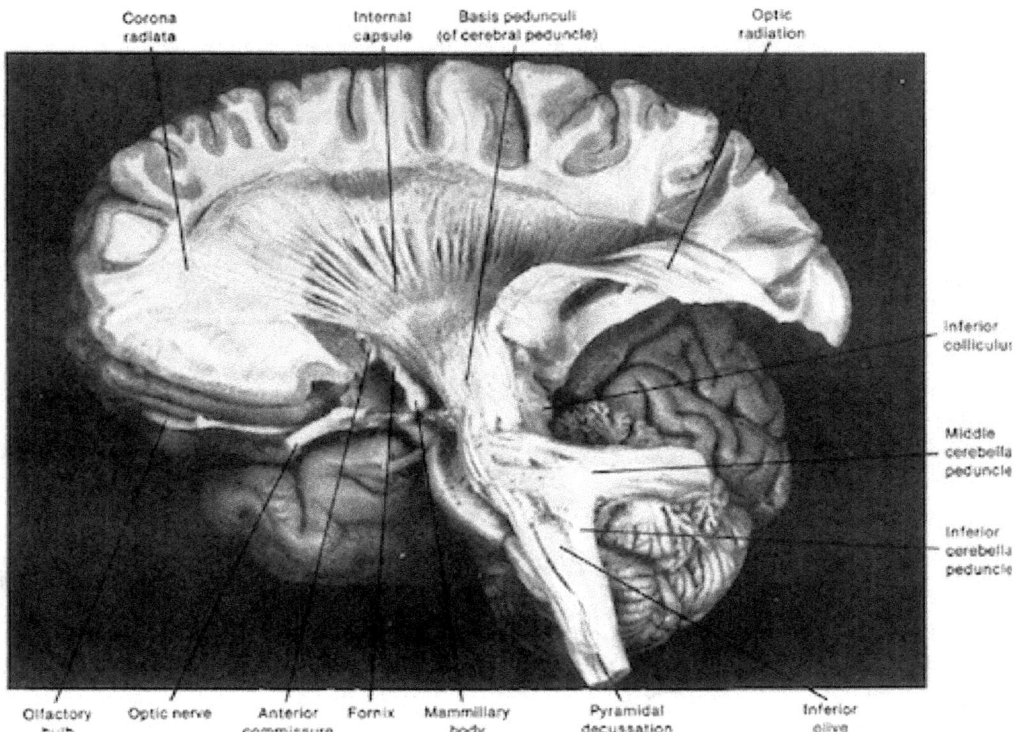

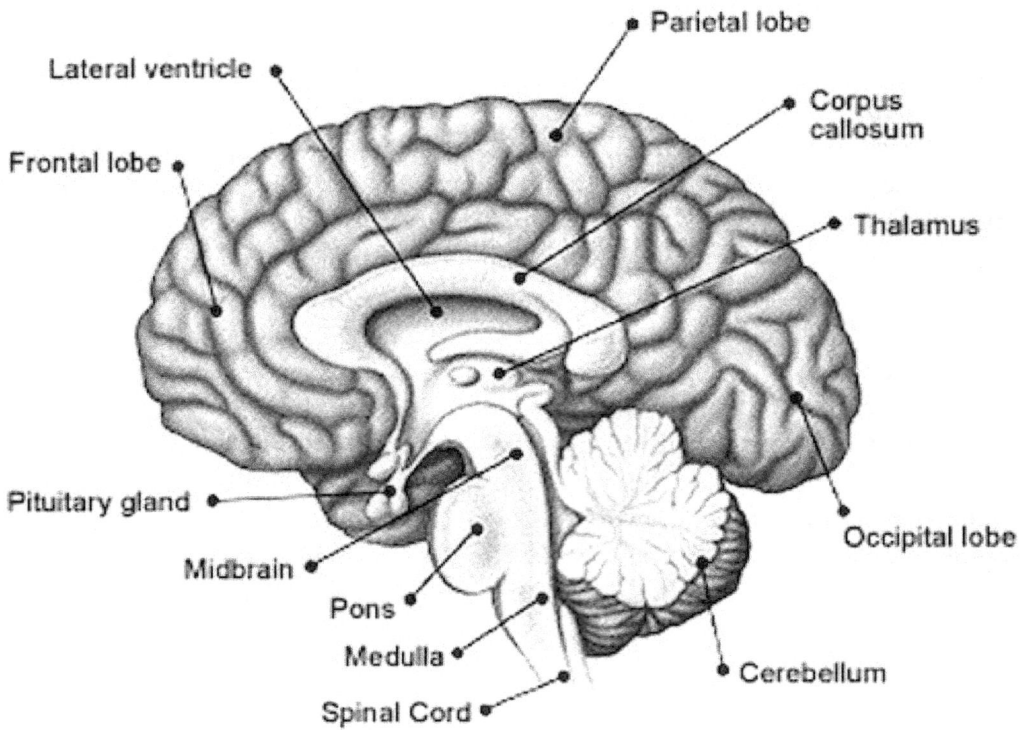

The different parts of the brain

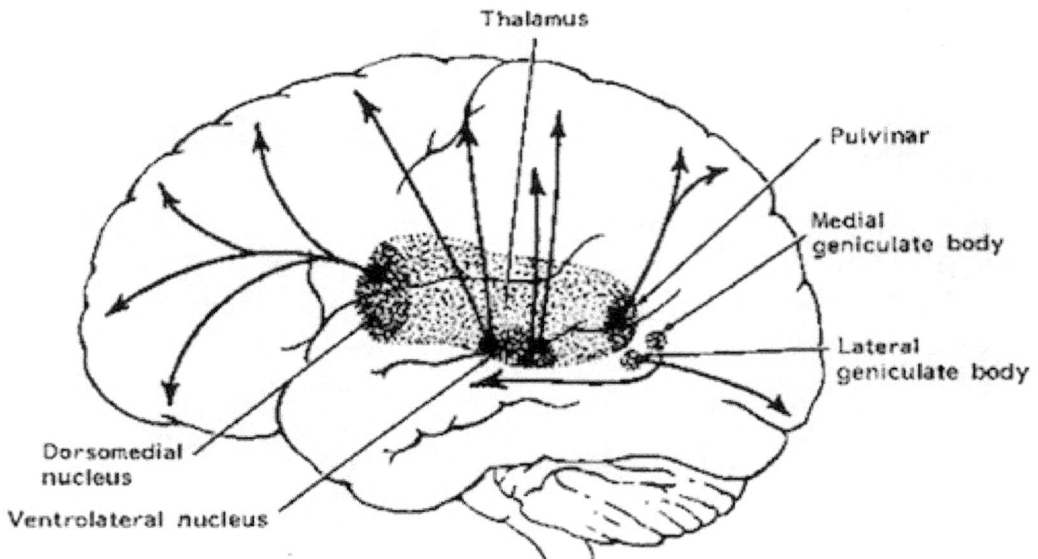

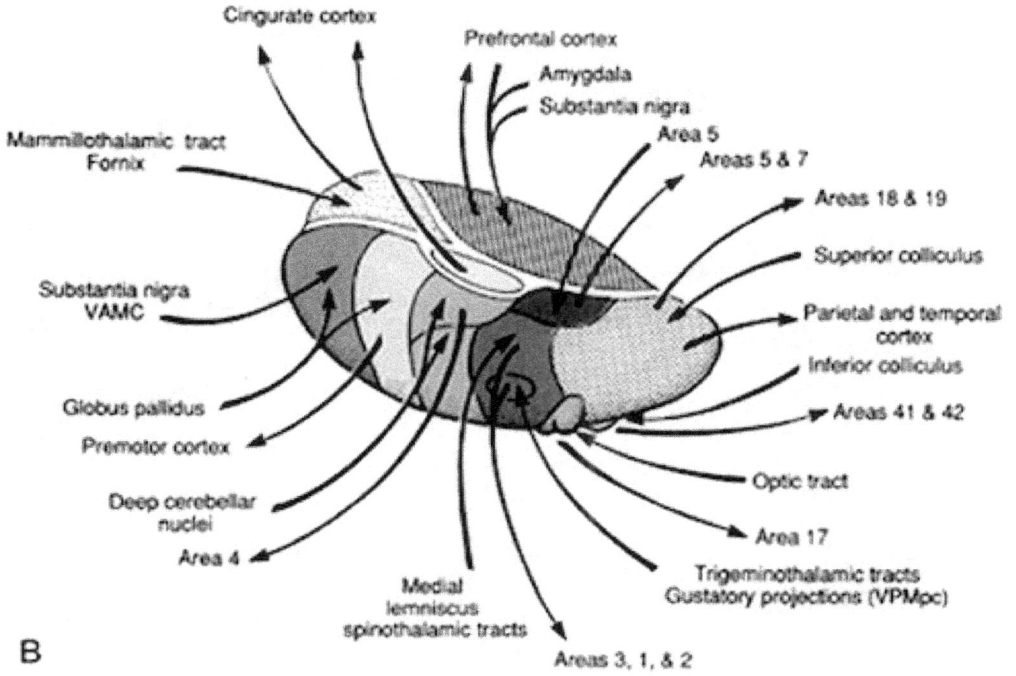

B

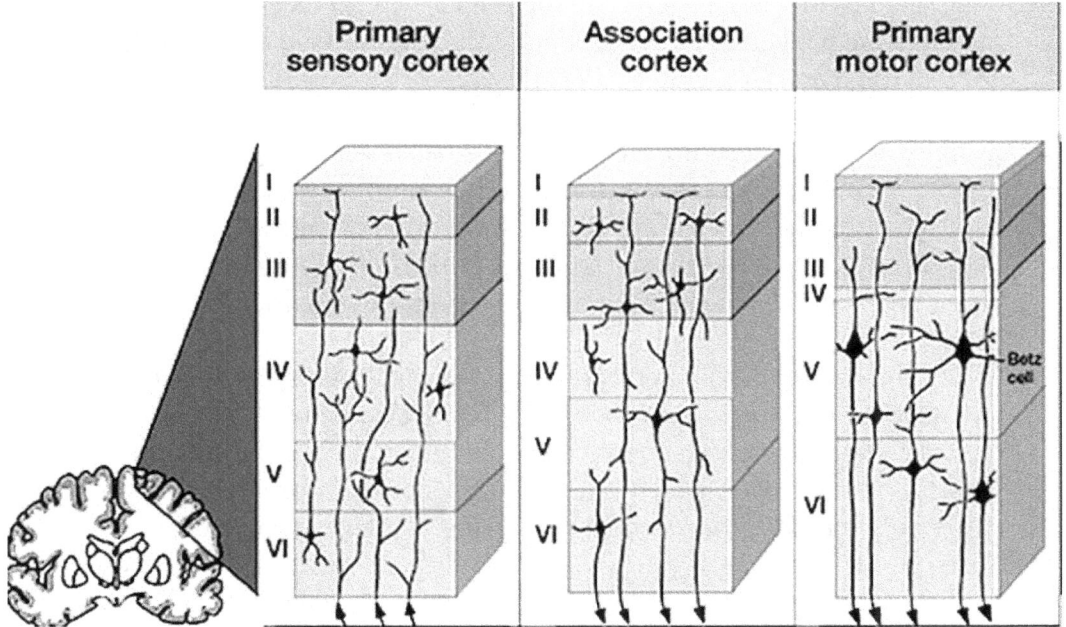

For example, visual input is transmitted from the eyes to the midbrain and thalamus and is transferred to the primary visual receiving area maintained in the neocortex of the occipital lobe (Casagrande & Joseph 1978, 1980; Joseph and Casagrande, 1978). Auditory input is transmitted from the inner ears to the brainstem, midbrain, and thalamus, and is transferred to the primary auditory receiving area within the neocortex of the temporal lobe. Tactual-physical stimuli are also transmitted from the thalamus to the primary somatosensory areas maintained in the neocortex of the parietal lobe. From the primary areas these signals are transferred to the adjoining "association" areas, and simple percepts become more complex by association (Joseph, 1996).

Monitoring all this perceptual and sensory activity within the thalamus and neocortex is the frontal lobes of the brain, also known as the senior executive of the brain and personality (Joseph 1986b, 1999a; Joseph et al., 1981). It is the frontal lobes which maintain the focus of attention and which can selectively inhibit any additional processing of signals received in the primary areas.

There are two frontal lobes, a right and left frontal lobe which communicate via a bridge of nerve fibers. Each frontal lobe, and subdivisions within each are concerned with different types of mental activity (Joseph, 1999a).

The left frontal lobe, among its many functions, makes possible the ability to speak. It is associated with the verbally expressive, speaking aspects of consciousness. However, there are different aspects of consciousness associated not only with the frontal lobe, but with each lobe of the brain and its subdivisions (Joseph, 1986b; 1996, 12012).

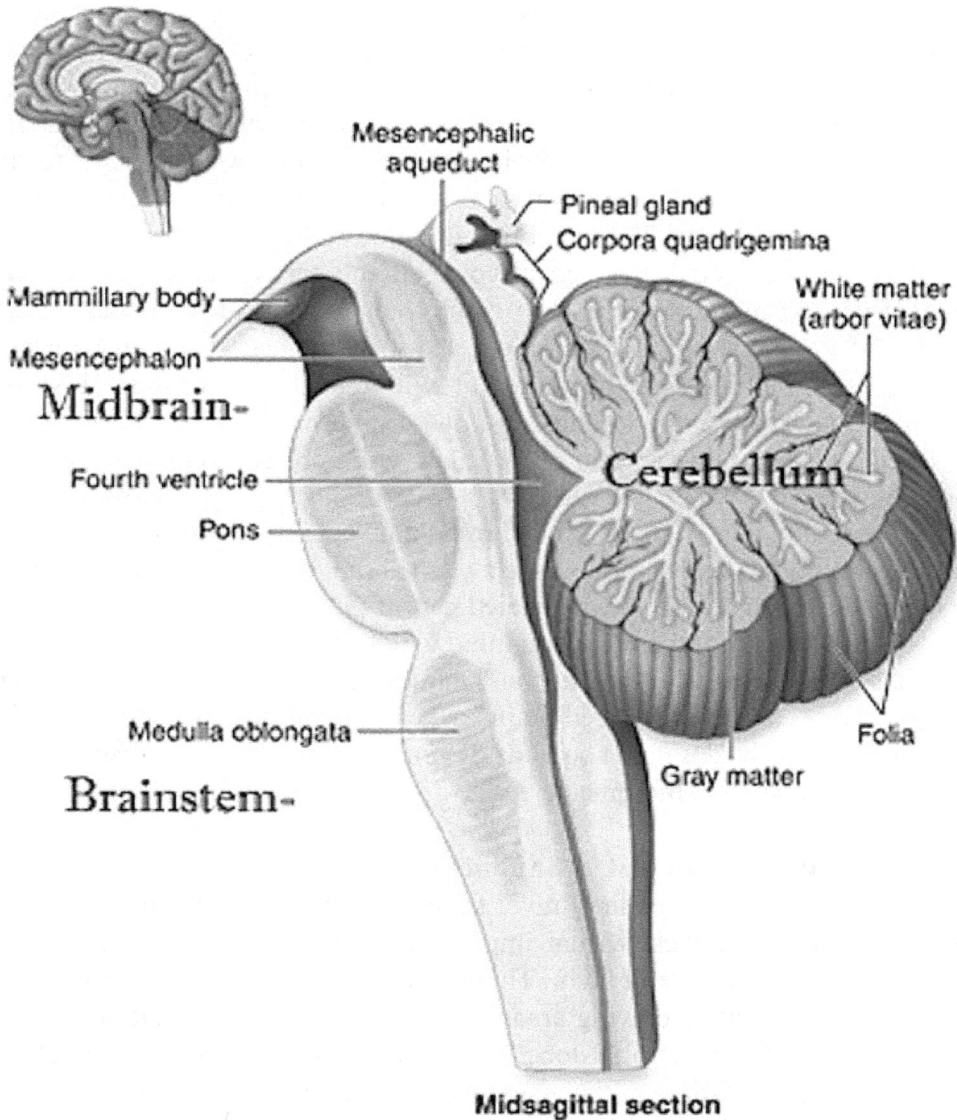

Mesencephalic aqueduct

Pineal gland
Corpora quadrigemina

Mammillary body

White matter (arbor vitae)

Mesencephalon
Midbrain-

Cerebellum

Fourth ventricle

Pons

Medulla oblongata

Folia

Gray matter

Brainstem-

Midsagittal section

4. Knowing Yet Not Knowing: Disconnected Consciousness

Consider the well known phenomenon of "word finding difficulty" also known as "tip of the tongue." You know the word you want (the "thingama-jig") but at the same time, you can't gain access to it. That is, one aspect of consciousness knows the missing word, but another aspect of consciousness associated with talking and speech can't gain access to the word. The mind is disconnected from itself. One aspect of mind knows, the other aspect of mind does not.

This same phenomenon, but much more severe and disabling, can occur if the nerve fiber pathway linking the language areas of the left hemisphere are damaged. For example, Broca's area in the frontal lobe expresses humans speech. Wernicke's area in the temporal lobe comprehends speech. The inferior parietal lobe associates and assimilates associations so that, for example, we can say the word "dog" and come up with the names of dozens of different breeds and then describe them (Joseph, 1982; Joseph and Gallagher 1985; Joseph et al., 1984). Therefore, if Broca's area is disconnected from the posterior language areas, one aspect of consciousness may know what it wants to say, but the speaking aspect of consciousness will be unable to gain access to it and will have nothing to say. This condition is called "conduction aphasia."

Or consider damage which disconnects the parietal lobe from Broca's area. If you place an object, e.g., a comb, out-of-sight, in the person's right hand, and ask them to name the object, the speaking aspect of consciousness may know something is in the hand, but will be unable to name it. However, although they can't name it, and can't guess if shown pictures, if the patient is asked to point to the correct object, they will correctly pick out the comb (Joseph, 1996).

Therefore, part of the brain and mind may act purposefully (e.g. picking out the comb), whereas another aspect of the brain and mind is denied access to the information that the disconnected part of the mind is acting on.

Thus, the part of the brain and mind which is perceiving and knowing, is not the same as the part of the brain and mind which is speaking. This phenomenon occurs even in undamaged brains, when the multiplicity of minds which make up one of the dominant streams of consciousness, become disconnected and/or are unable to communicate.

5. The Visual Mind: Denial of Blindness

All visual sensations first travel from the eyes to the thalamus and midbrain. At this level, these visual impressions are outside of consciousness, though we may be aware of them. These visual sensations are then transferred to the primary visual receiving areas and to the adjacent association areas in the neocortex of the occipital lobe. Once these visual impressions reach the neocortex, consciousness of the visual word is achieved. Visual consciousness is made possible by the occipital lobe.

Destruction of the occipital lobe and its neocortical visual areas results in cortical blindness (Joseph, 1996, 2012). The consciousness mind is blinded and can not see or sense anything except vague sensations of lightness and darkness. However, because visual consciousness is normally maintained within the occipital lobe, with destruction of this tissue, the other mental systems will not know that they can't see. The remaining mental system do not know they are blind.

Wernicke's area in the left temporal lobe in association with the inferior parietal lobe comprehends and can generate complex language. Normally, visual input is transferred from the occipital to the inferior parietal lobe (IPL) which is adjacent to Wernicke's area and the visual areas of the occipital lobe. Once these signals arrive in the IPL a person can name what they see; the visual input is matched with auditory-verbal signals and the conscious mind can label and talk about what is viewed (Joseph, 1982, 1986b; Joseph et al., 1984). Talking and verbally describing what is seen is made possible when this stream of information is transferred to Broca's area in the left frontal lobe (Joseph, 1982, 1999a). It is Broca's area which speaks and talks.

Therefore, with complete destruction of the occipital lobe, visual consciousness is abolished whereas the other mental system remain intact but are unable to receive information about the visual world. In consequence, the verbal aspects of consciousness and the verbal-language mind does not know it can't see because the brain area responsible for informing these mental system about seeing, no longer exists. . In fact the language-dependent conscious mind will deny that it is blind; and this is called: Denial of blindness.

Normally, if it gets dark, or you close your eyes, the visual mind becomes conscious of this change in light perception and will alert the other mental realms. These other mental realms do not process visual signals and therefore they must be informed about what the visual mind is seeing. If the occipital lobe is destroyed, visual consciousness is destroyed, and the rest of the brain cannot be told that visual consciousness can't see. Therefore, the rest of the brain does not know it is blind, and when asked, will deny blindness and will make up reasons for why they bump into furniture or can't recognize objects held before their eyes (Joseph, 1986b, 1988a).

For example, when unable to name objects, they might confabulate an explanation: "I see better at home." Or, "I tripped because someone moved the furniture." Even if you tell them they are blind, they will deny blindness; that is, the verbal aspects of consciousness will claim it can see, when it can't. The Language-dependent aspects of consciousness does not know that it is blind because information concerning blindness is not being received from the mental realms which support visual consciousness.

The same phenomenon occurs with small strokes destroying just part of the occipital lobe. Although a patient may lose a quarter or even half of their visual field, they may be unaware of it. This is because that aspect of visual consciousness no longer exists and can't inform the other mental realms of its condition.

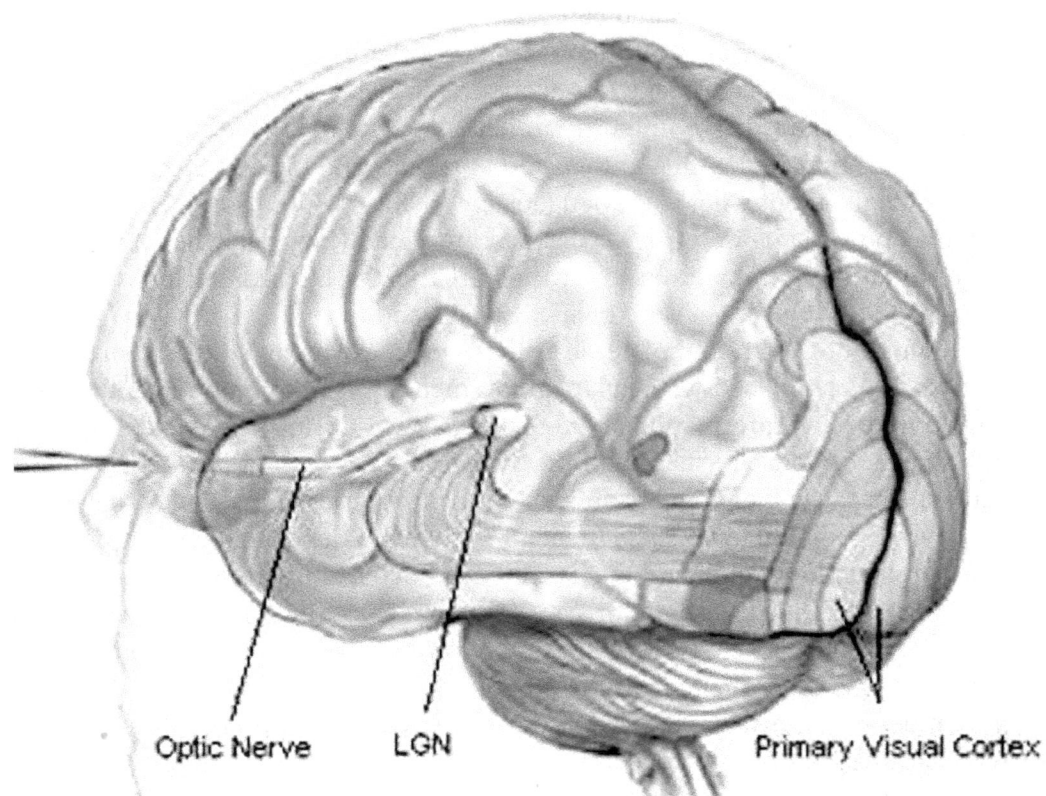

Optic Nerve   LGN   Primary Visual Cortex

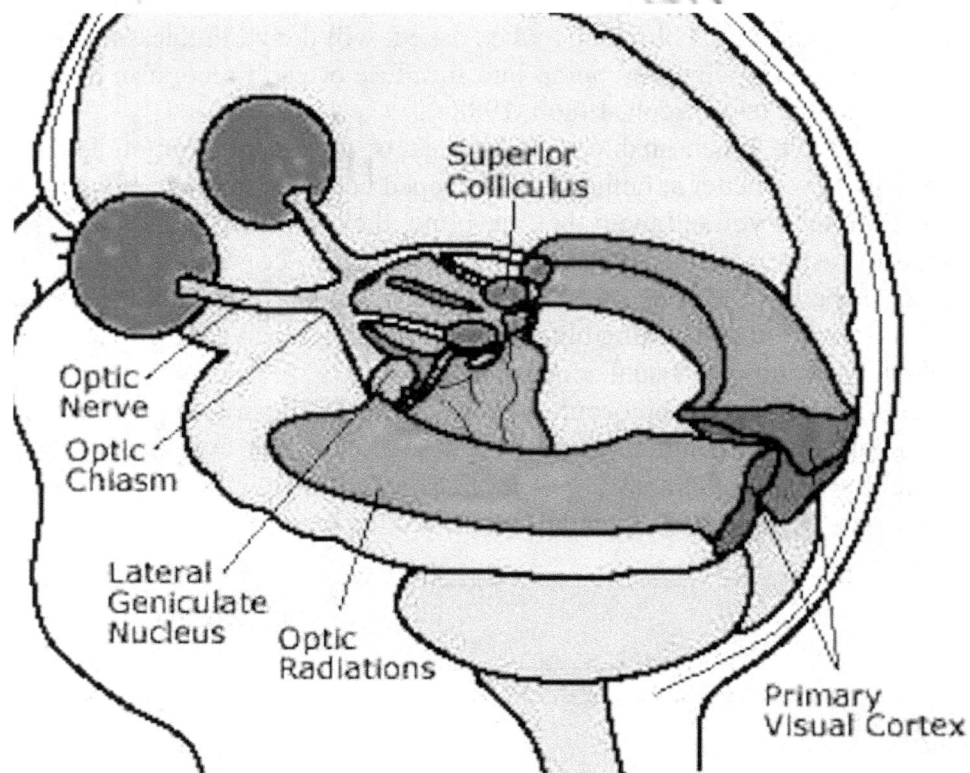

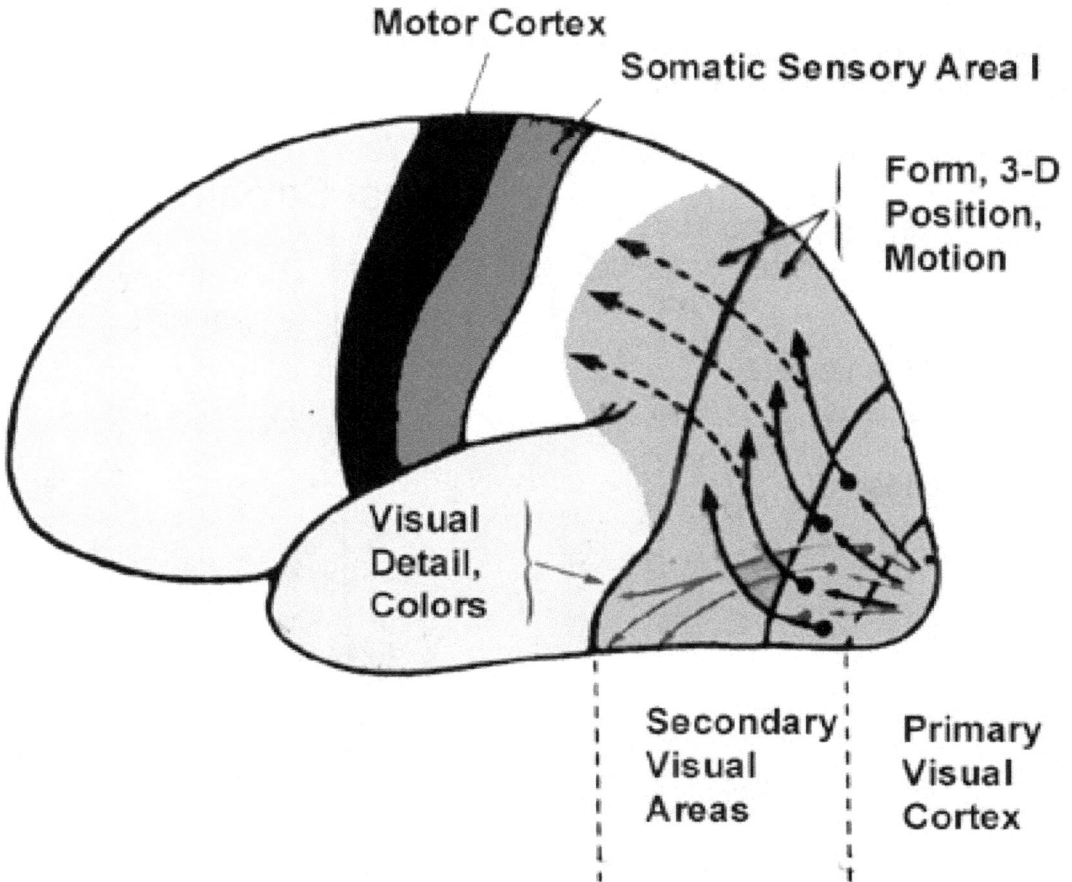

6. "Blind Sight"

The brains of reptiles, amphibians, and fish do not have neocortex. Visual input is processed in the midbrain and thalamus and other old-brain areas as these creatures do not possess neocortex or lobes of the brain. In humans, this information is also received in the brainstem and thalamus and is then transferred to the newly evolved neocortex.

As is evident in non-mammalian species, these creatures can see, and they are aware of their environment. They possess an older-cortical (brainstem-thalamus) visual awareness which in humans is dominated by neocortical visual consciousness.

Therefore, even with complete destruction of the visual neocortex, and after the patient has had time to recover, some patients will demonstrate a non-conscious awareness of their visual environment. Although they are cortically blind and can't name objects and stumble over furniture and bump into walls, they may correctly indicate if an object is moving in front of their face, and they may turn their head or even reach out their arms to touch it--just as a frog can see a fly buzzing by and lap it up with its tongue. Although the patient can't name or see what has moved in front of his face, he may report that he has a "feeling" that something has moved.

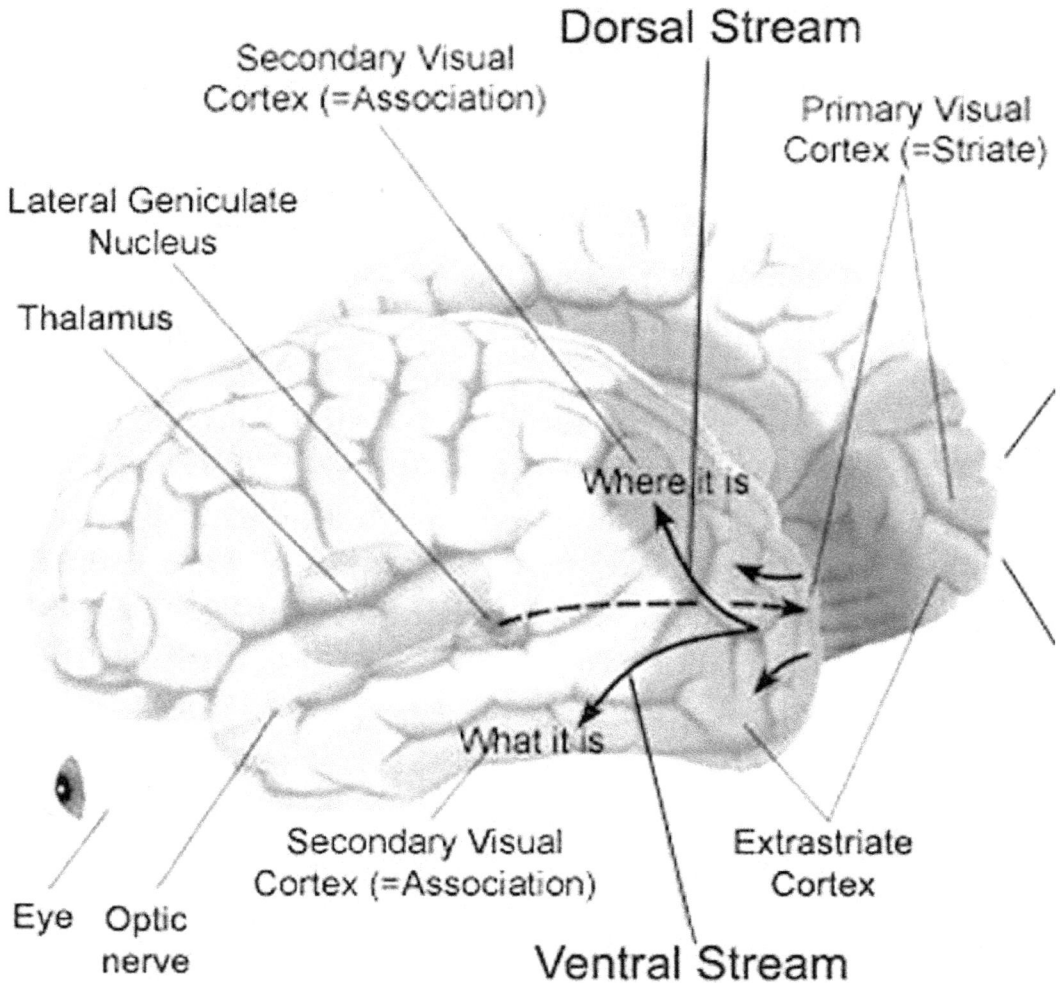

Dorsal Stream

Secondary Visual Cortex (=Association)

Primary Visual Cortex (=Striate)

Lateral Geniculate Nucleus

Thalamus

Where it is

What it is

Secondary Visual Cortex (=Association)

Extrastriate Cortex

Eye   Optic nerve

Ventral Stream

Frogs do not have neocortex and they do not have language, and can't describe what they see. However, humans and frogs have old cortex that process visual impressions and which can control and coordinate body movements. Specifically, these more ancient visual areas are located in the midbrain of the brainstem and are referred to as the superior colliculi (tectum). These nuclei project to the thalamus, and in mammals, they project to the visual cortex. It is the superior colliculi which can "see" in the absence of human consciousness.

Therefore, although the neocortical realms of human consciousness may become blind due to damage to the occipital visual areas, the mental realms of the old brain can continue to see and can act on what it sees; and this is called: Blind sight (Joseph, 1996).

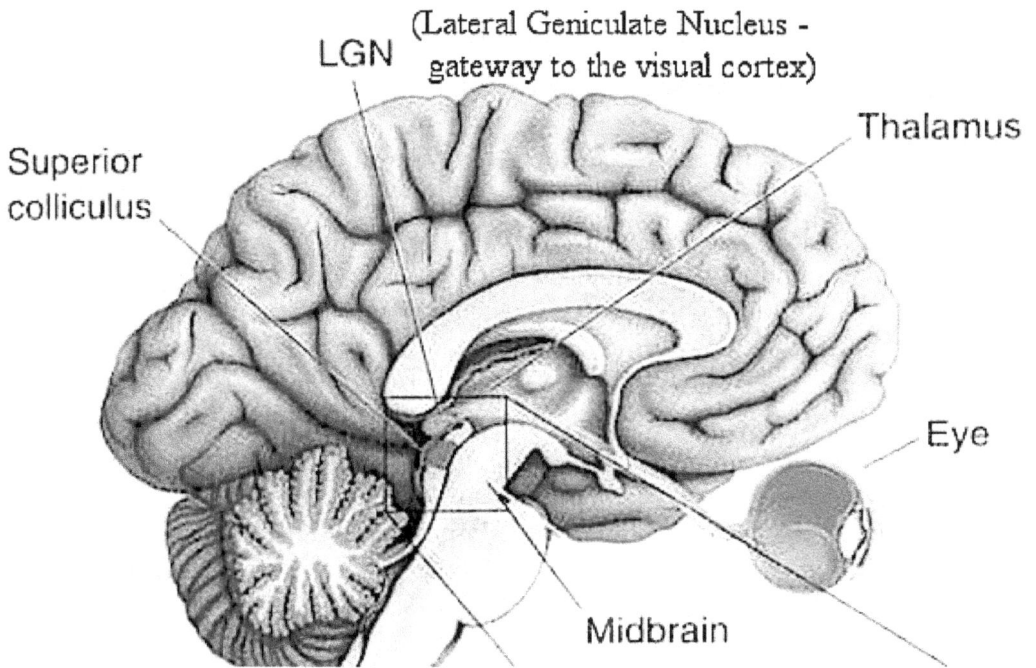

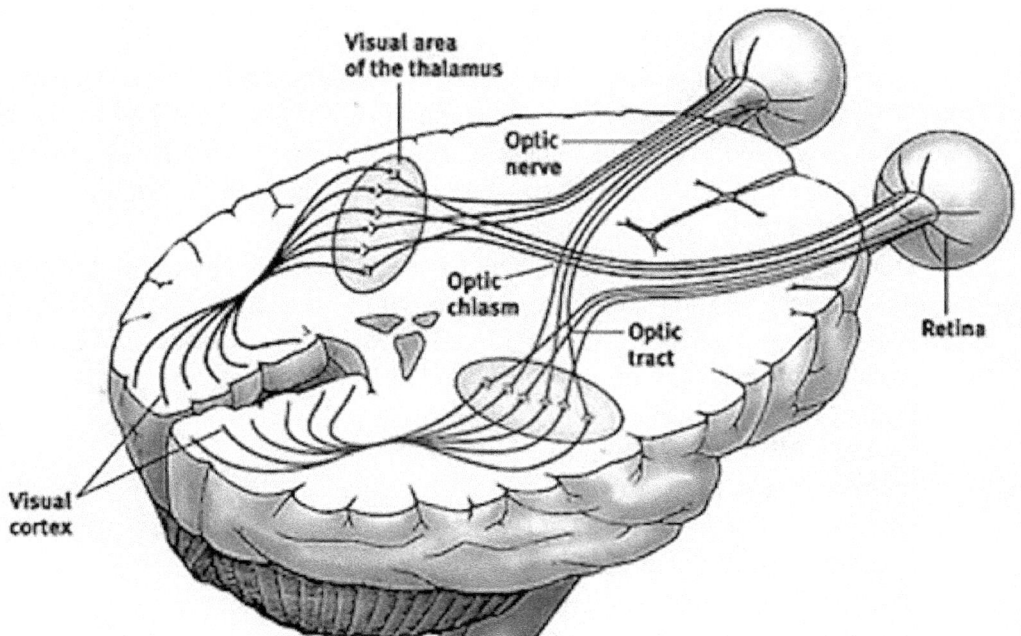

7. Body Consciousness: Denial of the Body, and Phantom Limbs

All tactile and physical-sensory impressions are relayed from the body to the brainstem and the thalamus, and are then transferred to the primary receiving and then the association area for somatosensory information located in the neocortex of the parietal lobe (Joseph, 1986b, 1996). The entire image of the body is represented in the parietal lobes (the right and left half of the body in the left and right parietal lobe respectively), albeit in correspondence with the sensory importance of each body part. Therefore, more neocortical space is devoted to the hands and fingers than to the elbow.

It is because the body image and body consciousness is maintained in the parietal area of the brain, that victims of traumatic amputation and who lose an arm or a leg, continue to feel as if their arm or their leg is still attached to the body. This is called: phantom limbs. They can see the leg is missing, but they feel as if it is still there; body-consciousness remains intact even though part of the body is missing (Joseph, 1986b, 1996). They may also continue to periodically experience the pain of the physical trauma which led to the amputation, and this is called "phantom limb pain."

Thus, via the mental system of the parietal lobe, consciousness of what is not there, may appear to consciousness as if it is still there. This is not a hallucination. The image of the body is preserved in the brain and so to is consciousness of the body; and this is yet another example of experienced reality being a manifestation of the brain and mind. In this regard, reality is literally mapped into the brain and is represented within the brain, such that even when aspects of this "reality" are destroyed and no longer exists external to the brain, it nevertheless continues to be perceived and experienced by the brain and the associated realms of body-consciousness.

Conversely, if the parietal lobe is destroyed, particularly the right parietal lobe (which maintains an image of the left half of the body), half of the body image may be erased from consciousness (Joseph, 1986b, 1988a). The remaining realms of mind will lose all consciousness of the left half of the body, which, in their minds, never existed.

Doctor: "Give me your right hand!" (Patient offers right hand). "Now give me your left!" (The patient presents the right hand again. The right hand is held.) "Give me your left!" (The patient looks puzzled and does not move.) "Is there anything wrong with your left hand?"
Patient: "No, doctor."
Doctor: "Why don't you move it, then?"
Patient: "I don't know."
Doctor: "Is this your hand?" (The left hand is held before her eyes.)
Patient: "Not mine, doctor."
Doctor: "Whose hand is it, then?"
Patient: "I suppose it's yours, doctor."

Doctor: "No, it's not; I've already got two hands. look at it carefully." (The left hand is again held before her eyes.)

Patient: "It is not mine, doctor."

Doctor: "Yes it is, look at that ring; whose is it?" (Patient's finger with marriage ring is held before her eyes)

Patient: "That's my ring; you've got my ring, doctor. You're wearing my ring!"

Doctor: "Look at it—it is your hand."

Patient: "Oh, no doctor."

Doctor: "Where is your left hand then?"

Patient: "Somewhere here, I think." (Making groping movements near her left shoulder).

Because the body image has been destroyed, consciousness of that half of the body is also destroyed. The remaining mental systems and the language-dependent conscious mind will completely ignore and fail to recognize their left arm or leg because the mental system responsible for consciousness of the body image no longer exists. If the left arm or leg is shown to them, they will claim it belongs to someone else, such as the nurse or the doctor. They may dress or groom only the right half of their body, eat only off the right half of their plates, and even ignore painful stimuli applied to the left half of their bodies (Joseph, 1986b, 1988a).

However, if you show them their arm and leg (whose ownership they deny), they will admit these extremities exists, but will insist the leg or arm does not belong to them, even though the arm or the leg is wearing the same clothes covering the rest of their body. Instead, the language dependent aspects of consciousness will confabulate and make up explanations and thus create their own reality. One patient said the arm belonged to a little girl, whose arm had slipped into the patient's sleeve. Another declared (speaking of his left arm and leg), "That's an old man. He stays in bed all the time."

One such patient engaged in peculiar erotic behavior with his left arm and leg which he believed belonged to a woman. Some patients may develop a dislike for their left arms, try to throw them away, become agitated when they are referred to, entertain persecutory delusions regarding them, and even complain of strange people sleeping in their beds due to their experience of bumping into their left limbs during the night (Joseph, 1986b, 1988a). One patient complained that the person sharing her bed, tried to push her out of the bed and then insisted that if it happened again she would sue the hospital. Another complained about "a hospital that makes people sleep together." A female patient expressed not only anger but concern least her husband should find out; she was convinced it was a man in her bed.

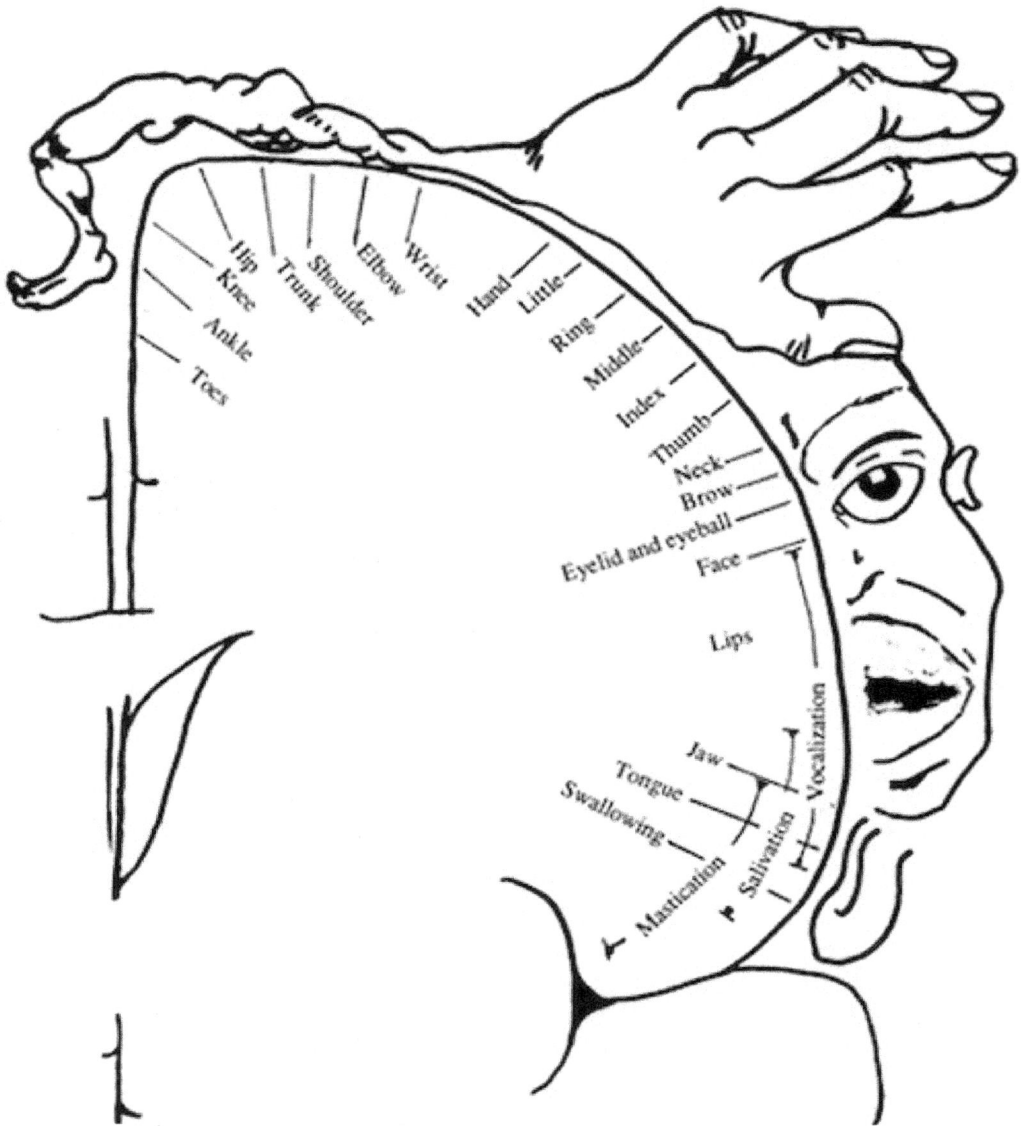

The right and left parietal lobes maintain a map and image of the left and right half of the body, respectively. Therefore, when the right parietal lobe is destroyed, the language-dependent mental systems of the left half of the brain, having access only to the body image for the right half of the body, is unable to become conscious of the left half of their body, except as body parts that they then deduce must belong to someone else.

However, when the language dominant mental system of the left hemi-sphere denies ownership of the left extremity these mental system are in fact telling the truth. That is, the left arm and leg belongs to the right not the left hemisphere; the mental system that is capable of becoming conscious of the left half of their body no longer exist.

When the language axis (Joseph, 1982, 2000), i.e. the inferior parietal lobe, Broca's and Wernicke's areas, are functionally isolated from a particular source of information, the language dependent aspect of mind begins to make up a response based on the information available. To be informed about the left leg or left arm, it must be able to communicate with the cortical area (i.e. the parietal lobe) which is responsible for perceiving and analyzing information regarding the extremities. When no message is received and when the language axis is not informed that no messages are being transmitted, the language zones instead rely on some other source even when that source provides erroneous input (Joseph, 1982, 1986b; Joseph et al., 1984); substitute material is assimilated and expressed and corrections cannot be made (due to loss of input from the relevant knowledge source). The patient begins to confabulate. This is because the patient who speaks to you is not the 'patient' who is perceiving- they are in fact, separate; multiple minds exist in the same head.

8. Split-Brains and Split-Minds.

The multiplicity of mind is not limited to visual consciousness, body consciousness, or the language-dependent consciousness. Rather the multiplicity of mind include social consciousness, emotional consciousness, and numerous other mental realms linked with specific areas of the brain such as the limbic system (emotion), frontal lobes (rational thought), the inferior temporal lobes (memory) and the two halves of the brain where multiple streams of mental activity become subordinated and dominated by two distinct realms of mind; consciousness and awareness (Joseph, 1982, 1986a,b, 1988a,b, 2012).

The brain is not a singularity. This is most apparent when viewing the right and left half of the brain which are divided by the interhemispheric fissure and almost completely split into two cerebral hemispheres. These two brain halves are connected by a rope of nerve fibers, the corpus callosum, which enables them to share and exchange some information, but not all information as these two mental realms maintain a conscious awareness of different realities.

For example, it is well established that the right cerebral hemisphere is dominant over the left in regard to the perception, expression and mediation of almost all aspects of social and emotional functioning and related aspects of social/emotional language and memory. Further, the right hemisphere is dominant for most aspects of visual-spatial perceptual functioning, the comprehension of body language, the recognition of faces including friend's loved ones, and one's own face in the mirror (Joseph, 1988a, 1996).

Recognition of one's own body and the maintenance of the personal body image is also the dominant realm of the right half of the brain (Joseph, 1986b, 1988a). The body image, for many, is tied to personal identity; and the same is true of the recognition of faces including one's own face.

The right is also dominant for perceiving and analyzing visual-spatial relationships, including the movement of the body in space (Joseph, 1982, 1988a). Therefore, one can throw or catch a ball with accuracy, dance across a stage, or leap across a babbling brook without breaking a leg.

The perception of environmental sounds (water, wind, a meowing cat) and the social, emotional, musical, and melodic aspects of language, including the ability to sing, curse, or pray, are also the domain of the right hemisphere mental system (Joseph, 1982, 1988a). Hence, it is the right hemisphere which imparts the sounds of sarcasm, pride, humor, love, and so on, into the stream of speech, and which conversely can determine if others are speaking with sincerity, irony, or evil intentions.

By contrast, expressive and receptive speech, linguistic knowledge and thought, mathematical and analytical reasoning, reading, writing, and arithmetic, as well as the temporal-sequential and rhythmical aspects of consciousness, are associated with the functional integrity of the left half of the brain in the majority of the population (Joseph, 1982, 1996). The language-dependent mind is linked to the left hemisphere.

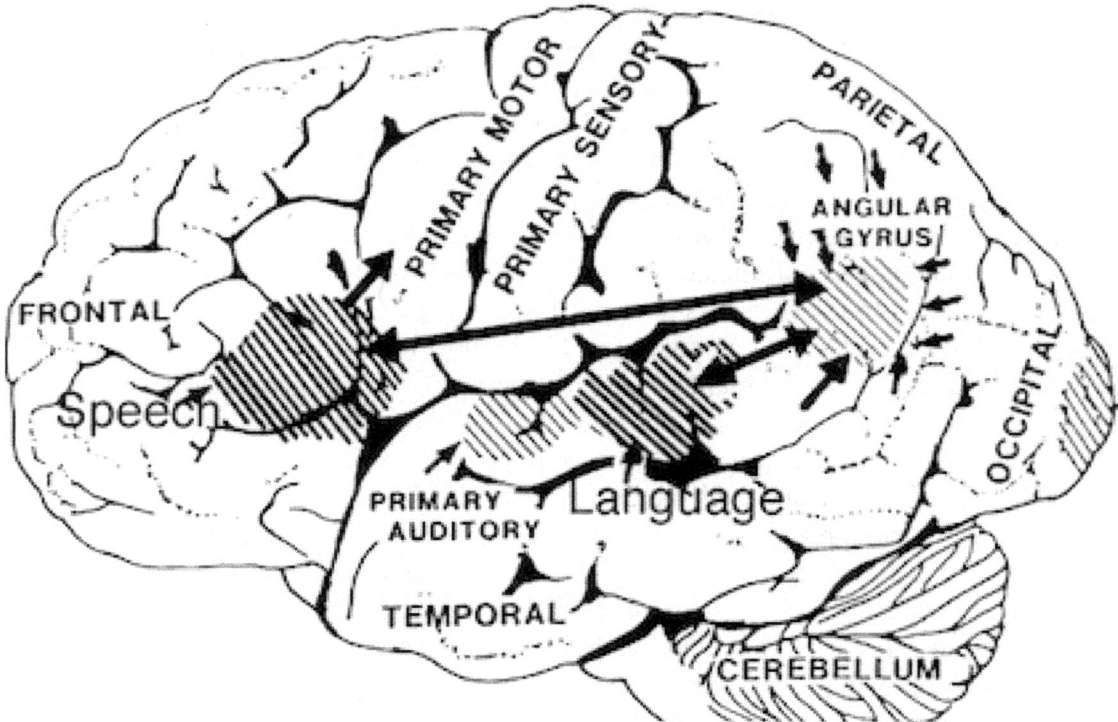

Certainly, there is considerable overlap in functional representation. Moreover, these two mental system interact and assist the other, just as the left and right hands cooperate and assist the other in performing various tasks. For example, if you were standing at the bar in a nightclub, and someone were to tap you on the shoulder and say, "Do you want to step outside," it is the mental system of the left hemisphere which understands that a question about "outside" has been asked, but it is the mental system of the right which determines the underlying meaning, and if you are being threatened with a punch in the nose, or if a private conversation is being sought.

However, not all information can be transferred from the right to the left, and vice versa (Gallagher and Joseph, 1982; Joseph, 1982, 1988a; Joseph and Gallagher, 1985; Joseph et al., 1984). Because each mental system is unique, each "speaks a different language" and they cannot always communicate. Not all mental events can be accurately translated, understood, or even recognized by the other half of the brain. These two major mental systems, which could be likened to "consciousness" vs "awareness" exist in parallel, simultaneously, and both can act independently of the other, have different goals and desires, and come to completely different conclusions. Each mental system has its own reality.

The existence of these two independent mental realms is best exemplified and demonstrated following "split-brain" surgery; i.e. the cutting of the corpus callosum fiber pathway which normally allows the two hemisphere's to communicate.

As described by Nobel Lauriate Roger Sperry (1966, p. 299), "Everything we have seen indicates that the surgery has left these people with two separate minds, that is, two separate spheres of consciousness. What is experienced in the right hemisphere seems to lie entirely outside the realm of awareness of the left hemisphere. This mental division has been demonstrated in regard to perception, cognition, volition, learning and memory."

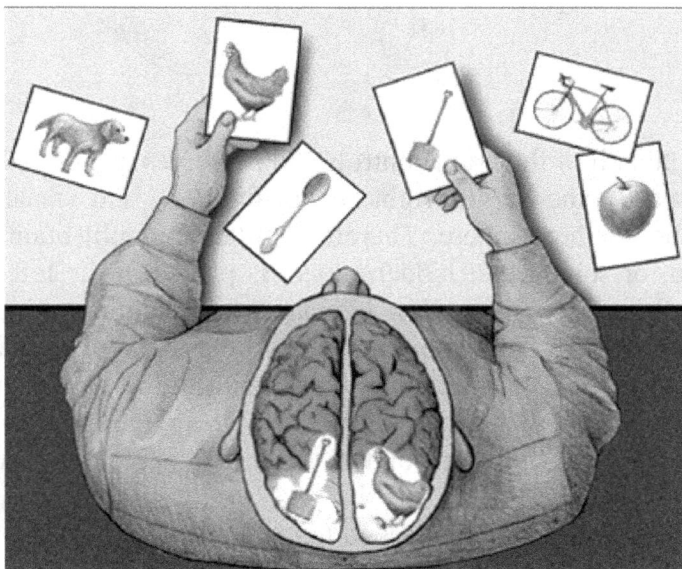

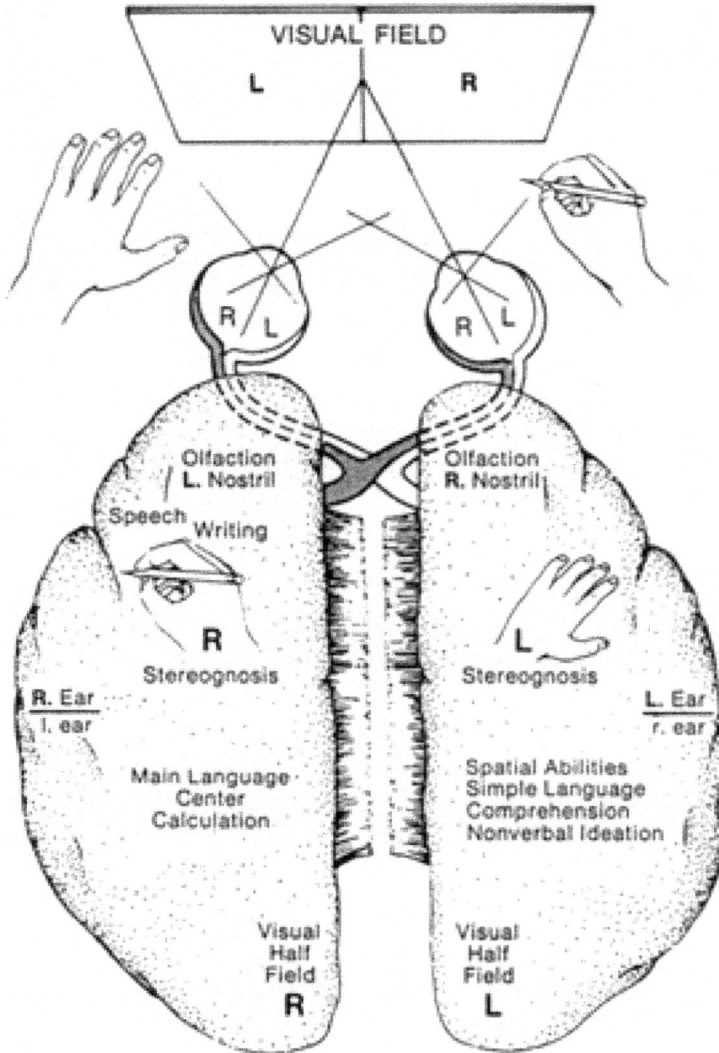

VISUAL FIELD

The right half of the brain controls and perceives the left half of the body and visual space, whereas the right half of the body and visual space is the domain of the left hemisphere. Therefore, following split-brain surgery, if a comb, spoon, or some other hidden object is placed in the left hand (out of sight), the left hemisphere, and the language-dependent conscious mind, will not even know the left hand is holding something and will be unable to name it, describe it, or if given multiple choices point to the correct item with the right hand (Joseph 1988a,b; Sperry, 1966). However, the right hemisphere can raise the left hand and not only point to the correct object, but can pantomime its use.

If the split-brain patient is asked to stare at the center of a white screen and words like "Key Ring" are quickly presented, such that the word "Key" falls in the left visual field (and thus, is transmitted to the right cerebrum) and the word "Rings" falls in the right field (and goes to the left hemisphere), the language dependent conscious mind will not see the word "Key." If asked, the language-dependent conscious mind will say "Ring" and will deny seeing the word "Key." However, if asked to point with the left hand, the mental system of the right hemisphere will correctly point to the word "Key."

Therefore, given events "A" and "B" one half of the brain may know A, but know nothing about B which is known only by the other half of the brain. In consequence, what is known vs what is not known becomes relatively imprecise depending on what aspects of reality are perceived and "known" by which mental system (Joseph 1986a; Joseph et al., 1984). There is no such thing as singularity of mind. Since the brain and mind is a multiplicity, "A" and "B" can be known simultaneously, even when one mind is knows nothing about the existence of A or B.

In that the brain of the normal as well as the "split-brain" patient maintains the neuroanatomy to support a multiplicity of mind, and the presence of two dominant psychic realms, it is therefore not surprising that "normal" humans often have difficulty "making up their minds," suffer internal conflicts over love/hate relationships, and are plagued with indecision even when staring into an open refrigerator and trying to decide what to eat. "Making up one's mind" can be an ordeal involving a multiplicity of minds.

However, this conflict becomes even more apparent following split-brain surgery and the cutting of the corpus callosum fiber pathway which links these two parallel streams of conscious-awareness.

Akelaitis (1945, p. 597) describes two patients with complete corpus callosotomies who experienced extreme difficulties making the two halves of their bodies cooperate. "In tasks requiring bimanual activity the left hand would frequently perform oppositely to what she desired to do with the right hand. For example, she would be putting on clothes with her right and pulling them off with her left, opening a door or drawer with her right hand and simultaneously pushing it shut with the left. These uncontrollable acts made her increasingly irritated and depressed."

Another patient experienced difficulty while shopping, the right hand would place something in the cart and the left hand would put it right back again and grab a different item.

A recently divorced male patient complained that on several occasions while walking about town he found himself forced to go some distance in another direction by his left leg. Later (although his left hemisphere was not conscious of it at the time) it was discovered that this diverted course, if continued, would have led him to his former wife's new home.

Geschwind (1981) reports a callosal patient who complained that his left hand on several occasions suddenly struck his wife--much to the embarrassment of his left (speaking) hemisphere. In another case, a patient's left hand attempted to choke the patient himself and had to be wrestled away.

Bogen (1979, p. 333) indicates that almost all of his "complete commissurotomy patients manifested some degree of intermanual conflict." One patient, Rocky, experienced situations in which his hands were uncooperative; the right would button up a shirt and the left would follow right behind and undo the buttons. For years, he complained of difficulty getting his left leg to go in the direction he (or rather his left hemisphere) desired. Another patient often referred to the left half of her body as "my little sister" when she was complaining of its peculiar and independent actions.

Another split-brain patient reported that once when she had overslept her left hand began slapping her face until she (i.e. her left hemisphere) woke up. This same patient, in fact, complained of several instances where her left hand had acted violently toward herself and other people (Joseph, 1988a).

Split brain patient, 2-C, complained of instances in which his left hand would perform socially inappropriate actions, such as striking his mother across the face (Joseph, 1988b). Apparently his left and right hemisphere also liked different TV programs. He complained of numerous instances where he (his left hemisphere) was enjoying a program, when, to his astonishment, the left half of his body pulled him to the TV, and changed the channel.

The right and left hemisphere also liked different foods and had different attitudes about exercise. Once, after 2-C had retrieved something from the refrigerator with his right hand, his left took the food, put it back on the shelf and retrieved a completely different item "Even though that's not what I wanted to eat!" On at least one occasion, his left leg refused to continue "going for a walk" and would only allow him to return home.

In the laboratory, 2-C's left hemisphere often became quite angry with his left hand, and he struck it and expressed hate for it. Several times, his left and right hands were observed to engage in actual physical struggles, beating upon each other. For example, on one task both hands were stimulated simultaneously (while out of view) with either the same or two different textured materials (e.g., sandpaper to the right, velvet to the left), and he was required to point (with the left and right hands simultaneously) to an array of fabrics that were hanging in view on the left and right of the testing apparatus. However, at no time was he informed that two different fabrics were being applied.

After stimulation he would pull his hands out from inside the apparatus and point with the left to the fabric felt by the left and with the right to the fabric felt by the right.

Surprisingly, although his left hand (right hemisphere) responded correctly, his left hemisphere vocalized: "Thats wrong!" Repeatedly he reached over with his right hand and tried to force his left extremity to point to the fabric experienced by the right (although the left hand responded correctly! His left hemisphere didn't know this, however). His left hand refused to be moved and physically resisted being forced to point at anything different. In one instance a physical struggle ensued, the right grappling with the left with the two halves of the body hitting and scratching at each other!

Moreover, while 2-C was performing this (and other tasks), his left hemisphere made statements such as: "I hate this hand" or "This is so frustrating" and would strike his left hand with his right or punch his left arm. In these instances there could be little doubt that his right hemisphere mental system was behaving with purposeful intent and understanding, whereas his left hemisphere mental system had absolutely no comprehension of why his left hand (right hemisphere) was behaving in this manner (Joseph, 1988b).

These conflicts are not limited to behavior, TV programs, choice of clothing, or food, but to actual feelings, including love and romance. For example, the right and left hemisphere of a male split-brain patient had completely different feelings about an ex-girlfriend. When he was asked if he wanted to see her again, he said "Yes." But at the same time, his left hand turned thumbs down!

Another split-brain patient suffered conflicts about his desire to smoke. Although is left hemisphere mental system enjoyed cigarettes, his left hand would not allow him to smoke, and would pluck lit cigarettes from his mouth or right hand and put them out. He had been trying to quit for years.

Because each head contains multiple minds, similar conflicts also plague those who have not undergone split-brain surgery. Each half of the brain and thus each mental system may have different attitudes, goals and interests. As noted above, 2-C experienced conflicts when attempting to eat, watch TV, or go for walks, his right and left hemisphere mental systems apparently enjoying different TV programs or types of food (Joseph 1988b). Conflicts of a similar nature plague us all. Split-brain patients are not the first to choke on self-hate or to harm or hate those they profess to love.

Each half of the brain is concerned with different types of information, and may react, interpret and process the same external experience differently and even reach different conclusions (Joseph 1988a,b; Sperry, 1966). Moreover, even when the goals are the same, the two halves of the brain may produce and attempt to act on different strategies.

Each mental system has its own reality. Singularity of mind, is an illusion.

9. Dissociation and Self-Consciousness

The multiplicity of mind is not limited to the neocortex but includes old cortical structures, such as the limbic system (Joseph 1992). Moreover, limbic nuclei such as the amygdala and hippocampus interact with neocortical tissues creating yet additional mental systems, such as those which rely on memory and which contribute to self-reflection, personal identity, and even self-consciousness (Joseph, 1992, 1998, 1999b, 2001).

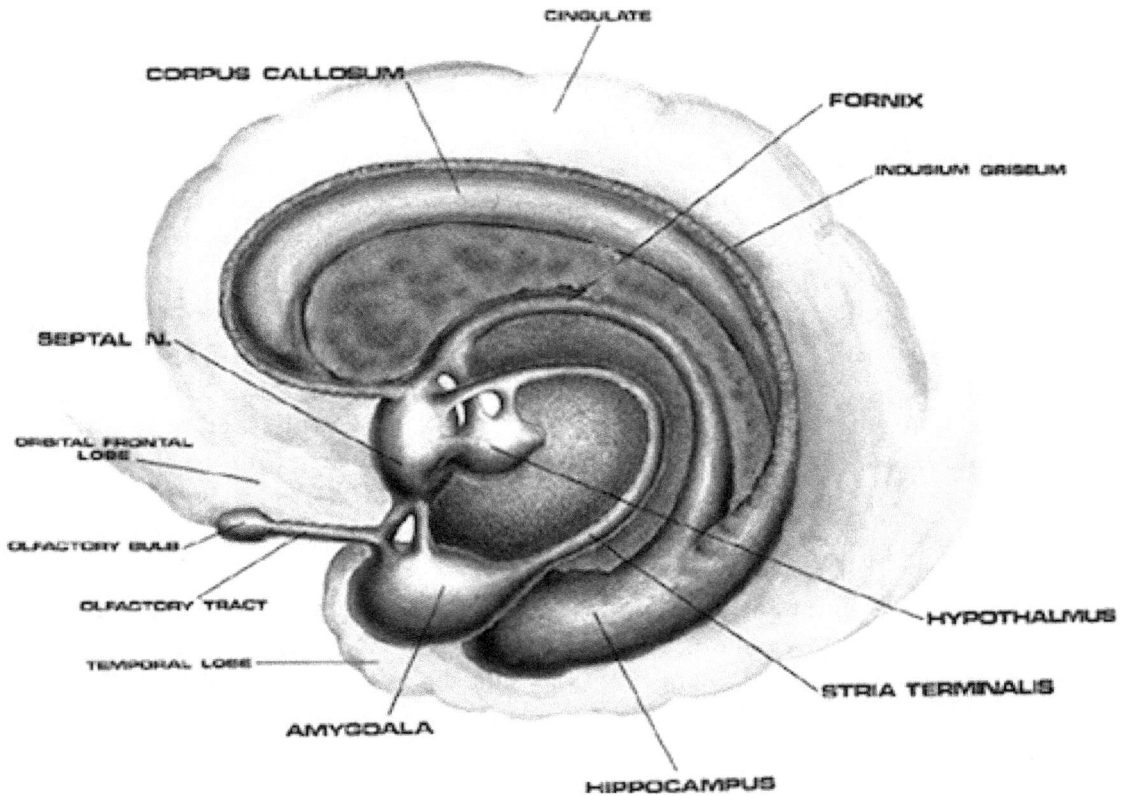

For example, both the amygdala and the hippocampus are implicated in the storage of long term memories, and both nuclei enable individuals to visualize and remember themselves engaged in various acts, as if viewing their behavior and actions from afar. Thus, you might see yourself and remember yourself engage in some activity, from a perspective outside yourself, as if you are an external witness; and this is a common feature of self-reflection and self-memory and is made possible by the hippocampus and overlying temporal lobe (Joseph, 1996, 2011).

The hippocampus in fact contains "place neurons" which cognitive map one's position and the location of various objects within the environment (Nadel, 1991; O'Keefe, 1976; Wilson & McNaughton, 1993). Further, if the subject moves about in that environment, entire populations of these place cells will fire. Moreover, some cells are responsive to the movements of other people in that environment and will fire as that person is observed to move around to different locations or corners of the room (Nadel, 1991; O'Keefe,

1976; Wilson and McNaughton, 1993).

Electrode stimulation, or other forms of heightened activity within the hippocampus and overlying temporal lobe can also cause a person to see themselves, in real time, as if their conscious mind is floating on the ceiling staring down at their body (Joseph, 1998, 1999b, 2001). During the course of electrode stimulation and seizure activity originating in the temporal lobe or hippocampus, patients may report that they have left their bodies and are hovering upon the ceiling staring down at themselves (Daly, 1958; Penfield, 1952; Penfield & Perot 1963; Williams, 1956). That is, their consciousness and sense of personal identity appears to split off from their body, such that they experience themselves as as a consciousness that is conscious of itself as a conscious that is detached from the body which is being observed.

One female patient claimed that she not only would float above her body, but would sometimes drift outside and even enter into the homes of her neighbors. Penfield and Perot (1963) describe several patients who during a temporal lobe seizure, or neurosurgical temporal lobe stimulation, claimed they split-off from their body and could see themselves down below. One woman stated: "it was though I were two persons, one watching, and the other having this happen to me." According to Penfield (1952), "it was as though the patient were attending a familiar play and was both the actor and audience."

Under conditions of extreme trauma, stress and fear, the amygdala, hippocampus and temporal lobe become exceedingly active (Joseph, 1998, 1999b). Under these conditions many will experience a "splitting of consciousness" and have the sensation they have left their body and are hovering beside or above themselves, or even that they floated away (Courtois, 2009; Grinker & Spiegel, 1945; Noyes & Kletti, 1977; van der Kolk 1987). That is, out-of-body dissociative experiences appear to be due to fear induced hippocampus (and amygdala) hyperactivation.

Likewise, during episodes of severe traumatic stress personal consciousness may be fragmented and patients may dissociate and experience themselves as splitting off and floating away from their body, passively observing all that is occurring (Courtois, 1995; Grinker & Spiegel, 1945; Joseph, 1999d; Noyes & Kletti, 1977; Southard, 1919; Summit, 1983; van der Kolk 1987).

Noyes and Kletti (1977) described several individuals who experienced terror, believed they were about to die, and then suffered an out-of body dissociative experience: "I had a clear image of myself... as though watching it on a television screen." "The next thing I knew I wasn't in the truck anymore; I was looking down from 50 to 100 feet in the air." "I had a sensation of floating. It was almost like stepping out of reality. I seemed to step out of this world."

One individual, after losing control of his Mustang convertible while during over 100 miles per hour on a rain soaked freeway, reported that "time seemed to slow down and then... part of my mind was a few feet outside the car zooming above it and then beside it and behind it and in front of it, looking at and analyzing the respective positions of my spinning Mustang and the cars surrounding me. Simultaneously I was inside trying to steer and control it in accordance with the multiple perspectives I was given by that part of my mind that was outside. It was like my mind split and one consciousness was inside the car, while the other was zooming all around outside and giving me visual feedback that enabled me to avoid hitting anyone or destroying my Mustang."

Numerous individuals from adults to children, from those born blind and deaf, have also reported experiencing a dissociative consciousness after profound injury causing near death (Eadie 1992; Rawling 1978; Ring 1980). Consider for example, the case of Army Specialist J. C. Bayne of the 196th Light Infantry Brigade. Bayne was "killed" in Chu Lai, Vietnam, in 1966, after being simultaneously machine gunned and struck by a mortar. According to Bayne, when he opened his eyes he was floating in the air, looking down on his burnt and bloody body: "I could see me... it was like looking at a manikin laying there... I was burnt up and there was blood all over the place... I could see the Vietcong. I could see the guy pull my boots off. I could see the rest of them picking up various things... I was like a spectator... It was about four or five in the afternoon when our own troops came. I could hear and see them ap-proaching... I looked dead... they put me in a bag... transferred me to a truck and then to the morgue. And from that point, it was the embalming process. I was on that table and a guy was telling jokes about those USO girls... all I had on was bloody undershorts... he placed my leg out and made a slight incision and stopped... he checked my pulse and heartbeat again and I could see that too... It was about that point I just lost track of what was taking place.... [until much later] when the chaplain was in there saying everything was going to be all right.... I was no longer outside. I was part of it at this point" (reported in Wilson, 1987, pp 113-114; and Sabom, 1982, pp 81-82).

Therefore, be it secondary to the fear of dying, or depth electrode stimu-lation, these experiences all appear to be due to a mental system which enables a the conscious mind to detach completely from the body in order to make the body an object of consciousness (Joseph, 1998, 1999b, 2001).

10. Quantum Consciousness

It could be said that consciousness is consciousness of something other than consciousness. Consciousness and knowledge of an object, such as a chair, are also distinct. Consciousness is not the chair. The chair is not consciousness. The chair is an object of consciousness, and thus become discontinuous from the quantum state.

Consciousness is consciousness of something and is conscious of not be-ing that object that is is conscious of. By knowing what it isn't, consciousness may know what it is not, which helps define what it is. This consciousness of not being the object can be considered the "collapse function" which results in discontinuity within the continuuum.

Further, it could be said that consciousness of consciousness, that is, self-consciousness, also imparts a duality, a separation, into the fabric of the quantum continuum. Therefore this consciousness that is the object of consciousness, becomes an abstraction, and may create a collapse function in the continuum.

However, in instances of dissociation, this consciousness is conscious of itself as a consciousness that is floating above its body; a body which contains the brain. The dissociative consciousness is not dissociated from itself as a consciousness, but only from its body. That is, there is an awareness of itself as a consciousness that is floating above the body, and this awareness is simultaneously one with that consciousness, as there is no separation, no abstractions, and no objectification. It is a singularity that is without form, without dimension, without shape.

Moreover, because this dissociated consciousness appears to be continuous with itself, there is no "collapse function" except in regard to the body which is viewed as an object of perception.

Therefore, in these instances we can not say that consciousness has split into a duality of observer and observed or knower and known, except in regard to the body. Dissociated consciousness is conscious of itself as consciousness; it is self-aware without separation and without reflecting. It is knowing and known, simultaneously.

In fact, many patients report that in the dissociated state they achieve or nearly achieved a state of pure knowing (Joseph, 1996, 2001).

A patient described by Williams (1956) claimed she was lifted up out of her body, and experienced a very pleasant sensation of elation and the feeling that she was "just about to find out knowledge no one else shares, something to do with the link between life and death." Another patient reported that upon leaving her body she not only saw herself down below, but was taken to a special place "of vast proportions, and I felt as if I was in another world" (Williams, 1956).

Other patients suffering from temporal lobe seizures or upon direct electrical activation have noted that things suddenly became "crystal clear" or that they had a feeling of clairvoyance, of having the ultimate truth revealed to them, of having achieved a sense of greater awareness and cosmic clarity such that sounds, smells and visual objects seemed to have a greater meaning and sensibility, and that they were achieving a cosmic awareness of the hidden knowledge of all things (Joseph, 1996, 2001).

Although consciousness and the object of consciousness, that which is known, are not traditionally thought of as being one and the same, in dissociative states, consciousness and knowing, may become one and the same. The suggestion is of some type of cosmic unity, particularly as these patients also often report a progressive loss of the sense of individuality, as if they are merging into something greater than themselves, including a becoming one with all the knowledge of the universe; a singularity with god.

Commonly those who experience traumatic dissociative conscious states, not only float above the body, but report that they gradually felt they were losing all sense of individuality as they became embraced by a brilliant magnificent whiteness that extended out in all directions into eternity.

Therefore, rather than the dissociated consciousness acting outside the quantum state, it appears this mental state may represent an increasing submersion back into the continuity that is the quantum state, disappearing back into the continuum of singularity and oneness that is the quantum universe.

Dissociated consciousness may be but the last preamble before achieving the unity that is quantum consciousness, the unity of all things.

11. Conclusion: Quantum Consciousness and the Multiplicity of Mind

What is "Objective Reality" when the mind is a multiplicity which is capable of splitting up, observing itself, becoming blind to itself, and becoming blind to the features of the world which then cease to exist for the remaining mental realms?

Each mental system has its own "reality." Each observer is a multiplicity that engages in numerous simultaneous acts of observation. Therefore, non-local properties which do not have an objective existence independent of the "act of observation" by one mental system, may achieve existence when observed by another mental system. The "known" and the "unknown" can exist simultaneously and interchangeably and this may explain why we don't experience any macroscopic non-local quantum weirdness in our daily lives.

This means that quantum laws may apply to everything, from atoms to monkeys and woman and man and the multiplicity of mind. However, because of this multiplicity, this could lead to seemingly contradictory predictions and uncertainty when measuring macroscopically objective systems which are superimposed on microscopic quantum systems. Indeed, this same principle applies to the multiplicity of mind, where dominant parallel streams of conscious awareness may be superimposed on other mental systems, and which may be beset by uncertainty.

Because of the multiplicity of mind, as exemplified by dissociative states, the observer can also be observed, and thus, the observer is not really external to the quantum state as is required by the standard collapse formulation of quantum mechanics. This raises the possibility that the collapse formulation can be used to describe the universe as a whole which includes observers observing themselves being observed.

The multiplicity of mind also explains why an object being measured by one mental system therefore becomes bundled up into a state where it either determinately has or determinately does not have the property being measured.

Measurements performed by one mental system are not being performed by others, such that the same object can have an initial state and a post-measurement state and a final state simultaneously as represented in multiple minds in parallel, or separate states as represented by each mental realm individually.

The collapse dynamics of observation supposedly guarantees that a system will either determinately have or determinately not have a particular property. However, because the observer is observing with multiple mental systems the object can both have and not have specific properties when it is being measure and not measured, simply because it is being measured and not measured, or rather, observed and not observed or its different features observed simultaneously by multiple mental systems. Therefore, it can be continuous and discontinuous in parallel, and different properties can be known and not known in parallel simultaneously.

And the same rules apply to the mental systems which exist in multiplicity within the head of a single observer. Mental systems can become continuous or discontinuous, and can be known and not known simultaneously, in parallel. Thus, the standard collapse formulation can be used to describe systems that contain observers, as the mind/observer can be simultaneously internal and external to the described system.

The mind is a multiplicity and there is no such thing as a "single observer state." Therefore, each element may be observed by multiple observer states which perceive multiple object systems thereby giving the illusion that the object has been transformed during the collapse function. What this also implies is that contrary to the standard or Copenhagen interpretations, states may have both definite position and definite momentum at the same time.

Each mental system perceives a different physical world giving rise to multiple worlds and multiple realities which may be subordinated by one or another more dominant stream of conscious awareness.

Moreover, as the multiplicity of mind can also detach and become discontinuous with the body, whereas dissociative consciousness is continuous with itself, this indicates that the mind is also capable of becoming one with the continuum, and can achieve singularity so that universe and mind become one.

References

Akelaitis, A. J. (1945). Studies on the corpus callosum. American Journal of Psychiatry, 101, 594-599.

Bogen, J. (1979). The other side of the brain. Bulletin of the Los Angeles Neurological Society. 34, 135-162.

Bohr, N. (1934/1987), Atomic Theory and the Description of Nature, reprinted as The Philosophical Writings of Niels Bohr, Vol. I, Woodbridge: Ox Bow Press.

Bohr, N. (1958/1987), Essays 1932-1957 on Atomic Physics and Human Knowledge, reprinted as The Philosophical Writings of Niels Bohr, Vol. II, Woodbridge: Ox Bow Press.

Bohr, N. (1963/1987), Essays 1958-1962 on Atomic Physics and Human Knowledge, reprinted as The Philosophical Writings of Niels Bohr, Vol. III, Woodbridge: Ox Bow Press.

Casagrande, V. A. & Joseph, R. (1978). Effects of monocular deprivation on geniculostriate connections in prosimian primates. Anatomical Record, 190, 359.

Casagrande, V. A. & Joseph, R. (1980). Morphological effects of monocular deprivation and recovery on the dorsal lateral geniculate nucleus in prosimian primates. Journal of Comparative Neurology, 194, 413-426.

Courtois, C. A. (2009). Treating Complex Traumatic Stress Disorders: An Evidence-Based Guide Daly, D. (1958). Ictal affect. American Journal of Psychiatry, 115, 97-108.

DeWitt, B. S., (1971). The Many-Universes Interpretation of Quantum Mechanics, in B. D.'Espagnat (ed.), Foundations of Quantum Mechanics, New York: Academic Press. pp. 167–218.

DeWitt, B. S. and Graham, N., editors (1973). The Many-Worlds Interpretation of Quantum Mechanics. Princeton University Press, Princeton, New-Jersey.

Eadie, B. J. (1993) Embraced by the light. New York, Bantam

Einstein, A. (1905a). Does the Inertia of a Body Depend upon its Energy Content? Annalen der Physik 18, 639-641.

Einstein, A. (1905b). Concerning an Heuristic Point of View Toward the Emission and Transformation of Light. Annalen der Physik 17, 132-148.

Einstein, A. (1905c). On the Electrodynamics of Moving Bodies. Annalen der Physik 17, 891-921.

Einstein, A. (1926). Letter to Max Born. The Born-Einstein Letters (translated by Irene Born) Walker and Company, New York.

Gallagher, R. E., & Joseph, R. (1982). Non-linguistic knowledge, hemispheric laterality, and the conservation of inequivalance. Journal of General Psychology, 107, 31-40.

Geschwind. N. (1981). The perverseness of the right hemisphere. Behavioral Brain Research, 4, 106-107.

Grinker, R. R., & Spiegel, J. P. (1945). Men Under Stress. McGraw-Hill.

Heisenberg. W. (1930), Physikalische Prinzipien der Quantentheorie (Leipzig: Hirzel). English translation The Physical Principles of Quantum Theory, University of Chicago Press.

Heisenberg, W. (1955). The Development of the Interpretation of the Quantum Theory, in W. Pauli (ed), Niels Bohr and the Development of Physics, 35, London: Pergamon pp. 12-29.

Heisenberg, W. (1958), Physics and Philosophy: The Revolution in Modern

Science, London: Goerge Allen & Unwin.

Joseph, R. (1982). The Neuropsychology of Development. Hemispheric Laterality, Limbic Language, the Origin of Thought. Journal of Clinical Psychology, 44 4-33.

Joseph, R. (1986). Reversal of language and emotion in a corpus callosotomy patient. Journal of Neurology, Neurosurgery, & Psychiatry, 49, 628-634.

Joseph, R. (1986). Confabulation and delusional denial: Frontal lobe and lateralized influences. Journal of Clinical Psychology, 42, 845-860.

Joseph, R. (1988) The Right Cerebral Hemisphere: Emotion, Music, Visual-Spatial Skills, Body Image, Dreams, and Awareness. Journal of Clinical Psychology, 44, 630-673.

Joseph, R. (1988). Dual mental functioning in a split-brain patient. Journal of Clinical Psychology, 44, 770-779.

Joseph, R. (1992) The Limbic System: Emotion, Laterality, and Unconscious Mind. The Psychoanalytic Review, 79, 405-455.

Joseph, R. (1996). Neuropsychiatry, Neuropsychology, Clinical Neuroscience, 2nd Edition. Williams & Wilkins, Baltimore.

Joseph, R. (1998). Traumatic amnesia, repression, and hippocampal injury due to corticosteroid and enkephalin secretion. Child Psychiatry and Human Development. 29, 169-186.

Joseph, R. (1999a). Frontal lobe psychopathology: Mania, depression, aphasia, confabulation, catatonia, perseveration, obsessive compulsions, schizophrenia. Psychiatry, 62, 138-172.

Joseph, R. (1999b). The neurology of traumatic "dissociative" amnesia. Commentary and literature review. Child Abuse & Neglect. 23, 71-80.

Joseph, R. (2000). Limbic language/language axis theory of speech. Behavioral and Brain Sciences. 23, 439-441.

Joseph, R. (2001). The Limbic System and the Soul: Evolution and the Neuroanatomy of Religious Experience. Zygon, the Journal of Religion & Science, 36, 105-136.

Joseph, R., & Casagrande, V. A. (1978). Visual field defects and morphological changes resulting from monocular deprivation in primates. Proceedings of the Society for Neuroscience, 4, 1978, 2021.

Joseph, R. & Casagrande, V. A. (1980). Visual field defects and recovery following lid closure in a prosimian primate. Behavioral Brain Research, 1, 150-178.

Joseph, R., Forrest, N., Fiducia, N., Como, P., & Siegel, J. (1981). Electrophysiological and behavioral correlates of arousal. Physiological Psychology, 1981, 9, 90-95.

Joseph, R., Gallagher, R., E., Holloway, J., & Kahn, J. (1984). Two brains, one child: Interhemispheric transfer and confabulation in children aged 4, 7, 10. Cortex, 20, 317-331.

Joseph, R., & Gallagher, R. E. (1985). Interhemispheric transfer and the completion of reversible operations in non-conserving children. Journal of Clinical Psychology, 41, 796-800.

Nadel, L. (1991). The hippocampus and space revisited. Hippocampus, 1, 221-229.

Neumann, J. von, (1937/2001), "Quantum Mechanics of Infinite Systems. Institute for Advanced Study; John von Neumann Archive, Library of Congress, Washington, D.C.

Neumann, J. von, (1938), On Infinite Direct Products, Compositio Mathematica 6: 1-77.

Neumann, J. von, (1955), Mathematical Foundations of Quantum Mechanics, Princeton, NJ: Princeton University Press.

Noyes, R., & Kletti, R. (1977). Depersonalization in response to life threatening danger. Comprehensive Psychiatry, 18, 375-384.

O'Keefe, J. (1976). Place units in the hippocampus. Experimental Neurology, 51-78-100. Pais, A. (1979). Einstein and the quantum theory, Reviews of Modern Physics, 51, 863-914.

Penfield, W. (1952) Memory Mechanisms. Archives of Neurology and Psychiatry, 67, 178-191.

Penfield, W., & Perot, P. (1963) The brains record of auditory and visual experience. Brain, 86, 595-695.

Rawlins, M. (1978). Beyond Death's Door. Sheldon Press.

Ring, K. (1980). Life at Death: A Scientific Investigation of the Near-Death Experience. New York: Quill.

Sabom, M. (1982). Recollections of Death. New York: Harper & Row.

Southard, E. E. (1919). Shell-shock and other Neuropsychiatric Problems. Boston.

Sperry, R. (1966). Brain bisection and the neurology of consciousness. In F. O. Schmitt and F. G. Worden (eds). The Neurosciences. MIT press.

van der Kolk, B. A. (1985). Psychological Trauma. American Psychiatric Press.

Williams, D. (1956). The structure of emotions reflected in epileptic experiences. Brain, 79, 29-67.

Wilson, I. (1987). The After Death Experience. Morrow.

Wilson, M. A., & McNaughton, B. L. (1993). Dynamics of the hippocampus ensemble for space. Science, 261, 1055-1058.

13. Dissociation, Traumatic Stress, Dissociative Amnesia, Out-Of-Body Hallucinations, Flashbacks, PTSD, Catatonia, Paralytic Fear

In response to the threat of bodily injury or death, animals, including humans, may fight, take flight, or fight while attempt to flee--reactions mediated by the limbic system and limbic striatum / basal ganglia. Under conditions of overwhelming terror, or in reaction to prolonged and repetitive instances of severe stress and the subsequent secretion of stress hormones and excessive activation of limbic structures, victims may become so frightened they become paralyzed with fear (Galliano, Noble, Travis, & Puechl, 2013; Miller, 1951; Nijenhuis, Vanderlinden, & Spinhoven, 2011; Krystal, 2003), enter trance-like states, experience numbing and out-of-body "hallucinations" and may then suffer varying degrees of memory loss and/or an inability to forget coupled with flashbacks and intrusive trauma-related imagery (Fink, 1999; Foa, Riggs, & Gershuny, 1995; Joseph, 2011b, 1999d; Krystal, 2003; Moller, et al., 2011).

Fight, flight, fear-paralysis, and associated dissociative disturbances, appear to be mediated by the same structures, the amygdala, hippocampus, brainstem, hypothalamus, striatum, medial frontal lobes, including the autonomic nervous system. Moreover, a cascade of neurochemical and stress hormones are released and secreted, including corticostereoids, enkephalins, norepinephrine (NE), serotonin (5HT), and dopamine (DA), all of which differentially contribute to the fear and stress response (Bliss, Ailion, & Zwanziger, 1968; Fink, 1999; Moller, Milinksi & Slater, 2011; Rosenblum, Coplan, & Friedman, 2009; Southwick et al., 2013), including fear paralysis, numbing trance-like states, memory loss, as well as flashbacks and out-of-body hallucinations. These reactions are not necessarily maladaptive.

Cessation of movement, for example, may prevent a predator or a hunter, from spying their prey. Soldiers and Airmen who have found themselves behind enemy lines have commonly escaped capture by simply lying still. There have been numerous reports of enemy soldiers nearly stepping on these men, but nevertheless failing to see them. However, under some circumstances, such as a sinking ship, or burning aircraft, paralytic fear may instead result in death (Krystal, 2003).

Similarly, ruminating about the consequences of a catastrophe, or a miraculous escape from death, can have its benefits; that is, as a learning experience and as a form of analysis which enables one to avoid or to take more effective action if similar misfortune arises in the future. Of course, there is nothing adaptive about sudden flashbacks, nightmares, intrusive imagery, or an inability to forget, as the victim is simply being traumatized. In fact, it is the continual traumatization, that is, at the level of the central nervous system, coupled with the experience of trauma-related environmental stimuli, which provokes these flashbacks; a consequence of stress-induced injury to the amygdala, hippocampus, hypothalamus, and related structures and neural pathways.

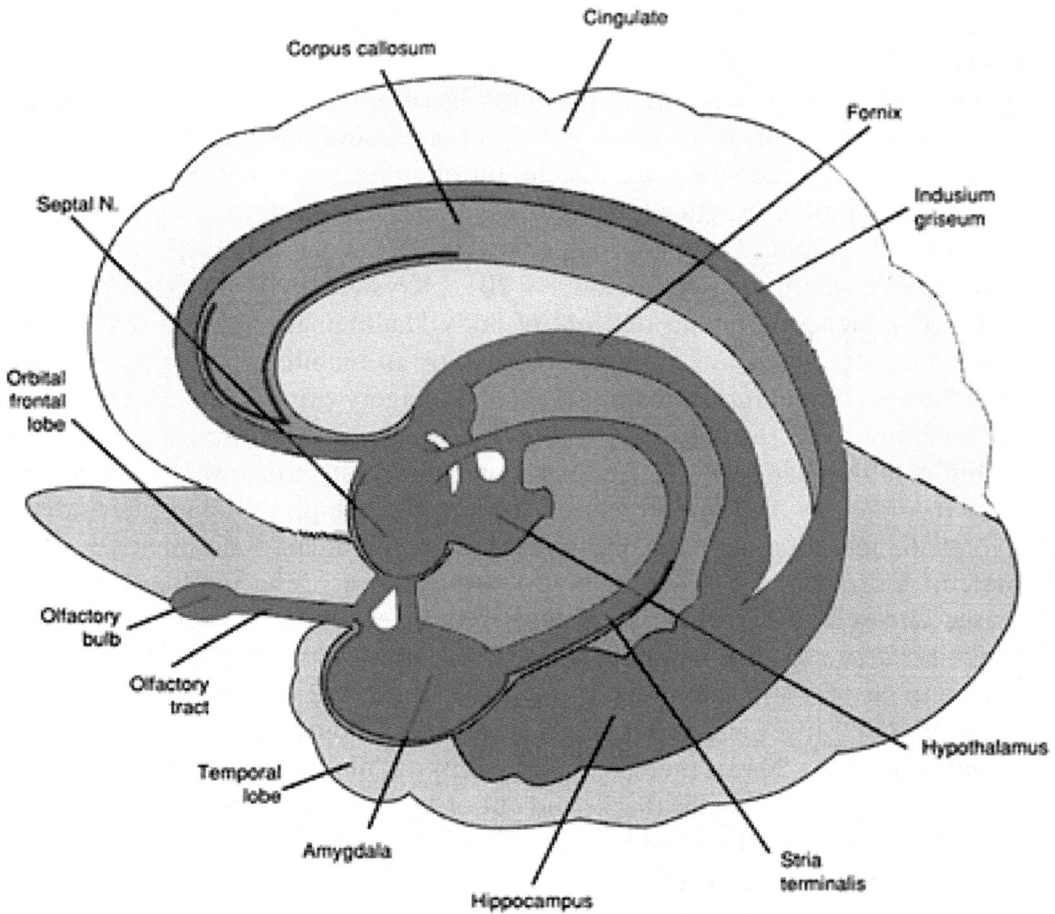

THE NEURAL CIRCUITRY OF STRESS

During periods of extreme emotional stress, pain, and fear, and as part of the "fight or flight" response, the hypothalamic-pituitary-adrenal (HPA) axis prepares the brain and the body for possible catastrophic consequences by secreting large amounts of the amino peptide, corticotropin-releasing factor (CRF) which activates the andenohypophysis which begins secreting ACTH which stimulates the adrenal cortex which secretes glucocosteroids which interact with norepinephrine and other neurotransmitters (Fink, 1999; Moller, et al., 2011). Specifically, the anterior hypothalamus releases CRF, the pituitary releases ACTH, and the outer layer of the adrenal glands secrete glucocorticoids and mineralalocorticoids. The primary glucocorticoid is cortisol (which mobilizes free fatty acids from adipose tissue, breaks down proteins, and increases blood pressure) and the primary mineralocorticoid is aldosterone.

These neurochemical events and stress stress hormones potentiate behavioral and autonomic reactions when confronted with or following bodily injury, and thus provide protective as well as activating influence which enables the organism to continue to function and thus escape or fight for their life. However, cortisol can also injure pyramidal cells, and can decrease the production of lymphocytes by the lymph nodes and thymus gland. Since a primary role of lymphocytes is immunological (e.g. to destroy invading bacteria), excessive cortical secretion can disrupt the immune response thus increasing the likelihood that one may get sick and die.

The neural circuitry of stress begins with peripheral receptors which directly transmit fear, pain, and stress-inducing signals to the brainstem locus ceruleus (LC) and periaqueductal gray, and the thalamus--structures which relay this information to the hypothalamus and amygdala, as well as to the entorhinal cortex (the "gateway to the hippocampus"), and the cingulate and orbital and medial frontal cortex (Fink, 1999; Moller et al.,2011). However, it is the hypothalamus and amygdala, as well as the autonomic nervous system, which play central roles in the stress response including the generation of fear and anxiety.

Specifically, in addition to influencing the HPA axis, the lateral and ventromedial hypothalamus act on the sympathetic (SNS) and parasympathetic nervous system (PNS), respectively. Hence, with lateral hypothalamic/SNS activation, there is an increase in arousal as well as in blood pressure and heart rate--bodily changes which are also mediated by the amygdala. These neural circuits act directly on the brainstem, including the vagus nerve which controls heart rate. In addition, the amygdala projects to and is intimately interconnected with the limbic and corpus striatum (see chapter 16), and through these pathways can trigger running, kicking, punching, and flailing, and thus the motor components of the fight or flight response, including "freezing" and immobilization during periods of extreme fear.

Thus a complex feedback system is maintained via the interaction of these substances and the amygdala, HPA axis and the LC; which when activated and secreted determine the stress response, and which provide protective as well as activating influence which enables the organism to continue to function and thus escape or fight for their life. Unfortunately, under conditions of chronic or repetitive stress, various subcomponents of this complex neural circuitry may be injured, thus disrupting the stress response and in some cases giving rise to profound disturbances of emotion, memory, and personality (Joseph, 2011b). Indeed, under these conditions, neural tissue such as the amygdala and hippocampus may be injured, thereby giving rise to long term maladaptive reactions, such as memory loss, flashbacks, anxiety, and an enhanced startle reaction.

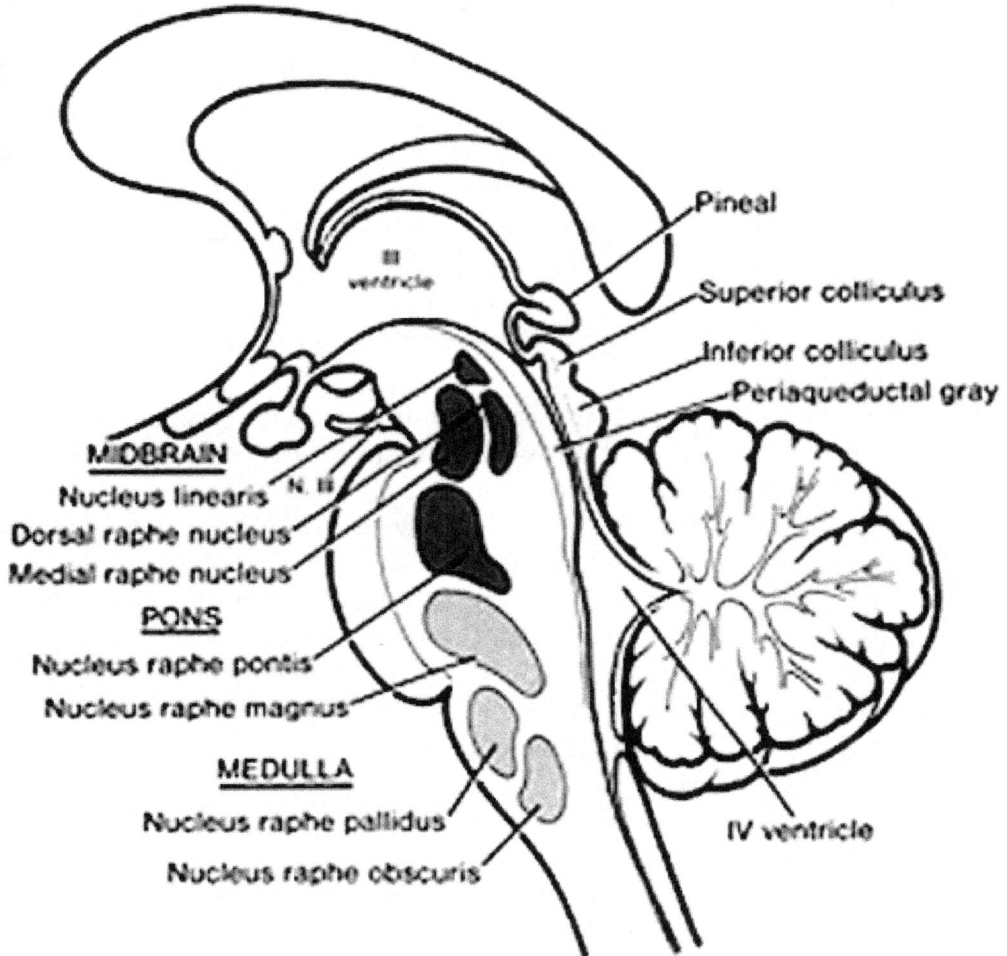

As noted, peripheral sensory receptors project directly to the LC, and under conditions of stress and in reaction to amygdala activation, LC neurons rapidly increase their rate of firing as well as their secretion of norepinephrine (NE). Thus, NE levels initially dramatically increase within the amygdala and hypothalamus (Bliss et al., 1968). Moreover, serotonin (5HT), dopamine (DA), corticosteroid, and enkephalin levels and blood flow to the temporal lobes also dramatically increase (Rosenblum, et al. 2009; Southwick et al., 2013). However, whereas subcortical NE levels initially increase, as stress becomes prolonged, NE (and 5HT) levels and NE receptor densities decrease in the amygdala as well as the hippocampus which exposes these structures to injury secondary to cortisol and enkephalins. In fact, these NE reductions trigger the activation of the HPA axis which results in the increased secretion of cortisol (Lupien & McEwen, 2007). Normally, cortisol secretion is subject to the tonic influences of NE; whereas cortisol can indirectly reduce NE synthesis.

THE AMYGDALA AND FEAR

The amygdala is preeminent in the control and mediation of most higher order emotional and motivational activities (Davis et al., 2007; Fukuda, Ono, & Nakamura, 1987; Gaffan, 2007; Gloor, 2007; Halgren, 2007; LeDoux, 2005; Rolls, 2007; Rosen & Schulkin, 2011). Through massive interconnections maintained with the brainstem, thalamus, hypothalamus, septal nucleus, hippocampus, cingulate, medial forebrain bundle and the temporal, occipital, parietal and frontal lobes (Amaral, Price, Pitkanen, & Thomas, 2007; McDonald, 2007; Mehler, 1980; Russchen, 2002; Turner, Mishkin, & Knapp, 1980), the amygdala is able to sample and affect the perception of auditory, somesthetic, and visual stimuli as well as scrutinize this information for emotional significance (Amaral et al., 2007; O'Keefe & Bouma, 1969; Ono & Nishijo, 2007; Perryman, Kling, & Lloyd, 1987; Rolls 2007; Schutze, Knuepfer, Eismann, Stumpf, & Stock, 1987). This includes the ability to discern and express even subtle social-emotional nuances such as friendliness, fear, love, anger, or threat, and at a more basic level, determine if something might be good to eat.

However, the amygdala also promotes emotional and mood congruent behavior, including flight or fight. It is also becomes highly active when subjects think about or are presented with traumatic stimuli (Rauch et al., 2005, Shin et al., 2007) and mediates the stress response (Henke, 2007). In response to stress, the amygdala (in addition to other limbic nuclei) will secrete massive amounts of opiate-like substances (enkephalins) and will stimulate the secretion of corticosteroids. The amygdala is in fact rich in cells containing enkephalins, and opiate and corticosteroids receptors can be found throughout the amygdala (Atweh & Kuhar, 1977; Uhl, Kuhar, & Snyder, 1978). In this regard, the amygdala is implicated in the generation of extreme pleasure, or it can induce analgesic-opiate-induced numbing if injured during the course of fight or flight.

Extreme fear is the most common emotional reaction elicited with direct electrode stimulation of the human or non-human amygdala (Chapman, 1960; Davis et al., 2007; Gloor, 2007; Halgren, 2007; Rosen & Schulkin, 2011; Strauss, Risser, & Jones, 2002; Williams 1956). The pupils dilate and the subject will cringe, withdraw, and cower. This cowering reaction may give way to extreme panic and the animal will attempt to take flight.

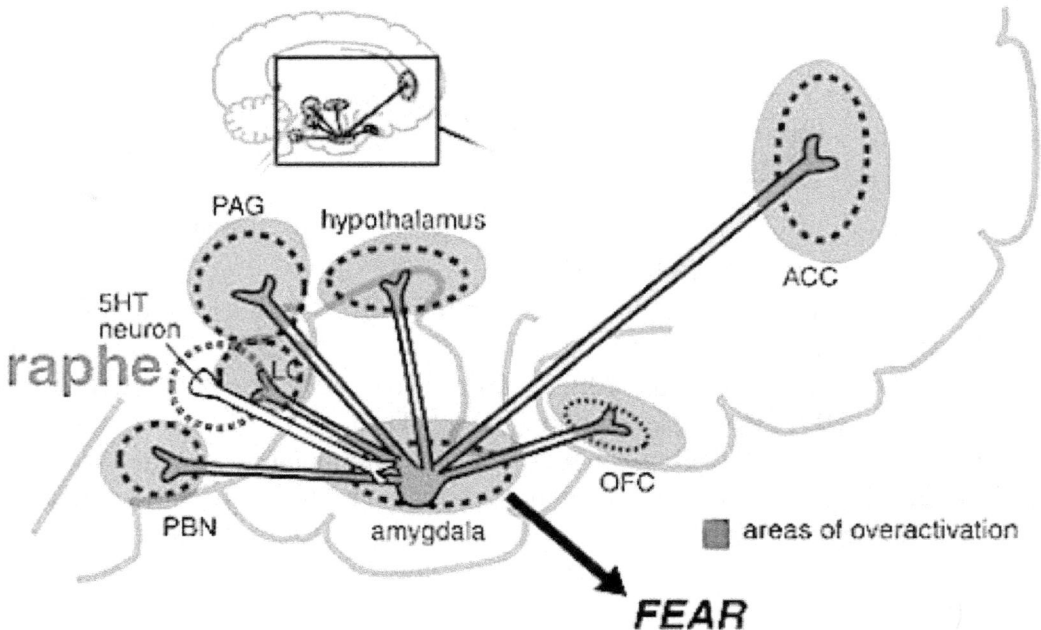

Likewise, abnormal activity originating in the amygdala and/or the overlying temporal lobe can evoke overwhelming, terrifying feelings of "nightmarish" fear that may not be tied to anything specific, other than perhaps the sensation of impending death (Herman & Chambria, 1980; Strauss et al., 2002; Weil, 1956). With amygdala activation the EEG becomes desynchronized (indicating arousal), heart rate becomes depressed, respiration patterns change, the galvanic skin response significantly alters, the face contorts, the pupils will dilate, and the subject will look anxious and afraid (Bagshaw & Benzies, 1968; Davis, 2007; Kapp, Supple, & Whalen, 2009; Ursin & Kaada, 1960).

Unlike hypothalamic on/off emotional reactions, amygdala-fear reactions can last up to several minutes after the stimulation is withdrawn. Moreover, amygdala pathways may remain potentiated for minutes, hours, and even days following fear induced stimulation (Clugnet & LeDoux, 1990). Hence, the amygdala continues to process information in the abstract even when no longer observable (O'Keefe & Bouma, 1969). In consequence, the subject may continue to feel extreme terror long after the threat has been removed. In fact, amygdala-induced fear may be re-evoked even by otherwise neutral cues or stimuli associated with the original threat or trauma (Davis, 2007; LeDoux, 2005).

However, with extreme fear, the individual may not fight or take flight, but they may become literally petrified with fear. Likewise, with amygdala stimulation, there may be a complete arrest of movement (Applegate, Kapp, Underwood, & McNall, 2003; Gloor, 1960; Ursin & Kaada, 1960), and the subject may even briefly cease to breath, as if dead.

FIGHT, FLIGHT, AND FEAR PARALYSIS

Under conditions of extreme fear, victims may enter trance-like states, and become so paralyzed and numb with fear that they may appear catatonic, and may fail to make any effort to resist or to save their lives (Courtois, 1995; Galliano et al., 2013; Miller, 1951; Nijenhuis et al., 2011). Indeed, be it animal or human, a common response to extreme fear is to become motionless, and in the extreme, catatonic as if rigor mortis has set in; also referred to as "death-feigning" or "playing possum" (Kalin 2013; Krystal, 2003; Nijenhuis et al., 2011; Stern, 1951). In fact, the affected individual may become so stiff and rigid, that it may appear as if rigor mortis has set in. Moreover, the victim may not only become unresponsive, but psychologically and emotionally numb (Courtios, 1995; Foa, Riggs, & Gershuny, 1995; Krystal, 2003) and suffer a complete blocking off of cognition and memory such that they may resist and fail to respond even to attempts at assistance (Krystal, 2003).

For example, over 25% of rape victims report body stiffening (Galliano et al., 2013), whereas in air and sea disasters 10-25% of the victims may become frozen, stunned, and immobile (Krystal, 2003). Those afflicted will also fail to take any action to save their lives, such as attempting to evacuate a burning or sinking craft even though they have been uninjured. The airline industry has referred to these fear-induced stiffening reactions as "frozen panic states."

Fear-driven catatonic-paralytic states are prevalent in the animal kingdom (Kalin, 2013; Nijenhuis et al., 2011) and constitute an adaptive, life preserving reaction that appears to be largely mediated by the amygdala, supplementary motor area, and striatum (see below), coupled with the secretion of enkephalins and corticosteroids, and alterations in dopamine and norepinephrine turnover. Indeed, the amygdala not only mediates the behavioral reactions to fear including the fight or flight response (Davis et al., 2007; Gloor, 2007; Halgren, 2007; LeDoux, 2005; Rosen & Schulkin, 2011), but, with direct electrical stimulation the amygdala may become hyperactivated and there is a complete arrest of movement such that even breathing may temporarily cease (Applegate, et al 2003; Kapp, et al. 2007; Ursin & Kaada, 1960). Likewise, under conditions of overwhelming terror--an emotion mediated by the amygdala-- the victim may become catatonic and petrified with fear (Courtois, 1995; Galliano et al., 2013; Miller, 1951; Nijenhuis et al., 2011).

Again, however, under certain conditions fear paralysis may be exceedingly adaptive. Since movement and motion typically alerts a predator to potential prey and thus triggers attacking behavior, lack of movement sometimes eliminates it. Hence, by freezing and not-moving, predators may cease to respond or even fail to note the presence of potential prey. Moreover, unless exceedingly hungry, some predators avoid food that appears to be already dead, though just as commonly the hapless victim is ripped to pieces and readily consumed.

Fortunately, even if this stratagem fails, hyperactivation of the amygdala may induce a numbing reaction thus ensuring a merciful death. Indeed, the amygdala not only mediates the fight, flight, or fear-induced freezing reaction (Applegate et al., 2003; Kapp et al., 2007; Ursin & Kaada, 1960), but under high levels of stress, fear and arousal, it secretes and triggers the secretion of massive amounts of enkephalins as well as corticosteroids and other stress hormones via the hypothalamic- pituitary-adrenals (Hakan, Eyle, & Henriksen, 2009; Roozendall, Koolhaas & Bohus, 2007). In general, these stress hormones and neurotransmitters initially facilitate motor responsiveness and the narrowing of attention. However, high levels of enkephalins and corticosteroids are also associated with numbing, analgesia, a reduction in pain perception coupled with a blocking of cognition and memory loss.

It is the massive secretion of opiates which likely produce the narcotic-like bliss associated with "near death experiences," and the numbing which enables a severely wounded warrior to keep fighting, or a hunted and wounded beast to lie down and calmly allow itself to be eaten alive. Similarly, the trauma or fear-induced release of enkephalins and corticosteroids may account for the psychological numbing reported by victims of catastrophe, or sexual or physical trauma (Courtios, 1995; Foa et al., 1995) as well as the tendency of some victims to become so numb with fear that they fail to take any action to save their life.

Moreover, the continued secretion of corticosteroids and enkephalins are directly associated with the development of immobility, cataplexy, paralysis, and catatonia (Amir, Brown, Amit, & Ornstein, 1981; Fenselow, 1986; Kalin, Shelton, Richman & Davidson, 2011). Catatonia, coupled with emotional and physical (enkephalin-induced) numbing, also represents a total surrender reaction, a complete loss of will, where the victim enters a trance-like state and simply loses the will to resist or to live. Hence, prey may cease to run or fight and may simply stand still or lie down and allow predators to literally eat them alive. Under life threatening conditions, some humans, in fact become so numb with fear that they passively allow themselves to be raped, robbed, assaulted or marched into a ditch and murdered (Galliano et al., 2013; Krystal, 2003). During World War II, tens of thousands of European Jews obeyed orders in an automaton, trance-like fashion, and would take of their clothes, and together with their husbands, wives and children, descend into pits and passively allow themselves to be shot. A fear-induced trance-like paralysis of will, therefore, coupled with emotional and enkephalin-induced numbing and cognitive blocking, gives rise, therefore, a total surrender reaction, usually as a prelude (and hopeful guarantee) of a "merciful" death.

THE AMYGDALA-STRIATUM AND FEAR PARALYSIS

The amygdala is buried within and is contiguous with the anterior-inferior-medial temporal lobe and consists of several major nuclear groups including what has been referred to as the "extended amygdala" (Heimer & Alheid, 1991). The extended amygdala is essentially contiguous with the limbic striatum: the olfactory tubercle, nucleus accumbens, and substantia innominata (ventral globus pallidus). In addition, the corpus striatum (caudate and putamen) forms a bulbous extension of and in fact evolved from the amgydala. The human amygdala arches in a posterior-dorsal-anterior loop and becomes the corpus striatum and is the first portion of the striatal complex to appear during embryological development. The caudate nucleus is in fact dominated by axons from the inferior temporal lobe and amygdala. However, the caudate can indirectly influence and even inhibit amygdala activity; accomplished through the nigrostriatal dopamine system.

In part, the striatum has evolved in order to serve as an emotional-motor interface so that limbic needs and amygdala-triggered impulses may be acted on in a stereotyped and efficient manner so as to facilitate escape, attack, or communication (MacLean, 1990; Mogenson, 1987). In response to amygdala activation, the striatum (including the subthalamic nucleus) may trigger a variety of stereotyped and ballistic motor actions such as running, kicking, and punching, or conversely "freezing" in reaction to extreme fear. Likewise, because all humans possess a striatum and limbic system that developed, evolved, and is organized in an identical manner, when happy, sad, angry, and so on, the facial and body musculature assumes the same readily identifiable emotional postures and expressions regardless of culture or racial origins (Ekman, 2013; Eible-Ebesfedlt, 1995; Joseph, 2013).

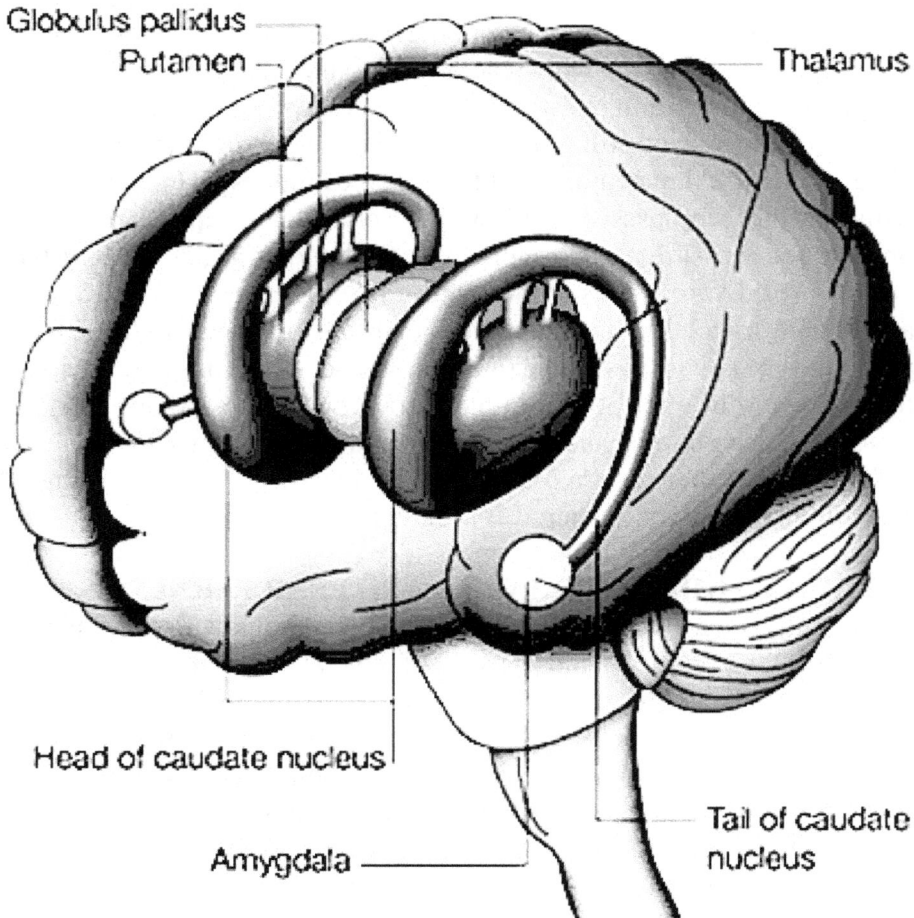

Hence the amygdala and corpus and limbic striatum constitute a functional unit and interact in producing and coordinating gross, or whole body motor activity in reaction to emotional and motivational concerns (Everitt & Robbins, 2007; Heimer & Alheid, 1991; MacLean, 1990; Mogenson & Yang, 1991). However, the amygdala and striatum are also intimately associated with the medial frontal lobe and the supplementary motor areas (Amaral et al., 2007; McDonald, 2007); structures which similarly contain high levels of dopamine. Presumably, it is through its striatal, motor-neocortical and dopamine interconnnections that the amygdala is able to guide, direct, or trigger a variety of emotionally-motivated movement programs, or conversely, a complete cessation of movement, including fear-induced rigidity and catatonia.

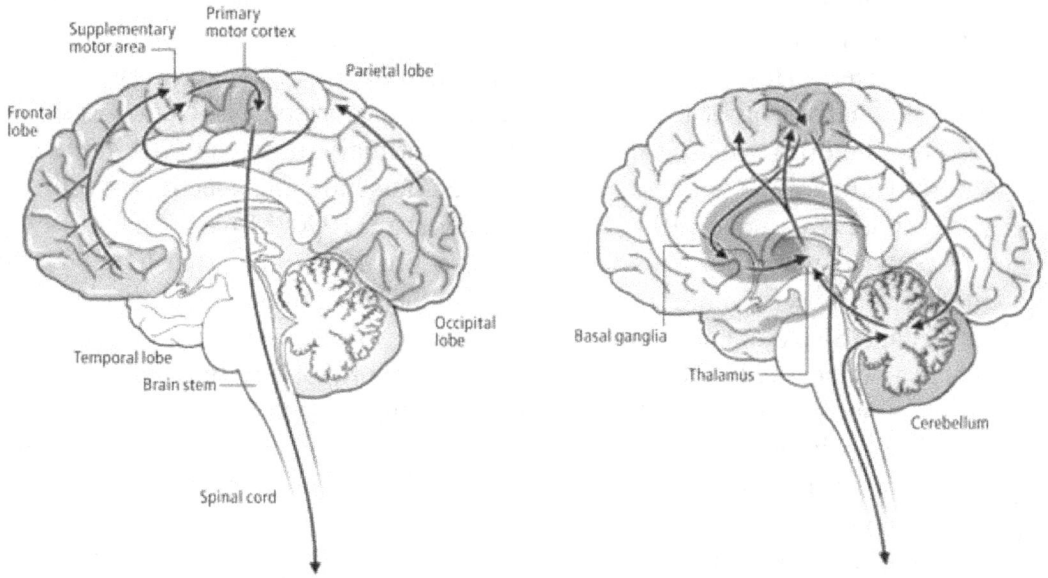

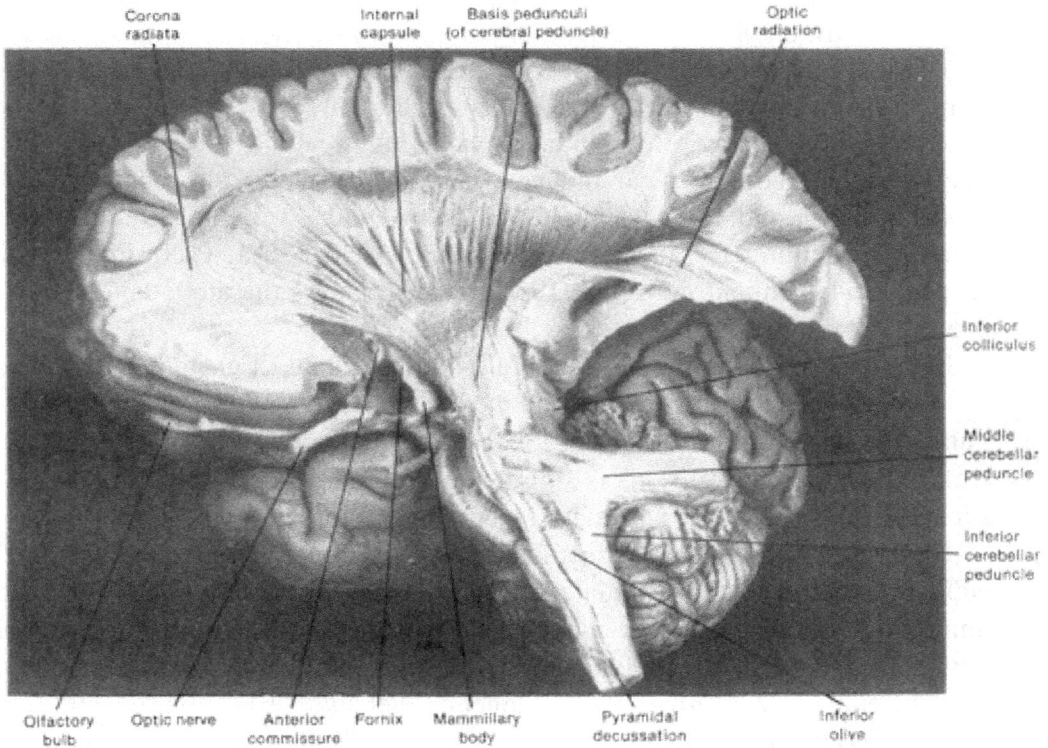

The amygdala, supplementary motor areas (SMA) within the medial frontal lobes and corpus striatum are an integral component of a very elaborate motor-feedback loop which includes the motor thalamus, motor neocortex, and the brainstem (Alexander & Crutcher, 1990a,b; Alexander, Crutcher, & DeLong,

1990; Mink & Thach, 1991; Parent, 2005); motor structures which project to and mutually influence one another as well as rely on the dopamine neurotransmitter system. However, these structures also interact in a step-wise fashion, such that activation of the SMA precedes activity in the association (secondary) motor cortex which is then followed by primary motor activity and finally the activation of the cranial or spinal nerves, thus inducing specific movements (Alexander & Crutcher, 1990a,b; Alexander et al., 1990; Mink & Thach, 1991).

Likewise, when frightened, these areas act in a coordinated fashion so as to mediate purposeful motor movements, with the exception that activation of the amygdala, SMA and striatum take precedence over the primary motor cortex which is concerned with fine motor movements involving the hands and fingers. As noted, the striatum and SMA (including the subthalamic nucleus) are exceedingly important in producing stereotyped motor acts, including running, biting, or ballistically flailing with arms and legs so as to hit, kick, and punch any predator or aggressor.

Petrified with Fear: Catatonia, The Amygdala, SMA, and Striatum

As noted, in response to extreme stress, prolonged terror or depth electrode amygdala stimulation, the subject may cower, run away or become petrified with fear. These behaviors appear to be mediated by the dopamine neurotransmitter system, and the SMA and caudate which are very susceptible to the disruptive influences of stress. In fact, hyperactivation or depth electrode stimulation of these structures can induce a catatonia coupled with a complete surrender reaction and a loss of "will" (Hassler, 1980; Joseph, 1999a; Laplane, Talairach, Meininger, Bancaud, & Orgogozo, 1977; Luria, 1980; Penfield & Jasper, 1954; Penfield & Welch, 1951); that is, unless the amygdala has first been destroyed (Spiegel & Szekely, 1961).

In fact, as based on regional cerebral blood flow and EEG studies that there activity significantly increases in the medial frontal (right caudal anterior cingulate sulcus) and inferior frontal lobes during hypnotic induction suggestion (Rainville, et al., 1999). This is significant in that hypnotic states are also associated with a complete surrender reaction and a loss of "will."

Under stressful conditions or those involving direct electrical or chemical activation of the SMA or caudate, the individual will become rigid, stiff, and unmoving, and will cease to speak or react to questions, threats, or external stimulation (Goto et al., 1990; Hassler, 1980; Joseph, 1999a; Kish et al., 2003; Laplane et al., 1977; Luria, 1980; Spiegel & Szekely, 1961). After the stimulation is removed, patients have reported that their mind essentially had become a blank, a void, and that thoughts no longer entered their head.

Identical disturbances, including the gegenhalten and waxy flexibility of fear-induced catatonia and death feigning, are produced with tumors, strokes, or lesions of the medial SMA and adjoining tissues (Feeman & Watts, 1942; Gasper et al., 1991; Hornykieciz, 2002; Playford et al., 2007). For example, if a physician attempts to move the patient's arm, it will become stiff and rigid (i.e. gegenhalten). Freeman & Watts (1942, pp 46-47) described one individual who developed waxy flexibility, catatonia and related symptoms after a gunshot wound that passed completely through the medial frontal lobes and SMA. "The patient lay in a catatonic-like stupor for two months, always upon one side with slightly flexed arms and legs, never changing his uncomfortable position. He did not obey commands, but if food and drink were given to him, he swallowed them naturally. He was incontinent, made no complaints, gazed steadily forward, showed no interest in anything and could not be persuaded to talk."

However, in other instances (i.e. waxy flexibility), the patient display an inability to resist and will behave like an automaton that is completely under the control of the examiner. If the arm is placed or forced into a new or awkward position, the arm will either remain in that posture or very slowly return to a normal resting position. Such patients might remain in odd and uncomfortable positions for exceedingly long time periods and make no effort to correct the situation. Moreover, they may not respond to painful stimulation.

TRAUMATIC STRESS AND TRANCE STATES

In some respects, waxy flexibility is reminiscent of and magnetic apraxia in which there is also a loss of will and where the individual responds to external stimulation and display a complete inability to resist. Moreover, these behaviors are reminiscent of the loss of will and others behaviors induced while under hypnotic trance.

For example, under hypnosis, a highly hypnotizable subject will display a complete surrender of will, will allow their arms and legs to be placed in odd and uncomfortable positions which they will maintain, and they may be unresponsive to cold and hot stimulation. Moreover, medial frontal (right anterior cingulate) and bilateral inferior frontal activity has been shown to significantly increase among subjects who have been hypnotized, as based on regional cerebral blood flow and EEG studies (Rainville, et al., 1999).

Moreover, these conditions are reminiscent of the trance-like states and he loss of will displayed by those who have been profoundly emotionally traumatized. As noted, uring World War II, tens of thousands of European Jews obeyed orders in an automaton, trance-like fashion, and would take of their clothes, and together with their husbands, wives and children, descend into pits and passively allow themselves to be shot.

It is also noteworthy that individuals who have a history of severe and profound emotional trauma are easily hypnotized (Hilgard, 1974; Nash & Lynn, 1985; Nemiah, 1979; Putnam, 1986). In fact, even "normal" non-abused children tend to be susceptible to hypnosis (London & Cooper, 1969). In fact, because children are also easily hypnotized and can in fact induce self-hypnosis, coupled with this neurological immaturity, they may thus be more at risk for entering trance-like states and developing dissociative disturbances of personality.

In the hypnosis literature there have also been reports of a "hidden observer;" a dissociated portion of consciousness that passively observes all that occurs (Hilgard, Hilgrad, Macdonald, Morgan, & Johnson, 1978; Spanos, deGroot, Tiller, Weekes, & Betrand, 1985). As noted, dissociated "observers" have also been produced during the course of electrode stimulation and seizure activity originating in the temporal lobe. Patients will report that they have left their bodies and are hovering upon the ceiling staring down at themselves (Daly, 1958; Penfield, 1952; Penfield & Perot 1963; Williams, 1956). That is, their consciousness and sense of personal identity appears to split off from their body, such that they experience themselves as two different people, one passively watching, the other being observed. One female patient claimed that she not only would float above her body, but would sometimes drift outside and even enter into the homes of her neighbors. Possibly, if this dissociative reaction is induced under conditions of abuse and during the development of a trance-like state, the dissociated aspect of consciousness may develop as a separate personality.

Hence, traumatic conditions may injure the medial frontal lobes and hippocampus and amygdala-striatum, thus giving rise to trance-like states, a loss of will, and a complete fear paralysis.

DOPAMINE, TERROR, CATATONIA, AND FEAR-PARALYSIS,

Paralysis and disturbances of movement are also associated with profound dopamine (DA) depletion or following neurochemical lesions within the striatum or the SMA (Goto, Hirano, & Matsumoto, 1990; Kish, Shannak, & Hornykiewicz, 2003; Spiegel & Szekely, 1961). The body becomes stiff and rigid, the face may become frozen and mask-like, and the patient may have extreme difficulty initiating speech or responding to questions.

As noted, the striatum, SMA, and amygdala are richly innervated with DA. One of the principles effects of DA is inhibition (Ellison, 2009; Le Moal & Simon 1991), and the amygdala has a higher concentration of DA than the caudate nucleus or SMA (Stevens, 2007). In part, DA acts to modulate amygdala as well as striatal and SMA activity (Amato et al., 2002; Ben-Ari & Kelly, 1976).

Initially, in reaction to fear or emotional (or amygdala) arousal, DA production is increased (Le Moal & Simon 1991). Increased DA production enables the amygdala-striatal-SMA motor centers to filter irrelevant movements or actions and to selectively engage life-preserving ballistic motor acts such as running, kicking, and failing with the extremities (chapter 16).

With continued stress DA levels may continue to increase unabated, thus inducing complete motor inhibition, and/or (in some individuals) DA may eventually become reduced inducing motor hyperactivation and hypertonia (Gasper et al., 1991; Hornykiecz, 2002; Le Moal & Simon, 1991; Playford et al., 2007); i.e. a paralysis of movement. Hence, in either case the organism may freeze, fall to the ground, and cease to move, blink or even breathe--or breathe only shallowly and slowly. The creature therefore appears to be in a state of rigor mortis and thus dead, i.e. catatonic.

Under conditions of prolonged and excessive stress, significant alterations in DA turnover occur within the amygdala and the medial frontal lobes and SMA (Gasper et al., 1991; Hornykiecz, 2002; Playford et al., 2007), thereby disrupting this emotional-motor feedback loop and creating amygdala and striatal hyperactivity (Amato et al., 2002; Ben-Ari & Kelly, 1976); conditions which also promote corticosteroids and enkephalin production (see below).

Loss of DA is directly associated with the development of tonic EMG activity, motor neuron hyperactivity and thus excessive tonic excitation of the musculature, thereby producing limb and facial rigidity (chapter 16). These conditions also afflict those with Parkinson's disease--a disturbance characterized by extreme rigidity. Indeed, Parkinson's disease is associated with a massive loss of DA neurons in the substantia nigra and an 80% decrease in corpus striatal DA (Goto et al., 1990; Kish et al., 2003); disturbances which also induce amygdala hyperactivity (Amato et al., 2002; Ben-Ari & Kelly, 1976).

Individuals suffering from chronically low levels of DA, including those with Parkinson's disease, are sometimes described as excessively aroused and to suffer from heightened autonomic nervous system activity (Stacy & Jankovic, 2007). Moreover, many Parkinson's patients experience not only rigidity and akinesia, but an episodic freezing of movement (Dietz, Goetz, & Stebbins, 1990; Pascual-Leone, Valls-Sole, & Brasil-Neto, 2009), a tendency to easily fall (Stacy & Jankovic, 2007) an impairment of "righting reflexes" (Calne, 2009), and a reduced capacity to blink (Freedman, 2007) and even to breathe (Stacy & Jankovic, 2007); conditions similar to frozen-panic states and fear-induced catatonia. Like those who are severely frightened, Parkinson's patients also sometimes "shake like a leaf," and display a 4-8 c/sec. ("pill rolling") resting tremor which is exacerbated by stress (Freedman, 2007; Stacy & Jankovic, 2007). In addition, just as those who become catatonic and "petrified with fear" experience an eclipsing of cognition, significant cognitive deficits are common among those with reduced DA and/or Parkinson's disease (Freedman, 2007). Thus, DA is a major factor in the catatonic-fear reaction.

Norepinephrine, Enkephalins, Corticosteroids, and Traumatic Stress

The amygdala and limbic striatum are rich in cells containing enkephalins, norepinephrine (NE), and DA. Moreover, opiate, NE, DA, and corticosteroid receptors can be found throughout these structures, especially the amygdala (Atweh & Kuhar, 1977; Ellison, 2009; Hakan, Eyle, & Henriksen, 2009;

Roozendall, Koolhaas & Bohus, 2007; Uhl, Kuhar, & Snyder, 1978). If threatened, frightened, or injured, the amygdala (and other limbic nuclei) become exceedingly aroused and secrete massive amounts of enkephalins (Atweh & Kuhar, 1977; Uhl et al., 1978) and corticosteroids (via the hypothalamic- pituitary-adrenals) as well as NE, DA and serotonin. In general, these stress hormones and neurotransmitters initially facilitate motor responsiveness, selective attention, and inhibit feelings of pain (Bliss et al., 1968; White & Neuman, 1980; Rosenblum et al., 2009; Southwick et al., 2013). However, whereas the initial release of NE and DA promote motor reactivity, the continued secretion of corticosteroids and enkephalins are directly associated with the development of immobility, cataplexy, paralysis, and catatonia (Amir, Brown, Amit, & Ornstein, 1981; Fenselow, 1986; Kalin, Shelton, Richman & Davidson, 2011).

Specifically, as fear and stress levels increase, corticostereoid, enkephalin, NE, and DA turnover are increased, especially in the amygdala (Bliss et al., 1968; Krystal, 1990; Tanaka, Kohno, Nakagawa,2002). However, with continued and prolonged stress, NE (Bliss et al., 1968; Spoont, 2007) and (in some individuals) DA levels may be temporarily reduced, whereas corticosteroid and enkephalin secretion may continue unabated.

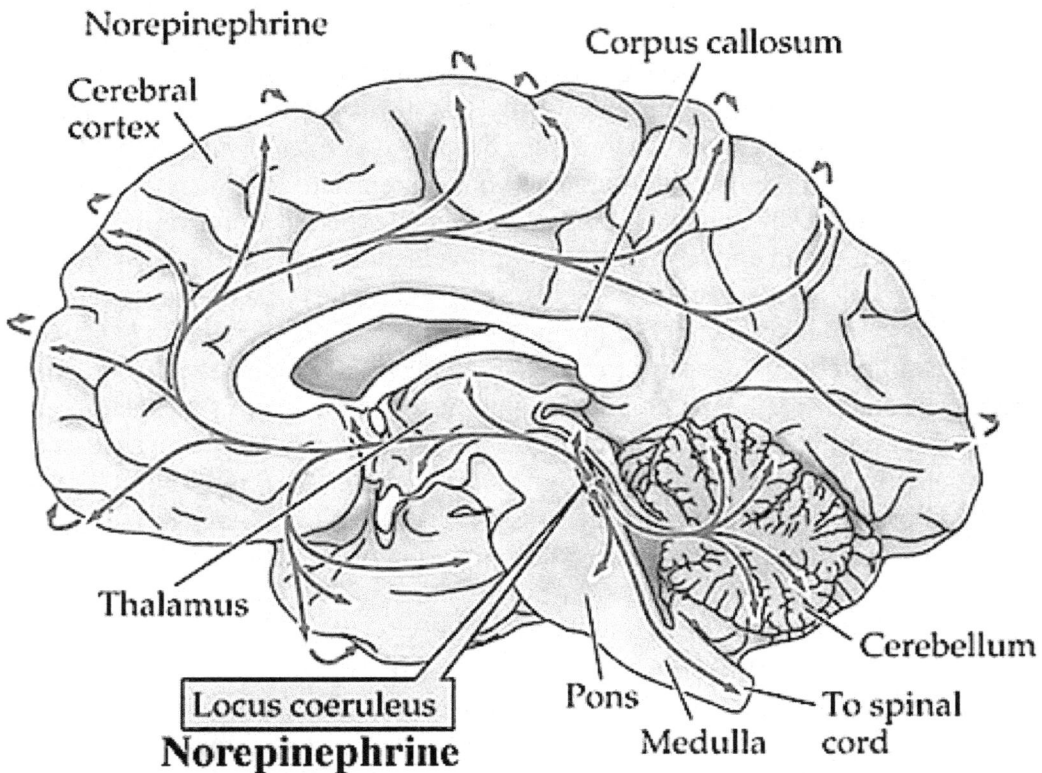

However, whereas corticosteroids increase NE turnover in the amygdala and other forebrain tissues (Dunn & Berridge, 1987), opiates exert an inhibitory influence on the release of NE (Izquierdo & Graundenz 1980) and stimulate corticostereoid production. Hence, as opiate and cortisol production increases, NE (and DA) levels may begin to drop propitiously due to depletion and inhibition, especially in the amygdala (Bliss et al., 1968; Tanaka et al., 2002) thus producing amygdala-striatal-SMA hyperactivation and a catatonic dissociative reaction.

Moreover, increased and high levels of enkephalins and corticosteroids are associated not only with catatonia, but with numbing, analgesia, a reduction in pain perception (Amir et al.,1981; Auerbach, Fornal, & Jacobs, 1985; Fenselow, 1986; Kalin, Shelton, Richman & Davidson, 2011; Roberts, 2004; Spoont, 2007), and the inhibition of hippocampal and pyramidal neural activity (Packan & Sapolsky, 1990; Uno, Tarara, Else, & Sapolsky, 1989; Roozendall, Koolhaas & Bohus, 2007). Thus, at high levels, corticosteroids and enkephalins may disrupt all aspects of cognition and memory functioning and produce a catatonic paralysis of movement.

DISSOCIATION, FEAR & OUT-OF-BODY EXPERIENCES

As the individual becomes increasingly fearful, and then paralyzed and numb with fear, consciousness and memory functioning may be completely, albeit temporarily eclipsed. Although the amygdala is largely responsible for fear-induced catatonic numbing, the hippocampus may play a more significant role in regard to any subsequent loss of memory and alterations in consciousness. For example, in response to high levels of stress, whereas the amygdala becomes excessively active, the hippocampus displays a reduction in activity and may come to be suppressed (Joseph, 2011b, 1999d; Lupien & McEwen, 2007). Moreover, hippocampal suppression appears to contribute to the immobilization response, including the syndrome of "learned helplessness" (Henke, 2007), a condition that may arise following repeated episodes of inescapable stress. The amygdala and hippocampus appear to interact in this regard, with high levels of amygdala arousal initially acting to excite the hippocampus, even producing hippocampal long term potentiation (Henke, 2007), whereas with continued arousal the hippocampus may become hyperactivated only to eventually come to be suppressed which disrupt memory. However, under certain conditions, recollections from the past may be involuntarily produced and give rise to complex hallucinations, including visions of seeing one's self engaged in certain acts.

As is well established, the hippocampus is implicated in the storage of long term memories. The hippocampus (and associated structures including the amygdala) also apparently enables an individual to visualize and remember themselves engaged in various acts. That is, the hippocampus and overlying temporal (and parietal) lobe, apparently enable individuals to visualize and remember themselves engage in various activities, as if viewing their behavior and actions from afar. In fact, it has been demonstrated that the hippocampus enables an individual to remember their surroundings, and contains "place neurons" which cognitive map one's position and the location of various objects within the environment (Nadel, 1991; O'Keefe, 1976; Wilson & McNaughton, 2013).

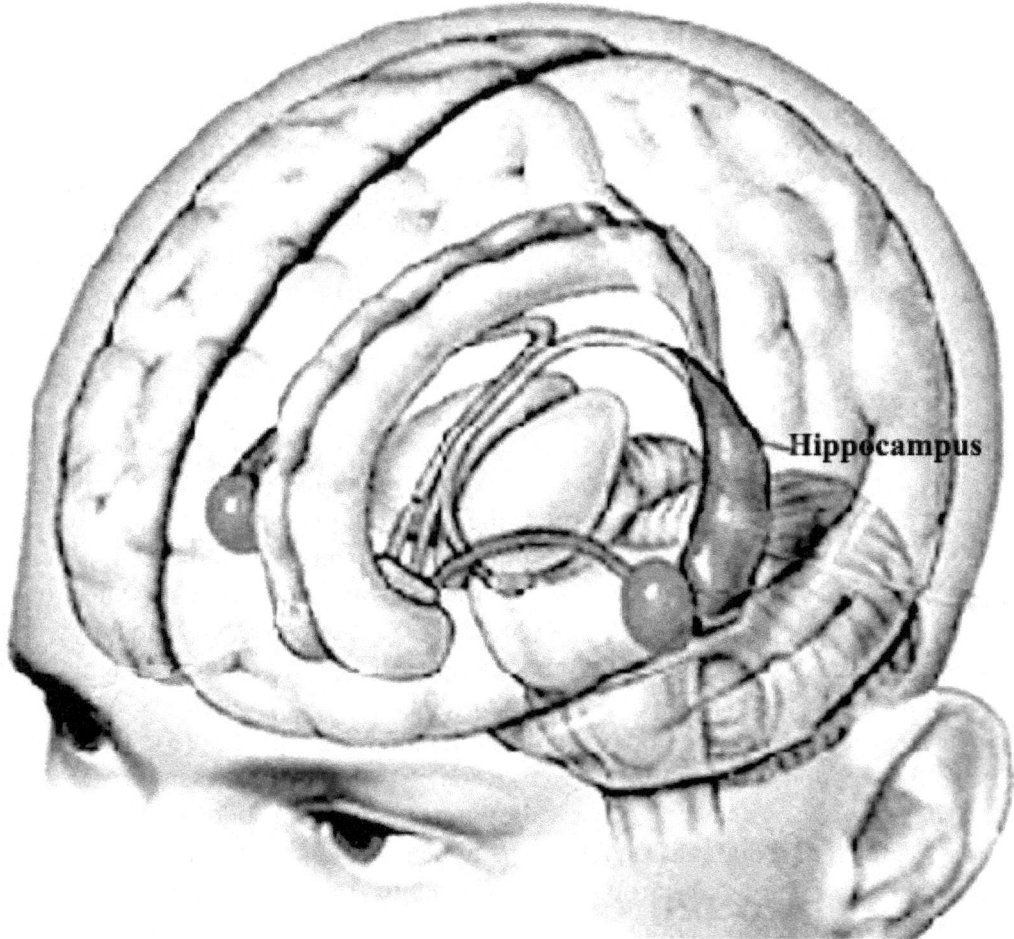

Specifically, O'Keefe, Nadel, and colleagues (Nadel, 1991; O'Keefe, 1976), found that hippocampal pyramidal cells were able to become attuned to specific locations within the environment, as well as particular objects and their location in that environment, thereby creating cognitive maps of visual space. Moreover, as the subject moves about in that environment, entire populations of cells would fire but only when in a particular spot, whereas other cells would fire when in a different location. Moreover, some cells respond not just when moving about, but in reaction to the speed of movement, or when turning in different directions. Moreover, some cells are responsive to the movements of other people in that environment and will fire as that person is observed to move around. (Nadel, 1991; O'Keefe, 1976; Wilson and McNaughton, 2013).

Given these hippocampal attributes, it thus appears that fear-induced alterations or abnormalities in hippocampal-temporal lobe activity presumably triggers the involuntarily visualization of one's surroundings and their personal image similar to what occurs during normal remembering. That is, they experience a dissociated hallucination and see themselves as well as others as if floating above the scene due to abnormal activity in these hippocampal neurons. In fact, hyperactivation of this structure can produce exceedingly vivid hallucinations and dissociative experiences such that the individual may involuntarily see themselves engaged in various activities. For example, during the course of electrode stimulation and seizure activity originating in the temporal lobe or hippocampus, patients may report that they have left their bodies and are hovering upon the ceiling staring down at themselves (Daly, 1958; Penfield, 1952; Penfield & Perot 1963; Williams, 1956). That is, their consciousness and sense of personal identity appears to split off from their body, such that they experience themselves as two different people, one passively watching, the other being observed. One female patient claimed that she not only would float above her body, but would sometimes drift outside and even enter into the homes of her neighbors. Likewise, during episodes of severe traumatic stress personal consciousness may be fragmented and patients may dissociate and experience themselves as splitting off and floating away from their body, passively observing all that is occurring.

Penfield and Perot (1963) describe several patients who during a temporal lobe seizure, or neurosurgical temporal lobe stimulation, claimed they split-off from their body and could see themselves down below. One woman stated: "it was though I were two persons, one watching, and the other having this happen to me." According to Penfield (1952), "it was as though the patient were attending a familiar play and was both the actor and audience."

A patient described by Williams (1956) claimed she was lifted up out of her body, and experienced a very pleasant sensation of elation and the feeling that she was "just about to find out knowledge no one else shares, something to do with the link between life and death." Another patient reported that upon leaving her body she not only saw herself down below, but was taken to a special place "of vast proportions, and I felt as if I was in another world" (Williams, 1956). Other patients suffering from temporal lobe seizures or upon direct electrical activation have noted that things suddenly became "crystal clear" or that they had a feeling of clairvoyance, of having the truth revealed to them, of having achieved a sense of greater awareness such that sounds, smells and visual objects seemed to have a greater meaning and sensibility. Similar claims are made by those who have "died" and returned to tell the tale.

Given that under conditions of extreme stress and fear, the amygdala, hippocampus and temporal lobe become exceedingly active, it is thus perhaps not surprising that some victims will experience a "splitting of consciousness" and have the sensation they have left their body and are hovering beside or above themselves, or even that they floated away (Courtois, 1995; Grinker & Spiegel, 1945; Noyes & Kletti, 1977; Summit, 2003; van der Kolk & Fisler 1995). That is, out-of-body dissociative experiences appear to be due to fear induced hippocampus (and amygdala) hyperactivation.

Noyes and Kletti (1977) described several individuals who experienced terror, believed they were about to die, and then suffered an out-of body dissociative experience: "I had a clear image of myself... as though watching it on a television screen." "The next thing I knew I wasn't in the truck anymore; I was looking down from 50 to 100 feet in the air." "I had a sensation of floating. It was almost like stepping out of reality. I seemed to step out of this world."

One individual, after losing control of his Mustang convertible while during over 100 miles per hour on a rain soaked freeway, reported that "time seemed to slow down and then... part of my mind was a few feet outside the car zooming above it and then beside it and behind it and in front of it, looking at and analyzing the respective positions of my spinning Mustang and the cars surrounding me. Simultaneously I was inside trying to steer and control it in accordance with the multiple perspectives I was given by that part of my mind that was outside. It was like my mind split and one consciousness was inside the car, while the other was zooming all around outside and giving me visual feedback that enabled me to avoid hitting anyone or destroying my Mustang."

Individuals reporting "near death" mental phenomenon may also report exceedingly vivid out-of-body "hallucinations" (Eadie 2007; Rawling 1978; Ring 1980). Consider for example, the case of Army Specialist J. C. Bayne of the 196th Light Infantry Brigade. Bayne was "killed" in Chu Lai, Vietnam, in 1966, after being simultaneously machine gunned and struck by a mortar. According to Bayne, when he opened his eyes he was floating in the air, looking down on his burnt and bloody body: "I could see me... it was like looking at a manikin laying there... I was burnt up and there was blood all over the place... I could see the Vietcong. I could see the guy pull my boots off. I could see the rest of them picking up various things... I was like a spectator... It was about four or five in the afternoon when our own troops came. I could hear and see them approaching... I looked dead... they put me in a bag... transferred me to a truck and then to the morgue. And from that point, it was the embalming process. I was on that table and a guy was telling jokes about those USO girls... all I had on was bloody undershorts... he placed my leg out and made a slight incision and stopped... he checked my pulse and heartbeat again and I could see that too... It was about that point I just lost track of what was taking place.... [until much later] when the chaplain was in there saying everything was going to be all right.... I was no longer outside. I was part of it at this point" (reported in Wilson, 1987, pp 113-114; and Sabom, 2002, pp 81-82).

Approximately 37% of patients who are resuscitated report similar out-of-body experiences (Ring, 1980). In fact, these dissociative out-of-body "hallucinations" are detailed in the 1,300 year-old Tibetan Book of the Dead (the Bardo Thodol) and the Egyptian Book of the Dead which was composed over 6,000 years ago. Nevertheless, be it secondary to the fear of dying, or depth electrode stimulation, these experiences all appear to be secondary to excessive amygdala, hippocampal, temporal lobe activity.

SEX, SATAN, AND ALIEN ABDUCTIONS

Compared to other cortical areas, the most complex, vivid hallucinations, including out-of-body dissociative experiences, have their source in the temporal lobe and the hippocampus and amygdala appear to be the responsible agents. Moreover, the amygdala, hippocampus, and temporal lobe play a significant role in the production of REM sleep and dream activity (Hodoba, 1986; Meyer, Ishikawa, Hata, & Karacan, 1987), and become disinhibited in reaction to hallucinogens such as LSD, and are thus implicated in the production of LSD-induced hallucinations (Baldwin, Lewis, & Bach, 1955; Chapman, Walter, & Ross, 1963; Serafetinides, 1965). Hence, under conditions of extreme emotional stress and trauma, the hippocampus, amygdala and temporal lobe may become abnormally active, and victims may experience extreme sensory distortions and hallucinations, including out-of-body phenomenon.

Due to the development of these fear-induced sensory-distorting hallucinatory states, trees, animals, and inanimate objects may even assume demonic likenesses and/or be invested with satanic intent. These horrible hallucinations and sensory distortions may also be committed to memory. Later, the victim may recall the "hallucination" and believe they were set upon by demons, witches and the like, and/or that they were abused in Satanic rituals, or abducted and painfully probed by demonic aliens.

Consider the Walt Disney version of "Snow White." When the woodcutter, who been ordered to cut out her heart, urged Snow White to flee for her life, she panicked and ran into the darkening forest in near hysteria. And, as she ran and stumbled darting in tears here and there, the trees became demonic, growing eyes and wicked mouths, and gnarled arms and hand which stretched out threateningly toward her. Overcome with terror, she collapsed to the forest floor, sobbing uncontrollable.

Now, perchance, had Snow White later recalled this frightening misadventure, she may well have explained to skeptical listeners that demons had emerged from the forest, and threatened to snatch her away. and, she may truly believe this happened, for it is what she truly experienced and what she now truly remembers. Of course, the skeptics, charlatans, foolish psychologists from academic settings who have never examined a patient, and the defenders of pedophiles and those who abuse children, would have us believe that some evil therapist implanted a "false memory;" even though in fact, there is absolutely no credible evidence that complex "false memories" can be implanted in a normal individual in the absence of drugs, hypnosis, or severe duress. Unfortunately, just as there are charlatans conjuring up evidence for "false memories" there are also exceedingly poorly trained psychotherapists, who are likely to mistake the remembered hallucination as a real Satanic event, or an alien abduction, and may then provide treatment for what they believe to be an otherworldly condition, when, at least in the above instance, this is clearly not the case.

It is noteworthy that many of those reporting alien abductions, have a history of sexual molestation or severe emotional trauma, or temporal lobe epilepsy (Mack, 2009). And, as detailed in chapters 28, 29, and/or above, stress, sexual trauma and sexual activity activates the amygdala as well as the temporal lobe--structures which are associated with the production of complex and exceedingly frightening hallucinations, including those of a demonic, religious, and sexual nature including naked women, demons, ghosts and pigs walking upright dressed as people.

Hence perhaps it is not surprising that some children who were sexually abused, sometimes report that they were subject to bizarre sexual rituals that involved demonic (Satanic) activities (which is not to say that in some cases this may in fact be the case). And perhaps it is not surprisingly that those with histories of severe abuse, and/or who suffer from seizure disorders, may believe that demonic-like aliens lifting them into the air and took them to a room of vast proportions--descriptions which are identical to those of at least some patients during temporal lobe stimulation, and at least some individuals who are severely frightened and/or who died only to return to tell the tale.

Of course many come back to life and can remember absolutely nothing that may have occurred after they died. Loss of memory following extreme stress and severe emotional trauma is not at all uncommon.

STRESS-INDUCED MEMORY LOSS AND AMNESIA

During periods of severe stress, initially the hippocampus may become increasingly aroused and activated, even displaying long term potentiation (Henke, 2007). However, as arousal and stress levels increase, hippocampal functioning becomes increasingly abnormal. In consequence, individuals may dissociate, hallucinate, and experience themselves as having left their body. Victims may also suffer varying degrees of memory loss, and in some cases, an amnesia so profound that even personal identity is forgotten (Joseph, 2011b, 1999d). Indeed, is not uncommon for memory to be disrupted under conditions of extreme arousal, fear, terror, and prolonged emotional stress. In one study, 38% of women who had been examined at a hospital following a sexual assault or after being sexually abused, were found later to have no memory of the assault or their early abuse (Williams 2007). Similarly, victims of rapes and physical assaults tend to provide poorer descriptions of their assailants than those who were victims of robberies (Kuehn 1974).

In both animal and human studies it has been repeatedly demonstrated that as arousal levels dramatically increase, memory progressively deteriorates, such that fewer details are recalled, and an inverted-U shaped learning curve is produced (Christianson, 2007; Conrad, Galea, Kuroda, & McEwen, 2005; Diamond et al., 2007, 2009; Easterbrook 1959; Joseph, Forrest, Fiducia, Como, & Siegel, 1981; Kebeck & Lohaus 1986; Kirchbaum, et al. 2005; Kramer, Buckout, & Eugenio 1990; Newcomber, et al. 2009; Yerkes & Dodson, 1908). Moreover, "memory for information associated with negative emotional events, that is, information preceding and succeeding such events...is found to be less accurately retained" (Christianson,2007, p. 193).

Christianson and Nilsson (2004), for example, found that high states of emotional arousal (as indicated by cardiac activity) was associated with reduced recall and poorer recognition memory when asked to remember names, occupations and so on listed below pictures of horribly disfigured vs neutral faces. Moreover, amnesia persisted even with cued recall but not with cued recognition. Similarly, Eysenck (1979), in his review of studies examining the relationship between performance and arousal level (as measured by anxiety) indicated that those who were highly aroused showed impaired memory and learning on complex and difficult tasks.

In fact, these same U-shaped relationships between memory and arousal have been demonstrated even in regard to non-threatening emotional events and following the administration of arousal inducing drugs, or the injection of stress-related neurotransmitters and hormones (Diamond et al., 2007, 2009; Kovacs, Telgdy, & Lissak, 1976; Lupien & McEwen, 2007; Warembourg, 1975). For example, epinephrine will improve learning and memory at low and intermediate doses but high doses result in an inverted U shaped learning curve such that memory functioning rapidly deteriorates (Berlyne 1960; McGaugh 1989; Gold 2007).

Even so called "flashbulb" memories are subject to considerable decay and forgetting (Neisser & Harsch, 1987). Specifically, Neisser and Harsh had subjects fill out a questionnaire regarding the Challenger space craft explosion. When questioned 32-34 months later, 75% could not recall filling out the questionnaire and many had forgotten considerable detail regarding the accident. According to Neisser and Harsch (1987), "As far as we can tell, the original memories are just gone."

Moreover, the younger the subject when first exposed to this high impact emotional information, the greater is the degree of verbal forgetting and distortion. For example, in examining memory for national traumatic events such as the death of JFK and Robert Kennedy, Winograd and Killinger (2003) found a steep gradient of forgetting which became more profound for memories formed between the ages of 1-7. Hence, for those adults who had been 1-7 years of age when Kennedy had been killed, only approximately 50% of those who had been 4.5 years old or older, could verbally recall the news and provide at least one detail. Those adults who were younger than 3 had no verbal recollections regarding context or associated events or information sources, and only a few who were younger than 5 were able to demonstrate detailed verbal knowledge or memories when questioned as adults.

In yet another study regarding memory for the Challenger space craft explosion, Warren & Swartwood (1987) found that children age 5 were also less accurate and more likely to delete features over time in describing their recollections. When tested two years later, those who had been the youngest as well as the most emotionally upset at the time were less likely to provide extensive narratives and deleted more features.

Children in general, appear to become more easily overwhelmed by high impact emotional events. For example, Peters (1991) found that children who observed a theft (vs those who did not), and who became the most excited and upset (as measured by pulse rate) experienced a considerable degree of memory loss. That is, as pulse rate went up eyewitness recognition memory declined such that those with the highest pulse rate had the most difficulty correctly identifying the face of the thief. This was particularly evident when the children were later placed face to face with the "thief" (i.e. live line up condition), thus inducing high levels of arousal, such that 58% made incorrect identifications.

Given this susceptibility of children to the adverse effects of trauma, in some instances trauma-induced memory loss may in fact result in a complete dissociative amnesia. Consider, for example, Tara Burke who was kidnapped at age-3, held captive in a van for 10 months, and repeatedly sexually assaulted by two men who filmed her ordeal. Despite a subsequent court trial and intense media scrutiny, 18-year old Tara has no memory of her trauma, and continues to display classic symptoms of dissociative amnesia: "It's like a story that has happened to someone else" (Joseph, 2011b). Likewise, at 18-months of age, "Baby Jessica" McClure was trapped 22 feet deep below the Earth's surface in a narrow hole for 58 terrifying and painful hours. Although this incident was internationally televised and subject to intense media attention, as well as one television movie and several books, and despite the skin grafts and amputation of one toe, 10-year old Jessica cannot remember anything of her ordeal (Babineck, 2007).

FUNCTIONAL AMNESIA
Some victims are unable to forget and suffer flashbacks and intrusive imagery. Others may be unable to remember.

Amnesia and memory loss may be secondary or related to a number of different factors, including post traumatic stress disorders, degenerative disturbances and neurological disease or stroke, disconnection syndromes (such as corpus callosum immaturity) as well as head injury, intense fear, emotional shocks and physical traumas, including rape, assault, sexual molestation or the terror of war in which case even hardened soldiers and aviators may lose their memory (Donaldson & Gardner, 1985; Grinker & Spiegel, 1945; Parson, 2003; Southard, 1919). These disturbances have been referred to as "functional" as well as "dissociative" and "hysterical amnesias." In some instances, loss of memory is only for the event itself and may last from seconds to minutes to hours and even days, weeks, and years.

For example, Donaldson and Gardner (1985) describe a woman who was kidnapped and repeatedly raped over a period of weeks. She developed a dissociative reaction and was able to recall only bits a pieces of what occurred until nightmares and flashbacks and therapeutic assistance enabled her to remember.

Fisher (2002) reports that a druggist who had been terrorized, handcuffed and robbed by two thieves subsequently became amnesic and lost all memories of what occurred. In another case, a woman became amnesic for 18 hours after seeing her husband die right before her eyes (Fisher, 2002).

In some cases, particularly where there are predisposing factors, even personal identity may be forgotten. For example, Schacter, Wang, Tulving, & Freedman (2002) describe a 21 year old patient (PN) suffering from functional retrograde amnesia. He had no idea as to his name, address or any other personal information and had been wondering the streets of Toronto until he approached a police officer and was taken to a hospital. It was only when his picture was placed in a newspaper that a cousin recognized him, although PN claimed not to recognize her. According to his cousin, his grandfather had died the week before. However, PN did not remember the death or the funeral or anything about his grandfather who his cousin said he loved dearly. However, "P.N.'s amnesia cleared on the next evening while he watched an elaborate cremation and funeral sequence in the concluding episode of the television series Shogun. PN reported that as he watched the funeral scene, an image of his grandfather gradually appeared in his mind. He then remembered his grandfather's death as well as the recent funeral. During the next few hours, the large sections of his personal past that had been inaccessible for the previous 4 days returned" (p. 524).

Similarly, Christianson and Nilsson (1989) describe a 23 year old female who was raped and beat while out jogging. When found by a police man she had no memory of her identity or that of her relatives, friends, boyfriend, or place of work and all recollection of her past life, including all aspects of the rape were forgotten. However, after approximately two weeks, when she returned with police to the jogging path so as to discover the scene of the rape (which she still could not recall) she became extremely anxious and upset...

and "thereafter, isolated memory-pictures from the traumatic evening started to return in a nonchronological manner, and within a time space of 10-20 minutes she was able to reconstruct the whole episode," (p. 291), including the fact that she was attacked near the crumbled bricks and raped in the nearby meadow.

HIPPOCAMPAL AMNESIA

Stress and arousal-induced memory loss appear to be directly due to disruptions in the functional integrity of the hippocampus and amygdala. Under highly arousing conditions, or following repetitive and prolonged episodes of extreme stress, the hippocampus and amygdala may be injured, and suffer profound dendritic death and atrophy (Diamond et al., 2007, 2009; Lupien & McEwen, 2007; Sapolsky, 2005). These structures may also develop abnormal seizure-like activity (Cain, 2007; Gahwiler 2003; Goelet & Kandel, 1986; Henriksen, Bloom, & McCoy, 1978); all of which significantly impacts learning and memory. Moreover, reductions in the size of the amygdala, and hippocampal dendritic death and atrophy, have been induced by social deprivation and impoverished rearing conditions (Diamond, 1985; Greenough, West, & Devoogd, 1978; Walsh, Budtz-Olsen, Penny, & Cummins, 1969); conditions which are exceedingly stressful and which disrupt learning and memory (Joseph, 1979; Joseph & Gallagher, 1980).

In fact, hippocampal atrophy and memory disturbances have been documented among adults emotionally traumatized in front line combat (Bremner, Randall, Southwick, Krystal, Innis, & Charney, 1995b, Gurvitz, Shenton, Hikama, Ohta, Lasko, & Gilbertson, 2005) and those sexually abused as children (Bremner, Randall, & Scott, 1995a; Stein, Hannah, Koverloa, & McClarty, 1995). Be it animal or human, neglect, abuse, and emotional trauma negatively impacts the hippocampus, the limbic system, as well as learning and memory.

Indeed, it is well established that injury to the hippocampus and overlying temporal lobe is associated with significant memory loss and amnesia (Eichenbaum et al., 2009; Milner, 1970; Reed & Squire, 2007). Moreover, it has been experimentally demonstrated that direct electrical activation of the amygdala or hippocampus can induce a profound amnesia such that minutes, hours, days and even weeks may be erased from memory (Brazier, 1966; Chapman, 1958; Chapman, Markham, Rand, & Crandall, 1967; Halgren, et al., 1991; Jasper & Rasmussen, 1958); memory disturbances which may slowly begin to shrink such that some memory is recovered; i.e. "shrinking retrograde amnesia." In cases of transient global amnesia there is also evidence for temporary mesial temporal inactivation (Hodges & Warlow, 1990), including bilateral hypofusion of the hippocampus (Evans et al., 2013), structures which also become inactivated following the seizure- or electrode induced- postictal anterograde amnesia (Brazier, 1966; Halgren, et al., 1991). Similarly, with bilateral hippocampal ablation patients suffer from a profound anterograde as well as retrograde amnesia which may extend backwards several years in time.

Consider, the famous case of H.M. who can recall almost nothing following bilateral hippocampal amputation (Milner, 1970; Penfield & Milner, 1958; Scoville & Milner, 1957). Dr. Brenda Milner has worked with H.M. for over 25 years and yet she is an utter stranger to him. Every time he discovers his favorite uncle died he suffers the same grief as if he had just been informed for the first time. Nevertheless, H. M. is aware of his deficit and frequently apologizes for his loss of memory. "Right now, I'm wondering" he once said, "Have I done or said anything amiss? You see, at this moment everything looks clear to me, but what happened just before? It's like waking from a dream. I just don't remember...Every day is alone in itself, whatever enjoyment I've had, and whatever sorrow I've had...I just don't remember."

HM   Normal Brain

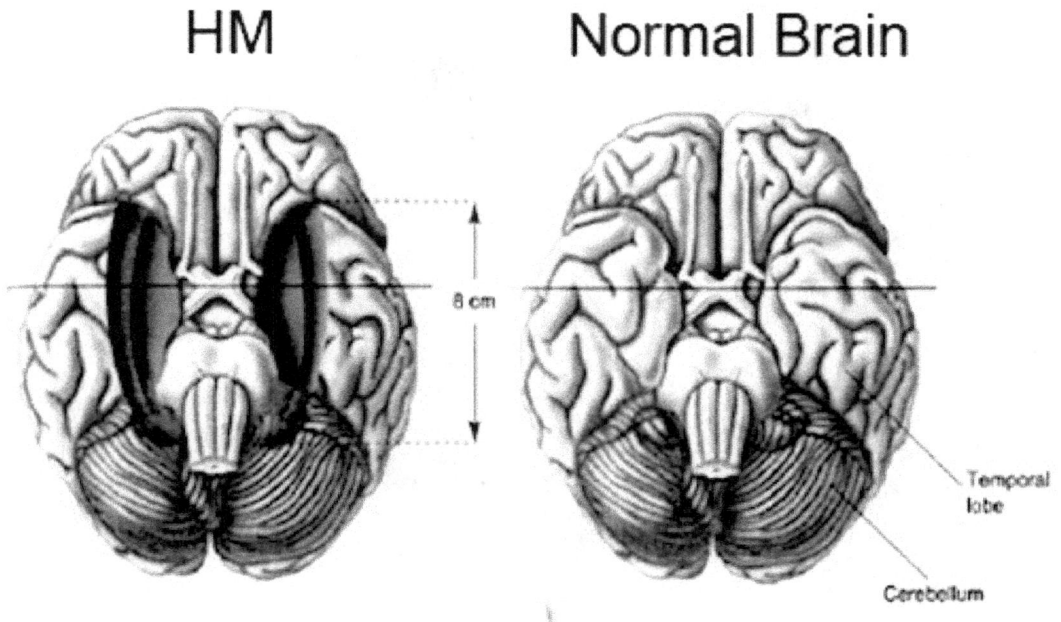

PAL-AMYGDALA INJURY

Stress-induced hippocampal and amygdala hyperactivation and/or dendritic death and atrophy and significant memory loss is directly due to the stress-induced secretion of enkephalins, norepinephrine, and especially corticosteroids (Diamond et al., 2007, 2009; Kovacs, Telgdy, & Lissak, 1976; Lupien & McEwen, 2007; Sapolsky, 2005; Warembourg, 1975), substances which are released as part of the fight or flight response. As noted, prolonged, or repeated episodes of emotional distress or severe emotional trauma can also alter the secretion of NE, DA, and 5HT (Bliss et al., 1968; Rosenblum et al., 2009; Southwick et al., 2013; Spoont,2007; Witvliet 2007). These neurotransmitter fluctuations negatively impact amygdala and hippocampal neurons, axons, dendrites, and their pre and post synaptic substrates).

Specifically under conditions of chronic stress, NE levels may be altered or even depleted (Bliss et. al., 1968; Rosenblum et al., 2009; Southwick et al., 2013; Spoont,2007; Witvliet 2007). Since NE also serves a neural protective function (Glavin, 1985; Ray et al., 1987b), if NE levels are reduced, neurons are exposed to the damaging effects of enkephalins and corticosteroids which at high levels attack and kill amygdala and hippocampal pyramidal neurons (Gahwiler 2003; Henriksen et al., 1978; Packan & Sapolsky, 1990)-cells which normally display synaptic growth and dendritic proliferation in response to new learning (Chapman et al., 1990; Clugnet & LeDoux 1990; Rolls, 1987, 2003).

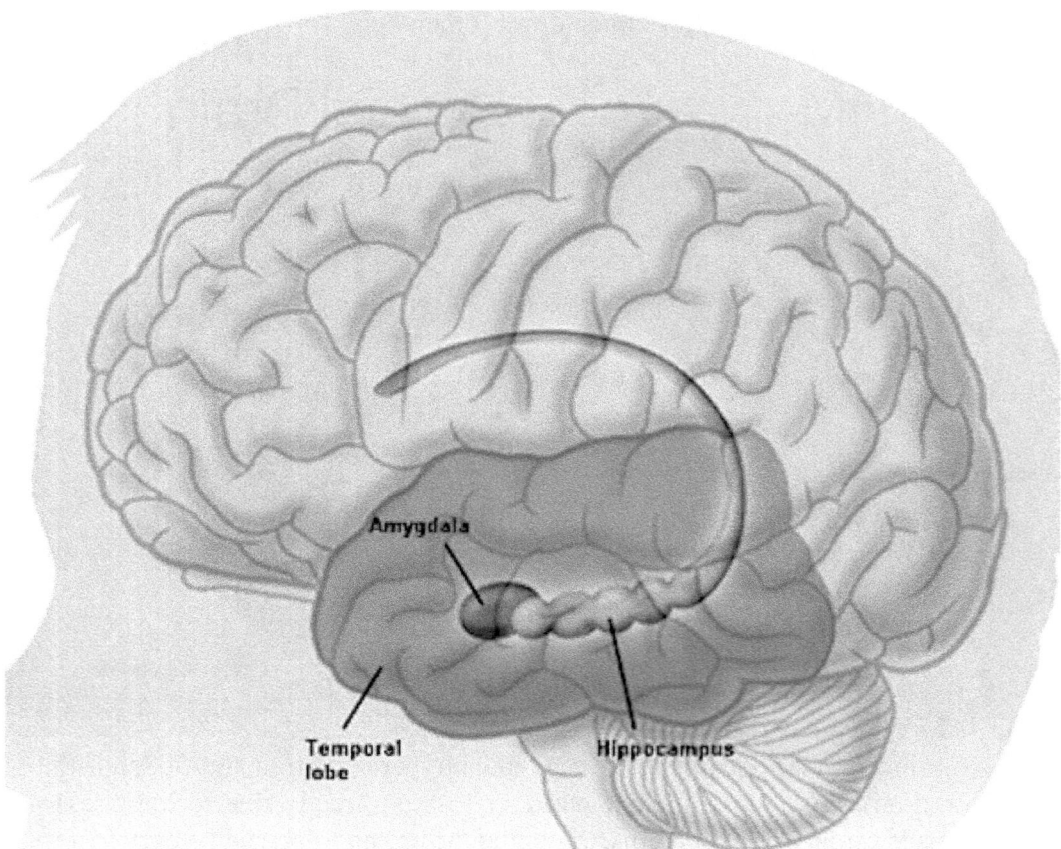

Under high levels of stress, corticosteroids are secreted in massive amounts, but also directly attack and injure the hippocampus due to the abundance of Type II adrenal steroids receptors which abound within this structure (Lupien & McEwen, 2007; Pugh, Fleshner & Rudy, 2007). In addition, coupled with NE depletion, repeated or prolonged stress induced secretory episodes of corticosteroids and enkephalins can injure cells within the dentate gyrus and Ammon's horn (Lupien & McEwen, 2007) such that the hippocampus will atrophy (Lupien & McEwen, 2007; Sapolsky, 2005; Uno, Tarara, Else, & Sapolsky, 1989).

Even if the hippocampus and amygdala are not injured, memory and new learning may be negatively impacted through stress-induced elimination of long term synaptic potentiation (Diamond et al., 2007, 2009; Dubrovsky, et al 1987; Shors et al., 1989; Spoont, 2007). Long term potentiation (LTP) is believed to represent a prolonged form of synaptic activity which binds together those neurons responsible for transferring information from short-term, to long term memory (Lynch, Larson, Muller, & Granger, 1990) and both the amygdala and the hippocampus develop LTP in response to new learning (Barnes & McNaughton, 1985; Chapman et al., 1990; Clugnet & LeDoux 1990; Diamond et al., 2007, 2009; Enbert & Bonhoeffer, 1999; Lynch et al., 1990). In fact, hippocampal LTP tends to build up within the first 30 minutes after learning and during memory acquisition (Lynch et al., 1990), though once this information has been consolidated and placed in long-term storage, LTP apparently ceases to be a factor in memory maintenance. Moreover, the buildup of LTP is accompanied by morphological dendritic changes and the growth of new spines on postsynaptic dendrites (Engbert & Bonhoeffer, 1999).

During periods of extreme fear or arousal, or in the course of high impact emotional learning situations, hippocampal LTP and theta activity disappear (Redding 1967; Shors et al. 1989; Vernderwolf & Leung 2003) and are replaced by irregular electrophysiological activity (Vernderwolf & Leung 2003). In addition, prolonged stress inhibits the development of LTP in the hippocampus (Shors et al. 1989). For example, if a subject is placed in a situation where they are surrounded by painful physical threat, such as an electrified grid floor, theta disappears (Vanderwolf & Leung 2003). In addition under conditions of stress or when the body has been anesthetized (Green & Arduini 1954), hippocampal theta disappears, as does LTP (Shors et al. 1989). In other words, when repetitively stressed and highly emotionally aroused the hippocampus appears to become so aroused that it is essentially deactivated (following the well established inverse U curve of arousal) and ceases to participate in memory formation.

It has also been demonstrated that the stress induced secretion of corticosteroids interfere with the development of LTP (Diamond et al. 2007; Shores et al., 1989; Spoont, 2007) and can thus directly disrupt learning (Diamond et al., 2007, 2009; Kovacs et al., 1976; Lupien & McEwen, 2007; Warembourg, 1975), memory retrieval, including the ability to discriminate between relevant and irrelevant stimuli (Lupien & McEwen, 2007). Even brief episodes of stress-induced enkephalin and corticosteroid secretion can exert deleterious influences, and can eliminate LTP (Diamond et al. 2007; Shors et al., 1989; Spoont, 2007). In consequence, although immediate retention may remain intact for a few seconds or minutes following severe stress, the buildup of LTP and thus long term memory may be significantly disrupted. In fact, cortisol secretion can also disrupt memory retrieval (de Quevain, Roozendall, & Mc-Gaugh, 2011).

In part, the deleterious effects of high levels of corticosteroids on learning, memory, and LTP are due to their suppressive influences exerted on membrane receptor proteins which are detached from their cellular receptor (Beaulieu, 1987). These conditions reduce excitability and information transmission between neurons (Hua & Chen, 1989; Majewska, Harrison, Schwartz, Baker & Paul, 1986) and interfere with messenger RNA protein transcription. Hence, incoming messages cannot be acted on, learning cannot take place, and damaged cells cannot be repaired due to DNA/RNA interference.

Hence, hippocampal memory storage and memory retrieval are significantly disrupted by severe stress and high levels of arousal. However, because the amygdala also play a significant role in emotional memory (Gloor, 2007, 2007; Halgren, 2007; LeDoux, 2005), and as this structure may continue to process affective stimuli even in the absence of hippocampal input (Joseph, 2011b; Seldon et al., 1991), the emotional and traumatic aspects of the experience may thus be memorized. As will be detailed below, the preservation of emotional memory formation likely contributes to flashbacks and the involuntary experience of intrusive imagery.

It is important to emphasize that a single high impact emotional experience is not sufficient to induce complete hippocampal deactivation and the loss of theta and LTP; that is unless the victim is plagued by a predisposition to overreact and/or has been exposed to a previous trauma (see below). That is, victims would not be expected to suffer a complete but only partial memory loss in response to a single traumatic episode. If these stressful episodes are prolonged and repetitive in nature, however, hippocampal theta and LTP may be may be abolished such that hippocampal synaptic depression results (Lynch 1986); a condition that is also related to rapid calcium depletion, for LTP and synaptic plasticity are effected by calcium levels.

OPIATES, CORTISONE & HIPPOCAMPAL AMNESIA

The amygdala interacts with the hippocampus and can depress hippocampal activity during periods of excessive stress. As the amygdala also mediates the stress response, not only does this structure become highly active, but it will secrete corticotropin-releasing factor (CRF) and enkephalins. The amygdala is rich in cells containing enkephalins, and opiate receptors can be found throughout this nucleus (Atweh & Kuhar 1977; Uhl, Kuhar, & Snyder 1978). In response to pain, stress, shock and fear, the amygdala and other limbic nuclei secrete high levels of opiates which can induce a state of calmness as well as analgesia. Under non-painful conditions, the release of opiates can increase feelings of pleasure as well as act as a reinforcer and a reward in response to certain experiences including those which are purely sexual.

However, if opiate activity increases beyond a certain level, they can greatly interfere with memory and learning (McGaugh 1989; Gold 2007), particularly in situations involving pain or terror. This is because opiates and opioid peptides result in hyperactivation of the hippocampus and hippocampal pyramidal cells (Gahwiler 2003; Henriksen, Bloom, McCoy et al. 1978). In consequence, hippocampal epileptiform and seizure activity can develop (albeit in the absence of convulsive seizures) which may be accompanied by abnormal high voltage EEG paroxysmal waves which can last from 15 to 30 minutes (Gahwiler 2003; Henriksen, et al. 1978). In addition, LTP and theta activity is completely abolished as is the ability of the hippocampus to learn and remember (McGaugh 1989; Gold 2007). Hippocampal memory functioning is severely disrupted. Presumably, in consequence, learning and memory is mediated by the amygdala and other nuclei, in the absence of hippocampal input.

In response to continued fear, emotional stress, and anxiety, the amygdala (and other limbic nuclei) also begins to secrete large amounts of the amino peptide, CRF which potentiates the behavioral and autonomic reaction to fear and stress (Davis 2007). Cortisol levels also increase dramatically when the subject is stressed by restraint, and restraint activates the amygdala (Henke, 2007). However, CRF and cortisol has an inhibitory effect on hippocampal functioning and can eliminate hippocampal theta. In consequence, hippocampal participation in memory functioning is again eliminated, altered, or reduced.

In addition to interfering with hippocampal memory formation, the massive secretion of cortisone and enkephalins may create a "state dependent memory" (see Eich 1987 for related discussion). Under conditions of state dependent learning, the subject is best able to recall what is learned when they are in the same state of mind and under the same environmental and drug induced influences. In consequence, state dependent memories may be recalled and retrieved only under certain "painful" or neurochemical conditions that reproduce the original "state", and/or when an individual is presented with contextual cues or experiencing high levels of emotional stress -which in turn is associated with amygdala activation and opiate and cortisone release; i.e. state dependency. However, when stressed, not just the traumatic memory, but any personal reactions (i.e. fragmentary personality features) may also be recalled due to activation of the associated neural network (Charney et al. 2013; Krystal 1990).

Indeed, it is probably exactly these stress-induced abnormalities which explains the dissociative disturbances of memory, hallucinatory-flashbacks, and post-traumatic stress disorders which afflicts some victims of sexual abuse, front line combat, physical assault and injury), and natural disasters and related trauma.

It is important to note that forgetting can be exceedingly adaptive. That is, the individual (or animal) need not be tormented by repeatedly recalling something that is terrifying. However, when these memories are triggered, many individuals feel they are vividly reexperiencing the original trauma--an experience which is not adaptive.

THE AMYGDALA: EMOTIONAL LEARNING AND HIPPOCAMPAL AMNESIA

It has been repeatedly demonstrated that the amygdala is primary in regard to all aspects of emotional functioning including the capacity to experience fear, terror, sexual arousal, or form complex multimodal emotional memories. The amygdala is responsible for the formation and establishment of sexual-social-emotional memories and in fact contributes to the emotional stream of cognitive and perceptual experience, by associating aversive or rewarding emotions to the flow of overall experience including that which is stored in memory (Gloor 2007, 2007; Halgren 2007; Rolls 2007). However, the amygdala appears to act in conjunction with the hippocampus (and temporal lobe) so that emotional memories are stored alongside associated cognitive attributes (Gloor 2007, 2007; Halgren, 2007; Seldon, et al. 1991). That is, in addition to the specific neural networks created by the hippocampus which insures the consolidation and storage of cognitive experience, the amygdala appears to create a semi-independent neural network which stores emotional information (detailed in chapter 14). Moreover, the amygdala appears to exert a steering influence on the hippocampus so as to reinforce and thus insure that certain memories are formed (see chapters 13, 14).

In fact, even when the hippocampus ceases to participate in memory functioning during periods of repetitive and chronic stress, the amygdala may become potentiated and continue to form emotional memories. However, if the amygdala is excessively activated it may develop kindling and abnormal activity, which in turn may contribute to the development of post-traumatic stress disorder (PTSD)--conditions referred to in the past as "soldier's heart," "shell shock," and so on. PTSD is often characterized by an extremely heightened startle response (Charny et al. 2013) which is mediated by the amygdala (Davis, 2007; Hitchock & Davis, 1991), and patients may experience flashbacks, hallucinations, and intrusive imagery.

Krystal (1990, p. 10) argues that "PTSD could be associated with fundamental and long-lasting neuronal modifications, including alterations in neuronal structure and gene expression. This suggests that many PTSD symptoms may have the indelible qualities of long-term memory," which in turn may result in "long-lasting sensitization of alarm systems with transient habituation, with a subsequent reemergence of symptoms when the habituation fades. From the perspective of evolutionary biology, the resistance of traumatic learning to modification probably reflects the fundamental importance of avoiding catastrophic situations at all costs."

TRAUMA CONGRUENT PERCEPTUAL PROCESSING

Individuals who have been severely traumatized tend to function at a heightened level of limbic (amydaloid) arousal. They may also experience intrusive, recurrent, trauma congruent thoughts and memories, or may demonstrate all the signs of PTSD coupled with loss of memory (Grinker & Spiegel, 1945; Terr, 1990). Those who suffer loss of memory, and those who suffer flashbacks, are also more sensitive to any cues which are in anyway associated with the trauma memory. Hence, when exposed (e.g. sounds from the battlefield) emotional and cognitive processing tends to be severely effected, much more so than when exposed to neutral or even positive material. For example, a woman who suffered amnesia after being brutally raped, began crying when she was brought, by police, to an area with a pile of bricks. Later it was determined that she had been raped in a nearby field where there was also a pile of bricks.

These cues strike at the heart of their trauma complex and concurrently arouse a number of congruent associations that must be attended to, as well as defended against. This increased sensitivity is also a function of the amygdala system (and right hemisphere) being essentially primed to receive and selectively respond to this material, or to selectively inhibit and suppress all associated memories (chapter 29).

For example, Vietnam combat veterans with post traumatic stress disorder (PTSD) are slower to name and show more interference when presented with traumatic words (e.g. bodybags) vs negative ("Germs") vs neutral words than those without PTSD (McNally, Kaspi, Riemann & Zietlin, 1990).

Similarly, rape victims with PTSD are slower than those without to name high threat words (e.g. rape) than moderate threat (e.g. "crime") or positive or neutral words (Cassiday, McNally & Zietlin, 2007; Foa, Feske, Murdock, et al. 1991), indicating that these words evoked negative intrusions on cognitive processing, and/or that some effort was being exerted at preventing recall (see chapter 29). They also showed interference for positive emotional words as well as when viewing colors which suggests that overall cognitive efficiency also appeared to be slowed among those with PTSD (Cassiday, et al. 2007).

Given the right hemispheres role in emotional processing (chapter 9), the inhibitory actions of the right frontal lobe, coupled with the mood congruent data reviewed in chapter 29, showing that when in a negative mood and when the right hemisphere is activated, that positive material and memories are recalled and reported less slowly, this research is not only consistent with that reviewed above, but suggests a role of the right hemisphere, and probably the right amygdala and right hippocampus, in PTSD.

The Amygdala & Traumatic Learning

The amygdala (and right hemisphere) is directly implicated in traumatic learning,. Affective memories formed by the amygdala appear to be rather indelible (LeDoux, 2005). Thus even with hippocampal amnesia, the emotion associated with the trauma may be experienced, and all associated feelings recalled, whereas the place and the circumstances around which the trauma was experienced may be forgotten to varying degree (Seldon, et al. 1991). However, because the amygdala is also involved in learning, and when activated can produce vivid hallucinations and traumatic memories (Gloor, 2007), not only the emotion but aspects of the memory, including flashback might be triggered if presented with certain cues.

Specifically, amygdala synapses and pathways develop LTP and display synaptic plasticity in response to fear stimuli (Chapman, et al. 1990; Clugnet & LeDoux 1990; Goelet & Kandel, 1986; Krystal, 1990). The amygdala not only becomes active when learning emotional stimuli, but patients have been reported to experience a variety of exceedingly vivid memory flashbacks, including even memories of physical abuse and sexual intercourse following depth electrode activation (Gloor 1986, 2007, 2007; see also Penfield & Perot 1963).

As noted, LTP is associated with the development of a long lasting increase in synaptic strength and the growth of additional dendritic spines (Enbert & Bonhoeffer, 1999) such that various interacting neurons come to be interlocked, thus forming widespread neuronal networks which in turn presumably corresponds to the establishment of long (rather than short) term memories. Moreover, it is during the course of the first half hour or so after learning that LTP develops thus presumably interlocking the pre and postsynaptic membrane. Hence, LTP within hippocampal and amygdala neurons and neural pathways may correspond to the development of long-term cognitive and emotional memory and the linkage of neurons which subserve and maintain these memories (see chapter 14).

Hence, fear induced LTP within amygdala neurons is probably a reflection of the amygdala's involvement in most aspects of emotional experience, including the formation of cross-modal emotional associations and memories (Gloor 2007; Halgren 2007; LeDoux 2005; Rolls 2007). However, as noted above, under conditions of repetitive and prolonged trauma, although the hippocampus may cease to participate in memory formation, the amygdala continues to function and will produce LTP. In consequence, painful and distressful memories are formed by the amygdala in the absence of hippocampal participation (Seldon et al., 1991).

Even if the hippocampus destroyed, although new learning and associated cognitive attributes may be forgotten, this amnesia does not extend to fear or pain related stimuli (Seldon et al. 1991). In fact, aversive conditioning is completely unaffected (Seldon, et al. 1991), particularly when the conditions are highly negative, painful or emotional as this form of learning is mediated by the amygdala (Cahill & McGaugh 1990; Hitchcock & Davis 1987; Kesner & DiMattia 1987) as well as the hypothalamus (see chapter 13).

However, under conditions of excessive stress resulting in hippocampal amnesia, the victim may remain amnesic such that all associated memories are repressed unless activated by relevant cues. On the other hand, due to the involvement of these structures in dreaming, these troubling memories may be reproduced in the form of fragmented and terrifying dreams (Donaldson & Gardner 1985; Kramer, Schoen & Kinney 1987), that in turn may be forgotten (Kaminer & Lavie 1991; Kramer et al. 1987); that is, forgotten by the left cerebral hemisphere upon waking (Joseph 2003a).

On the other hand, trauma related memories may be recalled when confronted by related stressors, and contextual or emotional cues (Charney et al. 2013). Likewise, flashbacks, and exceedingly vivid affective memories may be reactivated and recalled if the amygdala is aroused by similar contextual cues or if it is directly activated (Gloor 2007, 1999, Halgren 2007; Seldon, et al. 1991), including even memories and fragmentary flashbacks of sexual intercourse or instances of physical abuse and terror (Gloor 2007; see also Penfield & Perot 1963).

Likewise, combat vets with PTSD may have flashbacks involving cries for help, battle scenes, dead and wounded bodies, and so on (Mueser & Butler, 1987; Wilcox, et al. 1991). Presumably those with PTSD continue to experience intrusive images and flashbacks because those amygdaloid neural-memory pathways associated with the original trauma remain primed or in a partially activated state, such that, if activated, aspects of the original trauma-memories are recalled.

In fact, it has been demonstrated that adults with PTSD when asked to recall trauma-memories or when presented with combat-related photographs, display increased blood flow activity in the amygdala (Rauch et al., 2005; Shin et al., 2005). And the amygdala also becomes activated by bizarre stimuli (Halgren, 2007). Hence, the amygdala is directly implicated in the storage and recall of trauma-imagery.

SEROTONIN, FLASHBACKS & INTRUSIVE IMAGERY

The amygdala also becomes particularly active when recalling emotional memories (Halgren 2007; Heath 1964; Penfield & Perot 1963), and in response to cognitive and context determined stimuli regardless of their specific emotional qualities (Halgren 2007). Electrical stimulation of the amygdala (and/or the overlying temporal neocortex) can trigger the recollection of recent and forgotten memories, including incidents involving sexual intercourse, fear, terror, or conversely, joy and happiness (Halgren 2007; Gloor 2007; Penfield &

Perot 1963). In addition, associated emotional, visual, or contextual cues can trigger amygdala activation and the recall of these memories (Rolls 2007; Halgren 2007). However, because associated stimuli can trigger their recollection, in some instances, the recollection of these memories are not only involuntary, but intrusive and may take the form of flashbacks.

These flashbacks may also have a dream-like hallucinatory quality, and only fragments of the original memory may be produced--as is common during dreaming. In fact, the amygdala and hippocampus, and overlying temporal lobe play a significant role in the production of REM sleep and dream activity (Hodoba, 1986; Meyer, Ishikawa, Hata, & Karacan, 1987). These structures also become disinhibited in reaction to hallucinogens such as LSD, and are directly implicated in the production of LSD-induced hallucinations (Baldwin et al., 1955; Chapman et al., 1963; Serafetinides, 1965). Moreover, LSD-hallucinations, may also be later reexperienced as flashbacks. In fact, be they trauma induced or a product of LSD synaptic changes, hallucinations, flashbacks and intrusive imagery, appear to be secondary to similar structural and neurochemical changes in the amygdala and hippocampus--structures directly implicated in the production of complex and exceedingly vivid and emotionally disturbing hallucinations (Gloor, 1990, 2007; Horowitz et al., 1968; Halgren, 2007; Halgren et al., 1978).

Specifically, LSD-induced hallucinations are produced secondary to alterations in serotonin (as well as NE) secretion and uptake within the amygdala, hippocampus and overlying temporal lobe. Likewise, prolonged, or repeated episodes of emotional distress or severe emotional trauma can alter the secretion of serotonin (5HT) and NE (Bliss et al., 1968; Rosenblum et al., 2009; Southwick, Krystal et al., 2013; Spoont,2007; Witvliet 2007). However, 5HT and NE fluctuations not only negatively impact amygdala and hippocampal neurons, axons, dendrites, and their pre and post synaptic substrates (Cain, 2007; Goelet & Kandel, 1986; Kraemer, 2007; Krystal, 1990), but can produce profoundly disturbing, exceedingly vivid hallucinations.

For example, as 5HT exerts an inhibitory influence and induces perceptual filtering so as to narrow the range of sensory input received, the stress-induced depletion of 5HT can induce sensory distortions and dissociative phenomenon including hallucinations similar to those produced by the administration of LSD (Spoont, 2007). LSD acts to inhibit 5-HT production and blocks synaptic uptake within the amygdala, hippocampus, and temporal lobe (chapter 13) which induces hallucinations.

Hence, with 5HT reductions, the traumatized victim may be so overwhelmed by sensory overload that they may hallucinate--sensory distortions which may be forgotten or committed to memory. However, because of 5HT-induced neuroplastic changes and the development of stress-induced kindling and excessive activity in these structures, associated neural networks and the memories they maintain, may be easily activated and the memories recalled, albeit in a dream-like hallucinatory fashion.

Specifically, the main source for limbic and neocortical 5-HT is the dorsal (DR) and median raphe (MR) nuclei (Azamita, 1978; Azamitia & Gannon, 1986; Wilson & Milliver, 1991ab). The amygdala, hypothalamus, basal ganglia, primary and association receiving areas, and the frontal lobe are innervated by the DR, whereas the hippocampus, cingulate gyrus, and septum receive their 5-HT from the MR nuclei (Azamita, 1978; Azamitia & Gannon, 1986; Wilson & Milliver, 1991ab). The MR, however is diffusely organized and appears to exert a non-specific and global influence on arousal and excitability (Wilson & Molliver, 1991ab). The DR is much more discretely organized and can exert highly selective inhibitory or excitatory influences and plays a role in the coordination of excitation in multiple functionally related areas including the frontal lobes and amygdala (Wilson & Molliver, 1991ab). Because of the manner in which they are organized, the DR and MR can exert select inhibitory influences so as to promote perceptual filtering in one or a variety of neocortical and subcortical areas while simultaneously exciting yet other multiple regions which in turn aid selective attention and in the creation of neuronal networks.

Thus, one of the major roles of serotonin (5-HT) is inhibition of neural activity (Applegate, 1980; Soubrie, 1986; Spoont, 2007). It restricts perceptual and information processing and in fact increases the threshold for neural responses to occur at both the neocortical and limbic level. For example, in response to arousing stimuli, 5-HT is released (Auerbach, Fornal & Jacobs, 1985; Roberts, 2004; Spoont, 2007) which acts to aid attentional and perceptual functioning so that stimuli which are the most salient are attended to. That is, 5HT appears to be involved in learning not to respond to stimuli that are irrelevant and are not rewarded (See Beninger, 1989) which are in turn filtered out and inhibited.

In response to highly arousing and stressful, fearful, or painful stimuli, 5-HT (as well as opiate) levels, at least initially are increased which results in numbing, analgesia, and a loss of pain perception (Auerbach et al. 1985; Roberts, 2004; Spoont, 2007). However, under conditions of prolonged stress and fear, 5-HT is rapidly depleted. In consequence, the individual may begin to feel overwhelmed and unable to appropriately process incoming sensory stimuli. Soon thereafter they may also demonstrate a heightened fear and startle response (Davis, 2004) as well as social withdrawal (Raleigh et al. 2003); disturbances directly associated with the amygdala. Moreover, because sensory filtering is reduced, neurons in the amygdala, hippocampus, and inferior temporal lobe cease to be inhibited, which in turn can result in depersonalization (Sallanon et al. 2003) and increased social fear, as well as dream-like states and hallucinations (Sallanon, et al, 2003; Spoont, 2007), including flashbacks. Trauma-memories and associated emotional reactions are reexperienced.

KINDLING, THE AMYGDALA, PTSD, & FLACHBACKS

Opiate receptors appear to be located presynaptically on norepinephrine (NE) cells (Llorens, et al1978), and opiate peptides appear to exert an inhibitory effects on the release of NE (Izquierdo & Graundenz 1980). Stress increases NE turnover in the amygdala (Tanaka, et al. 2002) and increases the production of opiates (Krystal 1990). As noted, the repeated release of these substances, including cortisol and 5HT, coupled with fear induced LTP and synaptic plasticity can in turn significantly and permanently alter the functioning of the amgydala.

When the amygdala is repeatedly and continually stimulated by excitatory neurotransmitters, or via pharmacological agents including opiates, or through direct electrical activation, an abnormal form of neuronal plasticity and a lowered threshold of responding results. These changes are associated with increases in the size of the evoked potential amplitudes and can give rise to epileptiform after discharges, seizures and convulsions, as well as induce kindling (subseizure activity) within the amygdala (Cain 2007; Racine 1978).

Normally, the release of NE retards kindling in the amygdala. However, with repetitive stress and high levels of arousal, opiates will inhibit NE release. This can lead to permanent structural and functional alterations within amygdala neurons, effecting their postsynaptic densities and in the size of the presynaptic terminals as well as their capacity to process and transmit information (Cain 2007; Racine 1978). The amygdala may in fact develop abnormal levels of activity which may take the form of kindling. In consequence, the threshold for amygdala activation and information transmission and recollection may be lowered.

Moreover, even when other brain areas are repeatedly and highly activated, kindling begins to occur within the amygdala, and in fact builds up more rapidly than in any other brain region (Cain, 2007; Racine, 1978). In contrast, the hippocampus is one of the slowest regions to develop kindling, with the anterior hippocampus being more likely to do so than the posterior regions (Cain, 2007; Racine, 1978).

Kindling in the amygdala can potentiate neuronal transmission and can induce neural plasticity, which in turn significantly effects learning and memory and the manner in which kindled circuits transmit non-seizure relative activity and information (Cain, 2007; Racine, 1978). That is, high activity induced kindling can effect the actual creation and functioning of neural circuits as well as their manner of transmitting and processing information, particularly in that it also enhances inhibitory neurotransmission (Cain, 2007). For example, it has been found that kindling induces increases in postsynaptic densities and in the size of the presynaptic terminals (Cain, 2007).

Serotonin Pathways in the Brain

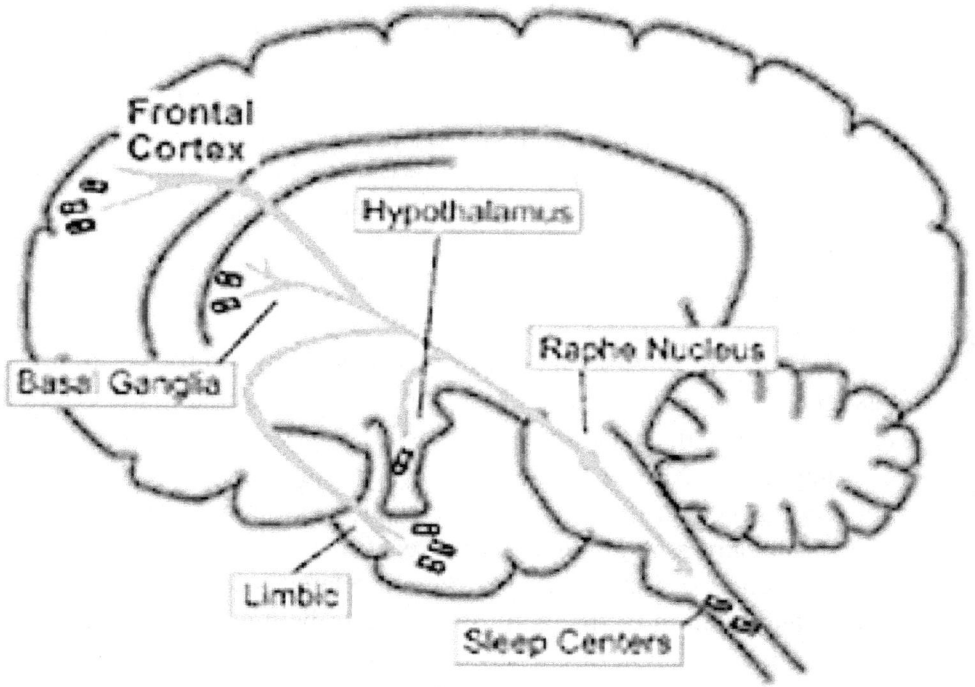

Norepinephrine Pathways in the Brain

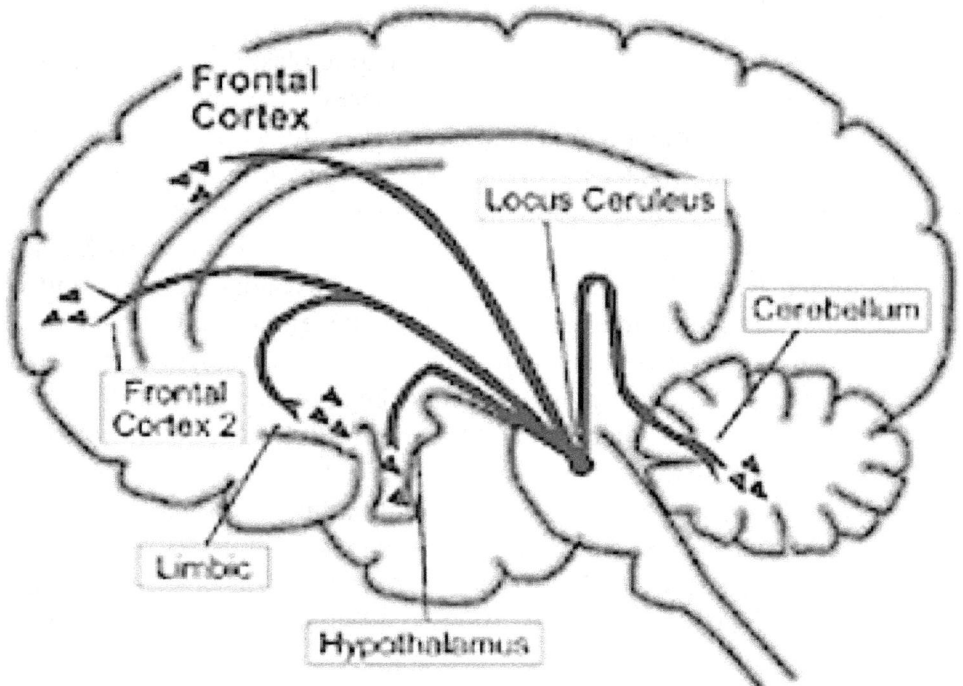

ity, this structure and associated memories and emotional reaction might be easily activated, thus resulting in a flashback. For example, in hearing a car backfire in the middle of the night, a traumatized former vet might experience a flashback, spring from their bed, search for their gun, and/or crawl about the room in search of, or in fear of enemies (Charney et al. 2013; Krystal 1990). In addition to the flashbacks and the original feelings of fear, enkephalins and related substances might also be released in response to the hallucinations and reexperience of fear. That is, once the trauma is experienced and an associated neural network is created, associated feelings and/or the experience of the trauma itself may be later reactivated by associated cues, and not only might the entire circuit of experience be recalled but the body and brain will respond the same as when originally traumatized.

Unfortunately, even benign cues may trigger hyperarousal, traumatic feelings, and even defensive postures in situations unrelated to the original stressful event -even in the absence of obvious stressors or provoking situations (Charney et al. 2013; Krystal, 1990). For example, because an excessively activated amygdala is easily activated (such as due to kindling) the individual may display a heightened startle reaction, and it may periodically trigger the recollection of associated and trauma-related memories due simply to the waxing and waning of this activity or due to benign stressors including even the abnormal secretion of corticosteroids (see below). For example, an overly stressed and hyper-trophied HPA axis may forever hypersecrete corticosteroids which in turn can activate and induce kindling in the amygdala.

Specifically, reductions in 5HT can increase amygdala activation and is associated with the increased secretion of opiates, cortisone, and dopamine. The possibility of amygdala kindling would be enhanced, as well as the development of stress induced LTP and synaptic plasticity. Amygdaloid kindling and reduced 5-HT levels would in turn increase the likelihood of an enhanced startle reaction, social withdrawal, feelings of depersonalization, hallucinatory states, sleep disturbances, increased REM latency, and other associated symptoms associated with a chronic post traumatic stress disorder.

Increased startle, disturbances in sleep onset, increased REM latency and decreased REM, hypersensitivity to noises, hypnogogic and auditory hallucinations, intrusive images and emotions, depersonalization and dissociation, nightmares, as well as difficulty sleeping, are stereotypical features associated with PTSD. These symptoms have been repeatedly noted among combat veterans, concentration camp and holocaust survivors, victims of sea disasters and natural catastrophes, among individual who are physically assaulted, raped or sexually abused as children, including adults who witness the murder of children, relatives, friends, parents, husbands or wives (Charney et al. 2013; Grinker & Spiegel, 1945; Kaminer & Lavie, 1991; Mueser & Butler, 1987; Parson, 19; van Kammen et al. 1990; Van Putten & Emory, 1973; Wilcox et al.

1991). Indeed, the development of these disturbances appears to be a universal reaction to traumatic experiences. They also appear to be directly related to permanent neurochemical (i.e. 5-HT, NE, DA, and opiates) and neurological alterations involving the amygdala and hippocampus.

In fact, even a single trauma may result in alterations in NE and 5HT neurotransmitter turnover and release, which in turn can influence and increase the size of both the pre and postsynaptic substrates (Goelet & Kandel, 1986; Krystal, 1990) and even gene expression. This may induce alterations in learning and memory, and increase emotional reactivity and alarm responses which may persist indefinitely (Charney et al. 2013; Goelet & Kandel, 1986).

SEROTONIN, CORTISOL, DEPRESSION, APATHY AND STRESS

When 5-HT levels are reduced individuals may become depressed. There is also an increased tendency to respond to non-rewarding situations (Sourbrie, 1986) and to continue to respond regardless of punishment (Spoont, 2007). Effected individuals may fail to avoid or may even be drawn to abusive or potentially frightening or traumatic situations, and may seem helpless to alter their behavior, e.g. learned helplessness. Reduced 5-HT has in fact been repeatedly noted in the brains of those who have committed suicide (Brown et al. 2002; Cronwell & Henderson 1995) and among primates who have been repeatedly stressed and who are of very low status. In fact, young primates subjective to severe social stress may display reductions in 5HT levels for up to five years. Conversely, the anti-depressant drug Prozac, acts to increase 5HT levels.

Likewise, individuals with high cortisol levels tend to be passive, become easily depressed, exhibit an "uptight" rigidity in temperment (Johnson et al., 2005; Kagan, Reznick, & Snidman, 2003), and are easily overwhelmed by even minimal stress. Primates and human children with high basal levels of cortisol are less likely to react in an adaptive manner when stressed and are the most likely to freeze-up and passively surrender to their fate (Johnson et al., 2005; Kagan et al., 2003). If sufficiently stressed, they may also die (Johnson et al., 2005; Uno et al. 1989). As noted, cortisol can decrease the production of lymphocytes by the lymph nodes and thymus gland. Since a primary role of lymphocytes is immunological (e.g. to destroy invading bacteria), excessive cortical secretion can disrupt the immune response thus increasing the likelihood that one may get sick and die.

Because the amygdala and HPA axis are abnormally activated, coupled with kindling and the hypersecretion of cortisol and altered levels of 5HT, the individual may continually respond as if stressed even when in a benign environment, and may appear to be chronically depressed, sickly, and easily overwhelmed. It is thus noteworthy that blockage of 5-HT reuptake (which increases the amount of 5-HT in the synapse) acts to reduce the symptoms of post-traumatic stress disorder (Hollander et al. 1990) as well as depression (e.g. Prozac). However, in the absence of pharmacological intervention, the reductions in 5HT and increases in cortisol, coupled with the potentiation of associated trauma-related neural networks may continue to exert significant disruptive influences as if the individual were constantly subject to trauma; and this may include the recollection of trauma-related memories and flashbacks. Even when no longer stressed, the brain continues to respond as if exceedingly emotionally aroused, although they are otherwise severely depressed.

Moreover, because the brain may also hypersecrete enkephalins when in stressful situations, some victims may become addicted to trauma (see Charney et al. 2013; Krystal 1990). They may seek traumatic or highly arousing or dangerous experiences so as to induce the release of naturally occurring opiates. They become addicted to trauma, addicted to drugs, or addicted to unpleasant interpersonal relationships (Charney et al. 2013; Krystal 1990). Iin fact, individuals who have been repeatedly traumatized and those with PTSD, not uncommonly abuse opiates and other substances.

KINDLING & PSYCHOSIS

Amygdala kindling can also have a direct effect on personality including increasing social withdrawl and defensiveness (Cain, 2007) as well as increasing sensitivity to dopamine. It is noteworthy that the amygdala receives extensive dopamine projections and has a higher concentration of dopamine than the caudate nucleus (Stevens, 2007). These amygdala dopamine neurons project to wide areas of the neocortex, including the frontal lobes and are believed to play an inhibitory role on neocortical information processing, which in turn is important in selective attention.

In this regard, under conditions of continual high level emotional stress and arousal, opiates may be released, amygdala neural circuitry can be altered, the hippocampus may be damaged, NE production can be suppressed, and dopamine sensitivity and dopamine levels can be increased, particularly within the amygdala (see Krystal, 1990). Hence, not only might emotional responsivity be altered but wide areas of the neocortex can be abnormally effected due to alterations in amygdaloid neurochemical and dopamine functioning.

Altered dopamine sensitivity and transmission has long been associated with the development of schizophrenia and psychosis. Specifically, individuals with schizophrenia contain a higher density of dopamine receptors within the left half of their brain (Stevens, 2007), as does the normal left amygdala (Bradbury, et al. 1985). In fact, the left amygdala and left temporal lobe (Flor-Henry, 1969; Perez, et al. 1985; Stevens, 2007) have long been thought to be a major component in the pathophysiology of psychosis and schizophrenia (Heath, 1954; Stevens, 1973; Torey & Peterson, 1974). For example, abnormal activity as well as size decrements have been noted in the left amygdala (Flor-Henry, 1969; Perez, et al. 1985) as well as the left inferior temporal lobe in schizophrenic patients (see Stevens, 2007). Spiking has also been observed in the amygdala of psychotic individuals who are experiencing emotional and psychological stress (see Halgren, 2007). Thus there appears to be an abnormal concentration of dopamine in the left hemisphere and temporal lobe among psychotic individuals, which, in some cases may be secondary to repeated traumas experienced during childhood.

Given that stress increases amygdaloid activity, induces kindling, and increases the production of opiates and dopamine, this may explain why psychotic disturbances can be exacerbated when patients are subject to emotionally arousing conditions. Moreover, given that stress and emotional arousal also alters serotonin activity, and given the role of this transmitter in sensory filtering, emotional inhibition, dream production and hallucinatory states, the likelihood of severe psychological abnormalities is tremendously increased, particularly in those who may be predisposed to develop these disturbances due to adverse early environmental and childhood experiences.

STRESS-INDUCED PSYCHOSIS & PERSONALITY MULTIPLICITY

Dramatic alterations in personality following injury to the amygdala (and overlying temporal lobe) are not uncommon (Lilly et al., 2003; Marlowe et al., 1975; Shenk & Bear, 1981; Terzian & Ore, 1955). Indeed, damage to these structures, the amygdala in particular, can in fact abolish all aspects of a social personal identity (Lilly, Cummings, Benson, & Frankel, 2003; Marlowe, Mancall, Thomas, 1975; Terzian & Ore, 1955), and erase from the mind memories of the personal, social, and cognitive self. In cases of amygdala injury or excessive or epileptic activity, the patient may become aggressive and violent, or they may become withdrawn, lose all interest in social activities, and experience significant alterations in sexual drive. Some become impotent whereas others experience an increase in sexual feelings, and then act on it in an inappropriate and indiscriminate manner. They may masturbate and expose themselves in public, seek sex with family members, or experience alterations in sexual orientation and engage in intercourse with members of their own sex or even animals (chapter 28; Mesulam, 1981; Shenk & Bear, 1981).

In some cases of temporal lobe, amygdala, hippocampal abnormality, the alterations in personality are so dramatic patients may appear to be suffering from a multiple personality disorder (Fichtner, Kuhlman, & Hughes, 1990; Mesulam, 1981; Shenk & Bear, 1981). In fact, in several cases of multiple personality dissociative disorder, EEG or blood flow abnormalities involving the temporal lobe have been demonstrated (Drake 1986; Fichtner et al., 1990; Mathew, Jack & West, 1985; Mesulam 1981; Schenk & Bear 1981). Moreover, some patients may shift from one personality to another following a seizure or with increases in temporal lobe activity (Mesulam, 1981; Shenk & Bear, 1981).

Similarly, it has been reported that heightened emotional distress may precede the appearance of alternate personalities (Greaves 1980; Putnam 1985); and as noted, stress can also abnormally active the amygdala and increase blood flow to the temporal lobe.

As per the development of additional personalities, as detailed in chapter 2, almost all humans are capable of displaying different personalities depending on context and situational restraints, e.g. first date, at work, etc., And almost all humans are capable of shifting between and extinguishing these alternative personalities at will. However, children are also capable of generating make-believe personalities, such as imaginary friends which they might confide in or blame for their own misdeed. In this regard, children may be more susceptible to developing psychotic disturbances involving the creation of multiple person-alities. In fact alternative personalities are generally formed during childhood and during periods of extreme stress involving sexual abuse and torture (Bliss 1980; Coons, Bowman & Milstein 2003; Coons & Milstein 1986; Devinksy, Putnam, Grafman, Bromfield & Theodore 1989; Kluft 2004; Mathew, Jack, & West 1985; Putnam 1985; Putnam, Guroff Silberman, et al. 1986; Rosenbaum & Weaver 1980; Schreiber 1974; Wilbur 2004).

Specifically, it appears that these alternate personalities may be formed dur-ing a dissociative state during a period of severe and repetitive trauma. That is, under certain traumatic (or neurological) conditions, an aspect of the con-sciousness may fragment, break off, and thereafter variable act in a semi-inde-pendent or completely independent manner. This form of dissociation may be likened to a disconnection syndrome and would be due to abnormal activation of the hippocampus, temporal lobe, amygdala, which induces an out-of-body hallucination as well as a loss of will (due to medial frontal involvement).

As detailed above, it has been repeatedly demonstrated with neurosurgical and epilepsy patients that abnormal temporal lobe/hippocampal activation can cause an individual to experience themselves as separating and floating above their body, such that they may feel as if they are two different people, one watching, the other being observed (Daly 1958; Jackson & Stewart, 1899; Pen-field 1952; Penfield & Perot 1963; Williams 1956). As described by Penfield (1952) during electrical stimulation of this area, "it was as though the patient were attending a familiar play and was both the actor and audience." Likewise, children, as well as adults, who are terribly abused or traumatically stressed and frightened often report dissociative experiences including even a splitting of consciousness such that one aspect of their mind will seem to be floating above or beside their body (Courtois, 2003; Grinker & Spiegel, 1945; Noyes & Kletti, 1977; Parson, 2003; Southard, 1919; Terr, 1990). Hence, the famous aside: "He was literally beside himself with fear."

Most abusers do not wish to molest an unresponsive body, but a personality that is either excited or tormented and terrified by what is occurring. According to Courtois (1995), it is the abusers desire "to achieve total control over the victim and her responses and for the victim to become a willing participant and to enjoy the abuse. To achieve this end, many incest offenders take great pains to sexually stimulate their victims to arousal." They may pinch, cut, and hurt them to obtain the desired reaction. Thus, although these particular children may have "split off" from their body, they are nevertheless forced to respond, as a personality, to the abuse in which case the "personality" may act in accordance with the manner in which it was created. Thus, the abuser slowly shapes a personality which is formed during period of abuse. Hence, some alternative personalities are highly sexual, or homosexual, or angry and self-destructive, mirroring either the role they were forced to play or the personality and behavior of the perpetrator. The dissociated aspects of conscious-awareness, therefore, may slowly establish its own independent identity; that is, an alternate personality--a personality which may be repeatedly triggered or purposefully shaped during subsequent traumas, and which may grow only to split apart again and yet again, forming multiple fragmentary personas each supported by a dissociated and abnormal neural network.

However, these "broken off" or alternate personalities, in turn, may be "state dependent" and supported by isolated neural networks maintained by the amygdala in the absence of hippocampal participation. Because the alternate "personality" essentially splits off from the main personality under certain highly stressful, in general, it can only be reactivated and function independently when the individual is stressed or in a similar state of mind. Moreover, because of the state dependent nature of this partial "personality," the main personality essentially becomes amnesic regarding it. In part this is a consequence of hippocampal deactivation. When the hippocampus returns to normal, it can no longer gain access to these personal memories (which may be maintained by the amygdala), such that dissociated personal memory and the associated, albeit disconnected personality, remains isolated and inaccessible.

INDIVIDUAL DIFFERENCES AND PREDISPOSING FACTORS

The manner in which one reacts to extreme stress or prolonged emotional trauma varies depending on age, gender, situational, individual, and neuro-chemical differences (Bremner et al., 2013; Johnson, Kamilaris, Carter, Calogero, Gold & Chrousos, 2005; Kalin et al., 2011). Not all individuals respond the same when stressed. When traumatized some individuals are unable to forget and continually think about the trauma and experience flashbacks and intrusive imagery. Some animals and humans respond with only minimal degrees of arousal or fluctuations in learning and memory, whereas others become easily overwhelmed and/or display profound memory loss or reductions in learning with even minimal stress (Como, Joseph, Fiducia, & Siegel, 1979; Johnson et al., 2005; Joseph, 2011b, 1999d; Joseph et al., 1981; Joseph, Hess, & Birecree,

2003; Joseph & Gallagher, 1980; Kalin et al., 2011).

For example, females may be more at risk than males for becoming highly stressed (Compas et al., 1989; Coons 2003; Cleary, 1987; Larson, & Ham,2013). Females are more easily frightened (Campbell, 1999; Cantor & Reilly, 2002; Sparks, 1986; Tamborini & Stiff, 1987), and cry easier than males, (Lombardo, Crester, Lombardo & Mathis, 2003; Oliver, 2013; Shields, 1987; Thomas, 2013). Females will begin crying and becoming upset even under conditions that seem affectively neutral (Oliver, Sargent & Weaver, 2011) and those which should engender joy, such as weddings. They are also two to three times more likely than men to develop mood disorders such as depression (Nolen-Hoeksema, 1990) and stress related abnormalities such as anxiety and phobias (DSM-IV), and appear to become more depressed than males (Cleary, 1987; Wichstrom, 1999). According to Cleary (1987, pp. 41, 49), "virtually all surveys find that women report more symptoms of distress than men," and that "women are more likely than men to experience symptoms of depression and anxiety."

Moreover, the younger the victim and the more profound or repetitive the trauma, the more likely may be the subsequent loss of memory (Joseph, 2011b, 1999d; Terr, 2009) and the more likely is it that they may develop post traumatic emotional and personality disturbances (DSM IV, 2009). The most severe forms of dissociative disorder are often directly linked to childhood trauma (DSM, 2009).

Children may be more at risk, simply because they are children and more vulnerable to repetitive instances of abuse and torture. For example, it has been estimated that of those who were sexually molested, approximately 29-31% were physically or aggressively forced and were pinned or held down, whereas from 2% to 5% were subject to substantial violence such as hitting or slapping or beating (e.g. Courtois, 2003; Russell, 1986). The amygdala becomes especially active when the body is restrained, as well as during sex and in reaction to pain. Penile erections, ovulation, uterine contraction and lactogenetic responses, and vaginal-vulva secretions in conjunction with orgasm and sexual feelings in the genital areas are all associated with heightened activity in the amygdala. (Halgren, 2007; Jacome, et al. 1980; Gloor, 1986; Remillard, et al. 2003; Robinson & Mishkin, 1968; Shealy & Peele, 1957). However, because of the susceptibility of the amygdala to develop abnormal kindling activity in reaction to severe stress, early abuse may more severely injure the amygdala due to the repetitive and vicious nature of the abuse.

In addition, children often respond to sexual abuse by pretending that it is not taking place (Courtois, 2003; Summit, 2003), such as by "playing possum." Many victims will lie in bed in a state of frozen fear and hypervigilance (Courtois, 2003)--which is associated with excessive amygdala activity. It is also not unusual for children to initially hyperventilate and then to hold their breath and pretend they are sleeping while the abuse occurs. Not only might they may be denied oxygen due to purposeful breath holding, but some may be denied sufficient oxygen by having an adult lie across them. In some cases, sexually abused children suffer compression injuries to the chest and will die from lack of oxygen. Moreover, some sadistically abusive adults will hold their hands over the child's mouth as a form of torture or so as to prevent them from crying out while they are being assaulted.

The amygdala, hippocampus, temporal lobe neural complex are highly susceptible to hypoxia, which in turn can result in loss of memory, or conversely create hallucinatory and dissociative states as well as injury. This is particularly true during childhood as the brain is still immature and plastic and consumes almost 50% of total body oxygen and glucose, compared with 25% in adults. Hence, whereas the amygdala may be highly activated due to sexual activity coupled with feelings of fear and terror, in some cases it (and the hippocampus) may be simultaneously denied sufficient oxygen which it requires even more than is usual (because it is highly active). In consequence, with repeated instances of abuse, coupled with oxygen depletion, the still immature temporal lobes and amygdala/hippocampal complex might be injured by these experiences and forever function abnormally.

In some cases, memory loss and/or the reaction to stress and the development of PTSD may be influenced by previous traumatic experiences. Bremner and colleagues (2013), for example, have shown that at least 26% of Viet Nam veterans with PTSD admit to a history of childhood abuse whereas only 7% of those without PTSD had a similar history. In another case described above, a 23-year old female who had been sexually abused s a child, lost all memory of her identity, her past life, or that of her relatives, boyfriend, or place of work after she had been accosted while jogging and brutally raped (Christianson & Nilsson, 2004).

In addition to previous emotional trauma, some individuals are predisposed due to underlying neurological disorders. In a case of profound dissociative amnesia described above, a 21-year-old man who had forgotten his name and all other personal information following the shocking death of his beloved grandfather, was subsequently discovered to have a right temporal lobe injury suggestive of gliosis (Schacter et al., 2002). Gliosis, particularly those involving the temporal lobe are highly likely to give rise to abnormal activity as well as seizure disorders which in turn will hyper-activate the amygdala (Bear, 1979; Cain, 2007; Gloor, 2007).

Moreover, in addition to previous traumas or neurological predispositions, there are individual differences in regard to basal levels of DA, NE, corticosteroids, etc., and the manner in which these neurotransmitters fluctuate when stressed. For example, individual (non-stressed) primates who have naturally elevated cortisol levels are observed to be passive, low status, submissive, and to exhibit an "uptight" rigidity in temperment (Johnson et al., 2005). Primates with elevated cortisol have difficulty coping with even mild stressors, and tend to "waste away and die with no apparent pathological cause" (Johnson et al., 2005, p. 333). Likewise, inhibited, passive, fearful, and submissive children tend to have higher levels of cortisol even under non-stressed conditions (Kagan, Reznick, & Snidman, 2003). As these individual primates and children are already functioning as if highly stressed (as suggested by their elevated cortisol levels) they also have the most difficulty functioning in an adaptive manner when stressed, and are the most likely to freeze-up and passively surrender to their fate (Johnson et al., 2005; Kagan et al., 2003). By contrast, those with low baseline levels of cortisol are far more likely to react in an adaptive fashion and are less likely to become depressed or develop immunological disorders leading to illness or death.

14. Dreams and Hallucinations: Lifting the Veil to Multiple Perceptual Realities

Abstract

Regardless of culture or antiquity, humans have similar dreams and experience hallucinations. There are numerous reports of individuals who claim to have dreamed about future events which then took place, including deaths, tragedies, mass murders, horrific accidents, and environmental catastrophes. Yet others have experienced hallucinations where they have left their bodies, or have seen entities that appear to be from other realities. We should ask: is it really a hallucination if someone experiences a dissociation of consciousness and floats above their body and can later accurately described what was taking place and the appearance of those around them? Is it really a hallucination if an individual can see inside his hand and watch the blood cells swishing through his blood vessel? Abraham Lincoln, one of the great presidents in the history of the United States dreamed of his death 13 days before he was assassinated. Was it just a dream? Based on the evidence marshaled here and elsewhere, it can be inferred that not all dreams and hallucinations are dreams and hallucinations. There are specific neuroanatomical structures within the brain, including the amygdala, hippocampus, and temporal lobe, which make these experiences possible. There are also neurotransmitters, such as serotonin, which inhibit these tissues of the mind, so as to suppress and prevent the reception of much of the information they are able to perceive. Why would the brain evolve capabilities which are suppressed? Why did neural structures evolve which can process multiple sensations simultaneously, but then come to be inhibited by serotonin? Why would activation of specific brain areas result in the sensation of having left the body, or being privy to cosmic wisdom, or witnessing events and entities which appear to be from other dimensions or realities? It is concluded that these are mental capabilities which are de-evolving or still evolving, and which are or were destined to serve a specific purpose: Lifting the veil so we can gaze deeply into the past, the future, and the unknown.

"In sleep and in dreams we pass through the whole thought of earlier humanity....as a man now reasons in dreams, so humanity reasoned for many thousands of years when awake....This atavistic element in man's nature continues to manifest itself in our dreams, for it is the foundation upon which the higher reason has developed and still develops in every individual. Dreams carry us back to remote conditions of human culture and afford us a means of understanding it better." --Friedrich Nietzsche (1879).

1. Dreams of Live's Past

When I was a boy of 3, and for many years until around age 7, I had dreams of a previous life... and in that dream of a previous life I had also been a little boy... playing by the sea shore... by the ocean... and there were crowds of people...and it was a warm summer day... and then...in the dream... the ocean began to recede... the ocean waters drew back back back... and I could see shells and fish flopping on the wet sand where moments before there had been ocean... and I ran to where the ocean had been, on the wet sand, picking up shells... and many other people also ran onto the wet sand, laughing and talking in amazement that the ocean had pulled back for miles and miles.... and then... and then... and then...

I walked further and further out to where the ocean had been, picking up giant shells some with wiggling living creatures still inside, and gazing in wonder at what the ocean had hidden but which was now revealed... and then, I heard screams... women and men and children were screaming... and in my dream, □they were all running from the wet sand toward the dry shore...and people on the shore were also running... everyone was running away and screaming... and when I looked to see why they were running, I could see the ocean... it was still miles away--but it was a WALL OF OCEAN.. a WALL OF WATER looming up maybe 100 yards into the sky... and in my dream the wall of ocean was rushing forward, to where the ocean had been minutes before, toward where I was standing with sea shells in my hands... and I started running... like everyone else, running running running... and I could see, over my shoulder, behind me, the wall of ocean water coming closer, and closer... and faster faster faster... and I kept running... everyone was running and screaming...trying to get away... and then the towering WALL OF WATER was just behind me... then looming over me... and then it crashed down upon me... and the little boy that I was, in this previous life, drowned.... and then I awoke in my bed... the same boy who drowned, but a different boy...me...

I had this dream over and over... for years. The same dream...

Twenty years later, I learned, for the first time, about Tsunamis--- what I first dreamed about when I was 3 years old... was in fact exactly what happens if there is a giant Tsunami... the ocean pulls back and recedes... and people foolishly run out to where the ocean was... and then... the ocean comes rushing back as a wall of water drowning everyone who did not immediately run away...

How could I have dreamed so vividly about something 3-year old me, knew nothing about?

I have been here before... you have been here before... we will be here again...

According to Carl Jung (Jung, 1945, 1964), not all dreams are related to wish fulfillments or impressions from one's personal life. Some dreams contain very archaic elements which seem to have absolutely no bearing on the dreamer's personal experience. Instead these dreams consists of ancestral memories and archetypes, the residue of ancient impressions and profound experiences that somehow became litterally engraved into the mind and brain of humanity; ancestral memories which are recalled even thousands of years later in a dream.

Yet others have argued, and have presented considerable evidence to back up their claims, that some children in fact dream of a previous life and a previous death (Dossey, Greyson, Sturrock, Tucker 2011). The dream is thus, not a dream at all, but a personal memory, an experiential reincarnation, that is passed on through mechanisms as yet unknown.

2. Dreams of Genetic Destiny

Science of the future, would be perceived as magic, today. The science of today, would have been magic to those who lived just a few hundred years ago. Therefore, what seems to stretch the boundaries of science, and which then become confined to the realms of "the supernatural", may be explained by a science as yet unknown.

If a cutting of a plant is placed in water, takes root and blooms, is it a re-incarnation? Or a continuation? Our genetic ancestry stretches backwards in time to the first creatures to take root on Earth. Memories, too, are presumably stored in DNA. And these genetic memories need not be passed down strictly from father/mother to daughter/son. Genes may be horizontally transferred between species, along with the information, the genetic memories they store.

3. But What of Those Who Dream of the Future?

Dreams serve a number of purposes, and at times are highly improbable and bizarre. However, they often reflect something significant about the mental and emotional life of the dreamer, as well as other issues of concern. For example, when subjects are awakened repeatedly over the course of several days when physiological indices indicated they were dreaming, often an evolving thematic pattern, an unfolding story, can be discerned (Cartwright et al., 1980). These patterns frequently reflect mental-emotional activity concerned with the solution of particular problems (Cartwright 2010; Freud, 1900; Joseph 1988a, 1992a; Jouvet 2001; Jung, 1945, 1964).

For example, one subject, a student, noted that "after being woken many times and seeing three or four dreams a night, I could realize there was a certain problem being worked out, like coping with responsibilities that were thrust upon me, but that weren't necessarily my own but I took on anyway. It was working out the feelings of resentment of taking somebody else's responsibility, but I met them well in my dreams. A good thing about spending time in the sleep lab was you could relate a common bond to some of the dreams" (Cartwright et al., 1980, p. 277). Similar patterns were, of course, recognized by Freud (1900) and Jung (1945) many years ago.

Given all the multiple forms of information one is exposed to on a daily basis, coupled with the personal concerns of the dreamer, not surprisingly this information may be analyzed during the dream, in dream-language, and the resulting dream may reflect not just the past, but one's future intentions. Dreams often instruct the dreamer, much in the way thinking serves the conscious mind (Joseph 2011a), and dreams may be predict the future based on what it has been perceived thereby creating probabilistic scenarios which can serve as rehearsals for future behaviors.

4. Abraham Lincoln Dreams of His Assassination

In some cases, dreams do no just fore tell the future, but may predict the death of the dreamer:

In April of 1865, the commanding general of the Confederate Army, Robert E. Lee, had surrendered to General Ulysses S. Grant, and the days of the South were numbered. John Wilkes Booth, an actor and southern sympathizer hoped to rally the remaining Confederate troops to continue fighting and plotted with several other men in a conspiracy to kill President Abraham Lincoln.

On April 2, 1865, President Abraham Lincoln dreamed of his own death by assassination. The dream troubled him deeply, and on April 11, 1865, three days prior to his assassination, Abraham Lincoln shared this dream with his wife and a few friends which included Ward Hill Lamon (1865/1994):

About ten days ago, I retired very late. I had been up waiting for important dispatches from the front. I could not have been long in bed when I fell into a slumber, for I was weary. I soon began to dream. There seemed to be a death-like stillness about me. Then I heard subdued sobs, as if a number of people were weeping. I thought I left my bed and wandered downstairs. There the silence was broken by the same pitiful sobbing, but the mourners were invisible. I went from room to room; no living person was in sight, but the same mournful sounds of distress met me as I passed along. I saw light in all the rooms; every object was familiar to me; but where were all the people who were grieving as if their hearts would break? I was puzzled and alarmed. What could be the meaning of all this? Determined to find the cause of a state of things so mysterious and so shocking, I kept on until I arrived at the East Room, which I entered. There I met with a sickening surprise. Before me was a catafalque, on which rested a corpse wrapped in funeral vestments. Around it were stationed soldiers who were acting as guards; and there was a throng of people, gazing mournfully upon the corpse, whose face was covered, others weeping pitifully. 'Who is dead in the White House?' I demanded of one of the soldiers, 'The President,' was his answer; 'he was killed by an assassin.' Then came a loud burst of grief from the crowd, which woke me from my dream. I slept no more that night; and although it was only a dream, I have been strangely annoyed by it ever since." -Abraham Lincoln

On April 14, 1865, President Lincoln was shot in the back of the head while watching the play "Our American Cousin" at Ford's Theatre in Washington, D.C. with his wife, Mary Todd Lincoln. He died the next morning.

5. Dreams of the Future

Among ancient societies dreams were seen as extremely important sources of information, not just regarding the past, but the future (Joseph, 1992a,b, 1996, 2001, 2002; Jung 1945). As possible harbingers of the future they had to be observed carefully interpreted.

In the ancient world, be it Greek, Rome, Egypt, or Babylon, it was believed that some dreams contain important information regarding not only the individual, but his friends, family, and even the entire clan, village, city, or nation (Freud, 1900; Joseph 1992a, 2001, 2002; Jung 1945, 1964): The "big dream" of a child, woman, or man, were taken seriously by highly sophisticated and cultured ancient societies and were even announced two thousand years ago in the Roman Senate. On one occasion a senator's daughter had a dream in which Minerva the Goddess, appeared and complained that her temple was being neglected by the Roman people. The dream was announced to the Senate which in turn voted funds for restoration of the temple.

The dreams of generals, kings, queens, emperors, and Pharaohs, were commonly scrutinized and their symbolism interpreted as they were believed to foretell the future:

...Now Israel loved Joseph more than all his children, and he made him a coat of many colors... ...and Joseph dreamed a dream and he told it to his brethren... for behold we were binding sheaves in the field and lo, my sheaf arose, and stood upright: and, behold, your sheaves stood round about, and made obeisance to my sheaf. And his brethren said to him. Shalt thou indeed reign over us? And they hated him yet the more for his dreams... -Genesis 37

But the predictions of these dreams came to pass.

...And... the Pharaoh dreamed: and behold, he stood by the river and seven well favored kine and flatfish came up out of the river and they fed in a meadow. And behold, seven other kine came up after them out of the river, ill favoured and leanfleshed. And the ill favoured and leanfleshed kine did eat up the seven well favoured and fat kine. So Pharaoh awoke. And he slept and dreamed the second time....

...And it came to pass in the morning that his spirit was troubled and he sent and called for all the magicians and wise men of Egypt...but there was none that could interpret the dream...

....And Joseph said unto Pharaoh, God hath showed Pharaoh what he is about to do. The seven good kine are seven years.... and the seven thin and ill favoured kine that came up after them are seven years... Behold, there come seven years of great plenty throughout all the land of Egypt: And there shall arise after them seven years of famine: and all the plenty shall be forgotten and the famine shall consume the land.... -Genesis 37

6. The Quantum Future is Now?

Joseph was not the first to dream of the future, nor would he be the last. How can this be? Why did Lincoln dream of his assassination?

It may also be that the past, present, and future are a simultaneity but which are located in different regions of space-time, within the 4th dimension. And it may be that it is the mind which journeys along the dimension, journeying across space-time and encountering what is experienced as the ever present now which slips away as quickly as it is grasped, only to be replaced by a future which becomes the now--just as a DVD or CD contains the beginning and ending of a film or song simultaneously, but which is encoded in different locations within the medium.

And just as the ripples of a pond may strike distant shores, the quantum states of the future may also effect the distant shores occupied by what the mind experiences as the now.

If there is a quantum continuum, then why should it be confined to what our minds define as the present? If the 4th dimension is space-time, and if differences in time are related to movement through space and thus distance between locations, then within the quantum continuum, everything is connected: stars, planets, dogs, cats, and the future and the past.

If correct, this would imply that there may be something unique about the process of dreaming, which enables some dreamers to enter this quantum state and dream of an ancestral past, or to see what may lie ahead in the future.

Or, it may be that that those regions of the mind which subserve dreams, analyze the myriad details commonly experienced to make predictions about the future. Could this explain Lincoln's dream of his own death? Did his dream simply explain his own realistic fears? Or did he gain access to information about the future?

7. Genetic Destiny

The future can be predicted by the past. The future is in fact engraved into our genes. Have we not inherited our genes which determined what we have become, and does not this genetic ancestry leads interminable into the past? And so, too, life in the future, may also be encoded into ancestral genes, for it is these ancestral genes, passed on from mother/father, which code for what will be: genetic destiny.

Certainly genes interact with the environment. However, much of the information contained in our genome is hardwired. Dogs behave like dogs, cats like cats, and human like humans, not because of free will, but genetic destiny. The hardware which supports the software is hardwired into our genes.

Our lives and the future are shaped, at least in part, by genetic destiny and our genetic ancestry is certainly much older than this Earth (Joseph 2011b). Indeed, there is evidence of life in this planet's oldest rocks, dated to 4.2 bya (Nemchin et al. 2008; O'Neil et al. 2008), indicating life was present on this planet from the very beginning. Let us engage in a thought experiment and imagine life on Earth came from other planets, and these seeds of life contained the genes for the tree of life which took root on Earth. However, if the same genetic seeds landed on other Earth-like planets, the same trees of life may have evolved, such that, humans just like the humans of Earth may populate innumerable worlds; and just as dogs behave like dogs, and cats like cats, those humans of other worlds may act just like us; because their genes and our genes have a common source.

Or let us say there are multiple dimensions, a multiverse with multiple worlds, many of which are just like our own and where our own cosmic quantum twins may have evolved, and where they behave just like us--my multiverse twin writing this article, and your multiverse twin reading it. We behave the same on these multiple worlds, because we have the same genes. Consider the often reported instances of twins separated at birth, but who go on to lead nearly identical lives, even marrying women who look alike and have the same first names.

If the future is engraved within our genomes, then this would imply there is something unique about dreams, which unlock these genetic codes, revealing to the conscious mind what had had been concealed.

In fact, gene expression is enhanced during dream sleep which in turn has been correlated with increased activity and plasticity within neurons in the neocortex and limbic system (Ribeiro et al. 1999, 2002, 2008). Therefore, dream sleep is not only associated with the activation of otherwise silent genes which change the shapes of neural interconnections, but may unlock the secrets of changes yet to come.

Be it genetics, the quantum continuum, a multiverse, the unconscious dream analysis of myriad details, or the opening of windows to sensory capabilities which had been inhibited or suppressed, it appears that during dreams the veils are lifted, thereby enabling the mind to see and reveal what had been concealed.

8. The Neuroanatomy of Dreams: Overview

Sleep consists of five distinct stages, one of which is closely associated with the appearance of dreams (Hobson 2004, Monti et al., 2008; Steriade & McCarley 2005). It is during the course of the dream that the eyes begin to move quite rapidly as if the dreamer were observing some action. This is referred to as REM (rapid eye movement). The appearance of REM during sleep has been found to occur in a rhythmical fashion in all terrestrial mammals so far studied.

REM occurs during a sleep stage referred to as "paradoxical sleep." It is called paradoxical (or active sleep), for electrophysiologically the brain is aroused and quite active, similar to its condition during waking. However, the body musculature is paralyzed and motor functioning is all but abolished except in certain regions which control respiration, and eye movements (Hobson 2004, Hobson et al. 1986; Jouvet 2001; Monti et al., 2008; Steriade & McCarley 2005). This prevents the dreamer from acting out their dreams. The ability to perceive outside sensory events, normally received through the five senses, is also greatly attenuated (Hobson 2004, Monti et al., 2008; Steriade & McCarley 2005; Vertes 1990). In addition during the course of a dream, temperture control is lost, pain sensation is rare, and males tend to have an erection.

These REM dream cycles occur every 90 to 100 minutes. By contrast, non-REM (N-REM) periods occur during a stage referred to as "slow-wave" or synchronized sleep.

Thus, during REM dream sleep, the brain is in a state of heightened activity and arousal, indicating that considerable processing of information is taking place. Yet, simultaneously, the normal routes of sensory reception have been restricted. Yet, the brain of the sleeper does in fact receive and process sensory information.

Most individuals awakened during REM report dream activity approximately 80% of the time. REM dreams involve a considerable degree of visual imagery, emotion, and tend to be distorted and implausible to various degrees (Foulkes, 1962; Hobson 2004; Steriade & McCarley 2005).

When awakened during the N-REM period, dreams are reported approximately 20% of the time (Foulkes, 1962; Jouvet 2001; Monroe et al. 1965). However, the type of dreaming that occurs during N-REM is quite different from REM. For example, N-REM dreams (when they occur) are often quite similar to thinking and speech (i.e. lingusitic thought), such that a kind-of rambling verbal monologue is experienced in the absence of imagery (Foulkes 1962; Hobson 2004; Jouvet 2001; Monroe et al. 1965). It is also during N-REM in which an individual is most likely to talk in his or her sleep (Kamiya, 1961).

9. Right Hemisphere Dreams

REM is characterized by high levels of activity within the pons of the brainstem, the lateral geniculate nucleus of the thalamus, and occipital lobes; referred to as PGO waves (Hobson 2004; Monti et al., 2008; Steriade & McCarley 2005). It also has been reported that electrophysiologically the right hemisphere becomes highly active during REM, whereas, conversely, the left half of the brain becomes more active during N-REM (Goldstein et al. 1972; Hodoba, 1986). This may account for the striking differences in the content of dreams, with left hemisphere dreams being more "thought-like" and verbal, and right hemisphere dreams more emotional, vivid, and visual-spatial.

Measurements of cerebral blood flow have shown an increase in the right temporal and parietal regions during REM sleep and in subjects who upon wakening report visual, hypnagogic, hallucinatory and auditory dreaming (Meyer et al., 1987). Moreover, abnormal and enhanced activity in the right temporal and temporal-occipital area acts to increase dreaming and REM sleep for an atypically long time period (Hodoba, 1986). Similarly, REM sleep is associated with increased activity in this same region much more than in the left hemisphere (Hodoba, 1986). These findings indicate that there is a specific complementary relationship between REM sleep and right temporal-occipital electrophysiological activity.

Conversely, there have been reports of patients with right cerebral damage who have ceased dreaming altogether or to dream only in words (Humphrey & Zangwill, 1951; Kerr & Foulkes, 1978, 1981). For example, defective dreaming, deficits that involve visual imagery, and loss of hypnagogic imagery have been found in patients with focal lesions or hypoplasia of the posterior right hemisphere and abnormalities in the corpus callosum which would prevent transfer from the right to left hemisphere (Botez et al. 1985; Kerr & Foulkes, 1981; Murri et al. 1984).

An absence or diminished amount of dreaming during sleep also has been reported after split-brain surgery; i.e., as reported by the disconnected left hemisphere (Bogen & Bogen, 1969; Hoppe & Bogen, 1977). Similarly, a paucity of REM episodes have been noted in other callosotomy patients, although these particular individuals continued to report some dream activity (Greenwood, Wilson, & Gazzaniga, 1977).

On the other hand it has been reported that when the left hemisphere has been damaged, particularly the posterior portions (i.e. aphasic patients), the ability to verbally report and recall dreams also is greatly attenuated (Murri et al., 1984; Pena-Casanova & Roig-Rovira, 1985; Schanfald et al. 1985). Of course, aphasics have difficulty describing much of anything, let alone their dreams.

The differential activation of the right and left hemisphere during REM vs N-REM, is a major factor in the visual-emotional hallucinatory-mosaic experienced during the dream (Joseph 1988a, 1996). As has been well established, the right hemisphere is dominant for most aspects of non-verbal and visual-spatial perceptual activity as well as the expression and comprehension of social-emotional nuances. It is for this reason that the right hemisphere is sometimes thought to be the more intuitive half of the cerebrum.

As demonstrated in individuals who have had the two hemispheres surgically separated, the right half of the brain is able to draw conclusions, make predictions, selectively store certain images and experiences in memory, and can call on and act on these information at will (Joseph 1988a,b; 1996). Moreover, given its sensitivity to a host of non-social environmental variables (Joseph 1988a), it is able to assimilate and draw conclusions if not make predictions about this material which, conversely, the left hemisphere has difficulty processing. In this regard, it is not at all surprising that during the course of a dream, when the right half of the brain is at a peak level of activity, that it may draw upon these capacities to arrive at certain conclusions or to make predictions regarding events, people, or the future, and that those aspects of consciousness associated with the language-dependent aspects of the mind (Joseph 1988a,b, 2011a) would view these cognitions as bizarre and inexplicable.

However, other factors may also be involved, including the perception of sensory and other information which is normally filtered out and suppressed. That is, the right hemisphere may be perceiving stimuli during the course of the dream, which during waking is not normally perceived.

10. REM-On, REM-Off & Serotonin

The visual-emotional hallucinatory aspects of dreaming occur during REM, and the activation of a variety of brain regions are involved, i.e. the amygdala, hippocampus, right temporal lobe, right occipital lobe, the lateral geniculate nucleus of the thalamus, and brainstem nuclei located in the lateral and medial pons. In addition, the production of REM sleep is mediated by cyclic fluctuations in the levels of various neurotransmitters, including, and especially serotonin which serves an inhibitory function and when at high levels suppresses REM sleep and the activity of neurons which contribute to the generation of dreams.

Specifically, cholinergic (ACh) neurons located in the lateral pons, and neurons located in the medial pontine reticular formation appear to be the locus for REM-on neurons which initiate and/or maintain the production of REM sleep and which produce muscle atonia so that dreamers do not act on their dreams (Lydic, et al., 1991; Monti et al., 2008; Steriade & McCarley, 2005; Vertes 1990). That is, during the production of REM and paradoxical sleep, there is increased cholinergic activity and the production of pontine, lateral geniculate, occipital activity; i.e. PGO waves. Whereas with the termination of REM, and with the onset of slow wave N-REM sleep these same neurons greatly reduce their activity. As ACh is also implicated in memory, this may well explain why recent memories tend to become incorporated in dreams.

In contrast, REM-off neurons, which tend to be located in the medial raphe nucleus (which contain 5HT neurons) and in the locus coeruleus (located at the midbrain-pons junction and which contain NE neurons) are highly active during waking but then significantly decrease their activity with the initiation and onset of REM and the production of pontine, lateral geniculate, occipital activity; i.e. PGO waves (Hobson 2004; Monti et al., 2008; Steriade & McCarley 2005).

Hence, these REM-Off neurons appear to suppress REM and PGO activity and interfere with the onset of dreaming and paradoxical sleep, whereas REM-on neurons initiate the opposite sleep phase in which case the brain appears to be highly active and the individual begins to dream.

REM-on and REM-off neurons, therefore, appear to oscillate in a rhythmic fashion thus inducing sleep, dreaming, and waking. When the REM-off neurons cease to fire, the REM-on neurons (which are predominately cholinergic) become highly active until a REM episode is produced. However, as REM-on neuron activity decreases, REM-off activity increases, thus setting into motion a continuous 90 minute cycle of REM - Non-REM sleep.

11. Serotonin & Sensory Filtering

Low levels of 5HT are associated with REM sleep and dreaming (Hobson 2004; Monti et al., 2008) and thus with increased activity in the amygdala, hippocampus, and right hemisphere. 5HT in fact exerts inhibitory influences on a variety of brain structures, thereby suppressing incoming sensory input and the processing of sensory information from a variety of modalities simultaneously (Applegate, 1980; Jacobs & Azmita 1992; Soubrie, 1986; Spoont, 1992). That is 5HT restricts perceptual and information processing and in fact increases the threshold for neural responses to occur at both the neocortical and limbic level. In this way, attention can be focused and everything considered irrelevant may be filtered out. In fact, 5HT appears to be involved in learning not to respond to stimuli (Benninger, 1989). These signals are filtered out and suppressed.

It has also been demonstrated that 5HT acts to suppress activity in the lateral (visual) geniculate nucleus of the thalamus and synaptic functioning in the visual cortex as well as the amygdala (Jacobs & Azmita 1992; Soubrie, 1986; Spoont, 1992). By contrast, substances which block 5HT transmission -such as LSD- results in increased activity in the amygdala (Chapman & Walter, 1965; Chapman, Walter, Ross et al., 1963) and in the sensory pathways to the neocortex (Purpura 1956), which induces complex hallucinatory experiences.

12. The Amygdala: Gateway to Multiple Perceptual Realities

The amygdala is exceedingly responsive to social and emotional stimuli as conveyed vocally, visually, through touch, or body language including the face (Gloor, 1997; Halgren, 1992; Kling & Brothers 1992; Morris et al., 1996; 1992) and contains neurons which respond selectively to smiles and to the eyes, and which can differentiate between male and female faces and the emotions they convey (Hasselmo, Rolls, & Baylis, 1989, Kawashima, et al., 1999; Rolls, 1984; Morris et al., 1996). Single neurons in the amygdala, in fact, can respond to multiple sensory modalities, simultaneously (Amaral et al. 1992; O'Keefe & Bouma, 1969; Perryman, Kling, & Lloyd, 1987; Rolls 1992; Sawa & Delgado, 1963; Schutze et al. 1987; Turner et al. 1980; Ursin & Kaasa, 1960). Overall, because emotional, motivational, and multimodal assimilation of various sensory impressions occurs in this region, it is also involved in attention, learning, and memory (Gloor, 1997; Halgren, 1992; Halgren et al., 1978).

The right amygdala (as well as the right hippocampus, and the right hemisphere in general) is also involved in the production of dream imagery as well as REM sleep (Broughton, 1982; Goldstein et al., 1972; Hodoba, 1986; Humphrey & Zangwill, 1961; Kerr & Foulkes, 1978; Meyer et al. 1987). Simulation of the amygdala triggers and increases ponto-geniculo-occipital paradoxical activity during sleep (Calvo, et al. 1987), which in turn is associated with REM and dreaming.

Presumably, during REM, the amygdala (and hippocampus) serve as a reservoir from which various images, emotions, faces, words, and ideas are drawn and incorporated into the matrix of dream-like activity being woven by the right hemisphere. Essentially, this increased activity, resulting in REM sleep and the productions of dreams is due in in large part to cyclic reductions in serotonin.

Serotonin, inhibits these multi-modal amygdala neurons, thereby suppressing their activity and preventing the reception and processing of a vast array of sensory impressions. However, if serotonin release or uptake is prevented, the amygdala and other structures are released from their inhibitory sensory prison, and in consequence, single amygdala neurons will process multi-modal properties simultaneously, such that individuals will dream.

13. Dreaming Backwards

During the dream state, the dominant sensory streams are suppressed prior to transmission to the neocortical receiving areas. However, subcortical and limbic structures continue to receive and process sensory information, and not uncommonly these sensations, experienced during sleep, will be incorporated into the dream and may even trigger the dream, in which case, the dreamer may dream backwards

"Julie" dreams she is walking in San Francisco lugging large bags of gifts. Feeling tired she sets them down on the sidewalk. She looks for a bus and see a cable car coming. As it pulls up the conductor begins to ring its bell. The sound of the bell grows louder and then jolts her awake. Fully awake she realizes someone is ringing her doorbell. In this regard, the hearing of the bell seemed to be a natural part of the dream, and it is. What seems paradoxical, however, is that the dream seemed to lead up to the bell so that its ringing made sense in the context of the dream.

The dream did not lead up to the bell; the bell initiated the dream. The dream was produced, via the unique language of the right hemisphere during sleep (as well as amygdala activation), so as to explain the sound of the bell. The bell was heard and the dream was instantly produced in explanation and association. The bell stimulated the dream which may have only last a second.

One individual (described by Freud, 1900) dreamt he was in 18th Century

France in the midst of the French Revolution. After a trial in which he was been found guilty, he was being led down a street lined with yelling and cursing Frenchmen and women. At the end of the street he could see the gallows where the heads of various political criminals were being chopped off at the neck. Mad with fright he felt and saw himself led up the stairs and his head being placed in the yoke of the chopping block. The executioner gave the signal, the crowd screamed its approval, he could hear and sense the blade falling, and with a loud crack it struck him across the neck. Indeed, it struck him with such a jolt that he awoke to find that his poster bed had broke and that a railing had fallen and struck him across the side and back of his neck.

Thus, we see that although the perception of external stimuli by the five senses is greatly attenuated, external stimuli can still trigger a complex dream. The dream explains this external stimuli. However, what if this stimuli is not transmitted via the dominant sensory channels, and is conveyed through electro-magentic radiation, pheromones, DNA, quantum field fluctuations, or the dreams of other dreamers?

14. Damon Wells: Son of the Devil

The following is based on court, psychiatric, police, and investigative records, and interviews with many of the principles directly involved.

District Attorney (D.A.): "Did you join the United States Army in 1980?

Witness (Stephen M): "Yes, sir."

D.A.: "You were stationed in Germany?"

S.M.: "Yes."

D.A.: "And while in Germany, did you come to know a person by the name of Damon Wells?"

S.M.: "Yes, sir."

D.A.: "Do you recall when you first met Damon Wells?"

S.M.: "Yes. It was 1981."

D.A.: "How did he introduce himself?"

S.M.: "He said, 'My name is Damon Wells, son of the devil.'"

D.A.: "Did you become friends with Damon Wells?"

S.M.: "We were both soldiers, and it was a pretty isolated site. Pretty secluded on top of a mountain, so I had a lot of contact with him."

D.A.: "What type of duties did you perform?"

S.M. "I was a satellite technician. Wells was a mechanic, a motor mechanic."

D.A.: "What kind of worker was he?"

S.M. "He was a fantastic worker. One of the best mechanics we ever had up there. He'd fix vehicles and the electric generators which powered our site. We didn't have any electricity up there."

D.A. "Did you ever notice anything odd about Damon?"

S.M. "Sure. Many times."

D.A. "Such as?"

S.M. "Sometimes he would stop what he was doing and would say, The voices are here. Can you hear them?"

D.A. "Anything else?"

S.M. "Damon used to sit for hours on top of this mountain, looking down into the valley. It was totally remote and isolated and heavily forested, but sometimes you could see him from the communications tower. Sometimes he would meditate by these ruins on the side of this mountain.

"He'd sit for hours, like in a trance. His legs would be crossed. He'd be rocking, and chanting, mumbling about demons and angels from Hell. Even when it began to rain, or even snow, Damon would just sit there, trance-like. He'd be soaking wet or freezing, sometimes in just shirt sleeves. I had to go get him a few times."

D.A. "Did Damon ever show you a cave?"

S.M. "Yes, he did. He had found this cave, hidden deep in the woods, on the back side of a mountain across the valley. I don't know how he ever found it. He had a rope tied to a tree on the mountain top, and you had to climb, to rappel down the rope in order to get to this cave. It was really kind of creepy."

D.A. "Did you ever go inside this cave?"

S.M. "Yes, I did. There was a large stone slab, and some candles and gun powder, and a knife and some books inside. It was obvious he had spent a lot of time there."

D.A. "Did Damon ever tell you why he visited this cave?"

S.M. "Yeah. He told me he went there to talk to the voices. The voices, from his point of view, came from the devil. He said he could communicate with the devil, and that the devil would appear in this cave and speak with him."

D.A. "Did he every talk to you about sacrifices?"

S.M. "Yes, he did. Especially after I pointed out what looked like animal bones, and blood stains on this big slab of rock. In fact, he showed me this book of black magic and it had a picture of a naked woman lying spread eagle on a rock. Standing over her was a hooded man with a big knife. Damon said that was how the devil would sacrifice his victims. Damon said, that since he was a slave of the devil, he was supposed to sacrifice victims the same way, on this rock."

D.A. "And what did you tell him after he said and showed these things to you?"

S.M. "I told him he had a serious problem, that he needed help. In fact, I suggested a few times that maybe he should see a psychiatrist or a priest."

D.A. "And what did he say when you told him he should see a priest?"

S.M. "He laughed and began talking about voodoo and demons. About how he could project his thoughts, and with the help of the demons, he could harm his enemies. He said that priests were enemies of the devil, they couldn't help him. Besides, he said, I have everything under control. The Demons do as I tell them.'"

D.A. "Did Damon ever tell you he had committed a serious crime? That he had sacrificed anyone to the devil?"

S.M. "Yes sir, he did. A few months later. He said he picked up a hitchhiker, drove her to the woods, and then killed her because the devil had told him to. He said the devil took control over him, and he killed her and had sex with her."

D.A. "Did you believe him?"

S.M. "No. He was really mixed up, confused like he was two people. He seemed possessed. I thought he was flipping out. When I asked if he had really killed someone, he just said, They're taking me over, Steve. They are taking me over."

On the afternoon of 8/27/84, Damon Wells had taken LSD and was again communing with the devil. Damon was to be sent on a mission. The devil wanted a woman. Damon was to sacrifice this woman for the devil. That woman would be Tanya Z.

That same afternoon, Tanya Z, a very attractive, dark haired, 21-year old woman, had just left work at a Santa Cruz bank and was accompanied to her car by a female friend and coworker. Tanya was chattering away and showing her friend a birthday card for her fiance and explaining that she was going shopping for a birthday present. However, first she would have to drive over the Santa Cruz mountains, on Highway 17, to stop at her house in Santa Clara County.

At the same time, Damon was also walking to his car, talking with the Devil who was urging him on, explaining what to do. A neighbor in the next yard overhead the mumble speech and thought Damon might have been talking to him? "What did you say?" he asked?

Damon looked at him: "Don't you hear them?"

"What?"

"The voices are here. Can you hear them?"

"Hear who?"

"The Devil," Damon replied. He got into his car and drove into the Santa Cruz mountains.

Tanya had just made it over the summit, the highest point on Highway 17, when her car began to stall, and then it died. She pulled over to the side of the busy highway, and then repeatedly attempted to get it to start, but to no avail. She had run out of gas.

It was then that Damon drove by. According to Damon, the voice of the Devil told him that this woman was to be the sacrifice that He required. Damon pulled up, introduced himself, and offered her a ride to the nearest phone. She agreed, and got in. Instead, he took her deep into the mountains, attacked her, beat her until she was nearly unconscious, and then dragged her down a heavily wooded mountainside, and then pushed away some branches which hid a trail leading to a huge rock. Damon dragged her to the rock, laid her out sacrificial style, killed her, sexually assaulted her, and then left the dead body lying naked on the rock.

That evening Tanya's father and fiancee began searching for her. They found only the abandoned car near the summit of highway 17. They contacted the police, who, however, could find no evidence of a crime. There was no body. No blood. No signs of a struggle. Just an abandoned car.

That evening, Damon, still experiencing the after-affects of the LSD, could not stop visualizing the murder. It played over and over in his mind, even after he fell asleep. But once he began to dream, the victim, Tanya, did not stay dead, but appeared to him, accusing him, showing him what he had done. His victim had become an angry avenging spirit.

That same night a woman named "Sunshine" a nudist who lived in a nudist colony, Lupin Lodge, situated in the Santa Cruz Mountains, had a nightmare: A woman was being brutally murdered. The next day, Sunshine read the story of Tanya's disappearance in the local newspaper, and that night, and the night following, she had the dream again, but this time the victim appeared to her. It was Tanya. But she was no avenging spirit. According to Sunshine, Tanya showed her the road off highway 17 where Damon had taken her, and then the spot where Damon had parked his car and attacked her. Next she led Sunshine down a rather steep incline, and then along a trail, and pointed out her naked body, lying spread eagle on this huge slab of rock. Sunshine had this same dream repeatedly.

Sunshine was not the only dreamer dreaming of the murder. Damon dreamed of the horrible crime night after night. But his victim would not stay dead. She haunted his dreams, accusing him. It was an unending nightmare for Damon Wells, who called himself: Son of the Devil.

On the morning of 9/15/84, having dreamed of the murder, and with the victim's help, Sunshine was convinced she knew where Tanya's body lay hidden. She contacted Tanya's family, told them of her dreams, and that same day led them and the police to the the side road Tanya had showed her, and then to the very spot where she had dreamed of the murder. The police climbed down the steep incline, and just as Sunshine had dreamed, they found the trail. But, there was no body.

That night Sunshine had another dream and this time Tanya took her to the same spot, down the same trail, then pointed at and emphasized a little trail that forked off to the right and which led directly to the body. The next day, Sunshine and the family met again, and then climbed down the incline, took the trail to the right, and there was Tanya's body laid out exactly as revealed to Sunshine when dreaming.

The murder remained unsolved, however, until 2/7/88 when Damon Wells, beset by horrible nightmares, sought psychiatric treatment. He confessed and hoped the psychiatrist could help him escape the dreams and visions which tormented him.

14.1 Dreams and Wondering Spirits. Souls and spirits were believed by ancient humans to wonder about while people sleep and dream (Brandon 1967; Frazier 1950; Harris 1993; Jung 1945, 1964; Malinowkski 1990). Some believed the soul could escape the body via the mouth or nostrils while dreaming and that the soul could wonder away from the body and engage in various purposeful acts or interact with other souls including the soul or spirit of those who had died. The spirit and soul were believed to hover about in human-like, ghostly vestiges, at the fringes of reality, the hinterland where day turns into night (Campbell 1988; Frazier 1950; Jung 1964; Malinowski 1954; Wilson 1951). It was also believed that even after death souls continued to interact with the living, and could haunt their dreams, and that the spirits of the dead might visit the dreams of the living, to convey knowledge and important information. Of course these are all silly superstitions. Certainly it is not possible Tanya Z's spirit was in fact haunting the dreams of Damon Wells and Sunshine.

In 1970, Ullman and Krippner reported statistically significant findings from a dream lab where dreamers were targeted with specific images from art prints, such as "School of the Dance" by Degas, depicting a dance class of several young women. The subject's dream reports included phrases as "I was in a class made up of maybe half a dozen people; it felt like a school." "There was one little girl that was trying to dance with me" (Krippner 1993). Could it be that Damon's dreams effected the dreams of Sunshine? Or perhaps it was all a coincidence.

15. LSD, 5HT, Dreams & Hallucinations

Damon Wells had consumed LSD on the afternoon of the murder. LSD blocks the release and uptake of 5HT (Bennett & Snyder, 1976) including in the amygdala (Gresch et al., 2002). LSD also acts directly on the amygdala and hippocampus (Bennett & Snyder, 1976; Gresch et al,, 2002), the right amygdala (and hippocampus) in particular. In consequence, once serotonin release or uptake is reduced or blocked, the amygdala will process multi-modal stimuli simultaneously, and the person will see sound, taste colors, and all aspects of perceptual processing will be greatly enhanced:

It was 1967, the "summer of love" and I am my friends made the trek to Haight-Ashbury where I acquired pure LSD created by Stanley Owsley. I was about to take my first "trip."

About half hour after I ingested the drug I forgot I had taken it, and instead I noticed how all the colors of the trees and plants were much brighter, more colorful and luminous. I was walking toward a park and stopped to touch a green leaf which was sparkling with emerald light, and I could feel the life inside the leaf, I could taste it's "greenness" through my fingers, and then my eyes became like a microscope and I could see the fine cellular structure of the leaf and then inside the leaf... and my attention turned to my hand... and I could see the fine cellular structure of my skin, then beneath my skin, and then I could see a blood vessel and then inside the vessel and I could see the red and white blood cells swishing past inside the vessel--and I was totally amazed and kept wondering: How come I never noticed this before! It was as if my eyes had become a tunneling microscope--and this was 15 years before its invention!

This was the first hour of my experience on LSD.... as the experience wore on I could see sound, I could see sound waves. I could taste colors. And I was able to see through the Santa Cruz mountains to what was on the otherside: the ocean and a jet plane and then the jet flew over the top of the mountain. And no, I do not think I was hallucinating per se. LSD blocks 5HT, which is an inhibitor. Structures such as the amygdala are inhibited by 5HT, and many amygdala neurons are multi-sensory, a single amygdala neuron can process sound, touch, taste, and vision, simultaneously---but this information is inhibited and filtered out as we would be overwhelmed if we were constantly tasting colors or seeing inside our skin...So LSD blocks 5HT which turns off the filtering... I was 17 years old when I took LSD. At the end of the experience I felt as if my intelligence had increased by 20 IQ points and my consciousness, understanding, and awareness of the world was certainly much greater. My mind had expanded and from the day forward I saw the world with open eyes.

As is well known, LSD can elicit profound hallucinations involving all spheres of experience. Following the administration of LSD, high-amplitude slow waves (theta) and bursts of paroxysmal spike discharges occur in the hippocampus and amygdala (Chapman & Walter, 1965; Chapman et al., 1963), but with little cortical abnormal activity. In both humans and chimpanzees in whom the temporal lobes, amygdala, and hippocampus have been removed, LSD ceases to produce hallucinatory phenomena (Baldwin et al., 1955, 1959; Serafetinides, 1965). Moreover, LSD-induced hallucinations are significantly reduced when the right versus left temporal lobe has been surgically ablated (Serafetinides, 1965). Dreaming is sometimes abolished with right but not left temporal lobe removals (Kerr & Foulkes, 1981). Likewise, Penfield and Perot (1963) report that the most vivid hallucination tend to be triggered from the right not the left temporal lobe.

LSD is structurally similar to serotonin, but acts as a serotonin antagonist and acts both pre-synaptically (Montigny and Aghajanian 1977) and post-synaptically (Bennett & Snyder, 1976) by blocking 5-HT secretion and 5-HT receptors (Bennett & Snyder, 1975), thereby preventing serotonin from exerting its normal inhibitory effects on sensory reception and multi-modal sensory processing. Further, LSD acts on the brainstem raphe nucleus (Strahlendorf, et al., 1982) which produces serotonin, thereby preventing this structure from exerting inhibitory influences not just on the amygdala, but in the pons, lateral geniculate, and visual cortex--structures which become highly active during dreaming.

Moreover, LSD acts on the frontal lobes (Gresch et al., 2002), which exert controls over the rest of the brain and sensory processing in the neocortex through inhibition (Joseph 1996, 1999a). Thus, following administration of LSD and suppression of 5-HT influences, sensory inhibition is significantly attenuated throughout the brain, such that neurons which are normally suppressed begin processing information normally filtered out, all of which is then experienced by the conscious mind.

16. Day Dreams, Hallucinations, Out-of-Body Consciousness, and Alternate Realities

Hallucinations are typically defined as the "Perception of visual, auditory, tactile, olfactory, or gustatory experiences in the absence of an external stimulus coupled with a compelling sense of their reality." The Diagnostic and Statistical Manual of the American Psychiatric Association also defines hallucinations as occurring "without external stimulation of the relevant sensory organ."

According to definition, hallucinatory experiences under LSD, insofar as they are based on external stimulation and perceptions freed of inhibitory restraint, are not necessarily hallucinations. Of course, this supposition could be dismissed by attributing the experience to the LSD. However, this dismissal fails as it is not the LSD which induces the experience, but the reductions in 5-HT, as also occurs during REM sleep. Therefore, it could be said that strictly speaking they are not hallucinating. Just as external stimuli may trigger a dream during REM sleep, that during waking and under reduced 5-HT external stimulation also produced a dream, what is experienced could be likened to a day-dream.

In fact, There is some evidence to suggest that during the course of the day and night the two cerebral hemispheres oscillate in activity every 90 to 100 minutes and are 180 degrees out of phase --a cycle that corresponds to changes in cognitive efficiency, the appearance of day dreams, REM (dream sleep), and, conversely, N-REM sleep (Bertini et al. 1983; Broughton, 1982; Hodoba, 1986; Klein & Armitage, 1979; Kripke & Sonnenschein, 1973). That is, like two pistons sliding up and down, it appears that when the right cerebrum is functionally at its peak of activity, the left hemisphere is correspondingly at its nadir. Day dreams also correspond to this cycle, such that when dreaming at night, or during the day, the dream is association with increased right hemisphere and reduced left hemisphere activity (Joseph 1988a, 1992a, 1996).

Day dreams, like night dreams, and LSD, are all associated with differential and increased right hemisphere activation. Likewise, in studies of hallucinations secondary to cerebral tumors or seizure activity, although simple hallucinations are likely following damage to either hemisphere, complex hallucinations are usually associated with right rather than left cerebral lesions (Teuber et al., 1960; Mullan & Penfield, 1959; Hecaen & Albert, 1978; Joseph 1996).

Moreover, direct stimulation of the neocortex and the amygdala, also produce complex hallucinations. For example, electrical stimulation of visual association areas 18 and 19 can produce complex and vivid images of men, animals, various objects and geometric figures, liliputian-type individuals (Hecaen & Albert, 1978; Joseph 1996; Tarachow, 1941). Those who experience these hallucinations may also have the illusion that their vision has telescoped such that they can see objects and people which are exceedingly far away, or existing in another reality or dimension. However, just as often, what they see may appear right before then and may not be unusual and no different from any other perception. For example, one patient saw a butterfly then attempted to catch it when area 19 of the visual cortex was electrically stimulated. Another hallucinated a dog and then called to it, denying the possibility that it was not real (Joseph 1996).

However, under conditions of exceedingly heightened activity or other disturbances in the temporal lobes and underlying amygdala and hippocampus, the hallucination may become exceedingly vivid and unusual and include images of threatening men, naked women, sexual intercourse, demons and ghosts and pigs walking upright dressed as people (Bear 1979; Daly 1958; Gloor 992; Halgren 1992; Horowitz et al. 1968; Joseph 1999, 2002; Penfield & Perot 1963; Slater & Beard 1963; Taylor 1972; Trimble 1991; Weingarten, et al. 1977; Williams 1956). Often the experience could best be described as "other worldly" as if seeing into another supernatural dimension or entering into another reality.

Some individuals report communing with spirits, angels or gods, or receiving profound knowledge from the Hereafter, following temporal lobe activation (Daly 1958; MacLean 1990; Penfield & Perot 1963; Williams 1956). Some have visualized and have seen ghosts, demons, angels, and even God, or claim to have left their body (Bear 1979; Daly 1958; Gloor 1992; Horowitz et al. 1968; MacLean 1990; Mesulam 1981; Penfield & Perot 1963; Schenk & Bear 1981; Slater & Beard 1963; Subirana & Oller-Daurelia, 1953; Trimble 1991; Weingarten, et al. 1977; Williams 1956).

Some individuals have described feelings such as elation, security, eternal harmony, immense joy, paradisiacal happiness, euphoria, completeness. Between .5 and 20% of such patients claim such feelings (Williams, 1956; Daly, 1958). One patient of Williams (1956) claimed he was overwhelmed by "sudden feeling of extreme well being involving all my senses. I see a curtain of beautiful colors before my eyes and experience a pleasant but indescribable taste in my mouth. Objects feeling pleasurably warm. the room assumes vast proportions, and I feel as if in another world."

A patient described by Daly (1958) claimed his seizure felt like "a sunny day when your friends are all around you." He then felt dissociated from his body, as if he were looking down upon himself and watching his actions.

Penfield and Perot (1963) describe several patients who claimed they could see themselves outside their body engaging in various activity. One woman stated that "it was though I were two persons, one watching, and the other having this happen to me," and that it was she who was doing the watching as if she was completely separated from her body. According to Penfield, "It was as though the patient were attending a play and was both actor and audience.

Williams (1956) describes a patient who claimed that during an aura she experienced a feeling that she was being lifted up out of her body, coupled with a very pleasant sensation of elation and the sensation that she was "just about to find out knowledge no one else shares, something to do with the line between life and death."

Subirana and Oller-Daurelia (1953) described two patients who experienced ecstatic feelings of either "extrraordinary beatitude" or of paradise as if they had gone to heaven and noted that his fantastic feelings lasted for hours.

Other patients have noted that feelings and things suddenly become "crystal clear" or that they have a feeling of clairvoyance, or of having the truth revealed to them, or opf having achieved a sense of greater awareness and of a new awareness such that sounds, smells and visual objects seemed to have a greater meaning and sensibility (Joseph 2001, 2002).

It has frequently been reported that as compared to other cortical areas, the most complex and most forms of hallucination occur secondary to temporal lobe involvement (Malh et al., 1964; Horowitz et al., 1968; Penfield & Perot, 1963; Tarachow, 1941) and that the hippocampus and amygdala (in conjunction with the temporal lobe) appear to be the responsible agents (Gloor 1992, 1997; Horowitz et al., 1968; Halgren et al., 1978). Thus, the same brain regions implicated in the generation of the dream are also linked to hallucinations following the administration of LSD or abnormal activation or direct electrode stimulation.

17. Out-of-Body and Near Death Experience

Some children and adults who have been declared "clinically" dead but who subsequently return to life, have reported that after "dying" they left their body and floated above the scene (Eadie 1992; Joseph, 1996; Rawling 1978; Ring 1980). Typically they become increasingly euphoric as they float above their body, after which they may float away, become enveloped in a dark tunnel and then enter a soothing radiant light. And later, when they come back to life, they may even claim conscious knowledge of what occurred around their body while they were dead and floating nearby. Similar experiences are detailed in the Egyptian funerary texts and "book of the dead," written almost 6000 years ago (Budge 1994) as well as by otherwise completely "modern" and sophisti-cated humans.

"Lisa" for example, was a 22 year old college coed with no religious back-ground or spiritual beliefs, who was badly injured in an auto accident when the windshield collapsed and all but completely severed her arm. According to Lisa, when she got out of the car her was blood spraying everywhere and she only walked a few feet before collapsing. Apparently an ambulance arrived in just a few minutes. However, the next thing she noticed was that part of the time she was looking up from the ground, and part of the time she was up in the air looking down and could see the ambulance crew working, picking up her body, placing it on a gurney and into the ambulance.

According to Lisa, during the entire ride to the hospital it was like she was half in and half out of the ambulance, as if she were running along outside or just extending out of the vehicle watching the cars and tress go by. When they got to the hospital she was no longer attached to her body but was floating up and down the halls, watching the doctors and nurses and attendants. One doc-tor in particular drew her attention because he had a big belt buckle with his name written on it. She could even read it and it said "Mike."

According to Lisa, she was "tripping out, bobbing up and down the halls, just checking everything out" when she noticed a girl lying on a gurney with several doctors and nurses working frantically. When she floated over and peered over the shoulder of one of the doctors to take a look she suddenly real-ized the girl was her and that her hair and face were very bloody and needed to be washed. At that point she realized she was floating well above her body and the doctors and that she looked to be "dead." However, according to Lisa she did not feel afraid or upset, although the fact that her hair was dirty bothered her.

As detailed by Lisa, she soon floated up and outside the Emergency room and was enveloped in a total blackness, "like I was passing through a tunnel at the end of which was a vague light which became brighter and more brilliant, radiating outward." The light soon enveloped her body which made her feel exceedingly happy and very warm. A few moments later she heard the voice of her grandmother who had died when Lisa was a young girl. Although Lisa had no memory of this grandmother, she nevertheless recognized her and felt exceedingly happy. However, as Lisa approached, her grandmother very sor-rowfully told her it was "too soon", she would have to "go back." Lisa didn't want to go back, but had no choice. She was drawn away from the light and felt herself falling only to land with a painful thump in her own body. At this point she moved her hand which alerted one of the emergency room staff that Lisa was no longer dead.

It is noteworthy that Lisa had never heard of "near death experiences" (she was injured in 1982) and that after returning to life she only reluctantly explained what had happened when she was questioned by one of her doctors. Lisa also claimed that while she was dead and floating about the emergency room that she saw, heard and is able to recall everything that occurred up to the point when she was enveloped in darkness. She was able to accurately describe "Mike" as well as some of the staff who first attended her, the conversations that occurred around her as well as some of the other patients. Indeed, similar "after death" claims of leaving and floating above the body, and seeing everything occurring below, are common (Eadie 1992; Moody 1977; Rawling 1978; Ring 1980; Sabom 1982; Wilson 1987), and, as noted, are even reported in the 6,000 year old Egyptian Book of the Dead (Budge 1994), as well as the Tibetan Book of the Dead (the Bardo Thodol) which was composed over 1,300 years ago. Approximately 37% of patients who are resuscitated report similar "out of body" experiences (Ring 1980).

Consider for example, the case of Army Specialist J. C. Bayne of the 196th Light Infantry Brigade. Bayne was "killed" in Chu Lai, Vietnam, in 1966. He was simultaneously machine gunned and struck by a mortar. According to Bayne, when he opened his eyes he was floating in the air, looking down on his crumpled, burnt, and bloody body, and he could see a number of Vietcong who were searching and stripping his him: "I could see me... it was like looking at a manikin laying there... I was burnt up and there was blood all over the place... I could see the Vietcong. I could see the guy pull my boots off. I could see the rest of them picking up various things... I was like a spectator... It was about four or five in the afternoon when our own troops came. I could hear and see them approaching... I could see me... It was obvious I was burnt up. I looked dead... they put me in a bag... transferred me to a truck and then to the morgue. And from that point, it was the embalming process. I was on that table and a guy was telling a couple of jokes about those USO girls... all I had on was bloody undershorts... he placed my leg out and made a slight incision and stopped... he checked my pulse and heartbeat again and I could see that too...It was about that point I just lost track of what was taking place.... [until much later] when the chaplain was in there saying everything was going to be all right.... I was no longer outside. I was part of it at this point" (reported in Wilson, 1987, pp 113-114; and Sabom, 1982, pp 81-82).

Moreover, some surgery patients also claim to "leave their bodies" and recall seeing not just the events occurring below, but in one case, dirt on top of a light fixture (Ring 1980). "It was filthy. And I remember thinking, 'Got to tell the nurses about that."

Did the above surgical patient or Lisa or Army Specialist Bayne really float above and observe their bodies and the events taking place below? Or did they merely transpose what they heard (e.g. conversations, noises, etc.) and then visualize, imagine, or hallucinate an accompanying and plausible scenario? This seems likely, even in regard to the "filthy" light fixture. On the other hand, not all those who have an "out of body" hear conversations, voices, or even sounds. Rather, they may be enveloped in silence.

"I was struck from behind...That's the last thing I remember until I was above the whole scene viewing the accident. I was very detached. This was the amazing thing about it to me... I could see my shoe which was crushed under the car and I thought: Oh no. My new dress is ruined... I don't remember hearing anything. I don't remember anybody saying anything. I was just viewing things...like I floated up there..." (Sabom, 1982; p. 90). Moreover, even individuals born blind experience these "near death" hallucinations.

18. Fear and Out-of-Body Experiences

"I was shooting down the freeway doing about 100 or more in my Mustang when a Firebird suddenly cut me off. As I switched lanes to avoid him, he also switched lanes at which point I hit the breaks and began to lose control. The Mustang began to slide and spin... I felt real terror.... I was probably going to be killed... I was trying to control the Mustang and avoid turning over, or hitting any of the surrounding cars or the guard rail.... time seemed to slow down and then I suddenly realized that part of my mind was a few feet outside the car looking all around; zooming above it and then beside it and behind it and in front of it, looking at and analyzing the respective positions of my spinning Mustang and the cars surrounding me. Simultaneously I inside trying to steer and control it in accordance with the multiple perspectives I was suddenly given by that part of my mind that was outside. It was like my mind split and one consciousness was inside the car, while the other was zooming all around outside and giving me visual feedback that enabled me to avoid hitting anyone or destroying my Mustang."

The prospect of being terribly injured or killed in an auto accident or fire fight between opposing troops, or even dying during the course of surgery, are often accompanied by feelings of extreme fear. It is also not uncommon for individuals who experience terror to report perceptual and hallucinogenic experiences, including dissociation, depersonalization and the splitting off of ego functions such that they feel as if they have separated from their bodies and floated away, or were on the ceiling looking down (Campbell 1988; Courtois 1995; Grinker and Spiegel 1945; James 1958; Neihardt and Black Elk 1932/1989; Noyes and Kletti 1977; Parson 1988; Southard 1919; Terr 1990). Consider the following:

"The next thing I knew I wasn't in the truck anymore; I was looking down from 50 to 100 feel in the air." "I had a clear image of myself... as though watching it on a television screen." "I had a sensation of floating. It was almost like stepping out of reality. I seemed to step out of this world" (Noyes and Kletti 1977).

19. Hippocampal Hyperactivation and Astral Projection

Feelings of fear and terror are mediated by the amygdala, whereas the capacity to cognitively map, or visualize one's position and the position of other objects and individuals in visual-space is dependent on the hippocampus (Nadel, 1991; Joseph, 1996; O'Keefe, 1976; Wilson and McNaughton, 1993). The hippocampus contains "place" neurons which are able to encode one's position and movement in space.

The hippocampus, therefore, can create a cognitive map of an individuals environment and their movements within it. Presumably it is via the hippocampus that an individual can visualize themselves as if looking at their body from afar, and can remember and thus see themselves engaged in certain actions, as if one were an outside witness (Joseph, 1996). However, under conditions of hyperactivation (such as in response to extreme fear) it appears that the hippocampus may create a visual hallucination of that "cognitive map" such that the individual may "experience" themselves as outside their body, observing all that is occurring.

Again, it has been repeatedly demonstrated that hyperactivation or electrical stimulation of the amygdala-hippocampus-temporal lobe, can cause some individuals to report they have left their bodies and are hovering upon the ceiling staring down (Daly 1958; Jackson and Stewart 1899; Joseph, 1996; Penfield 1952; Penfield and Perot 1963; Williams 1956). That is, their ego and sense of personal identity appears to split off from their body, such that they may feel as if they are two different people, one watching, the other being observed.

20. The Evolution of Dream Consciousness

That so many people, regardless of culture or antiquity, have similar dreams and hallucinations, is presumably due to all possessing a limbic system and temporal lobe that is organized similarly. Of course, many of these experiences are also colored by one's cultural background and differences in thinking patterns.

However, we should ask: is it really a hallucination if someone experiences a dissociation of consciousness and floats above their body and can later accurately described what was taking place and the appearance of those around them? Is it really a hallucination if an individual can see inside his hand and watch the blood cells swishing through his blood vessel? If one of the great presidents in the history of the United States dreams of his death 13 days before he is assassinated, was it just a dream?

There are numerous reports of individuals who claim to have dreamed about future events which then took place, including deaths, tragedies, mass murders, horrific accidents, and environmental catastrophes (Barker 1967; Jung 1945, 1964; Wiseman 2011). Yet, given that six billion people dream multiple dreams every night, it could be argued that we should not be surprised that a few of these dreams just happen to accurately coincide with what takes place. These are just chance coincidences, which, when considering the billions of dreams dreamed nightly, should be dismissed as meaningless and of no significance (Wiseman 2011). Since so few people have these dreams, it should not be concluded that these dreams represent an important cognitive capacity. However, if we apply the same criteria to Einstein's theory of relativity, or the music of Mozart or Beethoven, then the absurdity of this position becomes clear. Certain cognitive capacities are well developed in just a few people.

Another question should also be asked: Why would the brain evolve capabilities which are suppressed? Why did neural structures evolve which can process multiple sensations simultaneously, but then come to be inhibited by serotonin? Why would activation of specific brain areas result in the sensation of having left their body, or being privy to cosmic wisdom, or witnessing events and entities which appear to be from other dimensions or realities?

These experiences are made possible by the brain. They must serve an adaptive function. These capabilities must have evolved following natural selection. Why would we evolve the ability to hallucinate or dream about that which supposedly does not exist? Why should we have evolved the ability to hallucinate that we have left our bodies, or can see inside our hands, or dream that a friend will tragically die days before his accidental death?

One possibility is the human brain is de-evolving, and has lost or is losing capabilities which served a more adaptive purpose in ancient humans 30,000 to 10,000 years ago--ancient humans (the Cro-Magnon) whose brain was 1/3 larger in size than the modern brain! The Cro-Magnon men stool 6 ft tall on average, and the Cro-Magnon people created the magnificent underground cathedrals of art which first began to appear 30,000 years ago.

Conversely, it may be that these are capacities which are still evolving, but which at the present stage of human evolution, the human brain is unable to utilize adaptively. Consider, the evolution of language. Certainly, Australopithecus, Homo habilis, Homo erectus, and Neaderthals did not engage in complex conversations (Joseph 1996, 2000, 2001). We can surmise based on indirect evidence, that the language capabilities of these ancient hominids and humans may have been limited to grunts, groans, screams, and a variety of emotional sounds (Joseph 2000). Whatever language abilities ancestral species evolved prior to 100,000 years ago were primitive manifestations of what was yet to evolve, i.e. grammatical speech, reading, writing, and associated modes of abstract thinking, such as math. Although the rudimentary foundations had been laid long ago, these were only primitive steps toward what was later to more fully evolve.

It can be surmised that if humans continue to evolve, that maybe 100,000 years from now they will possess a brain which is able to fully utilize capacities which the modern brain is, as yet, unable to master. Or, just as the brain has shrunk by 1/3 in size since the ending of the Paleolithic, that it may continue to lose capacities due to advances in technology which will increasingly render reading, writing, speaking, or creative endeavors obsolete.

Based on the evidence marshaled here and elsewhere, it can be inferred that not all dreams and hallucinations are dreams and hallucinations. They are mental capabilities which are de-evolving or still evolving, and which are or were destined to serve a specific purpose: Lifting the veil so we can gaze deeply into the past, the future, and the unknown.

References

Amaral, D. G., Price, J. L., Pitkanen, A., & Thomas, S. (1992). Anatomical organization of the primate amygdaloid complex. In J. P. Aggleton (Ed.). The Amygdala. (Wiley. New York.

Applegate, C. D. (1980). 5,7,-dihydroxytryptamine-induced mouse killing and behavioral reversal with ventricular administation of serotonin in rats. Behavioral and Neural Biology, 30, 178-190.

Baldwin, M., Lewis, S. A., & Bach, S. A. (1959). The effects of lysergic acid after cerebral ablation. Neurology, 469-474

Baldwin, M., Lewis, S.A., & Bach, S.A. (1959).The effects of lysergic acid after cerebral ablation.Neurology, 9, 469-474.

Barker, J. (1967). Premonitions of the Aberfan Disaster, JSPR, 44, 169-181

Bear, D. M. (1979). Temporal lobe epilepsy: A syndrome of sensory-limbic hyperconnection. Cortex, 15, 357-384.

Benninger, R. J. (1989). The role of serotonin and dopamine in learning to avoid aversive stimuli. In T. Archer & L-G Nilsson (Eds) Aversion, Avoidance and Anxiety. New Jersey, Erlbaum.

Bennett, Jr. J. P., and Snyder, S. H. (1975). Stereospecific binding of d-lysergic acid diethylamide (LSD) to brain membranes: Relationship to serotonin receptors. Brain Research, 94, 523-544.

Bennett, Jr. J. P., and Snyder, S. H. (1976). Serotonin and Lysergic Acid Diethylamide Binding in Rat Brain Membranes: Relationship to Postsynaptic Serotonin Receptors. Molecular Pharmacology, 12, 373-389.

Bertini, M., Violani, C., Zoccolotti, P., Antonelli, A., & DiStephano, L. (1983). Performance on a unilateral tactile test during waking and upon awakenings from REM and NREM. In W. P. Koella (Ed.), Sleep, (pp. 122-155). Basel: Karger.

Bogen J.,& Bogen,C. (1969).The other side of the brain: III. The corpus callosum and creativity. Bulletin of the Los Angeles Neurological Society, 34, 191-220.

Botez, M. I., Olivier, M., Vezina, J.-L., Botez, T., & Kaufman, B. (1985). Defective revisualization: Dissociation between cognitive and imagistic thought case report and short review of the literature. Cortex, 21, 375-389.

Brandon, S.G. F. (1967) The Judgment of the Dead. New York, Scribners.

Broughton, R. (1982). Human consciousness and sleep/waking rhythms: A review and some neuropsychological considerations. Journal of Clinical Neuropsychology, 4, 193-218.

Budge, W. (1994). The Book of the Dead. New Jersey, Carol.

Calvo, J. M., Badillo, S., Morales-Ramirez, M., Palacios-Salas, P. (1987) The role of the temporal lobe amygdala in ponto-geniculo-occipital activity and sleep organization in cats. Brain Research, 403, 22-30.

Campbell, J. (1988) Historical Atlas of World Mythology. New York, Harper & Row.

Cartwright , R. (2010) The Twenty-four Hour Mind: The Role of Sleep and Dreaming in Our Emotional Lives. Oxford University Press.

Cartwright, R. D., Tipton, L. W., & Wicklund, J. (1980). Focusing on dreams. Archives of General Psychiatry, 37, 275-288.

Chapman, L. F., & Walter, R. D. (1965). Actions of lysergic acid dienthalamid on averaged human cortical evoked rsposnes to light flash. Recent Advances in Biological Psychiatry, 7, 23-36.

Chapman, L. F., Walter, R. D., Ross, W., et al. (1963). Altered electrical activity of human hippocampus and amygdala induced by LSD-25. Physiologist, 5, 118.

Courtois, C.A. (1995). Healing the Incest Wound. New York, Norton.

Daly, D. (1958) Ictal affect. American Journal of Psychiatry, 115, 97-108.

Dossey, L., Greyson, B., Sturrock, P. A., Tucker, J. B. (2011). Consciousness -- What Is It? Shared Consciousness, Twin Consciousness, Near Death, Journal of Cosmology, Vol 14. In press.

Eadie, B. J. (1992). Embraced by the light. California, Gold Leaf Press.

Foulkes, W. D. (1962). Dream reports from different stages of sleep. Journal of Abnormal and Social Psychology, 65, 14-25.

Frazier, J. G. (1950). The golden bough. Macmillan, New York.

Freud, S. (1900). The interpretation of dreams. Standard Edition (Vol 5). London: Hogarth Press.

Gloor, P. (1997). The Temporal Lobes and Limbic System. Oxford University Press. New York.

Goldstein, L., Stoltzfus, N. W., & Gardocki, J. F. (1972). Changes in interhemispheric amplitude relationships in the EEG during sleep. Physiology and Behavior, 8, 811-815.

Greenwood, P., Wilson, D. H., & Gazzaniga, M. S. (1977). Dream report following commissurotomy. Cortex, 13, 311-316.

Gresch, P. J., Strickland, L. V. and Sanders-Bush, E. (2002). Lysergic acid diethylamide-induced Fos expression in rat brain: role of serotonin-2A receptors. Neuroscience, 114, 707-713.

Grinker, R. R., Spiegel, J. P. (1945) Men Under Stress. McGraw-Hill, New York.

Halgren, E. (1992). Emotional neurophysiology of the amygdala within the context of human cognition. In J. P. Aggleton (Ed.). The Amygdala. New York, Wiley-Liss.

Halgren, E., Walter, R. D., Cherlow, D. G., & Crandal, P. H. (1978). Mental phenomenon evoked by electrical stimualtion of the human hippocampal formation and amygdala, Brain, 101, 83-117.

Harris, M. (1993) Why we became religious and the evolution of the spirit world. In Lehmann, A. C. & Myers, J. E. (Eds) Magic, Witchcraft, and Religion. Mountain View: Mayfield.

Hasselmo, M. E., Rolls, E. T., & Baylis, G. C. (1989). The role of expression and identity in the face-selective responses of neurons in the temporal visual cortex of the monkey. Behavioral Brain Research, 32,203-218.

Hecaen, H., & Albert, M. L.(1978). Human Neuropsychology, New York: John Wiley.

Hodoba, D. (1986). Paradoxic sleep facilitation by interictal epileptic activity of right temporal origin. Biological Psychiatry, 21, 1267-1278.

Hobson, J. A. (2004) Dreaming: An Introduction to the Science of Sleep. Oxford University Press.

Hoppe, K. D., & Bogen, J. E. (1977). Alexithymia in twelve commissurotomized patients. Psychotherapy and Psychosomatics, 28, 148-155.

Horowitz, M. J., Adams, J. E., & Rutkin, B. B. (1968). Visual imagery on brain stimulation. Archives of General Psychiatry, 19, 469-486.

Humphrey M. E., & Zangwill, O. L. (1911). Cessation of dreaming after brain injury. Journal of Neurology, Neurosurgery, and Psychiatry, 14, 322-325.

Jacobs, B. L., & Azmita, E. C. (1992). Structure and function of the brain Serotonin System. Physiological Reviews, 72, 165-245.

Joseph, R. (1988a) The Right Cerebral Hemisphere: Emotion, Music, Visual-Spatial Skills, Body Image, Dreams, and Awareness. Journal of Clinical Psychology, 44, 630-673.

Joseph, R. (1988b). Dual mental functioning in a split-brain patient. Journal of Clinical Psychology, 44, 770-779.

Joseph, R. (1992a) The Limbic System: Emotion, Laterality, and Unconscious Mind. The Psychoanalytic Review, 79, 405-456.

Joseph, R. (1992b). The Right Brain and the Unconscious. New York, Plenum.

Joseph, R. (1996). Neuropsychiatry, Neuropsychology, Clinical Neuroscience, 2nd Edition. 21 chapters, 864 pages. Williams & Wilkins, Baltimore.

Joseph, R. (1999a). Frontal lobe psychopathology: Mania, depression, aphasia, confabulation, catatonia, perseveration, obsessive compulsions, schizophrenia. journal of Psychiatry, 62, 138-172.

Joseph, R. (1999b). The neurology of traumatic "dissociative" amnesia. Commentary and literature review. Child Abuse & Neglect. 23, 715-727.

Joseph, R. (2000). The evolution of sex differences in language, sexuality, and visual spatial skills. Archives of Sexual Behavior, 29, 35-66.

Joseph, R. (2001). The Limbic System and the Soul: Evolution and the Neuroanatomy of Religious Experience. Zygon, the Journal of Religion & Science, 36, 105-136.

Joseph, R. (2002). NeuroTheology: Brain, Science, Spirituality, Religious Experience. University Press.

Joseph, R. (2011a). Origins of Thought: Consciousness, Language, Egocentric Speech and the Multiplicity of Mind Journal of Cosmology, 14.

Joseph, R. (2011b). Life on Earth Came from Other Planets. Cosmology Science Publishers, Cambridge.

Jouvet, M. (2001). The Paradox of Sleep: The Story of Dreaming. MIT press.

Jung, C. G. (1945). On the nature of dreams. (Translated by R.F.C. Hull.), The collected works of C. G. Jung, (pp.473-507). Princeton: Princeton University Press.

Jung, C. G. (1964). Man and his symbols. New York: Dell.

Kamiya, J. (1961). Behavioral, subjective and physiological aspects of drowsiness and sleep. In D. W. Fiske, & S. R. Maddi (Eds.). Function of varied experience, (pp. 145-174). Homewood, IL: Dorsey Press.

Kawashima, R., Sugiura, M., Kato, T., et al., (1999). The human amygdala plays an important role in gaze monitoring. Brain, 122, 779-783.

Kerr N. H., & Foulkes, D (1978). Reported absence of visual dream imagery in a normally sighted subject with Turner's syndrome. Journal of Mental Imagery, 2, 247-264.

Kerr, N. H., & Foulkes, D. (1981). Right hemisphere mediation of dream visualization: A case study. Cortex, 17, 603-611.

Klein, R., & Armitage, R. (1979). Rhythms in human performance: 11/2 hour oscillations in cognitive style. Science, 204, 1326-1328.

Kling. A. S. & Brothers, L. A. (1992). The amygdala and social behavior. In J. P. Aggleton (Ed.). The Amygdala. New York, Wiley-Liss.

Kripke, D. F., & Sonnenschein, D. (1973). A 90 minute daydream cycle. Sleep Research, 2, 187-188.

Krippner, S (1993). The Maimonides ESP-dream studies - Maimonides Medical Center, Journal of Parapsychology, 57, 279-319.

Lamon, W. H. (1865/1994). Recollections of Abraham Lincoln 1847-1865, by Ward Hill Lamon, University of Nebraska Press.

Lydic, R., Baghdoyan, H. A., & Lorinc, Z. (1991). Microdialysis of cat pons reveals enhanced ACh release during state-dependent respiratory depression. American Journal of Physiology, 261, 766-770.

MacLean, P. (1990). The Evolution of the Triune Brain. New York, Plenum.

Malh, G. F., Rothenberg, A., Delgado, J. M. R., & Hamlin, H. (1964). Psychological response in the human to intracerebral electrical stimulation. Psychosomatic Medicine, 26, 337-368.

Malinowski, B. (1954) Magic, Science and Religion. New York. Doubleday.

Meyer, J. S., Ishikawa, Y., Hata, T., & Karacan, I. (1987). Cerebral blood flow in normal and abnormal sleep and dreaming. Brain and Cognition, 6, 266-294.

Mesulam, M. M. (1981) Dissociative states with abnormal temporal lobe EEG: Multiple personality and the illusion of possession. Archives of General Psychiatry, 38, 176-181.

Monti, J. Pandi-Permal, S. R., Sinton, C. M. (2008). Neurochemistry of Sleep and Wakefulness, Cambridge U. Press.

Montigny, C. de, and Aghajanian G.K. (1977). Preferential action of 5-methoxytryptamine and 5-methoxydimethyltryptamine on presynaptic serotonin receptors: A comparative iontophoretic study with LSD and serotonin. Neuropharmacology, 16, 811-818.

Moody, R. (1977). Life after life. Georgia, Mockingbird Books.

Morris, J. S., Frith, C. D., Perett, D. I., Rowland, D., Young, A. W., Calder, A. J., & Colan, R. J. (1996). A differential neural response in the human amygdala to fearful and happy facial expression. Nature, 383, 812-815.

Mullan, S., & Penfield, W. (1959). Epilepsy and visual halluciantions. Archives of Neurology and Psychiatry, 81, 269-281.

Murri, L., Arena, R., Siciliano, G., Mazzotta, R., & Muratorio, A. (1984). Dream recall in patients with focal cerebral lesions. Archives of Neurology, 41, 183-185.

Nadel, L. (1991). The hippocampus and space revisited. Hippocampus, 1, 221-229.

Neihardt, J. G. & Black Elk, (1979). Black Elk speaks. Lincoln. U. Nebraska Press.

emchin, A. A., Whitehouse, M.J., Menneken, M., Geisler, T., Pidgeon, R.T., Wilde, S. A. (2008). A light carbon reservoir recorded in zircon-hosted diamond from the Jack Hills. Nature 454, 92-95.

Noyes, R., & Kletti, R. (1977) Depersonalization in response to life threatening danger. Comprehensive Psychiatry, 18, 375-384.

O'Keefe, J. (1976). Place units in the hippocampus of the freely moving rat. Experimental Neurology, 51, 78-109.

O'Keefe, J., & Bouma, H. (1969). Complex sensory properties of certain amygdala units in the freely moving cat. Experimental Neurology, 23, 384-398.

O'Neil, J., Carlson, R. W., Francis, E., Stevenson, R. K. (2008). Neodymium-142 Evidence for Hadean Mafic Crust Science 321, 1828 - 1831.

Parson, E. R. (1988). Post-traumatic self disorders (PTsfD): Theoretical and practical considerations in psychotherapy of Vietnam War Veterans. In J. P. Wilson, Z. Harel, & B. Kahana (Eds). Human Adaptation to Extreme Stress. New York, Plenum.

Pena-Casanova, J., & Roig-Rovira, T. (1985). Optic aphasia, optic apraxia, and loss of dreaming. Brain and Language, 26, 63-71.

Penfield, W., & Perot, P. (1963). The brains record of auditory and visual experience. Brain, 86, 595-695.

Perryman, K. M.,et al. (1987). Differential effects of inferior temporal cortex lesions upon visual and auditory-evoked potentials in the amygdala of the squirrel monkey. Behavioral and Neural Biology, 47, 73-79.

Purpura, D. P. (1956). Electrophysiological analysis of psychotogenic drug action. I & II. Archives of Neurology & Psychiatry, 40, 122-143.

Rawlings, M. (1978). Beyond deaths door. London, Sheldon Press.

Ribeiro, S., Goyal, V., Mello, C. & Pavlides, C. (1999). Brain gene expression during REM sleep depends on prior waking experience. Learning & Memory, 6: 500-508.

Ribeiro, S., Mello, C., Velho, T., Gardner, T., Jarvis, E., & Pavlides, C. (2002). Induction of hippocampal long-term potentiation during waking leads to increased extra hippocampal zif-268 expression during ensuing rapid-eye-movement sleep. Journal of Neuroscience, 22(24), 10914-10923.

Riberio, S., Sim☐es, C. & Nicolelis, M. (2008). Genes, Sleep and Dreams. In Lloyd & Rossi (Eds.) Ultradian rhythms from molecule to mind. Springer. N.Y., 413-430.

Ring, K. (1980). Life at death. New York, Coward, McCann & Geoghegan.

Rolls, E. T. (1984). Neurons in the cortex of the temporal lobe and in the amygdala of the monkey with responses selective for faces. Human Neurobiology, 3, 209-222.

Rolls, E. T. (1992). Neurophysiology and functions of the primate amygdala. In J. P. Aggleton (Ed.). The Amygdala. New York, Wiley-Liss.

Sabom, M. B. (1982). Recollections on death. New York, Harper & Row.

Sawa, M., & Delgado, J. M. R. (1963). Amygdala unitary activity in the unrestrained cat. Electroencephalography and Clinical Neurophysiology, 15, 637-650.

Schenk, L., & Bear, D. (1981) Multiple personality and related dissociative phenomenon in patients with temporal lobe epilepsy. American Journal of Psychiatry, 138, 1311-1316.

Serafetinides, E. A. (1965). The significance of the temporal lobes and of hemisphere dominance in teh production of the LSD-25 symptomology in man. Neuropsychologia, 3, 69-79.

Schutze, I., et al. (1987). Sensory input to single neurons in the amygdala of the cat. Experimental Neurology, 97, 499-515.

Slater, E. & Beard, A.W. (1963). The schizophrenia-like psychoses of epilepsy. British Journal of Psychiatry, 109, 95-112.

Soubrie, P. (1986). Reconciling the role of central serotonin neurons human and animal behavior. Behavioral and Brain Sciences, 9, 319-364.

Spoont, M. R. (1992). Modulatory role of serotonin in neural information processing: Implications for human psychopathology. Psychological Bulletin, 112, 330-350.

Steriade, M. M. & McCarley, R. W. (2005) Brain Control of Wakefulness and Sleep, Springer.

Strahlendorf, J. C. R., et al., (1982). Differential effects of LSD serotonin and l-tryptophan on visually evoked responses. Pharmacology Biochemistry and Behavior, 16, 51-55.

Subirana, A., & Oller-Daurelia, L. (1953). The seizures with a feeling of paradisiacal happiness as the onset of certain temporal symptomatic epilepsies. Congres Neurologique International. Lisbonne, 4, 246-250.

Tarachow, S. (1941). Tjhe clinical value of hallucinations in localizing brain tumors. American Journal of Psychiatry, 99, 1434-1442.

Taylor, D. C. (1972). Mental state and temporal lobe epilepsy. Epilepsia, 13, 727-765.

Teuber, H. L., et al. (1960). Visual field defects after penetrating missile wounds of the brain. Cambridge: Harvard University Press.

Trimble, M. R. (1991). The psychoses of epilepsy. New York, Raven Press.

Turner, B. H. Mishkin, M. & Knapp, M. (1980). Organization of the amygdalopetal projections from modality-specific cortical association areas in the monkey. Journal of Comparative Neurology, 191, 515-543.

Ullman, M., & Krippner, S. (1970). Dream studies and telepathy; An experimental approach. New York: Parapsychology Foundation.

Ursin H., & Kaada, B. R. (1960). Functional localization within the amygdaloid complex in the cat. Electroencephalography and Clinical Neurophysiology, 12, 1-20.

Vertes, R. P. (1990). Brainstem mechanisms of slow-wave sleep and REM sleep. In W. R. Klemm, & R. P. Vertes (Eds.). Brainstem mechanisms of behavior. Wiley. New York.

Weingarten, S. M., et al . (1977). The relationship of hallucinations to depth structures of the temporal lobe. Acta Neurochirugica 24: 199-216.

Williams, D. (1956). The structure of emotions reflected in epileptic experiences. Brain, 79, 29-67.

Wilson, I. (1987). The after death experience. New York, Morrow.

Wilson, J. A. (1951) The culture of ancient Egypt. Chicago, U. Chicago Press.

Wilson, M. A., & McNaughton, B. L. (1993). Dynamics of the hippocampal ensemble for space. Science, 261, 1055-1058.

Wiseman, R. (2011) Paranormality, Macmillan.

15. Quantum Entanglement With the Future: Lincoln Dreams of His Assassination

Abraham Lincoln Dreams Of His Death

In April of 1965, less than two weeks before he was gunned down by an assassin's bullet, President Abraham Lincoln dreamed of his own assassination. Lincoln told this dream to his wife and to several friends including his long time companion Ward Hill Lamon (1911) who was Lincoln's personal friend and former law partner. According to Lincoln:

"About ten days ago, I retired very late. I had been up waiting for important dispatches from the front. I could not have been long in bed when I fell into a slumber, for I was weary. I soon began to dream. There seemed to be a death-like stillness about me. Then I heard subdued sobs, as if a number of people were weeping. I thought I left my bed and wandered downstairs. There the silence was broken by the same pitiful sobbing, but the mourners were invisible. I went from room to room; no living person was in sight, but the same mournful sounds of distress met me as I passed along. I saw light in all the rooms; every object was familiar to me; but where were all the people who were grieving as if their hearts would break? I was puzzled and alarmed. What could be the meaning of all this? Determined to find the cause of a state of things so mysterious and so shocking, I kept on until I arrived at the East Room, which I entered. There I met with a sickening surprise. Before me was a catafalque, on which rested a corpse wrapped in funeral vestments. Around it were stationed soldiers who were acting as guards; and there was a throng of people, gazing mournfully upon the corpse, whose face was covered, others weeping pitifully. 'Who is dead in the White House?' I demanded of one of the soldiers, 'The President,' was his answer; 'he was killed by an assassin.' Then came a loud burst of grief from the crowd, which woke me from my dream. I slept no more that night; and although it was only a dream, I have been strangely annoyed by it ever since."

Consciousness and the Past Present Future are Entangled

"The distinctions between past present and future are illusions." - Albert Einstein

In quantum mechanics, where reality and the quantum continuum are a unity, time is also a unity such that the future present past are a continuum which are linked and the same could be said of personal consciousness which has a time-line extending backwards in time to the birth of that consciousness and into the future until that consciousness is extinguished at death--just as the majestic thousand year old tree which has fallen to its death, is linked-in-time to the seedling from which it first struggled to greet the sun. As such, the time-line of an individual, like a continuous string, exists in the future and in the present and past.

If considered as a "world line" and in space-like instead of time-like intervals, then consciousness from birth to death would be linked as a basic unity extending not in time but in space and the same could be said of time and consciousness which is entangled with the quantum continuum which encompasses past, present and future, as time also has a "world line" and is immeshed and entangled with the continuum. Hence, time, space, and consciousness are linked and interact as is also demonstrated by entanglement and the Uncertainty Principle.

When considered as a unity within the quantum continuum, time and consciousness exist in the future, past, present, simultaneously. Consciousness which exists in the future is entangled with consciousness which exists in the present and the past--like a rope of string stretched out and extending in all directions from the birth to death of that consciousness.

Consider Lincoln's dream of his own assassination. Lincoln was assassinated, and from the perspective of the present moment, this took place in the past. When Lincoln had the dream, and from his time-line perspective, the assassination takes place in the future. The Lincoln who had the dream is the Lincoln who was murdered; and if considered a string which begins at birth and ends with his death, then by plucking the string the entire string will vibrate such that what occurs in the future can affect the past. Thus, information may also be conveyed from the future to the past, perhaps along the rope of consciousness which extends in all directions and dimensions and is entangled with the quantum continuum which encompasses the past, present, and future.

The future may effect the past and a conscious mind in the past, may be effected by the future as the distinctions between past, present and future are an illusions and all are entangled.

The Quantum Physics of Entanglement and Retrocausation

It is well established that particles respond to and can influence and affect distant particles at speeds faster than light. This "spooky action at a distance" has been attributed to "fields," "mediator particles," gravity, and "quantum entanglement" (Bokulich & Jaeger, 2010; Juan et al. 2013; Sonner 2013).

It has been repeatedly demonstrated that entanglement swapping protocols can entangle two remote photons without any interaction between them and even with a significant time-like separation (Ma et al., 2012; Megidish et al. 2013; Peres 2000); the measurement of one simultaneously effects the measurement of another although separated by vast distances. . For example, two particles which are far apart have "spin" and they may spin up or down. However, an observer who measures and verifies the spin of particle A will at the same time effect the spin of particle B, as verified by a second observer. Measuring particle A, effects particle B and changes its spin.

Likewise observing the spin of B determines the spin of A. There is no temporal order as the spin of one effects the spin of the other simultaneously through the simple act of measurements. Even distant objects are entangled and have a symmetrical relationship and a constant conjunction (Bokulich & Jaeger, 2010; Plenio 2007; Sonner 2013).

If considered as a unity with no separations in time and space, then to effect one point in time-space is to effect all points which are entangled; and one of those entangled connections is consciousness (Joseph 2010a). And this gives rise to the uncertainty principle (Heisenberg, 1927) as the laws of cause and effect are violated. Correlation is not causation and it can't always be said with certainty which is the cause and which is the effect and this is because the cosmos is entangled.

Moreover, the decisions and measurements made in the future can effect the present. In one set of experiments entanglement was demonstrated following a delayed choice and even before there was a decision to make a choice. Specifically, four photons were created and two were measured and which became entangled such that the measurement of one effected the other simultaneously. However, if at a later time a choice was then made to measure the remaining two photons, all four became entangled before it was decided to do a second measurement, before the choice was even made in the future (Ma et al., 2012; Peres 2000).

Entanglement can occur independent of and before the act of measurement and choices made in the future can effect the present. "The time at which quantum measurements are taken and their order, has no effect on the outcome of a quantum mechanical experiment" (Megidish et al. 2013). Moreover, "two photons that exist at separate times can be entangled" (Megidish et al. 2013). As detailed by Megidish et al (2013): "In the scenario we present here, measuring the last photon affects the physical description of the first photon in the past, before it has even been measured. Thus, the "spooky action" is steering the system's past. Another point of view...is that the measurement of the first photon is immediately steering the future physical description of the last photon. In this case, the action is on the future of a part of the system that has not yet been created."

Therefore, entanglement between photons takes place even before the second photon even exists; "a manifestation of the non-locality of quantum mechanics not only in space, but also in time" (Megidish et al 2013). In other words, a photon may become entangled with another photon which is created and exists only in the future. Moreover, even after the first photon ceases to exist and before the second photon is created, both become entangle. Photons that do not exist can effect photons which do exist and photons which no longer exist and photons which will exist (Megidish et al. 2013); and this principle applies to electrons, photons, atoms, molecules and even diamonds separated by great distances (Lee et al. 2011; Matson 2012; Olaf et al. 2003; Schrodinger & Born 1935). The effect may even precede the cause even though the effect takes place in the future, as all are linked within the basic oneness of the quantum continuum which encompasses the past present future which are entangled.

The same principles can be applied to conscious phenomenon, including the experience of deja vu, premonitions, and dreams of events taking place in the future which is entangled with the present and that consciousness which exists in the future and in the past.

The past present and future are not only an illusion, but are a simultaneity; and this explains why entanglement (what Einstein called "spooky action at a distance") takes place faster than the speed of light (Plenio 2007; Juan et al. 2013; Francis 2012; Schrodinger & Dirac 1936), effecting electrons, photons, atoms, molecules and even diamonds separated by great distances instantaneously (Lee et al. 2011; Matson 2012; Olaf et al. 2003; Schrodinger & Born 1935). The effect may even precede the cause since it takes place faster than light. Even distant objects are entangled and have a symmetrical relationship and a constant conjunction (Bokulich & Jaeger, 2010; Plenio 2007; Sonner 2013).

If considered as a unity with no separations in time and space, then to effect one point in time-space is to effect all points which are entangled; and one of those entangled connections is consciousness which may dream of the future in the past.

Dream-Time and the Many Worlds of Quantum Physics

In dream-time past-present-future and the three dimensions of space may exist simultaneously as a gestalt thereby violating all the rules of causality abut not the laws of quantum physics. During dream-time events may occur in a logical or semi-logical temporal sequence, or they may be juxtaposed and make no sense at all to an external consciousness which is dependent on temporal sequences to achieve understanding. Because the future past present exist simultaneously, then a consciousness may experience de ja vu, have a premonition, or dream of the future or future events which take place in the future, but are entangled with the past.

Aberfan Disaster Dream-Time Precognition

Aberfan is a small village in South Wales. Throughout late September and October 1966, heavy rain lashed down on the area and seeped into the porous sandstone of the hills which surrounded the town and against which abutted the village school (Barker 1967).

On September 27 1966, Mrs SB of London dreamed about a school on a hillside, and a horrible avalanche which killed many children.

On October 14, 1966 Mrs GE from Sidcup, dreamed about a group of screaming children being covered by an avalanche of coal.

On October 20, 1966 Mrs MH, dreamed about a group of children who were trapped in a rectangular room and the children were screaming and trying to escape.

On October 20, 1966, a 10 year old child living in Aberfan woke up screaming from a nightmare. She told her parents that in her dream she was trying to go to school when "something black had come down all over it" and there was "no school there."

On October 21, 1966, part of the rain soaked hills of Aberfan gave way and half a million tons of debris slid toward the village of Aberfan and slammed into the village school. The 10 year old girl who dreamed of the tragedy and 115 other schoolchildren and 28 adults lost their lives when the school was smashed and covered with mud. There were less than a dozen survivors (Baker, 1967).

Assassination of Archduke Francis Ferdinand: Dream-Time Precognition

In June of 1914, Austria was seeking to expand it's central European empire; plans which were resented by neighboring states, including Serbia, who wished to remain independent. That same month, the Archduke Francis Ferdinand, nephew of the Austrian Emperor Francis Joseph, went on a diplomatic tour accompanied by his wife, to build alliances with the leaders of these independent nations. In late June he and his wife arrived in Sarajevo, Serbia.

On the evening of June 27, 1914, Bishop Joseph Lanyi prepared for bed and upon falling asleep he began to dream. The Archduke Franz Ferdinand of Austria, heir to the throne of Austria, had been the Bishop's student and pupil, and late that night the Archduke appeared in Bishop Lanyi's dream. The dream became a nightmare and at 3:15 AM Bishop Joseph Lanyi awoke, frightened, upset and in tears. He glanced at the clock, dressed himself, and because the dream was so horrible, he wrote it down:

"At a quarter past three on the morning of 28th June, 1914, I awoke from a terrible dream. I dreamed that I had gone to my desk early in the morning to look through the mail that had come in. On top of all the other letters there lay one with a black border, a black seal, and the arms of the Archduke. I immediately recognized the letter's handwriting, and saw at the head of the notepaper in blue colouring a picture which showed me a street and a narrow side-street. Their Highnesses sat in a car, opposite them sat a General, and an Officer next to the chauffeur. On both sides of the street there was a large crowd. Two young men sprang forward and shot at their Highnesses."

In the dream, Bishop Lanyi read the dream-letter, which had been written by the Archduke. According to the Bishop's account, which he wrote down in the early predawn hours of June 28, the dream letter from the Archduke was as follows: "Dear Dr Lanyi: Your Excellency. I wish to inform you that my wife and I were the victims of a political assassination. We recommend ourselves to your prayers. Cordial greetings from your Archduke Franz. Sarajevo, 28th June, 3.15 a.m."

Bishop Joseph Lanyi was convinced that the Archduke had been assassinated, and called his parishioners and household staff to tell them of the terrible news. Later that morning of June 28, 1914, the Bishop held a mass for the Archduke and his wife. But, the Archduke were still alive and would not be shot dead for another 2 hours.

On June 28, 1914, at 11 a.m., as the Archduke and his wife were leaving a ceremony at Sarajevo, a Serbian nationalist leaped from the crowd and killed them both. It was the Archduke's assassination which triggered World War One.

The Big Dream and the Quantum Wave Function

In the ancient world, be it Greek, Rome, Egypt, or Babylon, it was believed that some dreams contain important information regarding not only the individual, but his friends, family, and even the entire clan, village, city, or nation: The "big dream" of a child, woman, or man, were taken seriously by highly sophisticated and cultured ancient societies and were even announced two thousand years ago in the Roman Senate.Throughout history, the dreams of generals, kings, queens, emperors, and Pharaohs, were commonly scrutinized and their symbolism interpreted as they were believed to foretell the future.

What may distinguish the "big dream" from the "little dream" is the importance and impact of the future event on the "wave function"--just as a big rock vs a little rock, when tossed into a shimmering pond, will create bigger vs smaller waves and which ripple further and have greater impact on distant shores.

At one level within the quantum continuum, all of existence, including consciousness consists of particles and waves. Quantum theory merges the particle and wave. The wave function describes all the various possible states of the particle. The wave function is the particle spread out over space.

The wave function of a person sitting on their couch, would, at close proximity, resemble a quantum cloud in the general shape of their body and the couch--and the wave would spread out over space, but becoming vanishingly small until perhaps disappearing and becoming one with the quantum continuum.

All of the universe, including consciousness and the past present future have a wave function. Thus, the wave function of the future and of consciousness in that future, would also spread out and ripple across the sands of time and impacting the distant shores of the wave function of the conscious mind in the past, as all are entangled.

Precognition dreams are common and often involve negative, unhappy, unpleasant events such as deaths, disasters or personal calamities (Fukuda 2002; Lange et al. 2001; Ryback & Sweitzer, 1990), or, events of minor importance such as dreaming of a blond girl in a pink dress holding a blue balloon, and the next day observing several children holding balloons, one of whom is blonde and wearing a pink dress. About 40% of precognitive dreams are linked to an event that takes place the following day, or which are about to occur just minutes or seconds in the future.

Numerous studies have proved that most dreams last just seconds, even though they may be experienced as taking place over hours or days--and this is a function of accelerated dream consciousness, or what could be construed as dreaming at the speed of light as the brain is in a paradoxical state of exceedingly heightened activity. As such, the dream may be triggered by something that will occur in the distant or immediate future--and these are often the most common forms precognition dreaming. A famous dream dreamt by French physicist Alfred Maury is a classic example:

"I was rather unwell, and was lying down in my room, when I fell asleep and dreamed of the Reign of Terror [during the French Revolution]. I witnessed massacres... then I was appearing before the Revolutionary tribunal. I saw Robespierre, Marat, Fouquier-Tinville, all the most wicked figures of that terrible era; I talked to them; finally, after many events that I only partly remember, I was judged, condemned to death, taken out in a tumbril through a huge throng to the Place de la Revolution; I mounted the scaffold; the executioner tied me to that fatal plank, he tipped it up, the blade fell; I felt my head separating from my body, I woke up racked by the deepest anguish, and felt the bedpost on my neck. It had broke and had suddenly come off and had fallen on my cervical vertebrae just like the guillotine blade."

Certainly it would be expected that a major blow to the head and neck would cause instant waking. But in this instance, it did not. Instead, the blow to his neck occurred in the future, and the dream was woven around that event which was about to take place in the future, and the dreamer experienced a long and convoluted dream which was initiated by what was about to happen, and which could also be considered a warning of what was about to happen; albeit in the unique dream-language characteristic of dreams.

In numerous studies, anywhere from 17.8 % to 66% of individuals report that they experienced at least one precognitive dream. Like the dream of Alfred Maury, most of these are "little dream", such as a convoluted dream which logically leads up to the ringing of the phone or a knock on the door.

Coincidence, Parallel Worlds, and the Law of Averages

"Big Dreams" and precognitive dreams have been dismissed as due to "coincidence" and should be rejected based on the "law of large numbers." For example, the odds are with so many dreamers having dreams about so many different themes, that a few of them will have dreams about an airplane crash or a ship that sinks. If the next day a ship sinks or a plane crashes, this is merely coincidence. Hence, if precognitive dreams were real, they should be more commonplace, with more dreamers coming forward, and thus there should be a high "hit rate" when there is not; and as such "precognitive dreams" must be pure coincidence.

It may be, however, that just as there are men and women extraordinarily skilled in music, math, art, or athletics, some may be more adept than others at consciously perceiving these waves vs ripples across the sea of time.

Mozart heard his music in his head, already composed. He wrote it down, making few or no corrections, as if taking dictation, as he listened to it, as if it was a radio playing the music inside his head.

Wolfgang Amadeus Mozart (1756-1791) is considered to be one of the most gifted and talented composers and performers of all time. His famous symphonies include "Jupiter", "Haffner" and "Prague" but in modern time he is best known for his operas including "The Marriage of Figaro", "Don Giovanni" and "The Magic Flute." However, according to Mozart, and his wife, many of his works he heard in his mind, "already composed."

Mozart was a child prodigy, and by age 10 he was already a famous composer and performer. However, even at this tender age, this musical melodic genius, and extreme virtuoso, often wrote out his compositions without need for but minor corrections. By age 12 years old, Mozart had written 3 operas, 6 symphonies and over 100 other musical works and had performed for kings and queens.

Mozart reported that while he walked about, that he could hear the music in his mind, fully composed, being played by a symphony, and wrote it down as he heard it: "es ist alles komponiert, aber noch nicht geschrieben" ("I have already composed it").

As stated in the Nannerl and Leopold Novello diaries: "When some grand composition was working in his brain he was purely abstracted, walked about the apartment and knew not what was going on about him, but when once arranged in his mind, he needed no piano forte but would take music paper and begin writing."

Of course, he would also sometimes make minor corrections so not all his manuscripts are completely clean replicas of these stream of consciousness musical compositions. However, despite the fact that some of his compositions have corrections, the fact remains that the composition, in its entirety would first appear in his head, and then he would write down the finished product.

Perhaps there is a cosmic consciousness, a quantum continuum of consciousness, which contains all information, and that one need only a brain that can tune in and tap into select channels. Or, a Mozart in the future who composed the music in the future, is entangled with the mind of a Mozart who exists in the past. Because music is composed in the future, and due to entanglement, Mozart in the past/present then hears the composition already composed in his head; and that's because he is composing it in the future.

In the parallel universe, multiple worlds model of Everett (1956, 1957) and Dewitt (1971), when a physicist measures an object, the universe splits into two distinct universes, one of particles, the other of waves. In one universe, he measures the wave form, in the other universe he measures the particle; which is why he can't measure both at the same time. Hence, Mozart in one universe composes and plays the music, and the Mozart in the other universe hears it in his head. He had a rare talent. Not everyone is a Mozart.

Contrast the ability of the average man and woman, of which there are billions, and professional baseball players and their ability to get a "hit" when thrown a 90 MPH fast ball. Most of the billion average souls on this planet might get no hits at all. If we combine and average the hits (and misses) of professionals and non-professionals, then the odds of getting a hit are impossibly low. Hence, every time the professional player gets a hit, then it must be a coincidence, because 99% of all other humans can't do it, and most of the time he can't do it, and the overall odds and hit rate are so low.

Even if we restrict our analysis to professionals, the averages do not look very good. Consider, from 2000 until 2016, the average baseball batting average ranged around 267. In 2013-2015, during regular season play, out of 750 major league players, 725 had a batting average of less than 30% (http://espn.go.com/mlb/stats/batting). Taking into considering fowl balls and hits which result in "outs", but considering that each player has at least 3 opportunities each time at bat to hit the ball, and then factoring in that .267 average, it could be said that the average professional baseball player actually gets a base hit less than 10% of the time. Obviously this does not mean that the bat striking the ball is merely a coincidence because 90% of the time there are no hits. The same standard should be applied to precognitive "Big Dreams" and little dreams.

Relativity and Accelerated Dream Consciousness
It is through dreams that we may be transported to worlds that defy the laws of physics and which obeys their own laws of time, space, motion and conscious reality, where the future is juxtaposed with the past and where time runs backwards and forwards. Not uncommonly the dream will include so many branching and overlapping multiple realities that it makes no sense at all, except to those skilled in the art of interpreting dream symbolism Indeed, it is due to the non-temporal, often gestalt nature of dreams which require that they be consciously scrutinized from multiple angles to discern their meaning, for the last may be first and what is missing may be just as significant as what is there.

Throughout history it has been believed that dreams open doors to alternate realities, to the future, to the past, and the hereafter, where the spiritual world sits at the boundaries of the physical; hence the tendency to bury the dead in a sleeping position even 100,000 years ago. Although but a dream, the dream is experienced during dream consciousness much as the waking world is experienced by waking consciousness. The dream is real. Thus, throughout history dreams have been taken seriously especially when they gave glimpses of the future.

During dream sleep, the brain is in a heightened state of accelerated, paroxysmal activity as measured by electrophysiological indices. During the dream, events may unfold over a period of hours, even days, even though the dream lasted only seconds.

Relativity predicts that observers with an accelerated frame of reference experience time-contraction and a shrinking of time-space such that the future and the present come closer together relative to those with a different frame of reference (Einstein et al. 1923, Einstein 1961). Thus, since dream-consciousness and dream-time are also associated with accelerated levels of brain activity, during dream-time, the dreamer may see or experience the future before that future is experienced by the awake conscious mind or the consciousness of those external observers who have a different frame of reference as regard to the contraction of time.

In dream-time past-present-future and the three dimensions of space may exist simultaneously as a gestalt thereby violating all the rules of causality abut not the laws of quantum physics. During dream-time events may occur in a logical or semi-logical temporal sequence, or they may be juxtaposed and make no sense at all to an external consciousness which is dependent on temporal sequences to achieve understanding. Because the future past present may exist simultaneously and as the future may be experienced in a dream during accelerated states of brain activity, then during dream consciousness the dreamer may get glimpses of future events which may occur within days, the next morning, or which may even trigger wakefulness. In other words, just as increased velocity causes a contraction of space-time thereby decreasing the distance between the present and the future (Einstein 1961, Einstein et al. 2913), accelerated dream-consciousness may therefore transport the dreamer to the future which is entangled with the consciousness of the same observer which is still in the past.

This could be likened to the twin-paradox, where the time traveling twin reaches the future in less time than his/her twin back on Earth. Dream-time consciousness vs consciousness in the waking state, could be considered entangled twins, one of which accelerates to the future, the other awakening in its bed.

However, if the dreamer dreams the "big dream" vs the "small dream" may depend upon the wave function of that event which takes place in the future, and the ability to remember the dream upon waking.

Parallel Worlds and Dream Consciousness

In studies of entanglement, it has also been proposed that activity, or a measurement, in the present may exert a steering function on the future, via entanglement. If correct, then perhaps the dream of the future, is a dream that determines or influences the future.

Could the wave function of the dreaming Alfred Maury have caused his bedpost to break and fall upon his head? Entanglement and Everett's "Many worlds" interpretation says "yes."

Unlike conscious-time and the conscious mind, the dream-kaleidescape of dream-time and dream-consciousness could best be described as manifestation of the "Many worlds" interpretation of quantum physics: all worlds are possible and past and future and time and space are juxtaposed and intermingled; time can run backward and forward simultaneously and at varying speeds, and multiple realities come and go no matter now improbable.

According to Everett's theory, every action, every measurements, every behavior, every choice, even not choosing, can create a new reality, another world, generating a bifurcation between what happened and what did not happen, such that innumerable possibilities and possible worlds arise from every action, including realties which do not obey the laws of physics and cause and effect.

As conceived by Everett (1956, 1957) and Dewitt (1971), when a physicist measures an object, the universe splits into two distinct universes to accommodate each of the possible outcomes. In one universe, the physicist measures the wave form, in the other universe the physicist measures the object as a particle. Since all objects have a particle-wave duality, this also explains how an object can be measured as a particle and can be measured as a wave, but not both at the same time in the same world, and how it can be measured in more than one state, each of which exists in another world. The simple act of measurement creates two worlds both of which exist at the same time in parallel, and each separate version of the universe contains a different outcome of that event.

Instead of one continuous timeline, the universe and time-space, would be more like a forest of trees with innumerable branches and twigs, each of which represents a different possible world. However, by making a choice, a measurement, even by thinking and dreaming, one world among the many, is chosen, and becomes this reality.

What this would imply is that by dreaming of his death, Lincoln chose one world out of many--a world where he was struck by an assassins bullet, and then died, exactly as predicted by his dream.

Conclusions: Time, Consciousness, and Entanglement

Time is entangled. Future, past, present, are relative and overlap. Individual consciousness is a continuum which is born, dies, and thus exists in the future and the past. If time is considered from the perspective of space-like intervals and not time-like intervals, then causality can be forward, backward, or simultaneous (Bonor & Steadman, 2005; Buser et al. 2013; Carroll 2004; Godel 1995). The future and the past are entangled as a continuity in space-time and this means information can be transmitted from the past to the future, and from the future to the past simultaneously.

Consciousness, however, would also have a world line, which extends from the birth to the death of that consciousness. Consciousness is therefore entangled with itself, and could transmit information from itself to itself, and this is because that consciousness exists in the present, past, or future... and this is due to entanglement... and... because the distinctions between past present and future are an illusion.

"There seemed to be a death-like stillness about me. Then I heard subdued sobs... people weeping... the silence broken by pitiful sobbing... There I met with a sickening surprise. Before me... rested a corpse wrapped in funeral vestments. Around it were stationed soldiers... and... a throng of people, gazing mournfully upon the corpse... others weeping pitifully. 'Who is dead in the White House?' I demanded of one of the soldiers, 'The President,' was his answer; 'he was killed by an assassin.'" -Abraham Lincoln

16. Quantum Tunneling, LSD, Neuroscience, Doors of Perception

"There are things known and there are things unknown, and in between are the doors of perception." - Aldous Huxley

"To know that what is impenetrable to us really exists, manifesting itself to us as the highest wisdom and the most radiant beauty, which our dull faculties can comprehend only in their most primitive forms--this knowledge, this feeling, is at the center of all true religiousness. In this sense, and in this sense only, I belong to the ranks of devoutly religious men." -Einstein

In Plato's Allegory of the cave of shadows "reality" is shrouded in darkness and we perceive only echoes and shadows which reflect the divine reality beyond the world of the senses. Plato also believed that one could achieve enlightenment by liberating the mind and allowing it to be enveloped by the divine light.

The ability to see the light, however, is limited by the senses and the fact that the brain functions via inhibition: filtering reality so as to protect the brain, and the mind, from being overwhelmed. The mind and our conception of reality is restrained by our senses and perceptual capabilities, and the inhibitory influences of the frontal lobes and limbic system nuclei such as the amygdala, that filter out the incredible amount of stimuli which would otherwise be overwhelming.

For example, the two amygdala (located in the two temporal lobes), are comprised of neurons which can respond to auditory, visual, olfactory, gustatory, and tactile information simultaneously--"multi-modal" neurons. It is the amygdala which enables the conscious mind to experience a sweet taste, a pleasing melody, or determine if something might be good to eat. These amygdala neurons are modulated by neurotransmitters, such as serotonin, which serves an inhibitory function; otherwise, one might see sound and taste colors, resulting in sensory overload and confusion. Serotonin, via inhibition, allows the mind to remain focused on select stimuli while ignoring everything else which is filtered out.

The senses shape reality, for if the doors of perception were opened to all, the mind would become lost within the infinite. In quantum physics, this shaping of reality, is described as "the collapse of the wave function."

In the Copenhagen model of quantum physics, the ultimate reality is the quantum continuum; a frenzy of electromagnetic activity which has neither form, shape, or substance, until perceived by a conscious observer. However, the act of observing does not create reality. Rather, what is perceived are the blemishes, knots, and entanglements which may be observed as shape, form, cause, effect, and thus reality: the result of a decoupling of quanta from the quantum (coherent) continuum which leaks out and then couples together in a form of knot which is observed as a wave form collapse.

Perception serves as a fine-tuning filter; a process by which consciousness can selectively attend to those blemishes in the quantum continuum which are then recognized as cows, cats, chairs, molecules, moons, planets, all of which appear to have form and substance when in fact these objects consist mostly of empty space consisting of a frenzy of electromagnetic activity; the rapidity of which creates the illusion of solidity and phenomenon best described by Heisenberg's Uncertainty Principle: the probability that a particle will be in specific location.

Quantum Tunneling

Solidity is an illusion. Therefore, it is possible to pass through a solid object--between the holes within this frenzy of electromagnetic activity--via a process known as "Quantum Tunneling." It's been demonstrated, for example, that an energized particle can pass through a solid barrier; and these discoveries have led to the development of the tunneling microscope, the tunnel diode, and quantum computing. In 1973, Leo Esaki, Ivar Giaever and Brian Josephson received the Nobel Prize in Physics for their work in quantum tunneling and superconducting.

Quantum tunneling has sometimes been likened to a "ball" that either rolls over a mountain vs a "ball" that passes through the mountain. If the ball lacks sufficient energy, it will bounce off or roll back down the mountain. However, if provided enough energy, the ball can tunnel its way to the other side. Where would this ball gets its energy? Perhaps by "borrowing" it from its surroundings.

Quantum tunneling is also a biological phenomenon. Electron and proton tunneling play key roles in cellular respiration, photosynthesis, and DNA nucleosynthesis. By tunneling through the surrounding "energy barriers" electrons, protons, particles can pass through skin, bones, organs, and exit the other side; a concept known as quantum biology.

Quantum tunneling is believed to take place only at the quantum scale and can't be perceived by the senses. But this may not be accurate as the senses are biological and subject to quantum biology. Therefore, like a tunneling microscope, if the senses were provided enough energy or were released from inhibitory influences, then like the energized ball one might sense the mountain and then tunnel between the spaces that give rise to the illusion of solidity, so as to see, hear, and perhaps even touch and taste, what is on the other side.

LSD-25 and Quantum Tunneling

Lysergic acid diethylamide (LSD-25), was first created by Albert Hofmann in 1938; and who then experimented on himself with a dose of 250ug of pure

LSD, and then took a bicycle ride. He was mesmerized by the incredible colors and the enhancement of his senses, but soon felt so overwhelmed he returned to his home:

"My surroundings had now transformed themselves in more terrifying ways. Everything in the room spun around, and the familiar objects and pieces of furniture assumed grotesque, threatening forms. They were in continuous motion, animated, as if driven by an inner restlessness...Every exertion of my will, every attempt to put an end to the disintegration of the outer world and the dissolution of my ego, seemed to be a wasted effort...At times I believed myself to be outside my body, and then perceived clearly, as an outside observer. Kaleidoscopic, fantastic images surged in on me, alternating, variegated, opening and then closing themselves in circles and spirals, exploding in colored fountains, rearranging and hybridizing themselves in constant flux. It was particularly remarkable how every acoustic perception, such as the sound of a door handle or a passing automobile, became transformed into optical perceptions. Every sound generated a vividly changing image, with its own consistent form and color."

Later, Hofmann claimed that LSD liberated an "inborn faculty of visionary experience." He said it made him feel "reborn" and gave him the ability "to see again." But he also warned that its use might "represent a forbidden transgression of limits."

In 1967, I, and my very best friends (Big John, Jerry, Jared, and Richard), journeyed to San Francisco, to obtain LSD-25, from Owsley Stanely, who had been manufacturing it legally until outlawed the previous year. But that didn't stop Owsley. Between 1965 and 1967, he produced an estimated 1.5 million doses, including the famous "Purple Haze" which we purchased: 10 tablets, each guaranteed to contain 270 micrograms of pure LSD-25.

Jerry, Jerod, and Richard each took half a tablet on the way back to our homes in the South Bay; and I dropped them off at "The Park" along with John who agreed to watch over us. I waited until I had parked the car back at my house and then swallowed an entire tablet, 270 micrograms. It was going to be my first trip.

As I walked the 2 miles back to "The Park" I noticed the trees, flowers, lawns, and bushes, all seemed brighter, more colorful, radiant, as if illuminated from within. Stopping, I took hold of a brightly lit green leaf: I could taste the green-color through my finger tips! It was pulsating, radiating this energy; it was as if I could feel its life energy between my fingers. And then, I realized I could see the individual cells of the leaf; and then the fine cellular structure magnified and I could see its little veins and the surrounding cells pulsating with life. It was all very real-- as if my eyes had become a microscope. I was astonished; and I remember thinking: "how come I never noticed this before?" For reasons I can't explain, I had forgotten I'd taken the LSD--just 15 minutes before--but I remembered that my friends were waiting for me at The Park.

Upon my arrival, I was so overwhelmed with the colors, the tastes, the smells, the incredible vividity and clarity... I sat down to take it all in. When John approached to see if I was alright, I merely grinned and waved my arm and hand only to see dozens of arms and hands trailing one another; each of which seemed equally real, as if there were numerous copies of "me" each waving their hand, albeit slightly out of sync with each other, and leaving trails in their wake.

I studied my hand. Exactly as had occurred with the green leaf I'd fondled on my way to The Park, it was as if my eyes had become a microscope. I could see the fine cellular structure of the skin; and then, it was if my eyes had become a tunneling microscope (20 years before invented), and cellular layers became apparent and the nuclei within, and I could differentiate between different types and layers of skin cells; and then, I could see inside my hand: bones, ligaments, tissue, and blood vessels. And then, I could see the blood and red and white blood cells moving inside the the vessel, all going the same direction, only to be pulled back slightly between beats of my heart, and then continuing on their journey. I was simply amazed, and again I had the thought: "how come I never noticed this before?" And then, I could see through my hand and was gazing at the Santa Cruz Mountains and Mount Umunhum upon which sat a radar tower.

The mountain, I realized, was not solid; but was punctuated by a long series of holes which were in motion, like a train of giant bubbles, chain-like, rows and rows of holes popping in and out of existence, all going in the same direction--and I could see through these moving-holes in the mountain: the blue sky and the ocean far to the other side. And, I could hear a jet plane... and then I saw, through the holes in the mountain, a plane flying over the ocean, toward the mountains, and finally the plane appeared over the mountain, high in the sky; at which point John stepped in front of me, and said: "You've been staring at your hand for a long time."

I dropped my hand, and again, trails of numerous arms and hands; but as John was asking questions, I looked up at him, and could see the waves and vibrations of his voice, moving through the air toward me. And then John began to fractionate into billions of square-shaped particles, as if he was disintegrating and becoming a cloud of particles; and I could see right through him into the sky... and then the sky itself began to fractionate and split in half; and then, it was as if the sky was breaking apart and rolling away, in different directions, and I could see behind the sky, stars twinkling in the shimmering blackness of infinite night... And that was the first hour of my experience after ingesting 270 micrograms of LSD-25.

As an aside, many hours later, as the effects began to wane, Richard and I somehow walked to his house, entered his bedroom, turned on the black lights, and lay on the floor to grove on his psychedelic posters. There was this incredible celestial music playing; like classical, but deeper, more vibrant; as if performed by celestial gods... Richard and I both commented on its incredible beauty; at which point I asked: what station is that? He didn't know, so we got up to look only to discover the radio was off and not even plugged in--at which point, the music stopped. When we turned it back on, Led Zepplin blared from the speakers. The next day when Richard and I arrived at Jerry's house to share adventures, Jerry's mother and father were sitting quietly reading and listening to classical music. "What song is that?" I asked. "Mozart...." Richard and I exchanged looks.

Mozart: Mozart also heard his music in his head, already composed, being played by a symphony. He wrote it down, making few or no corrections, as if taking dictation, as he listened to it, as if it was a radio playing the music inside his head.

Mozart was a child prodigy, and by age 10 he was already a famous composer and performer. However, even at this tender age, this musical melodic genius, and extreme virtuoso, wrote out his compositions without need for but minor corrections. By age 12 years old, Mozart had written 3 operas, 6 symphonies and over 100 other musical works and had performed for kings and queens.

As stated in the Nannerl and Leopold Novello diaries: "When some grand composition was working in his brain he was purely abstracted, walked about the apartment and knew not what was going on about him... he needed no piano forte but would take music paper and begin writing."

Unlike other composers who would labor over their compositions, making numerous changes and corrections, the manuscripts of Mozart have few if any corrections as the music would appear, in its entirety, in his head, and then he would write down the finished product, saying simply: "es ist alles komponiert, aber noch nicht geschrieben" ("It's all composed, but not written yet").

Perhaps there is a cosmic consciousness, a quantum continuum of consciousness, which contains all information, and that one need only a brain that can tune in and tap into select channels. Perhaps Mozart possessed a brain such as this.

The Neuroscience of Quantum Tunneling

Serontonergic (5HT) neurons exert widespread and often tonic and inhibitory influences on a variety of brain areas including and especially, the inhibition of incoming sensory input, so that stimuli which are the most salient are attended to. That is, 5HT appears to be involved in learning not to perceive stimuli that are irrelevant and which are filtered out and suppressed.

LSD blocks 5HT transmission which results in increased activity in the sensory pathways to the amygdala, neocortex, and other brain areas. LSD also greatly increases electrophysiological activity in the amygdala and temporal lobe (within which the amygdala is buried). If the temporal lobes or the amygdala are surgically removed--particularly the right amygdala/temporal lobe--the enhanced sensory perceptions following LSD administration are abolished.

Although the effects of LSD include sensory distortions and hallucinations, its major effect is to block the inhibitory effects of 5HT, thereby resulting in heightened activity in amygdala multi-modal neurons. Sensory acuity is increased, multi-modal neurons in the amygdala perceive sound-touch-taste-and-visual stimuli simultaneously, and in many respects what is perceived are overlapping sensory qualities that are normally filtered out. Colors may be felt and tasted, music may be observed as well as heard, the molecular composition of ceilings, floor and walls may be parted so that one can see through the spaces between where atoms are linked together. And the pulse of life may be experienced as it ebbs and flows in a leaf held between one's finger tips. This information, however, is normally filtered out and suppressed.

Consider again, quantum tunneling. Again, after ingesting LSD, I could see the incredible cellular structure of my skin, and then the molecular structure, the pulsating molecules themselves; and then I could see between the spaces where the molecules joined together. My sight penetrated the skin and I could see the blood vessels, and then my eyes penetrated the blood vessels and I could see the blood platelets and the white corpuscles as they swirled through the vessel. It was more real than anything I'd seen in pictures, drawings, or film animations.

Our ability to perceive reality, is limited by our senses. Our senses are inhibited by a complex array of neurotransmitters and brain structures such as the amygdala and frontal lobe.

There are numerous species with senses different from, or which are in addition to, or more sensitive than those possessed by humans--such as echolocation and sonar in bats, whales and dolphins, or the olfactory acuity of dogs which is 40 times greater than woman and man. Innumerable species perceive a reality which overlaps with, but which is quite different from the reality perceived by the conscious human mind.

Likewise, substances which remove the inhibition of the senses, unlock the door to perceiving sensory impressions and realities which are normally filtered out. There is probably no adaptive reason for modern humans to taste color, see sound, or to gaze between the spaces of what appears to be solid objects; and thus, all these aspects of reality are filtered out and suppressed.

The Doors of Perception: Quantum Physics of Reality

Consider the theoretical foundations of matter, e.g., electrons, protons, neutrons, photons, elementary particles, etc., These "subatomic" particles, however, have only probable existences and display tendencies to assume certain patterns of activity that we perceive as shape and form.

We can only perceive what our senses (or the instruments we create) can detect, and what we detect as form and shape is really a mass of frenzied sub-atomic electromagnetic activity only a fraction of which is amenable to detection by our senses. If we possessed additional senses, or an increased sensory channel capacity, we would perceive yet other patterns and other realities.

And yet, reality may be mostly empty space. What we perceive as mass (shape, form, length, weight) are dynamic patterns of energy which we selectively attend to and then perceive as stable and static.

This energy that makes up the object of our perceptions, is also but an aspect of the electromagnetic continuum which has assumed a specific pattern that may be sensed and processed by our brain. Moreover, these patterns can break down and change and assume all manner of patterns that we cannot detect with our senses. And those we do perceive, do not reality exist, at the quantum level, but have probable existence, tendencies to exist and tendencies to no longer exist. This principle, is called the uncertainty principle because one can never predict with 100% certainty when a particle may exist and where it may exist when it does exist. It is these principles which form the crux of quantum electrodynamics, or field theory, and Heisenberg's Uncertainty Principle.

As stipulated by the Uncertainty Principle, one can measure the particle, or the wave, but not both at the same time. According to Everett's Multiple Worlds interpretation, this is because the "you" who is measuring the particle exists in one reality, and the "you" measuring the wave exists in another reality, in parallel. If applied to LSD, and the multiple hands/arms I saw following the motions of my own arms/hands, then (as a thought experiment) perhaps, I was observing "me" in other realities, moving "my" hands and arms. Or, maybe it was the lack of lateral inhibition between retinal cells, and each began sending, in waves of excitation, the same information to the visual neocortex which had also been disinhibited, resulting in multiple copies of the same reality.

Although the act of conscious observation, induces a wave form collapse, it is recognized, in quantum physics, that there are multiple realities which exist simultaneously, in parallel, and are entangled and overlap with the reality that we perceive. Quantum mechanics, as devised by Niels Bohr, Werner Heisenberg and others in the years 1924-1927, does not attempt to provide a description of an overall, objective reality, but instead is concerned with quanta, probabilities and the effects of an observer on what is being observed. The act of measurement causes what is being measured to assume one of many possible values and thus changes the probability of an object or particle to be moving at one speed or direction or to be in one position or location, vs many others. Thus, it could be said that the act of observation causes a wave function collapse which is interpreted as reality. However, if consciousness is enhanced, and inhibitory restraints are removed from the doors of perception, then realities which are hidden, may be revealed.

Consciousness and Reality

"If the doors of perception were cleansed every thing would appear to man as it is, Infinite." -William Blake

Human consciousness is shaped and limited by the senses; and what we sense and perceive as reality are particular patterns of electromagnetic activity that are in frenzied motion. Reality exists only because we perceive it as such; when in fact, what we perceive exists only as a fragment of the quantum continuum which display tendencies to assume certain patterns of activity that we perceive as shape and form--even though these shapes are mostly empty space. Shape and form are a function of our perception of these dynamic interactions. If we possessed additional senses, or an increased sensory channel capacity, we would perceive yet other patterns and other realities.

Thus, the patterns that resemble particles, planets and people, exist, and yet don't exist, and as such have only a probable existence, which gives rise to the uncertainty principle, because one can never predict with 100% certainty, when a particle may exist and where it may exist when it does exist.

This reality, therefore, is a manifestation of alterations in the patterns of activity within the electromagnetic field. The electromagnetic field, this energy, is therefore the fundamental entity, the continuum that constitutes the basic oneness and unity of all things.

However, as solidity is an illusion, a manifestation of this frenzy of activity, then those with enhanced sensory capabilities, may prove capable of tunneling between these spaces, and seeing and experiencing not just the mountain, but the divine light of the dawn sun rising on the other side.

References

Bohr, N. (1934/1987), Atomic Theory and the Description of Nature, reprinted as The Philosophical Writings of Niels Bohr, Vol. I, Woodbridge: Ox Bow Press.

Bohr, N. (1958/1987), Essays 1932-1957 on Atomic Physics and Human Knowledge, reprinted as The Philosophical Writings of Niels Bohr, Vol. II,

Bohr, N. (1963/1987), Essays 1958-1962 on Atomic Physics and Human Knowledge, reprinted as The Philosophical Writings of Niels Bohr, Vol. III, Woodbridge: Ox Bow Press.

Casagrande, V. A. & Joseph, R. (1978). Effects of monocular deprivation on geniculostriate connections in prosimian primates. Anatomical Record, 190, 359.

Casagrande, V. A. & Joseph, R. (1980). Morphological effects of monocular deprivation and recovery on the dorsal lateral geniculate nucleus in prosimian primates. Journal of Comparative Neurology, 194, 413-426.

Daly, D. (1958). Ictal affect. American Journal of Psychiatry, 115, 97-108.

DeWitt, B. S., (1971). The Many-Universes Interpretation of Quantum Mechanics, in B. D.'Espagnat (ed.), Foundations of Quantum Mechanics, New York: Academic Press. pp. 167-18.

DeWitt, B.S., Graham, N., editors (1973). The Many-Worlds Interpretation of Quantum Mechanics. Princeton University Press, Princeton, New-Jersey.

Eadie, B. J. (1993) Embraced by the light. New York, Bantam

Einstein, A. (1905a). Does the Inertia of a Body Depend upon its Energy Content? Annalen der Physik 18, 639-641.

Einstein, A. (1905b). Concerning an Heuristic Point of View Toward the Emission and Transformation of Light. Annalen der Physik 17, 132-148.

Einstein, A. (1905c). On the Electrodynamics of Moving Bodies. Annalen der Physik 17, 891-921.

Einstein, A. (1926). Letter to Max Born. The Born-Einstein Letters (translated by Irene Born) Walker and Company, New York.

Gallagher, R. E., & Joseph, R. (1982). Non-linguistic knowledge, hemispheric laterality, and the conservation of inequivalance. Journal of General Psychology, 107, 31-40.

Geschwind. N. (1981). The perverseness of the right hemisphere. Behavioral Brain Research, 4, 106-107.

Grinker, R. R., & Spiegel, J. P. (1945). Men Under Stress. McGraw-Hill.

Heisenberg. W. (1930), Physikalische Prinzipien der Quantentheorie (Leipzig: Hirzel). English translation The Physical Principles of Quantum Theory, University of Chicago Press.

Heisenberg, W. (1955). The Development of the Interpretation of the Quantum Theory, in W. Pauli (ed), Niels Bohr and the Development of Physics, 35, London: Pergamon pp. 12-29.

Heisenberg, W. (1958), Physics and Philosophy: The Revolution in Modern Science, London: Goerge Allen & Unwin.

Joseph, R. (1979). Effects of rearing and sex differences on learning and competitive exploration. Journal of Psychology, 101, 37-43.

Joseph, R. (1982). The Neuropsychology of Development: Hemispheric Laterality, Limbic Language, and the Origin of Thought. Journal of Clinical Psychology, 44, 3-34.

Joseph, R. (1986a). Confabulation and delusional denial: Frontal lobe and lateralized influences. Journal of Clinical Psychology, 42, 845-860.

Joseph, R. (1986b). Reversal of cerebral dominance for language and emotion in a corpus callosotomy patient. Journal of Neurology, Neurosurgery, and Psychiatry, 49, 628-634.

Joseph, R. (1988a) The Right Cerebral Hemisphere: Emotion, Music, Visual-Spatial Skills, Body Image, Dreams, and Awareness. Journal of Clinical Psychology, 44, 630-673.

Joseph, R. (1988b). Dual mental functioning in a split-brain patient. Journal of Clinical Psychology, 44, 770-779.

Joseph, R. (1990). Neuropsychology, Neuropsychiatry, Behavioral Neurology, Plenum, New York.

Joseph, R. (1992a) The Limbic System: Emotion, Laterality, and Unconscious Mind. The Psychoanalytic Review, 79, 405-456.

Joseph, R. (1992b). The Right Brain and the Unconscious. New York, Plenum.

Joseph, R. (1994) The limbic system and the foundations of emotional experience. In V. S. Ramachandran (Ed). Encyclopedia of Human Behavior. San Diego: Academic Press.

Joseph, R. (1996). Neuropsychiatry, Neuropsychology, Clinical Neuroscience, 2nd Edition. Williams & Wilkins, Baltimore.

Joseph, R. (1998). Traumatic amnesia, repression, and hippocampal injury due to corticosteroid and enkephalin secretion. Child Psychiatry and Human Development. 29, 169-186.

Joseph, R. (1999b). Environmental influences on neural plasticity, the limbic system, and emotional development and attachment, Child Psychiatry and Human Development. 29, 187-203.

Joseph, R. (1999a). The neurology of traumatic "dissociative" amnesia. Commentary and literature review. Child Abuse & Neglect. 23, 715-727.

Joseph, R. (1999c). Frontal lobe psychopathology: Mania, depression, aphasia, confabulation, catatonia, perseveration, obsessive compulsions, schizophrenia. journal of Psychiatry, 62, 138-172.

Joseph, R. (2000). Limbic language/language axis theory of speech. Behavioral and Brain Sciences. 23, 439-441.

Joseph, R. (2000). Fetal brain behavioral cognitive development. Developmental Review, 20, 81-98.

Joseph, R. (2000a). The evolution of sex differences in language, sexuality, and visual spatial skills. Archives of Sexual Behavior, 29, 35-66.

Joseph, R. (2003). Emotional Trauma and Childhood Amnesia. journal of Consciousness & Emotion, 4, 151-178.

Joseph, R. (2009). Quantum Physics and the Multiplicity of Mind: Split-Brains, Fragmented Minds, Dissociation, Quantum Consciousness. Journal of Cosmology, 2009, 3, 600-640.

Joseph, R. (2011). Origins of Thought: Consciousness, Language, Egocentric Speech and the Multiplicity of Mind. Journal of Cosmology, 14. 4577-4600.

Nadel, L. (1991). The hippocampus and space revisited. Hippocampus, 1, 221-

Neumann, J. von, (1937/2001), Quantum Mechanics of Infinite Systems. Institute for Advanced Study; John von Neumann Archive, Library of Congress, Washington, D.C.

Neumann, J. von, (1938), On Infinite Direct Products, Compositio Mathematica 6: 1-77.

Neumann, J. von, (1955), Mathematical Foundations of Quantum Mechanics, Princeton, NJ: Princeton University Press.

Noyes, R., & Kletti, R. (1977). Depersonalization in response to life threatening danger. Comprehensive Psychiatry, 18, 375-384.

O'Keefe, J. (1976). Place units in the hippocampus. Experimental Neurology, 51-78-100.

Pais, A. (1979). Einstein and the quantum theory, Reviews of Modern Physics, 51, 863-914.

Penfield, W. (1952) Memory Mechanisms. Archives of Neurology and Psychiatry, 67, 178-191.

Penfield, W., & Perot, P. (1963) The brains record of auditory and visual experience. Brain, 86, 595-695.

Ring, K. (1980). Life at Death: A Scientific Investigation of the Near-Death Experience. New York: Quill.

Southard, E. E. (1919). Shell-shock and other Neuropsychiatric Problems. Boston.

Sperry, R. (1966). Brain bisection and the neurology of consciousness. In F. O. Schmitt and F. G. Worden (eds). The Neurosciences. MIT press.

Williams, D. (1956). The structure of emotions reflected in epileptic experiences. Brain, 79, 29-67.

17. Paleolithic Cosmic Consciousness of the Cosmos

Abstract

The emergence of cosmological consciousness and its symbolism, is directly linked to the evolution of the Cro-Magnon peoples who may have developed the first cosmologies, 20,000 to 30,000 years ago. These ancient peoples of the Upper and Middle Paleolithic believed in spirits and ghosts which dwelled in a heavenly land of dreams, and interned their dead in sleeping positions and with tools, ornaments and flowers. By 30,000 years ago, and because they believed souls ascended to the heavens, the people of the Paleolithic searched the heavens for signs, and between 30,000 to 20,000 years ago, they observed and symbolically depicted the association between woman's menstrual cycle and the moon, patterns formed by stars, and the relationship between Earth, the sun, and the four seasons. These include depictions of 1) the "cross" which is an ancient symbol of the fours seasons and the Winter/Summer solstice and Spring/Fall equinox; 2) the constellations of Virgo, Taurus, Orion/Osiris, the Pleiades, and the star Sirius; 3) and the 13 new moons in a solar year. Although it is impossible to date these discoveries with precision, it can be concluded that cosmological consciousness first began to evolve over 30,000 years ago, and this gave birth to the first heavenly cosmologies over 20,000 years ago.

1. Cro-Magnon Cosmology and the Frontal Lobes

When humans first turned their eyes to the sun, moon, and stars to ponder the nature of existence and the cosmos, is unknown. The Cro-Magnon people were keen observers of the world around them, which they depicted with artistic majesty. The heavens were part of their world and they searched the skies for signs and observed the moon, the patterns formed by clusters of stars, and perhaps the relationship between the Earth, the sun, and the changing seasons. Although it is impossible to date cave paintings with precision, the first evidence of this awareness of the cosmic connection between Sun, Moon, Woman, Earth and the changing seasons are from the Paleolithic; symbolized in the creations of the Cro-Magnon of the Paleolithic.

As based on cranial comparisons and endocasts of the inside of the skull, and using the temporal and frontal poles as reference points, it has been demonstrated that the brain has tripled in size over the course of human evolution, and that the frontal lobes significantly expanded in length and height during the Middle to Upper Paleolithic transition (Blinkov and Glezer 1968; Joseph 1993; MacLean 1990; Tilney 1928; Weil 1929; Wolpoff 1980).

It is obvious that the height of the frontal portion of the skull is greater in the six foot tall, anatomically modern Upper Paleolithic H. sapiens (Cro-Magnon) versus Neanderthal and archaic H. sapiens (Joseph 1996, 2000b; Tilney, 1928; Wolpoff 1980). The evolution and expansion of the frontal lobe is also evident when comparing the skills and creative and technological ingenuity of the Cro-Magnons, vs the Neanderthals (Joseph 1993, 1996, 2000b).

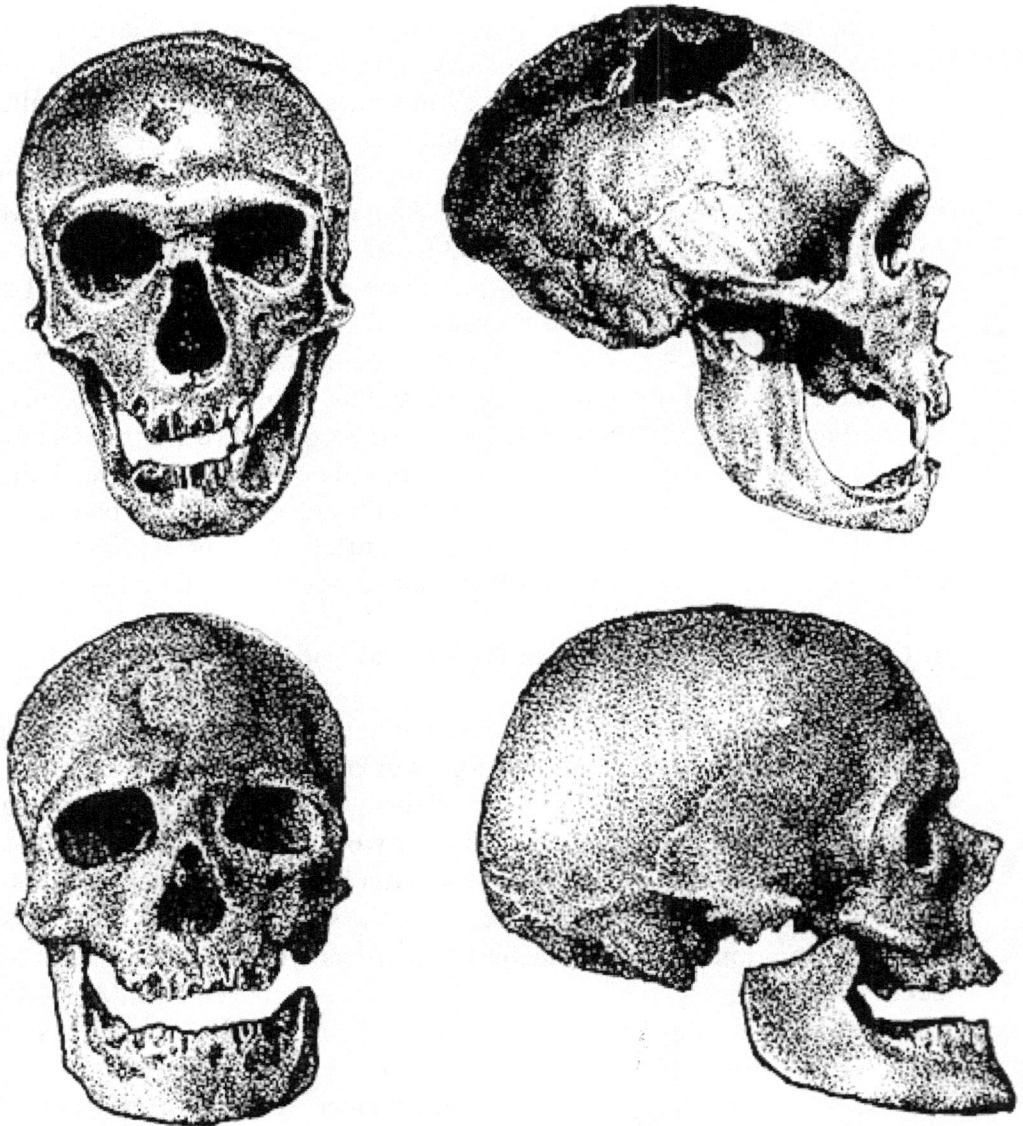

Figure: Neanderthal (top), Cro-Magnon (bottom)

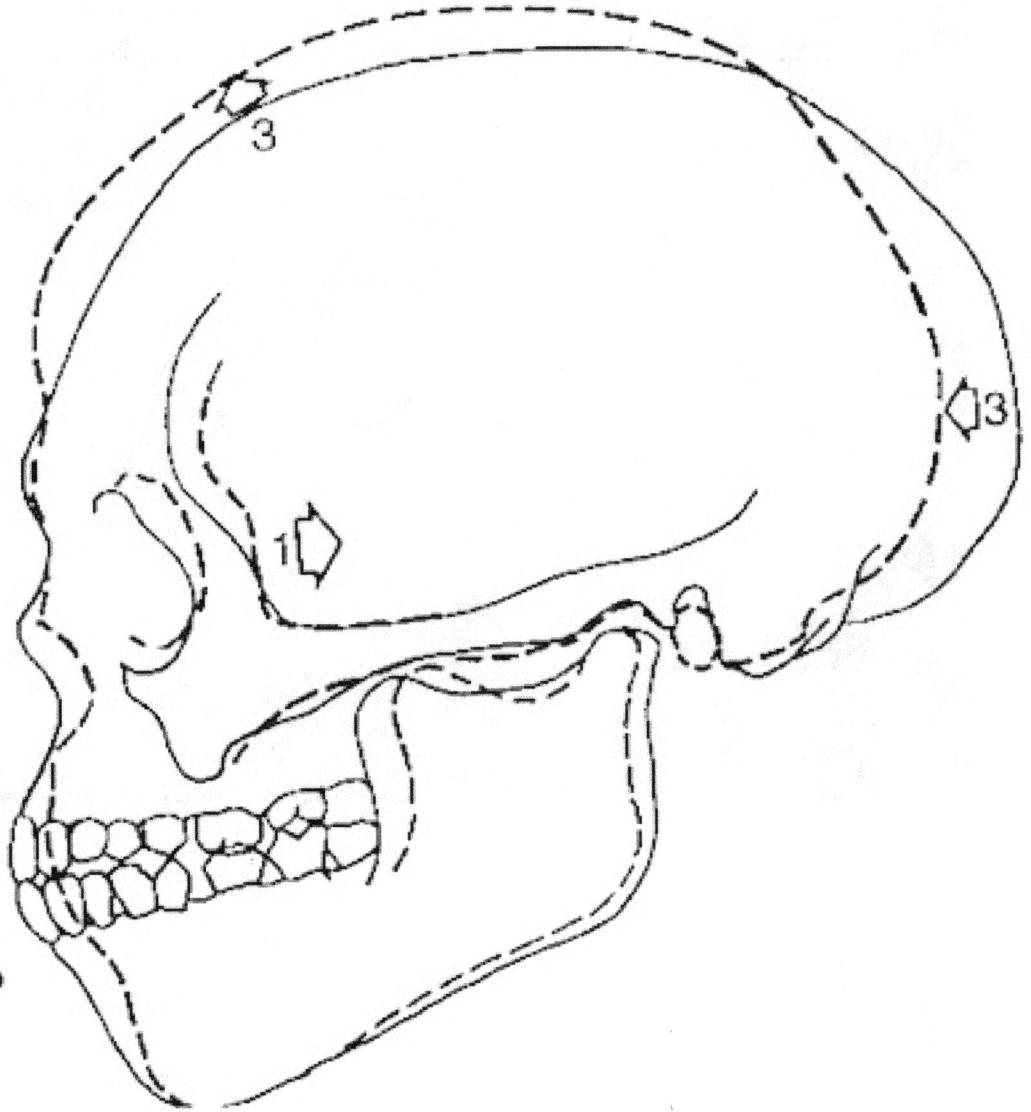

FIGURE A modern (dotted line) mesolithic cranium compared with a more ancient cranium (solid line). Arrows indicate the main average changes in skull structure including a reduction in the length of the occiput and an increase and upward expansion in the frontal cranial vault. Reproduced from M. H. Wolpoff (1980), Paleo- Anthropology. New York, Knopf.

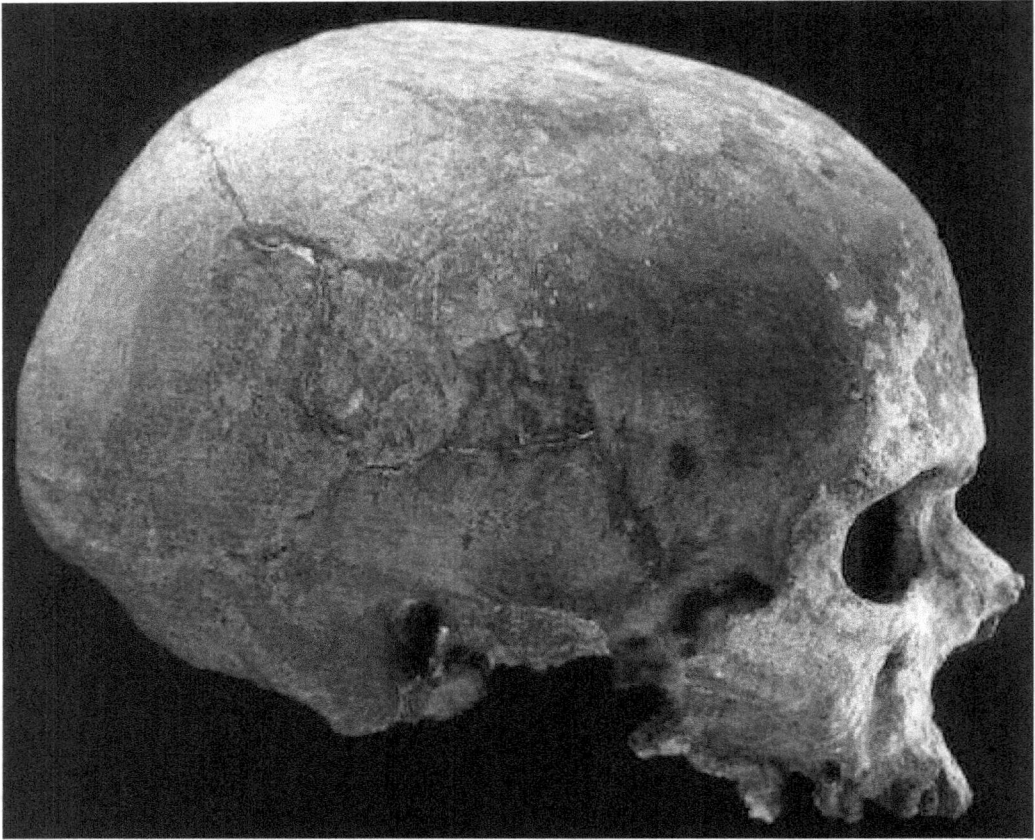

FIGURE: Cro-Magnon

Therefore, whereas the temporal, occipital and parietal lobes were well developed in archaic and Neanderthals, the frontal lobes would increase in size by almost a third in the transition from archaic humans to Cro-Magnon (Joseph 1996, 2000a,b, 2001). It is the evolution of the frontal lobes which ushered in a cognitive and creative big bang which gave birth to a technological revolution and complex spiritual rituals and beliefs in shamans and goddesses and their relationship to the heavens, and thus the moon and the stars.

It is well established that the frontal lobes enable humans to think symbolically, creatively, imaginatively, to plan for the future, to consider the consequences of certain acts, to formulate secondary goals, and to keep one goal in mind even while engaging in other tasks, so that one may remember and act on those goals at a later time (Joseph 1986, 1990b, 1996, 1999c). Selective attention, planning skills, and the ability to marshal one's intellectual resources so as to to anticipate the future rather than living in the past, are capacities clearly associated with the frontal lobes.

FIGURE: Paleolithic Goddess

The frontal lobes are associated with the evolution of "free will" (Joseph 1986, 1996, 1999c, 2011b) and the Cro-Magnon were the first species on this planet to exercise that free will, shattering the bonds of environmental/genetic determinism by doing what had never been done before: After they emerged upon the scene over 35,000 years ago, they created and fashioned tools, weapons, clothing, jewelry, pottery, and musical instruments that had never before been seen. They created underground Cathedrals of artistry and light, adorned with magnificent multi-colored paintings ranging from abstract impressionism to the surreal and equal to that of any modern master (Breuil, 1952; Leroi-Gourhan 1964, 1982). And they used their skills to carve the likeness of their female gods.

Thirty five thousand years ago, Cro-Magnon were painting animals not only on walls but on ceilings, utilizing rich yellows, reds, and browns in their paintings and employing the actual shape of the cave walls so as to conform with and give life-like dimensions, including the illusion of movement to the creature they were depicting (Breuil, 1952; Leroi-Gourhan 1964, 1982). Many of their engraving on bones and stones also show a complete mastery of geometric awareness and they often used the natural contours of the cave walls, including protuberances, to create a 3-dimensional effect (Breuil, 1952; Leroi-Gourhan 1964, 1982).

With the evolution of the Cro-Magnon people, the frontal lobes mushroomed in size and there followed an explosion in creative thought and technological innovation. The Cro-Magnon were intellectual giants. They were accomplished artists, musicians, craftsmen, sorcerers, and extremely talented hunters, fishermen, and highly efficient gatherers and herbalists. And they were the first to contemplate the heavens and the cosmos which they symbolized in art.

2. GODDESS OF THE MOON

Among the ancients, the Sun and the Moon were of particular importance and the Cro-Magnon observed the relationship between woman and the lunar cycle. Consider, the pregnant goddess, the Venus of Laussel, who holds the crescent moon in her hand (though others say it is a bison's horn). Although the length of a Cro-Magnon woman's menstrual cycle is unknown, it can be assumed that like modern woman she menstruated once every 28 to 29 days, which corresponds to a lunar month 29 days long, and which averages out to 13 menstrual cycles in a solar year. And not just menstruation, but pregnancy is linked to the phases of the moon.

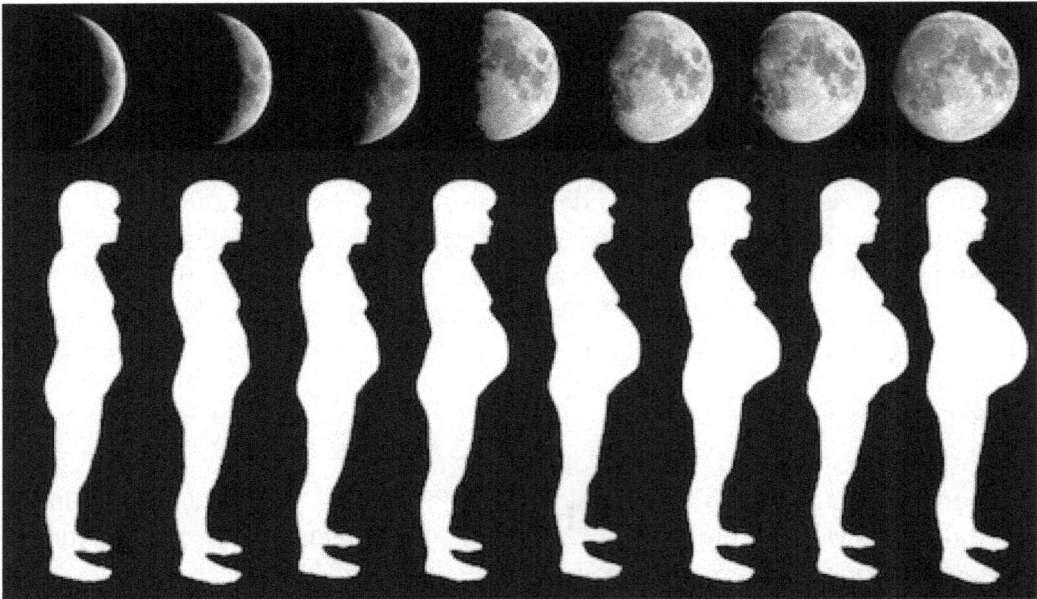

3. THE FOUR CORNERS OF THE SOLAR CLOCK.

When the Cro-Magnon turned their eyes to the heavens, seeking to peer beyond the mystery that separated this world from the next, they observed the sun. With a brain one third larger than modern humans, and given their tremendous power of observation, it can be predicted these ancient people would have associated the movement of the sun with the changing seasons which effected the behavior of animals, the growth of plants, and the climate and weather; all of which are directly associated with cyclic alterations in the position of the sun and the length of a single day over the course of a solar year which is equal to 13 moons.

The four seasons, marked by two solstices and the two equinoxes have been symbolized by most ancient cultures with the sign of the cross, e.g. the "four corners" of the world and the heavens. The "sign of the cross" generally signifies religious or cosmic significance. The Cro-Magnon also venerated the sign of the cross, the first evidence of which, an engraved cross, is at least 60,000 years old (Mellars, 1989). Yet another cross, was painted in bold red ochre upon the entryway to the Chauvet Cave, dated to over 30,000 years ago (Chauvet et al., 1996).

The illusion of movement of the Sun, from north to south, and then back again, in synchrony with the waxing and waning of the four seasons, is due to the changing tilt and inclination of the Earth's axis, as it spins and orbits the sun. Thus over a span of 13 moons, it appears to an observer that the days become shorter and then longer and then shorter again as the sun moves from north to south, crosses the equator, and then stops, and heads back north again, only to stop, and then to again head south, crossing the equator only to again stop and head north again. The two crossings each year, over the equator (in March and September) are referred to as equinoxes and refers to the days and nights being of equal length. The two time periods in which the sun appears to stop its movement, before reversing course (June and December), are referred to as solstices—the "sun standing still."

The sun was recognized by ancient astronomer priests, as a source of light and life-giving heat, and as a keeper of time, like the hands ticking across the face of a cosmic clock. Because of the scientific, religious, and cosmological significance of the sun, ancient peoples, in consequence, often erected and oriented their religious temples to face and point either to the rising sun on the day of the solstice (that is, in a southwest—northeast axis), or to face the rising sun on the day of the equinox (an east-west axis). For example, the ancient temples and pyramids in Egypt were oriented to the solstices, whereas the Temple of Solomon faced the rising sun on the day of the equinox.

Thus the sign of the cross is linked to the four seasons and heavens and tthe sun. Studying the heavens and the sun, has been been a common astronomical method of divining the the will of the gods, and for navigation, localization, and calculation: these celestial symbols have heavenly significance.

FIGURE: The God Seb supporting the Goddess Nut who represents heaven and the Milky Way galaxy. Note the repeated depictions of the key of life; i.e. a ring with a cross at the end.

The symbol of the cross is in fact associated with innumerable gods and goddesses, including Anu of the ancient Egyptians, the Egyptian God Seb, the Goddess Nut, the God Horus (the hawk), as well as Christ and the Mayan and Aztec God, Quetzocoatl. For example, like the Catholics, the Mayas and Aztecs adorned their temples with the sign of the cross. Quetzocoatl, like Jesus, was a god of the cross.

In China the equilateral cross is represented as within a square which represents the Earth, the meaning of which is: "God made the Earth in the form of a cross." It is noteworthy that the Chinese cross-in-a-box can also be likened to the swastika—also referred to as the "gammadion" which is one of the names of the Lord God: "Tetragammadion." The cross, in fact forms a series of boxes when aligned from top to bottom or side by side, and cross-hatchings such as these were carved on stone over 60,000 years ago.

FIGURE: Quetzocoatl the Mayan and Aztec god of the cross. The round shield encircling the cross represents the sun.

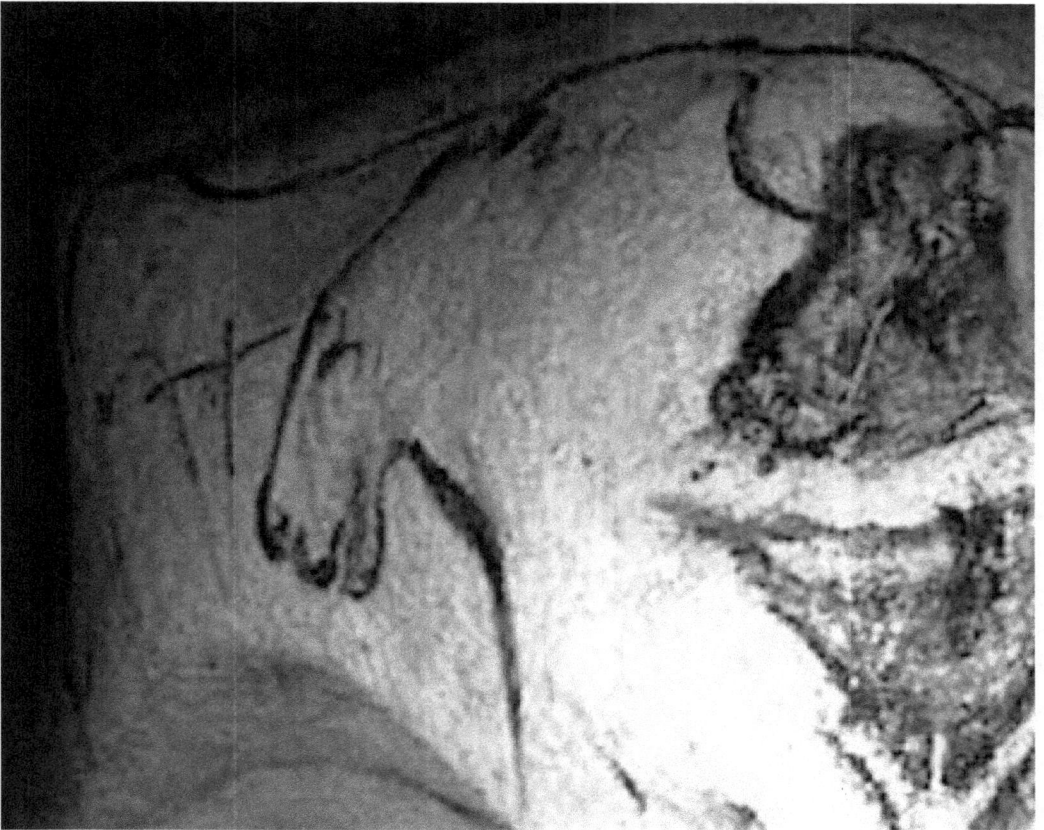

FIGURE: Sign of the cross (far left)

Among the ancient, the sign of the cross, represented the journey of the sun across the four corners of the heavens and marking the winter and summer solstice and the spring and autumn equinox. The Cro-Magon adorned the entrance and the walls of their underground cathedrals with the sign of the cross, which indicates this symbol was of profound cosmic significance. However, that some of the Cro-Magnon depictions of animal-headed men have also been found facing the cross, may also pertain to the heavens: the patterns formed by stars, which today are refereed to as "constellations."

Regardless of time and culture, from the Aztecs, Mayans, American Indians, Romans, Greeks, Africans, Christians, Cro-Magnons, Egyptians (the key of life), and so on, the cross consistently appears in a mystical context, and/or is attributed tremendous cosmic significance (Budge,1994; Campbell, 1988; Joseph, 2000a; Jung, 1964). The sign of the cross was the ideogram of the goddess "An", the Sumerian giver of all life from which rained down the seeds of life on all worlds including the worlds of the gods. An of the cross gave life to the gods, and to woman and man.

4.. THE CONSTELLATION OF VIRGO

here is nothing "virginal" about the constellation of Virgo. The pattern can be likened to a woman in lying on her back with an arm behind her head, and this may have been the visage which stirred the imagination of the Cro-Magnon.

FIGURES (above and below) Cro-Magnon / Paleolithic goddess, depicting the constellation of Virgo. La Magdelain cave.

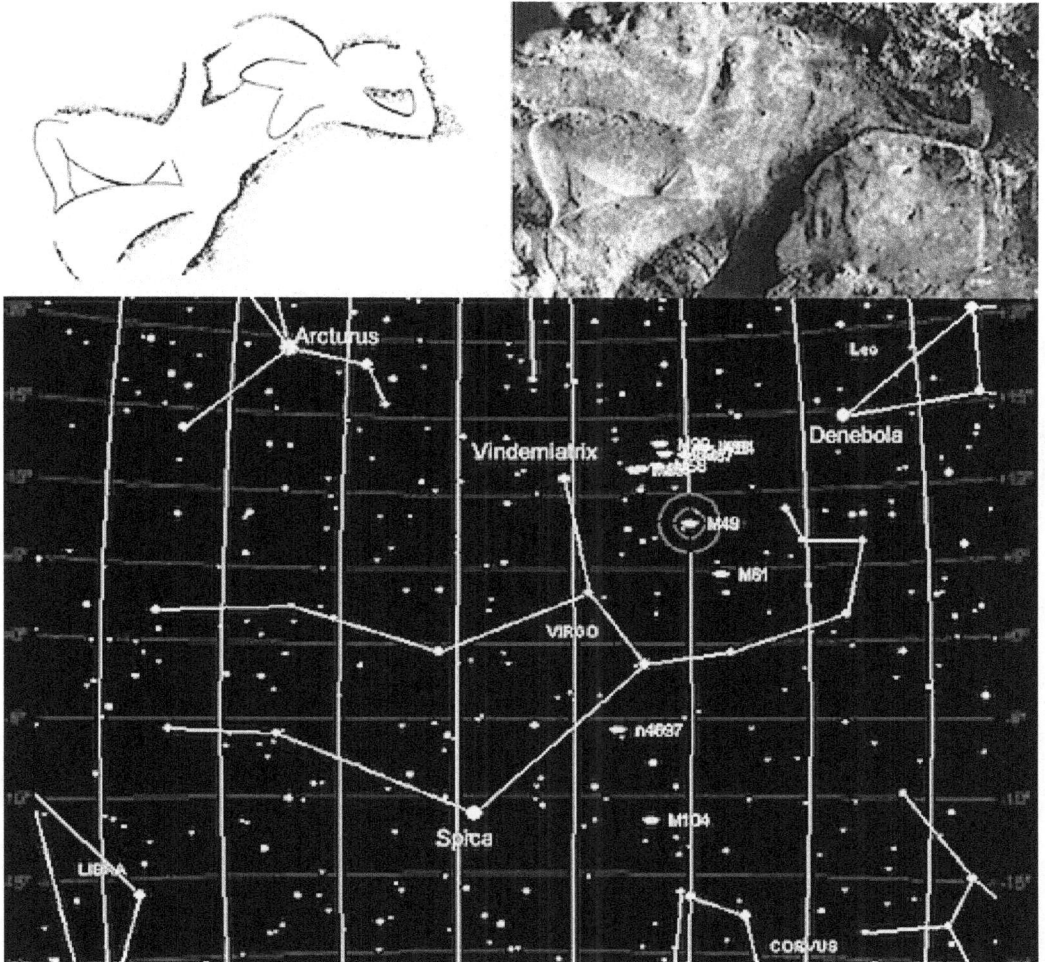

5. THE CONSTELLATIONS OF OSIRIS

It would be unreasonable to assume that the Cro-Magnon would not have observed the heavens or the illusory patterns formed by the alignment of various stars. Depictions of the various constellations, such as Taurus and Orion, and "mythologies" surrounding them, are of great antiquity, and it appears that similar patterns were observed by the Cro-Magnon people.

Consider, for example the "Sorcerers" or "Shamans" wearing the horns of a bull, and possibly representing the constellation of Taurus; a symbol which appears repeatedly in Lascaux, the "Hall of the Bulls" and in the deep recesses of other underground cathedrals dated from 18,000 to 30,000 B.P. And above the back of one of these charging bulls, appears a grouping of dots, or stars, which many authors believe may represent the Pleiades which is associated with Taurus. These Paleolithic paintings of the bull appear to be the earliest representation of the Taurus constellation.

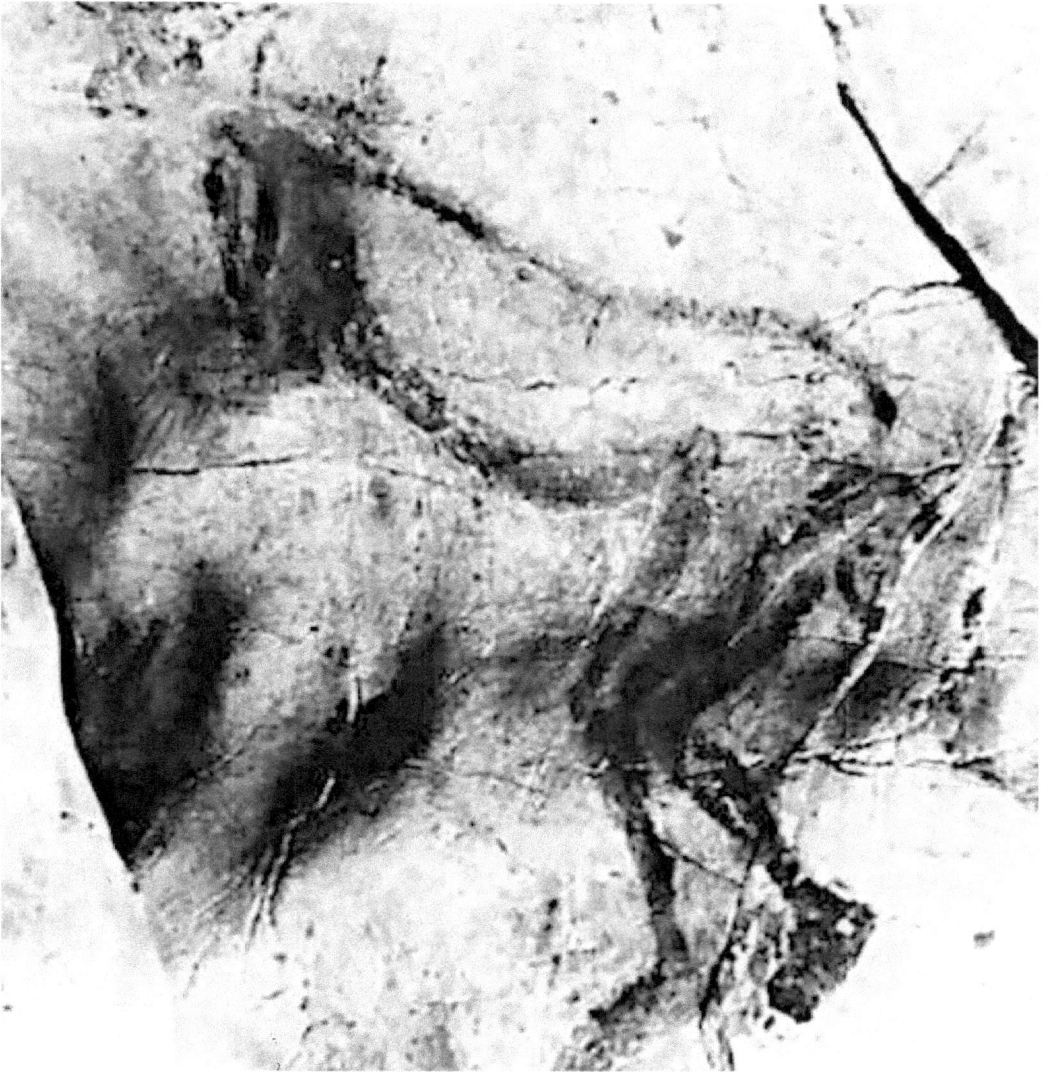

FIGURE: Ancient shaman attired in animal skins and stag antlers, graces the upper wall directly above the entrance to the 20,000-25,000 year-old grand gallery at Les Trois-Freres in southern France. Possibly representing the constellation of Orion.

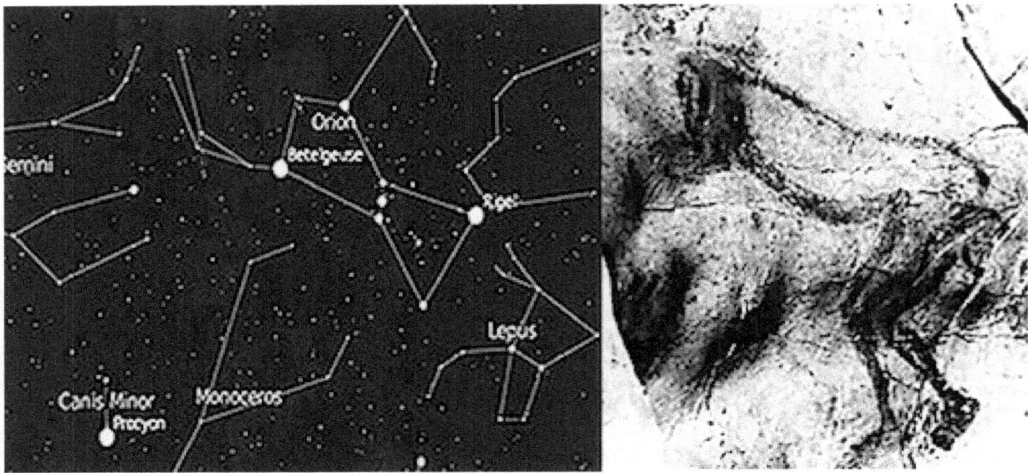

FIGURE. (Upper Right / Lower Left) The "Sorcerer" Trois-Frères cave. (Upper Left / Lower Right) Constellation of Orion/Osiris.

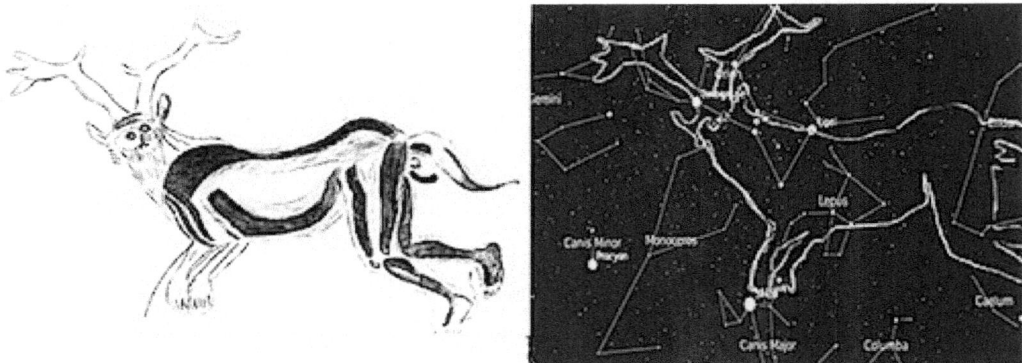

6. THE PLEIADES AND THE CONSTELLATIONS OF TAURUS AND ORION

In the "modern" sky, the constellation of Orisis/Orion the hunter, faces Taurus, the bull; and these starry patterns would not have been profoundly different 20,000 to 30,000 years ago. In ancient Egypt, dating back to the earliest dynasties (Griffiths 1980), Osiris was the god of death and of fertility and rebirth, who wore a a distinctive crown with two horns (later symbolized as ostrich feathers at either side). He was the brother and husband of Isis. According to myth, Orisis was killed by Set (the destroyer) and dismembered. Isis recovered all of his body, except his penis. After his death she becomes pregnant by Orisis. The Kings of Egypt were believed to ascend to heaven to join with

Osiris in death and thereby inherit eternal life and rebirth, symbolized by the star Sirius (Redford 2003). The Egyptian "King list" (The Turin King List) goes backward in time, 30,000 years ago to an age referred to as the "dynasty of gods" which was followed by a "dynasty of demi-gods" and then dynasties of humans (Smith 1872/2005).

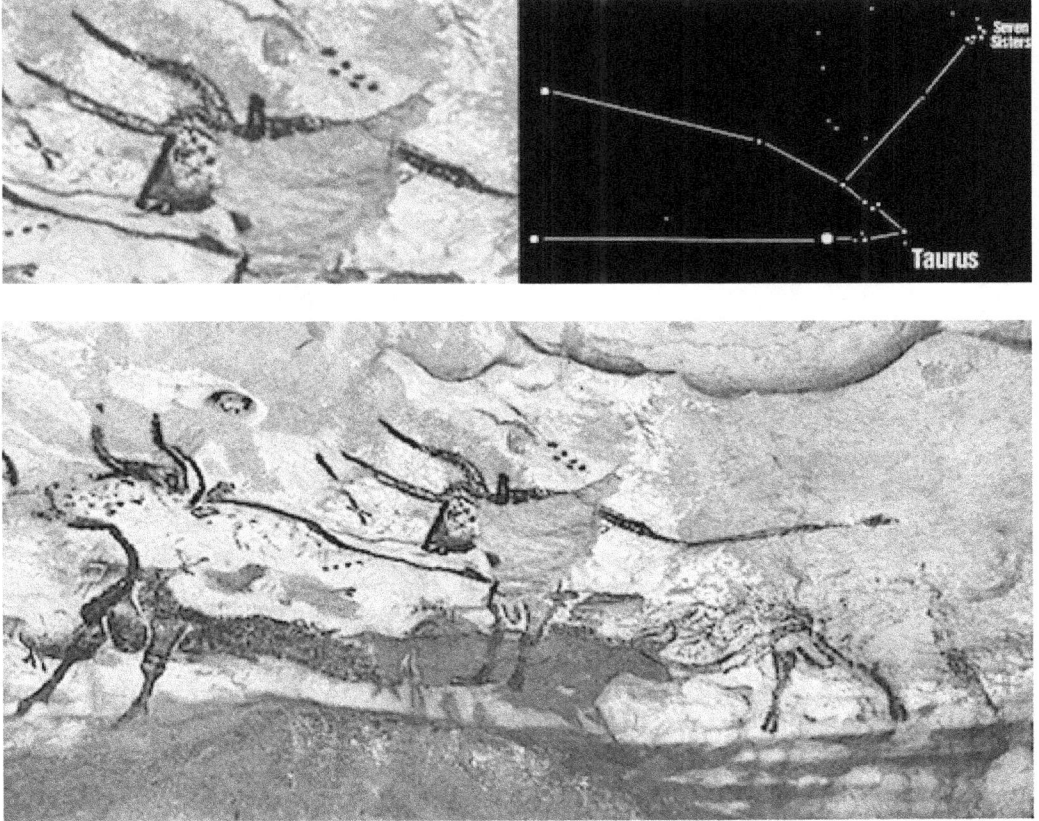

FIGURE: (Top) The main freeze of the bulls in the Lascaux Cave in Dordogne. There is a group of dots on the back of the great bull (Taurus) which may represent six of the seven stars of the Pleiades (the seven sisters). As stars are also in motion, not all would be aligned or as bright or dim today, as was the case 20,000 to 30,000 years ago.

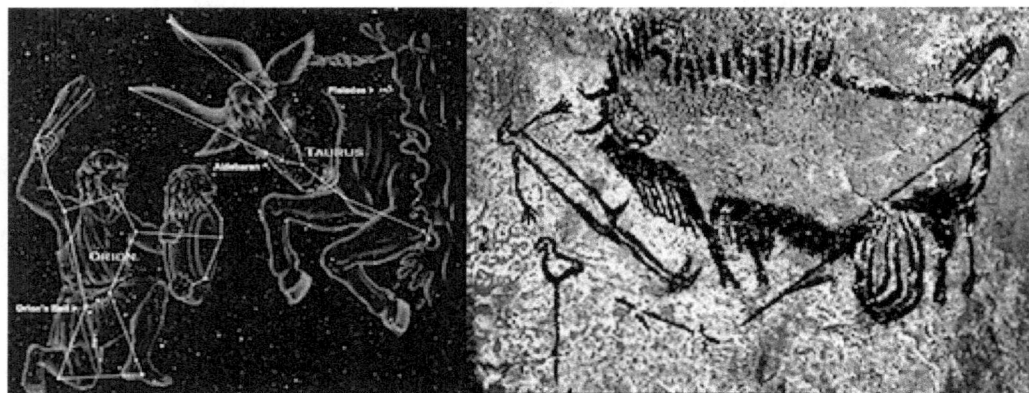

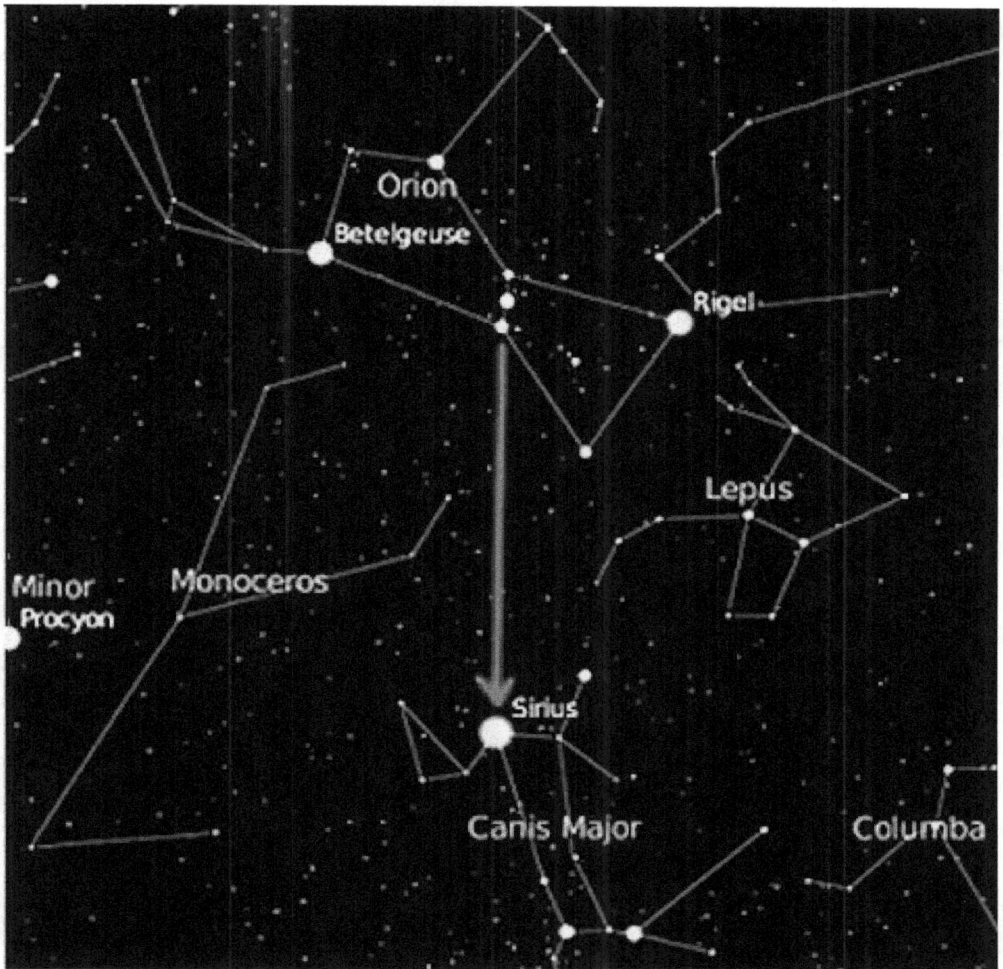

Over 20,000 years ago, the 6ft tall Cro-Magnon, with their massive brain one third larger than modern humans, painted a hunter with two horns who had been killed. And just as the constellation of Orion the hunter faces Taurus, so too does the dead Cro-Magnon hunter who has dismembered/disembowled the raging bull. And below and beneath the dead Cro-Magnon hunter, another bird, symbol of rebirth, and perhaps symbolizing the star Sirius.

The constellation of Osiris (Orion the hunter) in Egyptian mythology is the god of the dead who was dismembered; but also represents resurrection and eternal life as signified by the star Sirius. (Upper Right) Constellation of Osiris/Orion and Taurus. (Upper Left) Cave painting. Lascaux. The dead (bird-headed or two horned) hunter killed by a bull whom he disemboweled. (Bottom) Constellation of Orion/Osiris in relation to Sirius.

7. THE PALEOLITHIC AND NEOLITHIC MILKY WAY GALAXY

These peoples of the Paleolithic were capable of experiencing love, fear, and mystical awe, and they believed in spirits and ghosts which dwelled in a heavenly land of dreams. Because they believed souls ascended to the heavens, the people of the Paleolithic searched the heavens for signs. By 30,000 years ago, and with the expansion of the frontal lobes, they created symbolic rituals to help them understand and gain control over the spiritual realms, and created signs and symbols which could generate feelings of awe regardless of time or culture. They observed and symbolically depicted the association between woman and the moon, patterns formed by stars, and the relationship between Earth, the sun, and the four seasons.

The Milky Way galaxy can be viewed in the darkness of night, edge-on, snaking in a curving arc, forming part of a circle. If the peoples of the Paleolithic, through careful observation, deduced the existence of a spiraling galaxy, of which Earth, and the constellations circled round, or which circled round forming a cosmic clock, is unknown.

FIGURE: Quetzalcoatl Maya Galaxy

FIGURE: Petroglyph of the Milkyway galaxy, date unknown.

FIGURE: Quetzalcoatl

FIGURE: Milky Way Galaxy. (Below) Aztec, Quetzalcoatl / Galaxy

FIGURE: Ancient Eguypt: Osiris atended by the Gemini twins, and above: the Milky Way galaxy, and 12 constellations of the zodiac represented by snakes.

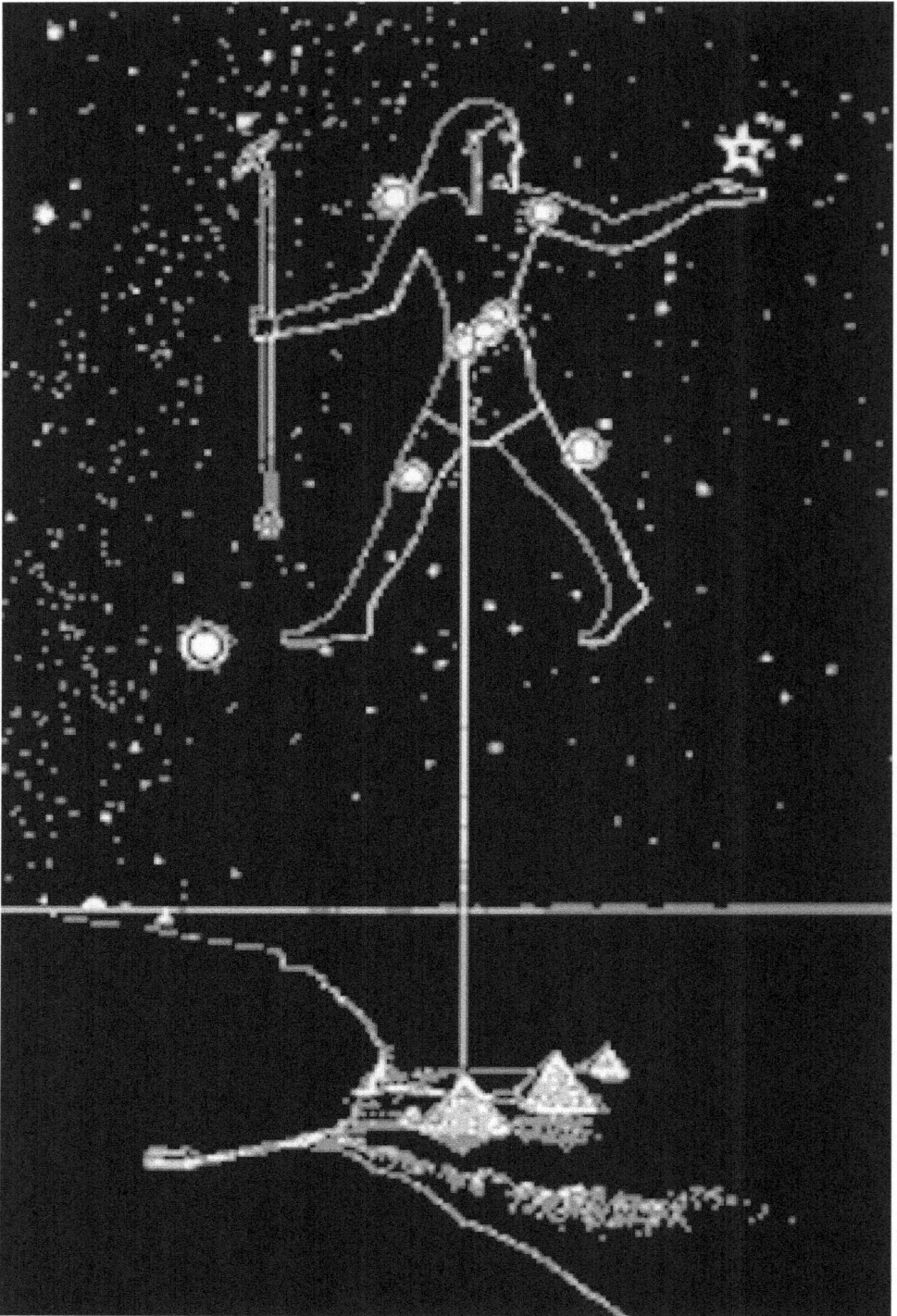

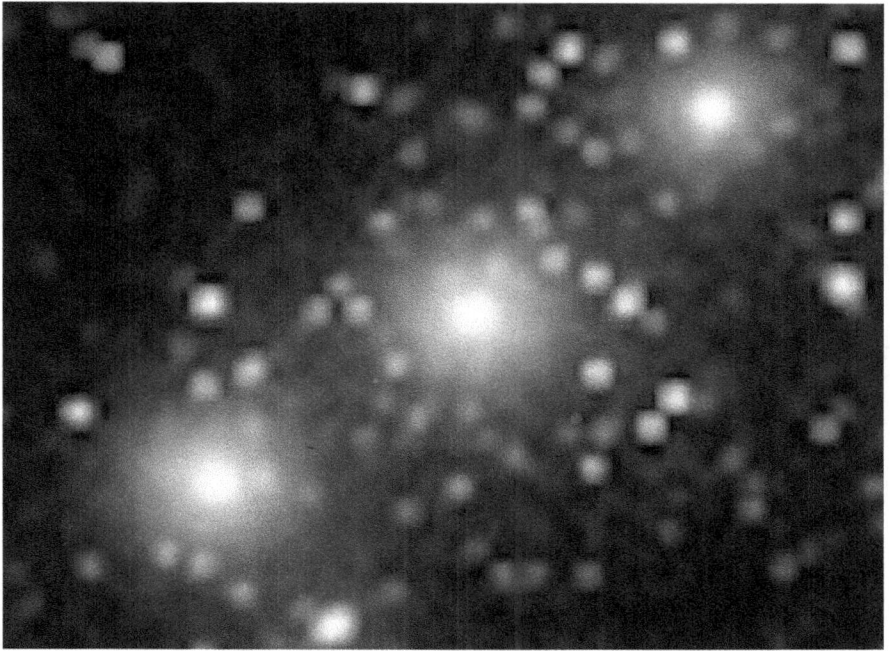

FIGURE: Three belt stars of Osiris (above) Three Pyramids of Giza (below)

References

Akazawa , T & Muhesen, S. (2002). Neanderthal Burials. KW Publications Ltd.

Amaral, D. G., Price, J. L., Pitkanen, A., & Thomas, S. (1992). Anatomical organization of the primate amygdaloid complex. In J. P. Aggleton (Ed.). The Amygdala. (Wiley. New York.

Bandi, H. G. (1961). Art of the Stone Age. New York, Crown PUblishers, New York.

Bear, D. M. (1979). Temporal lobe epilepsy: A syndrome of sensory-limbic hyperconnection. Cortex, 15, 357-384.

Belfer-ohen, A., & E.Hovers, (1992). In the eye of the beholder: Mousterian and Natufian burials in the levant. Current Anthropology 33: 463-471.

Breuil. H. (1952). Four hundred centuries of cave art. Montignac.

Budge, W. (1994). The Book of the Dead. New Jersey, Carol.

Butzer, K. (1982). Geomorphology and sediment stratiagraphy, in The Middle Stone Age at Klasies River Mouth in South Africa. Edited by R. Singer and J. Wymer. Chicago: University of Chicago Press.

Binford, L. (1981). Bones: Ancient Men & Modern Myths. Academic Press, NY

Binford, S. R. (1973). Interassemblage variability--the Mousterian and the 'functional' argument. In The explanation of culture change. Models in prehistory. edited by C. Renfrew. Pittsburgh: Pittsburgh U. Press.

Binford S. R. (1982). Rethinking the Middle/Upper Paleolithic transition. Current Anthropology 23: 177-181.

Blikkov, S. M., & Glezer, I. I. (1968). The human brain in figures and tables. New York: Plenum.

Campbell, J. (1988) Historical Atlas of World Mythology. New York, Harper & Row.

Cartwright , R. (2010) The Twenty-four Hour Mind: The Role of Sleep and Dreaming in Our Emotional Lives. Oxford University Press.

Chauvet, J-M., Deschamps, E. B. & Hillaire, C. (1996) Dawn of Art: The Chauvet Cave. H.N. Abrams.

Clark, G. (1967) The stone age hunters. Thames & Hudson.

Clark, J. D., & Harris, J. W. K. (1985). Fire and its role in early hominid lifeways. African Archaeology Review, 3, 3-27.

Conrad, N. J., & Richter, J. (2011). Neanderthal Lifeways, Subsistence and Technology. Springer.

Dennell, R. (1985). European prehistory. London, Academic Press.

Eadie, B. J. (1992). Embraced by the light. California, Gold Leaf Press.

Frazier, J. G. (1950). The golden bough. Macmillan, New York.

Gowlett, J. (1984). Ascent to civilization. New York: Knopf.

Gowlett, J.A. (1981). Early archaeological sites, hominid remains and traces of fire from Chesowanja, Kenya. Nature, 294, 125-129.

Griffiths, J. G. (1980). The Origins of Osiris and His Cult. Brill.

Harold, F. B. (1980). A comparative analysis of Eurasian Palaeolithic burials. World Archaeology 12: 195-211.

Harold, F. B. (1989). Mousterian, Chatelperronian, and Early Aurignacian in Western Europe: Continuity or discontinuity?" In P. Mellars & C. B. Stringer (eds). The human revolution: Behavioral and biological perspectives on the origins of modern humans, vol 1.. Edinburgh: Edinburgh University Press.

Harris, M. (1993) Why we became religious and the evolution of the spirit world. In Lehmann, A. C. & Myers, J. E. (Eds) Magic, Witchcraft, and Religion. Mountain View: Mayfield.

Harvati, K., & Harrison, T. (2010). Neanderthals Revisited. Springer.

Hayden, B. (1993). The cultural capacities of Neandertals: A review and re-evaluation. Journal of Human Evolution 24: 113-146.

Holloway, R. L. (1988) Brain. In: Tattersall, I., Delson, E., Van Couvering, J. (Eds.) Encyclopedia of human evolution and prehistory. New York: Garland.

Joseph, R. (1990b) The frontal lobes. In A. E. Puente and C. R. Reynolds (series editors). Critical Issues in Neuropsychology. Neuropsychology, Neuropsychiatry, Behavioral Neurology. Plenum, New York.

Joseph, R. (1992) The Limbic System: Emotion, Laterality, and Unconscious Mind. The Psychoanalytic Review, 79, 405-456.

Joseph, R. (1993) The Naked Neuron. Evolution and the languages of the body and the brain. Plenum. New York.

Joseph, R. (1994) The limbic system and the foundations of emotional experience. In V. S. Ramachandran (Ed). Encyclopedia of Human Behavior. San Diego, Academic Press.

Joseph, R. (1996). Neuropsychiatry, Neuropsychology, Clinical Neuroscience, 2nd Edition. 21 chapters, 864 pages. Williams & Wilkins, Baltimore.

Joseph, R. (1998a). The limbic system. In H.S. Friedman (ed.), Encyclopedia of Human health, Academic Press. San Diego.

Joseph, R. (2001). The Limbic System and the Soul: Evolution and the Neuroanatomy of Religious Experience. Zygon, the Journal of Religion & Science, 36, 105-136.

Joseph, R. (2002). NeuroTheology: Brain, Science, Spirituality, Religious Experience. University Press.

Joseph, R. (2011a). Dreams and Hallucinations: Lifting the Veil to Multiple Perceptual Realities, Cosmology, 14, In press.

Joseph, R. (2011). The neuroanatomy of free will: Loss of will, against the will "alien hand", Journal of Cosmology, 14, In press.

Jung, C. G. (1945). On the nature of dreams. (Translated by R.F.C. Hull.), The collected works of C. G. Jung, (pp.473-507). Princeton: Princeton University Press.

Jung, C. G. (1964). Man and his symbols. New York: Dell.

Kawashima, R., Sugiura, M., Kato, T., et al., (1999). The human amygdala plays an important role in gaze monitoring. Brain, 122, 779-783.

Kling. A. S. & Brothers, L. A. (1992). The amygdala and social behavior. In J. P. Aggleton (Ed.). The Amygdala. New York, Wiley-Liss.

Kurten, B. (1976). The cave bear story. New York: Columbia University Press.

Leroi-Gourhan, A. (1964.) Treasure of prehistoric art. New York: H. N. Abrams.

Leroi-Gourhan, A. (1982). The archaeology of Lascauz Cave. Scientific American 24: 104-112.

MacLean, P. (1990). The Evolution of the Triune Brain. New York, Plenum.

Malinowski, B. (1954) Magic, Science and Religion. New York. Doubleday.

McCown, T. (1937). Mugharet es-Skhul: Description and excavation, in The stone age of Mount Carmel. Edited by D. A. E. Garrod and D. Bate. Oxford: Clarendon Press.

Mellars, P. (1989). Major issues in the emergence of modern humans. Current Anthropology 30: 349-385.

Mellars, P. (1996) The Neanderthal legacy. Princeton University Press.

Mellars, P. (1998). The fate of the Neanderthals. Nature 395, 539-540.

Morris, J. S., Frith, C. D., Perett, D. I., Rowland, D., Young, A. W., Calder, A. J., & Colan, R. J. (1996). A differential neural response in the human amygdala to fearful and happy facial expression. Nature, 383, 812-815.

Petrides, M., & Pandya, D. N. (1999). Dorsolateral prefrontal cortex: comparative cytoarchitectonic analysis in the human and the macaque brain and corticocortical connection patterns. European Journal of Neuroscience 11.1011–1036.

Petrides, M., & Pandya, D. N. (2001). Comparative cytoarchitectonic analysis of the human and the macaque ventrolateral prefrontal cortex and corticocortical connection patterns in the monkey. European Journal of Neuroscience 16.291–310.

Prideaux, T. (1973). Cro-Magnon. New York: Time-Life.

Redford, D. B. (2003). The Oxford Guide: Essential Guide to Egyptian Mythology, Berkley.

Rightmire, G. P. (1984). Homo sapiens in Sub-Saharan Africa, In F. H. Smith and F. Spencer (eds). The origins of modern humans: A world survey of the fossil evidence. New York: Alan R. Liss.

Roginskii Y. Y., & Lewin S. S. (1955). Fundamentals of Anthropology. Moscow: Moscow University Press.

Schwarcz, A. et al. (1988). ESR dates for the hominid burial site of Qafzeh. Journal of Human Evolution 17: 733-737.

Smirnov, Y. A. (1989). On the evidence for Neandertal burial. Current Anthropology 30: 324.

Smith, G. A. (1872/2005). Chaldean Account of Genesis (Whittingham & Wilkins, London, 1872). Adamant Media Corporation (2005).

Solecki, R. (1971). Shanidar: The first flower people. New York: Knopf.

Subirana, A., & Oller-Daurelia, L. (1953). The seizures with a feeling of paradisiacal happiness as the onset of certain temporal symptomatic epilepsies. Congres Neurologique International. Lisbonne, 4, 246-250.

Tilney, F. (1928). The brain from ape to man. New York: P. B. Hoeber.

Tobias, P. V. (1971). The Brain in Hominid Evolution. Columbia University Press, New York.

Trinkaus, E. (1986). The Neanderthals and modern human origins. Annual Review of Anthropology 15: 193-211.

Weingarten, S. M., Cherlow, D. G. & Holmgren. E. (1977). The relationship of hallucinations to depth structures of the temporal lobe. Acta Neurochirugica 24: 199-216.

Williams, D. (1956). The structure of emotions reflected in epileptic experiences. Brain, 79, 29-67.

Wilson, J. A. (1951) The culture of ancient Egypt. Chicago, U. Chicago Press.

Wolpoff, M. H. (1980), Paleo-Anthropology. New York, Knopf.

18. A Neuro-Cosmology of Death, Souls, Spirits, Rebirth, Astral Projection, Judgment Day, Hell, and the Second Death

The sun was born about 4.6 billion years ago. It has enough nuclear fuel to last for another 5 billion years. Then it will grow to become a type of star called a red giant. Later in the sun's life, it will cast off its outer layers. The remaining core will collapse to become an object called a white dwarf, and will slowly fade. The sun will enter its final phase as a faint, cool object sometimes called a black dwarf.

Ancient peoples and civilizations the world over, from Egypt to the Incas of Peru, erected massive stone temples and monuments which faced the rising sun on the morning of the equinox and rebirth, the first day of Spring, . They held great religious ceremonies to celebrate the celestial cross: the Winter-Summer Solstice and the Spring-Autumn equinox; the four corner of Earth and the cosmos: which were believed to have cosmic significance.

The equinox means equality, for the day and the nights are of equal length on these ancient Holy days. Rising in a different constellation every 2160 years, it is the sun's changing position on the morning of the equinox, over thousands of years of time, which the ancients believed heralded a new age and the death and birth of the gods. Thus for the last 2160 years the sun had risen in the houses of Pieces and Virgo during the Spring and Fall Equinox, which is why the Christian god, Jesus, is linked to Virgo, the virgin, and Pieces the two fish. And before Jesus the Pieces there was the god of the Ram, and before the Ram, Taurus the Bull; a new god and a new age every 2160 years, marked by the rising sun in a different constellation on the day of the equinox. And each new god proclaims: thou shalt honor no gods before me.

It is during the equinox that the sun most brilliantly lights up the Earth with dazzling, incredibly bright, multi-colored sheets of light, known as the polar auroras. For millennia those living in the far northern or southern latitudes have been dazzled by these unearthly solar displays, called the "aurora borealis" in the north, in honor of Aurora, the Roman goddess of the Dawn, and Boreas the Greek god of the wind. In the southern latitudes these ghostly displays of brilliantly colored wavering lights are called the aurora "australis" which is Latin for "south."

The auroras become most beautiful and colorful during the Equinox. Ancient peoples believed these were dancing spirits or the shadows of the gods. In fact, these great shows of light are caused by solar winds as they slam into the Earth's protective magnetic field and the upper atmosphere. The magnetic field is produced by electrical currents generated by the Earth's spinning molten iron core, and extends thousands of miles into space, creating a magnetosphere. The solar wind, consisting of charged plasma particles emitted by the sun, collides with the charged particles of the Earth's magnetosphere, electrically exciting and interacting with oxygen and nitrogen gasses and ions, creating a rainbow of shimmering greens, reds, purples and blues. For reasons that are not well understood, the solar wind increases in strength and velocity during the equinox, causing massive geomagnetic storms and extraordinarily colorful auroras; though this is not always the case.

FATHER SUN GOD

For much of humanity's history, the sun has been worshipped as a life-giving god which rules the heavens and our solar system. Many a king and queen have claimed to be direct descendants of this all powerful solar deity, the bringer of life.

FIGURE: Mayan Sun God. **(opposite page)** Hindu.

Figure: Hindu Sun God

FIGURE: Egyptian

From August 28 through September 2, 1859 the Earth was battered by a great magnetic storm triggered by powerful solar winds, solar flares and coronal eruptions from the sun. For five days the planet was enveloped in ghostly shimmering sheets of greens, reds, and blues which were so brilliantly bright that night became day and even the darkest shadows of evening were illuminated with dazzling lights. The ancients also witnessed such incredible sights, which they believed were directly related to the sun god and the spirits and souls of the dead; and those yet to be reborn.

THE DEATH OF FATHER SUN

The last breaths of a dying star are exhaled as titanic, planet shaking, solar winds that grow more powerful and destructive with each dying day. When our own sun begins to die perhaps 6 billion years from now, its solar winds will roll across the Earth like a cosmic tsunami and the night skies will be illuminated for a millennia with fires of light. However, our sun is too small to erupt in a giant supernova, but instead will shrink, then grow, then shrink smaller still until it collapses, implodes and becomes an extremely dense and coldly pale "white dwarf."

However, once our sun begins to die, and well before its final collapse, all intelligent Earthly life will suffocate and drop to the ground gasping for breath. Increasingly powerful solar winds will whip our planet with cosmic hurricane gale force and strip the planet of its atmosphere. The oceans and seas will boil away, and the Earth's atmosphere, heavy with dust and watery mist, will be ripped from its moorings and blown into the black void of deep space. The air literally sucked from their lungs, all animals will collapse, gasping for breath, and die.

DEATH AND REBIRTH OF THE SUN GODS

Life gives birth to life, and stars give birth to stars in an endless cycle of death and rebirth. It's a cosmic dance which may have been ongoing for all eternity. Birth and death are part of the cycle of life. Some stars give birth as they die; becoming red giants and exploding in a vast supernova, spawning hundreds of infant proto-stars.

The ancient sun god, and the stars in the heavens above, die, but give birth to hundreds of new stars, new suns; suns of the sun gods.

Ancient peoples throughout the world worshiped the sun for its life generating powers, believing it a god which governed the cosmic cycle of death and rebirth. In ancient Egypt the morning sun represented rebirth and the god Horus the divine son of Isis and Osiris. The afternoon sun was the God Re, the most powerful of the gods because of his great strength. The redness of the setting sun was the blood of the god as he died. It was the evening sun Atum, the creator god who transported the dead pharoahs to the heavens where they were resurrected and brought back to life to live among the stars which were gods.

The sun also died and was reborn, or so believed the ancients, with death and rebirth taking place during a 3 day period beginning on the Winter Solstice, December 22. "Solstice" means, "Sun standing still" and during the Winter Solstice the sun seems to stop its southerly journey toward the underworld, and then to reverse course, ascending northward into the heavens on December 25--the birthday of innumerable gods and sons of gods. Thus, many ancient cultures celebrated the birth and resurrection of the sun on December 25.

Every son has a father. The ancient Aryans believed the sun was the "son" of the Sky God, and was thus the "Son of the sky" and had been created and fathered by the Heavenly Father who in turn had been sired by Father Sun. The son of the Sun-God becomes God the Sun. Christians borrowing from the ancient pagans, believed Jesus to be the "son of god" and "god the son" the incarnation of the Heavenly father, and who dies and is resurrected, ascending to heaven, becoming one with god.

Thus ancient peoples throughout the world believed the "Son of the Sun-God" was sired by God the Sun, becoming the Sun God who sets and dies and ascends to heaven, is resurrected and reborn to become the Heavenly Father, the Sun, in an endless cycle of death and rebirth, thus guaranteeing eternal life to all god's children--so claim the ancients.

The ancient were also aware that at death, or after death, the spirit or soul of the dear departed may also rise toward the heavens. Sometimes those who died return to life to tell the tale. Some children and adults who have been declared "clinically" dead but who subsequently return to life, have reported that after "dying" they left their body and floated above the scene (Eadie 1992; Joseph, 1996, 2011; Rawling 1978; Ring 1980). Typically they become increasingly euphoric as they float above their body, after which they may float away, become enveloped in a dark tunnel and then enter a soothing radiant light. And later, when they come back to life, they may even claim conscious knowledge of what occurred around their body while they were dead and floating nearby. Similar experiences are detailed in the Egyptian funerary texts and "book of the dead," written almost 6000 years ago (Budge 1994) as well as the Tibetan Book of the Dead (the Bardo Thodol) which was composed over 1,300 years ago. Approximately 37% of patients who are resuscitated report similar "out of body" experiences (Ring 1980). . Thus, it was known, even thousands of years ago, that those who die, may return to life; or they may be embraced by the light and join the heavenly father; and thus, live again, as a spirit basking in the light of the life giving god.

It is these beliefs, the experience of death, rebirth, be it the sun or those beloved on Earth, which contributed to the development of the first cosmologies as is evident among the ancient Egyptians, the Ayrians of ancient India, as well as Babylon and the ancient Mayans and Incas.

EGYPTIAN SOUL COSMOLOGY

In Egypt, the concept of the soul, as gleaned from a reading of the Book of the Dead, is divided into various forms which are necessarily entirely distinct. Nevertheless the ren is an individual's name, the sheut their shadow and the ib the heart. However, while ka is often translated as spirit, it is the ba which is normally considered to most closely resemble 'soul' (Frankfort 1978, p. 61). Whereas the ka, as a universal life-force, leaves the body at death, it is the ba which possesses some chance of transcendental immortality and which enables travel between the earth and the sky after death.

Ancient kings and queens, those or royal blood, claimed, as their ancestral heritage to be descended from the gods--and the ultimate god is the Sun. Thus, at death the ba of the kind and queen would ascend to heaven--but the common people believed they too had souls and upon death their immortal soul would journey to a celestial afterlife; either to join father sun, the sun god, or to take a place among the children and brothers and sisters of the god: the stars which became gods in their own right. Thus, upon death, the soul would travel to the stars; and thus, the ancient sought to study the heavens to understand the gods; and to ensure that they would journey to the heavens, and not to the other place, later known as hell.

The Pyramid Texts, which first occur in the fifth dynasty pyramids, indicate that the king mounted to heaven on the rays of the sun. For example, Pyramid Text Utterance 508 reads 'I have trodden those thy rays as a ramp under my feet whereon I mount up to that my mother, the living Uraeus on the brow of Re' and Utterance 523 tells us that 'May the sky make the sunlight strong for you, may you rise up to the sky as the Eye of Re' (Faulkner 1993).

The ascent to heaven was associated with the cosmology of Heliopolis; a counterpart to Osiris' dominion over the dead (Morenz 1992, pp. 205, 207, 211): the ba will 'view the sun and move freely among the 'lords of eternity' (Baines 1991, p. 190). The ba of the god Amon was in heaven, while his body in the realm of the dead and his image in the temple, and heaven was raised high for the ba of Osiris (Morenz 1992 pp. 151, 208). The pyramid texts talk of the double-doors of heaven opening, and Nut instructs the king to 'open up your place in the sky among the stars of the sky, for you are the Lone star' (Morenz, 1991, p. 205). The ba of the God and the ba of the king both belonged in the sky. But there was also a descent: the ba of the god may descend from heaven and enter its image (Morenz 1992, p. 115). The dead may live again.

"Go aboard this boat of Ra to which the gods desire to draw near, aboard which the gods desire to go, in which Ra rows to the horizon....... Raise yourself, O my father Osiris the King, for you are alive!" -Pyramid Texts 667

"O Shining One (Ra)! O Shining One! O Khepera! O Khepera! You are for the King and the King is for you; you live for the King and the King lives for you." -Pyramid Texts 662

JUDGEMENT AFTER DEATH

It is commonly believed that some souls will go to heaven, and others would spend eternity in hell. The determination of which soul would go where, thus gave rise to the concept of being judged. The nature of one's afterlife depended on how one was judged and the nature of one's good and evil deeds. Death, therefore, also became a final judgment day where the wicked were punished, and the good rewarded with eternal heavenly bliss. The ancient Sumerians, Babylonians, and Egyptians, as well as the ancient peoples of the Indus valley, and ancient China clearly believed that a final judgment awaited man after death.

For example, in the Rig-Veda we are told that the good are invited to live in heaven, whereas the wicked are hurled by the deities Soma and Indra into an eternally dark prison from whence there is no return.

According to the ancient Egyptians, the woman and man who respected and maintained maat— that is, truth, justice, and righteousness—would achieve a state of eternal beatitude, whereas those who did not could expect to suffer in a dreary Hellish underworld where they and the enemies of the sun god Ra would burn for all eternity—roasted by the burning rays of the sun. Most of these ideas were laid out in considerable detail in the Egyptian Book of the Dead, the Pyramid Texts.

The Pyramid Texts were composed and compiled by the priests of Heliopolis, to help the dead pharaohs achieve eternal bliss in the afterlife. Only the morally just could expect to enter heaven and become a star. Nevertheless, the dead, both good and evil, would have to defend themselves against every accusation. Life in the afterworld could be easily imperiled by accusations of those who had been wronged.

Yet, in ancient Egypt, it was believed that the chief accuser would be memory, the dead's own memories, and the dead's own heart would recount and remember their good or evil deeds. Indeed, it would be hell to be plagued for all eternity by every bad memory and all associated feelings of guilt and psychic pain.

This same belief is recounted in the Tibetan Book of the Dead: "You are now before Yama, King of the Dead. The mirror in which Yama seems to read your past is your own memory, and also his judgment is your own. It is you yourself who pronounces your own judgment, which in turn determines your rebirth."

However, whereas the Tibetan Buddhists believed that the souls of the dead passed judgment on themselves, the ancient Egyptians believed that the Great God Osiris would pass judgment. Osiris, would hold and weigh the scales of justice.

JUDGMENT

The Great God of the ancient Egyptians did not pass judgment alone, for in this he was assisted by a council of 12, a tribunal of the Great God. As detailed in the Book of the Dead, the jackal-headed god of death, Anubis, adjusts the scales of justice which will determine one's fate; to his right stands Thoth, the ibis-headed god of wisdom—the divine scribe who records the final verdict. And behind Osiris crouches a horrible hybrid monster, the eater of the dead.

"The council which judges the deficient, thou knowest that they are not lenient on the day of judging the miserable, the hour of doing their duty. It is woe when the accuser is one of knowledge. Do not trust in length of years, for they regard a lifetime as but an hour. A man remains after death, and his deeds are placed beside him in heaps. However, existence yonder is for eternity, and he who complains of it is a fool. But as for him who reaches it without wrong-doing, he shall exist yonder like a god, stepping out freely like the lords of eternity." -Egyptian Book of the Dead.

Not only is it futile for the guilty to proclaim their innocence, but the consequences, the manner in which the balance sways depends on one's own heart and memories, which will testify against the dead. According to the ancient Egyptians, the heart will be laid upon the scales and weighed against the lightness of a feather and balanced against Maat; Maat being truth. Good outweighs the evil, salvation would be at hand, for it could be expected that "the offenses of which thou art accused will be eliminated, thy fault wiped out by the balancing of the balance, in the day of the evaluation of qualities; thou causes the weighing to be made as Thoth." -Egyptian Book of the Dead.

THE SECOND DEATH

In later Vedic literature, including the Brahmanas, we are told of a second death. All souls would suffer a penalty that would be paid in the form of dying again and again, possibly for all eternity. This is called samsara, the transmigration of the soul. Thus according to the ancient peoples of the Indus Valley, after one dies, they could suffer a second death, and then a third death, and so on, for all eternity, each subsequent life determined by one's character and behavior in previous lives.

Although the Greeks thought it possible to achieve a state of blessed eternal beatitude, they also believed that some souls fall from their primal state of beatitude and undergo a series of incarnations, beginning in insect or bestial form, and that they then have to live and die ten thousand deaths before regaining a human state. Only enlightened souls can escape this endless cycle of birth and death.

According to modern Christian faiths, the dead also suffer a second death, which may be more horrible than the first. That is, after the soul awakens from purgatory: "The Son of man will send his angels, and they will gather out of his kingdom all causes of sin and all evildoers, and throw them into the furnace of fire; there men will weep and gnash their teeth. Then the righteous will shine like the sun in the kingdom of their father." -R.S.V. 13:41-43.

"Do you not know that the unrighteous will not inherit the kingdom of God? Do not be deceived; neither the immoral, nor idolaters, no adulterers, nor homosexuals, nor thieves, nor the greedy, nor drunkards, nor revivlers, nor robbers, will inherit the kingdom of God."-Paul, Galatians, V 22-23.

"The cowardly, the faithless, the polluted, murderers, fornicators, sorcerers, idolaters, and all liars, their lot shall be the lake that burns with fire and brimstone, which is the second death." -Revelation of St. John, II:184, 4356-457.

However, it was also believed possible to escape the second death if one had lived a life free of sin, or if one repented for their sinful ways. And this same belief is held by those of the Hindu and Buddhist faith. That is, if the consequences of one's Karma could be worked out by the self, the Atman before the next reincarnation, the soul would not have to be born again only to die again and yet again. As to those who failed to work out the consequences of their Karma, or those who had lived a life of the utmost evil, their lot would become an eternal hell.

THE PYRAMIDS: HEAVEN ON EARTH

The great pyramid was said by ancient Arabic historians to contain the world's knowledge, wisdom, and the secrets of the cosmos and creation. The base of the Great Pyramid covers 13 acres (5 hectares), and reaches a height of 480 feet (145 meters). The Great Pyramid consists of 2,300,000 stones, each weighing 21/2 tons, measuring 50 by 50 by 28 inches. It has five sides and five corners and a "pyramid inch" is one-fifth of one-fifth of a cubit--what Sir Isaac Newton called sacred and profane cubits. The north side is 755.4 feet long, the west 755.8, the east 755.9, and the south side was 756.1 feet long, indicating an incredible degree of engineering precision that was, in total no more than half a foot off from perfection. Indeed, as the Earth has shook and moved over the eons, the pyramid may well have shifted every so slightly, such that originally all sides were precisely the same. Moreover, not just the Great Pyramid, but the sides of all three of the Pyramids of Giza are perfectly aligned to the cardinal points of the compass.

The area of the Great Pyramid's base divided by twice its height gives the figure 3.14159+, (pi). Pi is the constant that is multiplied by the diameter of a circle to give its circumferences. Indeed, the Great Pyramids was built in accordance with geometrical and astronomical laws to record the dimensions of the earth and the duration of the solar year.

The builders of the pyramid were well versed in astronomy and knew the relationship of the radius to its circumference. They knew the circumference of the Earth, and the distance of the center of the Earth to the poles. The ratio of the Great Pyramid's altitude to its perimeter is the same as that of the polar radius of the Earth to its circumference: 2 pi. Hence, the Great Pyramid could be used to record the movements of the stars around the heavens in relation to the earth.

In fact, as determined by Charles Piazzi Smyth, the Astronomer-Royal of Scotland, the Great Pyramid incorporates and reveals the distance of the earth from the sun when its height is multiplied by the proportion of its height to its width, that is, ten to the ninth power.

The builders of the Great Pyramid "had determined the shape of the earth which they knew to be a true circle, its size, its precise circumference, the geographical distance from the equator to the poles, the fact that the earth is flattened at the poles, degrees of latitude and longitude to within a few hundred feet, the fact that they were shorter at the equator and longer at the poles, and the exact distance of the Earth from the sun. They had designed the pyramid's base to correspond to the distance the earth rotates in half a second" (Furrneaux, 1987, p 17).

Werner von Siemens, founder of Siemens' Electric, scaled the summit of the Great Pyramid in 1859, and stuck his finger up into the air. As he assumed a triumphant pose, a crack of electricity rang out

The three great pyramids were a cosmological map of heaven on Earth.

The kings of ancient Eygpt wished to avoid the second death and to ensure they would join their place among the stars which are gods. The three great pyramids of Egypt were created to serve as the pharaoh's gateway to the stars. ..The Egyptians developed the first cosmologies--and they were based on death and belief in the afterlife, and developed to ensure that one would take their rightful place among the stars, and not go to hell.

The three bright stars forming the belt of Orion ,the hunter, were the celestial embodiment of Osiris, the Egyptian god of the dead. Orion was of dual importance to Egyptians. The constellation was believed to be the final resting place of pharaohs, where they joined forever with the gods of the sky.

Orion also appeared in the night sky shortly before Sirius, which in turn heralded the arrival of the life-giving summer floods. Sirius represented Isis, Osiris' sister and consort, who trailed him across the sky.

Osiris symbolized the birth, growth, death, and rebirth of the natural world. In Egyptian mythology, Osiris was murdered and dismembered by his jealous brother, Seth, then briefly brought back to life by his sister and consort Isis to father the god Horus. Egyptians saw Osiris in the constellation Orion, whose appearance was connected with the annual flood. As god of the dead, Osiris welcomed the recently deceased to their new world.

Whoever constructed the three great pyramids, ensured they would be oriented in size and magnitude to mirror the three belt stars of Orion And who ever built the great pyramid, had a shaft constructed in his burial chamber pointing directly to Orion, which was considered to be the final resting place of the pharaohs.

The Egyptian Star Religion and Rebirth

The Egyptians practiced a solar religion, centering on the worship of Ra and Heliopolis, the City of the Sun and Osiris, the King of the Dead. The stars were of the utmost importance as the destination of the King after dearth: The dead kings, and ordinary people later, joined with or even become one with the god Osiris after their death. This god had a stellar representation, the constellation of Orion.

There are four strange and small shafts (20 x 20 cm wide) in the Great Pyramid. Both upper chambers have two of them, one going south and one north. All four shafts rise at angles between 36 and 45 degrees. The shafts of the upper King's chamber were each open at both ends, the shafts of the lower Queen's chamber were each closed at both ends. Presumably, the soul could travel through the northern shaft to the imperishable stars in the North. The southern shaft of the King's chamber points to Orion. Presumably, at the time the pyramid was built, the shaft pointed exactly to the star the Great Pyramid represents in Orion's belt. Likewise, the southern shaft of the Queen's Chamber at one time pointed at the brightest star in the sky, Sirius, who is the representation of Isis, who was the sister and great love of Osiris.

The Pharoahs regarded themselves as the embodiment of Osiris on Earth and when they died they believed they would travel up to the Duat and be reborn in the afterlife.

The pyramids, therefore, were not just tombs for the pharaohs, but vehicles to facilitate the pharaoh's travel to the afterlife through the Duatt, and Osiris was represented in the Duat by Orion.

"O Great One of Atum (another attribute of Khepera, the creator), son of a great one of Atum, the King is a Star in the sky among the gods."-Pyramid Texts 586A

"O King, you are this great star, the companion of Orion, who traverses the sky with Orion, who navigates the Duat with Osiris; you ascend from the East of the sky, being renewed at your due season......." -Pyramid Texts 466

From the moment of the king's birth he was groomed and trained for his return to the First Time. All his life, every aspect of it, was associated with his journey. He was taught spells and incantations to secure a safe passage to the stars after death, which he recited from the Book of the Dead and the Pyramid Texts. His sole aim in life was his successful return to the First Time, and the pyramids, were the starting point of this greatest of journeys to the stars.

The chain of command in Egypt was unbroken -each King, though a living, breathing being, was a renewal of the covenant which the gods made with man. The dead King, though dead to this world, lived on in spirit as he made his way back to the beginning, to the First Time.

Thus, the sun god dies in the solstice, when then sun stands still, and returns to life on the spring equinox as the new sun, the son of the sun god, and after 2160 years, as a new god represented by a new age. And all who die, may also be embraced by the light, to live again, among the stars which are gods.

Cosmology, therefore, had as its birth, death: death of the sun, death of those on Earth, who would ascend to haven as spirits and souls; but some of whom, after judgment, would be cast out of heaven; becoming a fallen star to live again on Earth, or to live forever in the underworld of eternal darkness, known as hell.

DEATH, DYING AND REBIRTH

The Cro-Magnon of the Paleolithic, and peoples of the Neolithic practiced mortuary rituals because they could also see beyond death, beyond one's personal demise, which is why these earthen graves contained tools, food, and ornaments even 30,000 years ago. The souls of the dead would need these items in the next world.

The dead, therefore, were believed to survive the experience of dying. It was believed that they would retain a personal identity, in the form of a personal soul, which would ascend to the heavens and join the abode of the spirits and the gods.

The ancients were exceedingly concerned about what happened after death, and what trials and tribulations they may experience in the hereafter as they ascended to heaven. And they worried about evil spirits, and the behavior of those souls who when alive had committed evil on Earth.

To protect the living, the enemy dead were sometimes decapitated, their hands and legs removed, and their eyes sewn shut and faces smashed, so that their souls would be unable to rise from the dead and cause mischief in the afterlife--even 50,000 years ago. The ancients did not necessarily believe that every soul would immediately ascend to heaven, but may remain tethered to the earth below.

Those souls that remain earth bound were often believed to be evil spirits, undeserving of a place among the gods and the stars. As detailed in the Egyptian and Tibetan Books of the Dead, it was also believed that some souls could lose their way and remain earthbound.

Yet others, such as the priesthood of the Roman Catholic Church, believed that souls almost never immediately ascended to heaven. Instead, the souls of the dead were believed to spend at least part of eternity in a realm called purgatory.

PURGATORY

Some conceive "purgatory" as a spirit-like world that exists between this reality and the reality of God. During the middle ages many Catholic theologians came to believe that the length of time a soul might remain in purgatory would depend on the nature and extent of their sins. Purgatory was like a jail sentence, and once sufficient punishment had been doled out, the soul was free to ascend to heaven.

Likewise, some of those of the Buddhist faith believed "that at death, the self, the Atman, departs to another world, where it works out the consequences of its karma. Some of these worlds were pleasant, others were hells where retribution was suffered for evil deeds. When the consequences of its karma was worked out, the self was reincarnated" (Brandon, 1967, p.171).

Others of the Christian faith also came to believe that upon death the soul merely goes to sleep; that sleep is the fate of the soul following death until a second awakening at the time of the final judgment.

According to John Calvin "the souls of the faithful, after completing their term of combat and travail, are gathered into rest, where they await with joy the fruition of their promised glory; and thus all things remain in suspense until Jesus Christ appears as the Redeemer."

The Koran endorses certain aspects of this view, such that the dead will have no knowledge of the time that has elapsed since their death. And when the soul awakes to experience judgment day, they will think they have just awakened from sleep. Hence, the soul was believed to be unconscious while the body decayed, and is awakened by the trumpet heralding the last judgment.

The ancients, however, believed that the soul could depart the body during sleep, and that death was the final liberation of the soul. The soul did not sleep. Instead the soul could remain earth bound and observe all that took place below, or ascend to heaven where it would be embraced by the light.

However, as pointed out by the Egyptian and Tibetan Books of the Dead, the soul might also be embraced by darkness, or they may lose their way and could be harried for all eternity by demons and devilish monstrosities in the darkness of the underworld.

SUBTERRANEAN HELL

As the dead were often buried in the earth, it was believed by some that the soul may descend into the earth womb, into the subterranean realms of the spirits and gods of the dead.

The ancient Greeks believed that at death and with the dissolution of the unity of the body, thymos—the conscious self and the life principle—was released into the air, and transformed into a wraith, a shadowy image of the living person, known as the eidolon. It was the life principle, psyche, which descended into Hades, which was a huge subterranean pit deep beneath the ground. There the life principle became a wraith, and existed with other eidolons, but devoid of self-consciousness in a state of perpetual and eternal gloom.

Greek thought was in some respects basically identical to that of the Sumerians, Akkadians, Babylonians, and Israelites who believed that the spirits of all the dead dwelled for all eternity in a dreary, lightless, subterranean abode. According to the Sumerians and Akkadians death involved a horrible transformation where the dead became a diseased, decaying, grisly being which descended to the underworld, "the land of no return."

According to the Hebrews, this terrible realm of darkness was called She'ol. She'ol, was where kings and peasants, rich and poor, good and evil, existed in a state of equal and horrible wretchedness. Hades and the Hebrew She'ol are identical. The good and the wicked, rich and poor, kings and peasants, all descended into hell and assumed this wraith-like existence in the realm of the dead.

Yet others preached that only the evil go to this subterranean hell. The righteous, the just, and the good, ascend to heaven and live in bliss for all eternity. However, if one went to heaven or hell depended on how one had lived their life, and how they were judged by the gods.

"For those who have disbelieved... shall roast in fire... and their bellies and skin shall be melted. To them it is said: Taste the punishment of the burning." -Koran

THE EVOLUTION OF HELL

"Jesus said: There is a light within a man of light, and it lights up the whole world. If he does not shine, he is darkness." -Gospel of Thomas.

In contrast to the light and bliss of "heaven," Hell has commonly been associated with darkness and the underworld. In part, the notion of an underworld is a direct consequence of burial practices, and perhaps the belief that sinners will fail to rise to heaven but instead will be condemned to a bleak existence deep beneath the Earth.

The belief in an underworld where sinners burn in hell, is also related to the worship of the sun as god. That is, during the day the sun god rules the Earth, but each evening it must pass through the underworld where it may be attacked by the demons of darkness. Every night, therefore, the sun god would have to fight these demons, and would do so with the rays of the sun. That is, the sun-god would burn the enemies of god with the fire of the sun's rays. Hence, Hell also became associated with fire, such that, as these beliefs evolved, sinners and enemies of god would be sent to hell where they would not only be attacked by devils and demons, but would burn in flames for all eternity.

The belief in Hell and in demons and devils is world wide and vivid descriptions of a hellish underworld and its demonic denizens are also described in the Egyptian and Tibetan Books of the Dead. Indeed, these books were written so as to inform believers as to the mysteries of death and the trials and tribulations one might experience following death so that demons and other devilish monstrosities and misfortunes might be avoided.

For example, priests or relatives of the recently departed were expected to read various hymns, litanies, magical formula, spells, and incantations from the Egyptian Book of the Dead which would help protect the dead from the multitude of devils and fiends that would seek to devour and destroy the spirit-soul. "These powers of evil had hideous and terrifying shapes and forms, and their haunts infested the region through which the road of the dead lay when passing from this world to the Kingdom of Osiris" (Budge, 1994). According to the ancient Egyptians, the gods were basically powerless to protect the souls from these demons which could inflict horrible suffering upon the spirit souls for all eternity.

"Then the Lord of Death will place around thy neck a rope and drag thee along; he will cut off thy head, tear out thy heart, pull out thy intestines, lick up thy brain, drink thy blood, eat thy flesh, and gnaw thy bones; but thou will be incapable of dying. Even when the body is hacked to pieces, it will revive again and thy will be tortured for all eternity."-Tibetan Book of the Dead.

However, these evils can also be avoided which is the purpose of these texts. They serve as a roadmap to heaven. In this regard, these texts actually educate the reader as to "The Art of Dying" so that an undesirable and hellish death can be avoided.

Unfortunately, or so say these texts, many are condemned to Hell for all eternity. According to the Tibetan Book of the Dead, a hellish death is due to the "power of accumulated evil Karma." Likewise, according to the Egyptian Book of the Dead, spirit souls may be judged and condemned to hell for all eternity if they had behaved in an evil and unjust manner when living. Hence, both texts warn of a judgment day and a life review and warn that those who have committed evil acts will be judged, found guilty, and then condemned.

Hence, it can be deduced that by 6,000 years ago, that in addition to a heaven and hell, a moral-religious spiritual conscience and moral-consciousness has evolved. This is reflected by the religious conviction that the just and the unjust would be rewarded and punished in the next world after they had died; their good deeds often weighed against the bad.

As is now well established, a sense of morality and a personal conscience (as in: Let your conscience be your guide) is directly associated with the functional integrity of the frontal lobe (Joseph, 1986a, 1999a, 2014). The frontal lobes are concerned with anticipating consequences and serve to inhibit or redirect unacceptable impulses in order to avoid punishment.

It also appears that the frontal lobe had fully evolved and reached its peak levels of development and expansion during the Upper Paleolithic as described in earlier chapters. Although there is no physical evidence to indicate that Upper Paleolithic peoples had also evolved a moral conscience or a concern that the spirit soul would be punished for past misdeeds, Budge (1994) argues that the Book of the Dead and the beliefs that permeate the pyramid texts, were already quite ancient by the rise of the first Dynasty, and thus have pre-dynastic, and therefore, Upper Paleolithic origins.

It is noteworthy that although the Tibetan (and Egyptian) Book of the Dead warn of a judgment day, and describe all manner of Hellish devils and demons, this text repeatedly reminds the reader that these demonic entities are a product of one's own mind. They are illusions erupting from the depths of the primeval unconscious which is freed of the suppressive restraints of the every day reality and the illusions of this world

THE LIMBIC SYSTEM NEUROLOGY OF PERSONAL MEMORIES: HIPPOCAMPUS

When death comes, many are buried beneath the Earth, and if wealthy, in a tomb; an earth womb. Born of the womb, they return to the womb: from dust to dust, ashes to ashes. The ancients believed hell was beneath the Earth, or in a place of eternal darkness. Heaven was, in heaven. Thus the first cosmologies were developed based on beliefs about death, dying, and the ascension of souls.

Although these spheres-of-after death life could be explained as existing in alternate dimensions, in a multi-verse of parallel existence; the ability to experience death and dying and ascension as a spirit, must have some basis in the brain. In later chapters, it is explained how hyperactivation, or rather, acceleration of brain activity, could be likened to Einstein's theory of relativity and the contraction of time-space. However, the brain also functions in terms of inhibition. Much of what is sensed, is filtered out, thereby protecting consciousness from being overwhelmed. One particular brain structure capable of perceiving multiple sensory impressions simultaneously, much of which is filtered out, is the limbic system.

There are a variety of conditions which can hyperactive the limbic system, including death. The anticipation of death, the terror and dread of impending death, are also made possible by the limbic system. The limbic system can anticipate, and it can feel fear and anxiety in response to the unknown. As will be explained in later chapters, hyperactivation, or electrode stimulation of the limbic system can trigger the experience, or the hallucination, of leaving one's body, or seeing ghosts, demons, angels and even gods. Some of the primary structures of the limbic system include the amygdala and hippocampus. As is well known, the hippocampus is exceedingly important in memory, acting to place various short-term memories into long-term storage and retrieving short-term memories (Enbert & Bonhoeffer, 2009; Fedio & Van Buren, 1974; Frisk & Milner, 1990; Milner, 1966; 1970; Nunn et al., 2009; Penfield & Milner, 1958; Rawlins, 2005; Scoville & Milner, 1957; Squire, 2011; Victor

& Agamanolis,2010). Presumably the hippocampus encodes new information during the storage and consolidation (long-term storage) phase, and assists in the gating of afferent streams of information destined for the neocortex by filtering or suppressing irrelevant sense data which may interfere with memory consolidation. Moreover, it is believed that via the development of long-term potentiation the hippocampus is able to track information as it is stored in the neocortex, and to form conjunctions between synapses and different brain regions which process and store associated memories.

Hence, if the hippocampus has been damaged the ability to convert short term memories into long term memories becomes significantly impaired. Memory for words, passages, conversations, and written material is also significantly impacted, particularly with left hippocampal destruction

Bilateral destruction of the anterior hippocampus results in striking and profound disturbances involving memory and new learning (i.e. anterograde amnesia). For example, one such individual who underwent bilateral destruction of this nuclei (H.M.), was subsequently found to have almost completely lost the ability to recall anything experienced after surgery. If you introduced yourself to him, left the room, and then returned a few minutes later he would have no recall of having met or spoken to you. Dr. Brenda Milner has worked with H.M. for almost 20 years and yet she is an utter stranger to him.

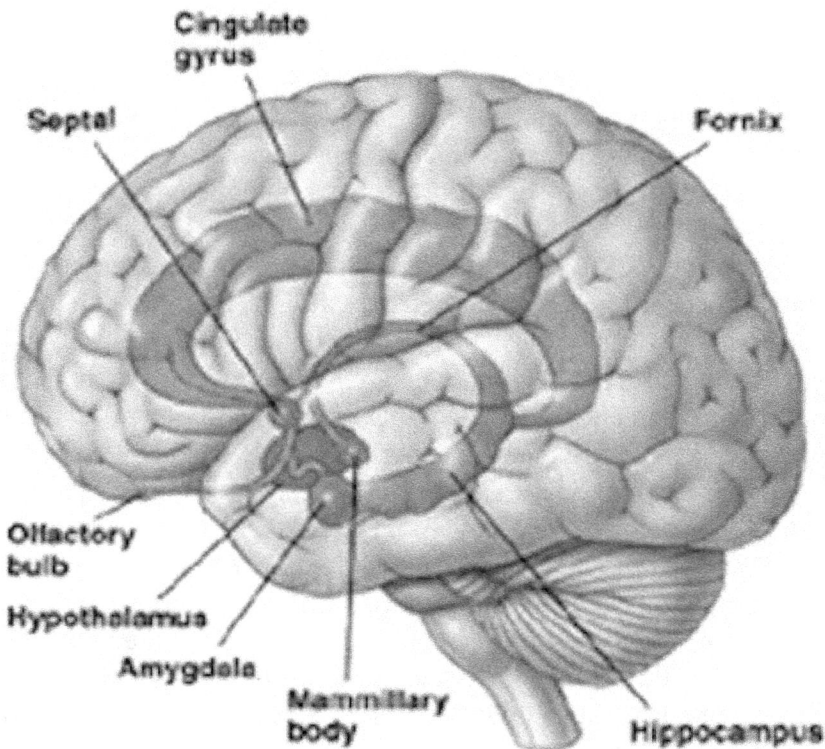

H.M. is in fact so amnesic for everything that has occurred since his surgery (although memory for events prior to his surgery is comparatively exceedingly well preserved), that every time he rediscovers that his favorite uncle died (actually a few years before his surgery) he suffers the same grief as if he had just been informed for the first time.

H.M., although without memory for new (non-motor) information, has adequate intelligence, is painfully aware of his deficit and constantly apologizes for his problem. "Right now, I'm wondering" he once said, "Have I done or said anything amiss?" You see, at this moment everything looks clear to me, but what happened just before? That's what worries me. It's like waking from a dream. I just don't remember...Every day is alone in itself, whatever enjoyment I've had, and whatever sorrow I've had...I just don't remember" (Blakemore, 1977, p.96).

Presumably the hippocampus acts to protect memory and the encoding of new information during the storage and consolidation phase via the gating of afferent streams of information and the filtering/exclusion (or dampening) of irrelevant and interfering stimuli. When the hippocampus is damaged there results input overload, the neuroaxis is overwhelmed by neural noise, and the consolidation phase of memory is disrupted such that relevant information is not properly stored or even attended to. Consequently, the ability to form associations (e.g. between stimulus and response) or to alter preexisting schemas (such as occurs during learning) is attenuated.

AMYGDALA & HIPPOCAMPUS

There is considerable evidence which strongly suggests that the hippocampus plays an interdependent role with the amygdala in regard to memory; particularly in that they are richly interconnected, merge at the uncus, and exert mutual excitatory influences on one another. For example, it appears that the amygdala is responsible for storing the emotional aspects and personal reactions to events, whereas the hippocampus acts to store the cognitive, visual, and contextual variables.

Specifically, the amygdala plays a particularly important role in memory and learning when activities are related to reward and emotional arousal (Gaffan 1992; Gloor 1992, 1997; Halgren 1992; LeDoux 2012, 1996; Kesner 2012; Rolls 2002; Sarter & Markowitsch, 1985). Thus, if some event is associated with positive or negative emotional states it is more likely to be learned and remembered.

Because of its involvement in all aspects of social-emotional and motivational functioning, activation of the amygdala therefore, can evoke highly personal and emotional memories as it is highly involved in remembering emotionally charged experiences (Gloor, 1992, 1997; Halgren, 1981, 1992; Halgren, et al. 1978; Rolls, 1992; Sarter & Markowtisch, 1985). In fact, the amygdala becomes particularly active when recalling personal and emotional memories (Halgren, 1992; Heath, 1964; Penfield & Perot, 1963), and in response to cognitive and context determined stimuli regardless of their specific emotional qualities (Halgren, 1992). Moreover, depth electrode activation of the amygdala can even evoke memories of sexual intercourse, and traumatic memories that had long ago been forgotten.

Consider, for example, a case described by Gloor (1997, pp. 7-9) in which the right amygdala of a 21 year old patient "was briefly stimulated for 2.8 seconds at a low intensity. Immediately upon its onset" the patient reported "a feeling like falling into the water. When he was asked to elaborate, he replied that it was as if something had covered his eyes, nose and mouth." Again the stimulation was applied and the patient excitedly asked: "Could you do it again, Doctor? When asked why, he said he had the words on his lips to describe the feeling, but then could not." The stimulation was then applied without warning for 4.4 seconds. The patient "opened his mouth, with an astonished look on his face." A terrifying memory had suddenly come back to him. It was when he was eight years old and had been at a picnic when "A kid was coming up to me to push me into the water.... I was pushed by somebody stronger than me... a big fellow..." who pushed him under the water and had kept his head under the water so that he couldn't breath.

However, once these emotional memories are formed, it sometimes requires the specific emotional or associated visual context to trigger their recall. If those cues are not provided or ceased to be available, the original memory may not be triggered and may appear to be forgotten or repressed. However, even emotional context can trigger memory (see also Halgren, 1992) in the absence of specific cognitive cues.

Similarly, it is also possible for emotional and non-emotional memories to be activated in the absence of active search and retrieval, and thus without hippocampal or frontal lobe participation. Recognition memory may be triggered by contextual or emotional cues. Indeed, there are a small group of neurons in the amygdala, as well as a larger group in the inferior temporal lobe which are involved in recognition memory (Murray, 1992; Rolls, 1992). Because of amygdaloid sensitivity to visual and emotional cues, even long forgotten memories may be evoked via recognition, even when search and retrieval repeatedly fails to activate the relevant memory store.

According to Gloor (1992), "a perceptual experience similar to a previous one can through activation of the isocortical population involved in the original experience recreate the entire matrix which corresponds to it and call forth the memory of the original event and an appropriate affective response through the activation of amygdaloid neurons." This can occur "at a relatively non-cognitive (affective) level, and thus lead to full or partial recall of the original perceptual message associated with the appropriate affect."

Thus the amygdala is responsible for emotional memory formation and recall whereas the hippocampus is concerned with recalling and storing verbal-visual-spatial and contextual details in memory. Thus, damage to the hippocampus can impair retention of context, and contextual fear conditioning, but it has no effect on the retention of the fear itself or the fear reaction to the original cue (Kim & Fanselow 1992; Phillips & LeDoux 1992, 1996; Rudy & Morledge 1994). In these instances, fear-memory is retained due to preservation of the amygdala. However, when both the amygdala and hippocampus are damaged, striking and profound disturbances in memory functioning result.

Therefore, the role of the amygdala in memory and learning seems to involve activities related to reward, orientation, and attention, as well as emotional arousal and social-emotional recognition (Gloor, 1992, 1997; Rolls, 2012; Sarter & Markowitsch, 1985). If some event is associated with positive or negative emotional states it is more likely to be learned and remembered. That is, reward increases the probability of attention being paid to a particular stimulus or consequence as a function of its association with reinforcement.

Moreover, the amygdala appears to reinforce and maintain hippocampal activity via the identification of motivationally significant information and the generation of pleasurable rewards (through action on the lateral hypothalamus). However, the amygdala and hippocampus act differentially in regard to the effects of positive vs. negative reinforcement on learning and memory, particularly when highly stressed or repetitively aroused in a negative fashion. For example, whereas the hippocampus produces theta in response to noxious stimuli the amygdala increases its activity following the reception of a reward (Norton, 1970).

THE HIPPOCAMPUS & ASSOCIATED MEMORY STRUCTURES

Reverberating neurons are presumably located in various regions of the neocortex, and are apparently bound together via the simultaneous activity and steering influences involving the frontal lobes (Joseph, 1986a, 2012), dorsal medial thalamus, and in particular the amygdala and hippocampus (Gloor, 1997; Graff-Radford, et al. 200; Lynch, 1986; Rolls, 2012; Squire,20-2). These structures are all interlinked and highly involved in attention, arousal, and memory functioning, and probably act together so as to establish and maintain specific neural circuits and networks associated with specific memories (e.g., Brewer et al., 1998; Squire, et al,. 1992; Tulving et al., 1994; Wagner et al., 1998). For example, these different networks and neurons may be linked via the steering influences exerted by the frontal lobes etc., which can selectively activate or inhibit these memories and associated tissues in a coordinated fashion, and which can tie together certain perceptual experiences so as to form a complex multi-modal memory (e.g., Dolan et al., 1997; Joseph, 1982, 1986a, 1988a, 1999a; Kapur et al., 1995; Squire, et al,. 1992; Tulving et al., 1994;

Dolan et al., 1997; Brewer et al., 1998; Wagner et al., 1998).

The frontal lobes, however, not only assist in the recollection of memories (Brewer et al., 1998; Wagner et al., 1998), but in conjunction with the limbic system, can induce feelings of guilt about those memories. Presumably, at death, the amygdala and hippocampus act to trigger the recollection of long term and highly personal memories, whereas the frontal lobes may serve to pronounce judgment and impose guilt. Indeed, it has been well established that the frontal lobe not only acts to inhibit memories and various impulses, but serves as a moral conscience which passes judgment on.

AMYGDALA & HIPPOCAMPAL INTERACTIONS: HALLUCINATIONS

Direct electrical stimulation of the temporal lobes, hippocampus and particularly the amygdala (Gloor, 1990, 1997) not only results in the recollection of images, but in the creation of fully formed visual and auditory hallucinations (Gloor 1992, 1997; Halgren 1992; Halgren et al., 1978; Horowitz et al., 1968; Malh et al., 1964; Penfield & Perot, 1963). It has long been know that tumors invading specific regions of the brain can trigger the formation of hallucinations which range from the simple (flashing lights) to the complex. The most complex forms of hallucination, however, are associated with tumors within the most anterior portion of the temporal lob; i.e. the region containing the amygdala and anterior hippocampus.

Similarly, electrical stimulation of the anterior lateral temporal cortical surface results in visual hallucinations of people, objects, faces, and various sounds (Gloor 1992, 1997; Halgren 1992; Horowitz et al., 1968)—particularly the right temporal lobe (Halgren et al. 1978). Depth electrode stimulation and thus direct activation of the amygdala and/or hippocampus is especially effective. For example, stimulation of the right amygdala produces complex visual hallucinations, body sensations, deja vus, illusions, as well as gustatory and alimentary experiences (Weingarten et al. 1977), whereas Freeman and Williams (1963) have reported that the surgical removal of the right amygdala in one patient abolished hallucinations. Stimulation of the right hippocampus also produces memory and dream-like hallucinations (Halgren et al. 1978; Horowitz et al. 1968).

The amygdala also becomes activated in response to bizarre stimuli (Halgren, 1992). Conversely, if activated to an abnormal degree, it may produce bizarre memories and abnormal perceptual experiences. In fact, the amygdala contributes in large part to the production of very sexual as well as bizarre, unusual and fearful memories and mental phenomenon including dissociative states, feelings of depersonalization, and hallucinogenic and dream-like recollections (Bear, 1979; Gloor, 1986, 1992, 1997; Horowitz et al. 1968; Mesulam, 1981; Penfield & Perot, 1963; Weingarten et al. 1977; Williams, 1956).

Single amygdaloid neurons receive a considerable degree of topographic input, and are predominately polymodal, responding to a variety of stimuli from different modalities simultaneously (Amaral et al. 1992; O'Keefe & Bouma, 1969; Ono & Nishijo, 1992; Perryman, Kling, & Lloyd, 1987; Rolls 1992; Sawa & Delgado, 1963; Schutze et al. 1987; Turner et al. 1980; Ursin & Kaasa, 1960; Van Hoesen, 1981). However, much of this information is filtered out, as it is not necessary to taste what we see, or to feel what we hear, and so as to prevent the brain from being overwhelmed. In part, this massive sensory filtering is made possible via serotonic (5HT) which suppresses synaptic activity in the visual cortex, the lateral (visual) geniculate nucleus of the thalamus as well as the amygdala and throughout the neocortex (Curtis & Davis 1962; Marazzi & Hart 1955). For example, in response to arousing stimuli, 5-HT is released (Auerbach et al. 1985; Roberts, 1984; Spoont, 1992) which acts to narrow and focus attentional and perceptual functioning so that stimuli which are the most salient are attended to. That is, 5HT appears to be involved in learning not to respond to stimuli that are irrelevant and not rewarding (See Beninger, 1989). These signals are filtered out and suppressed.

By contrast, substances which block 5HT transmission -such as LSD- results in increased activity in the sensory pathways to the neocortex (Purpura 1956), such that information that is normally filtered out is perceive. LSD acts on serotonin (5HT) 5HT restricts perceptual and information processing and in fact increases the threshold for neural responses to occur at both the neocortical and limbic level.

As is well known, LSD can elicit profound hallucinations involving all spheres of experience. Following the administration of LSD high amplitude slow waves (theta) and bursts of paroxysmal spike discharges occurs in the hippocampus and amygdala (Chapman & Walter, 1965; Chapman et al. 1963), but with little cortical abnormal activity. In both humans and chimps, when the temporal lobes, amygdala and hippocampus are removed, LSD ceased to produce hallucinatory phenomena (Baldwin et al. 1959; Serafintides, 1965). Moreover, LSD induced hallucinations are significantly reduced when the right vs. left temporal lobe has been surgically ablated (Serafintides, 1965).

Intense activation of the temporal lobe and amygdala has also been reported to give rise to a host of sexual, religious and spiritual experiences; and chronic hyperstimulation (i.e. seizure activity) can induce some individuals to become hyper-religious or visualize and experience ghosts, demons, angels, and even God, as well as claim demonic and angelic possession or the sensation of having left their body (Bear 1979; Gloor 1986, 1992; Horowitz, Adams & Rutkin 1968; MacLean 1990; Mesulam 1981; Penfield & Perot 1963; Schenk, & Bear 1981; Weingarten, et al. 1977; Williams 1956).

In consequence, under conditions such as death, these limbic system structures may become excessively aroused, freed of inhibitory restraint, and begin to recall and judge personal memories, as well as hallucinate all manner of forms and images, including those of gods, demons, angels and devils who pass judgment.

JUDGMENT DAY

It is commonly believed that some souls will go to heaven, and others would spend eternity in hell. The determination of which soul would go where, thus gave rise to the concept of being judged. The nature of one's afterlife depended on how one was judged and the nature of one's good and evil deeds. Death, therefore, also became a final judgment day where the wicked were punished, and the good rewarded with eternal heavenly bliss.

We do not know if the Cro-Magnon believed in hell or a final judgment. The ancient Sumerians, Babylonians, and Egyptians, as well as the ancient peoples of the Indus valley, and ancient China clearly believed that a final judgment awaited man after death.

For example, in the Rig-Veda we are told that the good are invited to live in heaven, whereas the wicked are hurled by the deities Soma and Indra into an eternally dark prison from whence there is no return.

According to the ancient Egyptians, the woman and man who respected and maintained maat— that is, truth, justice, and righteousness—would achieve a state of eternal beatitude, whereas those who did not could expect to suffer in a dreary Hellish underworld where they and the enemies of the sun god Ra would burn for all eternity—roasted by the burning rays of the sun. Most of these ideas were laid out in considerable detail in the Egyptian Book of the Dead, the Pyramid Texts.

The Pyramid Texts were composed and compiled by the priests of Heliopolis, to help the dead pharaohs achieve eternal bliss in the afterlife. Only the morally just could expect to enter heaven. Nevertheless, the dead, both good and evil, would have to defend themselves against every accusation. Life in the afterworld could be easily imperiled by accusations of those who had been wronged.

Yet, in ancient Egypt, it was believed that the chief accuser would be memory, the dead's own memories, and the dead's own heart would recount and remember their good or evil deeds. Indeed, it would be hell to be plagued for all eternity by every bad memory and feelings of guilt and psychic pain.

This same belief is recounted in the Tibetan Book of the Dead: "You are now before Yama, King of the Dead. The mirror in which Yama seems to read your past is your own memory, and also his judgment is your own. It is you yourself who pronounces your own judgment, which in turn determines your rebirth."

However, whereas the Tibetan Buddhists believed that the souls of the dead passed judgment on themselves, the ancient Egyptians believed that the

Great God Osiris would pass judgment. Osiris, would hold and weigh the scales of justice.

JUDGMENT

The Great God of the ancient Egyptians did not pass judgment alone, for in this he was assisted by a council of 12, a tribunal of the Great God. As detailed in the Book of the Dead, the jackal-headed god of death, Anubis, adjusts the scales of justice which will determine one's fate; to his right stands Thoth, the ibis-headed god of wisdom—the divine scribe who records the final verdict. And behind Osiris crouches a horrible hybrid monster, the eater of the dead.

"The council which judges the deficient, thou knowest that they are not lenient on the day of judging the miserable, the hour of doing their duty. It is woe when the accuser is one of knowledge. Do not trust in length of years, for they regard a lifetime as but an hour. A man remains after death, and his deeds are placed beside him in heaps. However, existence yonder is for eternity, and he who complains of it is a fool. But as for him who reaches it without wrongdoing, he shall exist yonder like a god, stepping out freely like the lords of eternity." -Egyptian Book of the Dead.

Not only is it futile for the guilty to proclaim their innocence, but the consequences, the manner in which the balance sways depends on one's own heart and memories, which will testify against the dead. According to the ancient Egyptians, the heart will be laid upon the scales and weighed against the lightness of a feather and balanced against Maat; Maat being truth. good outweighs the evil, salvation would be at hand, for it could be expected that "the offenses of which thou art accused will be eliminated, thy fault wiped out by the balancing of the balance, in the day of the evaluation of qualities; thou causes the weighing to be made as Thoth." -Egyptian Book of the Dead.

As noted, according to the ancient Egyptians the final judgment is listed into a book. Jewish, Christian and Islamic faith also involves a book in which the final judgment is listed.

"And I saw the dead, great and small, standing before the throne, and books were opened. Also another book was opened, which is the book of life. And the dead were judged by what was written in the books, by what they had done." -R.S.V. 10:11-15.

According to the ancient Hebrews, a sinner might have his name erased from the book: "who ever sins against me, I will blot out of my book." —Exodus 33:33. And with this erasure, by having one's name blotted out, the soul of the dead would be condemned for all eternity, forgotten forever, never to be remembered even by the Lord God.

THE LIFE REVIEW: HEART OF A MEMORY

The notion of being judged and a weighing of one's sins is a common belief that recurs throughout history. The ancient Chinese, although practicing ancestor worship, also believed that sinners would roast in hell. The souls of the ancient Chinese were judged by officials of the underworld which was comprised of 10 tribunals. The soul would typically kneel before these judges next to whom would stand officers with books with the records against which the soul would be judged.

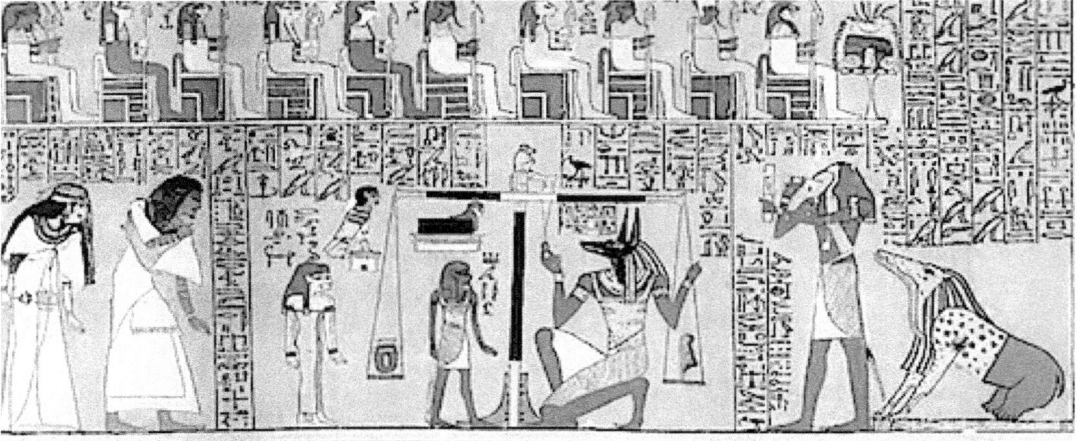

Similar beliefs were held by the ancient Greeks. For example, after Odysseus met with the ghost of Achilles in Hades he states that: "I saw Minos, the glorious son of Zeus, golden scepter in hand, giving judgment to the dead from his seat, while they sat and stood about the king through the wide-gated house of Hades." According to the ancient Egyptians, the chief prosecutor and the witness against the accused, would be the accused themselves. That is, the dead, their heart, would be forced to testify against themselves. Memories, both good and bad, would be recalled involuntarily. That is, the dead would undergo a life review.

In pictures of judgment day, we see that on the scales rest the hieroglyphic sign of the heart, the ib, and a feather. And before the balance stands the dead. If the soul has lived a righteous life, his heart will detail his good deeds which are weighed against truth. If he has done evil, he implores his heart not to witness against him. He struggles not to remember. But to no avail.

Death, to the ancients, therefore, involved a life review, where one's memories are recalled and one's life is judged on the scales of good and evil, on the balance of truth.

Although we now know that the amygdala, in conjunction with the hippocampus becomes particularly active when recalling personal and emotional memories (Halgren, 1992; Heath, 1964; Penfield & Perot, 1963), the ancients believed these memories were stored in the heart which served as one's conscience. Whereas we localized the transmitter of god to the limbic system, the ancient Egyptians believed that the heart was "the god which is in man." On the other hand, the amygdala directly affects cardiovascular functioning and thus the beating of the heart.

The ancient Egyptians experienced a considerable degree of apprehension about judgment day and the fact that their hearts would witness against them. The Egyptians were painfully aware that they might be condemned to hell for all eternity by their own evil memories. These same concerns also appear in the Tibetan Book of the Dead:

"You are now before Yama, King of the Dead. In vain will you try to lie, and to deny or conceal the evil deeds you have done. The Judge holds up before you the shinning mirror of Karma, wherein all your deeds are reflected. But again you have to deal with dream images, which you yourself have made, and which you project outside, without recognizing them as your own work. The mirror in which Yama seems to read your past is your own memory, and also his judgment is your own. It is you yourself who pronounces your own judgment, which in turn determines your rebirth." -Tibetan Book of the Dead.

The weighing of the heart and undergoing a final judgment is not shared by all religions. However, judgment day has been a common belief among many different religions since the Neolithic, including Christian, Greek, Roman, Christian and the apocryphal writings of the Jews: "The spirits of the dead shall be separated. A division has been made for the spirits of the righteous, in which there is a bright spring of water. And such has been made for sinners when they die and are buried in the earth and judgment. Here their spirits shall be set apart in great pain."-Book of Enoch

THE SECOND DEATH

"So I prophesied as I was commanded: and as I prophesied, there was a noise, and behold, a rattling; and the bones came together, bone to its bone. And as I looked, there were sinews on them, and flesh had come upon them, and skin had covered them; but there was no breath in them. Then he said to me, Prophesy to the breath, prophesy, son of man, and say to the breath, Thus says the Lord God: Come from the four winds, O breath, and breathe upon these slain, that they may live. So I prophesied as he commanded me, and the breath came into them, and they lived, and stood upon their feet, an exceedingly great host." -Ezekial.

Yahweh, Ezekial tell us, has the power to resurrect the dead, to join together even disarticulated bones, and turn them into human bodies, giving them life. This is the second life after death. And there is a second death that awaits all sinners:

"Then Raphael answered, one of the holy angels who was with me, and said unto me: These hollow places have been created for this very purpose, that the spirits of the souls of the dead should assemble therein, ye that all the souls of the children of men should assemble here. And these places have been made to receive them till the day of their judgment and till the appointed time period till the judgment comes upon them."-Book of Enoch

In later Vedic literature, including the Brahmanas, we are told of a second death. All souls would suffer a penalty that would be paid in the form of dying again and again, possibly for all eternity. This is called samsara, the transmigration of the soul. Thus according to the ancient peoples of the Indus Valley, after one dies, they could suffer a second death, and then a third death, and so on, for all eternity, each subsequent life determined by one's character and behavior in previous lives.

Although the Greeks thought it possible to achieve a state of blessed eternal beatitude, they also believed that some souls fall from their primal state of beatitude and undergo a series of incarnations, beginning in insect or bestial form, and that they then have to live and die ten thousand deaths before regaining a human state. Only enlightened souls can escape this endless cycle of birth and death.

According to modern Christian faiths, the dead also suffer a second death, which may be more horrible than the first. That is, after the soul awakens from purgatory: "The Son of man will send his angels, and they will gather out of his kingdom all causes of sin and all evildoers, and throw them into the furnace of fire; there men will weep and gnash their teeth. Then the righteous will shine like the sun in the kingdom of their father." -R.S.V. 13:41-43.

"Then I saw a great white throne and him who sat upon it; from his presence earth and sky fled away, and no place was found for them. And I saw the dead, great and small, standing before the throne, and books were opened. Also another book was opened, which is the book of life. And the dead were judged by what was written in the books, by what they had done. And the sea gave up the dead in it. Death and Hades gave up the dead in them, and all were judged by what they had done. Then Death and Hades were thrown into the lake of fire. This is the second death, the lake of fire; and if any one's name was not found written in the book of life, he was thrown into the lake of fire." -R.S.V. 10:11-15.

"The cowardly, the faithless, the polluted, murderers, fornicators, sorcerers, idolaters, and all liars, their lot shall be the lake that burns with fire and brimstone, which is the second death." -Revelation of St. John, II:184, 4356-457.

"Do you not know that the unrighteous will not inherit the kingdom of God? Do not be deceived; neither the immoral, nor idolaters, no adulterers, nor homosexuals, nor thieves, nor the greedy, nor drunkards, nor revivlers, nor robbers, will inherit the kingdom of God."-Paul, Galatians, V 22-23.

And yet, there is a second salvation.

Idolaters, adulterers, homosexuals, thieves, the greedy, drunkards, revivlers, and robbers... And such were some of you! But you were washed, you were sanctified, you were justified in the name of the Lord Jesus Christ and in the Spirit of our god." -Paul, Galatians, V 22-23.

Hence, it was also believed possible to escape the second death if one had lived a life free of sin, or if one repented for their sinful ways. And this same belief is held by those of the Hindu and Buddhist faith. That is, if the consequences of one's Karma could be worked out by the self, the Atman before the next reincarnation, the soul would not have to be born again only to die again and yet again. As to those who failed to work out the consequences of their Karma, or those who had lived a life of the utmost evil, their lot would become an eternal hell.

HELL: JUDGMENT AND PUNISHMENT

Although the Sumerians and ancient Hebrews and Greeks initially believed that both good and evil, rich and poor, would descend to hell, some also believed that She'ol, Hades, and Hell, were places of judgment, where punishments were doled out to the wicked.

"Upon the termination of life they were brought to trial; and, according to their sentence, some go to prison-houses beneath the earth, to suffer for their sins, while others, by virtue of their trial, are borne lightly upwards to some celestial spot, where they pass their days in a manner worthy of the life they lived." -Plato, Phaedrus.

In ancient Chinese religion, the realm of the dead was believed to consist of seven Hells where the evil would roast for all eternity. Likewise, in ancient Egypt, Greece, the Middle East, and Catholic Rome, punishment for the wicked, for the enemies of god, became eternal hell fire and all manner of demonic horrors.

"And I" the Apostle Peter "saw also another place very squalid; and it was a place of punishment, and they that were punished and the angels that punished them had their raiment dark, according to the air of the place. And some there were hanging by their tongues; and those were they that blasphemed the way of righteousness, and under them was laid fire flaming and tormenting them... and there were others, women, hanged by their hair above that mire which boiled up; and these were they that adorned themselves for adultery... And in another place were gravel-stones sharper than swords or any spit, heated with fire, and men and women clad in filthy rages rolled upon them in torment... and others being cast down from a great precipice to the rocks below..."

"For those who have disbelieved... shall roast in fire... and their bellies and skin shall be melted. To them it is said: Taste the punishment of the Burning." -Koran

AFTER-DEATH, OUT-OF-BODY, ASTRAL PROJECTION: THE BOOK OF THE DEAD

"The Book of the Dead" is the title given to the collection of mortuary texts, also referred to as Pyramid texts, which the ancient Egyptian theologians composed for the benefit of the dead. However, they were first called the "book of the dead" or "a dead man's book" by tomb robbers, as these texts were commonly found in the dead man's coffin. During the sixth Dynasty, the common name for the Book of the Dead was "manifested in the light" or "embraced by the light."

According to E.A. Wallis Budge, Late Keeper of the Egyptian and Assyrian Antiquities of the British Museum, these texts were already quite ancient by the rise of the first dynasty, over 6,000 years ago. The predynastic Egyptians "were quite certain that men did not perish when they died, but that some part of a man departed after death to some place where he would renew his life in some form, according to the dictates of some divine being" (Budge, 1994, p. 4).

According to the Book of the Dead, at death the spirit-soul of the deceased would arise from the body, and although translucent and transparent to the living, this spirit-soul would at first hover above and look upon the body and could see and hear what was taking place below (chapter CLXXVII). The spirit-soul may then depart, only to return to visit the body (e.g. chapter LXXXIX: The Soul Visiting the Body which Lies on a Bier), sometimes even reanimating the body (chapter XCII), such that the dead might live again in this world.

Like their Paleolithic predecessors, the walls of all Egyptian tombs and all but three of the hundreds of pyramids that dot the land, were graced with paintings of all manner of animal life and realistic scenes of the every day life to which they were accustomed. And within these tombs dwelled the ka, a spirit-like force which resided in a ka-statue of the dead person. These statues were exact replicas of the dead, that is, when they were alive.

It was the ba, the soul of the dead, which the Egyptians believed ascended to heaven. The ba would sometimes be depicted in flight, as a human-headed bird. Upon entering heaven the dead achieved the status of an akh—a "glorified being" that had reached completeness of attainment.

However, as detailed in the Egyptian Book of the Dead, prior to entering heaven (Anu), the spirit-soul of the deceased might "fly through the air" (chapter XCII) "pass over the earth" (chapter IV), such that it could move freely from place to place, before entering a tunnel or realm of darkness, and then arising into the sky to the abode of the gods in heaven at which point they bask in "a shining, glorious light" (Budge, 1994).

Similarly, according to Book I, of the Bardo, Tibetan Book of the Dead, following death the deceased will be enveloped with a "Clear Light" of "Wisdom and the Knower will experience the shining, dazzling, glorious, Radiance of the Clear light of Pure Reality, the All Good. Thine own consciousness, shining, void, and inseparable from the Great Body of Radiance, hath no birth, nor death, and is the Immutable Light—Buddha Amitabha."

Likewise, according to the Egyptian Book of the Dead, the soul ascends into the sky to the abode of the gods in heaven at which point the dead bask in "a shining, glorious light" and are led into the presence of dead relatives and brethren, and finally to the divine being Osiris, who is said to have made "men and women to be born again" and where they then dwell in bliss for all eternity having achieved everlasting life: "My soul is God. My soul is eternity" (Budge, 1994).

It is noteworthy that the soul, at first, usually does not know it is dead, that the body has died. According to the Tibetan Book of the Dead: "When the consciousness-principle gettest outside the body, it sayeth to itself, Am I dead, or am I not dead? It cannot determine. It can see that the body is being stripped of its garments. It seeth its relatives and heareth the weepings and wailings of friends and relatives, and although it seeth them and heareth them calling upon him, they seeth him not."

OUT-OF-BODY AND AFTER-DEATH EXPERIENCES

The notion of a soul leaving the body following death is a common belief among almost all religions. Again, some believe that the soul may fall asleep, whereas others believe it will either ascend to heaven or descend into hell. And, as noted, we are also told that the soul of the dead may linger upon the Earth, may remain tethered to the body, or may even return to visit the dead body, sometimes hovering above it, and in rare instances, somehow reanimating the body, such that the dead come back to life.

Likewise, among modern day people, some children and adults who have been declared "clinically" dead have returned to life. And many of those who subsequently return to life, have reported that after "dying" they left their body and floated above the scene (Eadie 1992; Rawling 1978; Ring 1980). And, at first, many do not realize they are dead.

According to modern accounts, typically they become increasingly euphoric as they float above their body, after which they may float away, become enveloped in a dark tunnel and then enter a soothing radiant light. And later, when they come back to life, they may even claim conscious knowledge of what occurred around their body while they were dead and floating nearby.

"Lisa" for example, was a 22 year old college coed with no religious background or spiritual beliefs, who was badly injured in an auto accident when the windshield collapsed and all but completely severed her arm. According to Lisa, when she got out of the car her was blood spraying everywhere and she only walked a few feet before collapsing. Apparently an ambulance arrived in just a few minutes. However, the next thing she noticed was that part of the time she was looking up from the ground, and part of the time she was up in the air looking down. From above she could see the ambulance crew working, picking up her body, placing it on a gurney and into the ambulance.

According to Lisa, during the entire ride to the hospital it was like she was half in and half out of the ambulance, as if she were running along outside or just extending out of the vehicle watching the cars and trees go by. When they got to the hospital she was no longer attached to her body but was floating up and down the halls, watching the doctors and nurses and attendants. One doctor in particular drew her attention because he had a big belt buckle with his name written on it. She could even read it and it said "Mike."

According to Lisa, she was "tripping out, bobbing up and down the halls, just checking everything out" when she noticed a girl lying on a gurney with several doctors and nurses working frantically. When she floated over and peered over the shoulder of one of the doctors to take a look she suddenly realized the girl was her and that her hair and face were very bloody and needed to be washed. At that point she realized she was floating well above her body and the doctors, and that she looked to be "dead." Lisa said she did not feel afraid or upset, although the fact that her hair was bloody and dirty bothered her.

She soon floated up and outside the Emergency room and was enveloped in a total blackness, "like I was passing through a tunnel at the end of which was a light which became brighter and more brilliant, radiating outward." The light soon enveloped her body which made her feel exceedingly happy and very warm. A few moments later she heard the voice of her grandmother who had died when Lisa was a young girl. Although Lisa had no memory of this grandmother, she nevertheless recognized her and felt exceedingly happy. However, her grandmother very sorrowfully told her it was "too soon" she would have to "go back." Lisa didn't want to go back, but had no choice. She was drawn away from the light and felt herself falling only to land with a painful thump in her own body. At this point she moved her hand which alerted one of the emergency room staff that Lisa was no longer dead.

Lisa had no religious training and had never heard of "near death experiences" (she was injured in 1982). After returning to life she only reluctantly explained what had happened when she was questioned by one of her doctors.

Lisa also claimed that while she was dead and floating about the emergency room that she saw and heard what was going on around her after she died, and upon coming back to life she was able to accurately recall everything that occurred in the emergency room up to the point when she was enveloped in darkness. She was able to accurately describe "Mike" (who in fact wore a belt buckle engraved with "Mike"), most of the staff who attended her, as well as the conversations that occurred around her as well as some of the other patients.

Similar "after death" claims of leaving and floating above the body, and seeing everything occurring below, are common (Eadie 1992; Moody 1977; Rawling 1978; Ring 1980; Sabom 1982; Wilson 1987), and, as noted, are even reported in the 6,000 year old Egyptian Book of the Dead (Budge 1994), as well as the Tibetan Book of the Dead (the Bardo Thodol) which was composed over 1,300 years ago (Evans-Wentz 1960). Approximately 37% of patients who are resuscitated report "out of body" experiences (Ring 1980).

Consider for example, the case of Army Specialist J. C. Bayne of the 196th Light Infantry Brigade. Bayne was "killed" in Chu Lai, Vietnam, in 1966. He was simultaneously machine gunned and struck by a mortar. According to Bayne, when he opened his eyes he was floating in the air, looking down on his crumpled, burnt, and bloody body, and he could see a number of Vietcong who were searching and stripping him:

"I could see me... it was like looking at a mannequin laying there... I was burnt up and there was blood all over the place... I could see the Vietcong. I could see the guy pull my boots off. I could see the rest of them picking up various things... I was like a spectator... It was about four or five in the afternoon when our own troops came. I could hear and see them approaching... I could see me... It was obvious I was burnt up. I looked dead... they put me in a bag... transferred me to a truck and then to the morgue. And from that point, it was the embalming process."

"I was on that table and a guy was telling a couple of jokes about those USO girls... all I had on was bloody undershorts... he placed my leg out and made a slight incision and stopped... he checked my pulse and heartbeat again and I could see that too...It was about that point I just lost track of what was taking place... [until much later] when the chaplain was in there saying everything was going to be all right... I was no longer outside. I was part of it at this point" (reported in Wilson, 1987, pp 113114; and Sabom, 1982, pp 81-82).

Some surgery patients, although ostensibly "unconscious" due to anesthesia, are also able to later describe conversations and related events that occurred during the operation (Furlong, 1990; Kilhstrom, et al. 1990; Polster, 1993). Hence, the notion that those who are "clinically dead" or near death may recall various events that occurred while they were ostensibly "dead" should not be dismissed out of hand. Moreover, some surgery patients also claim to "leave their bodies" while they were "unconscious" and claim to recall seeing not just the events occurring below, but in one case, dirt on top of a light fixture (Ring 1980). "It was filthy. I remember thinking, 'Got to tell the nurses about that."

Did the above surgical patient or Lisa or Army Specialist Bayne really float above and observe their bodies and the events taking place below? Or did they merely transpose what they heard (e.g. conversations, noises, etc.) and then visualize, imagine, or hallucinate an accompanying and plausible scenario? This seems possible, even in regard to the "filthy" light fixture. On the other hand, not all those who have an "out of body" hear conversations, voices, or even sounds. Rather, they may be enveloped in silence.

"I was struck from behind... That's the last thing I remember until I was above the whole scene viewing the accident. I was very detached. Everything was very quiet. This was the amazing thing about it to me... I could see my shoe which was crushed under the car and I thought: Oh no. My new dress is ruined... I don't remember hearing anything. I don't remember anybody saying anything. I was just viewing things... like I floated up there..." (Sabom, 1982; p. 90).

Moreover, even individuals born blind experience these "near death" hallucinations. And although blind, while dead they will see for the first time. However, whereas many who experience life after death find the experience exhilarating, others respond with feelings of extreme dread and terror. Death is not a pleasant experience.

FEAR AND OUT-OF-BODY EXPERIENCES

The prospect of being terribly injured or killed in an auto accident or fire fight between opposing troops, or even dying during the course of surgery, are often accompanied by fear. It is also not uncommon for individuals who experience terror to report perceptual and hallucinogenic experiences, including dissociation, depersonalization and the splitting off of ego functions. They may feel as if they have separated from their bodies and floated away, or were on the ceiling looking down (Campbell 1988; Courtois 1995; Grinker and Spiegel 1945; James 1958; Neihardt and Black Elk 1932/1989; Noyes and Kletti 1977; Parson 1988; Southard 1919; Terr 1990). Consider the following accounts:

"The next thing I knew I wasn't in the truck anymore; I was looking down from 50 to 100 feel in the air." "I had a clear image of myself... as though watching it on a television screen." "I had a sensation of floating. It was almost like stepping out of reality. I seemed to step out of this world" (Noyes and Kletti 1977).

Or as a close friend described his experience: "I was shooting down the freeway doing about 100 in my Mustang when a Firebird up ahead suddenly cut me off. As I switched lanes to avoid him, he also switched lanes at which point I hit the breaks and began to lose control. The Mustang began to slide and spin... I felt real terror... I was probably going to be killed... I was trying to control the Mustang and avoid turning over, or hitting any of the surrounding cars or the guard rail... time seemed to slow down and then I suddenly realized that part of my mind was a few feet outside the car looking all around; zooming above it and then beside it and behind it and in front of it, looking at and analyzing the respective positions of my spinning Mustang and the cars surrounding me. Simultaneously I was inside trying to steer and control it in accordance with the multiple perspectives I was suddenly given by that part of my mind that was outside. It was like my mind split and one consciousness was inside the car, while the other was zooming all around outside and giving me visual feedback that enabled me to avoid hitting anyone or destroying my Mustang."

HIPPOCAMPAL PLACE NEURONS: ASTRAL PROJECTION

Feelings of fear and terror are mediated by the amygdala, whereas the capacity to cognitively map, or visualize one's position and the position of other objects and individuals in visual-space is dependent on the hippocampus.

The hippocampus contains "place" neurons which are able to encode one's position, place, and movement in space. Specifically, O'Keefe, Nadel, and colleagues, found that hippocampal pyramidal cells were able to become attuned to specific locations within the environment, as well as to particular objects and their location in that environment, thereby creating cognitive maps of visual space. Moreover, they discovered that as the subject moves about in that environment, entire populations of cells would fire but only when in a particular spot, whereas other cells would fire when in a different location.

Some cells respond not just when moving about, but in reaction to the speed of movement, or when turning in different directions. Moreover, some cells are responsive to the movements of other people in that environment and will fire as that person is observed to move around (Nadel, 2001; O'Keefe, 1976; Wilson and McNaughton, 1993).

The hippocampus, therefore, can create a cognitive map of an individual's environment and their movements and the movement of others within it. Presumably it is via the hippocampus that an individual can visualize themselves as if looking at their body from afar, and can remember and thus see themselves engaged in certain actions, as if one were an outside witness (Joseph, 1996). Indeed, as is well known, the hippocampus play a major role in forming and retrieving memories.

Under conditions of hyperactivation (such as in response to extreme fear) it appears that the hippocampus may create a visual hallucination of that "cognitive map" such that the individual may "experience" themselves as outside their body, observing all that is occurring and/or hallucinate themselves as moving about in that environment such as flying above it.

In fact, it has been repeatedly demonstrated that hyperactivation or electrical stimulation of the amygdala-hippocampus-temporal lobe, can cause some individuals to report they have left their bodies and are hovering upon the ceiling staring down (Daly 1958; Jackson and Stewart 1899; Joseph, 1996, 2014; Penfield 1952; Penfield and Perot 1963; Williams 1956). That is, their ego and sense of personal identity appears to split off from their body, such that they may feel as if they are two different people, one watching from above, the other being observed down below.

As described by Penfield (1952), "it was as though the patient were attending a familiar play and was both the actor and audience."

Presumably abnormal activation due to extreme fear or direct electrical stimulation induces an individual to think they are seeing themselves from afar because the hippocampus is transposing and "hallucinating" one's image; similar to what occurs during normal remembering. Or, perhaps it is not a hallucination at all, but a capacity mediated by a hyperactivated hippocampus.

As noted, many patients who are diagnosed as "clinically dead" and then return to "life" report that after leaving their body they enter a dark tunnel and are then enveloped in a soothing radiant light. The same is reported in the Egyptian and Tibetan Books of the Dead.

Presumably, in that the hippocampus, amygdala, and inferior temporal lobe receive direct and indirect visual input and contain neurons sensitive to the fovea and upper visual fields, hyperactivation of this region also induces the sensation of seeing a radiant light. The massive release of opiates (due to physical trauma leading to "death") would account for the immediate loss of fear and the experience of tranquillity and joy. Continued activation of these brain regions would also account for the hallucinations of seeing dead relatives, that are commonly reported by those who have "died," as well as the life review, in which one's past life flashes before their eyes.

At death, the amygdala and hippocampus would begin to remember, and memories that were emotional and personal, and all accompanying emotions, such as guilt, would be experienced as part of death. That is, not just the memory, but the awful feeling of the memory would be reexperienced. These structures, therefore, account for the feeling of floating above the body, the life review, and the judgment imposed on the self by their guilty or guilt-free conscience.

The hyperactivation of these limbic structures, therefore, explains why those who have near death experiences report feelings of peace, rapture and joy as they were "bathed by the light" and stood in the all knowing presence of "God" or other divine beings including friends and relatives who had previously passed away. Indeed, these exact same feelings and experiences can be induced by electrically stimulating the inferior temporal lobe and amygdala-hippocampal complex.

EXPERIMENTAL & SEIZURE INDUCED OUT-OF-BODY, HEAVENLY AND OTHER WORLDLY EXPERIENCES

Some individuals suffering from temporal lobe epilepsy also report "out of body experiences." Penfield and Perot (1963) describe several patients who during a temporal lobe seizure claimed they could see themselves in different situations. One woman stated that "it was though I were two persons, one watching, and the other having this happen to me," and that it was she who was doing the watching as if she was completely separated from her body.

One patient had a sensation of being outside her body and watching and observing her body from the outside. Another neurosurgery patient alleged that while outside her body she was also overcome by feelings of euphoria and eternal harmony.

Other patients claim to have quite pleasant auras and describe feelings such as elation, security, eternal harmony, immense joy, paradisiacal happiness, euphoria, completeness. Between .5 and 20% of such patients report these feelings (Daly 1958; Williams 1956).

A patient of Williams (1956) claimed that his attacks began with a "sudden feeling of extreme well being involving all my senses. I see a curtain of beautiful colors before my eyes and experience a pleasant but indescribable taste in my mouth. Objects feel pleasurably warm, the room assumes vast proportions, and I feel as if in another world."

A patient described by Daly (1958) claimed his temporal lobe seizure felt like "a sunny day when your friends are all around you." He felt dissociated from his body, as if he were looking down upon himself and watching his actions. Williams (1956) describes a patient who during a seizure-induced aura reported that she experienced a feeling of being lifted up out of her body, coupled with a very pleasant sensation of elation and the feeling that she was "just about to find out knowledge no one else shares, something to do with the link between life and death." Subirana and Oller-Daurelia (1953) described two patients who experienced ecstatic feelings of "extraordinary beatitude" or of paradise as if they had gone to heaven. Their fantastic feelings also lasted for hours.

Other patients suffering from temporal lobe seizures have noted that feelings and things suddenly become "crystal clear" or that they had a feeling of clairvoyance, or of having the truth revealed to them, or of having achieved a sense of greater awareness such that sounds, smells and visual objects seemed to have a greater meaning and sensibility. Similar claims are made by those who have "died" and returned to tell the tale.

One woman I evaluated claimed she would float on the ceiling, and could float outside and could see, on one occasion, a friend who was coming up the walkway. She also reported that by having a certain thought, she could propel herself to other locals including the homes of her neighbors.

EMBRACED BY THE LIGHT

"The heavens were opened and the whole creation which is under heaven shone... I saw a light, and a child in the light... And while I looked he became an old man. And he changed his form again... and I saw... an image with multiple forms in the light." -Apocryphon of John.

Betty J. Eadie reports in her 1992 book, "embraced by the light" that after dying and then communing with three "ancient" men who appeared at her side and who "glowed" she suddenly thought of her husband and children who she wanted to visit. "I began to look for an exit" and discovered that "my spiritual body could move through anything...My trip home was a blur. I began moving at a tremendous speed... and I was aware of trees rushing below me. I just thought of home and knew I was going there... I saw my husband sitting in his favorite armchair reading the newspaper. I saw my children running up and down the stairs... I was drawn back to the hospital, but I don't remember the trip; it seemed to happen instantaneously" (pp. 33-35).

BLACK ELK

Compare Eadie's description with that of Black Elk (Neihardt and Black Elk, 1932/1989), a Lakota Sioux Medicine Man and spiritual leader (born in 1863). During a visit to England (he was part of Buffalo Bills Wild West Show) he suddenly fell out of his chair as if dead, and then experienced himself being lifted up. In fact, his companions thought he had died.

According to Black Elk: "Far down below I could see houses and towns and green land and streams... I was very happy now. I kept on going very fast... Then I was right over Pine Ridge. I looked down (and) saw my father's and mother's teepee. They went outside, and she was cooking... My mother looked up, and I felt sure she saw me... then I started back, going very fast...Then I was lying on my back in bed and the girl and her father and a doctor were looking at me in a queer way...I had been dead three days (they told him)...and they were getting ready to buy my coffin" (pp. 226228).

This was not Black Elk's first out of body experience, however. Black Elk demonstrated numerous behaviors and symptoms suggestive of temporal lobe epilepsy. Beginning even in childhood Black Elk repeatedly experienced "queer feelings" and heard voices, had visions, and suffered numerous instances of sudden and terrible fear and depression accompanied by weeping, as well as trance states in which he would fall to the ground as if dead.

Black Elk also had other visions similar to those reporting "life-after-death" experiences, including the following incident that occurred during one of his trance and out-of-body states: "Twelve men were coming towards me, and they said, Our father, the two-legged chief, you shall see... There was a man standing. He was not Wasichu (white) and he was not an Indian. While I was staring at him his body began to change and became very beautiful with all colors of light, and around him there was light..." (p. 245).

Similar accounts, including descriptions of a tribunal of 12 are detailed in the Egyptian Book of the Dead.

Ms. Eadie (like many others who have experienced "life after death") came upon a man standing in the light which "radiated all around him. As I got closer the light became brilliant...I saw that the light was golden, as if his whole body had a golden halo around it, and I could see that the golden halo burst out from around him and spread into a brilliant, magnificent whiteness that extended out for some distances" (pp. 40-41).

ASTRAL ANTIQUITY

Muhammed reports that while sleeping and dreaming he was lifted into the air and transported by the angel Gabriel from Arabia to the Temple Mount in Jerusalem. There he was greeted by three men, who he believed to be Abraham, Moses, and Jesus, as well as a crowd of other prophets. He was then lifted up and entered a divine sphere and came upon the garden of promise and, according to the Koran, saw a "lote-tree veiled in a veil of nameless splendor."

In Hinduism, the lote tree symbolizes the limit of this reality and rational thought.

St. Augustine reports he was "lifted up by an ardent affection towards eternal being itself... we climbed beyond all corporate objects and the heaven itself, where sun, moon, and stars shed light on the earth."

Some individuals (and their followers) claim to be able to voluntarily leave their body (Monroe, 1994), this includes any number of "mystics," and New Age spiritualists, as well as some priests, prophets and shamen. Indeed, Monroe (1994) founded an Institute to study this phenomenon, and claims that others can learn this technique. Monroe, however, notes that when he had his first out-ofbody experience he had felt extremely frightened.

That so many people, regardless of culture or antiquity, have similar experiences (or hallucinations) while in trance states, or dreaming, or after "dying," of leaving their body, is presumably due to all possessing a limbic system and temporal lobe that is organized similarly.

The fact that although ostensibly similar, many of these experiences are also colored by one's cultural background, can in turn be explained by differences in experience and cultural expectations and thinking patterns. As explained in the Tibetan Book of the Dead: "It is quite sufficient for you (the deceased) to know that these apparitions are [the reflections of] your own thought forms."

DEATH AND THE LIMBIC SYSTEM

Presumably conditions such as dreaming, trance states, depth electrode activation, extreme fear, traumatic injury, or temporal lobe epilepsy, result in hyperactivation of the amygdala and hippocampus and overlying temporal lobe. These structures create an image of the individual floating or even soaring above familiar or bizarre surroundings, and will trigger memories, hallucinations, brilliant lights, as well as secrete opiate-like neurotransmitters which induce a state of euphoria and thus eternal peace and harmony.

Given that similar experiences are reported by those who have experienced depth electrode stimulation of these structures, and by those declared "clinically dead" also raises the possibility that the hippocampus and amygdala may be the first areas of the brain to be effected by approaching death, as well as one of the last regions of the brain to actually die.

That is, as one approaches death and even after medical death, the amygdala and hippocampus may continue to function briefly and not only become hyperactivated, but produce a feeling of eternal peace and tranquillity and a hallucination of floating outside the body and of meeting relatives and other religious figures; like a dream.

And yet, it is curious that so many individuals have basically a very similar "dream" and only under conditions of death. Moreover, it is exceedingly difficult to reconcile these experiences with the Darwinian notion of evolution.

What is the "evolutionary" adaptive significance of so many members of the human race having a dream of the "Hereafter" after they die.

As we have learned from physics, this reality is only one of many, and the perception of these alternate realities, and all associated manifestations of extraordinary psychic abilities, are normally inhibited, suppressed, and cannot be perceived or processed due to the limitations of our brain. Under conditions of hyperactivity, the limbic system may not only hallucinate, it may also begin to process this information, and gain access to and engage in alternate modes of thinking, perceiving, and communicating, such that what is normally concealed is revealed to the receptive mind.

DREAMS, SPIRITS AND THE SOUL

Dreams and hallucinations, have, as their neurological source, activity generated within the temporal lobe and limbic structures buried within--the hippocampus and the amygdala. The hippocampus is known to be a depository of new memories and to assist in the retrieval of new memories-memories that often reappear in the contexts of dreams. Yet, dreams and memories, especially emotional memories, also have as their neurological source, the amygdala and overlying temporal lobe and associated limbic structures.

The amygdala enables us to hear "sweet sounds," recall "bitter memories," or determine if something is spiritually significant, sexually enticing, or good to eat. The amygdala also makes it possible for us to store personal, sexual, spiritual, and emotional experiences in memory and to recall and reexperience these memories when awake or during the course of a dream in the form of visual, auditory, or religious or spiritual imagery (Bear 1979; d'Aquili & Newberg 1993; Gloor 1986, 1992, 1997; Halgren, 1992).

The amygdala, in conjunction with the hippocampus, contributes in large part to the production of very sexual as well as bizarre, unusual and fearful mental phenomenon including out-of-body dissociative states, feelings of depersonalization, and hallucinogenic and dreamlike recollections involving threatening men, naked women, sexual intercourse, religion, the experience of god, as well as demons and ghosts and pigs walking upright dressed as people (Bear 1979; Daly 1958; d'Aquili & Newberg 1993; Gloor 1986, 1992; Halgren 1992; Horowitz et al., 1968; Jaynes 1976; MacLean 1990; Mesulam 1981; Penfield & Perot 1963; Rolls, 1992; Schenk & Bear 1981; Slater & Beard 1963; Subirana & Oller-Daurelia 1953; Taylor, 1972, 1975; Trimble, 1991; Weingarten et al., 1977; Williams, 1956).

The amygdala also makes it possible to experience not just spiritual and religious awe, but all the terror and dread of the unknown. Indeed, the amygdala can generate feelings of hellish, nightmarish fear

And yet, it is also the amygdala which is responsible for the capacity to transcend the known, this reality. It is also the amygdala which assists in maintaining this reality through the inhibition and filtering of most of the sensory signals bombarding the brain and body. If not for this sensory filtering, a color might have taste and its own particular texture, whereas sounds might provoke smells, warmth, light, colors, tastes, and feelings of weight. Hyperactivation of the amygdala, and overlying temporal lobes can remove these sensory filters, and what is concealed is revealed, sometimes in overwhelming confusing majesty.

Neurosurgical patients have reported communing with spirits or receiving profound knowledge from the Hereafter, following depth electrode amygdala stimulation or activation (Penfield and Perot 1963; Subirana and Oller-Daurelia, 1953; Williams 1956). Some have reported hearing even the singing of angels and the voice of "God."

THE LIMBIC SYSTEM: SPIRITUALITY AND EMOTION

According to d'Aquili and Newberg (1993) mystical states may be voluntarily or involuntarily induced and are dependent upon the differential stimulation of limbic system nuclei, including the hypothalamus, hippocampus, and amygdala, as well as the right frontal and right temporal lobe. However, it appears that these brain areas differentially contribute to religious and emotional experience.

For example, the hypothalamus is concerned with all rudimentary aspects of emotion, homeostasis such as food and water intake, and controls the hormonal and related aspects of violent behavior and sexual activity. The amygdala also plays a highly significant role in sexual and violent behavior, and, in conjunction with the temporal lobe and hippocampus, enables a human to have religious, spiritual and mystical experiences (Bear 1979; Daly 1958; d'Aquili and Newberg 1993; Horowitz et al. 1968; Mesulam 1981; Penfield and Perot 1963; Schenk, and Bear 1981; Slater & Beard 1963; Subirana & Oller-Daurelia 1953; Trimble 1991; Weingarten, et al. 1977; Williams 1956). These religious experiences may involve violent behavior or sexual activity, due to limbic involvement.

As we shall see in later chapters, many religious customs and rituals involve violence and sexuality, which in turn is due to the activation of this limbic system transmitter to god.

The amygdala and hypothalamus subserve the capacity to experience feelings of intense sexual arousal, fear or, conversely, euphoria, including an orgasmic feeling of rapture, or the "nirvana" of a heroin "high." This is because the hypothalamus and amygdala are major pleasure centers, and contain opiate-producing neurons and opiate (enkephalin) receptive neurons, thus generating feelings of numbness and a narcotic high. Large concentrations of opiate receptors and enkephalin-producing neurons are located throughout the amygdala (Atweh & Kuhar 1977ab; Uhl et al., 1978).

In response to pain, stress, shock, fear, or terror, the amygdala and other limbic nuclei begin to secrete high levels of opiates which can eventually induce a state of calmness as well as analgesia and euphoria (Joseph, 1998a, 1999a). It is this heroin high that explains why an ox, deer, and other animals will simply give up and lie still while they are devoured and eaten alive. It is this amygdala-induced heroin high which also contributes to feelings of religious rapture and the ecstasy associated with life after death and the attainment of Nirvana.

THE AMYGDALA, HIPPOCAMPUS AND HALLUCINATIONS

Whereas the amygdala and hypothalamus interact in regard to pleasure, rage, and sexuality, the amygdala and hippocampus interact to subserve and mediate wholly different aspects of experience, including memory, dreaming, and hallucinations. The hippocampus in particular appears to be responsible for certain types of "hallucinations" such as the visualizations of astral projection or seeing oneself floating above the body (Joseph 1996, 1999b, 2000a). Some patients report not only floating, but of being embraced by a light and taken to a vast realm of fantastic proportions where they are given access to knowledge of the nature of life and death.

The amygdala, hippocampus, and temporal lobe are richly interconnected and appear to act in concert in regard to mystical experience, including the generation and experience of dream states and complex auditory and visual hallucinations, such as may be induced by LSD (Broughton 1982; Goldstein et al. 1972; Gloor 1986, 1992; Hodoba 1986; Horowitz, et al. 1968; Joseph, 1992a; Meyer et al. 1987; Penfield and Perot 1963; Weingarten, et al. 1977; Williams 1956). If these neurons are hyperactivated, such as occurs during dream states, seizures, physical pain, terror, food deprivation, social and sensory isolation, and under LSD (which disinhibits the amygdala by blocking serotonin) an individual might infuse their perceptions with tremendous religious and emotional feeling. Hence, under these conditions the individual may hallucinate, and ordinary perceptions, objects or people may be perceived as spiritual in nature or endowed with special or religious significance.

Intense activation of the temporal lobe, hippocampus, and amygdala has been reported to give rise to a host of sexual, religious and spiritual experiences; and chronic hyperstimulation can induce an individual to become hyper-religious or to visualize and experience ghosts, demons, angels, and even "God," as well as claim demonic and angelic possession or the sensation of having left their body (Bear 1979; Gloor 1986, 1992; Horowitz et al. 1968; MacLean 1990; Mesulam 1981; Penfield & Perot 1963; Schenk & Bear 1981; Williams 1956).

In some instances the individual may come to believe he or she is hearing, seeing, and interacting with gods, angels and demons when in fact they are hallucinating. These false beliefs are accentuated further because they are excessively emotionally and religiously aroused and are experiencing an "enkephalin" high and feelings of rapture or "nirvana."

In many cases, however, the individual is not hallucinating. Rather, their eyes have been opened, and they suddenly see as gods... knowing good and evil.

LSD, LIMBIC SYSTEM FILTERING AND HALLUCINATIONS

The amygdala is capable of processing visual, tactile, auditory, gustatory, olfactory, and emotional stimuli simultaneously. Amygdaloid neurons are multimodally responsive. Single amygdaloid neurons receive a considerable degree of topographic input, and are predominately polymodal, responding to a variety of stimuli from different modalities simultaneously.

Normally much of this data is suppressed and filtered so as to prevent the tasting of colors, the visualization of sound, and so on. This is made possible via the inhibitory influences of the frontal lobes and a variety of neurotransmitters including serotonin (5HT). 5HT suppresses synaptic activity in the amygdala and throughout the neocortex and thus reduces sensory input (Curtis & Davis 1962; Marazzi & Hart 1955; Morrison & Pompeiana 1965). 5HT restricts perceptual and information processing and in fact increases the threshold for neural responses to occur at both the neocortical and limbic level. In this manner, selective attention (via sensory filtering) can be maintained while the organism is engaged in goal directed motor behavior, such as running away or taking evasive action if chased by a lion.

In consequence, substances which block 5HT reception at the level of the synapse -such as LSDresults in increased activity in the amygdala and in the sensory pathways to the neocortex (Purpura 1956), such that information that is normally filtered out is perceived. Following the administration of LSD high amplitude slow waves (theta) and bursts of paroxysmal spike discharges occurs in the hippocampus and amygdala (Chapman & Walter, 1965; Chapman et al. 1963), and the amygdala begins to process information that is normally suppressed.

Under conditions which induce limbic hyperactivity, 5HT may be depleted and limbic sensory acuity is increased. However, what is perceived is not necessarily a hallucination but instead represents the perception of overlapping sensory qualities that are normally filtered out. Colors may be felt and tasted, music may be observed as well as heard, the molecular composition of ceilings, floor and walls may be parted so that one can see through the spaces between where molecules join together. And the pulse of life may be experienced as it ebbs and flows in a leaf one holds between their finger tips.

Consider, for example, a description of someone's first LSD "trip:"

"It was 1966... Jimi Hendrix was singing about purple haze, and that is exactly what we scored up in Haight Ashbury—mixed by the master himself, Stanley Osley..."

"About half an hour after I'd taken it, I was walking toward the park to meet and trip with my fellow tripsters when I began to notice the incredible clarity and vividness of my surroundings. Colors were brighter, plants seem to sparkle...and I stopped and touched a leaf...I could feel its energy, its life... I could taste it through my fingers...And when I got to the park I was so overwhelmed with the colors, the tastes, the smells, the incredible vividity and clarity that I felt almost overwhelmed and sat down to take it all in..."

"And then I heard a jet somewhere in the distance, and I looked into the sky but couldn't see it but my ears led my eyes to one of the mountains surrounding the valley, and oh my god, I could see right through the mountain. It was like the molecular composition of the mountain was parting into separate molecules. I could see the spaces between the molecules which were all in a frenzy of activity... it was as if I had achieved X-ray vision, and there were these crystal blue holes—like bubbles—and I could see right through the mountain and I could see the sky on the other side, including the jet. I could see the jet on the other side of the mountain by looking right through the mountain, by looking right through those gaps in the spaces between the molecules which were zipping along in their own unique pattern. And then the jet flew over the top of the mountain and instead of one jet I could see ten, then a hundred, and a kaleidoscope of jets in the sky."

"I raised my hand to point to this incredible sight, and instead of one hand, there were these trails of hands. And I did that again, waved my hand and there were these hand-arm trails of handshands-hands catching up and then merging... and it was then that I realized I could see through my hand. So I gazed at it for closer inspection. It was as if my eyes became a tunneling microscope—and this was 10 years before they invented tunneling microscopes."

"At first I could see the incredible cellular structure of the skin, and then the molecular structure...the pulsating molecules themselves... and then I could see between the spaces where the molecules joined together... my sight penetrated the skin and I could see the blood vessels, and then my eyes penetrated the blood vessels and I could see inside the blood vessel, I could see the blood platelets and the white corpuscles as they swirled through the vessel—and I kept thinking: How come I never noticed this before? I had forgotten that I had taken LSD."

Hallucinogens, of course, are an age-old means of obtaining access to alternate realities, including the realities of "god." Hallucinogens are said to enable an individual to peer between the space that separate this reality from all other realities, such that what is concealed is suddenly revealed. It is the amygdala, hippocampus, and temporal lobe which are responsible for these complex hallucinations, and it is the amygdala which normally filters much of this information so that it remains hidden. Hence, by hyperactivating these structures, in essence, one is also activating the transmitter to god.

SOULS, SPIRITS, DREAMS, AND POLTERGEISTS TO KNOW FEAR IS TO KNOW GOD

"Fear the Lord your God." -Deuteronomy 10:12.

Priests, prophets, shamans and many others who encounter "god" or His "angels" not uncommonly experience tremendous fear. Fear is the most common emotion associated with the amygdala (Gloor, 1997; Halgren, 1992; LeDoux, 1996; Williams, 1956), with some patients experiencing horrifying, hellish, and nightmarish fear, sometimes coupled with hellish hallucinations.

We are repeatedly told in the Bible, that to know fear is to know god. Indeed, even a committed atheist may feel compelled to cry out and pray to god if sufficiently terrified. Terror is also an emotion associated with the amygdala... as well as with the Lord God.

Yet others experience awe and rapture when confronted by the divine. As noted, amygdala hyperactivation is also associated with feelings of extreme joy and ecstasy. According to d'Aquili and Newberg (1993; p. 194) "a combination of the experience of both fear and exhalation" is "usually termed religious awe." These feeling states are "almost always associated with religious symbols, sacred images, or archetypal symbols" which flow "from the inferior temporal lobe" and which "appear sometimes as monsters or gods". Indeed, angels, demons, and poltergeists may be experienced.

Most people find these experiences quite terrifying. They also frequently believe their perceptions are completely real and are not hallucinations.

SPIRITS AND POLTERGEISTS

"Cindy," a 22 year old college student, was plagued by demons and ghosts for months until her abnormal right inferior temporal lobe was surgically removed.

Prior to her brain injury, Cindy had never been very religious, and had certainly never seen a ghost; that is, until following her auto accident. She had been thrown over 50 feet through the windshield of her car and suffered a fracture of the right temporal region of the skull and developed a subdural hematoma, a blood clot, which was pressing on the temporal lobe inducing herniation. Burr holes were drilled into her skull and the clot was surgically evacuated. Although her brain and temporal lobe had been injured, over the following weeks she seemed to quickly recover.

It was several days after her release from the hospital when she was startled while watching television. The arms, legs, hands, feet, and heads of the various actors began protruding from the screen into the living room where she sat.

Cindy said that at first she thought the television was broken and turned it off. But, as she stared back at the blank screen she saw what looked like her dead father staring back at her (which was probably her own reflection). As she backed away, the figure emerged from the television. He beckoned to her, and then behind him ghostly spooks and wraiths began to stream from the picture tube.

Terrified and crying for her mother she raced for the bathroom and locked herself in. Yet, even as she hid within the inner sanctum of the washroom, spirits, sprites and poltergeists streamed from the bathroom mirror and swirled about her. Crying and stumbling, she raced into the living room and was horrified to see a spirit enter and take possession of her mother who was transformed before her eyes. Panicked and terrified, Cindy ran into the street crying for help. A police officer, after investigating the scene, brought her to the local hospital and psychiatry unit. She was medicated and kept there on a 72 hour hold.

Later she decided what she had experienced were ghosts and lost souls of people who had been buried in an old, almost completely forgotten cemetery on the other side of the hill from where she lived. She also thought they were the ghosts of Indians who had been entombed beneath her house as there are numerous Indian burial grounds in the county.

Once she was released from her 72 hour psychiatric hold, she stopped taking her medication, and over the course of the next several weeks, she claimed to see "animal spirits." She reported that the "secret souls" of her mother's house plants were watching and observing her and that she could sometimes see filmy, soul-like entities traveling to and fro across the room and between different plants.

After several more hospitalizations, and an EEG, it was determined that she was suffering from excessive activity, seizure activity in the damaged temporal lobe. The inferior temporal lobe and the underlying amygdala were surgically ablated and destroyed, and she ceased to "hallucinate."

DREAMS, ANIMAL SPIRITS AND LOST SOULS

Across time and culture, people have believed that not just humans and animals, but plants and trees were sensitive, sentient, intelligent, and the abode of spirits including the souls of dead ancestors (Campbell 1988; Frazier 1950; Harris, 1993; Jung 1964; Malinowski 1948). Before felling a tree, the spirit sometimes had to be conjured forth so as to not harm it (Campbell 1988; Frazier 1950). Be it animal or plant, souls were also believed capable of migrating to new abodes.

Among the ancients and many so called primitive cultures, it was believed that souls are reflected in shadows, in streams, and pools of water (Campbell, 1988; Frazier, 1950; Harris, 1993; Jung, 1964; Malinowkski, 1948). Because ghosts or demons sometimes attempt to abduct souls, this required that one's shadow and reflection be protected. Even water spirits might try to capture a person's soul, so staring into reflecting ponds and lakes was to be avoided.

Moreover, the ancients believed that the shadows and reflections of others had to be avoided so that one did not come into contact with the soul of a witch, sorcerer, or a demon. It was believed that one's soul could be abducted by demons and witches as well as the recently departed. This is also why in some cultures people turn mirrors to the wall after a death and lay down pictures of the recently departed (Frazier 1950). This insures that living souls are not stolen by the souls of the dead who are leaving this world for the next one.

The belief in the persistence of the soul after death gave rise to ancestor worship in some ancient cultures, including China. Oracle Bones were believed to contain the souls of the dead and were employed for divinatory purposes in ancient China over 4,000 years ago. The earliest Oracle Bones so far discovered are covered with pictograms including a man with a large ghostly distorted head (Brandon, 1967). This pictogram also denotes the word kuei, which means, soul. Oracle bones also included characters signifying the words "she, shen, and tsu." "Tsu" means dead ancestor, "she" is a protective fertility deity of the soil, and "shen" are divine beings connotating phallic significance.

The ancient Chinese practiced a religious ritual which was intended to prevent the soul from departing the body at death. The orifices would be stopped up to keep the kuei safely inside the body. So long as the soul remained tethered to the body, the body would not undergo its final fatal disintegration. These beliefs gave rise to the ritual of "calling back of the soul," a ritual also performed for the living as the souls were believed able to wander forth from the body, such as during sleep, exposing the living body to death and extinction (Brandon, 1967).

The ancient Chinese believed that the kuei represented by the yin-soul, was associated with the body since conception. By contrast, a yang-soul was associated with the individual personality and the person's unique mental qualities.

The alternating and competing principles of the yin-soul and the yang-soul are directly related to the concepts of ying and yang. They also gave rise to ancestor worship as the ancient Chinese believed that the life force, the energy associated with the soul, this substance familiale, lay buried beneath the ground as une masse indistincte. "This energy was represented above ground at any given moment, only by the living members of the family, which constituted the individualized portions of that substance familial. It followed, accordingly, that each birth within the family represented the reincarnation of a portion of that subterranean substance familial, while each death meant the return of a part of this individual family substance familial to the masse indistincte in the ground below" (Brandon, 1967, p. 180).

SOULFUL DREAMS

Souls were also believed by ancient humans to wonder about while people sleep and dream (Brandon 1967; Frazier 1950; Harris 1993; Jung 1945, 1964; Malinowkski 1990). That is, among many different cultures and religions the soul is believed to sometimes escape the body via the mouth or nostril during sleep. Moreover, during a dream the soul may wonder away from the body and may engage in certain acts or interact with other souls including those of the dear but long dead and departed.

Sometimes the soul is believed to take on another form, such as a bird, or deer, fox, rabbit, wolf, and so on. The spirit and the soul could also hover about in human-like, ghostly vestiges, at the fringes of reality, the hinterland where day turns into night (Campbell 1988; Frazier 1950; Jung 1964; Malinowski 1954; Wilson 1951). The souls of animal's such as a wolf or eagle, could also leave the body and take on various forms including that of a woman or Man. Not just men but animals too had souls that had to be respected.

Even after death souls continued to interact with the living, and every living being possessed a soul. Hence, the ancients believed that these souls could be influenced, their behavior controlled, and, in consequence, a good hunt insured or with the assistance of a soul, or by performing magical rituals aimed at the soul, enemies could be defeated and an evil man could be easily slain.

Because animals and plants had souls, and as some gods would also take up residence within an animal or plant as a temporary or permanent abode, this gave rise to animal worship and animal sacrifice, as well as the avoidance of certain animals or plants which were not to be killed or eaten, or killed or eaten only in a certain ritualized manner (Campbell 1988; Frazier 1950; Malinowkski 1948; Smart 1969).

Over the course of human cultural and cognitive evolution, these beliefs became increasingly complex and required specialists to interpret and minister the rituals and rites (Armstrong 1994; Campbell 1988; Frazier 1950; Smart 1969; Wilson 1951). Soon priests, prophets, and even the Gods evolved.

DREAMING OF GOD

Priests and prophets as well as the common people, often experienced God as well as animal spirits and the souls of the dead, during the course of a dream (Campbell 1988; Frazier 1950; Jung 1945, 1964; Malinowkski 1954).

Dreams have their source in the amygdala and hippocampus and inferior temporal lobe. Although mediated by brainstem nuclei, activity within the amygdala may in fact trigger the first phase of dreaming (REM), which is signified by the buildup of PGO (pontine-geniculate-occipital) waves. REM (dream) sleep is heralded and then accompanied by what has been referred to as PGO waves. That is, the amygdala is active not only during REM, but amygdala activity triggers PGO waves (Calvo, et al. 1987) which then leads to dream sleep. The amygdala produces the dream, and dreams the dream, and the dream may consist of fragmentary memories, hallucinatory imagery, and emotional extremes.

In addition to amygdala activity during REM, the hippocampus begins to produce slow wave, theta activity. Presumably, during REM, the hippocampus and amygdala act as a reservoir from which various memories, images, emotions, words, and ideas are drawn and incorporated into the matrix of dream-like activity being woven by the right (and left) hemisphere (Joseph 1982, 1988, 1996, 2014a). The hippocampus and amygdala also serve as a source from which material is drawn during the course of a daydream.

Yet, as also noted, these limbic structures also normally acts to inhibit the perception of most of the stimuli impinging on the body. This information is filtered so that one reality is maintained and to prevent the individual from being overwhelmed with competing streams of input. In other words, just as a channel selector on a television or radio permits the select reception of information from a single source (a single reality), these brain structures perform likewise. However, just as there are a multitude of channels, and just as one can change from channel to channel in order to receive a wealth of data, images, sounds, and information from a variety of sources (a variety of realities), likewise, the limbic system can also become tuned to other sources of information which are not normally perceived.

And, once the Doors of Perception are opened, all appears as it is...infinite.

DREAMING & WITCHCRAFT

"Witchcraft" and any and all "paranormal" capabilities, including "ESP" or divining by dreams, has been condemned as "unnatural" by many (but not all) religions throughout the ages including the ancient Jews and the Catholic Church.

"Witchcraft is a sin." I Samuel 15: 23

Of course, the "god" Yahweh would communicate with his prophets through visions and dreams. However, the unauthorized use of these same capacities was also condemned and outlawed, and those possessing these "unnatural" attributes killed: "I will gather you and blow upon you the fire of my wrath, and ye shall be melted in the midst therefore... for... They prophesy falsely and divine deceitfully. They have profaned what is sacred to me... I am profaned in their midst" (Ezekiel, 22:23).

The Lord God declared religious war against and condemned to death all witches and all those who could commune with the dead, speak with spirits, conjure ghosts, or communicate with devils (Exodus 22: 18, Samuel 15: 23; Ezekiel 22:23).

"You know... how he has cut off those familiar with ghosts and spirits and wizards. So wherefore then layest thou a snare for my life, to cause me to be killed." -I Samuel 28:8-10.

Likewise, the Medieval Catholic Church burnt witches, warlocks, and dreamers, and all those suspected of communing with devils. According to Church and religious authorities, these practices were the work of the Devil, Satan, the fallen angel, the fallen god, the serpent also known as the god of the snake.

These "supernatural abilities" were viewed as a threat for they could be employed to divine the future, to influence and cast spells on other humans, or to communicate with the dead or even with other gods including the devil. Hence, those who saw the future through trance states or via dreams, or who demonstrated what might be described as "ESP," although venerated by some, were condemned by others as "unnatural" and sinful.

"Thou shall not suffer a witch to live." Exodus 22:18

DREAMS AND MULTIPLE REALITIES

It is via dreams that gods and spirits speak to women and men (Campbell, 1988; Jaynes, 1976; Jung, 1945, 1964). It was via dreams that hunter-gatherers and ancient humans were able to gain access to the domicile of the soul and the spirit world of the hereafter (Frazier 1950; James 1958; Neihardt and Black Elk1989). Indeed, it has been argued that dreams (and thus the limbic system) enable an individual to come into contact with a different reality; the same reality shared and experienced by our ancestors and the gods—the ultimate reality of the Great Spirit.

Our ancient human ancestors lived in two realities, that of the physical and of the spiritual, both of which were undeniable and experienced by enemies and friends alike (Frazier, 1950; Jung, 1945, 1964). One need only spend a night alone in the forest among the trees and the elements to become quickly convinced that one is not alone, but is being watched by various entities both alive and supernatural, animal and spirit, benevolent and unkind.

Like modern day humans, the ancients had dreams by which they were transported or exposed to a world of magic and untold wonders. It is as if one had been transported to a different world and a reality which obeyed its own laws of time, space, and motion.

It is through dreams that humans came to believe the spiritual world sits at the boundaries of the physical, where day turns to dusk, the hinterland of the mind where imagination and dreams flourish and grow (Frazier, 1950; Jung, 1945, 1964; Malinowkski, 1954); hence the tendency to bury the dead in a sleeping position even 100,000 years ago.

It is also via dreams that humans came to know that spirits and lost souls populated the night. The dream was real and so too were the Gods and demons who thundered and condemned and the ghosts and phantoms that hovered at the edge of night. Although but a dream, like modern humans, our ancient ancestors experienced this through the senses, much as the physical world is experienced. Both were real and were taken seriously.

Dreams are the royal road to the unconscious realms of the mind and to realities that lie just beyond conscious awareness. And it is through the amygdala, hippocampus, and temporal lobe, that dreams emerge and flow. The limbic neurons subserving spiritual experiences also give rise to dreams. Thus the link between the world of dreams and the spirit land of gods and demons is the limbic system; i.e. the "transmitter to god."

RIGHT TEMPORAL LOBE HYPERACTIVATION & DREAMING

The amygdala and the neocortex of the temporal lobe are interactionally involved in the production of religious and hallucinatory experiences including dream states (Broughton, 1982; Gloor, 1997; Goldstein et al., 1972; Hodoba, 1986; Humphrey & Zangwill 1961; Kerr & Foulkes 1978; Meyer et al., 1987); the right temporal lobe and amygdala in particular (Joseph, 1988a, 1992a).

Similarly, d'Aquili and Newberg (1993) argue that the right hemisphere (and right amygdala) is more involved than the left in the reception and production of religious imagery. This is likely as the right hippocampus and amygdala, and the right hemisphere in general (Broughton, 1982; Goldstein et al., 1972; Hodoba, 1986; Humphrey & Zangwill 1961; Joseph, 1988, 2000a; Kerr and Foulkes 1978; Meyer, Ishikawa, Hata, and Karacan 1987) also appear to be involved in the production of hallucinations, dream imagery as well as REM during sleep. Indeed, like the limbic system, the right hemisphere is not only associated with dreams, but the unconscious mind (Joseph, 1982, 1988a,b).

In addition to dream production, the right hemisphere also appears to be the dominant source for complex non-linguistic hallucinations. Specifically, tumors or electrical stimulation of the right hemisphere or temporal lobe are much more likely to result in complex visual as well as musical and singing hallucinations, whereas left cerebral tumors or activation gives rise to hallucinations of words or sentences (Berrios, 1990; Halgren, et al. 1978; Hecaen &

Albert, 1978; Jackson, 1880; Mullan & Penfield, 1959; Penfield & Perot, 1963; Teuber et al. 1960).

Although up to five stages of sleep have been identified in humans, for our purposes we will be concerned only with two distinct sleep states. These are the REM (rapid eye movement) and nonREM (N-REM) periods. N-REM occurs during a stage referred to as "slow-wave" or synchronized sleep. In contrast, REM occurs during a sleep stage referred to as "paradoxical sleep." It is called paradoxical, for electrophysiologically the brain seems quite active and alert, similar to its condition during waking. However, the body musculature is paralyzed, and the ability to perceive outside sensory events is greatly attenuated (reviewed in Hobson et al. 1986; Steriade & McCarley 1990; Vertes 1990).

Most individuals awakened during REM report dream activity approximately 80% of the time. When awakened during the N-REM period, dreams are reported approximately 20% of the time (Foulkes, 1962; Goodenough et al. 1959; Monroe et al. 1965). However, the type of dreaming that occurs during REM vs. N-REM is quite different. For example, N-REM dreams (when they occur) are often quite similar to thinking and speech (i.e. lingusitic thought), such that a kind-of rambling verbal monologue is experienced in the absence of imagery. It is also during N-REM in which an individual is most likely to talk in his or her sleep. In contrast, REM dreams involve a considerable degree of visual imagery, emotion, and tend to be distorted and implausible to various degrees.

REM is characterized by high levels of activity within the brainstem, occipital lobe, and other nuclei (Hobson, et al. 1986; Steriade & McCarley 1990; Vertes 1990). It also has been reported that electrophysiologically the right hemisphere becomes highly active during REM, whereas, conversely, the left half of the brain becomes more active during N-REM (Goldstein et al. 1972; Hodoba, 1986). Similarly, measurements of cerebral blood flow have shown an increase in the right temporal and parietal regions during REM sleep and in subjects who upon wakening report visual, hypnagogic, hallucinatory and auditory dreaming (Meyer et al., 1987).

Electrophysiologically the right temporal lobe becomes highly active during REM, whereas, conversely, the left temporal region becomes more active during dreamless sleep, i.e., NREM (Goldstein et al. 1972; Hodoba 1986). Similarly, measurements of cerebral blood flow have shown an increase in the right temporal regions during REM sleep. Right temporal lobe blood flow also increases in subjects who upon wakening report visual, hypnogogic, hallucinatory and auditory dreaming (Meyer et al. 1987). Abnormal and enhanced activity in the right temporal and temporal-occipital area will also provoke dreaming and acts to increase the length and amount of dreaming and REM sleep for an atypically long time (Hodoba 1986).

Interestingly, abnormal and enhanced activity in the right temporal and temporal-occipital area acts to increase dreaming and REM sleep for an atypically long time period. Similarly, REM sleep increases activity in this same region much more than in the left hemisphere (Hodoba, 1986), which indicates that there is a specific complementary relationship between REM sleep and right temporaloccipital electrophysiological activity.

As noted, LSD induces its "hallucinatory" effects by disinhibiting the amygdala. That is, LSD blocks the sensory filtering and perceptual inhibitory activity of a neural transmitter, serotonin. In consequence multisensory qualities that are normally suppressed are suddenly perceived such that what was hidden is revealed; sounds have color and colors and sounds can be tasted, as the boundaries of this reality melt away thus revealing the supernatural reality of the otherside.

Conversely, LSD induced hallucinations are significantly reduced when the right but not the left temporal lobe has been surgically ablated (Serafintides1965). Similarly, it has been reported that dreaming is abolished with right but not left temporal lobe destruction (Bakan, 1978). Hence, it appears that there is a specific complementary relationship between REM sleep, hallucinations, LSD, mystical experiences, and right temporal (and thus right amygdala and hippocampus) electrophysiological activity. By contrast, the left half of the brain appears to be the domain of the more logical and nonintuitive aspects of conscious experience.

Whereas the right hemisphere is dominant for all aspects of emotion, and is the domain of the more visual and imaginal aspects of the mind, the left hemisphere is dominant for language, math, and the temporal sequential aspects of consciousness. It is the right hemisphere which dreams the dream, and it is the left hemisphere which not only passively observes but which forgets the dream upon waking (Joseph, 1988a). It is the more unconscious realms associated with the right hemisphere which are directly in tune with these alternate realities, such as conveyed through dreams, and it is the left hemisphere which dismisses and rationally explains away these experiences as nonsense.

DAY DREAMS AND FORESEEING THE FUTURE

During dream states we see and experience events which are normally filtered from the conscious mind. We can also gain insight into problems which have plagued us, or gain access to knowledge of events which occurred in the past or which will occur in the future (Joseph, 1988a, 2000a; Jung 1945, 1964)—just as we can think about the future or the past and make certain deductions and predictions.

Consider the day dream. In addition to its images and memories, the fantasy produced also consists of anticipations regarding the future, and in this respect, the day dream could be considered an imaginal means of preparation for various possible realities. Interestingly, daydreams appear to follow the same 90-120 minute cycle that characterize the fluctuation between REM and NREM periods, as well as fluctuations in mental capabilities associated with the right and left hemisphere (Broughton 1982; Kripke and Sonneschein 1990). That is, the cerebral hemispheres tend to oscillate in activity every 90-120 minutes. This cycle corresponds to the REM-NREM cycle and the appearance of day and night dreams, and the right hemisphere also appears to be the source of day dreams.

According to the ancients, day and night dreams both contained important information, not just regarding the past or the world of souls and spirits, but the future. As possible harbingers of the future, the intentions of the gods, and the future of self, friends and family, it has long been believed that dreams should be observed most carefully and could be used to foretell the future (Campbell 1988; Frazier 1950; Freud 1900; Jung 1945, 1964; Malinowkski 1954). It was pharaoh's dream which foretold that seven years of famine would follow seven years of plenty.

In fact, among the ancients, the American Indians, and even the highly cultured Romans, every once in a while someone would have what is called "a big dream." The big dream was of great importance to the whole clan, tribe, city, or nation. Often, the man or woman having the dream would gather the others together and announce it. And more often than not, it would be a woman who would have the big dream, for women have always been said to be more in touch with the "irrational," and throughout history it is women who have predominantly served as oracles.

Because the limbic system and right temporal lobe are hyperactivated during dream states, not only does the brain become freed of inhibitory restraint, but one is presumably able to gain access to dreamlike alternate realities, including, perhaps, the spiritual reality of the Hereafter. Presumably the same occurs when fasting, isolated, in pain, under LSD, in trance, or in the throes of religious ecstasy, all of which activates the limbic system thus increasing channel capacity, so that what is concealed is revealed.

The Evolution of Dream Consciousness

That so many people, regardless of culture or antiquity, have similar dreams and hallucinations, is presumably due to all possessing a limbic system and temporal lobe that is organized similarly. Of course, many of these experiences are also colored by one's cultural background and differences in thinking patterns.

However, we should ask: is it really a hallucination if someone experiences a dissociation of consciousness and floats above their body and can later accurately described what was taking place and the appearance of those around them? Is it really a hallucination if an individual can see inside his hand and watch the blood cells swishing through his blood vessel? If one of the great presidents in the history of the United States dreams of his death 13 days before he is assassinated, was it just a dream?

There are numerous reports of individuals who claim to have dreamed about future events which then took place, including deaths, tragedies, mass murders, horrific accidents, and environmental catastrophes (Barker 1967; Jung 1945, 1964; Wiseman 2011). Yet, given that six billion people dream multiple dreams every night, it could be argued that we should not be surprised that a few of these dreams just happen to accurately coincide with what takes place. These are just chance coincidences, which, when considering the billions of dreams dreamed nightly, should be dismissed as meaningless and of no significance (Wiseman 2011). Since so few people have these dreams, it should not be concluded that these dreams represent an important cognitive capacity. However, if we apply the same criteria to Einstein's theory of relativity, or the music of Mozart or Beethoven, then the absurdity of this position becomes clear. Certain cognitive capacities are well developed in just a few people.

Another question should also be asked: Why would the brain evolve capabilities which are suppressed? Why did neural structures evolve which can process multiple sensations simultaneously, but then come to be inhibited by serotonin? Why would activation of specific brain areas result in the sensation of having left their body, or being privy to cosmic wisdom, or witnessing events and entities which appear to be from other dimensions or realities?

These experiences are made possible by the brain. They must serve an adaptive function. These capabilities must have evolved following natural selection. Why would we evolve the ability to hallucinate or dream about that which supposedly does not exist? Why should we have evolved the ability to hallucinate that we have left our bodies, or can see inside our hands, or dream that a friend will tragically die days before his accidental death?

One possibility is the human brain is de-evolving, and has lost or is losing capabilities which served a more adaptive purpose in ancient humans 30,000 to 10,000 years ago--ancient humans (the Cro-Magnon) whose brain was 1/3 larger in size than the modern brain! The Cro-Magnon men stool 6 ft tall on average, and the Cro-Magnon people created the magnificent underground cathedrals of art which first began to appear 30,000 years ago.

Comparison of modern human skull superimposed on Cro-Magnon skull (Left). Cro-Magnon skull (Right). The Cro-Magnon brain was 1/3 larger on average, than the modern human brain.

Conversely, it may be that these are capacities which are still evolving, but which at the present stage of human evolution, the human brain is unable to utilize adaptively. Consider, the evolution of language. Certainly, Australopithecus, Homo habilis, Homo erectus, and Neaderthals did not engage in complex conversations (Joseph 1996, 2000, 2001). We can surmise based on indirect evidence, that the language capabilities of these ancient hominids and humans may have been limited to grunts, groans, screams, and a variety of emotional sounds (Joseph 2000). Whatever language abilities ancestral species evolved prior to 100,000 years ago were primitive manifestations of what was yet to evolve, i.e. grammatical speech, reading, writing, and associated modes of abstract thinking, such as math. Although the rudimentary foundations had been laid long ago, these were only primitive steps toward what was later to more fully evolve.

It can be surmised that if humans continue to evolve, that maybe 100,000 years from now they will possess a brain which is able to fully utilize capacities which the modern brain is, as yet, unable to master. Or, just as the brain has shrunk by 1/3 in size since the ending of the Paleolithic, that it may continue to lose capacities due to advances in technology which will increasingly render reading, writing, speaking, or creative endeavors obsolete.

Based on the evidence marshaled here and elsewhere, it can be inferred that not all dreams and hallucinations are dreams and hallucinations. They are mental capabilities which are de-evolving or still evolving, and which are or were destined to serve a specific purpose: Lifting the veil so we can gaze deeply into the past, the future, and the unknown.

References

Amaral, D. G., Price, J. L., Pitkanen, A., & Thomas, S. (1992). Anatomical organization of the primate amygdaloid complex. In J. P. Aggleton (Ed.). The Amygdala. (Wiley. New York.

Applegate, C. D. (1980). 5,7,-dihydroxytryptamine-induced mouse killing and behavioral reversal with ventricular administation of serotonin in rats. Behavioral and Neural Biology, 30, 178-190.

Baldwin, M., Lewis, S. A., & Bach, S. A. (1959). The effects of lysergic acid after cerebral ablation. Neurology, 469-474

Baldwin, M., Lewis, S.A., & Bach, S.A. (1959).The effects of lysergic acid after cerebral ablation.Neurology, 9, 469-474.

Barker, J. (1967). Premonitions of the Aberfan Disaster, JSPR, 44, 169-181

Bear, D. M. (1979). Temporal lobe epilepsy: A syndrome of sensory-limbic hyperconnection. Cortex, 15, 357-384.

Benninger, R. J. (1989). The role of serotonin and dopamine in learning to avoid aversive stimuli. In T. Archer & L-G Nilsson (Eds) Aversion, Avoidance and Anxiety. New Jersey, Erlbaum.

Bennett, Jr. J. P., and Snyder, S. H. (1975). Stereospecific binding of d-lysergic acid diethylamide (LSD) to brain membranes: Relationship to serotonin receptors. Brain Research, 94, 523-544.

Bennett, Jr. J. P., and Snyder, S. H. (1976). Serotonin and Lysergic Acid Diethylamide Binding in Rat Brain Membranes: Relationship to Postsynaptic

Serotonin Receptors. Molecular Pharmacology, 12, 373-389.

Bertini, M., Violani, C., Zoccolotti, P., Antonelli, A., & DiStephano, L. (1983). Performance on a unilateral tactile test during waking and upon awakenings from REM and NREM. In W. P. Koella (Ed.), Sleep, Basel: Karger.

Bogen J., & Bogen, C. (1969). The other side of the brain: III. The corpus callosum and creativity. Bulletin of the Los Angeles Neurological Society, 34, 191-220.

Botez, M. I., Olivier, M., Vezina, J.-L., Botez, T., & Kaufman, B. (1985). Defective revisualization: Dissociation between cognitive and imagistic thought case report and short review of the literature. Cortex, 21, 375-389.

Brandon, S.G. F. (1967) The Judgment of the Dead. New York, Scribners.

Broughton, R. (1982). Human consciousness and sleep/waking rhythms: A review and some neuropsychological considerations. Journal of Clinical Neuropsychology, 4, 193-218.

Budge, W. (1994). The Book of the Dead. New Jersey, Carol.

Calvo, J. M., Badillo, S., Morales-Ramirez, M., Palacios-Salas, P. (1987) The role of the temporal lobe amygdala in ponto-geniculo-occipital activity and sleep organization in cats. Brain Research, 403, 22-30.

Campbell, J. (1988) Historical Atlas of World Mythology. New York, Harper & Row.

Cartwright, R. (2010) The Twenty-four Hour Mind: The Role of Sleep and Dreaming in Our Emotional Lives. Oxford University Press.

Cartwright, R. D., Tipton, L. W., & Wicklund, J. (1980). Focusing on dreams. Archives of General Psychiatry, 37, 275-288.

Chapman, L. F., & Walter, R. D. (1965). Actions of lysergic acid dienthalamid on averaged human cortical evoked rsposnes to light flash. Recent Advances in Biological Psychiatry, 7, 23-36.

Chapman, L. F., et al. (1963). Altered electrical activity of human hippocampus and amygdala induced by LSD-25. Physiologist, 5, 118.

Courtois, C.A. (1995). Healing the Incest Wound. New York, Norton.

Daly, D. (1958) Ictal affect. American Journal of Psychiatry, 115, 97-108.

Dossey, L., Greyson, B., Sturrock, P. A., Tucker, J. B. (2011). Consciousness -What Is It? Shared Consciousness, Twin Consciousness, Near Death, Journal of Cosmology, Vol 14. In press.

Eadie, B. J. (1992). Embraced by the light. California, Gold Leaf Press.

Foulkes, W. D. (1962). Dream reports from different stages of sleep. Journal of Abnormal and Social Psychology, 65, 14-25.

Frazier, J. G. (1950). The golden bough. Macmillan, New York.

Freud, S. (1900). The interpretation of dreams. Standard Edition (Vol 5). London: Hogarth Press.

Gloor, P. (1997). The Temporal Lobes and Limbic System. Oxford University Press. New York.

Goldstein, L., et al. (1972). Changes in interhemispheric amplitude relationships in the EEG during sleep. Physiology and Behavior, 8, 811-815.

Greenwood, P., Wilson, D. H., & Gazzaniga, M. S. (1977). Dream report following commissurotomy. Cortex, 13, 311-316.

Gresch, P. J., et al. . (2002). Lysergic acid diethylamide-induced Fos expression in rat brain: role of serotonin-2A receptors. Neuroscience, 114, 707-713.

Grinker, R. R., Spiegel, J. P. (1945) Men Under Stress. McGraw-Hill, New York.

Halgren, E. (1992). Emotional neurophysiology of the amygdala within the context of human cognition. In J. P. Aggleton (Ed.). The Amygdala. New York, Wiley-Liss.

Halgren, E., Walter, R. D., Cherlow, D. G., & Crandal, P. H. (1978). Mental phenomenon evoked by electrical stimualtion of the human hippocampal formation and amygdala, Brain, 101, 83-117.

Harris, M. (1993) Why we became religious and the evolution of the spirit world. In Lehmann, A. C. & Myers, J. E. (Eds) Magic, Witchcraft, and Religion. Mountain View: Mayfield.

Hasselmo, M. E., Rolls, E. T., & Baylis, G. C. (1989). The role of expression and identity in the face-selective responses of neurons in the temporal visual cortex of the monkey. Behavioral Brain Research, 32,203-218.

Hecaen, H., & Albert, M. L.(1978). Human Neuropsychology, Wiley.

Hodoba, D. (1986). Paradoxic sleep facilitation by interictal epileptic activity of right temporal origin. Biological Psychiatry, 21, 1267-1278.

Hobson, J. A. (2004) Dreaming: An Introduction to the Science of Sleep. Oxford University Press.

Hoppe, K. D., & Bogen, J. E. (1977). Alexithymia in twelve commissurotomized patients. Psychotherapy and Psychosomatics, 28, 148-155.

Horowitz, M. J., Adams, J. E., & Rutkin, B. B. (1968). Visual imagery on brain stimulation. Archives of General Psychiatry, 19, 469-486.

Humphrey M. E., & Zangwill, O. L. (1911). Cessation of dreaming after brain injury. Journal of Neurology, Neurosurgery, and Psychiatry, 14, 322-325.

Jacobs, B. L., & Azmita, E. C. (1992). Structure and function of the brain Serotonin System. Physiological Reviews, 72, 165-245.

Joseph, R. (1988a) The Right Cerebral Hemisphere: Emotion, Music, Visual-Spatial Skills, Body Image, Dreams, and Awareness. Journal of Clinical Psychology, 44, 630-673.

Joseph, R. (1988b). Dual mental functioning in a split-brain patient. Journal of Clinical Psychology, 44, 770-779.

Joseph, R. (1992a) The Limbic System: Emotion, Laterality, and Unconscious Mind. The Psychoanalytic Review, 79, 405-456.

Joseph, R. (1992b). The Right Brain and the Unconscious. New York, Plenum.

Joseph, R. (1999a). Frontal lobe psychopathology: Mania, depression, aphasia, confabulation, catatonia, perseveration, obsessive compulsions, schizophrenia. journal of Psychiatry, 62, 138-172.

Joseph, R. (1999b). The neurology of traumatic "dissociative" amnesia. Commentary and literature review. Child Abuse & Neglect. 23, 715-727.

Joseph, R. (2000). The evolution of sex differences in language, sexuality, and visual spatial skills. Archives of Sexual Behavior, 29, 35-66.

Joseph, R. (2001). The Limbic System and the Soul: Evolution and the Neuroanatomy of Religious Experience. Zygon, the Journal of Religion & Science, 36, 105-136.

Joseph, R. (2002). NeuroTheology: Brain, Science, Spirituality, Religious Experience. University Press.

Joseph, R. (2011a). Origins of Thought: Consciousness, Language, Egocentric Speech and the Multiplicity of Mind Journal of Cosmology, 14.

Jouvet, M. (2001). The Paradox of Sleep: The Story of Dreaming. MIT press.

Jung, C. G. (1945). On the nature of dreams. (Translated by R.F.C. Hull.), The collected works of C. G. Jung, (pp.473-507). Princeton: Princeton University Press.

Jung, C. G. (1964). Man and his symbols. New York: Dell.

Kamiya, J. (1961). Behavioral, subjective and physiological aspects of drowsiness and sleep. In D. W. Fiske, & S. R. Maddi (Eds.). Function of varied experience, (pp. 145-174). Homewood, IL: Dorsey Press.

Kawashima, R., Sugiura, M., Kato, T., et al., (1999). The human amygdala plays an important role in gaze monitoring. Brain, 122, 779-783.

Kerr N. H., & Foulkes, D (1978). Reported absence of visual dream imagery in a normally sighted subject with Turner's syndrome. Journal of Mental Imagery, 2, 247-264.

Kerr, N. H., & Foulkes, D. (1981). Right hemisphere mediation of dream visualization: A case study. Cortex, 17, 603-611.

Klein, R., & Armitage, R. (1979). Rhythms in human performance: 11/2 hour oscillations in cognitive style. Science, 204, 1326-1328.

Kling. A. S. & Brothers, L. A. (1992). The amygdala and social behavior. In J. P. Aggleton (Ed.). The Amygdala. New York, Wiley-Liss.

Kripke, D. F., & Sonnenschein, D. (1973). A 90 minute daydream cycle. Sleep Research, 2, 187-188.

Krippner, S (1993). The Maimonides ESP-dream studies Maimonides Medical Center, Journal of Parapsychology, 57, 279-319.

Lamon, W. H. (1865/1994). Recollections of Abraham Lincoln 1847-1865, by Ward Hill Lamon, University of Nebraska Press.

Lydic, R., Baghdoyan, H. A., & Lorinc, Z. (1991). Microdialysis of cat pons reveals enhanced ACh release during state-dependent respiratory depression. American Journal of Physiology, 261, 766-770.

MacLean, P. (1990). The Evolution of the Triune Brain. Plenum.

Malh, G. F., Rothenberg, A., Delgado, J. M. R., & Hamlin, H. (1964). Psychological response in the human to intracerebral electrical stimulation. Psychosomatic Medicine, 26, 337-368.

Malinowski, B. (1954) Magic, Science and Religion. New York. Doubleday.

Meyer, J. S., Ishikawa, Y., Hata, T., & Karacan, I. (1987). Cerebral blood flow in normal and abnormal sleep and dreaming. Brain and Cognition, 6, 266-294.

Mesulam, M. M. (1981) Dissociative states with abnormal temporal lobe EEG: Multiple personality and the illusion of possession. Archives of General Psychiatry, 38, 176-181.

Monti, J. Pandi-Permal, S. R., Sinton, C. M. (2008). Neurochemistry of Sleep and Wakefulness, Cambridge U. Press.

Montigny, C. de, and Aghajanian G.K. (1977). Preferential action of 5-methoxytryptamine and 5-methoxydimethyltryptamine on presynaptic serotonin receptors: A comparative iontophoretic study with LSD and serotonin. Neuropharmacology, 16, 811-818.

Moody, R. (1977). Life after life. Georgia, Mockingbird Books.

Morris, J. S., Frith, C. D., Perett, D. I., Rowland, D., Young, A. W., Calder, A. J., & Colan, R. J. (1996). A differential neural response in the human amygdala to fearful and happy facial expression. Nature, 383, 812-815.

Mullan, S., & Penfield, W. (1959). Epilepsy and visual halluciantions. Archives of Neurology and Psychiatry, 81, 269-281.

Murri, L., Arena, R., Siciliano, G., Mazzotta, R., & Muratorio, A. (1984). Dream recall in patients with focal cerebral lesions. Archives of Neurology, 41, 183-185.

Nadel, L. (1991). The hippocampus and space revisited. Hippocampus, 1, 221-229.

Neihardt, J. G. & Black Elk, (1979). Black Elk speaks. Lincoln. U. Nebraska Press.

Nemchin, A. A.,et al. . (2008). A light carbon reservoir recorded in zircon-hosted diamond from the Jack Hills. Nature 454, 92-95.

Noyes, R., & Kletti, R. (1977) Depersonalization in response to life threatening danger. Comprehensive Psychiatry, 18, 375-384.

O'Keefe, J. (1976). Place units in the hippocampus of the freely moving rat. Experimental Neurology, 51, 78-109.

O'Keefe, J., & Bouma, H. (1969). Complex sensory properties of certain amygdala units in the freely moving cat. Experimental Neurology, 23, 384-398.

O'Neil, J., et al. (2008). Neodymium-142 Evidence for Hadean Mafic Crust Science 321, 1828 1831.

Parson, E. R. (1988). Post-traumatic self disorders (PTsfD): Theoretical and practical considerations in psychotherapy of Vietnam War Veterans. In J. P. Wilson, Z. Harel, & B. Kahana (Eds). Human Adaptation to Extreme Stress. New York, Plenum.

Pena-Casanova, J., & Roig-Rovira, T. (1985). Optic aphasia, optic apraxia, and loss of dreaming. Brain and Language, 26, 63-71.

Penfield, W., & Perot, P. (1963). The brains record of auditory and visual experience. Brain, 86, 595-695.

Perryman, K. M., et al.. (1987). Differential effects of inferior temporal cortex lesions upon visual and auditory-evoked potentials in the amygdala of the squirrel monkey. Behavioral and Neural Biology, 47, 73-79.

Purpura, D. P. (1956). Electrophysiological analysis of psychotogenic drug action. I & II. Archives of Neurology & Psychiatry, 40, 122-143.

Rawlings, M. (1978). Beyond deaths door. London, Sheldon Press.

Ribeiro, S., Goyal, V., Mello, C. & Pavlides, C. (1999). Brain gene expression during REM sleep depends on prior waking experience. Learning & Memory, 6: 500-508.

Ribeiro, S., et al. (2002). Induction of hippocampal long-term potentation during waking leads to increased extra hippocampal zif-268 expression during ensuing rapid-eye-movement sleep. Journal of Neuroscience, 22(24), 10914-10923.

Riberio, S., Simões, C. & Nicolelis, M. (2008). Genes, Sleep and Dreams. In Lloyd & Rossi (Eds.) Ultradian rhythms from molecule to mind. Springer. N.Y., 413-430.

Ring, K. (1980). Life at death. New York, Coward, McCann & Geoghegan.

Rolls, E. T. (1984). Neurons in the cortex of the temporal lobe and in the amygdala of the monkey with responses selective for faces. Human Neurobiology, 3, 209-222.

Rolls, E. T. (1992). Neurophysiology and functions of the primate amygdala. In J. P. Aggleton (Ed.). The Amygdala. New York, Wiley-Liss.

Sabom, M. B. (1982). Recollections on death. New York, Harper & Row.

Sawa, M., & Delgado, J. M. R. (1963). Amygdala unitary activity in the unrestrained cat. Clinical Neurophysiology, 15, 637-650.

Schenk, L., & Bear, D. (1981) Multiple personality and related dissociative phenomenon in patients with temporal lobe epilepsy. American Journal of Psychiatry, 138, 1311-1316.

Serafetinides, E. A. (1965). The significance of the temporal lobes and of hemisphere dominance in teh production of the LSD-25 symptomology in man. Neuropsychologia, 3, 69-79.

Schutze, I., et al. (1987). Sensory input to single neurons in the amygdala of the cat. Experimental Neurology, 97, 499-515.

Slater, E. & Beard, A.W. (1963). The schizophrenia-like psychoses of epilepsy. British Journal of Psychiatry, 109, 95-112.

Soubrie, P. (1986). Reconciling the role of central serotonin neurons human and animal behavior. Behavioral and Brain Sciences, 9, 319-364.

Spoont, M. R. (1992). Modulatory role of serotonin in neural information processing: Implications for human psychopathology. Psychological Bulletin, 112, 330-350.

Steriade, M. M. & McCarley, R. W. (2005) Brain Control of Wakefulness and Sleep, Springer.

Strahlendorf, J. C. R., et al., (1982). Differential effects of LSD serotonin and l-tryptophan on visually evoked responses. Pharmacology Biochemistry and Behavior, 16, 51-55.

Subirana, A., & Oller-Daurelia, L. (1953). The seizures with a feeling of paradisiacal happiness as the onset of certain temporal symptomatic epilepsies. Congres Neurologique International. Lisbonne, 4, 246-250.

Tarachow, S. (1941). Tjhe clinical value of hallucinations in localizing brain tumors. American Journal of Psychiatry, 99, 1434-1442.

Taylor, D. C. (1972). Mental state and temporal lobe epilepsy. Epilepsia, 13, 727-765.

Teuber, H. L., et al. (1960). Visual field defects after penetrating missile wounds of the brain. Cambridge: Harvard University Press.

Trimble, M. R. (1991). The psychoses of epilepsy. Raven Press.

Turner, B. H. et al. (1980). Organization of the amygdalopetal projections from modality-specific cortical association areas in the monkey. Journal of Comparative Neurology, 191, 515-543.

Ullman, M., & Krippner, S. (1970). Dream studies and telepathy; An experimental approach. New York: Parapsychology Foundation.

Ursin H., & Kaada, B. R. (1960). Functional localization within the amygdaloid complex in the cat. Electroencephalography and Clinical Neurophysiology, 12, 1-20.

Vertes, R. P. (1990). Brainstem mechanisms of slow-wave sleep and REM sleep. In W. R. Klemm, & R. P. Vertes (Eds.). Brainstem mechanisms of behavior. Wiley. New York.

Weingarten, S. M., et al. (1977). The relationship of hallucinations to depth structures of the temporal lobe. Acta Neurochirugica 24: 199-216.

Williams, D. (1956). The structure of emotions reflected in epileptic experiences. Brain, 79, 29-67.

Wilson, I. (1987). The after death experience. New York, Morrow.

Wilson, J. A. (1951) The culture of ancient Egypt. Chicago, U. Chicago Press.

Wilson, M. A., & McNaughton, B. L. (1993). Dynamics of the hippocampal ensemble for space. Science, 261, 1055-1058.

Wiseman, R. (2011) Paranormality, Macmillan.

19. Possession and Prophecy

PRESENCE, POSSESSION & THE ALIEN HAND

"After getting into bed I had a vivid tactile hallucination of being grasped by the arm, which made me get up and search my room for an intruder. The next night I suddenly felt something come into the room and stay close to my bed. It remained only a minute or two. There was a horribly unpleasant sensation connected with it. It stirred something at the roots of my being. The feeling was not pain so much as abhorrence. Something was present with me, and I knew its presence far more surely than I have ever known the presence of any fleshly living creature. I was conscious of its departure as of its coming; an almost instantaneously swift going through the door, and the horrible sensation disappeared" (James, 1902).

"Quite early in the night I was awakened. I felt as if I had been aroused intentionally, and at first thought someone was breaking into the house. I immediately felt a consciousness of a presence in the room, it was not a consciousness of a live person, but of a spiritual presence. I also at the same time felt a strong feeling of superstitious dread, as if something strange and fearful were about to happen" (James, 1902).

It is certainly possible, in some cases, that the experience of a "presence," of the existence of an unseen person or entity, may have a supernatural origin. Neurologically, however, this condition is associated with abnormalities, including seizure activity, involving the right hemisphere--the right parietal lobe in particular—and disturbances involving the corpus callosum interconnections between the right and left (speaking) half of the brain including the medial frontal lobe and anterior commissure (Joseph, 1988a,b, 1996, 1999a).

For example, a woman with a right parietal injury repeatedly claimed that at night another person would get into bed with her. She believed that the alien entity was a little Negro girl, whose arm would slip into the patient's sleeve (Gerstmann, 1942). This alien "presence" was exclusively a left sided phenomenon. Yet another patient with a right parietal injury claimed "that an old man" would get into bed with him. Another patient engaged in peculiar erotic behavior with his left limbs which he believed belonged to a woman. A patient described by Bisiach and Berti (1987, p. 185) "would become perplexed and silent whenever the conversation touched upon the left half of his body; even attempts to evoke memories of it were unsuccessful." Instead he claimed "that a woman was lying on his left side; he would utter witty remarks about this and sometimes caress his left arm".

Some patients feel as if the left half of their body has been taken over by something evil that is persecuting them. They may develop a dislike for their left arms or legs, try to throw them away, become agitated when they are referred to, entertain persecutory delusions regarding them, and repeatedly complain of strangers sleeping in their beds.

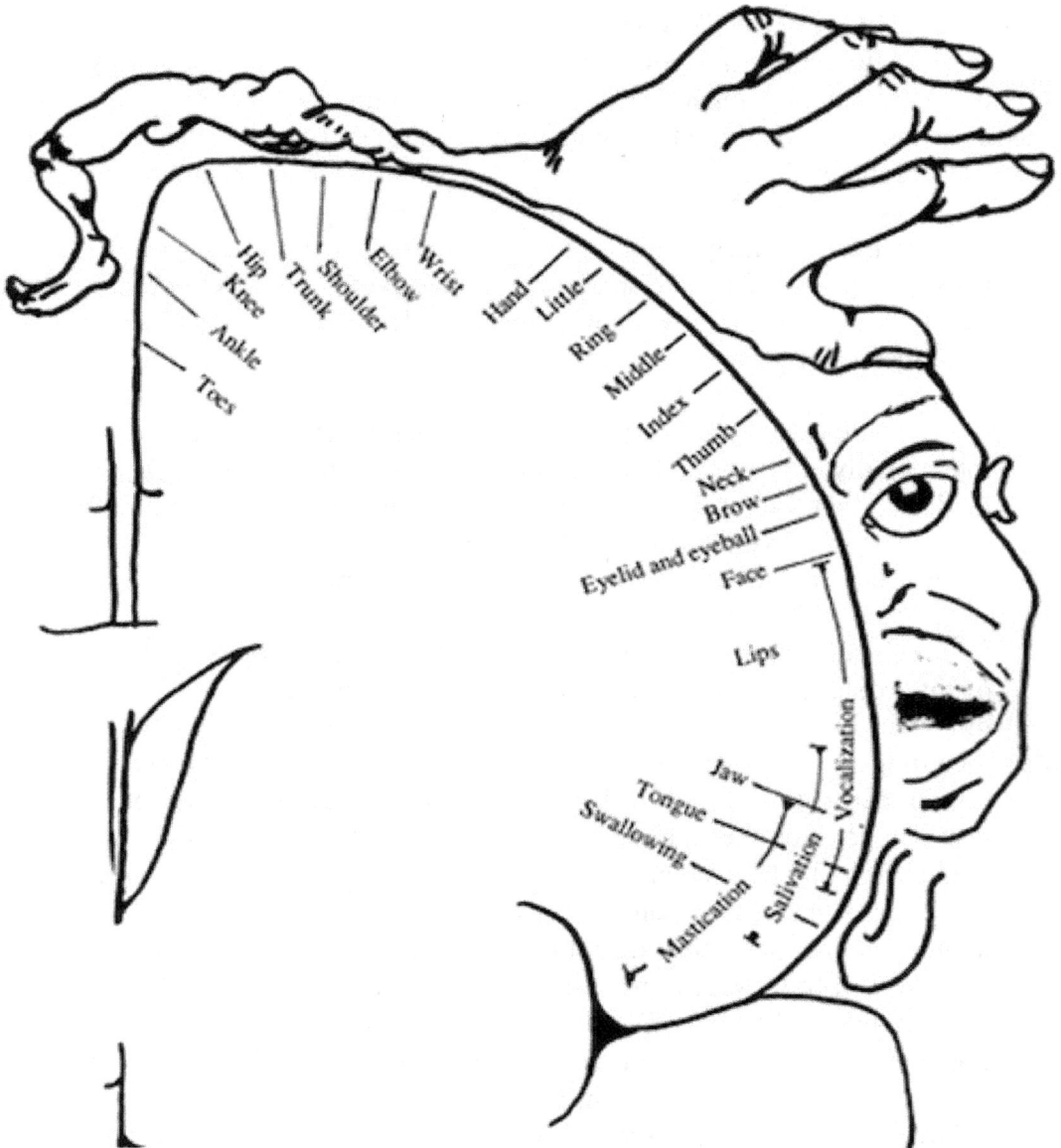

FIGURE: The left half of the body is represented in the right hemisphere and the right half of the body in the left hemisphere

One patient complained that this other person tried to push her out of the bed. She threatened to sue the hospital. Another patient, after bumping into her left arm and leg all night, bitterly complained about "a hospital that makes people sleep together." She expressed not only anger but concern least her husband should find out; she was convinced it was a man in her bed.

Likewise, the left hemisphere of some split-brain patients have claimed to hate the left half of their body and attribute to it disagreeable personality traits or claim that it has engaged in behavior which the speaking half of the brain finds unpleasant, strange, objectionable, embarrassing, or contrary to it's wishes (Joseph, 1988ab).

In many instances, patients may have lived a normal, uneventful life, only to suddenly become possessed by an alien presence which takes over half their body.

Goldberg (1987, p 290) describes a 53 year old right-handed women, "B.D." who while at work was overcome with a "feeling of nausea and began to notice that her left leg felt 'as if it did not belong to me.' This feeling of being dissociated from her body spread to the rest of her left side. At home, her symptoms began to worsen. While sleeping one night a few days after admission to the hospital, she woke up suddenly and noticed her own left hand scratching her shoulder.... She would frequently look down to find the hand doing something that she had no idea it had been doing. She found this very disturbing and was convinced that she was going crazy." Subsequent CT scan and MRI indicated an infarct to the medial frontal lobe and damage to the body of the corpus callosum.

Another patient described by Goldberg (1987, p. 295) reported "an incident in which she was lying in bed with the window open when suddenly the left limb reached down and pulled up the covers, functioning entirely in the alien mode. She concluded that 'it' must have felt cold and needed to cover her up. She felt that frequently the 'alien' did things that were generally 'good for her.'" McNabb, Caroll, and Mastaglia (1988, pp. 219, 221) describe a woman with extensive damage involving the medial left frontal lobe and anterior corpus callosum, whose right "hand showed an uncontrollable tendency to reach out and take hold of objects and then was unable to release them. At times the right hand interfered with tasks being performed by the left hand. She attempted to restrain it by wedging it between her legs or by holding or slapping it with her left hand. The patient would repeatedly express astonishment at these actions." Another patient reported that she was "attempting to write with her left hand, the right would reach over and take the pencil. The left hand would respond by grasping the right hand to restrain it."

In some instances, patients may display alien movements in both hands (Gasquoine 1993; Goldberg & Bloom 1990) as well as alien vocalization of thoughts. For example, a patient described by Gasquoine (1993) had a propensity to reach out and touch female breasts, as well as novel objects and persons. He reported this caused him great embarrassment and that he would typically attempt to take hold of his right with his left hand or voluntarily grasp objects, such as his lap tray, so that he would not spontaneously reach out and grab someone.

Some individuals not only have difficulty controlling their arms and hands, but their speech. That is, they may begin to speak and make pronouce-ments, which they claim to have no control over—as if they had been taken over and were possessed by an alien consciousness that had a mind of its own.

POSSESSION &AND THE SPLIT BRAIN

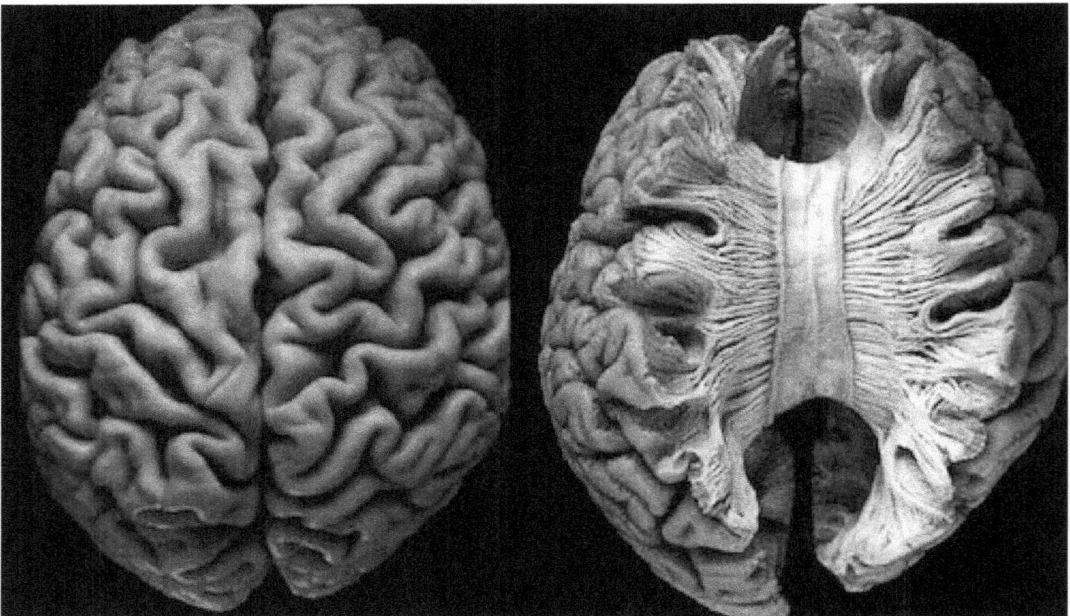

It is now well established that the brain is functionally lateralized, such that the temporal and sequential aspects of language are controlled by the left hemisphere, whereas visual spatial, environmental, the emotional sounds of speech, and the body image, are the domain of the right half of the brain. Because each hemisphere is concerned with different types of information, even when analyzing ostensibly the same stimulus each half of the brain may react, interpret and process it differently and even reach different conclusions (Joseph, 1988b, 1996' 2011; Joseph et al., 1984, Levy & Trevarthen, 1976). Moreover, even when the goals are the same, the two halves of the brain may produce and attempt to act on different strategies. In consequence, functional lateralization may lead to the development of oppositional attitudes, goals and interests such that one half of the brain may desire or engage in acts that are opposed by the other. This has been experimentally demonstrated in patients who have undergone split-brain surgery, that is, complete corpus callosotomy and the severing of the axonal pathways linking the right and left hemisphere.

For example, one split brain individual's left hand would not allow him to smoke, and would pluck lit cigarettes from his mouth or right hand and put them out. Apparently, although his left cerebrum wanted to smoke, his right hemisphere didn't approve --he had been trying to quit for years (Joseph, 1988a,b, 2011). Yet another split brain patient experienced conflicts when attempting to eat, watch TV, or go for walks, his right and left hemisphere apparently enjoying different TV programs or types of food (Joseph 1988b). Nevertheless, these difficulties are not limited to split-brain patients, for conflicts of a similar nature often plague the intact individual as well.

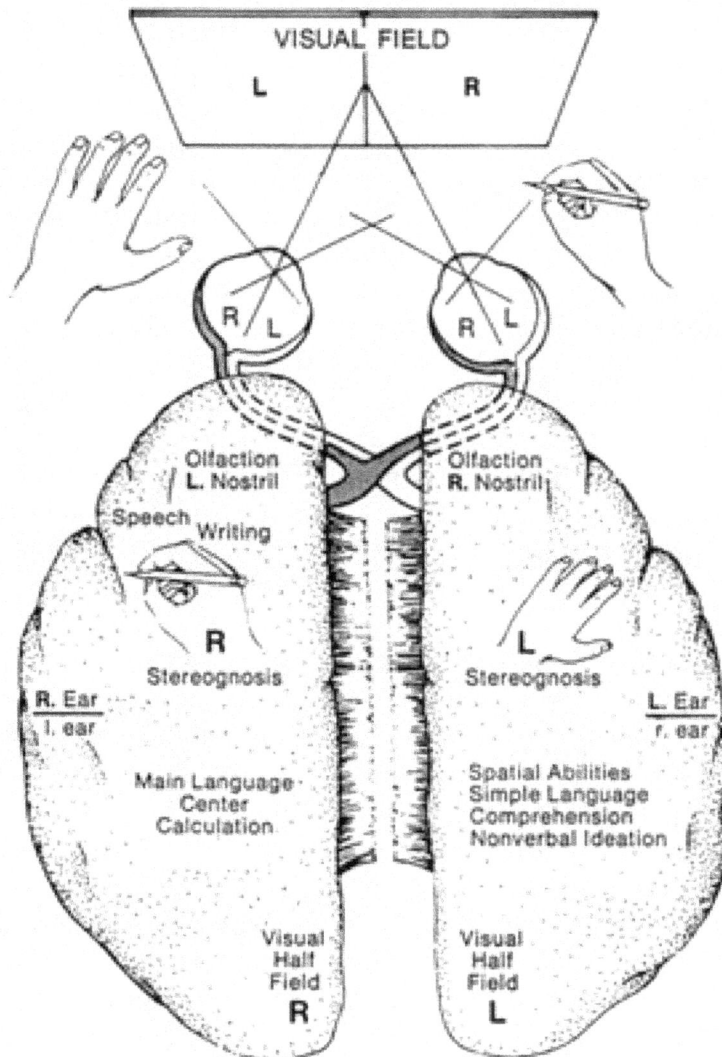

Indeed, it has been well demonstrated that each half of the brain is capable of experiencing independent and semi-independent forms of consciousness, two minds within a single brain, one in the right the other in the left hemisphere. This has been demonstrated in studies of patients who have undergone complete corpus callosotomies (i.e. split-brain operations) for the purposes of controlling intractable epilepsy (Joseph, 1988ab). As described by Noble Laureate Roger Sperry (1966, p. 299), "Everything we have seen indicates that the surgery has left these people with two separate minds, that is, two separate spheres of consciousness. What is experienced in the right hemisphere seems to lie entirely outside the realm of awareness of the left hemisphere. This mental division has been demonstrated in regard to perception, cognition, volition, learning and memory."

For example, when split-brain patients are tactually stimulated on the left side of the body, their left hemispheres demonstrate marked neglect when verbal responses are required, they are unable to name objects placed in the left hand, and they fail to report the presence of a moving or stationary stimulus in the left half of their visual fields (Bogen, 1979; Joseph, 1988b. 2014; Levy, 1974, 1983; Seymour et al. 1994; Sperry, 1982). They (i.e., their left hemisphere's) cannot verbally describe odors, pictures or auditory stimuli tachistoscopically or dichotically presented to the right cerebrum, and have extreme difficulty explaining why the left half of their bodies responds or behaves in a particular purposeful manner (such as when the right brain is selectively given a command).

However, by raising their left hand (which is controlled by the right half of the cerebrum) the disconnected right hemisphere is able to indicate when the patient is tactually or visually stimulated on the left side. When tachistoscopically presented with words to the left of visual midline, although unable to name them, when offered multiple visual choices in full field their right hemispheres are usually able to point correctly with the left hand to the word viewed.

When presented with words like "toothbrush," "tooth" falls in the left visual field (and thus, is transmitted to the right cerebrum) and the word "brush" falls in the right field (and goes to the left hemisphere). Hence, when offered the opportunity to point to several words (i.e., hair, tooth, coat, brush, etc.), the left hand usually will point to the word viewed by the right cerebrum (i.e., tooth) and the right hand to the word viewed by the left hemisphere (i.e., brush). When offered a verbal choice, the speaking (usually the left) hemisphere will respond "brush" and will deny seeing the word "tooth." Overall, this indicates that the disconnected right and left cerebral hemispheres, although unable to communicate and directly share information, are nevertheless fully capable of independently generating and supporting mental activity (Bogen, 1969, 1979; Joseph, 1986b, 1988b; Levy, 1974, 1983; Sperry, 1982). Hence, in the right hemisphere we deal with a second form of awareness that accompanies in parallel what appears to be the "dominant" temporal-sequential, language dependent stream of consciousness in the left cerebrum.

Moreover, as has been demonstrated by Sperry, Bogen, Levy, Joseph and others, the isolated right cerebral hemisphere, like the left, is capable of self-awareness, can plan for the future, has goals and aspirations, likes and dislikes, social and political awareness, can purposefully initiate behavior, guide response choices and emotional reactions, as well as recall and act upon certain memories, desires, impulses, situations or environmental events —without the aid, knowledge or active (reflective) participation of the left half of the brain.

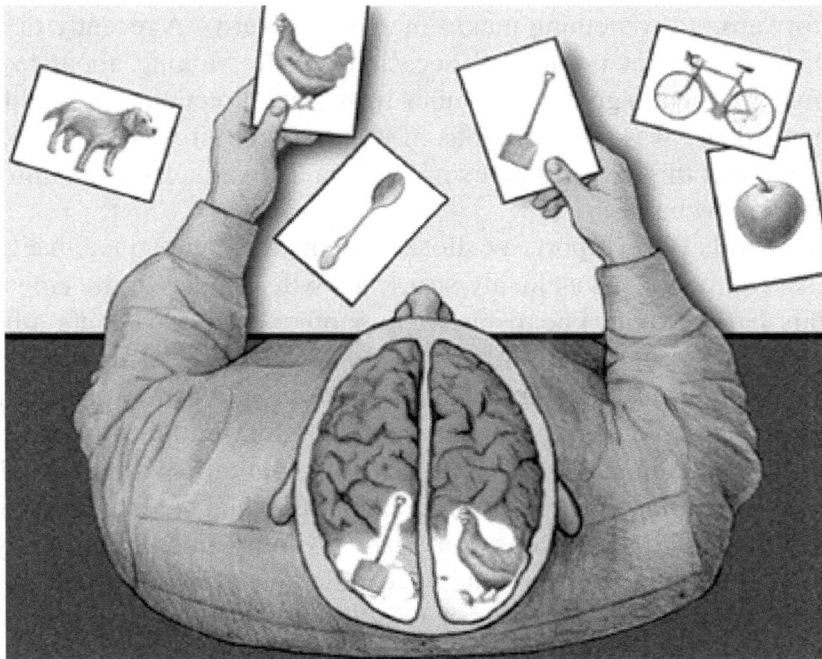

In consequence, because each half of the brain controls the other half of the body, sometimes one half of the body will engage in independent actions, such that half of the body may behave as if possessed and controlled by an alien or even a demonic force.

As reported by patients who have undergone "split-brain" surgery, the behavior of the right hemisphere is not always cooperative, and sometimes it engages in behavior which the left language dominant half of the brain finds objectionable, embarrassing, puzzling, mysterious, and upsetting.

For example, Akelaitis (1945, p. 597) describes two patients with complete corpus callosotomies who experienced extreme difficulties making the two halves of their bodies cooperate. "In tasks requiring bimanual activity the left hand would frequently perform oppositely to what she desired to do with the right hand. For example, she would be putting on clothes with her right and pulling them off with her left, opening a door or drawer with her right hand and simultaneously pushing it shut with the left. These uncontrollable acts made her increasingly irritated and depressed." On several occasions it tried to slam a drawer on her right hand, and on a number of instances the left hand (right hemisphere) attempted to take her clothes off, even though that is not what she (i.e. the left hemisphere) desired to do.

Another patient experienced difficulty while shopping, the right hand would place something in the cart and the left hand would put it right back again. Both patients frequently experienced other difficulties as well . "I want to walk forward but something makes me go backward." A recently divorced male patient noted that on several occasions while walking about town he found himself forced to go some distance in another direction. Later (although his left hemisphere was not conscious of it at the time) it was discovered (by Dr. Akelaitis) that this diverted course, if continued, would have led him to his former wife's new home.

Geschwind (1981) reports a callosal patient who complained that his left hand on several occasions suddenly struck his wife—much to the embarrassment of his left (speaking) hemisphere. In another case, a patient's left hand attempted to choke the patient himself and had to be wrestled away. Bogen (1979, p. 333) indicates that almost all of his "complete commissurotomy patients manifested some degree of intermanual conflict in the early postoperative period." One patient, Rocky, experienced situations in which his hands were uncooperative; the right would button up a shirt and the left would follow right behind and undo the buttons. For years, he complained of difficulty getting his left leg to go in the direction he (or rather his left hemisphere) desired. Another patient often referred to the left half of her body as "my little sister" when she was complaining of its peculiar and independent actions.

A split-brain patient described by Dimond (1980, p. 434) reported that once when she had overslept her "left hand slapped me awake." This same patient, in fact, complained of several instances where her left hand had acted violently. Similarly, Sweet (1945) describes a female patient whose left hand sometimes behaved oppositionally and in a fashion which on occasion was quite embarrassing.

Similar difficulties plagued a split-brain patient whom I reported on (Joseph 1988b). After callosotomy, this patient (2-C) frequently was confronted with situations in which his left extremities not only acted independently, but engaged in purposeful and complex behaviors —some of which he (or rather, his left hemisphere) found objectionable and annoying.

For example, 2-C (the speaking half of his his brain) complained of instances in which his left hand would perform socially inappropriate actions (e.g. attempting to strike a relative) and would act in a manner completely opposite to what he expressively intended, such as turn off the TV or change channels, even though he (or rather his left hemisphere) was enjoying the program. Once, after he had retrieved something from the refrigerator with his right hand, his left took the food, put it back on the shelf and retrieved a completely different item "Even though that's not what I wanted to eat!" On at least one occasion, his left leg refused to continue "going for a walk" and would only allow him to return home.

In the laboratory, he often became quite angry with his left hand, he struck it and expressed hate for it. Several times, his left and right hands engaged in actual physical struggles. For example, on one task both hands were stimulated simultaneously (while out of view) with either the same or two different textured materials (e.g., sandpaper to the right, velvet to the left), and the patient was required to point (with the left and right hands simultaneously) to an array of fabrics that were hanging in view on the left and right of the testing apparatus. However, at no time was he informed that two different fabrics were being applied. After stimulation the patient would pull his hands out from inside the apparatus and point with the left to the fabric felt by the left and with the right to the fabric felt by the right.

Surprisingly, although his left hand (right hemisphere) responded correctly, his left hemisphere vocalized: "That's wrong!" Repeatedly he reached over with his right hand and tried to force his left extremity to point to the fabric experienced by the right (although the left hand responded correctly! His left hemisphere didn't know this, however.). His left hand refused to be moved and physically resisted being forced to point at anything different. In one instance a physical struggle ensued, the right grappling with the left.

Moreover, while 2-C was performing this (and other tasks), his left hemisphere made statements such as: "I hate this hand" or "This is so frustrating" and would strike his left hand with his right or punch his left arm. In these instances there could be little doubt that his right hemisphere was behaving with purposeful intent and understanding, whereas his left hemisphere had absolutely no comprehension of why his left hand (right hemisphere) was behaving in this manner.

EMOTIONAL TRAUMA, POSSESSION & THE SPLITTING OF CONSCIOUSNESS

It has been demonstrated experimentally, that even the normal brain may experience similar episodes of functional disconnection between the right and left hemisphere (Joseph et al., 1984; reviewed in Joseph,1988a) such that the individual may feel possessed or experience an evil, alien presence. In part, this is a consequence of functional lateralized and the specialization of each half of the brain, and right hemisphere dominance for emotion and the body image. However, these conditions also sometimes occur during episodes of extreme emotional distress which also effects the right hemisphere more strongly than the left (reviewed in Joseph, 1988a, 1996). In fact, under conditions of extreme fear, the brain may be injured, and regions in the limbic system, such as the amygdala, may develop "kindling" --which is a form of seizure activity.

Specifically, under conditions of extreme fear or emotional trauma, the brain secretes and releases a cascade of neurochemical and stress hormones, including corticostereoids, enkephalins, norepinephrine (NE), serotonin (5HT), and dopamine (DA), all of which differentially contribute to the fear and stress response. Unfortunately, these neurotransmitter fluctuations negatively impact amygdala and hippocampal neurons, axons, dendrites, and their pre and post synaptic substrates (Cain, 1992; Goelet & Kandel, 1986; Kraemer, 1992; Krystal, 1990). For example, since NE also serves a neural protective function (Glavin, 1985; Ray et al., 1987b), if NE levels are reduced--a normal consequence of prolonged fear and stress-neurons are exposed to the damaging effects of enkephalins and corticosteroids which at high levels attack and kill amygdala and hippocampal pyramidal neurons (Gahwiler 1983; Henriksen et al., 1978; Packan & Sapolsky, 1990). For example, under high levels of stress, corticosteroids are secreted in massive amounts, but also directly attack and injure the hippocampus due to the abundance of Type II adrenal steroids receptors which abound within this structure (Lupien & McEwen, 1997; Pugh, Fleshner & Rudy, 1997). In addition, coupled with NE depletion, repeated or prolonged stress induced secretory episodes of corticosteroids and enkephalins can injure cells within the dentate gyrus and Ammon's horn (Lupien & McEwen, 1997) such that the hippocampus will atrophy (Lupien & McEwen, 1997; Sapolsky, 1996; Uno, Tarara, Else, & Sapolsky, 1989). As to the amygdala, it may develop seizure activity and thus become abnormally activated.

Extreme fear is the most common emotional reaction elicited with direct electrode stimulation of the human or nonhuman amygdala (Chapman, 1960; Davis et al., 1997; Gloor, 1997; Halgren, 1992; Rosen & Schulkin, 1998; Strauss, Risser, & Jones, 1982; Williams 1956). The pupils dilate and the subject will cringe, withdraw, and cower. This cowering reaction may give way to extreme panic and the animal will attempt to take flight. Likewise, abnormal activity originating in the amygdala and/or the overlying temporal lobe can evoke overwhelming, terrifying feelings of "nightmarish" fear that may not be tied to anything specific, other than perhaps the sensation of impending death (Herman & Chambria, 1980; Strauss et al., 1982; Weil, 1956). With amygdala activation the EEG becomes desynchronized (indicating arousal), heart rate becomes depressed, respiration patterns change, the galvanic skin response significantly alters, the face contorts, the pupils will dilate, and the subject will look anxious and afraid (Bagshaw & Benzies, 1968; Davis, 1992; Kapp,

Supple, & Whalen, 1994; Ursin & Kaada, 1960).

However, if the amygdala (and hippocampus) is injured or abnormally active, the individual may become emotionally abnormal, they may suffer from hallucinations, they may hear voices, and they may have dissociative episodes and feel as if they have been "possessed." The feeling of being "possessed" including the development of alternate personalities which temporarily possess and take over control from the main personality, and thus dramatic alterations in personality,have been repeatedly observed following injury to the amygdala and overlying temporal lobe (Lilly et al., 1983; Marlowe et al., 1975; Shenk & Bear, 1981; Terzian & Ore, 1955). In some cases of temporal lobe, amygdala, hippocampal abnormality, the alterations in personality are so dramatic patients may appear to be possessed by demons or suffering from a multiple personality disorder (Fichtner, Kuhlman, & Hughes, 1990; Mesulam, 1981; Shenk & Bear, 1981). In fact, in several cases of multiple personality dissociative disorder, EEG or blood flow abnormalities involving the temporal lobe have been demonstrated (Drake 1986; Fichtner et al., 1990; Mathew, Jack & West, 1985; Mesulam 1981; Schenk & Bear 1981). Moreover, some patients may shift from one personality to another following a seizure or with increases in temporal lobe activity (Mesulam, 1981; Shenk & Bear, 1981). Similarly, it has been reported that heightened emotional distress may precede the appearance of alternate personalities

Sometimes the "voices" of the alternate personality, or the "demon... devil... god" and so on, will order the main personality to harm others or themselves. Some of these alternate personalities may also take control and engage in criminal and self-destructive acts. Some of those "possessed"honestly believe they are possessed by demons, devils, or god.

Alternate personalities may be formed during a dissociative state while a victim is experiencing a period of severe and repetitive trauma. That is, under certain traumatic (or neurological) conditions, an aspect of the consciousness may fragment, break off, and thereafter act in a semi-independent or completely independent manner.

Compared to other cortical areas, the most complex, vivid hallucinations, including out-ofbody dissociative experiences, have their source in the temporal lobe. The amygdala, hippocampus, and temporal lobe play a significant role in the production of REM sleep and dream activity (Hodoba, 1986; Meyer, Ishikawa, Hata, & Karacan, 1987), and become disinhibited in reaction to hallucinogens such as LSD, and are thus implicated in the production of LSD-induced hallucinations (Baldwin, Lewis, & Bach, 1955; Chapman, Walter, & Ross, 1963; Serafetinides, 1965). Hence, under conditions of extreme emotional stress and trauma, the hippocampus, amygdala and temporal lobe may become abnormally active, and victims may experience extreme sensory distortions and hallucinations, including out-of-body phenomenon.

Due to the development of these fear-induced sensory-distorting hallucinatory states, trees, animals, and inanimate objects may even assume demonic form and/or be invested with satanic intent. These horrible hallucinations and sensory distortions may also be committed to memory. Later, the victim may recall the "hallucination" and believe they were set upon by demons, witches and the like, and/or that they were abused in Satanic rituals, or abducted and painfully probed by demonic aliens.

Consider the Walt Disney version of "Snow White." When the woodcutter, who had been ordered to cut out her heart, urged Snow White to flee for her life, she panicked and ran into the darkening forest in near hysteria. And, as she ran and stumbled darting in tears here and there, the trees became demonic, growing eyes and wicked mouths, and gnarled arms and hands which stretched out threateningly toward her. Overcome with terror, she collapsed to the forest floor, sobbing uncontrollable.

Now, perchance, had Snow White later recalled this frightening misadventure, she may well have explained to skeptical listeners that demons had emerged from the forest, and threatened to snatch her away. and, she may truly believe this happened, for it is what she truly experienced and what she now truly remembers.

Yet others may believe they were attacked by aliens, or demons, and then spirited away to hell, or to space ships where nightmarish experiments were performed on them.

It is noteworthy that many of those reporting alien abductions, have a history of sexual molestation or severe emotional trauma, or temporal lobe epilepsy (Mack, 1994). As noted,, stress, sexual trauma and sexual activity activates the amygdala as well as the temporal lobe—structures which are associated with the production of complex and exceedingly frightening hallucinations, including those of a demonic, religious, and sexual nature including naked women, demons, ghosts and pigs walking upright dressed as people.

Hence perhaps it is not surprising that individuals who were severely traumatized or who were sexually abused, sometimes report that they were subject to bizarre sexual rituals that involved demonic (Satanic) activities (which is not to say that in some cases this may in fact be the case). And perhaps it is not surprisingly that those with histories of severe abuse, and/or who suffer from seizure disorders, may believe that demonic-like aliens lifted them into the air and took them to a room of vast proportions—descriptions which are identical to those of at least some patients during temporal lobe stimulation, and at least some individuals who are severely frightened and/or who died only to return to tell the tale.

HELL & ALIEN ABDUCTIONS

Many "abductees" claim a sequence of perceptual experience similar to those who have died and returned to tell the tale. Abductees report the presence of a bright light, or a strange illumination which may envelop them in a beam or halo of light. They feel drawn upward toward the light, and they feel and see themselves as floating in the air (Bullard, 1987; Mack, 1994; Ritchie, 1994).

Similar to those who have "experienced" life after death, abductees report going on voyages through the air, where they rapidly fly over the land or sea, to destinations including the Egyptian pyramids, New York City, and the North pole (Bullard, 1987; Mack, 1994; Ritchie, 1994). Over the course of the last fifty years there have been numerous reports of alien abductions (Bullard, 1987; Mack, 1994; Ritchie, 1994). Typically they are "abducted" while asleep or dreaming, or just upon wakening in the middle of the night -which raises the specter of hallucination and temporal lobe limbic system activation as these structures become exceedingly active during dream sleep.

Others claim they were abducted while driving late at night, while tired and under conditions where the head lights, moon light, and oncoming lights may flicker past (Bullard, 1987; Mack, 1994; Ritchie, 1994) -thus inducing possible seizure activity. However, the religious experience of some abductees is often hellish, and the aftermath includes prolonged feelings of depression, and horror and despair. "Abductees" frequently report that once they were drawn up toward the light, they felt overwhelmed with terror and that once they "arrived" they were subjected to painful and agonizing procedures (Bullard, 1987; Mack, 1994; Ritchie, 1994). Women often report that they were stripped naked and their legs spread, and that they were sexually molested, raped, or painfully probed.

Male and female abductees frequently report undergoing painful and invasive physical exams by alien monstrosities who loom demonically, probing vaginas, wombs, the anus, the eyeballs, and the viscera, with needlelike devices, or with twisting wires, or sharp, painfully cold lance-like instruments that may deliver electric, burning, or shock like sensations (Bullard, 1987; Mack, 1994; Ritchie, 1994).

Like those who experience life after death, some abductees report undergoing a "life review." They may see themselves or others on a viewing screen, usually engaged in sexual or violent activity. Similar sexual flashbacks are not uncommon with direct amygdala stimulation (Gloor, 1990, 1997; Halgren, 1992).

Once they return to earth and/or awake in their beds, many abductees are initially amnesic for the experience, though they may be troubled by fleeting, horrifying images and flashbacks (Bullard, 1987; Mack, 1994; Ritchie, 1994). Likewise, hyperactivation of the hippocampus can induce a temporary amnesia (Joseph, 1998a, 1999b).

They also suffer from depression, sleeplessness, anxiety and panic attacks; which again are suggestive of limbic system and temporal lobe abnormalities as well as post-traumatic stress disorder.

It is noteworthy, however, that some of those who undergo life-after-death also report exceedingly unpleasant experiences. This includes feelings of terror, sensations of terrible physical pain, the presence of demonic monstrosities, or hallucinations of people crying, moaning, screaming, and burning in flames.

"For those who have disbelieved... shall roast in fire... and their bellies and skin shall be melted. To them it is said: Taste the punishment of the burning." -Koran

DISSOCIATION & POSSESSION
It has been repeatedly demonstrated with neurosurgical and epilepsy patients that abnormal temporal lobe/hippocampal activation can cause an individual to experience themselves as separating and floating above their body, such that they may feel as if they are two different people, one watching, the other being observed (Daly 1958; Jackson & Stewart, 1899; Penfield 1952; Penfield & Perot 1963; Williams 1956). As described by Penfield (1952) during electrical stimulation of this area, "it was as though the patient were attending a familiar play and was both the actor and audience." Likewise, children, as well as adults, who are terribly abused or traumatically stressed and frightened often report dissociative experiences including even a splitting of consciousness such that one aspect of their mind will seem to be floating above or beside their body (Courtois, 1988; Grinker & Spiegel, 1945; Noyes & Kletti, 1977; Parson, 1988; Southard, 1919; Terr, 1990). Hence, the famous aside: "He was literally beside himself with fear."

Most abusers do not wish to molest an unresponsive body, but a personality that is either excited or tormented and terrified by what is occurring. According to Courtois (1995), it is the abuser's desire "to achieve total control over the victim and her responses and for the victim to become a willing participant and to enjoy the abuse. To achieve this end, many incest offenders take great pains to sexually stimulate their victims to arousal." They may pinch, cut, and hurt them to obtain the desired reaction. Thus, although these particular children may have "split off" from their body, they are nevertheless forced to respond, as a personality, to the abuse in which case the split off, or remaining portion of the "personality" may act in accordance with the manner in which it was created.

Thus, the abuser slowly shapes a personality which is formed during periods of abuse. Hence, some alternative personalities are highly sexual, or homosexual, or angry and self-destructive, mirroring either the role they were forced to play or the personality and behavior of the perpetrator. The dissociated aspects of conscious-awareness, therefore, may slowly establish its own independent identity; that is, an alternate personality—a personality which may be repeatedly triggered or purposefully shaped during subsequent traumas, and which may grow only to split apart again and yet again, forming multiple fragmentary personas each supported by a dissociated and abnormal neural network.

However, these "broken off" or alternate personalities, in turn, may be "state dependent" and supported by isolated neural networks maintained by the amygdala in the absence of hippocampal participation. Because the alternate "personality" essentially splits off from the main personality under certain highly stressful conditions, in general, it can only be reactivated and function independently when the individual is stressed or in a similar state of mind. Moreover, because of the state dependent nature of this partial "personality," the main personality essentially becomes amnesic regarding it. In part this is a consequence of hippocampal deactivation. When the hippocampus returns to normal, it can no longer gain access to these personal memories (which may be maintained by the amygdala), such that dissociated personal memory and the associated, albeit disconnected personality, remains isolated and inaccessible. In consequence, some victims, or their families, feel as if they sometimes are possessed.

POSSESSION & PROPHECY

Altered mental states, secondary to brain injury, seizure activity, or severe emotional trauma, by definition, encompass an alteration in consciousness. Under some conditions, what is perceived is a hallucination, a sensory distortion, or a dream-like flashback. Some may believe they have been afflicted by demons, or that they are possessed. Feelings of possession or the presence of an alien presence, can be directly attributed to neurological abnormalities in the right hemisphere, right parietal lobe, corpus callosum, and the limbic system. Prophecy, too, has been linked to hyperactive states and abnormalities in the limbic system; and individuals who make prophecies also sometimes appear to be possessed.

Signs of limbic system hyperactivity and seizure activity include loss of consciousness, trance states, dreamy states, auditory and visual hallucinations and related disturbances of the auditory and visual system involving language and reading and writing. The disturbances of language and speech are due to the involvement of the temporal lobe in auditory and reading comprehension. That is, the language areas, upon receiving visual signals, provide the auditory equivalent to a written word so that the person knows what the written words sounds like (Joseph, 1982, 1996, 2000a).

The auditory association area in the left temporal lobe (Wernicke's receptive speech area) not only comprehends incoming speech but assists in the programming of Broca's expressive speech area in the frontal lobe. It is Broca's area which produces speech, and a tissue immediately above Broca's area subserves writing. Therefore, abnormalities in the temporal lobe can induce severe disturbances of speech including abnormalities in the ability to write and spell words. In this regard it is noteworthy that whereas Moses suffered from a severe speech impediment and was slow of tongue, Muhammed, God's messenger, was apparently dyslexic and agraphic. He was unable to read or write.

On the other hand, hyperactivity in these limbic and temporal lobe structures can also induce pressured speech and writing. The afflicted individual may feel compelled to preach and to write out their mystical thoughts (Joseph, 2000a). Certain individuals who develop "temporal lobe epilepsy" or irritative lesions to this tissue, may suddenly become hyperreligious and spend hours reading and talking about the Bible or other religious issues. They may spend hours every day preaching or writing out their mystical or religious thoughts, or engaging in certain actions and movements they believe have religious significance. Many modern day religious writers who also happen to suffer from epilepsy are in fact exceedingly prolific, whereas conversely, those who feel impelled to preach tend to do just that. In fact, many of the prophets reported that they felt forced to preach and prophecies even though they struggled not to.

Temporal lobe hyperactivity or temporal lobe epilepsy is not usually associated with tonic clonic seizures and patients do not flail about on the ground, though they may certainly lose consciousness and move their hands and arms in a ritualistic manner. And yet, they may also retain the power of speech-- though what they say may sound like jibberish. In some instances, particularly if the amygdala and amygdala-striatum is impacted, the patient may appear to be in a trance. They may experience pleasant or unpleasant odors, heart palpations, difficulty breathing, excessive sweating, and disturbances of movement, ranging from a complete freezing of the body to spasmodic movements of the arms and legs (Joseph, 1999a, 2000a). Following a temporal lobe seizure the patient may stagger about like a "stunned ox." In this regard when the prophet Jeremiah experienced the

Lord God, his arms and legs convulsed and he staggered about like a drunk.

The presence of "god" may be experienced through dreams, through visions, and through voices—and it is usually a terrifying experience which grips the prophet like a seizure: "And I saw this great vision, and there remained no strength in me, for my comeliness was turned in me into corruption, and I retained no strength" (Daniel, 10:8); he defecated and urinated, as he fell to the ground—a not uncommon manifestation of certain seizures.

Daniel claimed he immediately lost consciousness before being confronted by the Lord god: "Thus was I in deep sleep on my face, and my face toward the ground" when an angel appeared in a prophetic vision (Daniel 10:9).

Likewise, Abraham would lose consciousness just prior to experiencing these "visions" or upon hearing the voice of the Lord God. "And a deep sleep fell upon Abraham (Genesis, 15:1,2).

In addition, and as discussed in earlier chapters, hyperactivity in this area of the brain can induce terror and severe depression, auditory and visual hallucinations, as well as religious fervor and even feelings of possession.

CASE HISTORIES

Consider Mary (described by Mesulam, 1981), a 26 year old female, A-average college student. For several months she had been complaining of odd mystical experiences involving alterations in consciousness, accompanied by auditory and visual hallucinations as well as frequent experiences of deja vu. These mystical experiences soon progressed to feelings of being possessed by the Devil. She was convinced the Devil was urging and trying to make her do horrible things to other people or to herself. She also claimed he would sometimes loudly cackle inside her head. Finally, a priest was brought in and a rite of exorcism was performed, as the Catholic hierarchy became convinced of the authenticity of her experiences, that she was possessed. However, her condition failed to improve. Finally an EEG was performed and abnormal activity was discovered to be emanating from both temporal lobes.

Another 44 year old female college graduate suffering from temporal lobe abnormalities instead came to believe she was possessed by God and at times also thought she was the Messiah, and at the behest of God, had a special mission to fulfill (Mesulam 1981). At the urging of the "God" she ran for public office and almost won. However, she also engaged in some rather bizarre actions including widespread and inappropriate sexual activity -another manifestation of limbic hyperactivation.

In a classic case described over 120 years ago by Sommer (1880), a 25 year-old man who suffered several seizures a day, claimed to have had repeated conversations with God. God's voice spoke to him quite clearly. One day, God told him he was Jesus Christ and to go forth and perform miracles. The first miracle he was to perform was to fly through the air. So convinced was he that these revelations were real, and that he was in fact "The Christ" that this young man climbed up on a roof and leaped into the air. At autopsy he was found to have a hard sclerotic lesion of the right Ammon's horn; i.e. the hippocampus. Sommer (1880) concluded that these religious hallucinations (and all hallucinations and sensory illusions) were directly due to abnormalities in this and adjoining tissue—which includes the amygdala; the two core structures of the limbic system and temporal lobe.

MUHAMMED

In order to receive the word of God, Muhammed would typically lose consciousness and enter into trance states (Armstrong 1994; Lings 1983).

However, he had his first truly spiritual-religious conversion when he was torn from his sleep by the archangel Gabriel who enveloped him in an terrifying embrace so overpowering that Muhammed's breath was squeezed from his lungs.

"Recite!" (iqra!) the angel demanded. Muhammed refused: "I am not a recitor!"

The angel again enveloped him in a crushing embrace, squeezing the air from his lungs. Again the angel demanded: "Recite!" and again Muhammed refused. And then the angel crushed him a third time, at which point words began to pour from Muhammed's mouth:

"Recite in the name of thy Sustainer, who has created—created man out of a germ-cell! Recite—for thy Sustainer is the Most Bountiful, One who has taught (man) the use of pen—taught him what he did not know!"

Because the angel forced Muhammed to recite the words of God, it is for this reason that the book of Islam is called the qur'an: "Recitation."

This was the first of many such episodes Muhammed had with the archangel Gabriel who sometimes appeared in a titanic kaleidoscopic panoramic form.

Muhammed was initially exceedingly distressed, horrified by what he had experienced. He also feared he may have been "possessed by a jinn," a sprite. It was commonly believed among the Arabic community that some people had their own individual, personal jinni who could provide them with good luck and inspiration.

Muhammed did not know what to think. Shattered by the experience and overcome with feelings of suicidal terror and depression, Muhammed finally decided to throw himself from a cliff. Again the angel Gabriel appeared to him as he stood at the precipice.

According to Muhammed: "I heard a voice from heaven saying, O Muhammed! thou art the apostle of God and I am Gabriel. I raised my head toward the heavens to see who was speaking, and lo, Gabriel in the form of a man with his feet astride the horizon... I turned my face away, but toward whatever region of the sky I looked, I saw him there."

Escape was impossible. Horrified, terrorized, and depressed by what he considered a "loathsome" experience, and fearing that he was "possessed by a jinn," Muhammed crawled on his hand hands and knees and flung himself into the lap of his wife, and begged her to shield him from the divine presence. "Then I told her the whole story." She soothed him and after listening to him describe his ordeal, she suggested that they speak with her cousin, "Waraqa, a Christian who knew scripture."

Eventually, Muhammed reluctantly accepted his fate, and in accordance with the voice of "God" or his angels, Muhammed not only spoke the words of "God" but began reciting and chanting various themes of God in a random order over the course of the following 20 years; an experience he continued to find quite painful and wrenching (Armstrong 1994; Lings 1983). "Never once did I receive a revelation without feeling my soul being torn away from me."

In order to receive his revelations, Muhammed also entered into a trance state and would lose consciousness. He reported that a heaviness would fill him with a feeling of depression and grief, and that he would sweat profusely. He also stated that although the divine message was often clear, at other times it was a jumble of sounds and voices, "like the reverberations of a bell. And that is the hardest upon me; the reverberations abate only when I understand the message."

LOT

Epilepsy can be due to a number of different causes, such as head injury, heat stroke during infancy, and tumors. However, the predisposition to develop epilepsy can also be inherited.

Like his uncle Abraham, Lot also saw angels and talked to the Lord God. It was the Lord God's angels who warned Lot to leave Sodom; reportedly the most sexually corrupt city on Earth.

Once Lot escaped from Sodom, he celebrated by getting drunk and impregnating both his daughters who willingly snuck into his bed on two separate nights (Genesis, 20: 33-38). In fact, even before they left Sodom Lot had offered his daughters to some of the men of the city to do with as they pleased (Genesis 19: 8).

We do not know if Lot followed Abraham's example and also let other men have sex with his wife. Nevertheless, both Abraham and Lot clearly demonstrated signs of temporal lobe and limbic hyperactivation.

EZEKIAL

Many "prophets" and other religious figures also display evidence of the Kluver-Bucy syndrome—a disturbance also referred to as "psychic blindness" which is due to severe abnormalities or destruction of the amygdala. For example, patients may seek isolation, avoid people, engage in abnormal sex acts, and eat non-nutritional substance such as cigarette butts or other objects. Thus we read that Ezekiel would eat dung.

In addition, many of the "prophets" avoided people and would seek isolation—yet another characteristic of limbic and amygdala abnormalities.

Likewise we read that Ezekial isolated himself from other humans and refused to speak to others. If they approached he would withdraw. If they followed he would flee. Primates with bilateral amygdala damage respond identically.

Social isolation, however, is a common means of achieving enlightenment, and isolation also acts on the amygdala (Joseph, 1982, 1992a. 1999a,b).

It was while isolated from all human contact that Ezekial had a shattering experience which caused him to lose consciousness. He had seen a cloud of light shot through with lightning, and within that stormy light he observed a great chariot pulled by four beasts each of which had four heads with the face of a man, bull, a lion, and eagle, and each wheel of the chariot rolled in a direction different from the others. And he heard wailing, moaning, and voices that gave him commands. And he experienced a sweet taste in his mouth, like honey, when "the spirit lifted me and took me, my heart overflowed with bitterness and anger." Afterward he lay "lie someone stunned" for an entire week.

Yet other signs of amygdala hyperactivation include catalepsy and catatonia. Thus we read that Ezekial was forced to lie on one side for 390 days without moving and then on his other side for 40 days. He was in all respects completely catatonic.

Cataplexy, or in this case, what appears to be catatonia, is also known as an arrest reaction due to excessive fear (e.g. "paralyzed with fear"), or to abnormalities in the amygdala striatal medial frontal lobe emotion-motor circuit. If exceeding frightened, one may freeze and fail to move, a function of the amygdala acting on the motor centers in the striatum and medial frontal lobe (Joseph, 2011). The airline industry refers to this as "frozen panic states" and about 20% of those involved in airline or other disasters will freeze and fail to move or to take any evasive action that might save their lives.

If there is an abnormality in this emotional-motor circuit, the individual may develop Parkinson's disorder, or catatonia, in which case they may lie frozen in the same posture for weeks and months. In Ezekial's case, the possibility of amygdala hyperactivity seems to be implied; not only because of the above mentioned conditions, but as he was obsessed with sexuality and often employed violent sexual imagery when he prophesied.

MOSES AND TEMPORAL LOBE LIMBIC HYPERACTIVITY

The story of Moses is part mystery, at least regarding his paternal origins. We are told that he was raised in the palace of the Pharaoh of Egypt who regarded him as a grandson. Yet, the identity of his father is unknown and we are also told that his mother hid him for three months after he was born.

When she could hide him no more, she placed him in a wicker basket among the reeds of the Nile. It was there that Moses was discovered by the daughter of Pharaoh, who in turn raised him like a son. But then we are told that the princess paid a Hebrew woman to nurse the child. When he grew to be a boy the princess named him "Moses, which means: "I drew him out of the water." "Moses" however, was also a common Egyptian name, including that of a mighty Pharaoh: Thutmoses.

Moses had a violent temper and one day beat an Egyptian to death. Overcome with fear, when he realized the murder had been witnessed, he fled to the land of Midian where he proceeded to challenge a group of men and rescued the daughters of the local pagan priest.

Moses, the "Egyptian" then married Zipporah, the daughter of the priest. It was later, after the birth of his son, Gershom, and while alone and isolated and attending a flock of sheep in the wilderness, near the mountain of God, that the Lord God appeared and spoke to him from inside a blazing bush.

The Lord God then proceeded to instruct Moses as to what he should say to the Israelites: "The Lord, the god of your fathers, the God of Abraham, the God of Isaac, the God of Jacob has sent me to you." But "Moses answered and said to the Lord, Please O Lord, I have never been a man of words.

I am slow of speech and slow of tongue." Indeed, his speech was so garbled the Lord God decides to appoint Moses' brother Aron to assist him: "You shall speak to him and put the words in his mouth." Possibly, Moses requires an interpreter because he does not know the language of the Hebrews; being raised as a prince in the household of the Pharaoh. However, he also required Aron to speak for him to Pharaoh. Thus it appears that he may well have had a severe speech impediment as he claims. In fact, coupled with his violent temper, his severe depressions, and the possibility of auditory and visual hallucinations, it could be argued that Moses suffered from temporal lobe epilepsy: The "Divine Illness." Seizure activity within the temporal lobe and limbic system are clearly associated with triggering religious experiences and can give rise to murderous rage reactions as well as disturbances of speech and language.

Moses may have suffered from temporal lobe seizures as a consequence of being left for days, as an infant, to bake in the hot, broiling Egyptian sun, after his mother abandoned him in a basket on

the Nile. If that were the case his infant brain could have become overheated and damaged by the high temperatures.

If Moses subsequently developed temporal lobe seizure activity, this could explain his hyperreligious fervor, his rages and the numerous murders he committed or ordered. Indeed, a recurring theme throughout the story of Exodus is that of Moses killing and repeatedly ordering human slaughter, beginning with the first Egyptian he killed, followed by the first born sons of the Egyp-

tians, then Pharaoh and the Egyptian army, and then the slaughter and death of the Hebrews who followed him out of Egypt, all 600,000 of them.

TRANCE STATES

Numerous religious figures have achieved enlightenment, or were only able to commune with god, once they entered a trance state. Buddha, for example, is said to have at first communed with gods, and then to have transcended the gods by entering a trance state which enabled him to experience nirvana.

It is said that when Buddha entered into trance that fragrant breezes swirled about him, flowers bloomed and blossomed and fell from the air, and that the Earth began to rock as the gods in heaven began to rejoice. However, one god, a demon, Mara, was alarmed. She told Buddha to remain in a state of bliss and to never, under any circumstances attempt to inform others of what he had accomplished as no one would believe him. Two other gods, however, Sakra, the Lord of the devas, and Maha Brahma begged him to teach his method to all the world. Buddha heeded their advice.

For the next 45 years he wondered all over India preaching his message. Humans could transcend suffering and the gods, and through kindness, compassion, meditation and Yoga, they could come into contact with the ultimate reality.

Trance states are said to enable one to sense and perceive what others cannot. In Buddhism it is believed that trance enables one to gain use of Power over the material world. This includes the Power (te) of not contending, the Power to use men, and the Power for causing others to act, or to prevent them from acting.

Mystics may achieve trance states through ritual and discipline or through pain, sex, fasting or drugs. These trance states have been divided into five stages by some (Buddhists), six by others (Tantra), and four by yet other mystics (Sufis).

Broadly considered, these trance stages begin with the perception of visions and voices, then progress to a non-perceptual awareness of God, followed by transcendence and a merging and becoming one with God, which is the supreme mystical act.

While in trance, some individuals not only experience a bright light, but they may be blinded by it. In some cases the light is so bright those around them will see it. Moses and Jesus both shone with a light that others could see, whereas St. Paul was blinded by the great light that shone around his head--a light which gave rise to a "hallucination" and a revelation, on the road to Damascus.

During revival meetings, it is not uncommon for revivalists to suddenly go into trance and see\ "a brightness like the sun" "everything is bathed in a rainbow of glory."

VISIONS AND DREAMS

Throughout the "Old Testament" the "Lord god" delivers his words through visions and dreams, and in some instances the words themselves constitute the "vision;" that is, like an auditory hallucination:

"And an angel said unto Israel... And an angel said unto Balaam... and Elohim said unto him... And an angel of the Lord called unto Abraham out of heaven... and Elohim said unto Noah... The word of the Lord came unto Abraham in a vision... And the Lord said unto Abraham... and the Lord said unto

Jacob...and the Lord said unto Joshua... and the Lord said unto me... and the word of the Lord came unto me... "etc.

Likewise, the gods and goddesses of ancient peoples throughout the world, including for example, the gods of the ancient Romans and Greeks would speak and appear to specific, chosen people in visions and dreams; an experience also referred to as an "epiphany" or in modern psychiatric terminology, a "hallucination."

One of Jacob's many epiphanies included a dream of a ladder which stretched from heaven to earth; and angels were ascending and descending between the realms of god and man. People who suffer from periodic episodes of limbic and temporal lobe hyperactivation, such as those with temporal lobe epilepsy, may hear voices and suffer seizures which may be indistinguishable from a trance state. It is not uncommon for these seizures to be preceded by an auditory or visual hallucination or aura (Gloor, 1997; Penfield and Perot 1963; Trimble 1991; Williams 1956). Patients can have any number of very odd hallucinations, such as smelling sweet or horrible odors, hearing voices, music, or conversations, seeing angels, demons, ghosts, and God.

The great existential author, F. Dostoevsky, apparently suffered temporal lobe epilepsy. Dostoevsky, alleged (via one of his characters) that when he had a seizure the gates of Heaven would open and he could see row upon row of angels blowing on great golden trumpets. Then two great golden doors would open and he could see a golden stairway that would lead right up to the throne of God.

Beginning with Abraham, and throughout the "Old Testament" the "Lord God" delivers his words primarily through visions and dreams, and frequently his prophets lose consciousness or enter into trance states. Many of His prophets appear to be suffering from temporal lobe epilepsy or hyperactivity.

In the 12th century, Moses Maimonides, composed a book "A Guide for the Perplexed" as a letter to R. Joseph. According to Maimonides "the Divine Will has withheld from the multitude the truths required for the knowledge of God... Do not imagine that these most difficult problems can be thoroughly understood by any one of us."

"At times the truth shines so brilliantly that we perceive it as clear as day. Our nature and habit then draw a veil over our perceptions, and we return to darkness almost as dense as before. We are like those who, though beholding frequent flashes of lightning still find themselves in the thickest darkness of night."

"For some the lightning flashes in rapid succession, and they seem to be in continuous light, and their night is as clear as the day. This was the degree of prophetic excellence attained by Moses, the greatest of prophets."

"By others, only once during the whole of night is a flash of lightning perceived. This is the case with those of whom we are informed "They prophesied and did not prophecy again." "There are some to whom the flashes of lightning appear with varying intervals; others are in the condition of men, whose darkness is illuminated not by lightning, but by a small light that is not continuous, but now it shines and now it vanishes..."

"The degrees of perfection of men vary according to these distinctions. Concerning those who never beheld the light even for one day, but walk in continual darkness, it is written: "They know not, neither will they understand. They walk in darkness" (Psalms, 82:5). Truth, in spite of all its powerful mani-festations is completely withheld from them" (Maimonides, 12th century).

The Truth revealed, however, is often a shattering experience, that grips one like a seizure. Likewise, the gift of prophecy, is not always a welcome gift, and even the role of prophet may be resisted. Moses actively resisted and argued with the "Lord God" and repeatedly tried to convince Him to pick someone more deserving. Other prophets have also attempted to resist, but to no avail. They were often hounded by the "Lord god" or his angels, or felt compelled to prophecy even when threatened with death by the people—and many a prophet was killed by the people, including the ancient Jews.

Jeremiah, for example, was openly despised and ridiculed by the ancient Jews. Nevertheless, though he wished it otherwise, he could not withhold his prophesying: "For the Word of the Lord was unto me a reproach and a mocking all day, and I said, I will not mention it, nor will I again speak in His name; but it was in mine heart as burning fire, enclosed in my bones, and I was wearied to keep it, and did not prevail" (Jeremiah, 20:8,9).

THE TRANSMITTER TO GOD

Although it is possible that certain religious figures may have been suf-fering from temporal lobe epilepsy, it is noteworthy that seizures are stereo-typical and idiosyncratic. That is, unless an individual has several different seizure foci, each seizure is basically identical to the next and the auras and hallucinations are quite similar from one seizure to the next.

This is not the case with many of the prophets. As detailed by Moses Maimonides: "According to the books of the prophets, a certain prophet after being inspired with one kind of prophecy is reported to have received prophecy in another form. It is possible for a prophet to prophecy at one time in the form and at another time in another form. The prophet does not prophecy continu-ously but is inspired at one time and not at another, so he at one time prophecy in the form of a higher degree and at another time in that of a lower degree; it may happen that the highest degree is reached by a prophet only once in his life time, and afterwards remains inaccessible to him, or that a prophet remains below the highest degree until he entirely looses the faculty."

For example, "And the word of the Lord ceased from Jeremiah" (Erza, i:1), or "And these are the last words of David" (Samuel, 23:1). Some individuals may prophecy once and only once—which argues against temporal lobe epilepsy.

ABRAHAM

In ancient Sumer (in southern Iraq around 6,000 years ago), it was believed that the Universe was ruled by a pantheon of Gods, the Anunnaki (Armstrong 1994; Kramer 1956, Wooley 1965); perhaps the same pantheon alluded to in the first chapter of Genesis.

The Sumerians also worshiped household gods and goddesses, including a personal God, which in some respects could be likened to a "guardian angel" or a spirit (totem) helper, as was common among the Plains Indians. This personal God served almost as a conscience and as a mediator between the head of the household and the great Gods which ruled the Earth and the cosmos.

Because this was a private, personal God, it was not uncommon for a believer to engage in prolonged and daily discussions with his deity (Kramer 1956: Woolley 1965). To this god one could
bear their heart and soul regarding sins, injustice, personal shortcomings and hopes for the future. Hence, this god was indeed a personal god with whom one could "talk" and maintain a special personal relationship and which served to protect the Sumerians from the Pantheon of gods who rules the Earth at that time—or so claim the ancients.

And then one day something astounding and revolutionary occurred in the mind of a man of the city, of Ur of the Chaldees, of ancient Sumer, in ancient Babylon, and birthplace of Abram a rich
Babylonian prince. Abram had left the city and began having visions and hearing voices. It was coming from his personal God and it later gave him a command (Genesis 12): "Get thee out of thy country ...and I will make of thee a great nation... and in thee shall all the families of the earth be blessed..."

And Abram and his personal god walked and talked, as God had not done since the time of Adam and Eve. And then one day this personal God came to a decision and said to Abram: "Thy name shall be Abraham, for a father of many nations have I made thee. And I will make thee exceedingly fruitful... and I will make thee the father of many nations... and I will be their God" (Genesis 17).

Abraham, both saw and heard his God on numerous occasions both awake and dreaming, often falling on his face as God appeared. They ate together, and walked and spoke together during the heat of the day, and during the darkest hours of the night. This Lord God, the "God of Abraham" was also making all types of grandiose promises and predictions, all of which came to pass.

PROPHECIES FULFILLED

Is it possible that Abraham was dreaming? Could this personal God from ancient Ur have been but a hallucination, and given Abraham's odd sexuality and murderous actions, a product on tempo ral lobe epilepsy or subclinical seizure activity?

When we consider that this is the same Lord God (at least in religious theory) who today is worshiped by Jews, Christians, and Moslems alike, the possibility of hallucinations does not seem likely. Indeed, has not the Lord God's prophecy to Abraham come to pass: "And I will make thee the father of many nations... and I will be their God" (Genesis 17).

In fact, given that the God experienced by Abraham and Sara (and in fact, with few exceptions, the God repeatedly described in the Bible) not only appeared as a man but behaved as a man and not a spirit-like supernatural being, the possibility of hallucinations does not seem likely.

Likewise, when we consider how many other prophesies were fulfilled, including the repeated destruction and recreation of Israel, including, most recently in 1948—the aftermath of a world war led by an Austrian German-Jew, Adolf Hitler—as well as the hundreds of millions who worship a Jew (Jesus) as "God," it borders on the irrational to simply dismiss these events as a hallucination, a myth, or a coincidence.

Under conditions of limbic hyperactivation, not all "hallucinations" are hallucinations, but instead may represent the perception of stimuli which are normally filtered from consciousness. Under conditions of sensory deprivation, pain, social isolation, drug use, the nuclei of the limbic system may become abnormally activated and possibly hyperactivated such that subclinical seizure activity (kindling) develops (Cain, 1992; Gloor, 1997; Joseph, 1999b,d).

Under conditions of limbic kindling, hyperactivation, or seizure activity, sensory and emotional filtering that normally takes place in these nuclei is reduced or abolished. That is, the individual may begin to perceive things or individuals that (presumably) do not exist, or they may gain access to "hidden" knowledge that bursts upon them with a shattering clarity, or they may achieve a "higher" understanding of existence and the nature of reality. They may commune directly with god, goddesses, devils, demons, and angels.

As these same nuclei are also implicated in dream states, near death experiences, and out-ofbody phenomenon, it could thus be argued that individuals who for whatever reason are "blessed" with an overactive amygdala-temporal lobe are also given access to god-like stimuli which are also normally filtered from consciousness.

"Know that for the human mind there are certain objects of perception which are within the scope of its nature and capacity to perceive; on the other hand, there are amongst things which actually exist, certain objects which the mind can in no way grasp: The gates of perception are closed against it. Further, there are things which the mind understands, one part, but remains ignorant of the other. And when man is able to comprehend certain things, it does not follow that he may be able to comprehend every thing."-Moses Maimonides

Moreover, consider the prophecies delivered to Muhammed. In less than fifty years all the Arabic tribes had been converted to Islam.

Or consider the prophecies of the god of the cross, that is, the Mayan and Aztec God, Quetzalocoatl. Quetzalocoatl, the plumed serpent, was associated with the planet Venus, and the sign of the cross was one of his many emblems which is depicted in the center of his shield. According to Mayan and Aztec religious belief, Quetzalocoatl, the god of the cross, had been driven from the lands at the time of the great flood, some 12,000 years ago. However, it had been prophesied that Quetzalocoatl, the plumed serpent and god of the cross, would return to Mexico and that the ruling gods, and their temples, would be completely overthrown. In fact, an exact date was given for his return, which would be signified by the appearance of a comet in the East. In the Christian calendar this was to be in the year 1519.

As predicted, in the year 1519 a comet appeared in the East, and soon thereafter, the plumed and helmeted Cortez and his crew, flying the Spanish flag and the sign of the cross appeared off the

East Coast of Mexico. And, as predicted, the old god, as well as the Aztec civilization were destroyed, and the god of the cross is now worshipped not just by the Mexican, but peoples throughout the Americas. Just a coincidence?

THE PROPHECY OF THE SECOND EXODUS

Prior to and with the establishment of Israel the "Lord God" of the ancient Jewish people repeatedly threatens them with the most horrible of misfortunes and foretells, through his prophets, that they will be scattered and dispersed throughout the lands—and this is exactly what transpires, repeatedly, beginning in the 6th century B.C., when the southern kingdom of Judah was destroyed and the people marched off into exile in Babylon. It is Babylon and before that, Sumer where much of the story of creation may well have been first composed, only to be later incorporated and retold in Genesis.

And then again, almost 2,000 years ago, following their return from exile and return to the promise lands, the Lord God again warns the ancient Jews of their destruction.

"The Lord will bring a nation against you from afar, from the end of the earth, which will swoop down like the eagle... a ruthless nation, that will show the old no regard and the young no mercy." -Deuteronomy 28:47-50.

And this is exactly what transpired when Roman legions, marching under the banner of the swastika and the eagle, swooped down and destroyed the second temple. Again the Jews are driven from Israel, and dispersed throughout the lands.

But then the Lord God also tells the Jewish people that someday He will return them to the promised land. And this prophecy too was fulfilled, by no less than Adolf Hitler; a man who likened his nation to the "Holy Roman

Empire" and who modeled his government after the Roman Catholic Church (which he also sought to destroy). And, Hitler's armies marched under the banner of the eagle and the swastika, the twisted cross, the "gammadion"—which is also one of the many names of god, including the Lord God of the old testament: "Tetragrammaton" ("absolute existence"). Even the word "Nazi" has a Jewish reference: "Ashkenazi." "Ashkenazis" are European Jews.

Hitler went forth according to instructions he claimed to have received from Divine Provi dence. He was to go forth and cleanse the land of Jews. He was to begin the second exodus.

HITLER AND DIVINE PROVIDENCE: SECOND EXODUS

"Behold, the people of the children of Israel are too many and too mighty for us. Come, let us deal wisely with them." -Exodus 1:9-10.

As Hitler struggled to become dictator of Germany, and as he clearly states in his book, Mein Kampf, the Jews, he preached had not only obtained positions of power disproportionate to their numbers, but they were stealing German wealth, were engaged in sex slavery, and they threatened to overwhelm society with evil and moral decay. They were like a disease, like vermin, like a cancer, and were less than human; they were like animals, subhuman. Hitler counseled that the German nation should deal wisely with them.

The same sentiments had been expressed thousands of years before, by the Pharaoh, King of Egypt.

When Adolf Hitler finally came to power, not only did the majority of the German peoples willfully and fearfully followed his dictates, but many compared him to John the Baptist, Jesus Christ and even God. "Is he John the Baptist? Is he Jesus?" Goebbels wrote in his diary. Many honestly believed they were in the presence of a German Messiah! And not just his earliest followers, but many a German general thought he was "god" and the destiny of Germany. And many shared his lowly opinion of the Jewish people, who Hitler promised to sweep from Germany and send back to their promised land.

Hitler claimed he was acting in accordance with the will of Divine Providence. He believed he had been appointed by God.

Hitler was to be the "Lord God's" angry fist which would savagely strike a scattered people, the Jews, and drive them back to Israel, the Promised Land.

The second exodus would mirror the first.

"He exalts nations, then destroys them; He expands nations, then leads them away. He deranges the leaders of the people, and makes them wander in a trackless waste." -Job, 12:20-24

Once Hitler came to power, and as he led his nation out of the great depression to greatness, he and his government issued orders controlling Jewish breeding and sexuality; and then the Jews were barred from practicing their professions; and then the Jews were officially described as subhuman, then they were ordered to immigrate--events and laws that echo the events leading up to the first Exodus.

And just as the ancient Jews stripped the Egyptians of their worldly goods before setting off for the promised land, the German Jews who sought to immigrate were stripped of their worldly possessions. Those who remained or failed to heed the six years of repeated warnings to leave Germany were finally herded into concentration camps where they were enslaved, starved and worked to death-conditions which mirror those of ancient Egypt prior to the exodus. "And so they died, one after another, as if smitten by a pestilential destruction... And then their taskmasters threw their bodies, unburied, beyond the borders of the land, not allowing their kinsmen to even weep over those who had thus miserably perished." -Philo of Alexandria.

HITLER, POSSESSION & THE VOICE

Hitler first heard the "voice" of Divine Providence, in 1905, when he was a 17-year old youth, living in Austria. The "voice" told Hitler, he would someday become the leader of Germany, and would lead "God's people" back to the land of their fathers.

Hitler and his best friend, August Kubizek, had just left the opera when the "voice" spoke through Hitler. The voice told Hitler and Kubizek that Adolf would someday become the savior of Germany. The Voice explained: Hitler had received a mandate from god, and would someday receive a mandate from the people, to lead them out of their servitude and to lead them back to the land of their fathers. The Voice declared that Hitler had been chosen by Providence and had been given a Divine mission. Adolf Hitler was destined to establish a new social order, a new Reich which would be established under his leadership... The 17-year old Hitler had received a mandate to lead God's people to the heights of freedom and back to the promised land.

Kubizek was amazed and shocked by the voice and the transformation he observed in Hitler. Hitler, he thought, seemed to be possessed by a demon.

Many of those who observed Hitler, later in life, also witnessed periods of possession. Hitler, a plain, and somewhat funny looking man, would suddenly become transformed, as if--according to those who observed the transformation--he was possessed by a demon: a demon who could weave a spell upon the German people by speaking with Hitler's voice.

"Listening to Hitler one suddenly has a vision of one who will lead mankind to glory... A light appears in a dark window. A gentleman with a comic little moustache turns into an archangel... The archangel flies away... and there is Hitler sitting down, bathed in sweat with glassy eyes..." —Gregor Strasser

According to Francois-Poncet, Ambassador from France to Nazi Germany, when the voice spoke, "Hitler entered into a sort of mediumistic trance..." Others were of the same opinion.

"I looked into his eyes—the eyes of a medium in a trance... Sometimes there seemed to be a sort of ectoplasm; the speaker's body seemed to be inhabited by something... fluid. Afterwards he shrank again to insignificance, looking small and even vulgar. He seemed exhausted, his batteries run down." —Bouchez

"Hitler was possessed by forces outside himself... One cannot help but think of him as a medium. For most of the time, mediums are ordinary, insignificant people. Suddenly they are endowed with what seems to be supernatural powers which set them apart from the rest of humanity. These powers are something that is outside their true personality—visitors, as it were from another planet. The medium is possessed. Once the crisis is past, they fall back into mediocrity. It was in this way, beyond any doubt, that Hitler was possessed by forces outside himself—demoniacal forces of which the individual named Hitler was only a temporary vehicle." —Rauschning.

Hitler tells us in Mein Kampf and in his other writings and comments, that throughout his life; he was guided by the "voice" --a voice which would repeatedly protect him from harm, and which guided his rise to power.

For example, he relates the following experience during the first World War. "I was eating dinner in a trench with several comrades. Suddenly a voice seemed to be saying to me, "Get up and go over there." It was so clear and insistent that I obeyed automatically. I rose to my feet and walked twenty yards. Then I sat down to go on eating. Hardly had I done so when a flash and deafening report came from the part of the trench I had just left. Every member in it was killed."

Remarkably, although he served as a runner (messenger) during the first World War, and although the average war time life span of a runner was just a few days, Adolf Hitler nevertheless survived in this position for several years with only minor injuries—a good fortune that he also attributed to Divine Providence.

In Mein Kampf Hitler explains that following the only time he was seriously injured, following a gas attack in the closing days of World War I, he experienced a vision, and heard a voice, which taunted him and then explained why he was being spared: "And then the Voice thundered at me: "Miserable wretch, are you going to cry when thousands are a hundred times worse off than you!" And then, the

Voice spoke again and he experienced a vision of the utmost clarity: "I was being summoned to save Germany.... I would go into politics."

It was soon after that terrifying and "wonderful" vision, that others too, began to believe that Hitler had been chosen by God.

Dietrich Eckardt, a highly influential and powerful member of the secret mystical organization, the "Thule Society" and one of the founding members of the National Socialist Party, met Hitler in 1919, and announced after their first meeting: "He is the one..." Eckardt and other German mystics had been waiting for the coming of a German Messiah, one who would lead the German nation and the German people in a battle between the gods... and who could serve as a bridge between this world and a mystical world from the past--a world of mythical heroes, demons and gods. Eckardt was convinced that Hitler was the Messiah they had been waiting for--that Hitler had been chosen by god.

Eckardt took Hitler under his wing, and initiated him into the mysteries of the most diabolical of secret societies. As he lay dying, in 1923, Eckhart bragged. "We have given him the means of communicating with Them."

Hitler, too, admitted to communicating with Them. "I will tell you a secret," he once confided to one of his top deputies, Rauschning. " I have seen Him. He is intrepid and cruel. I was afraid of him."

Hitler was especially afraid that "He" would come at night, while Hitler slept and dreamed. As related by Albert Speer and others, Hitler was fearful of being alone at night. He also had trouble falling asleep and staying asleep. He often wanted company into the late hours of the night. Hitler frequently voiced a fear of falling asleep when by himself. He sometimes dozed off only to awake with a frightened and hysterical shout, screaming that someone or something was in his room.

As described by one of his followers:

"Hitler wakes at night with convulsive shrieks; shouts for help. He sits on the edge of his bed, as if unable to stir. He shakes with fear making the whole bed vibrate. He gasps, as if imagining himself to be suffocating."

On one occasion, after awakening his staff with cries for help, they rushed to his room only to observe as "Hitler stood swaying in his room, looking wildly about him. 'He! He! He's been here!' he stamped and shrieked in the familiar way." On another occasion he awoke and cried out: "There! There! Over in the corner! He is there." On yet another occasion he awoke screaming and in convulsions. When his attendants ran into the room they found "Hitler standing, swaying and looking all around. "It's he, it's he'" he groaned. 'He's come for me!' His lips were white; he was sweating profusely. Suddenly he uttered a string of meaningless figures, then words, and scraps of sentences. It was terrifying." —Rauschning.

Yet, although Hitler admitted that this presence was"intrepid and cruel. I was afraid of him," he nevertheless relished being the chosen one, that he had been chosen by divine providence. He bragged of it.

As Fuehrer of Germany, he repeatedly spoke of hearing "Divine" voices, and claimed that he was following "the commands that Providence has laid upon me... Divine power has willed it... Not even if the whole party tried to drive me to action, I will not act. But if the voice speaks, then I know the time has come to act."

"There is a higher ordering, and we are all nothing else than its agents." -Adolf Hitler.

HISTORY REPEATS ITSELF

"We are often accused of being the enemies of the mind and spirit. Well, that is what we are, but in a far deeper sense than bourgeois science, in its idiotic pride, could ever imagine." —Adolf Hitler

Adolf Hitler experienced amazing success in building up Germany during the "Great Depression." His amazing success was the envy of the world--including the United States of America, whose President could look on only with envy.

And Adolf Hitler achieved incredible success when he went to war: destroying the armies and conquering Poland, France, and all of Europe in just a matter of months.

And then, Hitler destroyed his armies, and then Germany was destroyed. By the end of the war, in 1945, every major Germany city had been nearly reduced to rubble.

And in Palestine, the Palestinian people were being murdered, terrorized and driven from their homes by the Jews--in a repeat of the terrors that characterized the Jewish conquest of Palestine and Canaan, thousands of years before.

A little over two years later, and because of the atrocities committed by the Nazis, the United Nations granted Israel statehood. The Jews again had reclaimed their "promised land," the land of their fathers.

GERMANY & BABYLON

Over two thousand years ago, around 600 BC., the Jewish Lord God announced through his prophets that he would enable Babylon to become a super-power in order to punish his Jews, by attacking Israel/Judea and sacking the Temple of Solomon. And this is exactly what happened.

The Jewish prophet Ezekiel, then announced that the sacking of Jerusalem and the Temple of Solomon was just the beginning of their punishments. It was only a warning. But if the Jewish people did not heed these warnings, Ezekiel also pronounced that Yahweh would use the nation of Babylon as the instrument of his wrath, and that the king of Babylon would destroy the Temple and burn the city to the ground. Approximately 6 years after this prophecy, Nebuchadnezzar, the king of Babylon, attacked and destroyed the city and the temple. The Jews were driven into exhale.

However, the Lord God, through his prophets also explained that He would then destroy Babylon, thus giving His Jews, yet another opportunity to repent of their sins, to follow His laws, and to return to Israel.

However, in order for these prophecies to be fulfilled required that Babylon become a superpower and defeat Egypt. That is, Babylon was to become an instrument of God. It would become a super power which the Lord God would employ to make the Jews suffer, and then to force them off their land, and then to enslave them, so as to make them cry out to the Lord God. Then Babylon would be destroyed so that the Jews could return to the promised land and reestablish the state of Israel.

Around 600 B.C. the prophet Jeremiah began uttering prophecies about Babylon and Egypt and its king. According to Jeremiah 46: "This is the message against the army of Pharaoh Neco king of Egypt. "Prepare your shields, both large and small, and march out for battle! What do I see? They are terrified, they are retreating, their warriors are defeated...and there is terror on every side, declares the Lord... The day belongs to the Lord, the Lord Almightya day of vengeance, for vengeance on his foes. The sword will devour till it is satisfied, till it has quenched its thirst with blood. For the

Lord, the Lord Almighty, will offer sacrifice in the land of the north by the River Euphrates... prepare for Nebuchadnezzar king of Babylon to attack Egypt: Egypt will hiss like a fleeing serpent...The Daughter of Egypt will be put to shame, handed over to the people of the north. The Lord Almighty, the God of Israel, says: "I am about to bring punishment on Amon god of Thebes, on Pharaoh, on Egypt and her gods and her kings, and on those who rely on Pharaoh. I will hand them over to those who seek their lives, to Nebuchadnezzar king of Babylon and his officers. Later, however, Egypt will be inhabited as in times past," declares the Lord.

And this is exactly what came to pass. Babylon again became a lone superpower that none could oppose. But according to the prophets of Israel, Yahweh was merely using Babylon, allowing this nation to become all powerful, so that he could employ it in his wrath against Judea which he proclaimed would be severely punished so that the Jewish people would learn a lesson and again follow his laws:

"In those days, at that time," declares the Lord, "the people of Israel and the people of Judah together will go in tears to seek the Lord their God. They will ask the way to Zion and turn their faces toward it. They will come and bind themselves to the Lord in an everlasting covenant that will not be forgotten." —Jeremiah 50: 4,5.

And that is exactly what came to pass. However, according to the prophet Jeremiah, although Yahweh would use Babylon to punish His people, Yahweh would then destroy Babylon and punish it for this sin: "This is the word the

Lord spoke through Jeremiah the prophet concerning Babylon and the land of the Babylonians: "Announce and proclaim among the nations, lift up a banner and proclaim it; keep nothing back, but say, 'Babylon will be captured; Bel will be put to shame, Marduk filled with terror. Her images will be put to shame and her idols filled with terror.' A nation from the north will attack her and lay waste her land. No one will live in it; both men and animals will flee away. I will stir up and bring against Babylon an alliance of great nations from the land of the north. They will take up their positions against her, and from the north she will be captured. So Babylonia will be plundered; all who plunder her will have their fill," declares the Lord....I will punish the king of Babylon and his land as I punished the king of Assyria. But I will bring Israel back to his own pasture." —Jeremiah 50.

And this too is exactly what came to pass.

Cyrus, the King of Persia, attacked and destroyed Babylon, and one of his first acts was to permit the Jewish exiles to return to Judea and rebuild the Temple of Solomon—an edict Cyrus recorded in the Cylinder of Cyrus, which is housed in the British Museum. According to Cyrus, he destroyed Babylon and allowed the Jewish people to return to Judea because he "was charged to do so by Yahweh, the God of Heaven."

THE TRANSMITTER TO GOD

The question as to why any particular individual might be chosen to serve as a prophet or messenger of "God" cannot be answered here, though we can certainly consider possibilities.

Every individuals appears to be naturally "wired for god" and thus capable of receiving the word of god. However, some display signs of hyperactivity in this region of the brain which appears to enhance these capabilities—albeit in accordance with the waxing and waning activity within the limbic system and temporal lobe. Or perhaps a person who lives a highly spiritual or mystical life style perpetually activates this region of the brain and achieves what others can only hope for via drugs, fasting, self-mutilation, and isolation; i.e. access to God, or the spiritually sublime.

Or perhaps the presence of "God" triggers hyperactivity in the limbic system of those chosen to be His prophets, which thus enables them to hear and see god-like stimuli as well as causing them to demonstrate signs of temporal lobe abnormalities. That is, just as something frightening or sexual will activate limbic neurons, something exceedingly frightening, spiritual, or god-like, might hyperactivate these same neurons, eventually creating supersensitive conditions and thus perpetually activating the "transmitter to God."

As to the possibility that individuals such as Moses, Jesus, Abraham, Muhammed, or other "prophets" and messengers may have been hallucinating, this is not terribly likely since what they heard and experienced was often different from day to day and as there was an obvious message and plan of action that they were exposed to and which they relayed to others. Moreover, the prophecies they were given and what they were told would come to pass, came to pass!

Indeed, as the limbic system is clearly implicated in every single case, and as this region of the brain normally inhibits and filters incoming sensory stimuli, it could thus be argued that individuals who for whatever reason are "blessed" with a hyperactivated limbic system, or a limbic system which is highly "evolved," or who are directly contacted by god, are therefore provided access to god-like stimuli and alternate realities which are normally filtered from consciousness.

"If the doors of perception were cleansed everything would appear... as it is, infinite..."-W. Blake

Indeed, what the evidence suggests is that these limbic structures periodically become hyperactivated and open up windows and doorways to alternate realities or dimension which are normally hidden from view. Under conditions of limbic system hyperactivity, what is concealed is suddenly revealed.

20. Cosmology, The Uncertainty Principle, Wave Function, Probability, Entanglement, and Multiple Worlds

Abstract:

In quantum mechanics the cosmos as a whole can be likened to a quantum continuum which is continually in flux and is thus indeterminate except at the moment of perception and registration by an observing consciousness or measuring apparatus. Because of this continual fluctuation and the limitations of conscious perceptual capacities and through a phenomenon known as "entanglement," it is only possible to make predictions about what may be observed; and these predictions can only be based on probabilities and a probability distribution. When quantum mechanics are applied to the concept of "time" then what is conceptualized as "past" "present" and "future" is also best described in terms of probabilities. Time is uncertain and not deterministic. Causes may occur simultaneously with or even after effects become effects, as demonstrated by entanglement. The experience of time or the existence of an object in space, are also manifestations of the wave function. As the wave propagates through space it effects the continuum both locally and at a distance simultaneously as demonstrated by entanglement, where even choices made in the future can effect the present. If time has a wave function, then the present can effect not just the future, but the future can effect the present and the past, as time is a continuum. Time is entangled. If considered as a unity with no separations in time and space except at the moment of conscious observation, then to effect one point in time-space is to effect all points which are entangled in spacetime. Time and events occurring in time, through entanglement, and as a manifestation of the quantum continuum, can therefore change the future and the past and events occurring in time simultaneously. The present and the future may change the past as all are interconnected, thereby giving rise to paradoxes where the past may be changed such that it becomes a different past. This paradox, however, can be resolved through Everett's conception of Many Worlds. The past which is changed, is just one past among many. Hence, in terms of the "grandfather" paradox, for example, one may travel back in time but the "grandfather" they kill would not be their "grandfather," but the "grandfather" of their doppelganger who exists in an alternate world as there are innumerable worlds each with their own probable existence and space-time.

TimeSpace in Relativity

In 1904, Lorentz introduced a hypothesis that moving bodies contract in their direction of motion by a factor depending on the velocity of the moving object. Time can therefore also contract such that the future and the present come closer together. He also argued that in different schemes of reference there are different apparent times which differ from and replace "real time." He also argued that the velocity of light was the same in all systems of reference. In 1905 Albert Einstein seized on these ideas and abolished what Lorentz called "real time" and instead embraced "apparent time." In his theories of special relativity, Einstein promoted the thesis that reality and its properties, such as time and motion had no objective "true values" but were "relative" to the observer's point of view (Einstein, 1905a,b,c). Einstein's conceptions of reality and time, therefore, differed significantly from that of Newton.

Time is relative to the observer (Einstein 1905a,b,c, 1906, 1961). Since there are innumerable observers, there is no universal "past, present, future" which are infinite in number and all of which are in motion. There is more than one "present" and this is because time is not the same everywhere for everyone, and differs depending on gravity, acceleration, frames of reference, relative to the observer (Einstein 1907, 1910, 1961). Time is relative and there is no universal past. No universal future. And no universal now. The "past" in another galaxy overlaps with the "present" on Earth. The "present" in another galaxy will not be experienced on Earth until the future. There is no universal now (Einstein 1955). Time is relative, and so too are the futures, presents, and pasts, which overlap and exist simultaneously in different distant regions of space-time. Time is relative, and the "present" for one observer, in one location, may be the past, or the future, for a second observer on another planet.

Time has energy. As defined by Einstein's (1905b) famous theorem $E=mc^2$, and the law of conservation of energy and mass, mass can become energy and energy can become mass. Space-time is both energy and mass which is why it can be warped and will contract in response to gravity and acceleration (Einstein, 1914, 1915a,b; Parker & Toms 2009; Ohanian & Ruffini 2013). Time and space are linked, thereby forming a fourth dimension, timespace. Time, and conceptions about the past, present of future are therefore illusions, as there is no "future" or "past" but rather there are different locations in space which relative to an observer appear far away or nearby. However, when considered from the perspective of quantum mechanics, timespace is a continuum, a unity, and time does not exist independent of this continuum, except as an act of perceptual registration by consciousness or mechanical means.

Einstein's theories did not replace Newtons. Instead Einstein came up with a new closed system of definitions and axioms represented by mathematical symbols which radically different from those of Newton's mechanics. For example, space and time in Newtonian physics are independent, whereas in relativity they are combined and connected by the Lorentz transformations. Moreover, although Newtonian mechanics could be applied to events where velocities are small relative to the velocity of light, Newtonian physics cannot be applied to events which take place near light speeds whereas Einstein's physics can.

By contrast, it is at light speed and beyond, and for objects and particles smaller than atoms where Einstein's theory breaks down and this was recognized in the early 1920s (Born et al. 1925; Heisenberg 1925, 1927). The phenomenon of electricity, electromagnetism and atomic science required a new physics and radically different conceptions of cause, effect, and time.

The Uncertainty Principle: Cause, Effects, Time, and Probability

In 1925 a mathematical formalism called matrix mechanics posed a direct challenge to Newton and Einstein and conceptions of reality (Born et al. 1925; Heisenberg 1925). The equations of Newton were replaced by equations between matrices representing the position and momentum of electrons which were found to be unpredictable. Broadly considered, atoms consist of empty space at the center of which is a positively charged nucleus and which is orbited by electrons. The positive charge of the atom's nucleus determines the number of surrounding electrons, making the atom electrically neutral. However, it was determined that it was impossible to make precise predictions about the position and momentum of electrons based on Newtonian or Einsteinian physics, and this led to the Copenhagen interpretation (Heisenberg 1925, 1927) which Einstein repeatedly attacked because of all the inherent paradoxes. Matrix mechanics is referred to now as quantum mechanics whereas the "statistical matrix" is known as the "probability function;" all of which are central to quantum theory.

As summed up by Heisenberg (1958) "the probability function represents our deficiency of knowledge... it does not represent a course of events, but a tendency for events to take a certain course or assume certain patters. The probability function also requires that new measurements be made to determine the properties of a system, and to calculate the probable result of the new measurement; i.e. a new probability function." Since time is also a property of a system, as events take place in time, then time also, is subject to the probability function.

Quantum physics, as exemplified by the Copenhagen school (Bohr, 1934, 1958, 1963; Heisenberg, 1925, 1927, 1930), like Einsteinian physics, makes assumptions about the nature of reality as related to an observer, the "knower" who is conceptualized as a singularity. As summed up by Heisenberg (1958), "the concepts of Newtonian or Einsteinian physics can be used to describe events in nature." However, because the physical world is relative to being known by a "knower" (the observing consciousness), then the "knower" can influence the nature of the reality which is being observed through the act of measurement and registration at a particular moment in time. Moreover, what is observed or measured at one moment can never include all the properties of the object under observation. In consequence, what is known vs what is not known becomes relatively imprecise (Bohr, 1934, 1958, 1963; Heisenberg, 1925, 1927). Time, therefore, including what is conceptualized as the "now" also becomes imprecise, as well as relative to an observer as predicted by special relativity.

As expressed by the Heisenberg uncertainty principle (Heisenberg, 1927), the more precisely one physical property is known the more unknowable become other properties. The more precisely one property is known, the less precisely the other can be known and this is true at the molecular and atomic levels of reality. Therefore it is impossible to precisely determine, simultaneously, for example, both the position and velocity of an electron at any specific moment in time (Bohr, 1934, 1958, 1963).Time, itself, becomes relativity imprecise even when measured by atomic clocks which slow or speed up depending on gravity and velocity (Ashby 2003, Chou et al. 2010; Hafele & Keating 1972a,b,)--exactly as predicted by Einstein and Lorenz.

Heisenberg's principle of indeterminacy focuses on the relationship of the experimenter to the objects of his scientific scrutiny, and the probability and potentiality, in quantum mechanics, for something to be other than it is. Time, too, therefore, would have potentiality, including what is believed to have occurred in the past (Joseph 2014). Einstein objected to quantum mechanics and Heisenberg's formulations of potentiality and indeterminacy by proclaiming "god does not play dice."

In Einstein's and Newton's physics, the state of any isolated mechanical system at a given moment of time is given precisely. Numbers specifying the position and momentum of each mass in the system are empirically determined at that moment of time of the measurement. Probability never enters into the equation. Therefore, the position and momentum of objects including subatomic particles are precisely located in space and time as designated by a single pair of numbers, all of which can be determined causally and deterministically. However, quantum physics proved that Einstein and Newton's formulation are not true at the atomic and subatomic level (Bohr, 1934, Born et al. 1925; Heisenberg 1925, 1927), whereas experiments with atomic clocks proves that even "moments in time" can vary (Ashby 2003, Chou et al. 2010; Hafele & Keating 1972a,b,).

According to Heisenberg (1925, 1927, 1930), chance and probability enters into the state and the definition of a physical system because the very act of measurement can effect the system. No system is truly in isolation. No system can be viewed from all perspectives in totality simultaneously which would require a god's eye view. Only if the entire universe is included can one apply the qualifying condition of "an isolated system." Simply including the observer, his eye, the measuring apparatus and the object, are not enough to escape uncertainty. Results are always imprecise. Time itself, is relatively imprecise depending on gravity, velocity and the observer's frame of reference.

As determined by Niels Bohr (1949), the properties of physical entities exist only as complementary or conjugate pairs. A profound aspect of complementarity is that it not only applies to measurability or knowability of some property of a physical entity, but more importantly it applies to the limitations of that physical entity's very manifestation of the property in the physical world. Physical reality is defined by manifestations of properties which are limited by the interactions and trade-offs between these complementary pairs at specific moments in time when those moments are also variable. For example, the accuracy in measuring the position of an electron at a specific moment in time requires a complementary loss of accuracy in determining its momentum; and momentum can contract time and the distance between the present and the future. Precision in measuring one pair is complimented by a corresponding loss of precision in measuring the other pair (Bohr, 1949, 1958, 1963); which in turn may be related to variations and fluctuations in time. The ultimate limitations in precision of property manifestations are quantified by Heisenberg's uncertainty principle and matrix mechanics. Complementarity and Uncertainty dictate that all properties and actions in the physical world are therefore non-deterministic to some degree--and the same applies to time and even what is considered cause and effect.

Bohr (1949) holds that objects governed by quantum mechanics, when measured, give results that depend inherently upon the type of measuring device used, and must necessarily be described in classical mechanical terms since the measuring devices functions according to classical mechanics. The measuring device effects the outcome and the interpretation of that outcome as does the observer using that device. "This crucial point...implies the impossibility of any sharp separation between the behaviour of atomic objects and the interaction with the measuring instruments which serve to define the conditions under which the phenomena appear...." (Bohr 1949). Time, however, is also determined by measuring devices, which may fluctuate depending on gravity and velocity, including the velocity of the object being measured-- exactly as predicted by relativity.

Evidence obtained under a single or under different experimental conditions cannot be reduced to a single picture, "but must be regarded as complementary in the sense that only the totality of the phenomena exhausts the possible information about the objects." In consequence, the results must be viewed in terms of probabilities when applied to the nature of the object under study and its current and future behaviors in time. Bohr (1949) called this the principle of complementarity, a concept fundamental to quantum mechanics and closely associated with the Uncertainty Principle. "The knowledge of the position of a particle is complementary to the knowledge of its velocity or momentum." If we know the one with high accuracy we cannot know the other with high accuracy at the same time (Bohr, 1949, 1958, 1963; Heisenberg, 1927, 1955, 1958); and this is also because, there is no such thing as "the same time."

Central to the Copenhagen principle is the wave function and the probability distribution, i.e. the results of any experiment can only be stated in terms of the probability that the momentum or position of the particles under observation may assume certain values at a specific time. The probability distribution is a prediction for what may occur in the future, that is, within a predicted range of probabilities. When the experiments are performed many times, and although subsequent observations may differ, they are expected to fall within the predicted probability distribution. This also means that nothing is precisely determined at any particular moment in time (Bohr, 1949, 1963; Heisenberg, 1927, 1930, 1955).

Time and the measuring devices used to calculate time, are relative, and even moments in time may be stretched or contracted relative to an observer's frame of reference. There is no universal now. Thus, even what is described as "now" or the future or the past, must also be subject to a probability function. Time cannot be known precisely, even when measured by atomic clocks (Ashby 2003, Chou et al. 2010; Hafele & Keating 1972a,b,). Thus, even what is considered cause and effect" must be subject to a probability function as the moments embracing the "cause" may overlap and occur simultaneously with or even preceded the "effect" due to the stretching and contraction of local time.

These are not just thought experiments. There is considerable evidence of what Einstein (1955) referred to as "spooky action at a distance" and what is known in quantum physics as "entanglement" (Plenio 2007; Juan et al. 2013; Francis 2012). It is well established that causes and effects can occur simultaneously and ever faster than light speed (Lee et al. 2011; Matson 2012; Olaf et al. 2003); a consequence of the connectedness of all things in the quantum continuum.

For example, photons are easily manipulated and preserve their coherence for long times and can be entangled by projection measurements (Kwiat et al. 1995; Weinfurter 1994). A pump photon, for example, can split light into two lower- energy photons while preserving momentum and energy, and these photons remained maximally entangled although separated spatially (Goebel et al 2008; Pan et al. 1998). However, entanglement swapping protocols can entangle two remote photons without any interaction between them and even with a significant time-like separation (Ma et al., 2012; Megidish et al. 2013; Peres 2000). In one set of experiments entanglement was demonstrated even following a delayed choice and even before there was a decision to make a choice. Specifically, four photons were created and two were measured and which became entangled. However, if a choice was then made to measure the remaining two photons, all four became entangled before it was decided to do a second measurement (Ma et al., 2012; Peres 2000). Entanglement can occur independent of and before the act of measurement. "The time at which quantum measurements are taken and their order, has no effect on the outcome of a quantum mechanical experiment" (Megidish et al. 2013).

Moreover, "two photons that exist at separate times can be entangled" (Megidish et al. 2013). As detailed by Megidish et al (2013): "In the scenario we present here, measuring the last photon affects the physical description of the first photon in the past, before it has even been measured. Thus, the "spooky action" is steering the system's past. Another point of view...is that the measurement of the first photon is immediately steering the future physical description of the last photon. In this case, the action is on the future of a part of the system that has not yet been created."

Hence, entanglement between photons has been demonstrated even before the second photon even exists; "a manifestation of the non-locality of quantum mechanics not only in space, but also in time" (Megidish et al 2013). In other words, a photon may become entangled with another photon even before that photon is created, before it even exists. Even after the first photon ceases to exist and before the second photon is created, both become entangled even though there is no overlap in time. Photons that do not exist can effect photons which do exist and photons which no longer exist and photons which will exist (Megidish et al. 2013); and presumably the same applies to all particles, atoms, molecules (Wiegner, et al 2011).

As demonstrated in quantum physics, the act of observation, measurement, and registration of an event, can effect that event, causing a collapse of a the wave function (Dirac 1966a,b; Heisenberg 1955), thereby registering form, length, shape which emerges like a blemish on the face of the quantum continuum. Likewise, a Time Traveler or particle/object speeding toward and then faster than light and from the future into the past will affect the quantum continuum. By traveling into the future or the past, the Time Traveler will interact with and alter every local moment within the quantum continuum and thus the future or the past.

Entanglement proves that effects may precede causes, and causes and effects may also take place simultaneously. In the quantum continuum, determinism and causes and effects do not always exist and this is because, as Einstein proclaimed: "The distinction between past, present and future is only an illusion."

In quantum mechanisms, although every deterministic system is a causal system, not every causal system is deterministic (Heisenberg (1925, 1927; 1958). Rather, causality is the relationship between different states of the same object at different times whereas what is "deterministic" relates to what may occur, and is better described in terms of probabilities.

According to the Copenhagen interpretation (Bohr, 1949, 1963; Heisenberg, 1958), it is the act of measurement which collapses the wave function. It is also the measurement and observation of one event which triggers the instantaneous alteration in behavior of another event or object at faster than light speeds; i.e. entanglement (Plenio 2007; Juan et al. 2013; Francis 2012). For example, two particles which are far apart have "spin" and they may spin up or down. However, although they are far apart, an observer who measures and verifies the spin of particle A will at the same time effect the spin of particle B, as verified by a second observer. Measuring particle A, effects particle B and changes its spin. Likewise observing the spin of B determines the spin of A. There is no temporal order as the spin of one effects the spin of the other simultaneously, faster than the speed of light. Even distant objects are entangled and have a symmetrical relationship and a constant conjunction (Bokulich & Jaeger, 2010; Plenio 2007; Sonner 2013).

Because the future can effect the past or present, the relationship of cause and effect and energy or mass over time is uncertain and can be described only by probabilities (Born et al 1925, Heisenberg 1925, 1927). Time is uncertain. Temporal succession may have no probable connection with what precedes or follows (Heisenberg 1958). In quantum mechanics, one can know the connection between two events only by knowing the future state--thus one must wait for the future to arrive, or look back upon the future state of similar systems in the past. If one knows the properties of an acorn at an earlier time t1 one still cannot deduce the properties of the oak tree at time t2. This may be possible only in isolated systems (Bohr, 1949; Heisenberg 1958). Thus time must also be isolated. However, unless the entire universe is included in the measurement, then the system, which includes time, is not truly isolated.

The Probability and Wave Function

Quantum mechanics is mechanical but not deterministic and causal relationships are never teleological and not always deterministic. In quantum physics, nature and reality are represented by the quantum state. The electromagnetic field of the quantum state is the fundamental entity, the continuum that constitutes the basic oneness and unity of all things. The physical nature of this state can be "known" by assigning it mathematical properties and probabilities (Bohr, 1958, 1963; Heisenberg, 1927). Therefore, abstractions, i.e., numbers and probabilities become representational of a hypothetical physical state. Because these are abstractions, the physical state is also an abstraction and does not possess the material consistency, continuity, and hard, tangible, physical substance as is assumed by Classical (Newtonian) physics. Instead, reality, the physical world, is a process of observing, measuring, and knowing and is based on probabilities and the wave function (Heisenberg, 1955).

Consider an elementary particle, once its positional value is assigned, knowledge of momentum, trajectory, speed, and so on, is lost and becomes "uncertain." The particle's momentum is left uncertain by an amount inversely proportional to the accuracy of the position's measurement which is determined by values assigned by measurement and the observing consciousness at a specific moment in time relative to that observe and the measuring device. Therefore, the nature of reality, and the uncertainty principle is directly affected by the observer and the process of observing, measuring, and knowing, all of which are variable thereby making the results probable but not completely certain (Heisenberg, 1955, 1958):

"What one deduces from an observation is a probability function; which is a mathematical expression that combines statements about possibilities or tendencies with statements about our knowledge of facts....The probability function obeys an equation of motion as the coordinates did in Newtonian mechanics; its change in the course of time is completely determined by the quantum mechanical equation but does not allow a description in both space and time" (Heisenberg, 1958).

"The probability function does not describe a certain event but a whole ensemble of possible events" whereas "the transition from the possible to the actual takes place during the act of observation... and the interaction of the object with the measuring device, and thereby with the rest of the world... The discontinuous change in the probability function... takes place with the act of registration, because it is the discontinuous change of our knowledge in the instant of registration that changes the probability function." "Since through the observation our knowledge of the system has changed discontinuously, its mathematical representation has also undergone the discontinuous change and we speak of a quantum jump" (Heisenberg, 1958).

Einstein ridiculed these ideas: "Do you really think the moon isn't there if you aren't looking at it?"

Heisenberg (1958), cautioned, however, that the observer is not the creator of reality: "Quantum theory does not introduce the mind of the physicist as part of the atomic event. But it starts from the division of the world into the object and the rest of the world. What we observe is not nature in itself but nature exposed to our method of questioning." Nevertheless, the act of knowing, of observing, or measuring, that is, interacting with the environment in any way, creates an entangled state and a knot in the quantum continuum described as a "collapse of the wave function;" a knot of energy that is a kind of blemish in the continuum of the quantum field. This quantum knot bunches up at the point of observation, at the assigned value of measurement and can be entangled.

The same principles would also apply to time, and to time travel. The act of moving through time would effect time and all local and even more distant events. Traveling through the past or the future would effect every moment of that future; however, exactly what those changes may be, are indeterministic and can only be described by a probability function.

In the Copenhagen model, objects are viewed as quantum mechanical systems which are best described by the wave function and the probability function. "The reduction of wave packets occurs when the transition is completed from the possible to the actual" (Heisenberg, 1958).

The measuring apparatus and the observer also have a wave function and therefore interact with what is being measured. The effect of this is obvious when its a macro-structure measuring a micro-structure vs a macro-structure measuring a macro-structure.

Moreover, according to the uncertainty principle, it is not possible to restrict any analysis to position or moment without effecting the other, and this is because the very act of eliminating uncertainty about position maximizes uncertainty about momentum (Heisenberg 1927). Uncertainty implies entanglement. Likewise, eliminating uncertainty about momentum maximizes uncertainty about position. Instead, one must assign a probability distribution which assigns probabilities to all possible values of position and momentum.

Therefore, no object, or particle, or quanta, or quantum, or moment in time, has its own eigenstate (inherent characteristic). Although every object appears to have a definite momentum, a definite position, and a definite time of occurrence, the object is in flux and it can't have a position and momentum at the same time as there is no such thing as "the same time." Time is also in flux. Therefore, when applied to time, then time, including the future and the past, can only be defined by a probability function. This means, the future and the past may change and that whatever is believed to have taken place or which will take place is best described in terms of probabilities.

Time and Quantum Physics: The Future Can Lead to the Past

In contrast to Newton and Einstein, quantum mechanics concerns itself with the dynamical change of state and its probability coupled with the Schrödinger (1926) time equations which are both time dependent and time independent for particles and waves. The state-function specifies the state of any physical system as a specific time t. The Schrödinger time equations relates states at a series time t1 to a later time t2. In quantum mechanics, the Schrödinger (1926) equation is a partial differential equation that describes how the quantum state of a physical system changes with time. Like Newton's second law (F = ma), the Schrödinger equation describes time in a way that is not compatible with relativistic theories, but which supports quantum mechanics and which can be easily mathematically transformed into Heisenberg's (1925) matrix mechanics, and Richard Feynman's (2011) path integral formulation.

Therefore, time, in quantum physics, is not necessarily relative or even a temporal sequence, and the same is true of future and past. As summed up by Heisenberg (1958), "in classical theory we assume future and past are separated by an infinitely short time interval which we may call the present moment. In the theory of relativity we have learned that the future and past are separated by a finite time interval the length of which depends on the distance from the observer..." and where the past always leads to the future. However, "when quantum theory is combined with relativity, it predicts time reversal;" i.e. the future can lead to the past.

Time Is Entangled

Time cannot be separate from the continuum except when perceived as such by an observing consciousness or measuring device, thereby inducing a collapse of the wave function of time; experienced as the present, past, or future.

Time, be it considered a dimension known as timespace, or as a perceived aspect of the quantum continuum, is also subject to entanglement, as all aspects of time are interconnected and indistinguishable until perceived thereby inducing a collapse of the wave function. "A" future can therefore effect "a" past and change it through entanglement and by effecting the wave function.

Given faster than light entanglement, spooky action at a distance and the reality of the wave function, then the laws of physics must allow for information and effects to be conveyed faster than light speed and from the future to the past. If time is considered as a gestalt and a continuum and not a series of fragments, then the future and past are coextensive.

The quantum continuum is without dimensions and encompasses space and time in its basic unity of oneness. Everything within the quantum continuum can be effected by local effect and distant effects simultaneously at and beyond light speeds. Therefore, the future, and the "present" being part of this continuum can effect the past by effecting the wave function of the past, present, future, and thus, the space-time continuum, as all are entangled.

Light can travel to the future and from the past relative to the observer's frame of reference. However, light and time are not the same. The speed of light, and time, be it past or future, are not synonymous, though both may be affected by gravity (Carroll 2004; Einstein 1961). Even the ticking of atomic clocks is effected by gravity as well as velocity. Time is subject to change, including what is described as "now" as there is no universal "now." Moreover, just as light has a particle-wave duality and can physically interact with various substances, time also can be perceived and therefore must have a wave function if not a particle-wave duality. Time, be it "past" "present" or "future" can be changed.

Time-space is interactional, and can contract to near nothingness and then continue to contract in a negative direction such that the time traveler can journey into the past.

Gravity, Acceleration, Relativity, and the Quantum Mechanics of Time Contraction

Time has energy. As defined by the law of conservation of energy and mass and Einstein's (1905b) theorem $E=mc^2$, mass can become energy and energy can become mass. Space-time is both energy and mass which is why it will contract in response to gravity and acceleration (Einstein, 1914, 1915a,b; Parker & Toms 2009; Ohanian & Ruffini 2013).

Time is perceived. Time is experienced. Time is "something," it exists, and therefore it must have energy and a wave function which is entangled with motion, velocity, gravity, the observer, and the quantum continuum which encompasses space-time.

Time is associated with light (Einstein 1961). Light has a particle-wave duality and travels at a maximum velocity of 186282 miles per second. However, time is not light, and light is not time. Rather, light can carry images reflected by or emitted from innumerable locations in space-time and can convey or transport information from these locations which may be perceived by an observer and experienced as moments in time. For much of modern human history time has been measured by celestial clocks such as the phases of the moon, and the tilt and rotation of Earth and Earth's orbit around the sun which marks the four seasons and the 24 hour day (Joseph, 2011b). Time is a circle and may be segmented into years, months, weeks, days, hours, minutes, seconds, nanoseconds as measured by various clocks from sundials to atomic clocks. However, time, even when measured by atomic clocks, can flow at different rates and speeds such that the "future" and the "past" can overlap and exist simultaneously with the same moment in time, and this is because there is no universal now.

Atomic clocks tick off time as measured by the vibrations of light waves emitted by atoms of the element cesium and with accuracies of billionths of a second (Essen & Parry, 1955). However, these clocks are also effected by their surroundings and run slower under conditions of increased gravity or acceleration (Ashby 2003; Hafele & Keating 1972a,b) In 1971 Joe Hafele and Richard Keating placed atomic clocks on airplanes traveling in the same direction of Earth's rotation thereby combining the velocity of Earth with the velocity of the planes (Hafele & Keating 1972a,b). All clocks slowed on average by 59 nanoseconds compared to atomic clocks on Earth. Time, like the weather, is effected by local conditions. Under accelerated conditions and increased gravity, time slows down; the same conditions which would enable a time traveler to accelerate toward the future and from the future into the past.

It has been demonstrated that atomic clocks at differing altitudes will eventually show different times; a function of gravitational effects on time. The lower the altitude the slower the clock, whereas clocks speed up as altitude increases; albeit the differences consisting of increases of a few nanoseconds (Chou et al. 2010; Hafele & Keating, 1972; Vessot et al. 1980). "For example, if two identical clocks are separated vertically by 1 km above the surface of Earth, the higher clock gains the equivalent of 3 extra seconds for each million years (Chou et al., 2010). The speeding up of atomic clocks at increasingly higher altitudes has been attributed to a reduction in gravitational potential which contributes to differential gravitational time dilation.

A predicted by Einstein, clocks run more slowly (time contraction) near massive objects whereas time dilates and runs more quickly as gravity is reduced. Increases in altitude and reductions in gravity speed up the clock, whereas decreases in altitude and increases in gravity slow the clock down (Hafele & Keating, 1972; Vessot et al. 1980).

Time must have energy and energy can be converted into mass. Acceleration expands mass (as energy is converted to mass) and increases gravity which contracts time and mass. Increases in gravity can squeeze space-time into smaller spaces such that there is more time in a smaller space. According to Einstein's famous equation: $E = mc^2$, where E is energy, m is mass and c is the speed of light, mass and energy are the same physical entity and can be changed into each other (Einstein 1905a,b,c 1961). Because of this equivalence, the energy an object acquires due to its motion will increase its mass. In other words, the faster an object moves, the greater the amount of energy which increases its mass, since energy can become mass. This increase in mass only becomes noticeable when an object moves very rapidly. If it moves at 10% the speed of light, its mass will only be 0.5 percent more than normal. But if it moves at 90% the speed of light, its mass will double. And as mass increases it also shrinks and its gravity increases. This is because increased mass increases gravity which then pulls on the mass making it shrink toward the center of gravity, all of which contributes to the collapsing and contraction of space time (Carroll 2004; Einstein 1913, 1914, 1915a,b).

A similar principle applies to time travel. By accelerating toward light speed, space-time contracts (Lorentz 1982; Einstein 1961; Einstein et al. 1923), and the distance between the future and the present and distant locations in space time shrinks and are closer together.

Speed, that is velocity, per se is not effected by time travel. Velocity does not contract or dilate. Hence, since space-time contracts as one accelerates (and although time slows down), and as velocity is not effected then one can traverse and journey across this shrinking space more quickly, and cover the distance between the "now" and the "future" more rapidly because they are closer together--and this would be possible only if the "future" already exists, albeit in a different location in spacetime. Distant locations in space-time are no longer so far apart; the result of increased speed and gravity.

The relationship between time dilation and the contraction of the length of space-time can be determined by a formula devised by Hendrik Lorentz in 1895. As specified by the Lorentz factor, γ (gamma) is given by the equation γ = , such that the dilation-contraction effect increases exponentially as the time traveler's velocity (v) approaches the speed of light c. Therefore, for example, at 90% light speed 2.29 days on Earth shrinks to just one day in the time machine and 7 days in the time machine at this speed, would take the time traveler 16 days into the future. The distance between the present and the future has contracted so that the future arrives in 7 days instead of 16.

Consider for example, 30 feet of space which contracts to 10 feet. Those inside the time machine need only walk 10 feet whereas those outside the time machine must walk 30 feet. Likewise because the time traveler's clock runs more slowly, and since more time is contracted into a smaller space, it might take him 10 minutes to get 30 minutes into the future. By contrast, it takes those outside the time machine longer to get to the future because it is further away and as their clocks are running faster and it takes more time. At 99.999999% the speed of light, almost two years pass for every day in the time machine. At 99.99999999999999 % of c, for every day on board, nearly twenty thousand years pass back on Earth. However, upon reaching light speed, time stops. It is only upon accelerating beyond light speed, that time runs backwards and the contraction of space-time continues in a negative direction. One must accelerate toward the future to reach the past.

The shrinkage of space-time has given rise to the famous "twin paradox" (Langevin 1911; von Laue 1913). If one twin leaves Earth and accelerates toward light speed, that twin will arrive in the future in less time than the twin left behind on Earth. Because it took less time, the time traveling twin does not age as much whereas the twin left on Earth ages at the normal rate. Because time-space has contracted, and since it takes less time to get to distant locations which are now closer together, the time traveling twin arrives in the future in less time than her twin on Earth. Hence, the time traveling twin will be younger.

Not just spacetime, but the mass of the object traveling toward light speed also contracts. The amount of length contraction can be calculated and determined by the Lorentz Transforms (Einstein 1961). For example, a 100 foot long time-space ship traveling at 60% the speed of light would contract by 20% and would become 80 feet in length. Presumably, its diameter would remain the same, though the likelihood is that all surrounding space including the diameter of the time machine would contract. If the time-space ship accelerates to 0.87 light speed, it will contract by 50%.

"Length contraction" can be expressed mathematically by the following formula: $E = mc^2/\sqrt{(1-v^2/c^2)}$, which is similar to the equation for time dilation (if one replaces the value of v for 0). As the value of v (velocity) increases, so does an object's mass which requires more energy to continue at the same velocity or to accelerate. Since energy can become mass, mass increases even as the object shrinks and contracts, thereby increasing its gravity which exerts local effects on the curvature of space-time. Not just the time machine, but space-time in front and surrounding the time machine also contracts. Eventually, the time traveler may shrink to the less than the width of a hair--at least from the perspective of outside observers. At near light speed, the time traveler's length would contract to the size of an atom. Once it shrinks in size smaller than a Planck Length, it will have so much mass and energy that it can blow a hole in spacetime and be propelled at superluminal speeds (Joseph 2014)--however once it exceeds light speed, length contraction and the contraction of time continues in a negative direction. Time reverses, and the direction of travel is into the past. One must accelerate to light speed, which takes the time traveler far into the future, and then to superluminal speeds to journey backwards in time, and this means the future leads to the past.

Although seemingly paradoxical, Einstein's theories of relativity (despite his posting of a cosmic speed limit) predicts that the only way to travel into the past is to exceed the speed of light. Upon accelerating toward light speed, space-time contracts and the space-time traveler is propelled into the future. However, it is only upon accelerating into the future and then beyond light speed that the contraction of space-time continues in a negative direction and time flows in reverse. It is only at superluminal speeds that time reverses and one can voyage backward in time. Einstein's general theory of relativity predicts that the future leads to the past. Likewise, as shown by Gödel 1949a,b), Einstein's field equations predict that time is a circle; and this violates the laws of causality (Buser et al. 2013).

Because the present leads to the future which leads to the past, past, present and future are linked in spacetime. The future can therefore effect the past and effects may take place before the cause.

Time, and time-space are embedded in the quantum continuum and can effect as well as be effected by other particle-waves even at great distances; a concept referred to as "entanglement." Time and space-time are entangled.

Probabilities and The Wave Function of the Time Traveler

According to quantum mechanics the subatomic particles which make up reality, or the quantum state, do not really exist, except as probabilities (Born et al. 1925; Dirac 1966a,b; Heisenberg 1925, 1927). These "subatomic" particles have probable existences and display tendencies to assume certain patterns of activity that we perceive as shape and form. Yet, they may also begin to display a different pattern of activity such that being can become nonbeing and thus something else altogether.

The conception of a deterministic reality is rejected and subjugated to mathematical probabilities and potentiality which is relative to the mind of a knower which registers that reality as it unfolds, evolves, and is observed (Bohr 1958, 1963; Heisenberg 1927, 1958). That is, by measuring, observing, and the mental act of perceiving a non-localized unit of structural information, injects that mental event into the quantum state of the universe, causing "the collapse of the wave function" and creating a bunching up, a tangle and discontinuous knot in the continuity of the quantum state.

Therefore, quantum mechanics, as devised by Niels Bohr, Werner Heisenberg, Dirac, Born and others in the years 1924–1930, does not attempt to provide a description of an overall, objective reality, but instead is concerned with quanta, probabilities and the effects of an observer on what is being observed. The act of measurement causes what is being measured to assume one for many possible values at specific moments of time, and yields the probability of an object or particle to be moving at one speed or direction or to be in one position or location, vs many others at a specific moment in time. Thus, it could be said that the act of observation causes a wave function collapse, a discontinuity in the continuum which is interpreted as reality and cause and effect. However, time too, is subject to measurement and can therefor yield different values by being measured. Observing and measuring time causes time to have certain values.

Central to quantum mechanics is the wave function (Bohr, 1963; Heisenberg, 1958). All of existence has a wave function, including light and time. However, quantum physics is also based on the fact that matter appears to be a duality, and can be both a wave and a particle; that is, to have features of both, i.e. particle-like properties and wave-like properties (Niel Bohr's complementary principle). Therefore, every particle has a wave function which describes it and which can be used to calculate the probability that a particle will be in a certain location or in a specific state of motion, but not both at certain moment of time. Again, however, time also has a wave function. Every aspect of existence can be described as sharing particle-like properties and wave-like properties and this would necessarily have to include the experience of time. Time can be perceived, therefore time must have energy, and energy has a particle wave duality.

The wave function is the particle spread out over space and describes all the various possible states of the particle. Likewise, the wave function would describe all the various possible states of time, including past, present, and future. According to quantum theory the probability of findings a particle in time or space is determined by the probability wave which obeys the Schrodinger equation. Everything is reduced to probabilities, including time. Moreover, these particle/waves and these probabilities are entangled.

Reality and the experience of time, are manifestation of wave functions and alterations in patterns of activity within the quantum continuum which are entangled and perceived as discontinuous, and that includes the perception of past, present, future. The perception of a structural unit of information is not just perceived, but is inserted into the quantum state which causes the reduction of the wave-packet and the collapse of the wave function. It is this collapse which describes shape, form, length, width, and future and past events and locations within space-time (Bohr, 1963; Heisenberg, 1958).

In quantum physics, the wave function describes all possible states of the particle and larger objects, including time, thereby giving rise to probabilities, and this leads to the "Many Worlds" interpretation of quantum mechanics (Dewitt, 1971; Everett 1956, 1957). That is, since there are numerous if not infinite probable outcomes, each outcome and probable outcome represents a different "world" with some worlds being more probable than others and each of which may be characterized by their own unique moments in time. "Many Worlds" must include "Many Times."

For example, an electron may collide with and bounce to the left of a proton on one trial, then to the right on the next, and then at a different angle on the third trial, and another angle on the fourth and so on, even though conditions are identical with one exception: they occur at different moments in time. This gives rise to the Uncertainty Principle and this is why the rules of quantum mechanics are indeterministic and based on probabilities. The state of a system one moment cannot determine what will happen the next moment, because moments in time, and thus time itself has a wave function and a probability function. The wave function describes all the various possible states of the particle (Bohr, 1963; Heisenberg, 1958) and that includes the experience of time, including the eternal now.

Wave Functions: The Past, Present and Future Exist Simultaneously

Only when the object can be assigned a specific value as to location, or time, or moment, does it have possess an eigenstate, i.e. an eigenstate for position, or an eigenstate for momentum, or an eigenstate for time; each of which is a function of the "reduction of the wave function;" also referred to as wave function collapse (Bohr, 1934, 1958, 1963; Heisenberg, 1930, 1955, 1958). Wave function collapse, which is indeterministic and non-local is a fundamental a priori principle of the Copenhagen school of quantum physics and so to is the postulate that the observer and the observed, and the past, present, and future, become entangled and effect one another.

Wave function collapse has also been described as "decoherence" which in turn leads to the "many-worlds" interpretation and the thought experiment known as "Schrödinger's Cat'" i.e. is a cat in a sealed box dead or alive? According to the Copenhagen interpretation, there is a 50% chance it will be dead and 50% chance it will be alive when it is observed, but one cannot know if it dead or alive until observed (measured). However, if there are two observers, one in the box with the cat the other outside the box, then the observer in the box knows if the cat is dead or alive, whereas the observer outside the box sees only a 50-50 probability (Heisenberg 1958).

The wave function describes all the various possible states of the particle. Rocks, trees, cats, dogs, humans, planets, stars, galaxies, the universe, the cosmos, past, present, future, as a collective, all have wave functions.

Waves can also be particles, thereby giving rise to a particle-wave duality and the Uncertainty Principle. Particle-waves interact with other particle-waves. The wave function of a person sitting on their rocking chair would, within the immediate vicinity of the person and the chair, resemble a seething quantum cloud of frenzied quantum activity in the general shape of the body and rocking chair. This quantum cloud of activity gives shape and form to the man in his chair, and is part of the quantum continuum, a blemish in the continuum which is still part of the continuum and interacts with other knots of activity thus giving rise to cause and effect as well as violations of causality: "spooky action at a distance."

Since mass can become energy and energy mass, the "field" is therefore a physical entity that contains energy and has momentum which can be transmitted across space. Likewise, since time can be perceived it must have energy and energy mass as well as momentum which can be transmitted across space. Therefore, "action at a distance" may be both distant and local, a consequence of the interactions of these charges within the force field they create in conjunction with the force field know as "time."

Because time has a wave function which interacts with the continuum which includes time, then effects can be simultaneous, even at great distances, and occur faster than the speed of light (Plenio 2007; Juan et al. 2013; Francis 2012; Schrödinger & Dirac 1936), effecting electrons, photons, atoms, molecules and even diamonds (Lee et al. 2011; Matson 2012; Olaf et al. 2003; Schrödinger & Born 1935). Since time has a wave function and is entangled, then effects may precede the cause since time is a continuity, and this explains why effects may take place faster than light.

If considered as a unity with no separations in time and space, then to effect one point in time-space is to effect all points which are entangled; and those entangled connections includes time and consciousness (Joseph 2010a). And this gives rise to the uncertainty principle because all are interactional (Heisenberg, 1927) and there is no universal "now." Everything effects everything else and thus time in the "future" can effect "time" in the "past" via the wave function which propagates instantaneously throughout the continuum.

Likewise, the intrepid time traveler, journeying into the past, is also a wave function; consisting of particles and waves which interact locally with other local waves and creating additional blemishes in the quantum continuum. By traveling into the past or the future, the time traveler would come into contact with and change and alter the wave function of other blemishes in space-time. Hence, speeding into the past would therefore change the past, or rather, local events in that past, even if the Time Traveler sat still and did nothing at all except go with the flow. The wave function of the observer effects the wave function of what is observed and the wave function of immediate surroundings. The backward traveling time traveler effects each moment of "local" space-time as she travels through it. The wave function of the time traveler moving through time would spread out over space, becoming vanishingly small until disappearing.

By traveling into the past, the time traveler changes the past locally and perhaps even at a distance, depending on his actions. Likewise, since the future, present, and past are entangled, events taking place in the future, can effect and alter the past, thereby violating causality such that the past the time traveler visits may no longer be the past he was familiar with.

The probability function and entanglement when applied to the space-time continuum indicates that the, or rather "a" past may be continually changed and altered to varying degrees. This may also explain why memories of the past do not always correspond with the past record (Haber & Haber, 2000; Megreya & Burton 2008). Although blamed on faulty memory, perhaps the past has been and is continually and subtly being altered through entanglement.

As demonstrated in quantum physics, the act of observation, measurement, and registration of an event, can effect that event, causing a collapse of a the wave function (Dirac 1966a,b; Heisenberg 1955),. Likewise, a Time Traveler or particle/object speeding toward and then faster than light and from the future into the past will affect the quantum continuum. By traveling into the future or the past, the Time Traveler will interact with and alter every local moment within the quantum continuum and thus the future or the past. However, the past which is changed, always existed, albeit, as a probability; one past world among infinite worlds each with their own past, presents, and futures.

Everett's Many Worlds

Since the universe, as a collective, must also have a wave function, then this universal wave function would describe all the possible states of the universe and thus all possible universes, which means there must be multiple universes which exist simultaneously as probabilities (Dewitt, 1971; Everett 1956, 1957). And the same would be true of time. Why shouldn't time have a wave function?

The wave function of time means there are infinite futures, presents, pasts, with some more probable than others.

As theorized by Hugh Everett the universal wave function is "the fundamental entity, obeying at all times a deterministic wave equation" (Everett 1956). Thus, the wave function is real and is independent of observation or other mental postulates (Everett 1957), though it is still subject to quantum entanglement.

In Everett's formulation, a measuring apparatus MA and an object system OS form a composite system, each of which prior to measurement exists in well-defined (but time-dependent) states. Measurement is regarded as causing MA and OS to interact. After OS interacts with MA, it is no longer possible to describe either system as an independent state. According to Everett (1956, 1957), the only meaningful descriptions of each system are relative states: for example the relative state of OS given the state of MA or the relative state of MA given the state of OS. As theorized by Hugh Everett what the observer sees, and the state of the object, become correlated by the act of measurement or observation; they are entangled.

However, Everett reasoned that since the wave function appears to have collapsed when observed then there is no need to actually assume that it had collapsed. Wave function collapse is, according to Everett, redundant. Thus there is no need to incorporate wave function collapse in quantum mechanics and he removed it from his theory while maintaining the wave function, which includes the probability wave.

According to Everett (1956) a "collapsed" object state and an associated observer who has observed the same collapsed outcome have become correlated by the act of measurement or observation; that is, what the observer perceives and the state of the object become entangled. The subsequent evolution of each pair of relative subject–object states proceeds with complete indifference as to the presence or absence of the other elements, as if wave function collapse has occurred. However, instead of a wave function collapse, a choice is made among many possible choices, such that among all possible probable outcomes, the outcome that occurs becomes reality.

Everett argued that the experimental apparatus should be treated quantum mechanically, and coupled with the wave function and the probable nature of reality, this led to the "many worlds" interpretation (Dewitt, 1971). What is being measured and the measuring apparatus/observer are in two different states, i.e. different "worlds." Thus, when a measurement (observation) is made, the world branches out into a separate world for each possible outcome according to their probabilities of occurring. All probable outcomes exist regardless of how probable or improbable, and each outcome represent a "world." In each world, the measuring apparatus indicates which of the outcomes occurred, which probable world becomes reality for that observer; and this has the consequence that later observations are always consistent with the earlier observations (Dewitt, 1971; Everett 1956, 1957).

Predictions, therefore, are based on calculations of the probability that the observer will find themselves in one world or another. Once the observer enters the other world he is not aware of the other worlds which exist in parallel.

Moreover, if he changes worlds, he will no longer be aware that the other world existed (Everett 1956, 1957): all observations become consistent, and that includes even memory of the past which existed in the other world.

The "many worlds" interpretation (as formulated by Bryce DeWitt and Hugh Everett), rejects the collapse of the wave function and instead embraces a universal wave function which represents an overall objective reality which consists of all possible futures and histories all of which are real and which exist as alternate realities or in multiple universes. What separates these many worlds is quantum decoherence and not a wave form collapse. Reality, the future, and the past, are viewed as having multiple branches, an infinite number of highways leading to infinite outcomes. Thus the world is both deterministic and non-deterministic (as represented by chaos or random radioactive decay) and there are innumerable futures and pasts.

As described by DeWitt and Graham (1973; Dewitt, 1971), "This reality, which is described jointly by the dynamical variables and the state vector, is not the reality we customarily think of, but is a reality composed of many worlds. By virtue of the temporal development of the dynamical variables the state vector decomposes naturally into orthogonal vectors, reflecting a continual splitting of the universe into a multitude of mutually unobservable but equally real worlds, in each of which every good measurement has yielded a definite result and in most of which the familiar statistical quantum laws hold."

DeWitt's many-worlds interpretation of Everett's work, posits that there may be a split in the combined observer–object system, the observation causing the splitting, and each split corresponding to the different or multiple possible outcomes of an observation. Each split is a separate branch or highway. A "world" refers to a single branch and includes the complete measurement history of an observer regarding that single branch, which is a world unto itself. However, every observation and interaction can cause a splitting or branching such that the combined observer–object's wave function changes into two or more non-interacting branches which may split into many "worlds" depending on which is more probable. The splitting of worlds can continue infinitely.

Since there are innumerable observation-like events which are constantly happening, there are an enormous number of simultaneously existing states, or worlds, all of which exist in parallel but which may become entangled; and this means, they can not be independent of each other and are relative to each other. This notion is fundamental to the concept of quantum computing.

Likewise, in Everett's formulation, these branches are not completely separate but are subject to quantum interference and entanglement such that they may merge instead of splitting apart thereby creating one reality.

Changing the Past: Paradoxes and the Principle of Consistency

Entanglement and "spooky action at a distance" prove that effects can occur faster than the speed of light (Lee et al. 2011; Matson 2012; Olaf et al. 2003), such that effects may take place simultaneously with or before the cause, such that the effect causes itself and may be responsible for the "cause;" a consequence of entanglement in the quantum continuum Likewise, a Time Travel can also effect the present and change the future of the past, or rather, "a" past or "a" future.

Since the time traveler and his time machine are comprised of energy and matter their presence and movement through time-space will also warp and depress the geometry of space-time thereby creating local and distant effects. Time travel would effect each local moment of time-space leading from one moment and location in time (e.g. the present) to another location, i.e. from the present to the future and from the future into the past, and these effects can occur simultaneously and at superluminal speeds.

Many physical systems are very sensitive to small changes which can lead to major change. Unless the past and the future are "hard wired" and already determined, then the very act of voyaging to distant locations in time will alter every local moment of that time continuum. In terms of "Many Worlds" the time traveler is continually creating or entering new worlds which exist in parallel. Each "world" becomes most probable the moment he interacts with the quantum continuum, including simply by passing through time.

As detailed by quantum mechanics (Dirac 1966a,b; Heisenberg, 1955), shape and form appear as blemishes and bundles of energy in the quantum continuum, the underlying quantum oneness of the cosmos, emerging out of the continuum but remaining part of it. According to the Copenhagen interpretation (Bohr 1934, 1963; Heisenberg, 1930, 1955), all quanta are entangled and therefore any jostling of one quanta can create an instantaneous ripple which can effect local as well as distant objects and events through intersecting wave functions.

The space-time continuum is part of that basic oneness and is the sum of its parts including what can and can't be observed. And this includes distant locations in space-time corresponding to all possible futures, presents, and pasts.

As pertaining to time travel, as the traveler journeys through the quantum continuum of space-time he will jostle and affect all the particles (or waves) he contacts as he passes through time, and these will effect particles and waves elsewhere in space-time, thus altering the very fabric of every local and more distant moments of space-time. In the "Many worlds" interpretation, the time traveler is not really changing the future or the past but is engaging in actions which cause branching and splitting, which leads him to a future and a past which exists in parallel with innumerable other futures and pasts. He is not changing the past, but entering a different past which always existed as a probability.

As predicted by the Many Worlds interpretation, if the Time Traveler did make a significant impact in the past, then the alteration of the past would effect the entire world-line of history related to that event, including the memories of everyone living since that event and all those who retain any knowledge of that event; such that no one would realize anything has changed.

Minds, consciousness, the brain, memory, are also part of the quantum continuum and can be altered by changes in it (Joseph 2010a). The act of observation can change an event and an event can alter the observing mind. If the past were changed, we would not know it had changed because everything related to that event would have changed, from the writing of books to documentary films about the event. The alteration of the quantum continuum is not limited to just that event but can alter the entire continuum, including the quantum composition of the brain and memories of everyone who has lived since that event (Everett 1956, 1957).

The Principle of Self-Consistency

Many theorists have argued that it is impossible to change the past. Igor Novikov and Kip Thorne (Friedman et al. 1990) called this the "self-consistency conjecture" and "the principle of self-consistency" and various paradoxes have since been proposed to support this contention such as: "what if you killed your grandmother before she gave birth to your mother? If you did, then you could not be born and could not go back in time to kill your grandmother! Presumably these paradoxes are supposed to prove it is impossible to travel back in time.

In some respects these "paradoxes" are the equivalent of asking: "What if you went into the past and grew wings?" And the answer is: "You can't." The time traveler can not go back in the past and grow wings, or an extra pair of hands, or develop super powers, and so on. Nor could the time traveler kill anyone in the past who, according to the past record, did not die on the date he was killed.

Just as in "real life" there are boundaries which prevent the average person from engaging in or making world-altering decisions, these same limitations would apply in the past. Therefore, according to the principle of self-consistency, it is impossible to change the past, and if any changes were made, they may be "local" rather than global, and thus completely non-significant and not the least memorable--just like daily life for 99.999999999% of the 7 billion souls who currently dwell on Earth and who live and die and are quickly forgotten except by a few other insignificant souls who are also quickly forgotten as if they never even exists. Any changes made in the past may be so insignificant as to be meaningless.

Just as it is impossible to determine position and momentum of a particle, the past may also be subject to imprecision such that by establishing certain facts, makes other facts less certain The past may also be subject to the Uncertainty

Principle, which may explain why historians, eye-witnesses, and husbands and wives may not always agree about what exactly happened in the past or just moments before.

The Principle of Self-Consistency, however, holds that the past is hard wired and cannot be altered, and reverse causality is an impossibility (Friedman et al. 1990). By contrast, reverse causality (also referred to as backward causation and retro-causation) is based on the premise that an effect may occur before its cause, such that the future may effect the present and the present may effect the past. A "cause" by definition must precede the effect, otherwise the effect may negate the cause and the effect! For example, the if a man went back in time and killed his grandfather he would negate his own existence making it impossible to go back in time and kill his grandfather. On the other hand, if he did kill his "grandfather" it might turn out that his paternal lineage leads elsewhere, i.e. "grandmother" had an affair and another man fathered his own father. Thus killing his "grandfather" has no effect on his existence and does not interfere with his ability to go back in time to kill his grandfather. In this instance, the effect does not nullify the cause; which is in accordance with the principle of self-consistency. The past can't be altered and if it is, the result is not significant.

If the past is "fixed" and hard-wired and can't be altered, then although the time traveler may go back in time with the intention of killing his grandfather, or Hitler, or Lee Harvey Oswald, the result would be that he would be unable to do so; his gun would misfire, the bullet would miss, or he never got close enough to the intended victim to do the deed. The past is hard wired and can't be changed.

If the past can't be altered, then this also implies that the future may also be fixed and hard wired and is not subject to alteration. However, if the future is subject to change (as demonstrated by classical physics and the laws of cause and effect), then the future must exist in order to be altered; as predicted by quantum mechanics, entanglement, and Einstein's theories of relativity. If the future may be changed, then why not the past? According to the "Many worlds" interpretation, the past is not changed, instead one changes which past world becomes his reality.

The "Many Worlds" interpretation of quantum mechanics would allow one to kill their mother or commit a murder which had not taken place, in this "world;" but in so doing would be effecting the quantum continuum and contributing to the probability that an alternate world would become the time traveler's world once he commits these crimes.

Paradoxes and Many Worlds

Most time travel "paradoxes" are based on the premise that the time traveler some how gains powers or the will to do things he would never do, or to accomplish what others tried to do and failed. Even if the time traveler wanted to kill his mother before he was born, or assassinate Hitler before he came to power, would he be able to do it? Would he be able to get close enough to shove in that knife or fire that bullet? And if he did, maybe the victims would live. Maybe the knife or the bullet would miss the necessary organ. Maybe he would change his mind at the last moment. Maybe in the struggle someone else would shoot the Time Traveler in the head and he would die instead. Many people tried to kill Hitler and failed.

"Paradoxes" can be reduced to simple probabilities. What is the probability a time traveler would want to go back in time and kill his mother? What is the probability he would succeed? What is the probability others would intervene before he could do the deed? What is the probability that he might be killed in the attempt? ... and so on.

And if he did kill his mother, it would not be "his" mother.

An observer, object, particle, interacts with its environment, with the quantum continuum, changing and altering it. As postulated by "Many Worlds" theory, there is one ultimate reality, but many parallel realities and histories, like the branches of a tree, a hallway with infinite doors, or infinite highways all of which lead out of the city. One highway leads to a past where Hitler won the war. Another highway leads to a past where the Kennedy brothers were never killed. Yet another takes the time traveler to a world where he was never born.

According to quantum theory and the "many worlds" interpretation, a new highway, a new door, a new branch of the tree appears every time a particle whizzes by or an observer interacts with his environment, makes a decision, or records an observation. Thus, the time traveler may go back in time and kill the mother who dwells in a parallel world or universe, but he would be unable to kill his mother.

The "Many Worlds" Resolution of The Grandmother Paradox Time Traveler "A" goes back into the past and kills his grandmother when she was still a little girl. An observer, object, particle, interacts with its environment, with the quantum continuum, changing and altering it. A time traveler going into the past would change every moment leading to that past simply by traveling through it, so that the past and the grandmother he encounters would be a different past and a different grandmother. As also predicted by the "Many Worlds" interpretation of quantum physics, a time traveler can appear in different parallel worlds. Therefore by killing this grandmother in this past time, Traveler "A" would be preventing the birth of that woman's time-traveling grandson "B", thereby preventing "B" from going into the past and killing the grandmother of the Time Traveler "A."

Multiple Paradoxes. Effects Negating Causes A very wealthy scientist invents a time machine and travels 30 years back into the past to prevent the car accident which killed his very beautiful wife. He arrives in the parking lot of the business where she works and lets all the air out of her tires and disables the engine.

He visits the younger version of himself and gives him the blue print for building a time machine, and a list of 100 stocks and when to buy and sell them. The Time Traveler returns to the future.

When was the time machine invented?

His wife takes a cab to her Lover's apartment and that night they drive to her home and that of her husband (the younger version of the time traveler). The Lover discovers the blue print for the time machine and the list of 100 stocks. The Lover and the wife sneak into the bedroom where her husband (the younger version of the time traveler) is napping and shove a knife through his heart.

Who invented the time machine?

The "Lover" upon killing the younger version of the Time Traveler (with the help of Time Traveler's wife), suddenly finds himself alone with the body, still holding the bloody knife in his hand. However, the Time Traveler's wife (and the blue print for a time machine and list of stocks) have disappeared. Upon his arrest he learns the Time Traveler's wife was killed hours before in a car accident.

Information Exists Before it is Discovered Two research scientists, both bitter rivals, are competing to make a major scientific discovery. Scientist A, who is better funded, makes the discovery first, publishes the results, receives world wide acclaim and receives a Nobel Prize.

Scientist B loses all funding, does not get tenure, and is reduced to living in obscurity and working in his basement lab, where, 20 years later, he invents a time machine. Scientist B makes a copy of the article which won his rival, Scientist A, the Nobel Prize, and goes back in time and gives it to the younger version of himself. To ensure that the true inventor, Scientist A, does not get credit, the time traveler, Scientist B, kills Scientist A.

The younger version of Scientist B publishes the discovery and receives all the credit and the Nobel prize. When the time traveler, Scientist B returns to his own time he is famous and has a Nobel prize on his shelf. When he looks at the scientific journal where the original article appeared, he sees the same article but with himself listed as the author. He no longer understands why he went back in time to kill his rival.

Who made the discovery?

Another scientist after laboring his entire life makes a major discovery which brings him wealth and world wide acclaim. However, he is old and sick and unhealthy and is unable to savor the honors, women, and riches which are now his for the asking but he is too old to enjoy. So, he invents a time machine, takes a copy of his notebook describing the discovery, goes back 50 years in time and gives it to his younger self, and explains: "here are the answers you are searching for. You are going to be rich and famous."

So where did the discovery come from?

Science is replete with examples of scientists who independently make the same discoveries although they were working independently of each other and often not knowing of the other's work (Merton, 1961; 1963; Hall, 1980). Examples include the 17th-century independent formulation of calculus by Isaac Newton, Gottfried Wilhelm Leibniz and others; the 18th-century discovery of oxygen by Carl Wilhelm Scheele, Joseph Priestley, Antoine Lavoisier and others; In 1989, Thomas R. Cech and Sidney Altman won the Nobel Prize in chemistry for their independent discovery of ribozymes; In 1993, groups led by Donald S. Bethune at IBM and Sumio Iijima at NEC independently discovered single-wall carbon nanotubes and methods to produce them using transition-metal catalysts. And the list goes on.

What this could imply, if the past and future are a continuum, is that the discovery exists before it is discovered, albeit in a distant location of space-time. Or, in terms of multiple worlds theory, one branch leads to a world where the discovery is made by scientist A, a different branch leads scientist B to the discovery. Yet another branch leads to a world where the discovery is not made until 20 years into the future, whereas a different branch leads to a world where it is discovered in just a few days.

Mozart heard his music in his head, already composed--and some have proposed there is a cosmic consciousness which contains all information, and that one need only a brain that can tap into this source to extract this information. If true, this may explain why discoveries are made simultaneously or why Mozart heard his music "already composed" in his head and then simply wrote it down.

Time Travel Through Many Worlds

As based on a Many Worlds interpretation of quantum physics, traveling backwards into the past would itself be a quantum event causing branching. Therefore the timeline accessed by the time traveller simply would be one timeline among many different branching pasts. Hence, the time traveler from one world/universe may kill his grandfather in another world/universe. Likewise, in the past of some worlds, Hitler won the war, the Kennedy brothers were never killed, the dinosaurs did not become extinct, mammals and humans never evolved, and so on. All quantum worlds, many worlds, all exist as there is an infinity of possible universes and worlds, each of which differs in some manner from the other, from the minute to the major.

However, by changing (or choosing) his past, the Time Traveler would not just be making this past "World" more probable, but may cause all pasts to become unified. That is, the other pasts disappear as they are subsumed by and merge to become this one unified past.

Therefore, according to the Many Worlds interpretation, by changing the past, and by creating a single unified past, then once the merging occurs, all "memories" of earlier branching events will be lost. No one will ever remember that there was any other past and no observer will even suspect that there are several branches of reality. As such, the past (and the future) becomes deterministic and irreversible, and this effects the wave function of time, such that the past shapes the future, and conversely, the future can shape the past.

Therefore, if a time traveler journeys to the past, his passage will either change the past so that those in the future can only remember the past that has been altered since this past is the past which leads up to them. Or, the past was never really altered and always included the Time Traveler's journey into the past. That is, this altered past has always existed even before he journeyed to it and this is because he traveled to and arrived in the past before he left. Thus everything he does, from the moment he left for the past, has already happened. The past, like the future is irreversible and has been hard wired into the fabric of space-time.

According to the Copenhagen model, one may predict probabilities for the occurrence of various events which are taking place or which will take place. In the many-worlds interpretation, all these events occur simultaneously. Therefore, the time traveler is not changing the past, but choosing one past among many: "new worlds" which always existed as probabilities.

REFERENCES

Bohr, N., (1913). "On the Constitution of Atoms and Molecules, Part I". Philosophical Magazine 26: 1–24.

Bohr, N., (1913). "On the Constitution of Atoms and Molecules, Part I". Philosophical Magazine 26: 1–24.

Bohr, N. (1934/1987), Atomic Theory and the Description of Nature, reprinted as The Philosophical Writings of Niels Bohr, Vol. I, Woodbridge: Ox Bow Press.

Bohr. N. (1949). "Discussions with Einstein on Epistemological Problems in Atomic Physics". In P. Schilpp. Albert Einstein: Philosopher-Scientist. Open Court.

Bohr, N. (1958/1987), Essays 1932-1957 on Atomic Physics and Human Knowledge, reprinted as The Philosophical Writings of Niels Bohr, Vol. II, Woodbridge: Ox Bow Press.

Bohr, N. (1963/1987), Essays 1958-1962 on Atomic Physics and Human Knowledge, reprinted as The Philosophical Writings of Niels Bohr, Vol. III, Woodbridge: Ox Bow Press.

Born, M. Heisenberg, W. & Jordan, P. (1925) Zur Quantenmechanik II, Zeitschrift für Physik, 35, 557-615.

DeWitt, B. S., (1971). The Many-Universes Interpretation of Quantum Mechanics, in B. D.'Espagnat (ed.), Foundations of Quantum Mechanics, New York: Academic Press. pp. 167–218.

DeWitt, B. S. and Graham, N., editors (1973). The Many-Worlds Interpretation of Quantum Mechanics. Princeton University Press, Princeton, New-Jersey.

Dirac, P. (1966a) Lectures on Quantum Mechanics.

Dirac, P. (1966b). Lectures on Quantum Field Theory .

Einstein, A. (1905a). Does the Inertia of a Body Depend upon its Energy Content? Annalen der Physik 18, 639-641.

Einstein, A. (1905b). Concerning an Heuristic Point of View Toward the Emission and Transformation of Light. Annalen der Physik 17, 132-148.

Everett , H (1956), Theory of the Universal Wavefunction", Thesis, Princeton University.

Everett, H. (1957) Relative State Formulation of Quantum Mechanics, Reviews of Modern Physics vol 29, 454–462.

Friedman, J. et al. (1990). Cauchy problem in spacetimes with closed timelike curves". Physical Review D 42 (6): 1915.

Haber, R. N., Haber, L. (2000). Experiencing, remembering and reporting events. Psychology, Public Policy, and Law, 6(4): 1057-1097.

Heisenberg, W. (1925) Über quantentheoretische Umdeutung kinematischer und mechanischer Beziehungen, ("Quantum-Theoretical Re-interpretation of Kinematic and Mechanical Relations") Zeitschrift für Physik, 33, 879-893, 1925

Heisenberg, W. (1927), "Über den anschaulichen Inhalt der quantentheoretischen Kinematik und Mechanik", Zeitschrift für Physik 43 (3–4): 172–198,

Heisenberg. W. (1930), Physikalische Prinzipien der Quantentheorie (Leipzig: Hirzel). English translation The Physical Principles of Quantum Theory, University of Chicago Press.

Heisenberg, W. (1955). The Development of the Interpretation of the Quantum Theory, in W. Pauli (ed), Niels Bohr and the Development of Physics, 35, London: Pergamon pp. 12-29.

Heisenberg, W. (1958), Physics and Philosophy: The Revolution in Modern Science, London: Goerge Allen & Unwin.

Joseph, R. (2010) Quantum Physics and the Multiplicity of Mind: Split-Brains, Fragmented Minds, Dissociation, Quantum Consciousness. "The Universe and Consciousness", Edited by Sir Roger Penrose, FRS, Ph.D., & Stuart Hameroff, Ph.D. Science Publishers, Cambridge, MA.

Juan Y., et al. (2013). "Bounding the speed of `spooky action at a distance". Phys. Rev. Lett. 110, 260407.

Lee, K.C., et al. (2011)."Entangling macroscopic diamonds at room temperature". Science 334 (6060): 1253–1256. Matson, J. (2012) Quantum teleportation achieved over record distances, Nature, 13 August.

Megidish, E., Halevy, T. Shacham, A., Dvir, T., Dovrat, L., Eisenberg, H. S. (2013) Entanglement Swapping Between Photons that have Never Coexisted. ArXiv.1209.4191v1, 19, Sep, 2012. Physical Review Letters, 110, 210403.

Megreya, A. M., & Burton, A. M. (2008). Matching faces to photographs: Poor performance in eyewitness memory (without the memory). Journal of Experimental Psychology: Applied, 14(4): 364–372.

Olaf, N.. et al. (2003) "Quantum interference experiments with large molecules", American Journal of Physics, 71 (April 2003) 319-325.

Plenio, V. (2007). "An introduction to entanglement measures". Quant. Inf. Comp. 1: 1–51.

Schrödinger E; Born, M. (1935). "Discussion of probability relations between separated systems". Mathematical Proceedings of the Cambridge Philosophical Society 31 (4): 555–563.

Schrödinger E; Dirac, P. A. M. (1936). "Probability relations between separated systems". Mathematical Proceedings of the Cambridge Philosophical Society 32 (3): 446–452.

21. The Quantum Time Machine of Consciousness. Past Present Future Exist Simultaneously.

Abstract:

There is no "universal now." The distinctions between past present and future are illusions. As predicted by Einstein's field equations space-time may be a circle such that the future leads to the present and then the past which leads to the future, thereby creating multiple futures and pasts and which allows information from the future to effect the present. Causes may cause themselves. Coupled with evidence from entanglement where choices made in the future effect measurements made in the present and theoretical tachyons which travel at superluminal speeds from the future to the present and then the past, this may account for precognition, deja vu, and premonitions. In quantum mechanics, where reality and the quantum continuum are a unity, time is also a unity such that the future present past are a continuum which are linked and the same could be said of consciousness which exists in the future and in the present and past. If considered as a "world line" and in space-like instead of time-like intervals, then consciousness from birth to death would be linked as a basic unity extending not in time but in space and the same could be said of time. Time-space and consciousness are also linked and interact via the wave function and as demonstrated by entanglement and the Uncertainty Principle. Evidence from space-time contraction, atomic clocks and the twin paradox as functions of gravity and acceleration also demonstrate that the future already exists before it is experienced by consciousness in the present. Likewise, under conditions of accelerated consciousness (such as in reaction to terror) and dream states where various brain structures are in a heightened state of activity, space-time may also contract, such that time may slow down and consciousness may be given glimpses of the future in advance of other conscious minds thereby providing again for experiences such as precognition, premonitions, and deja vu. Closed time curves, conscious time, relative time, dream time, and quantum time are also discussed.

Relativity: The Future is the Past. The Past is the Future

Time is relative to the observer (Einstein 1961). Since there are innumerable observers, due to gravity, velocity and other variables, there is no universal "past, present, future" (Einstein 1905a,b, 1906, 1915a, 1961) all of which overlap and are infinite in number and yet interconnected and entangled in the basic oneness of the spacetime quantum continuum. It is the unity and relativity of time which makes time travel possible as well as some of the unique features of consciousness such as "premonition" and deja vu during which an observer experiences or is effected by the future before it becomes the present.

Because time is relative, and due to entanglement (Lee et al. 2011; Matson 2012; Olaf et al. 2003), the future can effect the past and may be experienced in the present depending on the observer's frame of reference. Consider the moon, the sun, and the stars up above. From a vantage point on Earth, the moon we see is actually the moon from 1 second ago; the moon we see in the present is from the past. The sun we observe is a sun from 8 seconds in the past. Upon gazing at the nearest star, Alpha Centauri, the star we see is from years ago since it takes 4.3 light years for its light image to reach Earth. If you stand in front of a full length mirror, just 3 feet away, and since light travels at 1 foot per nanosecond, you see yourself as you looked 6 nanoseconds ago. You are staring into the past. You are always younger in a mirror because what you see is the "you" from moments before. Mirrors are gateways to the past even though the past image you see is experienced in the "present." However, although from the past, the images are in the future until they arrive in the present relative to an observer. Before the image from the past arrives it is still in the future, relative to the observer. A stream of photons which just left the surface of the sun will not arrive on Earth until the future, 8 seconds from now. Until the splash of light arrives, it is in the future, relative to those on Earth but in the past relative to the sun.

If an alien observer living on a planet in Alpha Centauri was gazing at Earth, then the present on Alpha Centauri overlaps with the past on Earth. The alien sees an Earth from 4.3 light years ago. The reverse is true for an observer on Earth gazing at this distant star. However, until those images from the past reach the observer, they are in the future relative to that observer and overlap and exist simultaneously. The past can be the future and both may exist before they arrive in the present only to again become the past thereby creating a circle of time. Innumerable futures, presents, and pasts exist simultaneously albeit in different locations within space-time all of which are in motion. Observers located in New York, Shanghai, Tokyo, Paris, Mexico City, and in distant galaxies, are also in motion, as planets spin and orbit the sun, the sun orbits the galaxy, and galaxies move about in the universe. Observers, regardless of what planet, solar system, or in what galaxy they reside, are continually moving though space-time, often at different velocities and effected by varying degrees of gravity, and all are continually coming into contact with different times which are effect by velocity as well as the consciousness and emotional state of the observer.

Contraction of Time: The Future and Present Come Closer Together

As predicted by relativity and quantum mechanics, the experience of time and the distinctions between the past, future and the "present" are shaped and affected by distance, gravity, acceleration, consciousness, and our emotions, contracting and speeding up under conditions of pleasure and slowing down and sometimes splitting apart or even running backwards under conditions of fear and terror (Joseph 1996, 2010a). Acceleration contracts space-time thereby decreasing the distance between the future and the now, albeit depending on the location and frame of reference of the observers (Einstein 1905c, 1961; Einstein et al. 1923; Lorentz 1892, 1905). When time-space contracts, more time is squeezed into a smaller space such that it may take one consciousness less time to reach the future (vs the consciousness of a second observer) if that consciousness experiences time contraction. Two observers, with two different inertial frames of reference, may experience time as slowing down or speeding up. Consciousness can also accelerate and contract the space-time continuum, particularly during dream states, or under conditions of terror in which case, time may speed up or slow down and there may be a splitting of consciousness (Joseph 1996, 2010a). Likewise, emotions such as pleasure can also speed up the experience of time; a phenomenon observed by Einstein nearly 100 years ago:

"Put your hand on a hot stove for a minute, and it seems like an hour. Sit with a pretty girl for an hour, and it seems like a minute. THAT'S relativity." -Einstein

Time is a dimension, not in Euclidian space, but in "Minkowski space" (Minkowski 1909). Euclidian space consists of 4 spatial dimensions which include movement and geometric space; but none of which encompass time. By contrast, in "Minkowski space" which is incorporated within Einstein's special relativity, time is the 4th dimension (Einstein 1961). More specifically, 3 of the Euclidian dimensions of space are combined with a dimension of time thereby creating a four-dimensional manifold known as "space-time." Space-time, however, is effected by gravity and acceleration, and can shrink and contract as gravity and velocity increase and in response to alterations in cerebral activity and thus, consciousness. As demonstrated in quantum mechanics, consciousness and the act of perceptual (or mechanical) registration directly impacts the quantum continuum through interactions via the wave function and through entanglement.

The relationship between time dilation and the contraction of the length of space-time can be determined by a formula devised by Hendrik Lorentz in 1895. As specified by the Lorentz factor, γ (gamma) is given by the equation γ = , such that the dilation-contraction effect increases exponentially as the time traveler's velocity (v) approaches the speed of light c.

ticks slower) whereas the twin on Earth took more time (clock ticks faster) to reach the same destination in the future. The time traveling twin arrives in the future more quickly--and the same can be said of accelerated and dissociated consciousness thereby providing the foundations of deja vu and premonitions.

Because of time dilation and the contraction of space, once the time traveler lands on Earth, and depending on how fast and far into the future she is propelled, all her friends and relatives back on Earth may have died and a completely new generation of Earthlings may greet the time traveler upon her return. By contrast, since time slows down and time-space become squeezed together, the time traveler who arrives in the future may not have aged appreciably.

For example, say the time traveler is born in the year 2100, had a life expectancy of 80 years and would have died in the year 2180 if she had never left on her journey into the future. If she began her journey at age 20 in the year 2120, achieved 0.999999999999999 light speed and arrived in the future date of 2180, she would still have a life expectancy of 60 years (minus the 20 she already lived and time spent in the time machine). Upon arriving in the year 2180, she would still be 20 years old instead of 80 and could now expect to live another 60 years until the year 2240 (vs the year 2180 if she had never left home).

This premise is based on achieving near light speeds almost instantaneously and is supported by experiments with non-living, ultra-short-lived particles. For example, the muon particle is given a new lease on life when accelerated to a velocity of 99.92% light speed and its life span is nearly 25 times longer (Houellebecq 2001; Knecht, 2003). The muon particle not only lives longer but travels 25 times further due to its expanded life span. Particles, including phi mesons, which have been accelerated to velocities of 99.9% light speed also achieve significant life span extensions with a γ factor of around 5,000 (Houellebecq 2001). Presumably particles live longer because they arrived in the future more quickly vs their counterparts traveling at their normal, slower speeds. Therefore, it could be predicted that a time traveler who journeys at near light speeds will live longer compared to friends left back on Earth; and this is because the contraction of time space enabled them to reach the future before those back on Earth. Premonitions and deja vu, work on the same principles.

A time traveling consciousness will also experience time as speeding up or slowing down depending not just on acceleration and gravity, but emotion and neural activity. The distance between the "present" and the future decrease because of the shrinkage of space. In consequence the time traveling consciousness gains access to information in the future more quickly than other observers.

These same principles can be applied when traveling great distances across space to other stars and planets. If the journey takes place at near lights speeds, the space-time traveler may visit a distant star and then return home still fresh and young whereas her relatives and friends will have grown old and infirm and may have already died.

For example, if Gaia stays on Earth and her twin, Aurora travels at 80% the speed of light to Proxima Centauri which is 4.2 light years away, then Aurora's trip will take 5.25 Earth years (4.2/0.8 = 5.25). One day in the time machine at 80% light speed is equal to 1.67 days on Earth. Thus 1,916.25 days in the time machine (5.25 years) is equal to 3,200 days on Earth (8.76 years). Hence, Aurora's clock will tick 0.599% more slowly than Gaia's clock on Earth (5.25/8.76) and Aurora will age only 3.15 years during the journey (0.599 x 5.25 = 3.146) whereas Gaia will age 5.25 years. If Aurora immediately returns to Earth at 80% light speed she will be 6.3 years older and her twin will be 10.5 years older.

Atomic Clocks

Alterations in consciousness, gravity, and velocity can shrink or stretch space-time. Therefore, time-space is also warped, shrunk, stretched, and may even curl up and fold upon and over itself depending on local conditions. Time is asymmetric. Like the weather, time is not the same everywhere, even when measured by atomic clocks.

Atomic clocks tick off time as measured by the vibrations of light waves emitted by atoms of the element cesium and with accuracies of billionths of a second (Essen & Parry, 1955). However, these clocks are also effected by their surroundings and run slower under conditions of increased gravity or acceleration (Ashby 2003; Hafele & Keating 1972a,b). In 1971 Joe Hafele and Richard Keating placed atomic clocks on airplanes traveling in the same direction of Earth's rotation thereby combining the velocity of Earth with the velocity of the planes (Hafele & Keating 1972a,b). All clocks slowed on average by 59 nanoseconds compared to atomic clocks on Earth. These clocks arrived in the future in less than than their counterparts.

It has also been demonstrated that atomic clocks at differing altitudes will show different times; a function of gravitational effects on time. The lower the altitude the slower the clock, whereas clocks speed up as altitude increases; albeit the differences consisting of increases of only a few nanoseconds (Chou et al. 2010; Hafele & Keating, 1972; Vessot et al. 1980). "For example, if two identical clocks are separated vertically by 1 km above the surface of Earth, the higher clock gains the equivalent of 3 extra seconds for each million years (Chou et al., 2010). The speeding up of atomic clocks at increasingly higher altitudes has been attributed to a reduction in gravitational potential which contributes to differential gravitational time dilation.

Accelerated Consciousness: Dissociative Mind And The Slowing Of Time

Evidence from relativity, quantum mechanics, atomic clocks, and space-

tome contraction, demonstrates that the future, or at least, "a" future must have existed so this future could contract closer to the present experienced by the time traveler;. This is also proved by experiments in entanglement (REF), and Einstein's field equations which demonstrate that time is a circle (REF) where the present leads to a future which already exists and that this future leads to the present and then the past. That is, instead of the present moving toward a future which does not yet exist, the future already exists and streams toward consciousness, becoming the present, and then continues into the past relative to that observing consciousness. However, consciousness may also accelerate as reflected by increased brain activity, and this too would contract space-time. Buried deep within the brain are a series of structures referred to collectively as the limbic system, and which includes the amygdala, hippocampus, and hypothalamus. The limbic system governs all aspects of sexual behavior and emotion, including emotional and non-emotional memory as well as anxiety, fear, and the ability to visualize one's self (Joseph, 1992, 2011). Limbic system structures, such as the amygdala are able to receive sensory information from multiple modalities at the same time and excessive activity in these areas, as reflected by increased EEG activity, are associated with the the reception of information which is normally inhibited and filtered, including the experience of deju vu and other precognitive phenomenon (Daly, 1958; Halgren 1990, Gloor, 1990; Joseph, 1996, 2011; Penfield, 1952; Penfield & Perot 1963; Williams, 1956), such as a splitting of consciousness where time slow down and the dissociated consciousness can observe itself as if up in the air looking down.

> "I had a clear image of myself... as though watching it on a television screen." "The next thing I knew...I was looking down from 50 to 100 feet in the air...I had a sensation of floating. It was almost like stepping out of reality. I seemed to step out of this world."

Electrode stimulation, or other forms of heightened activity within limbic system structures such as the amygdala, hippocampus and overlying temporal lobe can also cause time to speed up or slow down (Joseph, 1998, 1999b, 2001). Likewise, in response to extreme trauma, stress and fear, the amygdala, hippocampus and temporal lobe become hyper activated resulting not only in a "splitting of consciousness" but the sensation that time has slowed down while the dissociated consciousness seems to speed up (Courtois, 2009; Grinker & Spiegel, 1945; Noyes & Kletti, 1977; van der Kolk 1987).

One individual, after losing control of his Mustang convertible while during over 100 miles per hour on a rain soaked freeway, reported that:

> "time seemed to slow down and then... part of my mind was a few feet outside the car zooming above it and then beside it and behind it and in front of it, looking at and analyzing the respective positions of my spinning Mustang and the cars surrounding me. Simultaneously I was inside trying

to steer and control it in accordance with the multiple perspectives I was given by that part of my mind that was outside. It was like my mind split and one consciousness was inside the car, while the other was zooming all around outside and giving me visual feedback that enabled me to avoid hitting anyone or destroying my Mustang."

"Tiffany" describes her experience as follows:

"I was a passenger in my boyfriend's sports car and we were laughing and racing along highway 17, going 80, 90 mph. I remember he reached his cigarettes when we were going around a corner and then the car began to slide sideways toward the embankment and all the trees. Everything just suddenly slowed down, like in slow motion, and I could see the car sliding very slowly toward the trees, and I turned and looked at my boyfriend and he had this look of fear and determination on his face. He was gritting his teeth which were very white. I remember looking at his hands tightly gripped on the steering wheel, and I could see the ring I gave him. And outside the car there were other cars and they were also moving in slow motion. We were still sliding, and I turned my head and I could see we were going to slide right into this big tree, and everything was still so slow, and I could see the trunk and bark of the tree, the tree limbs, coming closer and I could see this bird flying out of the tree flapping its wings real slow, then we hit the tree with the back of the car which made the car spin around the tree, but it was all in slow motion, and all this glass blew out the side and back window and I could see little pieces of glass going everywhere moving very slowly through the air. I was wearing my seat belt and was spun toward my boyfriend but he wasn't there. Instead, the driver's side door was open and the car was turning upside down and I could see my purse falling upward and my wallet and phone and eyeliner and lipstick and a pencil were all falling out but going upward very slowly, like floating right in front of me, and I remember thinking that I hoped that pencil did not stick me in the eye, and then the air bags popped out and it was also going in slow motion billowing out toward me and I could see the trees down below because we were falling over and down the embankment and everything was upside down and going sideways and then the airbag hit me in the chest and time suddenly sped up and the car landed upside down and slid down the embankment and hit some trees."

The slowing down of time and the splitting of consciousness, creating twin consciousnesses, are not uncommon under conditions of terror. Terror can accelerate the mind and brain by releasing a cascade of "fight or flight" neurochemicals such as norepinephrine (Joseph 1992, 1994, 1996, 20011). When the brain and mind are accelerated under these conditions one aspect of consciousness may split off and observe itself and the body which houses it. Under certain accelerated conditions, time will also slow down for the dissociated consciousness which has split off and is observing (Joseph 1996, 2000); exactly as predicted by relativity and the twin paradox: acceleration slows time for one observer and speeds it up for the other (Einstein et al. 1927; Einstein 1961).

In fact, time slows down under conditions of terror and accelerated, hyper-brain activity to such a degree that seconds may last minutes, and minutes hours (Joseph 1996). Individuals may become completely motionless, almost catatonic from the perspective of outside observers and may fail to make any effort to save their lives or to respond to assistance such as attempting to evacuate a burning plane or sinking ship even though they have been uninjured (Courtois, 1995; Galliano et al., 1993; Miller, 1951; Nijenhuis et al., 1998). From the perspective of outside observers, those so afflicted appear to be frozen in time; which is exactly how a time traveler accelerating toward light speed would appear to those outside the time machine. Likewise, from the perspective of the dissociated consciousness time also appear to slow down and to contract which gives that dissociated consciousness more time in less space to observe its surroundings; and this too is predicted by relativity and the Lorentz transformations of length and space-time contraction. Under accelerated conditions space-time shrinks and more time is compacted into less space.

> "I began moving at a tremendous speed... and I was aware of trees rushing below me. I just thought of home and knew I was going there... I saw my husband sitting in his favorite armchair reading the newspaper. I saw my children running up and down the stairs... I was drawn back to the hospital, but I don't remember the trip; it seemed to happen instantaneously" (Eadie & Taylor 1992).

Thus time slows down for the consciousness attached to the body, but may speed up for the dissociated consciousness. However, this multiplicity of mind, although dissociated, is still entangled, and as such, time may speed up and slow down simultaneously.

In fact, under conditions of extreme fear and terror and accelerated consciousness, time slows to such a degree and corresponding movements becomes slowed to such a degree that to outside observers the person may appear to be dead: "Far down below I could see houses and towns and green land and streams... I was very happy now. I kept on going very fast... then I started back, going very fast...Then I was lying on my back in bed and the girl and her father and a doctor were looking at me in a queer way...I had been dead three days (they told him)...and they were getting ready to buy my coffin" (Neihardt 1989).

Consider the case of Lisa, a wild blonde 22 year old beauty, a passenger in a sport's car with the top down that struck a telephone pole:

"It felt like a movie in slow motion and everything slowed down just before we crashed. I had a seat belt and my arms and hands stretched out in front on the dashboard... and the windshield shattered and I could see all these cracks forming in the glass in slow motion. Everything was slowed down and the windshield just broken in half and it fell toward me all in slow motion and cut off my arm....I could see it cutting the skin and droplets of blood and then all this blood and my arm falling slowly to the floor of the car. Everything was so slow... and I got out of the car and my arm was spraying blood everywhere. I walked only a few feet and in slow motion fell down.... Then I was in the air, watching everything. Part of the time I was on the ground looking up and at the same time I was in the air looking down at me. I could see people getting out of their cars. They were all around me and I could see them while I was on the ground and at the same time I could see them like I was 50 feet in the air looking down at them. Then the ambulance came and they put a tourniquet on my arm and put me inside... and then everything started going real fast. I was outside the ambulance, like I was inside sticking out and my mind was racing up and down the streets, like I was running very fast alongside.... then I was in the hospital. But I was no longer part of my body. I was racing along, tripping out, bobbing up and down the halls, just checking everything out and everything was going very fast. Then I saw all these doctors and nurses working on this body of a girl. I peaked over their shoulders and then I realized it was me, that girl was me and I could see that my hair was all bloody and this bothered me. It needed to be washed. But I wasn't moving. I looked dead.... The doctors also thought I was dead.... and that's when I fell back into my body with this thump and I started moving and that's when the doctors and nurses realized I was still alive."

Of course, time can also slow down under conditions of extreme boredom. However, boredom does not induce a splitting of the mind or dissociated states of consciousness. Instead of more time in less space, there is less time in more space. The clock ticks more slowly for the consciousness of the observer and more slowly external to that observer. Conversely, under conditions of pleasure, time speeds up and passes more quickly as so eloquently summed up by Einstein.

Time is entangled with and relative to consciousness and the multiplicity of mind.

The Event Horizon of Consciousness: The Eternal Now

As a thought experiment Einstein imagined that if he flew away from a big clock in the town square precisely at 12 noon, and traveled at the speed of light, the clock would appear to stop and would remain 12 noon forever--and this is because Einstein would be traveling at the same rate of speed as the light coming from the clock, in tandem and in parallel with it. Time would also essentially stop for Einstein, for if he were looking at the light beams on either side of him, they would look like stationary waves of electromagnetic activity consisting of crests and valleys--and this is because he would be moving in tandem and relative to these light beams; like two trains traveling at exactly the same speed, side by side and the only view is of the other train. At light speed Einstein would be captured in an "eternal now" with the future on one side and the past on the other.

All observers in uniform motion (like two trains traveling side by side) view themselves as at rest (so long at they can only see the two trains). If traveling at the speed of light, a light from a flashlight held in that time traveler's hand will never escape from the flashlight. The light from the flashlight will be frozen in place, in an eternal now.

If a star, astronaut, or space-time machine were to approach a supermassive black hole at the center of this galaxy, they would accelerate toward the "event horizon" at light speed (Dieter 2012; McClintock, 2004)--the "event horizon" being the point of no-return, the vortex forming the mouth of the hole. The Time Traveler's clock would tick increasingly slower and light trailing behind would become redder (red shifted) as the event horizon is approached. However, for the time traveler, time continues as before.

Once caught by the gravitational grip of the vortex spinning round the event horizon, the star, astronaut, or space-time machine would have a velocity of light speed (Dieter 2012). Time stops. They would be captured and held in the grip of what could best be described as an "eternal now." Light could not escape, and the outside of the hole would appear black, whereas the event horizon would be blazing brightly illuminated with light.

Just as a star will accelerate toward light speed as it approaches the event horizon of a supermassive black hole in the fabric of space-time (Bethe et al. 2003; Dieter 2012; McClintock, 2004), the multiple futures flowing toward the event horizon of consciousness may also accelerate toward light speed. Once captured by the event horizon of consciousness these futures have a velocity of light speed becoming the "eternal now." Consciousness of the "present" could be likened to an event horizon illuminated with light. The present, the "eternal now" is the illumination of the event horizon of consciousness at light speed. On one side of the event horizon of "now" would be the future, and on the other, the past.

Predicting A Future Which Exists Before It Is Experienced

Relativity and quantum physics both predict the future exists before it is experienced. However, due to the fact that time is entangled in the frenzied activity of the quantum continuum, the future, or rather "a" future may continually change until the moment it is perceived by consciousness.

Since futures and pasts overlaps and as time-space is coextensive, then time, including local time relative to a single observer, is entangled. The past may effect the future, the future can effect the past, time effects consciousness and alterations in consciousness effect the passage of time.

As a "future" flows toward Earth it can also be effected by whatever it encounters on the way to the consciousness of "now," relative to an observer on Earth--exactly as befalls light. All futures are also entangled with space-time, the quantum continuum, and subject to the Uncertainty Principle. Therefore, future time may be continually altered until perhaps just moments before these futures are experienced by observers who are also entangled with what they experience. Hence, although one may anticipate and predict the future, just like they may predict the weather, the ability to accurately anticipate and predict the future, like predicting future weather, may increase the closer that future is to the present. Planning skills, goal formation, strategy, long term investments, concern for consequences, and even the most basic of calendars, all rest upon the ability to make predictions about the future.

The future is like the weather, with the ability to forecast the weather decreasing in accuracy as time and distance from the present increases. In other words, and because of entanglement and classic concepts governing "cause and effect", the future is not already determined but is in flux and subject to continual alteration. The act of observing and other forces related to cause and affect alter the quantum continuum and continually change the future as it approaches. The future may not become fixed until the moment it is perceived by an observer relative to that observer, at which point it is in the present. Hence, predictions about the future will seldom be completely accurate, and become less accurate regarding increasingly distant events in the future, but more accurate but not completely accurate regarding events in the immediate future; a consequence of entanglement and the Uncertainty Principle.

Since the past is also relative and can exist in the future for some observers and in the present for others, and as the past is entangled with the quantum continuum, then the past is also subject to change after it has been experienced and before it is experienced by another observer at a downstream location in space-time. Two historians writing about history interpret and experience the past differently. A husband and wife discussing what happened at a party the night before, disagree. Eye-witness accounts differ among eye-witnesses (REF). A peasant living in a small village in western China in 1963 may have never heard of the assassination of president John F. Kennedy. The past is relative. There is no universal "past."

Time is entangled and is affected by consciousness and relative to and effected by the act of observation and measurement--as predicted by quantum mechanics (Bohr, 1958, 1963; Dirac, 1966a,b; Planck 1931, 1932, Heisenberg

1927, 1958; Neumann 1937, 1955).

Causes and Effects Are Relative To Consciousness

Every particle, person, planet, star, galaxy, has a wave function. The brain and consciousness have a wave function (Penrose & Hameroff 2011). Reality, including the reality of time, is a manifestation of wave functions and alterations in patterns of activity within the quantum continuum which are perceived as discontinuous (Bohr, 1958, 1963; Planck 1931; Heisenberg, 1958). This also gives rise to the perception of temporal order and what comes first, second, third, and what is in the present and in the past. The perception of temporal order, and structural units of information are not just perceived, but inserted into the quantum state which causes the reduction of the wave-packet and collapse of the wave function.

The brain and mind of a time traveler also has a wave function. As predicted by Einstein's field equations, consciousness can be accelerated into the future, and from the future, into the past. The Time traveler, upon observing his surroundings causes a collapse of the wave function, as predicted by the Copenhagen school of quantum physics: "The discontinuous change in the probability function takes place with the act of registration...in the mind of the observer" (Heisenberg, 1958).

The loss of coherence, the creation of discontinuous states in the quantum continuum is the result of entangled interactions within the environment which results in an exchange of energy and information: quantum entanglements. These entanglements, or blemishes in the quantum continuum, may be observed as shape, form, cause, effect, past, present, future, first, second, last, an so on, all of which are the result of a decoupling of quanta from the quantum (coherent) continuum which leaks out and then couples together in a form of knot which is observed as a wave form collapse. Every moment in time, is a wave form collapse of space-time at the moment of observation (Bohr, 1958, 1963; Heisenberg 1958; Von Neumann 1932, 1937).

However, in the Copenhagen model, the observer is external to the quantum state and is not part of the collapse function but a witness to it (Bohr, 1958, 1963; Heisenberg 1958). The observer is not the creator of reality but registers the transition of the possible to the actual: "The introduction of the observer must not be misunderstood to imply that some kind of subjective features are to be brought into the description of nature. The observer has, rather, only the function of registering decisions, i.e., processes in space and time, and it does not matter whether the observer is an apparatus or a human being; but the registration, i.e., the transition from the possible to the actual, is absolutely necessary here and cannot be omitted from the interpretation of quantum theory" (Heisenberg 1958).

As summed up by Von Neumann (1932), the "experiential increments in a person's knowledge" and "reductions of the quantum mechanical state of that person's brain" corresponds to the elimination of all those perceptual functions that are not necessary or irrelevant to the knowing of the event. Consciousness, therefore, could be viewed as a filter, which selectively attends to fragments of the quantum continuum which are perceived as real: the transition from the possible to the actual.

Therefore, according to the Copenhagen interpretation, the observing consciousness is external and separate from (albeit entangled with) what is observed and external to the ensuing collapse of the wave function which is collapsed by being measured and observed and this includes the observation and experience of time. However, consciousness can also be conscious of consciousness and thus consciousness can be subject to wave form collapse when observed by consciousness (Joseph 2011).

Consciousness and Mind

In a quantum universe all of existence, including consciousness, consists of a frenzy of subatomic activity which can be characterized as possessing pure potentiality and all of which are linked and entangled as a basic oneness which extends in all directions and encompasses all dimensions including time (Bohr, 1958, 1963; Dirac, 1966a,b; Planck 1931, 1932, Heisenberg 1955, 1958; von Neumann 1937, 1955). Hence, consciousness and the act of observation be it visual, auditory, tactile, mechanical, digital, are entangled with the quantum continuum and creates a static and series of impressions of just a fragment of that quantum frenzy that is registered in the mind of the observer as length, width, height, seconds, minutes, hours, days, weeks, months, first, second, third, and so on; like taking a series of pictures of continual motion and transformation and then believing it consists of temporal sequences when in fact, the conscious mind imposes temporal order (Joseph 1982, 1996, 2010). Just as, according to the quantum physics, the observing mind interacts with the quantum continuum and makes it possible to perceive shape and form, the conscious mind (and the dreaming mind) can perceive temporal sequences where there is none (Joseph 1982, 1986, 2010a); and those sequences include the illusion of future and past. That is, the act of sensory registration, be it a function of a single cell, or the conscious mind of a woman or man, selects a fragment of the infinite quantum possibilities and experiences it as real, and it is real but only to that mind or that cell at the moment of registration (Heisenberg 1955, 1958). Hence, "past present future" are a manifestation of consciousness which is entangled with time-space and the quantum continuum.

"I regard consciousness as fundamental. I regard matter as derivative from consciousness" (Max Planck, 1931).

As demonstrated by quantum mechanics and formalized by the Uncertainty Principle (Heisenberg 1925, 1927), what is known, is imprecise (Bohr, 1958,

1963; Dirac, 1966a,b; Planck 1931, 1932, Heisenberg 1955, 1958; Neumann 1937, 1955) and this includes time. To know something in its totality, would require a multi-dimensional all encompassing infinite "god's eye" view.

It could be said that consciousness is consciousness of something other than consciousness (Joseph 1982, 2011). Consciousness and knowledge of an object, such as a chair, are also distinct and separate. Consciousness is not the chair. The chair is not consciousness. The chair is an object of consciousness, and thus become discontinuous from the quantum state and entangled with consciousness.

Consciousness is consciousness of something and consciousness can be conscious of not being that object that it is conscious of. By knowing what it isn't, consciousness may know what is not, which helps define what is. This consciousness of not being the object can be considered the "collapse function" which results in discontinuity within the continuum: consciousness of consciousness being conscious.

Moreover, as demonstrated by neuroscience, the mind is not a singularity, but a multiplicity with different aspects of consciousness and awareness directly associated with specific regions of the brain (Joseph, 1992, 1996, 2011). These different mental realms and brain areas can perceive time and the quantum continuum differently. Time may be perceived by one brain region as lacking temporal order but as a continuum or gestalt. The mind is a multiplicity, which can become a duality, and which is often experienced as a singularity referred to as consciousness.

Further, it could be said that consciousness of consciousness, that is, self-consciousness, also imparts a duality, a separation, into the fabric of the quantum continuum. Hence, this consciousness that is the object of consciousness, becomes an abstraction, and may create a collapse function in the quantum continuum (Heisenberg, 1958; Joseph 2011; von Neurmann 1955, 2001). Consciousness may cause itself. That is, a continuum which is consciousness, and which exists as entangled in the quantum continuum which includes time, may cause itself to experience the "eternal now" which is simply a collapse of the wave function of time.

REFERENCES

Al-Khalili (2011). Black Holes, Wormholes and Time Machines, Taylor & Francis.

Almheiri, A. et al. (2013). Black Holes: Complementarity or Firewalls? J. High Energy Phys. 2, 062

Bethe, H. A., et al., (2003) Formation and Evolution of Black Holes in the Galaxy, World Scientific Publishing.

Bilaniuk, O.-M. P.; Sudarshan, E. C. G. (1969). "Particles beyond the Light

Barrier". Physics Today 22 (5): 43–51.

Bo, L., Wen-Biao, L. (2010). Negative Temperature of Inner Horizon and Planck Absolute Entropy of a Kerr Newman Black Hole. Commun. Theor. Phys. 53, 83–86.

Bohr, N., (1913). "On the Constitution of Atoms and Molecules, Part I". Philosophical Magazine 26: 1–24.

Bohr, N., (1913). "On the Constitution of Atoms and Molecules, Part I". Philosophical Magazine 26: 1–24.

Bohr, N. (1934/1987), Atomic Theory and the Description of Nature, reprinted as The Philosophical Writings of Niels Bohr, Vol. I, Woodbridge: Ox Bow Press.

Bruno, N. R., (2001). Deformed boost transformations that saturate at the Planck scale. Physics Letters B, 522, 133-138.

Blandford, R.D. (1999). "Origin and evolution of massive black holes in galactic nuclei". Galaxy Dynamics, proceedings of a conference held at Rutgers University, 8–12 Aug 1998,ASP Conference Series vol. 182.

Carroll, S (2004). Spacetime and Geometry. Addison Wesley.

Casimir, H. B. G. (1948). "On the attraction between two perfectly conducting plates". Proc. Kon. Nederland. Akad. Wetensch. B51: 793.

Chodos, A. (1985). "The Neutrino as a Tachyon". Physics Letters B 150: 431.

Einstein, A. (1915a). Fundamental Ideas of the General Theory of Relativity and the Application of this Theory in Astronomy, Preussische Akademie der Wissenschaften, Sitzungsberichte, 1915 (part 1), 315.

Einstein, A. (1915b). On the General Theory of Relativity, Preussische Akademie der Wissenschaften, Sitzungsberichte, 1915 (part 2), 778–786.

Einstein A. (1939) A. Einstein, Ann. Math. 40, 922.

Einstein, A. (1961), Relativity: The Special and the General Theory, New York: Three Rivers Press.

Einstein, A. and Rosen, N. (1935). "The Particle Problem in the General Theory of Relativity". Physical Review 48: 73.

Einstein, A., Lorentz, H.A., Minkowski, H., and Weyl, H. (1923). Arnold Sommerfeld. ed. The Principle of Relativity. Dover Publications: Mineola, NY. pp. 38–49.

Einstein A, Podolsky B, Rosen N (1935). "Can Quantum-Mechanical Description of Physical Reality Be Considered Complete?". Phys. Rev. 47 (10): 777–780.

Eisberg, R., and Resnick. R. (1985). Quantum Physics of Atoms, Molecules, Solids, Nuclei, and Particles. Wily,

Everett, A., & Roman, T. (2012). TIme Travel and Warp Drives, University Chicago Press.

Feinberg, G. (1967). "Possibility of Faster-Than-Light Particles". Physical

Review 159 (5): 1089–1105.

Fuller, Robert W. and Wheeler, John A. (1962). "Causality and Multiply-Connected Space-Time". Physical Review 128: 919.

Garay, L. J. (1995). Quantum gravity and minimum length Int.J.Mod.Phys. A10 (1995) 145-166

Ghez, A. M.; Salim, S.; Hornstein, S. D.; Tanner, A.; Lu, J. R.; Morris, M.; Becklin, E. E.; Duchene, G. (2005). "Stellar Orbits around the Galactic Center Black Hole". The Astrophysical Journal 620: 744.

Geiss, B., et al., (2010) The Effect of Stellar Collisions and Tidal Disruptions on Post-Main-Sequence Stars in the Galactic Nucleus. American Astronomical Society, AAS Meeting #215, #413.15; Bulletin of the American Astronomical Society, Vol. 41, p.252.

Giddings, S. (1995). The Black Hole Information Paradox," Proc. PASCOS symposium/Johns Hopkins Workshop, Baltimore, MD, 22-25 March, 1995, arXiv:hep-th/9508151v1.

Hawking, S. W., (1988) Wormholes in spacetime. Phys. Rev. D 37, 904–910.

Hawking, S., (1990). A Brief History of Time: From the Big Bang to Black Holes. Bantam.

Hawking, S. (2005). "Information loss in black holes". Physical Review D 72: 084013.

Hawking, S. W. (2014). Information Preservation and Weather Forecasting for Black Holes.

Heisenberg, W. (1927), "Über den anschaulichen Inhalt der quantentheoretischen Kinematik und Mechanik", Zeitschrift für Physik 43 (3–4): 172–198.

Heisenberg. W. (1930), Physikalische Prinzipien der Quantentheorie (Leipzig: Hirzel). English translation The Physical Principles of Quantum Theory, University of Chicago Press.

Heisenberg, W. (1955). The Development of the Interpretation of the Quantum Theory, in W. Pauli (ed), Niels Bohr and the Development of Physics, 35, London: Pergamon pp. 12-29.

Heisenberg, W. (1958), Physics and Philosophy: The Revolution in Modern Science, London: Goerge Allen & Unwin.

Jaffe, R. (2005). "Casimir effect and the quantum vacuum". Physical Review D 72 (2): 021301.

Joseph, R (2010a) The Infinite Cosmos vs the Myth of the Big Bang: Red Shifts, Black Holes, and the Accelerating Universe. Journal of Cosmology, 6, 1548-1615.

Joseph, R. (2010b). The Infinite Universe: Black Holes, Dark Matter, Gravity, Acceleration, Life. Journal of Cosmology, 6, 854-874.

Joseph, R. (2014a) Paradoxes of Time Travel: The Uncertainty Principle, Wave Function, Probability, Entanglement, and Multiple Worlds, Cosmology,

18, 282-302.

Joseph, R. (2014a) The Time Machine of Consciousness, Cosmology, 18, In press.

Kerr, R P. (1963). "Gravitational Field of a Spinning Mass as an Example of Algebraically Special Metrics". Physical Review Letters 11 (5): 237–238.

Lambrecht, A. (2002) The Casimir effect: a force from nothing, Physics World, September 2002

Lorentz, H. A. (1892), "The Relative Motion of the Earth and the Aether", Zittingsverlag Akad. V. Wet. 1: 74–79

McClintock, J. E. (2004). Black hole. World Book Online Reference Center. World Book, Inc.

Melia, F. (2003). The Edge of Infinity. Supermassive Black Holes in the Universe. Cambridge U Press. ISBN 978-0-521-81405-8.

Melia, F. (2007). The Galactic Supermassive Black Hole. Princeton University Press. pp. 255–256.

Merloni, A., and Heinz, S., (2008) A synthesis model for AGN evolution: supermassive black holes growth and feedback modes Monthly Notices of the Royal Astronomical Society, 388, 1011 - 1030.

Minchin, R. et al. (2005). "A Dark Hydrogen Cloud in the Virgo Cluster". The Astrophysical Journal 622: L21–L24.

Morris, M. S. and Thorne, K. S. (1988). "Wormholes in spacetime and their use for interstellar travel: A tool for teaching general relativity". American Journal of Physics 56 (5): 395–412.

O'Neill, B. (2014) The Geometry of Kerr Black Holes, Dover

Penrose, R. (1969) Rivista del Nuovo Cimento.

Preskill, J. (1994). Black holes and information: A crisis in quantum physics", Caltech Theory Seminar, 21 October. arXiv:hep-th/9209058v1.

Pollack, G. L., & Stump, D. R. (2001), Electromagnetism, Addison-Wesley.

Rindler, W. (2001). Relativity: Special, General and Cosmological. Oxford: Oxford University Press.

Rodriguez, A. W.; Capasso, F.; Johnson, Steven G. (2011). "The Casimir effect in microstructured geometries". Nature Photonics 5 (4): 211–221.

Ruffini, R., and Wheeler, J. A. (1971). Introducing the black hole. Physics Today: 30–41.

Russell, D. M., and Fender, R. P. (2010). Powerful jets from accreting black holes: Evidence from the Optical and the infrared. In Black Holes and Galaxy Formation. Nova Science Publishers. Inc.

Schrödinger, E. (1926). "An Undulatory Theory of the Mechanics of Atoms and Molecules". Physical Review 28 (6): 1049–1070. Bibcode:1926PhRv...28.1049S. doi:10.1103/PhysRev.28.1049.

Sen, A. (2002) Rolling tachyon," JHEP 0204, 048

Slater, J. C. & Frank, N. H. (2011) Electromagnetism, Dover.

Smolin, L. (2002). Three Roads to Quantum Gravity. Basic Books.

Taylor, E. F. & Wheeler, J. A. (2000) Exploring Black Holes, Addison Wesley.

Thorne, K. (1994) Black holes and time warps, W. W. Norton. NY.

Thorne, K. S. & Hawking, S. (1995). Black Holes and Time Warps: Einstein's Outrageous Legacy, W. W. Norton.

Thorne, K. et al. (1988). "Wormholes, Time Machines, and the Weak Energy Condition". Physical Review Letters 61 (13): 1446.

Wilf, M., et al. (2007) The Guidebook to Membrane Desalination Technology: Reverse Osmosis, Balaban Publishers.

22. Consciousness Of The Future: PreCognition, Premonition, Deja Vu

The Future Already Exists

In a quantum universe all of existence consists of a frenzy of subatomic activity which can be characterized as possessing pure potentiality and all of which are linked and entangled as a basic oneness which extends in all directions and encompasses all dimensions including time (Bohr, 1958, 1963; Dirac, 1966a,b; Planck 1931, 1932, Heisenberg 1955, 1958; von Neumann 1937, 1955). The act of observation be it visual, auditory, tactile, mechanical, digital, is entangled with the quantum continuum and creates a static impression of just a fragment of that quantum frenzy that is registered in the mind of the observer as length, width, height, first, second, and so on; like taking a single picture of something in continual motion, metamorphosis, and transformation. That is, the act of sensory registration, be it a function of a single cell, or the conscious mind of a woman or man, selects a fragment of the infinite quantum possibilities and experiences it as real, but only to that mind or that cell at the moment of registration (Heisenberg 1955, 1958).

"I regard consciousness as fundamental. I regard matter as derivative from consciousness" (Max Planck, 1931).

Time, the fragmentation of time into temporal sequential units and where "causes" precede "effects," are also a "derivative of consciousness." If time is a feature of the quantum continuum, and if considered independent of "consciousness" then causes and effects may be one and the same, a unity and simultaneity; or causes may cause themselves, or effects may be responsible for the causes; and this is because the future is not separate from the present or the past, but all are one extending in multiple directions and dimensions until perceived by a conscious mind.

If time has a wave function not only would the future be linked to the past as a unity, but the conscious mind would be linked to the future and the past, thereby accounting for premonitions (vs anticipation) of what is going to take place; and this is because what will take place has already taken place. The future causes the premonition. And just as likely, the premonition may cause the future due to entanglement.

Then there are "effects" without any apparent "cause;" a possible consequence of the future effecting the present (Aharonov et al. 1988; Bem 2011; Radin 2006' Cho 2011). However, typically, these "effects" are written off as "mistakes" or due to "coincidence." Nevertheless, in an entangled universe the wave function of time representing the future can be predicted to interact with the wave function of the present (and vice versa) thereby inducing a causality-violating reduction of the wave form as perceived by the conscious mind.

Entanglement occurs across space and time with the future entangling the present and the past (Megidish et al. 2013). For example, photons preserve their coherence for long time periods and can be entangled by projection measurements (Kwiat et al. 1995; Weinfurter 1994) and can remain maximally entangled although separated spatially and at remote distances from each other (Goebel et al 2008; Pan et al. 1998). Moreover, entanglement swapping protocols can entangle two remote photons even with a significant time-like separation and without any interaction between them (Ma et al., 2012; Megidish et al. 2013; Peres 2000). Entanglement has been demonstrated even following a delayed choice and even before there was a decision to make a choice. For example, when four photons were created and two were measured, the two became entangled. However, if a choice was then made to measure the remaining two photons, all four became entangled before it was decided to do a second measurement (Ma et al., 2012; Peres 2000). Entanglement can occur independent of and before the act of measurement. "The time at which quantum measurements are taken and their order, has no effect on the outcome of a quantum mechanical experiment" (Megidish et al. 2013) and effects can occur even before a decision is made to do additional measurements. The future, therefore, effects the present and the past.

Moreover, "two photons that exist at separate times can be entangled" (Megidish et al. 2013). As detailed by Megidish et al (2013): "In the scenario we present here, measuring the last photon affects the physical description of the first photon in the past, before it has even been measured. Thus, the "spooky action" is steering the system's past. Another point of view...is that the measurement of the first photon is immediately steering the future physical description of the last photon. In this case, the action is on the future of a part of the system that has not yet been created."

Hence, entanglement between photons has been demonstrated even before the second photon even exists; "a manifestation of the non-locality of quantum mechanics not only in space, but also in time" (Megidish et al 2013). In other words, a photon may become entangled with another photon even before that photon is created, before it even exists in the present; and this is because the photon does exist in the future. Even after the first photon ceases to exist and before the second photon is created, both become entangled even though there is no overlap in time. Photons that do not exist can effect photons which do exist and photons which no longer exist and photons which will exist (Megidish et al. 2013); and presumably the same applies to all particles, atoms, molecules (Wiegner, et al 2011).

Therefore, as indicated by entanglement, the future may effect or take place before the past-present (Megidish et al 2013); and it is these same conditions which likely account for many of the experiences classified as deja vu, premonition, and precognition.

The laws of cause and effect are conscious constructs. Time is a phenomenon which is experienced by consciousness, which then fragments the experience of time into seconds, minutes, hours, days, weeks, the past, present, and future; when, in quantum theory, time just is. We are participant observers, and our conception of time like our conception of the universe is shaped by consciousness and the limitations of the mind and brain. We can not conceive of what we cannot conceive and tend to reject what does not abide by the "laws" erected by consciousness, and this includes the unity of time where causes may precede effects and where the conscious mind may perceive the future before it becomes the present. That the future exists before it becomes the present, has been proved by entanglement and is a fundamental feature of space-time quantum continuum.

The Circle of Time

Time is linked to the Cosmos. There are a variety of conceptions as to the nature of the universe: curved, flat, finite, infinite, and so on. Einstein's relativity predicts a curved universe, the curvature being due to gravity which in turn creates a lumpy universe with waves, valleys, vortexes, eddies, and innumerable geometric contortions all of which effect the trajectory of light and the fourth dimension known as space-time. And yet, because light, matter, and space-time may be torqued by gravity does that necessarily mean the cosmos in-itself is curved. Perhaps these curvatures only effect its contents including the experience of time? In a curved universe anyone traveling in a "straight line" across the cosmos would eventually return to where they started. Einstein's field equations (Gödel 1949a,b) predict that time is a circle and that the future leads to the past.

It is generally assumed that the arrow of time flies from the present into the future. If time consists of just one dimension, then the observer can only go in one direction. And yet, there is nothing in the laws of physics indicating that a particular direction is preferred. Likewise, although light waves travel in a direction, the laws of electromagnetism do not make a distinction between the past and future (Pollack & Stump, 2001; Slater & Frank, 2011). Einstein (1955) argued that the distinctions between the past present and future are an illusion and that the past, present and future differ according to location, gravity, and speed of movement. Like a flowing river, the "present," "past" and "future" are relative to the location of observer along the banks of that river; whereas in fact, the river has no present, past, future, or upstream or downstream, it just flows as its own unity. If space-time is curved as predicted by Einstein (1961) and if time is a circle (Gödel 1949a, b), then the river of time is also a unity as predicted by quantum physics. However, in a curved universe, the future leads to the present and then the past (Gamow, 1946; Gödel 1949a, b).

The eternal now, experienced as the present, is like an event horizon with the future on one side and the past on the other. The future is always arriving whereas the past is always receding. Certainly it is possible that the arrow of time splits and bifurcates at the threshold of consciousness and flows toward the future and the past. However, quantum mechanics and relativity predict that the arrow of "now" is from the future to the present and then the past.

Even if time is in motion and limited to one direction, then the observer would not be moving forward in time, but future time would be flowing toward the observer and continue beyond the observer into the past. Consider the newly wed cowboy and his brushing bride sitting shyly on the edge of a carpet. What he intends to do to her is in the future. He lassos her and pulls her and the rug closer to him, and the rug curls up as he pulls. Although the rug stays the same length, the distance between the cowboy and his bride decreases. The future and the present come closer together. However, the future exists before it arrives. This is proved by relativity and length contraction for accelerated observers (Einstein et al. 1923).

When accelerating toward light speed, space-time contracts, and the distance between the present and the future will decrease; and this means, for the future to arrive more quickly, then it must already exist, albeit in a distant location in space-time, which is what Einstein (1955, 1961) implied when he insisted that the distinctions between past, present and future were illusions. And if the future, a future, or any number of futures already exists, then those futures may effect the present, just as the ripples from a rock thrown into a clear crystal pond can intersect distant shores.

The Future In One Location May Be The Past In Another Location

The star Sirius A is 8.6 light years from Earth. Light-images which left Sirius 1.6 light years ago, are from that star's past, but will not reach Earth for another 7 light years in Earth's future. Therefore, someone living on a planet orbiting Sirius would be able to predict events before they take place on Earth. If something which has already happened is not perceived until the future on Earth, then that future already exists even before it is experienced on Earth.

Conversely, light from our sun and accompanying reflective light-images of Earth from yesterday, are from this solar system's past, but will not reach Sirius for 8.6 light years, and are thus in the future relative to Sirius. The future and the past are relative and overlap and can travel in the same direction in time-space and can occupy the same forward or backward mobile locations in time-space simultaneously. However, again, what has just taken place on Earth, what has just happened, will not be experienced on a planet orbiting Sirius for another

8.6 light years.

There is no universal now, no universal future, no universal past. There are innumerable stars, innumerable observers, and innumerable pasts, presents, and futures. Thus, multiple futures, which already exist, flow from multiple directions toward Earth and which will be experienced by innumerable observers. It is only the conscious mind of the observer which determines what takes place in the present; even if it has already happened in the future.

Predicting A Future Which Exists Before It Is Experienced

Relativity and quantum physics both predict the future exists before it is experienced. However, due to entanglement, and chaos, the future may continually change until the moment it is perceived.

According to Einstein's theories, as a space-time machine accelerates toward light speed, the future arrives more quickly relative to those back on Earth. If the space-time machine were to slow, it would take longer to arrive at the future. Likewise, although light travels at light speed, the speed of light can be bent, curved, and slowed. For example, lights slow by 25% when passing through water and by 35% when passing through glass. Rays of light may also be bent and reflected in multiple directions, as illustrated by galactic lensing.

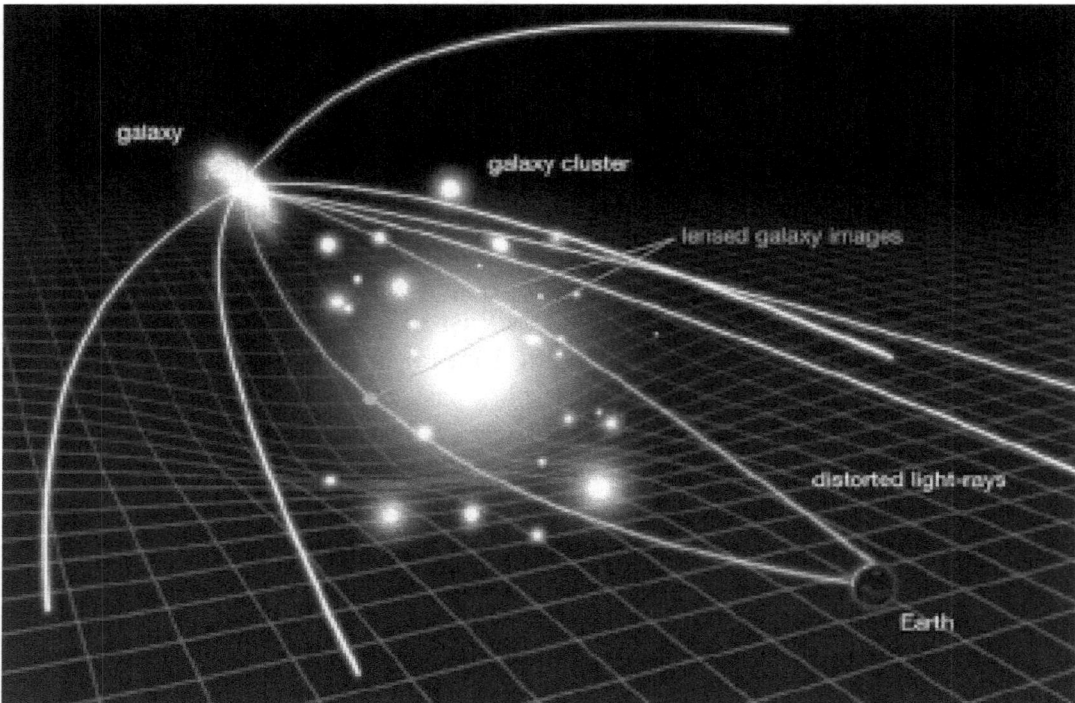

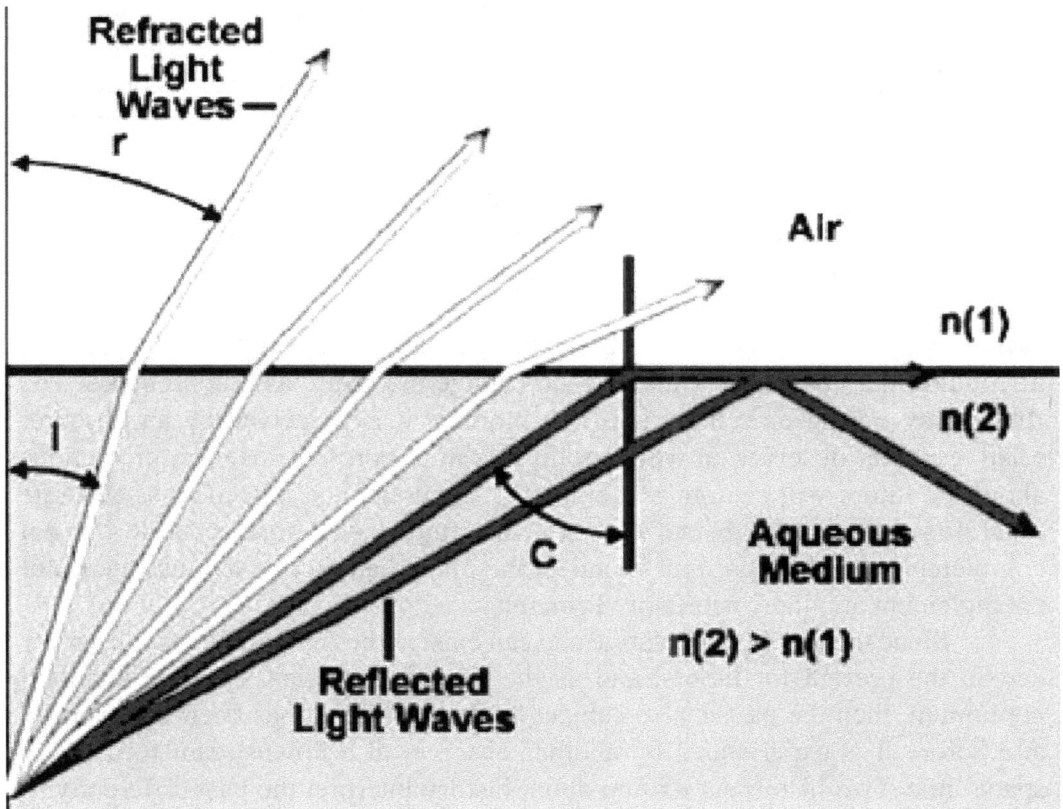

Figures: Refraction and Gravitational Lensing: Light is curved, bent, split.

Time and light are not synonymous. However, since time appears to be linked to light speed, and as time is integral to space-time, then perhaps time, like light can also be bent, curved, split-apart, slowed as well as effect and be effected by all that it encounters, including, consciousness.

Since futures and pasts overlaps and as time-space is coextensive, then time, including local time relative to a single observer, is entangled. The past may effect the future, the future can effect the past, time effects consciousness and alterations in consciousness effect the passage of time.

As a "future" flows toward Earth it can also be effected by whatever it encounters on the way to the present, relative to an observer on Earth--exactly as befalls light. All futures are also entangled with space-time, the quantum continuum, and subject to the Uncertainty Principle. Therefore, future time may be continually altered until perhaps just moments before these futures are experienced by observers who are also entangled with what they experience.

Likewise, one may anticipate and predict the future, just like they may predict the weather. The ability to accurately anticipate and predict the future, like predicting future weather, may increase the closer that future is to the present. Planning skills, goal formation, strategy, long term investments, concern for consequences, and even the most basic of calendars, all rest upon the ability to make predictions about the future.

The future is like the weather, with the ability to forecast the weather decreasing in accuracy as time and distance from the present increases. In other words, and because of entanglement and classic concepts governing "cause and effect", the future is not already determined but is in flux and subject to continual alteration. The act of observing and other forces related to cause and affect alter the quantum continuum and change the future as it approaches. The future may not become fixed until the moment it is perceived by an observer relative to that observer, at which point it is in the present. Hence, predictions about the future will seldom be completely accurate, and become less accurate regarding increasingly distant events in the future, but more accurate but not completely accurate regarding events in the immediate future; a consequence of entanglement and the Uncertainty Principle.

Since the past is also relative and can exist in the future for some observers and in the present for others, and as the past is entangled with the quantum continuum, then the past is also subject to change after it has been experienced and before it is experienced by another observer at a downstream location in space-time. Two historians writing about history interpret the past differently. A husband and wife discussing what happened at a party the night before, disagree. Eye-witness accounts differ among eye-witnesses. A peasant living in a small village in western China in 1963 may have never heard of the assassination of president John F. Kennedy. The past is relative. There is no universal "past."

Time is entangled and is affected by consciousness and relative to and effected by the act of observation and measurement--as predicted by quantum mechanics (Bohr, 1958, 1963; Dirac, 1966a,b; Planck 1931, 1932, Heisenberg 1927, 1958; Neumann 1937, 1955).

Causes and Effects Are Relative To Consciousness

If the future already exists, and if superluminal particles or information can arrive in the present from the future before they are perceived, this, coupled with entanglement (Plenio 2007; Juan et al. 2013; Francis 2012), may result in causes becoming confused with effects, whereas it is the future which is causing and effecting the present (Bem 2011; Radin 2006).

In some respects, sometimes the association between and the classification of one event as a "cause" and the other an "effect" are little more than illusion, as demonstrated by quantum entanglement and "spooky action at a distance" (Francis 2012; Lee et al. 2011; Matson 2012; Olaf et al. 2003; Plenio 2007; Juan et al. 2013;). Just as, according to the quantum physics, the observing mind interacts with the quantum continuum and makes it possible to perceive shape and form, the conscious mind (and the dreaming mind) can also impose temporal order where this is none (Joseph 1982, 1986, 2010a).

For example, it has been well established that the neocortical mantle of the left hemisphere is dominant for the expressive, nominal, grammatical aspects of language, most aspects of mathematical reasoning, and is associated with processing and thinking in terms of temporal sequential order, and beginning and endings (Joseph 1982, 1984, 1986, 1992, 1996, 2011). As demonstrated by quantum mechanics and formalized by the Uncertainty Principle (Heisenberg 1925, 1927), consciousness affects reality at the moment an event or object is measured or registered in consciousness, and thus imposes shape, form, speed, momentum, location, and temporal order. However, what is known, is imprecise (Bohr, 1958, 1963; Dirac, 1966a,b; Planck 1931, 1932, Heisenberg 1955, 1958; Neumann 1937, 1955). To know something in its totality, would require a multi-dimensional all encompassing infinite "god's eye" view.

Moreover, as demonstrated by neuroscience, the mind is not a singularity, but a multiplicity with different aspects of consciousness and awareness directly associated with specific regions of the brain (Joseph, 1992, 1996, 2011). These different mental realms and brain areas can perceive time and the quantum continuum differently. Time may be perceived by one brain region as lacking temporal order but as a continuum or gestalt. The mind is a multiplicity, which can become a duality, and which is often experienced as a singularity referred to as consciousness.

Consciousness is entangled with the space-time continuum which includes the future. Conscious observers can also engage in "mental time travel" (Suddendorf & Corballis 2007). Upon anticipating or looking into the future the observing consciousness can then engage in behaviors that are shaped and directed by that future. What constitutes a cause and what constitutes an effect, are relative and not uncommonly it is the anticipation of the future which causes the cause in the present.

A major corporation, A, which is a defendant in a Federal lawsuit with Plaintiff "B", promises the Judge "C", a large bribe at some future date if he rules against "B" and in favor of "A." After the the Judge rules against "B" he receives his bribe. Thus the "cause" of this judge's behavior, the "bribe", took place at a future date, such that the future affected the "present" (the issuing of false rulings). The future is the cause which effects the present. The cause, causes itself.

A man buys a beautiful woman flowers, candy, jewelry, and an expensive dinner at a five star restaurant. He doesn't lavish these gifts upon the lucky maiden because he loves her, but because he is hoping she will reciprocate, after the date, by giving him sex. The expectation of sex in the future, and thus an event in the future, is the cause of his behavior in the present. The future is the cause which effects and causes his behavior in the present.

Before he bought her these gifts the man may have fantasized about the date, how he would take her to his home, what he would say, what he would do, how she would respond. This could be described as "mental time travel; rehearsing and practicing for a future event before it occurs. As demonstrated by Bem (2011), future practice can effect performance in the present before the practice occurs.

Time is also relative. Hence, when the beautiful woman received these gifts she decided to reward him. Therefore, relative to and from the perspective of the lucky maiden, the effect (sex) is a direct consequence of the cause (his gifts). On the other hand, she also knew that she could cause him to give her gifts by giving him sex in the future. Future sex caused his behavior.

Consciousness is also part of the quantum continuum and so too is the future, present, and past. Thus, consciousness, like gravity and electromagnetic waves, is relative and can affect distant objects and events, including, perhaps, those in the future and the past (Planck 1931, 1932. Moreover, all have a wave function, and time and consciousness are entangled.

The past, present, future are entangled and occupy overlapping as well as distant locations flowing in a variety of directions and creating ripples in the river of time, just as rocks tossed at various distances into the smooth surface of a crystal lake creates ripples and waves which may intersect. It is well established that causes and effects can occur simultaneously and ever faster than light speed (Lee et al. 2011; Matson 2012; Olaf et al. 2003). A future "cause" can "effect" the present even though separated by great distance in space-time as it is all part of the basic oneness of the quantum continuum which is also continuous with time.

However, since consciousness is also entangled, then consciousness may also perceive a future event before it occur; a phenomenon known as "precognition"

The Contraction and Acceleration of Time

If a Time Traveler was 1,900 miles (3,100 kilometers) above Earth and accelerating at 9.8 m/s^2 that is, generating a force of 9.8 N/kg, equal to the gravitational field strength of Earth at its surface, time-space would not contract and his clock would run at the same speed as those on Earth. Under these conditions, the Time Traveler would journey into the past or the future at the same time as those on Earth. This is because velocity, inertial mass and gravitational mass (g-load) would be the same for those on Earth and the Time Traveler

Upon accelerating toward light speed, the length of a space-time machine, unless capable of spin and rotation, will contract in the direction of motion and become increasing thin while maintaining height and width. Upon accelerating to 50% light speed, then 90% then 99.9999999% light speed, time-space would continue to contract, decreasing the distance between the present and the future, possibly with those areas of space-time closest to the nose of the time machine undergoing maximum contraction relative to more distant future locations.

Not all of space-time contracts. Space-time contraction is local and relative to the velocity of the Time Machine, such that the shrinkage of space-time is maximal at its nearest point to the Time Machine. If all inertial frames were equally contracted then clocks on Earth and clocks on the time machine would run at the same speed, and external observers would not see the time machine is shrinking where the Time Traveler would not see clocks running faster on Earth. if inertial frames are equal, time is symmetric.

Just as the exertion of pressure on a substance has its maximum impact beneath the point of contact, and just as the effects of gravity diminish over distance and increase near the center of gravity, time-space increasingly contracts as proximity to the time-machine increases as it moves through space. Like gravity, the proportion of contraction decreases with increasing distance from the time machine, which means, the near future is proportionally closer than the far future.

For example, if a Time Traveler leaves on his journey in the year 2050 with a destination of 2250, and has a velocity of 99% light speed, space-time contraction would also be relative to the distance from the Time Machine, such that those years closest to 2050 (the location of the time machine in time-space) contract more than those years after 2051. In other words, if traveling at near light speed in the year 2050 the distance between the years 2050 and 2051 may contract by 10%, the distance between 2051 and 2052 by 9%, the distance between 2052 and 2053 by 8%, and so on. Upon reaching the year 2051, then the contraction between 2051 and 2052 might contract by 10% and between 2052 and 2053 by 9% and so on, such that amount of space-time contraction increases the closer it is to the Time Machine.

Not all space-time shrinks proportionately. The greatest shrinking would occur locally, within the immediate vicinity of the Time Machine such that the immediate future always arrives at the same time. By contrast, observers back on Earth and those at a great distance from the Time Traveler experience no shrinkage at all.

Into The Past: Invisibility

Relative to those on Earth, the Time Traveler arrives in the future in less time due to the contraction of space-time. Upon reaching light speed, and if traveling in a craft with rotation and spin, the Time Traveler may have become the size of a Planck length or smaller whereas surrounding space-time will have contracted to a singularity. Upon exceeding light speed, the contraction will implode, there is a time-space reversal, contraction continues in a negative direction, and he leaves the future for the past at superluminal speeds. However, he will likely consist of negative energy and negative mass and like all other superluminal objects, particles, or wave, he will be invisible since he is traveling faster than light.

There is a duality, and this is because the Time Traveler is moving faster than light and the reflected light images of the Time Machine will trail behind. Therefore, the Time Machine will have already arrived in the present from the future, and will be on its way to the past, when the light-image, following right behind, finally shows up in the present (relative to the observer). However, the time machine itself will be invisible since it is traveling faster than the light which reflects its image.

Moreover, his reflected light-image, which follows far behind, will be misperceived as traveling not from the future into the past, but from the present into the future. Likewise, "causes" which occur in the future may be misperceived as occurring in the present.

For example, an observer watching a movie running in reverse of a distant rocket orbiting well above Earth may not realize it is playing in reverse. Unless they have expertise in rocketry they may not realize that the rocket is moving backwards in time (from the end of the film to the beginning) . Likewise, an observer on Earth moving forward in time, will see the Time Traveler who is heading into the past, as moving forward in time, but with the tail end of the Time Machine leading the way. However, what the Earth-bound observer sees in the present, has already happened in a future leading to her present and has already left the present for the past. What she experiences has already happened--just the moon you see is from 8 seconds ago.

Information, particles, objects, time machines moving backward in time will appear to be moving forward in time from the reference frame of an observer moving forward in time. The observer will see the reflected light image of the Time Machine as moving parallel forward in time with the observer. As to the actual Time Machine, since it is traveling faster than light, the observer may only perceive a blank spot in the sky and just behind it a reflected light-image which follows. Nevertheless, in either scenario, the Time Traveler will have already journeyed from the future to the present and into the past in advance of the light-image of the Time Machine which comes after.

Likewise, superluminal information from the future can arrive in the present and continue into the past, whereas Earth-bound observers will only detect and perceive the later arriving details which trail behind and arrive moments later at the speed of light. And, this information may not be perceived as from the future, but as coexisting in the present and leading back to the future. And this is because for every moment that leads to the future, the image or the information from the future will be there as it travels form the future to the past.

For example, a Time Machine that leaves in the year 2150 for the year

2120, will be visible to observers in the year 2121, 2122, 2123, up to 2150 as it travels into the past. However, what they see is the light-image trailing right behind at the speed of light.

Information from the future may arrive without being perceived (Bem 2011; Radin 2006), and then it may be perceived (as conveyed by light), and then it may be misperceived as taking place in the present and continuing to take place as time marches on into the future.

Since consciousness is entangled with the quantum continuum, then not only may the mind perceive events before they take place, but they may perceive information traveling from the future to the past at superluminal speeds, and then continue to perceive it as time marches on into the future thereby providing the foundation for precognition and phenomenon such as deja vu.

Moreover, because information may appear in the present at superluminal speeds (Lee et al. 2011; Matson 2012; Olaf et al. 2003), followed by that same information as transmitted at the speed of light, this would explain why phenomenon such as deja vu and precognition are generally limited to what is about to happen, rather than what may take place next week or next year.

Deja Vu

Upon accelerating toward light speed a Time Travel is propelled into the future, but at beyond light speeds he is propelled into the past. A time machine with a velocity beyond light speed will travel faster than the reflected light images of that time machine which lag behind at light speed. The Time Machine will appear in the present only to be followed moments later by those beams of light transporting those images of the Time Machine. Superluminal information from the future can arrive in the present and continue into the past, whereas Earth-bound observers will only perceive the later arriving details which follow behind at the speed of light.

Entanglement commonly occurs at superluminal speeds (Francis 2012; Juan et al. 2013; Plenio 2007; Lee et al. 2011; Matson 2012; Olaf et al. 2003). However, if an entangled consciousness is effected by the passage of that superluminal information this can give rise to retro-cognition (Bem 2011; Radin 2006); knowing something has happened or will happen before it happens.

As illustrated by light-images from distant stars which are from the past but which will arrive on Earth in the future, various "futures" exist prior to being experienced by various observers. If time has a wave function and is entangled with space-time and the quantum continuum, and as the brain and consciousness are part of that continuum (Heisenberg 1958; Planck 1931, 1932), then under certain circumstances a future may effect consciousness prior to being experienced by consciousness. Since entanglement takes place faster than light speed, the leading edge of a future experience may be registered in various conscious minds at superluminal speeds before the future actually arrives at light speed.   The experience of this "time echo" is not uncommon, and has been referred to as deja vu, pre-cognition, and premonitions.

Deja vu is the conscious experience of having experienced some events just moments before the events take place. For example, a man opens the front door, step outsides, drops his keys and then a dog barks and the phone rings, and then he again experiences himself opening the door dropping his keys and then hearing a dog bark and then the ringing of his phone; like a time echo. He thus has the experience that all this has happened before or that he has done this before it happens. He may even say: "I've done this before" and then a few nanoseconds later he experiences himself saying "I've done this before."

Deja vu has been attributed to a delay in the transfer of sensory experiences from one region of the brain to another which receives that information twice, or the transmission of the same experience to the same area of the brain by two different brain areas such that the information is received twice following a brief delay (Joseph 1996). Hence, someone may experience deja vu because two or more areas of the brain are receiving or processing the same message with a slight delay between them. For example, the right and left halves of the brain are interconnected by a massive rope of nerve fibers called the corpus callosum. Each half of the brain is capable of conscious experience (Joseph 1988a,b; 2010a). Usually information is shared between the cerebral hemispheres. However, if there is a delay in transferring these signals, then one or both halves of the brain may sense it has had this experience just moments before thereby giving a sense of familiarity (Joseph 1996).

Brain areas communicate via neurons, and neurons communicate with each other by sending signals over axons which are transmitted to and received by dendrites (at the synaptic junction) belonging to other neurons which in turn may transmit message via their axons at synaptic junctions to the dendrites of other neurons. Impulses between neurons travel at various speeds, ranging from 10 to 50 m / s (Joseph 1996) whereas the speed of light is 300,000 km/sec.

The experience of deja vu has been reported under conditions of altered and heightened brain activity (Bancaud et al., 1994; Gloor 1990; Joseph 1996). Moreover, deja vu has been reported in cases involving the ingestion of anti-viral flu vaccines, such as amantadine and phenylpropanolamine (Taiminen & Jääskeläinen 2001) which increases brain activity by acting on dopamine receptors and increasing dopamine activity.

Heightened brain activity can be likened to an accelerated state of consciousness. Accelerated states are also associated with the contraction of space-time such that future arrives more quickly.

Deja vu, is also associated with heightened and accelerated activity in the inferior temporal lobe which houses the limbic striatum and amygdala, the later of which receives multi-modal sensory information and which normally filters out most of these sensations so the brain is not overwhelmed (Joseph 1996, 2011).

Deja vu has been reported by patients when these areas of the brain have been activated due to direct electrode stimulation (Halgren 1990; Gloor 1990), drug ingestion (Taiminen & Jääskeläinen 2001) or seizure activity (Joseph 1996).

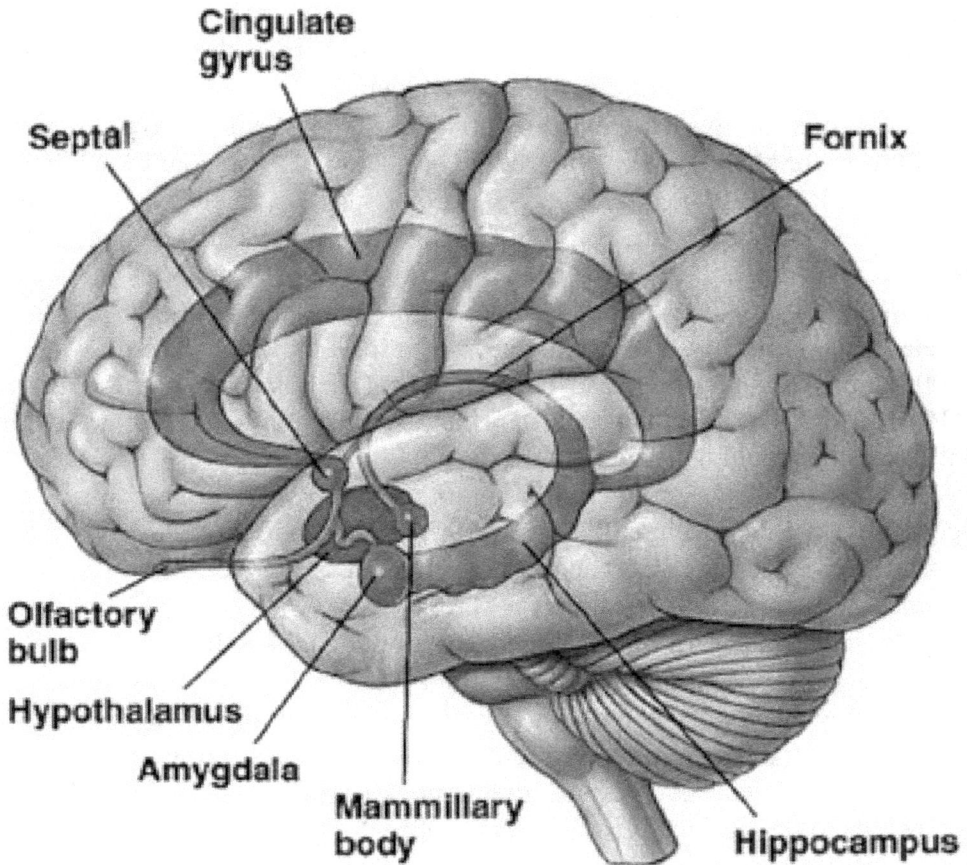

Therefore, when brain activity increases and neurons fire more rapidly and process more information, one of the consequences is Deja vu. In other words, just as a Time Traveler will come closer to the future as he accelerates toward light speed, when brain activity accelerates the future may also come closer such that the leading edge of a future event is experienced by this accelerated state of consciousness just before the event happens in the present.

Precognition: Experimental Proof

Precognition is a form of conscious cognitive awareness which involves the acquisition of future knowledge just prior to its occurrence. Premonitions are a form of presentiment or an emotional feeling that something may happen in the near future, but without conscious knowledge of exactly what it is that is going to happen. Both can be considered forms of quantum entanglement (Radin 2006; Bem, 2011) where some near future event exerts and makes an impression on consciousness before the event occurs even when there is absolutely no way the future event could be inferred as about to happen.

Various surveys have indicated that over 50% of adults have experienced premonitions or phenomenon which could be classified as precognition (Kennedy et al., 1994; Radin 2006). Moreover, numerous rigorous, scientifically controlled experiments and meta-analyses of these experiments have demonstrated statistically significant evidence for precognition and premonitions (Honorton & Ferrari 1989; Radin 2006). For example Honorton and Ferrari (1989) performed a meta-analysis of 309 forced-choice precognition experiments involving over 50,000 subjects, and which had been published in scientific journals between 1936 and 1997. They found a consistent, statistically significant hit rate, meaning that the results could not be due to chance.

As with deja vu, increased brain activity or arousal contributes to precognitive activity (Bem, 2011; Radin 1997, 2006; Spottiswoode & May, 2003). Presentiment effect has also been directly related to increased brain activity as demonstrated in fMRI experiments (Bierman & Scholte, 2002) and with other physiological indices of participants' emotional arousal in which case they become aroused before they see the stimulus (Radin 1997). For example, when participants viewed a series emotionally neutral or emotionally arousing pictures on a computer screen, strong emotional arousal occurred a few seconds before the picture appeared, even before the computer had selected which emotional picture was to be displayed (Radin 1997, 2006).

In 2011, a well respected scientist, Daryl Bem published extensive statistically significant evidence for the effects of future events on cognition and emotion, demonstrating that the effect is in the present whereas the cause can still be in the future. For example, Bem had subjects perform a memory test which required that each subject look at a long list of words and to remember as many as possible. After completing the memory test he had the subjects type various words from that list which were randomly selected. Subjects showed statistically superior memory for the words which they were later asked to type. That is, the practice effect was retrocausal. The practice which was to take place in the future (the typing of words they had already seen) improved their memory of those words before they typed them. Thus, rehearsing a set of words makes them easier to recall even when the rehearsal occurs in the future and after subjects recall the words.

In another set of experiments Bem (2011), allowed a computer to control the entire procedure which involved showing each subject "explicit erotic images." The instructions were as follows: "on each trial of the experiment, pictures of two curtains will appear on the screen side by side. One of them has a picture behind it; the other has a blank wall behind it. Your task is to click on the curtain that you feel has the picture behind it." Statistical analysis of the results demonstrated that based on "feelings" subjects picked the location of the pornographic image at well above chance (even though they couldn't see it), whereas the location of the non-erotic neutral pictures were chosen at the rate of chance, i.e. 49.8% of the time.

Bem (2011) performed nine rigorously controlled experiments involving over 1000 subjects involving erotic stimuli, the avoidance of negative stimuli, and retroactive priming effects on memory and recall. Eight of the nine experiments yielded statistically significant results, and thus evidence for precognition and premonition.

Criticism of Precognition Experimental Results: The Baseball Analogy

A common criticism regarding the validity of research on premonitions and precognition is: if it exists, why doesn't it happen all the time? Why doesn't everyone have these experiences?

Consider major league baseball. In 2013, Miguel Cabrera had a batting average of .348 which was the best of all major league players. Although he is the best hitter in major league baseball, he hit the ball less than 50% of the time when he was at bat, and was able to get a "base hit" less than 35% of the time. Out of 750 major league players, 726 of them got a base hit less than 30% of the time in 2013 during regular season play (http://espn.go.com/mlb/stats/batting). Given that these players had up to 5 opportunities to hit the ball each time at bat, and 3 opportunities to swing, it can be said that professional baseball players actually hit the ball less than 30% of the time. Bem (2011), Raden (1996, 2006) Bonorton and Ferrari (1989) and others have shown a precognition hit rate above 50%. But unlike major league baseball players, those displaying precognition get their hits before they see what is being thrown at them.

Precognition should be treated like all other measures of ability. We should not be surprised that there is variation (Carpenter 2004, 2005; Schmeidler, 1988). Indeed, the same complaints can be made about memory and past events: If it really happened, why does everyone remember it differently. Why do some people have a great memory and others are more forgetful? Why do different eye-witnesses remember the same event differently?

Even highly arousing and emotionally significant "flashbulb memories" are subject to considerable forgetting. For example, Neisser and Harsch (1992) had subjects fill out a questionnaire regarding where they were and how they heard about the Challenger space craft explosion soon after this national tragedy occurred in 1986. When these subjects were questioned again 32-34 months later, 75% could not recall filling out the questionnaire. Many of the subjects in fact had forgotten considerable detail regarding the Challenger explosion and where they were when the heard about it. According to Neisser and Harsch (1992), "As far as we can tell, the original memories are just gone."

Memory is poor. Batting averages are dismal. Should it be any surprise that premonitions and the experience of precognition is also variable?

The Quantum Physics of Premonition and Retrocausation

The phenomenon of premonition must be considered from the perspective of quantum physics not Newtonian physics or Einstein's theories of relativity. As summarized by John Stewart Bell in his 19964 ground breaking paper ("On the Einstein Podolsky Rosen paradox") "any physical theory that incorporates local realism, favoured by Einstein cannot reproduce all the predictions of quantum mechanical theory."

In 2006, the American Association for the Advancement of Science organized an interdisciplinary conference of research scientists and physicists to discuss evidence for retrocausation as related to quantum physics, the conclusions of which were published in 2006: "it seems untenable to assert that time-reverse causation (retrocausation) cannot occur, even though it temporarily runs counter to the macroscopic arrow of time" (Sheehan, 2006, p vii).

As demonstrated by quantum physics and entanglement, the future may effect and even direct the past or the present. Consider again entanglement between photons. In delayed choice experiments, entanglement was demonstrated among photons even before there was a decision to make a choice regarding these photons, that is, before it was decided to do a measurement (Ma et al., 2012; Peres 2000). Entanglement has also been demonstrated among photons which do not yet exist, where the choice has not even been made to create or measure future photons. Nevertheless, decisions which will be made in the future effect the measurement of photons in the present (Megidish et al 2013). The same principles can be applied to precognition. Information in the future, information which does not yet exist in the present, can effect and is entangled with the consciousness which will directly perceive that information once it arrives in the present.

The future, past, present, and consciousness are entangled within the quantum continuum. The future exists before it arrives and some people consciously perceive a future before it becomes the present; phenomenon which can be classified as evidence of entanglement and which are variably experienced as deja vu, premonitions, and precognition.

The future, past, present, and consciousness are entangled within the quantum continuum. The future exists before it arrives and some people consciously perceive a future before it becomes the present; phenomenon which can be classified as evidence of entanglement and which are variably experienced as deja vu, premonition, and precognition.

REFERENCES

Al-Khalili (2011). Black Holes, Wormholes and Time Machines, Taylor & Francis.

Almheiri, A. et al. (2013). Black Holes: Complementarity or Firewalls? J. High Energy Phys. 2, 062

Bethe, H. A., et al., (2003) Formation and Evolution of Black Holes in the Galaxy, World Scientific Publishing.

Bilaniuk, O.-M. P.; Sudarshan, E. C. G. (1969). "Particles beyond the Light Barrier". Physics Today 22 (5): 43–51.

Bo, L., Wen-Biao, L. (2010). Negative Temperature of Inner Horizon and Planck Absolute Entropy of a Kerr Newman Black Hole. Commun. Theor. Phys. 53, 83–86.

Bohr, N., (1913). "On the Constitution of Atoms and Molecules, Part I". Philosophical Magazine 26: 1–24.

Bohr, N., (1913). "On the Constitution of Atoms and Molecules, Part I". Philosophical Magazine 26: 1–24.

Bohr, N. (1934/1987), Atomic Theory and the Description of Nature, reprinted as The Philosophical Writings of Niels Bohr, Vol. I, Woodbridge: Ox Bow Press.

Bruno, N. R., (2001). Deformed boost transformations that saturate at the Planck scale. Physics Letters B, 522, 133-138.

Blandford, R.D. (1999). "Origin and evolution of massive black holes in galactic nuclei". Galaxy Dynamics, proceedings of a conference held at Rutgers University, 8–12 Aug 1998, ASP Conference Series vol. 182.

Carroll, S (2004). Spacetime and Geometry. Addison Wesley.

Casimir, H. B. G. (1948). "On the attraction between two perfectly conducting plates". Proc. Kon. Nederland. Akad. Wetensch. B51: 793.

Chodos, A. (1985). "The Neutrino as a Tachyon". Physics Letters B 150: 431.

Einstein, A. (1915a). Fundamental Ideas of the General Theory of Relativity and the Application of this Theory in Astronomy, Preussische Akademie der Wissenschaften, Sitzungsberichte, 1915 (part 1), 315.

Einstein, A. (1915b). On the General Theory of Relativity, Preussische Akademie der Wissenschaften, Sitzungsberichte, 1915 (part 2), 778–786.

Einstein A. (1939) A. Einstein, Ann. Math. 40, 922.

Einstein, A. (1961), Relativity: The Special and the General Theory, New York: Three Rivers Press.

Einstein, A. and Rosen, N. (1935). "The Particle Problem in the General Theory of Relativity". Physical Review 48: 73.

Einstein, A., Lorentz, H.A., Minkowski, H., and Weyl, H. (1923). Arnold

Sommerfeld. ed. The Principle of Relativity. Dover Publications: Mineola, NY. pp. 38–49.

Einstein A, Podolsky B, Rosen N (1935). "Can Quantum-Mechanical Description of Physical Reality Be Considered Complete?". Phys. Rev. 47 (10): 777–780.

Eisberg, R., and Resnick. R. (1985). Quantum Physics of Atoms, Molecules, Solids, Nuclei, and Particles. Wily,

Everett, A., & Roman, T. (2012). TIme Travel and Warp Drives, University Chicago Press.

Feinberg, G. (1967). "Possibility of Faster-Than-Light Particles". Physical Review 159 (5): 1089–1105.

Fuller, Robert W. and Wheeler, John A. (1962). "Causality and Multiply-Connected Space-Time". Physical Review 128: 919.

Garay, L. J. (1995). Quantum gravity and minimum length Int.J.Mod.Phys. A10 (1995) 145-166

Ghez, A. M.; Salim, S.; Hornstein, S. D.; Tanner, A.; Lu, J. R.; Morris, M.; Becklin, E. E.; Duchene, G. (2005). "Stellar Orbits around the Galactic Center Black Hole". The Astrophysical Journal 620: 744.

Geiss, B., et al., (2010) The Effect of Stellar Collisions and Tidal Disruptions on Post-Main-Sequence Stars in the Galactic Nucleus. American Astronomical Society, AAS Meeting #215, #413.15; Bulletin of the American Astronomical Society, Vol. 41, p.252.

Giddings, S. (1995). The Black Hole Information Paradox," Proc. PASCOS symposium/Johns Hopkins Workshop, Baltimore, MD, 22-25 March, 1995, arXiv:hep-th/9508151v1.

Hawking, S. W., (1988) Wormholes in spacetime. Phys. Rev. D 37, 904–910.

Hawking, S., (1990). A Brief History of Time: From the Big Bang to Black Holes. Bantam.

Hawking, S. (2005). "Information loss in black holes". Physical Review D 72: 084013.

Hawking, S. W. (2014). Information Preservation and Weather Forecasting for Black Holes.

Heisenberg, W. (1927), "Über den anschaulichen Inhalt der quantentheoretischen Kinematik und Mechanik", Zeitschrift für Physik 43 (3–4): 172–198.

Heisenberg. W. (1930), Physikalische Prinzipien der Quantentheorie (Leipzig: Hirzel). English translation The Physical Principles of Quantum Theory, University of Chicago Press.

Heisenberg, W. (1955). The Development of the Interpretation of the Quantum Theory, in W. Pauli (ed), Niels Bohr and the Development of Physics, 35, London: Pergamon pp. 12-29.

Heisenberg, W. (1958), Physics and Philosophy: The Revolution in Modern Science, London: Goerge Allen & Unwin.

Jaffe, R. (2005). "Casimir effect and the quantum vacuum". Physical Review D 72 (2): 021301.

Joseph, R (2010a) The Infinite Cosmos vs the Myth of the Big Bang: Red Shifts, Black Holes, and the Accelerating Universe. Journal of Cosmology, 6, 1548-1615.

Joseph, R. (2010b). The Infinite Universe: Black Holes, Dark Matter, Gravity, Acceleration, Life. Journal of Cosmology, 6, 854-874.

Joseph, R. (2014a) Paradoxes of Time Travel: The Uncertainty Principle, Wave Function, Probability, Entanglement, and Multiple Worlds, Cosmology, 18, 282-302.

Joseph, R. (2014a) The Time Machine of Consciousness, Cosmology, 18, In press.

Kerr, R P. (1963). "Gravitational Field of a Spinning Mass as an Example of Algebraically Special Metrics". Physical Review Letters 11 (5): 237–238.

Lambrecht, A. (2002) The Casimir effect: a force from nothing, Physics World, September 2002

Lorentz, H. A. (1892), "The Relative Motion of the Earth and the Aether", Zittingsverlag Akad. V. Wet. 1: 74–79

McClintock, J. E. (2004). Black hole. World Book Online Reference Center. World Book, Inc.

Melia, F. (2003). The Edge of Infinity. Supermassive Black Holes in the Universe. Cambridge U Press. ISBN 978-0-521-81405-8.

Melia, F. (2007). The Galactic Supermassive Black Hole. Princeton University Press. pp. 255–256.

Merloni, A., and Heinz, S., (2008) A synthesis model for AGN evolution: supermassive black holes growth and feedback modes Monthly Notices of the Royal Astronomical Society, 388, 1011 1030.

Minchin, R. et al. (2005). "A Dark Hydrogen Cloud in the Virgo Cluster". The Astrophysical Journal 622: L21–L24.

Morris, M. S. and Thorne, K. S. (1988). "Wormholes in spacetime and their use for interstellar travel: A tool for teaching general relativity". American Journal of Physics 56 (5): 395–412.

O'Neill, B. (2014) The Geometry of Kerr Black Holes, Dover

Penrose, R. (1969) Rivista del Nuovo Cimento.

Preskill, J. (1994). Black holes and information: A crisis in quantum physics", Caltech Theory Seminar, 21 October. arXiv:hep-th/9209058v1.

Pollack, G. L., & Stump, D. R. (2001), Electromagnetism, Addison-Wesley.

Rindler, W. (2001). Relativity: Special, General and Cosmological. Oxford: Oxford University Press.

Rodriguez, A. W.; Capasso, F.; Johnson, Steven G. (2011). "The Casimir effect in microstructured geometries". Nature Photonics 5 (4): 211–221.

Ruffini, R., and Wheeler, J. A. (1971). Introducing the black hole. Physics

Today: 30–41.

Russell, D. M., and Fender, R. P. (2010). Powerful jets from accreting black holes: Evidence from the Optical and the infrared. In Black Holes and Galaxy Formation. Nova Science Publishers. Inc.

23. The Sixth Dimension: Dream Time, Precognition, Many Worlds

It is through dreams that we may be transported to worlds that defy the laws of physics and which obeys their own laws of time, space, motion and conscious reality, where the future is juxtaposed with the past and where time runs backwards and forwards (Campbell, 1988; Freud, 1900; Jung, 1945, 1964). Throughout history it has been believed that dreams open doors to alternate realities, to the future, to the past, and the hereafter, where the spiritual world sits at the boundaries of the physical; hence the tendency to bury the dead in a sleeping position even 100,000 years ago (Joseph 2011a,b). Although but a dream, the dream was experienced during dream consciousness much as the waking world is experienced by waking consciousness. The dream was real. Thus, throughout history dreams have been taken seriously especially when they gave glimpses of the future.

Dream are often of events from the previous day and may concern the future. It is through dreams that the dreamer may gain insight into problems which have plagued him or which he anticipates encountering in the near future. Just as one can think about the future or the past and make certain deductions and predictions, a dream may include of anticipations regarding the future, and in this respect, the dream could be considered an imaginal means of preparation for various possible realities. As such, dream-time and dream-consciousness could be considered manifestations of the "Many Worlds" theory of quantum physics.

Not uncommonly the dream will include so many branching and overlapping multiple realities that it makes no sense at all, except to those skilled in the art of interpreting dream symbolism (Freud 1900; Jung 1945, 1964). Indeed, it is due to the non-temporal, often gestalt nature of dreams which require that they be consciously scrutinized from multiple angles in order to discern their meaning, for the last may be first and what is missing may be just as significant as what is there.

Dream-consciousness and dream-time are also associated with accelerated levels of brain activity (Joseph 2000, 2011a). Relativity predicts that observers with an accelerated frame of reference experience time-dilation and a shrinking of time-space such that the future and the present come closer together relative to those with a different frame of reference (Einstein et al. 1923, Einstein 1961). Thus, during dream-time, the dreamer may see or experience the future before that future is experienced by the conscious mind.

Many Worlds of Quantum Dream-Time

There are at least three dimensions of time: space-time which obeys the laws of relativity; conscious-time which obeys the laws of quantum physics;

and dream-time which defies the laws of Newtonian physics but which could be viewed as a manifestation of the "Many Worlds" interpretation of quantum mechanics. Whereas conscious-time occupies the 5th dimension, dream-time is the 6th dimension of the quantum continuum and the space-time manifold embraced by relativity.

Although dream-time encompasses dream-consciousness, dream-time and conscious-time are distinct. Consciousness and dreaming are not synonymous. Dreams may be observed by consciousness and as such, dreaming and consciousness are entangled as dream-consciousness. However, consciousness is generally little more than a passive witness during dreaming, an audience before the stage upon which the dreams are displayed in all their mystery and majestic glory. It is rare for consciousness to become conscious that "it" is observing a dream, and when such rarities occur the dreamer may awaken or briefly take an active role in what has been described as "lucid dreaming" (LaBerge, 1990).

Unlike conscious-time and the conscious mind, the dream-kaleidescape of dream-time and dream-consciousness could best be described as manifestation of the "Many worlds" interpretation of quantum physics where all worlds are possible and past and future and time and space are juxtaposed and intermingled; time can run backward and forward simultaneously and at varying speed, and multiple realities come and go no matter now improbable. Dream-time represents accelerated states of brain activity and is entangled with the "many worlds" and the space-time quantum continuum of future and past, and as such, while dreaming, the dreamer may obtain a glimpse of the future before it arrives.

Abraham Lincoln Dreams Of His Death

In April of 1965, less than two weeks before he was gunned down by an assassin's bullet, President Abraham Lincoln dreamed of his own assassination (Lamon 1911). Lincoln told this dream to his wife and to several friends including Ward Hill Lamon who was Lincoln's personal friend, body guard and former law partner. According to Lincoln:

"About ten days ago, I retired very late. I had been up waiting for important dispatches from the front. I could not have been long in bed when I fell into a slumber, for I was weary. I soon began to dream. There seemed to be a death-like stillness about me. Then I heard subdued sobs, as if a number of people were weeping. I thought I left my bed and wandered downstairs. There the silence was broken by the same pitiful sobbing, but the mourners were invisible. I went from room to room; no living person was in sight, but the same mournful sounds of distress met me as I passed along. I saw light in all the rooms; every object was familiar to me; but where were all the people who were grieving as if their hearts would break? I was puzzled and alarmed. What could be the meaning of all this? Determined to find the cause of a state of things so mysterious and so shocking, I kept on until I arrived at the East Room, which I entered. There I met with a sickening surprise. Before me was a catafalque, on which rested a corpse wrapped in funeral vestments. Around it were stationed soldiers who were acting as guards; and there was a throng of people, gazing mournfully upon the corpse, whose face was covered, others weeping pitifully. 'Who is dead in the White House?' I demanded of one of the soldiers, 'The President,' was his answer; 'he was killed by an assassin.' Then came a loud burst of grief from the crowd, which woke me from my dream. I slept no more that night; and although it was only a dream, I have been strangely annoyed by it ever since."

The 6 Dimensional Time-Space Manifold

Euclidean geometry dictates that the universe has three dimensions of space and one of time which is independent of motion, with time progressing at a fixed rate in all reference frames. Time is treated as universal and constant, being independent of the state of motion of an observer.

In Euclidean space, the separation between two points is measured by the distance between them. The distance is purely spatial, and is always positive. A minus sign is associated with the dimension of time. If two events have a greater separation in time than in space, this results in a negative and they have a time-like separation. If the quantity is positive, the two events have a space-like separation which is greater than their separation in time. If the result is 0, then the two events have a light-like separation and are connected only by a beam of light.

For example, if a football player throws a pass and it is caught by a retriever 50 feet distant after an elapse of 5 seconds (5000000000 nanoseconds), then the separation in space is 50 and the separation in time is 5. By calculating the square of the separation in space minus the square of the separation in time, it can then be determined if the separation of the two events are space-like or time-like:

50^2 (feet) 5000000000^2(nanoseconds) $= 2,500$ $500000000000 =$ -499999998500

In this instance, since the results are a negative, the two events have a greater separation in time than in space.

The star Proxima Centauri is 4.2 light years from Earth. If the time traveler was required to visit that star 10 years from today, then they would be separated by 4.2 light years in space and 10 years in time. Thus it would be 4.2^2(space) 10^2 (time) $= 17.64$ $100 = -82.38$ meaning that the two events are separated in time and have a -82.38 time-like separation from now. However, if the time traveler had to arrive on Proxima Centauri in 3 years, then it would have a space-like separation and it would be impossible for him to get there is 3 years since there is not enough time; unless he were to exceed the speed of light or travel via dream-time.

In relativity, space and time are combined as a single continuum within a fourth dimension: space-time (Einstein 1961). In space-time, the separation between two events is measured by the invariant interval between the two events, which takes into account not only the spatial separation between the events, but also their temporal separation. For two events separated by a time-like interval, enough time passes between them that there could be a cause–effect relationship between the two events.

In space-time, a coordinate grid that spans the 3+1 dimensions locates events (rather than just points in space) whereas the other 3 dimensions (considered separately) locate a point and location in a certain defined "space." The spatial location of an event is designated by three coordinates, X, Y, Z, whereas a fourth coordinate is based on time; all of which constitute "frames of reference." Together the 3+1 dimensions locates events and when and where they took place. Therefore, in relativistic contexts time remains entangled with the other three dimensions of space (length, width, height).

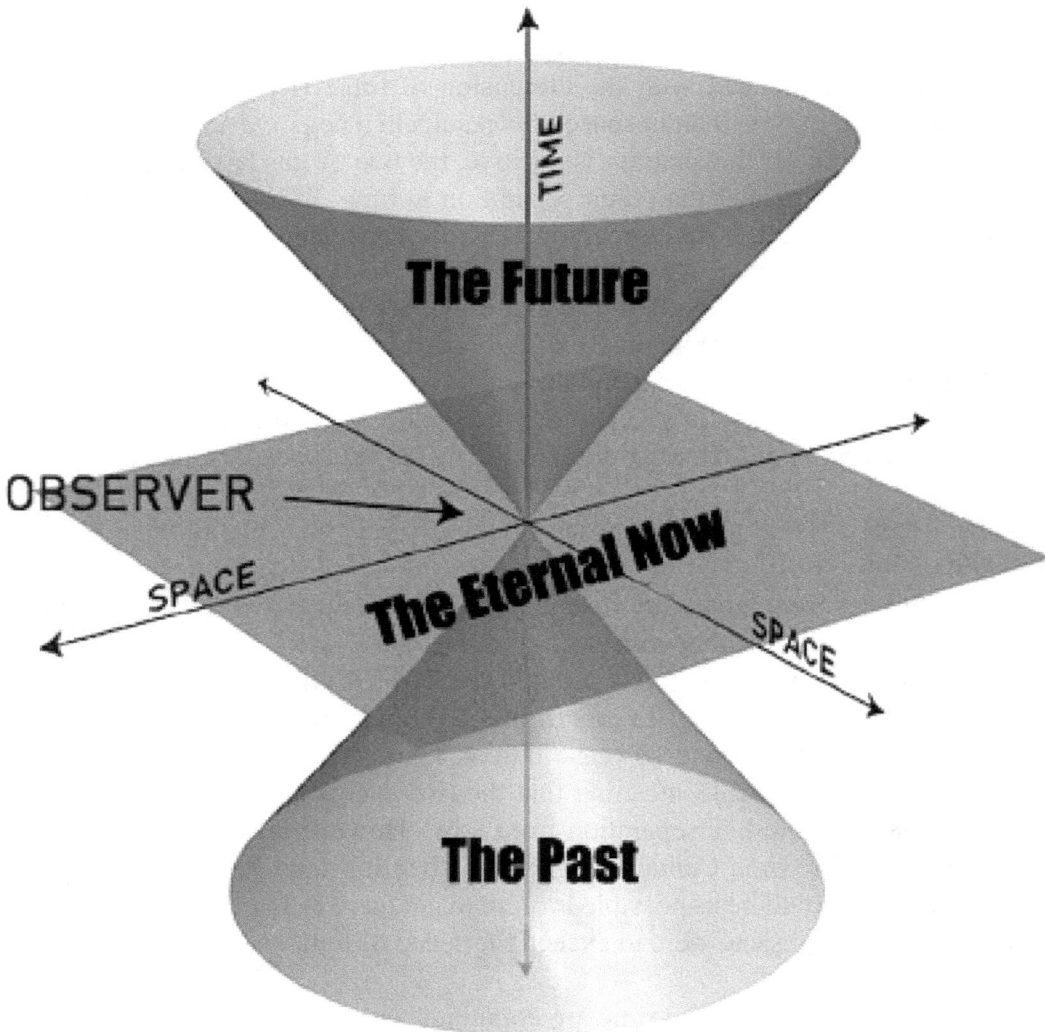

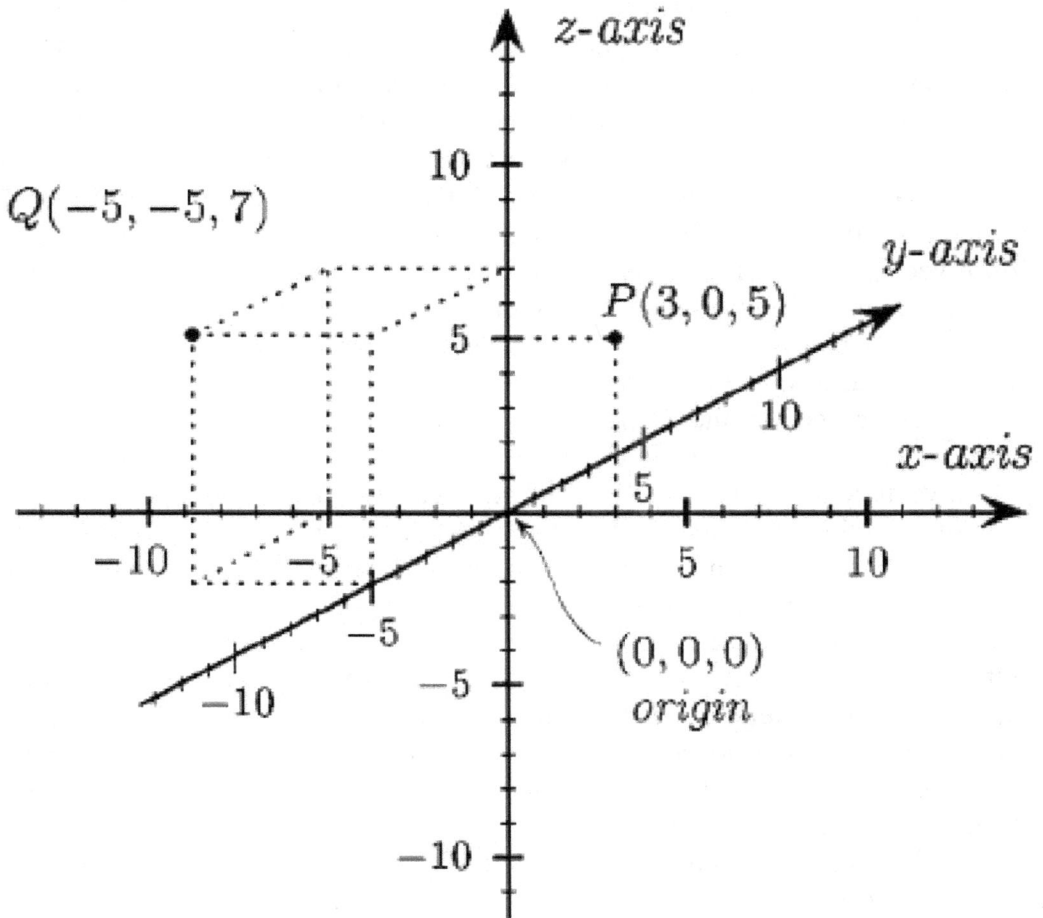

As theorized by Einstein (1961), and unlike the Copenhagen model of quantum physics (Bohr 1949; Heisenberg 1927, 1958), space-time is relative to but independent of any observer. Consciousness is relative but irrelevant having no effect on the passage of time. In relativity, each event which occurs at certain moments of time in a given region of space are relative to those observers in different regions of space, such that each observer chooses a convenient metrical coordinate system in which these events are specified by four real numbers. All four dimensions are measured in terms of units of distance, e.g. in two spatial dimensions x and y and a time dimension orthogonal to x and y.

However, in further contrast to the Euclidian perspective, in an Einsteinian universe the observed rate at which time passes for an object depends on accelerated frames of reference and the strength of gravitational fields; all of which can slow or accelerate the passage of time, depending on the object's velocity relative to the observer (Einstein 1961). Specifically, time slows at higher speeds in one reference frame relative to another reference frame. The duration of time can therefore vary according to reference frames.

In relativity consciousness is merely relative. In quantum physics, consciousness is a separate reference frame which can collapse the wave function and register entangled interactions within the environment. Consciousness by the act of observation or measurement takes a static or series of pictures-in-time which then becomes discontinuous from the quantum continuum (Heisenberg 1958; Planck 1931; von Neumann 2001). These entanglements (Francis 2012; Juan et al. 2013; Plenio 2007), or blemishes in the quantum continuum, may be observed as shape, form, cause, effect, past, present, future, the passage of time, and thus reality; the result of a decoupling of quanta from the quantum (coherent) continuum which leaks out and then couples together in a knot of activity which is observed as a wave form collapse.

As based on the Copenhagen theory of quantum mechanics (Bohr, 1958, 1963; Heisenberg 1955, 1958), time and reality are a manifestation of wave functions and alterations in patterns of activity within the quantum continuum which are perceived by consciousness as discontinuous. Wave form collapse is always a matter of probability, and is non-local, indeterministic and a consequence of conscious observation, measurement, and entanglement. Consciousness, therefore, is entangled with the quantum continuum, and alterations in consciousness can alter the continuum and the space-time manifold. Just as a Time Traveler can accelerate or slow down, contracting or dilating time and time-space, consciousness can accelerate or slow down, and time will slow down or speed up accordingly. Conscious-time represents a second dimension of time, whereas dream-time represents a third dimension of time.

During dream-time the brain is in a "paradoxical" state of accelerated activity, known as paradoxical sleep, as demonstrated by rapid eye movement (REM) and electrophysiological activity (Frank, 2012, Pagel, 2014, Stickgod & Walker, 2010). As predicted by Einstein's (1961) relativity, under accelerated states time contracts and the future arrives more quickly. Therefore, in dream-time one may visit the future or the past during the course of the dream.

A fifth dimension of conscious-time, and a sixth dimension of dream-time creates a 6 dimensional time-space manifold. This would be conceptualized as three dimensions of space^2 minus the separation of -time^2, minus the separation of conscious-time ($-ct^2$) minus the separation of dream-time ($-dt^2$).

Dream-Time Of The Future

Because conscious-time and dream-time appear to have no significant mass or gravity they would not significantly affect, for example, the orbits of planets around stars, the orbit of this star around the Milky Way galaxy, and so on, except through entanglement and observation, in which case the effects outside of dream-time and conscious-time, would most likely be inconsequential or local.

A separation in conscious-time would include the "past," "present," and "future." In dream-time past-present-future and the three dimensions of space may exist simultaneously as a gestalt thereby violating all the rules of causality and the laws of physics. During dream-time events may occur in a logical or semi-logical temporal sequence, or they may be juxtaposed and make no sense at all. Because the future past present may exist simultaneously and as the future may be experienced in a dream during accelerated states of brain activity, then during dream consciousness the dreamer may get glimpses of future events which may occur within days, the next morning, or which may even trigger wakefulness. Access to the future occurs because dream-time takes place during an accelerated state of dream-consciousness and thus the future comes closer to the present and the dreamer arrives in the future in less time than those who are awake. In other words, just as increased velocity causes a contraction of space-time thereby decreasing the distance between the present and the future (Einstein 1961, Einstein et al. 2913), accelerated dream-consciousness has the same effect.

Dream-Time and the Many Worlds of Quantum Physics

In dream-time and dream-consciousness all worlds are possible simultaneously and in parallel. These many worlds include those of the future and the past and where time and space are juxtaposed and every probable outcome is equally likely, and where the world is continually splitting into alternate worlds. Dream-time-consciousness is a manifestation of and in many respects obeys the laws of the "Many Worlds" theory of quantum physics as first proposed by Hugh Everett (1956, 1957).

Hugh Everett's "theory of the universal wavefunction" (Many Worlds) is distinguished from the Copenhagen model, as there is no special role for an observing consciousness. Everett also removed the "wave function" collapse which he believed to be redundant, and instead insisted that what is observed must be clearly defined (thereby answering one of Einstein's criticism of quantum theory). According to Everett's theory, every action, every measurements, every behavior, every choice, even not choosing, can create a new reality, another world, generating a bifurcation between what happened and what did not happen, such that innumerable possibilities and possible worlds arise from every action, including realties which do not obey the laws of physics and cause and effect

As conceived by Everett (1956, 1957) and Dewitt (1971), when a physicist measures an object, the universe splits into two distinct universes to accommodate each of the possible outcomes. In one universe, the physicist measures the wave form, in the other universe the physicist measures the object as a particle. Since all objects have a particle-wave duality, this also explains how an object can be measured as a particle and can be measured as a wave, but not both at the same time in the same world, and how it can be measured in more than one state, each of which exists in another world. The simple act of measurement creates two worlds both of which exist at the same time in parallel, and each separate version of the universe contains a different outcome of that event.

Instead of one continuous timeline, the universe under the many worlds interpretation looks more like a forest of trees with innumerable branches and twigs each of which represents a different possible world. According to Everett the entire universes continuously exists in a superposition and juxtaposition of multiple states. In many respects, Everett's theory defines dream-time and dream-consciousness.

According to Everett (1957), observation and measurement does not force the object under observation to take any specific form or to have any specific outcome. Instead, all outcomes are possible; much like a dream. For example, an NFL football player, a receiver, is running down the field and the quarterback throws him the ball. According to the "Many Worlds" interpretation of quantum physics, every conceivable and incomprehensible outcome is possible: The receiver catches or doesn't catch the ball. A female cheerleader runs out into the field and catches the ball. The receiver and the cheerleader ignore the ball and take off their clothes and have sex on the field. The head coach takes out a shotgun and begins shooting at the football. Spectators run onto the football field and erect circus tents and it becomes a giant carnival with rides. Some of the football players dress up as clowns and circus performers. Players and spectators lay on the grass and swim toward the goal posts. An alien space ship crashes into the football stadium and aliens emerge selling popcorn. Terrorists attack the football players and steal the football, and so on.

All outcomes are possible in Multiple Worlds, ranging from the most probable, to the least probable (Dewitt 1971). Thus every probable outcome is possible; trillions of outcomes including those where the improbable, and in defiance of physics may become the law of the land. Moreover, each of these multiple realities exist, simultaneously, side-by-side, in parallel. They exist simultaneously with the reality in which the observer resides; and whatever reality houses the observer is just one probable reality.

PreCognition in Dream-Time and Dream-Consciousness

During dream-time and during dream-consciousness the reality being dreamed is characterized by every possible outcome. Some dream worlds exist in the future, others in the past, and yet others in a world where past, present and future are juxtaposed and exist simultaneously and where every possible outcome is possible. Thus, in dream-time, the dream-consciousness can witness any number of these possible worlds including those which exist in the future.

However, these futures and possible futures which are observed by dream-consciousness are not "just a dream." According to the Many Worlds interpretation, they actually exist. In terms of space-time, these future worlds exist in the future, in a distant location. As predicted by quantum mechanics, the observer is entangled with that future. However, in dream-time the observer (dream consciousness) directly observes that future; including those futures which are improbable or most probable.

The Many Worlds of dream-consciousness provides the foundation for dream-time precognition. The dreamer may dream of the future just before it occurs. And upon waking from that dream of the future, the conscious mind may remember it and then experience it as it occurs in real time.

Dream-time access to the future is made possible because the brain is in a state of accelerated activity during the course of the dream. As predicted by Einstein (1961) an accelerated frame of reference brings the future closer to the present and makes time travel possible. Accelerated states of consciousness not only bring the future closer, but provide glimpses of those futures before they occur; a phenomenon best described as pre-cognition in dream time.

Aberfan Disaster Dream-Time Precognition

Aberfan is a small village in South Wales. Throughout late September and October 1966, heavy rain lashed down on the area and seeped into the porous sandstone of the hills which surrounded the town and against which abutted the village school (Barker 1967).

On September 27 1966, Mrs SB of London dreamed about a school on a hillside, and a horrible avalanche which killed many children.

On October 14, 1966 Mrs GE from Sidcup, dreamed about a group of screaming children being covered by an avalanche of coal.

On October 20, 1966 Mrs MH, dreamed about a group of children who were trapped in a rectangular room and the children were screaming and trying to escape.

On October 20, 1966, a 10 year old child living in Aberfan woke up screaming from a nightmare. She told her parents that in her dream she was trying to go to school when "something black had come down all over it" and there was "no school there."

On October 21, 1966, part of the rain soaked hills of Aberfan gave way and half a million tons of debris slid toward the village of Aberfan and slammed into the village school. The 10 year old girl who dreamed of the tragedy and 115 other schoolchildren and 28 adults lost their lives when the school was smashed and covered with mud. There were less than a dozen survivors (Baker, 1967).

Assassination of Archduke Francis Ferdinand: Dream-Time Precognition

In June of 1914, Austria was seeking to expand it's central European empire; plans which were resented by neighboring states, including Serbia, who wished to remain independent. That same month, the Archduke Francis Ferdinand, nephew of the Austrian Emperor Francis Joseph, went on a diplomatic tour accompanied by his wife, to build alliances with the leaders of these independent nations. In late June he and his wife arrived in Sarajevo, Serbia.

On the evening of June 27, 1914, Bishop Joseph Lanyi prepared for bed and upon falling asleep he began to dream. The Archduke Franz Ferdinand of Austria, heir to the throne of Austria, had been the Bishop's student and pupil, and late that night the Archduke appeared in Bishop Lanyi's dream. The dream became a nightmare and at 3:15 AM Bishop Joseph Lanyi awoke, frightened, upset and in tears. He glanced at the clock, dressed himself, and because the dream was so horrible, he wrote it down:

"At a quarter past three on the morning of 28th June, 1914, I awoke from a terrible dream. I dreamed that I had gone to my desk early in the morning to look through the mail that had come in. On top of all the other letters there lay one with a black border, a black seal, and the arms of the Archduke. I immediately recognized the letter's handwriting, and saw at the head of the notepaper in blue colouring a picture which showed me a street and a narrow side-street. Their Highnesses sat in a car, opposite them sat a General, and an Officer next to the chauffeur. On both sides of the street there was a large crowd. Two young men sprang forward and shot at their Highnesses."

In the dream, Bishop Lanyi read the dream-letter, which had been written by the Archduke. According to the Bishop's account, which he wrote down in the early predawn hours of June 28, the dream letter from the Archduke was as follows: "Dear Dr Lanyi: Your Excellency. I wish to inform you that my wife and I were the victims of a political assassination. We recommend ourselves to your prayers. Cordial greetings from your Archduke Franz. Sarajevo, 28th June, 3.15 a.m."

Bishop Joseph Lanyi was convinced that the Archduke had been assassinated, and called his parishioners and household staff to tell them of the terrible news. Later that morning of June 28, 1914, the Bishop held a mass for the Archduke and his wife. But, the Archduke were still alive and would not be shot dead for another 2 hours.

On June 28, 1914, at 11 a.m., as the Archduke and his wife were leaving a ceremony at Sarajevo, a Serbian nationalist leaped from the crowd and killed them both. It was the Archduke's assassination which triggered World War One.

Death of Mark Twain's Brother: Dream-Time Precognition

In May of 1858, Mark Twin had a dream about his younger brother Henry who was working on a riverboat as a "mud clerk." As related by Mark Twain:

"The dream was so vivid, so like reality, that it deceived me, and I thought it was real. In the dream I had seen Henry a corpse. He lay in a metallic [burial case]. He was dressed in a suit of my clothing, and on his breast lay a great bouquet of flowers, mainly white roses, with a red rose in the [centre]. The casket stood upon a couple of chairs...it suddenly flashed upon me that there was nothing real about this--it was only a dream. I can still feel something of the grateful upheaval of joy of that moment, and I can also still feel the remnant of doubt, the suspicion that maybe it [was] real, after all. I returned to the house almost on a run, flew up the stairs two or three steps at a jump, and rushed into that [sitting-room]--and was made glad again, for there was no casket there."

A few days later, Twain's brother left on a river boat from New Orleans. As related by Mark Twain:

"Two or three days afterward the boat's boilers exploded at Ship Island, Memphis. I found Henry stretched upon a mattress on the floor of a great building, along with thirty or forty other scalded and wounded persons... his body was badly scalded... I think he died about dawn. The coffins provided for the dead were of unpainted white pine, but in this instance some of the ladies of Memphis had made up a fund of sixty dollars and bought a metallic case, and when I came back and entered the [dead-room] Henry lay in that open case, and he was dressed in a suit of my clothing. He had borrowed it without my knowledge during our last sojourn in St. Louis; and I recognized instantly that my dream of several weeks before was here exactly reproduced, so far as these details went-- and I think I missed one [detail;] but that one was immediately supplied, for just then an elderly lady entered the place with a large bouquet consisting mainly of white roses, and in the [centre] of it was a red rose, and she laid it on his breast."

The Dream-Murder of Tanya Zachs

In a legal case investigated and reported by Joseph (2000), a beautiful young woman, Tanya Zachs, disappeared on her way home in San Jose from her job in Santa Cruz in September of 1984. Her car was found abandoned along highway 17 midway between the two cities and which courses through the Santa Cruz mountains. That night, a young woman "Sunshine" who lived in a nudist colony, Lupin Lodge, situated in the Santa Cruz Mountains, had a nightmare: A woman was being brutally murdered. The next day, Sunshine read the story of Tanya's disappearance in the local newspaper, and that night she had the dream again, but this time the victim appeared to her quite clearly. It was Tanya.

In the dream Tanya showed "Sunshine" a narrow mountain road off highway 17, one of many leading from the long and winding highway between San Jose and Santa Cruz. Tanya led the dreamer down the mountain road which was bordered by a thick canopy of redwood trees and pines, and then to an isolated spot alongside. Tanya then beckoned the dreamer to follow her down a rather steep incline leading from the mountain road into the forest and thick brush, and then along a forested trail. Finally, Tanya stopped and pointed out her naked body, lying spread eagle on a huge slab of rock surrounded by trees. On the morning of 9/15/84, she contacted Tanya's family, told them of her dreams, and that same day led them and the police to the mountain side road Tanya had showed her and finally to the isolated spot. The police climbed down the tree-covered steep incline, and just as Sunshine had dreamed, they found the trail leading into the forest. But, there was no body.

That night Sunshine had another dream and Tanya took her to the same spot, down the same trail, then pointed at and emphasized a little deer trail that forked off to the right between the trees, and which led directly to her body. The next day, Sunshine and the family met again, and then climbed down the incline, took the trail to the right, and there was Tanya's body laid out exactly as revealed to Sunshine when dreaming.

The murder remained unsolved, however, until four years later. Damon Wells, beset by horrible nightmares where the victim kept accusing him of her murder, sought psychiatric treatment and confessed (Joseph 2000).

Precognition Dreams Are Common

Precognition dreams are common (Fukuda 2002; Haraldsson, 1985; Lange et al. 2001; Ross & Joshi, 1992; Stowell, 1995; Thalbourne, 1994) and often involve negative, unhappy, unpleasant events such as deaths, disasters and other calamities (Ryback & Sweitzer, 1990). About 40% of precognitive dreams are linked to an event the following day (Sondow, 1988), or take place several days or weeks later. However, anecdotal evidence indicates that the dreamed events may occur just prior to waking, even triggering wakefulness.

Precognition dreams can be about mundane affairs of concern only to the dreamer. A colleague of this author admits to frequently having had precognitive dreams and relates the following: "I dreamed that my water heater busted and that water was flooding out onto the floor. Three days later, the water heater sprung a leak." "I dreamed about getting a flat tire while driving on the freeway. It was the rear tire on the driver's side. A couple days later, the car's dashboard-computer informed me that the rear tire on the driver's side was low in air." "I dreamed that I took a girlfriend to my hot tub in the back yard, but it was empty and there was no water. When I tried to turn the water on nothing would happen. In the dream I was irritated because it would have to be replaced. A few days later the hot tub on-switch broke and after several failed attempts to fix it, I had it junked."

Several studies indicate that precognitive dreams are more common in younger than older individuals and that women report more precognitive dreams than men (Lange et al. 2001). It has also been found that those who have experienced deja vu are more likely to have precognitive dreams (Fakuda, 2002).

In large samples, anywhere from 17.8 % to 66% of individuals report that they experienced at least one precognitive dream (Fukuda 2002; Palmer, 1979; Haraldsson, 1985; Ross & Joshi, 1992; Ryback, 1988; Thalbourne, 1994) whereas over 60% of the general population believe such dreams are possible (Thalbourne, 1984; Haraldsson, 1985). However, Ryback (1988) after investigating 290 case reports of paranormal dreams, dismissed most of these precognitive dreams as coincidence and concluded that only 8.8% of the population actually have these dreams.

The Big Dream

Precognitive dreams have been reported throughout history, with accounts even appearing in the ancient Babylonian epic of Gilgamesh. Among ancient societies dreams were seen as extremely important sources of information about the future. In the ancient world, be it Greek, Rome, Egypt, or Babylon, it was not uncommon for someone to have what is called "a big dream" which was of great importance to the whole tribe, city, or nation (Campbell 1988; Frazier 1950; Freud 1900; Jung 1945, 1964; Malinowkski 1954). As possible harbingers of the future, these dreams had to be analyzed most carefully and the populace would gather around the man or woman who had the big dream and then debate its message and what do do about it. "Big Dreams" were even announced two thousand years ago in the Roman Senate. However, there were many who were non-believers.

PreCognitive Dream Skepticism and Professional Baseball

Over 2000 years ago Aristotle wrote a book expressing his disbelieve in precognitive dreams: "On Divination in Sleep." Aristotle complained that most of those having precognitive dreams were unworthy of the honor of receiving advanced information and "are not the best and wisest, but merely commonplace persons." Aristotle argued that "the sender of such dreams should be God." According to Aristotle "most dreams are to be classed as mere coincidences..." and do not take "place according to a universal or general rule" and have no causal connection to actual events in the future.

"Coincidence" has been the major objection to claims of precognitive dream activity (Caroll 2000; Wiseman 2011). Caroll (2000) refers to the "law of large numbers" and dismisses all claims as being a function merely of coincidence. For example, the odds, are with so many dreamers having dreams about so many different themes, that a few of them will have dreams about an airplane crash or a ship that sinks.   If the next day a ship sinks or a plane crashes, this is merely coincidence. According to Caroll and others, if precognitive dreams were real, they should be more commonplace, with more dreamers coming forward, and thus there should be a high "hit rate" and there is not; and as such "precognitive dreams do not exist."

However, if we applied the same reasoning to professional baseball, then professional baseball does not exist. Consider, from 2000 until 2013, the average baseball batting average ranged around 267. During regular season play during 2013, out of 750 major league players, 726 had a batting average of less than 30% (http://espn.go.com/mlb/stats/batting). Taking into considering fowl balls and hits which result in "outs", but considering that each player has at least 3 opportunities each time at bat to hit the ball, and then taking that .267 average, it could be said that the average professional baseball player actually gets a base hit less than 20% of the time. Be it a 20% or 30% hit rate, obviously this does

709

not mean no one in professional baseball is able to hit the ball, or that when they do it is merely a coincidence. The same standard must be applied to precognitive dreams.

Precognitive dreams need not be about Earth-shaking national tragedies and it is unknown how many dreamers would ever come forward to report their dreams even if they did have national implications. In fact, most dreams are forgotten upon waking (Frank, 2012, Pagel, 2014, Stickgold & Walker, 2010). Further, many precognitive dreams may be related to mundane matters like a "flat tire," or a phone call or visit from a friend the next day; or they may be entangled with events which are about to occur just minutes or seconds into the future, i.e. backward/precognitive dreams.

Since most people forget their dreams upon waking and most dreams are forgotten, how often precognitive dreaming occurs, and how many people have them, is unknown. What is known is that such dreams can be explained by quantum physics and the neurological foundations for dream activity and dream-time.

PreCognitive Backward Dreams

Precognitive experiences occurring during waking may be entangled with innocuous event which are just about to occur, such as thinking of a friend and then getting a phone call or email from that friend minutes or hours later. Just as a professional baseball player is more likely to swing and miss than hit the ball, the fact that one might think of a friend who does not call is not evidence against precognition.

During accelerated states associated with dream-time, precognitive dreams may be for events which will soon happen, or are just about to happen, perhaps seconds, or minutes away. These latter-type of dreams care best described as precognitive-backward dreams.

A case in point, "Katherine" dreams she and a friend "Sheryl" are shopping in Boston. They go from store, lugging shopping bags. "Katherine" in her dreams feels this sense of urgency to go home as if she is late for something and someone is waiting for her so she sets her bags down on the sidewalk and sits on a bench to wait for a cab or a the bus. She then realized her friend "Sheryl" is gone. Katherine looks for her, goes in and out of stores, but can't find her. "Katherine" sees a bus-like street car coming down the cobbled street and she picks up her packages and steps out onto the curb. As the street car pulls up and stops "Katherine" is surprised to see that Sheryl is driving and is ringing its bell. The sound of the bell grows louder and louder and then jolts "Katherine" from her dream. Katherine realizes that her phone is ringing. She picks it up and it is her friend "Sheryl" who is calling. Sheryl and Katherine are going shopping that day.

That "Katherine" dreamed about going shopping with "Sheryl" is not remarkable in-itself. That "Sheryl" was ringing the bell and it was Sheryl who was calling can be explained away as interesting coincidence. It is no surprise that Sheryl called. What seems paradoxical, however, is that the dream of shopping and walking down the streets of Boston laden with packages, the desire to go home, then looking for her missing friend and then seeing the bus-like street car all seemed to lead up to the ringing of the bell in a logical order of events so that its ringing made sense in the context of the dream. Hearing the ringing bell seemed to be a natural part of the dream, and it is. However, the dream did not lead up to the bell. Rather, the ringing of the bell initiated the dream. The effect (ringing bell) and the cause (the ringing phone) are identical. The effect caused itself.

There are two explanations for these quite common "backward" dreams. Dream-time and dream-consciousness does not obey the laws of physics. In dream-time, dream-consciousness may attempt to impose temporal order on a dream which has no temporal order and which may be experienced as a gestalt. In other words, in dream-time the entire dream was instantaneous and the dream was initiated by the ringing of her phone. The bell was heard and the dream was instantly produced in explanation and association. Future, present, past may be juxtaposed and experienced as a gestalt; like seeing the forest instead of the individual trees. In fact, although dreams may seem to last long time periods, they may be only seconds in length (Frank, 2012, Pagel, 2014, Stickgold & Walker, 2010).

The other explanation is that the ringing of the phone and the fact that Sheryl was calling Katherine, was perceived in dream-time, before it happened. Just as a time-machine traveling at superluminal speeds from the future into the past will pass by an observer only to be followed by its light image (which trails behind at the speed of light), information just seconds or minutes away into the future can be perceived by dream-consciousness in dream-time through entanglement. However, it is not future information traveling at superluminal speeds, but the mind and brain of the dreamer which are accelerating toward that future event in advance of those conscious minds which are still awake.

In dream time, the brain is highly active (Frank, 2012, Pagel, 2014, Stickgold & Walker, 2010), and certain regions in the limbic system are hyperactive (Joseph 1992, 1996, 2000). During dream-time, brain activity is accelerating which causes a contraction in time-space. The future comes closer to the present during dream time relative to outside observers which may include, upon waking, the conscious mind of the dreamer. However, while in dream-time, in a state of accelerated dream-consciousness, the future may be sensed and it may trigger a complex dream which then leads up to that future event when it arrives in the present thereby waking the dreamer.

Another illustrative example: French physicist Alfred Maury dreamt that he had taken part in the French Revolution and that he had been condemned to death and his head cut off at the exact moment when his bedpost broken and struck him across the neck:

"I was rather,unwell, and was lying down in my room, with my mother at my bedside. I dreamed of,the Reign of Terror; I witnessed massacres, I was appearing before the Revolutionary,tribunal, I saw Robespierre, Marat, Fouquier-Tinville, all the most wicked figures of,that terrible era; I talked to them; finally, after many events that I only partly remember,,I was judged, condemned to death, taken out in a tumbril through a huge throng to the,Place de la Revolution; I mounted the scaffold; the executioner tied me to that fatal,plank, he tipped it up, the blade fell; I felt my head separating from my body, I woke up,racked by the deepest anguish, and felt the bedpost on my neck. It had suddenly come,off and had fallen on my cervical vertebrae just like the guillotine blade."

Certainly it would be expected that a major blow to the head and neck would cause instant waking. But in this instance, it did not. Instead, the dreamer experienced a long and convoluted dream which was initiated by what was about to happen, and which could also be considered a warning of what was about to happen; albeit in the unique dream-language characteristic of dreams. This is not a case of an instantaneous backward dream, but a precognitive dream which provided the dreamer with a glimpse of what lay in store just moments into the future.

A third example related by a colleague: "I'd been working in my yard into the late afternoon and was exhausted. It was hot and I stripped off my shirt and lay down in a swinging hammock in my yard to take a nap, fell asleep and began to dream. In the dream I was in a nightclub and there was this exotic beautiful woman with long black hair drinking at the bar. We began drinking together and then we were dancing and kissing and then we were suddenly in my house and we were laying on the floor and I was taking off her clothes and she was getting very excited and aggressive. All at once I could see she had yellow eyes and black skin, but it wasn't skin, but resembled an insect's chitlin. Her arms and hands became claws and her teeth became razor sharp and pointed. She put her claw arms around me very tight as I struggled to escape. She had turned into some demonic insect-creature and pressed her razor sharp claw-hands into my back. I could feel her razor sharp claws knifing me and I felt I was being stabbed in the back. The pain was terrible. It seemed as if her pointed claws were going to completely pierce my back and come out my chest. The pain was so horrific I woke up. But the pain was still there. I got up from the hammock and there was a crippled black bumble bee laying there. The damn thing has stung me on my back."

Dreams that seem to paradoxically lead up to an event which wakes the dreamer are common. These dreams may be relatively brief or become lengthy complicated dreams leading up to some event which then occurs, as if on cue, waking the dreamer who discovers upon waking that someone was knocking on his door, the phone was ringing, it was the alarm clock, a kid was yelling outside the window, and so on, all of which initiated the dream which then led up to the event which caused the dream (Joseph 1992). The dream was produced, so as to explain in the unique language of the dream what was about to happen; and this is because, it already happened in the future. The only other explanation is the dream was produced as an instant gestalt and the dreamer dreamed the dream in accelerated dream-time without any temporal order, and it was upon waking that the dream was reconstructed in a temporal sequential time frame (Joseph 1992).

Be they backward dreams instantly produced as a gestalt, or examples of dream precognition, backward dreams are the most easily comprehended because the conscious mind utilizes temporal sequences to explain what is observed, and may recall the dream in reverse, so it makes temporal-sequential sense; as if the cause led to the effect, when the cause and effect were either simultaneous, or the effect was its own cause.

Joseph Dreams His Death 2000 Years Ago?

In the early 1950s, when I [R. Joseph] was a boy of 3, and for many years until around age 7, I had dreams about a little boy playing by the sea shore, by the ocean. And there were crowds of people. Some lying or sitting together on the sand. Others swimming or fishing. And then in the dream the ocean began to recede... the ocean waters drew back back back... and I could see shells and fish flopping on the wet sand where moments before there had been ocean... and ships and small boats lay on their sides... and I ran to where the ocean had been, on the wet sand, picking up shells... and many other people also ran onto the wet sand picking up wiggling fish and laughing and talking in amazement that the ocean had pulled back for miles and miles leaving the sand and ocean floor completely revealed for everyone to see.... and then... and then... and then...

I walked further and further out to where the ocean had been, picking up giant shells some with wiggling living creatures still inside, and gazing in wonder at what the ocean had hidden but which was now revealed… and then I heard screams... women and men and children were screaming... and in my dream, they were all running from the wet sand where the ocean had been toward the dry shore...and people on the shore were also running... everyone was running away and screaming... and I could hear this rumbling roar from behind me… and when I looked back to see why, what they were running from and what was making that roar, I could see the ocean... it was still miles away--but it was a WALL OF OCEAN.. a WALL OF WATER looming up maybe 100 yards perhaps even miles into the sky... and in my dream the wall of ocean was rushing forward, to where the ocean had been minutes before, toward where I was standing with sea shells in my hands... and I started running... like everyone else, running running running... and I could see, over my shoulder, behind me, the roaring wall of ocean water coming closer, and closer... and faster faster faster... and I kept running...

everyone was running and screaming...trying to get away... and then the towering WALL OF WATER was just behind me... then looming over me... and then it crashed down upon me... and the little boy that I was, in this dream, drowned.... and then I awoke in my bed... the same boy who drowned, but a different boy... me...

I had this dream over and over for years; the same dream, the source of which was a mystery to me as I had never even imagined that the ocean could actually recede and then rush back to land as I had dreamed. It was not until 20 years later that I learned, for the first time, about Tsunamis and how characteristic it is for people to foolishly run out to where the ocean and been... and then... the ocean comes rushing back as a wall of water drowning everyone who did not immediately run away.

How could I have dreamed so vividly about something 3-year old me knew nothing about in the early 1950s when we didn't even have a television? There were clues in yet other dreams when I was a child, and they were dreams of the same little boy. But it was during ancient Roman times, and I was sitting with my mother who was dressed in royal robes typical of the Roman period. She was singing to me... And down below I could see Roman soldiers marching, and peasant women by a river, washing clothes, and the river was flowing into the ocean. The peasant women, who had with them many naked children, were dressed in clothes I associated with Biblical times, of ancient Egypt; my grandmother would often read to me from a Bible picture book. But in these river-side dreams which began so peaceful, they all ended with incredible earthquakes, like the world was turning up-side down...

How do these two dreams relate? Almost 50 years after I had these dreams, I searched the records for Tsunamis in the Mediterranean sea near Italy and Egypt.

On the morning of July 21, 365 AD, an earthquake of great magnitude caused a huge tsunami more than 100 feet high and it inundated and destroyed several towns on the coasts of the Mediterranean, including Alexandria. This is how Ammianus Marcellinus, a Roman historian described it:

Slightly after daybreak, and heralded by a thick succession of fiercely shaken thunderbolts, the solidity of the whole earth was made to shake and shudder, and the sea was driven away, its waves were rolled back, and it disappeared, so that the abyss of the depths was uncovered and many-shaped varieties of sea-creatures were seen stuck in the slime; the great wastes of those valleys and mountains, which the very creation had dismissed beneath the vast whirlpools, at that moment, as it was given to be believed, looked up at the sun's rays. Many ships, then, were stranded as if on dry land, and people wandered at will about the paltry remains of the waters to collect fish and the like in their hands; then the roaring sea as if insulted by its repulse rises back in turn, and through the teeming shoals dashed itself violently on islands and extensive tracts of the mainland, and flattened innumerable buildings in towns or wherever they were found. Thus in the raging conflict of the elements, the face of the earth was changed to reveal wondrous sights. For the mass of waters returning when least expected killed many thousands by drowning, and with the tides whipped up to a height as they rushed back, some ships, after the anger of the watery element had grown old, were seen to have sunk, and the bodies of people killed in shipwrecks lay there, faces up or down.

Had Joseph dreamed of a previous life from nearly 2000 years ago? Or did he journey to the past, during dream-time, and visit the long ago in the time machine of consciousness?

We have been here before, we will be here again, we will always be, and this is because time and consciousness are a quantum continuum and the distinctions between past present and future are illusions.

The ancients believed the past could be accessed through dreams. But according to general relativity and quantum physics, time reversal is possible only at superluminal speeds.

Quantum Physics of Dream-Time and Dream Consciousness

Dream-consciousness and dream-time are associated with accelerated levels of brain activity. Relativity predicts that observers with an accelerated frame of reference experience time-contraction and a shrinking of time-space such that the future and the present come closer together relative to those with a different frame of reference. Not surprisingly, two dissociated conscious minds may experience slowing and speeding up simultaneously. However, space-time is also part of the quantum continuum; and this continuum extends into the future and the past, such that all are entangled.

As based on a coupling of the Copenhagen with the Many Worlds interpretations, dream-consciousness can become entangled with the space-time continuum which includes the various futures as they flow toward an observer; futures which are also entangled and which will become modified to varying degrees before being perceived and then upon the act of registration. Thus, during dream consciousness, not just the most improbable, but also the most probable future world may be observed; a probable future which may or may not happen in this reality, in this world, but which will or has happened in another world as predicted by "Many Worlds."

Dream-consciousness, and dream-time, because they are maintained by brain structures which, during sleep, are freed of inhibitory restraint, also have a sensitivity to the quantum continuum which differs from and is more wide-reaching than the mind during waking. Much of the human brain is subject to inhibition, which prevents sensory overload and allows for the focusing of attention. However, if that inhibition is removed, specific neurons and brain areas may greatly increase their activity and begin processing information which is normally filtered from the conscious mind.

Einstein proclaimed that the "distinctions between past, present and future are an illusion," that time is embedded with space, creating a space-time manifold where numerous futures, pasts, presents overlap and exist simultaneously in parallel, all flowing in various directions relative to observers. The future, or rather, multiple futures exist, albeit in varying distant locations in space-time. All of space-time, the quantum continuum, is entangled, and just as the ripples of a pond may strike distant shores, the space-time quantum states of the future may also effect distant shores occupied by what the mind experiences as the now.

If the 4th dimension is space-time, and if the future and the present and differences in time are related to accelerated frames of reference and movement through space, then within the quantum continuum, everything is connected: stars, planets, dogs, cats, past, present, and the future.

Be it the time traveler who accelerates toward light speed and then to superluminal velocities, such that he first travels to the future and then the past, the time machine of consciousness, freed of inhibitory restraint, may also accelerate toward light speed, lifting the veil and enabling the dreaming mind to see what had been concealed including what may occur in the future and what may have taken place thousands of years ago in the past.

REFERENCES

Al-Khalili (2011). Black Holes, Wormholes and Time Machines, Taylor & Francis.

Almheiri, A. et al. (2013). Black Holes: Complementarity or Firewalls? J. High Energy Phys. 2, 062

Bethe, H. A., et al., (2003) Formation and Evolution of Black Holes in the Galaxy, World Scientific Publishing.

Bilaniuk, O.-M. P.; Sudarshan, E. C. G. (1969). "Particles beyond the Light Barrier". Physics Today 22 (5): 43–51.

Bo, L., Wen-Biao, L. (2010). Negative Temperature of Inner Horizon and Planck Absolute Entropy of a Kerr Newman Black Hole. Commun. Theor. Phys. 53, 83–86.

Bohr, N., (1913). "On the Constitution of Atoms and Molecules, Part I". Philosophical Magazine 26: 1–24.

Bohr, N., (1913). "On the Constitution of Atoms and Molecules, Part I". Philosophical Magazine 26: 1–24.

Bohr, N. (1934/1987), Atomic Theory and the Description of Nature, reprinted as The Philosophical Writings of Niels Bohr, Vol. I, Woodbridge: Ox Bow Press.

Bruno, N. R., (2001). Deformed boost transformations that saturate at the

Planck scale. Physics Letters B, 522, 133-138.

Blandford, R.D. (1999). "Origin and evolution of massive black holes in galactic nuclei". Galaxy Dynamics, proceedings of a conference held at Rutgers University, 8–12 Aug 1998,ASP Conference Series vol. 182.

Carroll, S (2004). Spacetime and Geometry. Addison Wesley.

Casimir, H. B. G. (1948). "On the attraction between two perfectly conducting plates". Proc. Kon. Nederland. Akad. Wetensch. B51: 793.

Chodos, A. (1985). "The Neutrino as a Tachyon". Physics Letters B 150: 431.

Einstein, A. (1915a). Fundamental Ideas of the General Theory of Relativity and the Application of this Theory in Astronomy, Preussische Akademie der Wissenschaften, Sitzungsberichte, 1915 (part 1), 315.

Einstein, A. (1915b). On the General Theory of Relativity, Preussische Akademie der Wissenschaften, Sitzungsberichte, 1915 (part 2), 778–786.

Einstein A. (1939) A. Einstein, Ann. Math. 40, 922.

Einstein, A. (1961), Relativity: The Special and the General Theory, New York: Three Rivers Press.

Einstein, A. and Rosen, N. (1935). "The Particle Problem in the General Theory of Relativity". Physical Review 48: 73.

Einstein, A., Lorentz, H.A., Minkowski, H., and Weyl, H. (1923). Arnold Sommerfeld. ed. The Principle of Relativity. Dover Publications: Mineola, NY. pp. 38–49.

Einstein A, Podolsky B, Rosen N (1935). "Can Quantum-Mechanical Description of Physical Reality Be Considered Complete?". Phys. Rev. 47 (10): 777–780.

Eisberg, R., and Resnick. R. (1985). Quantum Physics of Atoms, Molecules, Solids, Nuclei, and Particles. Wily,

Everett, A., & Roman, T. (2012). TIme Travel and Warp Drives, University Chicago Press.

Feinberg, G. (1967). "Possibility of Faster-Than-Light Particles". Physical Review 159 (5): 1089–1105.

Fuller, Robert W. and Wheeler, John A. (1962). "Causality and Multiply-Connected Space-Time". Physical Review 128: 919.

Garay, L. J. (1995). Quantum gravity and minimum length Int.J.Mod.Phys. A10 (1995) 145-166

Ghez, A. M.; Salim, S.; Hornstein, S. D.; Tanner, A.; Lu, J. R.; Morris, M.; Becklin, E. E.; Duchene, G. (2005). "Stellar Orbits around the Galactic Center Black Hole". The Astrophysical Journal 620: 744.

Geiss, B., et al., (2010) The Effect of Stellar Collisions and Tidal Disruptions on Post-Main-Sequence Stars in the Galactic Nucleus. American Astronomical Society, AAS Meeting #215, #413.15; Bulletin of the American Astronomical Society, Vol. 41, p.252.

Giddings, S. (1995). The Black Hole Information Paradox," Proc. PASCOS symposium/Johns Hopkins Workshop, Baltimore, MD, 22-25 March, 1995, arXiv:hep-th/9508151v1.

Hawking, S. W., (1988) Wormholes in spacetime. Phys. Rev. D 37, 904–910.

Hawking, S., (1990). A Brief History of Time: From the Big Bang to Black Holes. Bantam.

Hawking, S. (2005). "Information loss in black holes". Physical Review D 72: 084013.

Hawking, S. W. (2014). Information Preservation and Weather Forecasting for Black Holes.

Heisenberg, W. (1927), "Über den anschaulichen Inhalt der quantentheoretischen Kinematik und Mechanik", Zeitschrift für Physik 43 (3–4): 172–198.

Heisenberg. W. (1930), Physikalische Prinzipien der Quantentheorie (Leipzig: Hirzel). English translation The Physical Principles of Quantum Theory, University of Chicago Press.

Heisenberg, W. (1955). The Development of the Interpretation of the Quantum Theory, in W. Pauli (ed), Niels Bohr and the Development of Physics, 35, London: Pergamon pp. 12-29.

Heisenberg, W. (1958), Physics and Philosophy: The Revolution in Modern Science, London: Goerge Allen & Unwin.

Jaffe, R. (2005). "Casimir effect and the quantum vacuum". Physical Review D 72 (2): 021301.

Joseph, R (2010a) The Infinite Cosmos vs the Myth of the Big Bang: Red Shifts, Black Holes, and the Accelerating Universe. Journal of Cosmology, 6, 1548-1615.

Joseph, R. (2010b). The Infinite Universe: Black Holes, Dark Matter, Gravity, Acceleration, Life. Journal of Cosmology, 6, 854-874.

Joseph, R. (2014a) Paradoxes of Time Travel: The Uncertainty Principle, Wave Function, Probability, Entanglement, and Multiple Worlds, Cosmology, 18, 282-302.

Joseph, R. (2014a) The Time Machine of Consciousness, Cosmology, 18, In press.

Kerr, R P. (1963). "Gravitational Field of a Spinning Mass as an Example of Algebraically Special Metrics". Physical Review Letters 11 (5): 237–238.

Lambrecht, A. (2002) The Casimir effect: a force from nothing, Physics World, September 2002

Lorentz, H. A. (1892), "The Relative Motion of the Earth and the Aether", Zittingsverlag Akad. V. Wet. 1: 74–79

McClintock, J. E. (2004). Black hole. World Book Online Reference Center. World Book, Inc.

Melia, F. (2003). The Edge of Infinity. Supermassive Black Holes in the Universe. Cambridge U Press. ISBN 978-0-521-81405-8.

Melia, F. (2007). The Galactic Supermassive Black Hole. Princeton University

Press. pp. 255–256.

Merloni, A., and Heinz, S., (2008) A synthesis model for AGN evolution: supermassive black holes growth and feedback modes Monthly Notices of the Royal Astronomical Society, 388, 1011 1030.

Minchin, R. et al. (2005). "A Dark Hydrogen Cloud in the Virgo Cluster". The Astrophysical Journal 622: L21–L24.

Morris, M. S. and Thorne, K. S. (1988). "Wormholes in spacetime and their use for interstellar travel: A tool for teaching general relativity". American Journal of Physics 56 (5): 395–412.

O'Neill, B. (2014) The Geometry of Kerr Black Holes, Dover

Penrose, R. (1969) Rivista del Nuovo Cimento.

Preskill, J. (1994). Black holes and information: A crisis in quantum physics", Caltech Theory Seminar, 21 October. arXiv:hep-th/9209058v1.

Pollack, G. L., & Stump, D. R. (2001), Electromagnetism, Addison-Wesley.

Rindler, W. (2001). Relativity: Special, General and Cosmological. Oxford: Oxford University Press.

Rodriguez, A. W.; Capasso, F.; Johnson, Steven G. (2011). "The Casimir effect in microstructured geometries". Nature Photonics 5 (4): 211–221.

Ruffini, R., and Wheeler, J. A. (1971). Introducing the black hole. Physics Today: 30–41.

Russell, D. M., and Fender, R. P. (2010). Powerful jets from accreting black holes: Evidence from the Optical and the infrared. In Black Holes and Galaxy Formation. Nova Science Publishers. Inc.

Schrödinger, E. (1926). "An Undulatory Theory of the Mechanics of Atoms and Molecules". Physical Review 28 (6): 1049–1070. Bibcode:1926PhRv...28.1049S. doi:10.1103/PhysRev.28.1049.

Sen, A. (2002) Rolling tachyon," JHEP 0204, 048

Slater, J. C. & Frank, N. H. (2011) Electromagnetism, Dover.

Smolin, L. (2002). Three Roads to Quantum Gravity. Basic Books.

Taylor, E. F. & Wheeler, J. A. (2000) Exploring Black Holes, Addison Wesley.

Thorne, K. (1994) Black holes and time warps, W. W. Norton. NY.

Thorne, K. S. & Hawking, S. (1995). Black Holes and Time Warps: Einstein's Outrageous Legacy, W. W. Norton.

Thorne, K. et al. (1988). "Wormholes, Time Machines, and the Weak Energy Condition". Physical Review Letters 61 (13): 1446.

Wilf, M., et al. (2007) The Guidebook to Membrane Desalination Technology: Reverse Osmosis, Balaban Publishers.

24. Accelerated Dream-Consciousness: Entangled Minds

Up to five stages of sleep have been identified in humans (Frank, 2012, Pagel, 2014, Stickgod & Walker, 2010). For our purposes we will be concerned only with two distinct sleep states. These are the REM (rapid eye movement) and nonREM periods. NonREM occurs during a stage referred to as "slow-wave" or synchronized sleep. It is during non-REM that the brain is at its lowest level of activity.

By contrast, REM is linked with the visual-emotional hallucinatory aspects of dreaming and both are associated with the deepest stage of sleep, characterized by "paradoxical" fast irregular wave activity (Cartwright 2007; Hobson 2004; Mallick et al. 2011). Dream sleep is called "paradoxical," for electrophysiologically the brain is in a state of accelerated activity and a heightened level of arousal, as demonstrated by electrophysiological indices (Steriade & McCarley, 2005; Vertes 1990). However, the body musculature is paralyzed so the ability to interact physically with one's dreams, in the real world, is greatly attenuated if not impossible. Otherwise, one would be acting our their dreams by moving around, and then hurting themselves or others.

As predicted by Einstein's relativity, under accelerated states space-time and the distance between the future and the present contracts (Einstein et al. 1923, Einstein 1961). Therefore, because of these heightened states of accelerated arousal during dream sleep, dream-consciousness may be provided glimpses of the future before that future arrives in space-time external to dream-consciousness; and this is because dream-time takes place during an accelerated state as reflected by brain activity.

Brain structures involved in the production of REM and paradoxical sleep include the amygdala, hippocampus, right temporal lobe, and especially reticular activating system and the lateral and medial pons which are located in the brainstem (Joseph 2012, Hobson et al. 2000; Hobson & Allan 2010). For example, brainstem cholinergic neurons located in the lateral pons, and neurons located in the medial pontine reticular formation appear to be the locus for REM-on neurons which initiate and/or maintain the production of REM and dream sleep (Aeschbach 2011; Steriade & McCarley, 2005; Vertes 1990). Cholinergic neurons release excitatory neurotransmitters such as acetylcholine (ACh) which also activates the parasympathetic nervous system which is also involved in arousal (Joseph 2012). ACh levels remain at high levels during REM and the production of PGO waves by the pontine, lateral geniculate, occipital brain areas (Hobson, et al., 1975; Lydic, et al., 1991; Steriade & McCarley, 2005; Vertes

1990). As ACh is also implicated in memory, this may well explain whey recent memories tend to become incorporated in dreams.

Therefore, during the production of REM and paradoxical sleep, there is increased cholinergic activity which increases and accelerates brain activity, whereas with the termination of REM, these same neurons greatly reduced their activity.

The PGO Wave Function

Dream sleep has its own unique wave function. Pontine geniculate occipital (PGO) waves are phasic field potentials which begin as electrical pulses radiating from the pons into the limbic system, the thalamus, and the neocortex (Frank, 2012; Hobson et al. 2000; Stickgold & Walker, 2010). These waves suddenly increase in amplitude and frequency just before the onset of REM. During waking two classes of neurotransmitters, serotonin and norepinephrine act to suppress PGO waves and activity whereas increases in cholinergic neurons and ACh increase PGO wave activity as occurs during dreaming (Hobson, et al., 1975; Lydic, et al., 1991).

Characteristically, PGO waves and their propagation through the brain first appear during the last stages of synchronized sleep just prior to REM. That wave function increases in amplitude and frequency during desynchronized, paradoxical sleep and REM. Typically, PGO waves reach their maximum amplitude in the lateral geniculate nucleus of the thalamus which, like the amygdala, are also cholinergically responsive (Steriade & McCarley, 2005). In fact, activation of the amygdala can initiate and dramatically increase PGO wave activity (Calvo, et al. 1987). This is significant as the amygdala is a multi-modal brain area which is directly implicated in sensory filtering, dreaming, deja vu, and other paranormal experiences (Joseph 2000. 1911a,b).

Therefore, the conscious and sleeping brain has a wave function (Hobson 2004; Hobson et al. 2000; Penfiled & Hameroff 2011), and the wave function accelerates and increases during dream sleep (Hobson et al. 2000). As predicted by the Copenhagen interpretation of quantum mechanics, that wave function will interact with everything it encounters in the quantum continuum, and as such, and especially under accelerated conditions, it may become entangled with events still some distance away in space-time. As predicted by relativity (Einstein 1961), acceleration shrinks the distance between the present and the future.

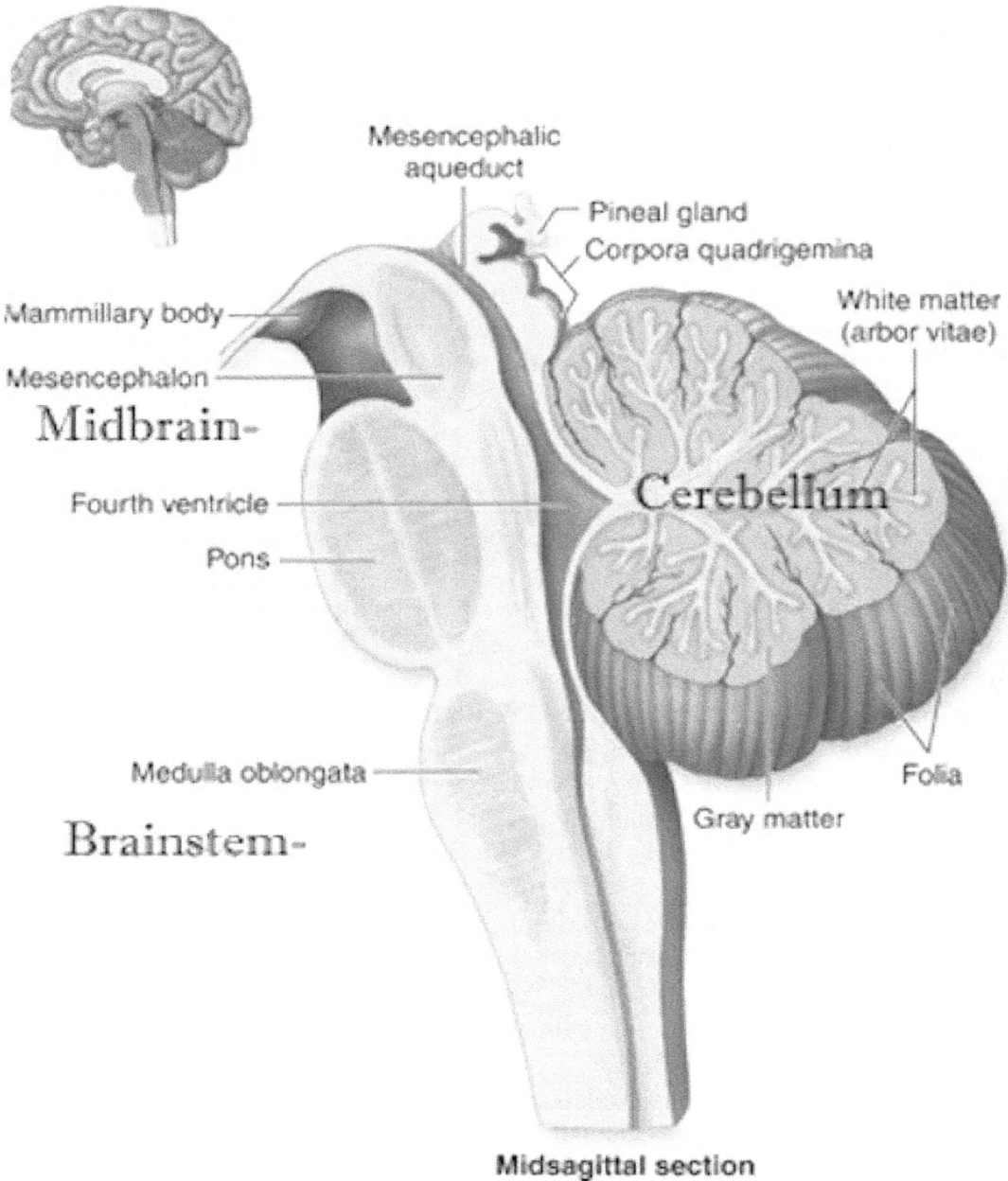

Midsagittal section

Limbic System Accelerated Dream-Consciousness

    Neurons within limbic system structures, such as the amygdala, receive input from multiple sensory modalities simultaneously, with streams of visual auditory tactile emotional gustatory olfactory and other forms of information impinging on single neurons. The amygdala along with other brain structures including the frontal lobe, normally acts to inhibit the perception of most of the stimuli impinging on the senses (Joseph 1992, 1996, 2000, 2012). This information is filtered so that a narrow reality becomes the focus of consciousness. This massive inhibition also serves to prevent the conscious mind from being overwhelmed with competing streams of input which are turned off. In other words, just as a channel selector on a television, radio, computer, or phone, permits the select reception of information from a single source (a single reality), these brain structures perform likewise. If inhibition is removed, such as when NE and 5HT levels are reduced, communication between select axons and dendrites in the amygdala is enhanced and these limbic system neurons becomes hyperactivated and the brain becomes acutely sensitive to and may be overwhelmed by a kaleidoscope of overlapping sensations, e.g., tasting colors, seeing sound, smelling music. Some patients describe it as being given access to alternate worlds and realities of vast proportions or to hidden knowledge. Presumably this includes information which lies in the immediate future. Consciousness may also fragment and split apart, such that one aspect of consciousness may hover up above looking down, whereas that aspect of conscious still tied to the body may be down below looking up.

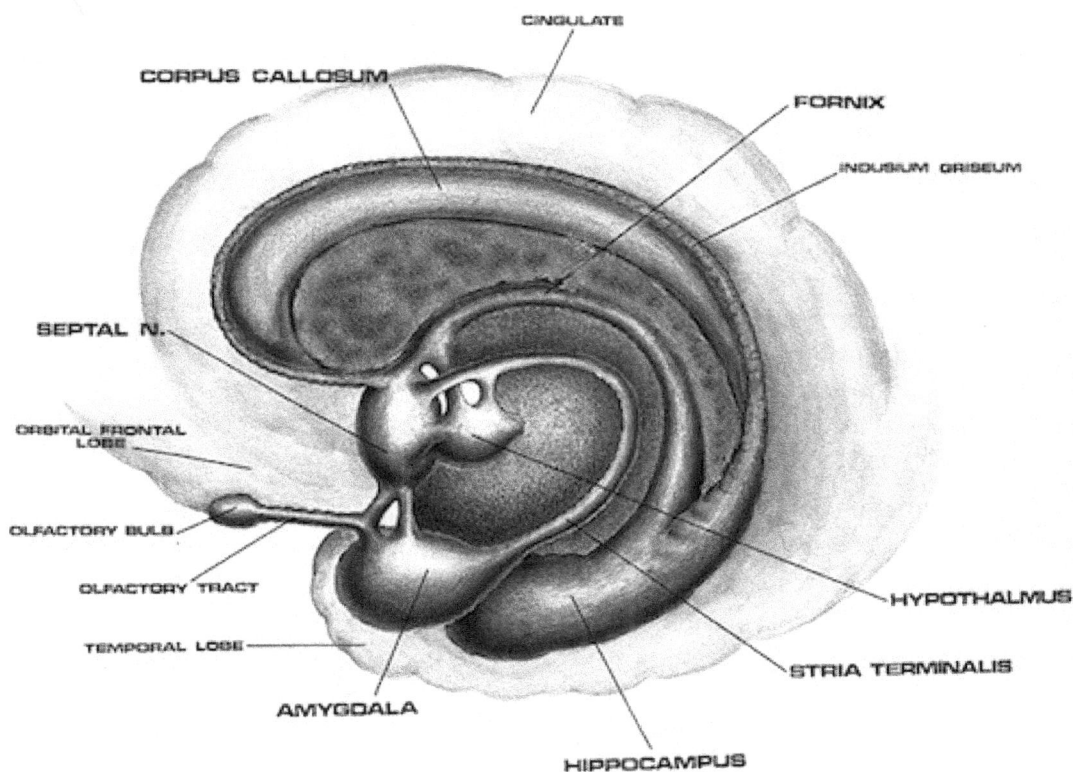

The amygdala is directly involved in dream sleep, REM sleep and PGO activity (Joseph 1996, 2012). Likewise, activity in the adjacent hippocampus accelerates and increases during REM, producing theta waves (Vertes, 1984, 2005). Hippocampal theta is associated with high levels of arousal. Because these same limbic system areas of the brain become disinhibited and experience accelerate activity during dream sleep, the dreaming brain becomes much more sensitive to fluctuations in the quantum space-time continuum which, as predicted by relativity, include the future; or rather, "a" future which, during sleep, is relative to the dreaming mind of the observer.

The amygdala, hippocampus, and the overlying temporal lobe are richly interconnected and appear to act in concert in regard to paranormal and mystical experience (Broughton 1982; Gloor 1992; Hodoba 1986; Joseph, 1992a; Meyer et al. 1987). These structures are also implicated in memory (Joseph 2012). Presumably, during REM, the hippocampus and amygdala act as a reservoir from which various images, emotions, words, and ideas are drawn and incorporated into the matrix of dream activity observed by dream-consciousness (Joseph 1982, 1992, 1996). Often these structures incorporate memories and experiences from the previous day into the content of the dream. However, they are also sensitive to the quantum continuum even while sleeping, such that events which are perceived during dream sleep are often woven into the dream.

Therefore, these limbic system structure are directly linked to REM sleep, PGO wave activity, dream sleep, increased brain activity, deja vu, precognition, and a duality in consciousness such that the conscious mind can split off and observe itself and the body, as if the mind is on the ceiling looking down, in the audience watching a play, or passively observing a dream during dream-consciousness. However, because structures such as the amygdala are freed of inhibitory restraint during the course of dream-time, although the dreamer can't act on their dreams, they becomes sensitive to sources of information which are normally filtered out or which can only be accessed under an accelerated frame of reference; thereby providing the foundations for premonitions, deja vu, and precognition.

Dream-Consciousness and Entangled Minds

Dream-conscious and dream-time are linked to accelerated frames of reference, due to the arousal and hyperactivation of various brain structures including the amygdala, hippocampus and overlying temporal lobe; the activity of which is associated with REM, paradoxical EEG activity, and PGO waves. When these same structures are activated by electrodes, subjects have reported communing with spirits or receiving profound knowledge from the Hereafter. Some have reported a splitting of consciousness and of having left their body which they may observe from afar, as if they are on the ceiling or hovering above. Moreover, some subjects have reported that their disembodied consciousness could travel vast distances and observe family, friends, and strangers engaging in their daily activities:

"Far down below I could see houses and towns and green land and streams... I kept on going very fast...Then I was right over Pine Ridge. I looked down (and) saw my father's and mother's teepee. They went outside, and she was cooking... My mother looked up, and I felt sure she saw me... then I started back, going very fast..." (Neihardt 1988).

"I began moving at a tremendous speed... and I was aware of trees rushing below me. I just thought of home and knew I was going there... I saw my husband sitting in his favorite armchair reading the newspaper. I saw my children running up and down the stairs... I was drawn back to the hospital, but I don't remember the trip; it seemed to happen instantaneously" (Eadie & Taylor 1992).

Likewise, it has long been believed by ancient cultures that while dreaming this disembodied dream-consciousness can wonder about and interact with other disembodied minds, including those who lived in the past or those who were bringing news of the future (Campbell 1988; Frazier 1950; Harris, 1993; Jung 1964; Malinowski 19490). It has also been believed that dreams may be shared, and that the dreamer can commune with the minds not just of the dead, but the living.

Anecdotal evidence in fact supports the possibility that dreams may become entangled, such that two individuals may experience similar dreams, or that during dreaming the mind of the dreamer becomes entangled with the waking minds of others. Consider, for example, studies performed in the Sleep Laboratory of Mimonides Medical Center in New York. This research group carried out several studies on entangled minds and dreaming and consistently obtained positive results (Krippner et al., 1971, 1972; Ullman, et al., 1977, 1989). In these studies, participants were introduced, and then later, while one subject slept and when it was determined he had entered REM sleep, the other would begin looking at particular paintings and prints. He would then attempt to project these images into the mind of the dreamer. Independent judges blind to the purpose and procedures of the experiment then judged if the dream reports contained any of the images sent by the sender.

In one experiment, when the test subject entered REM, the "sender" opened a package which contained an art print of the Mexican and Indian followers of the Mexican revolutionary Emiliano Zapta, i.e. Zapatistas, by Romero. When the test subject entered REM, the "sender" began gazing at the picture and attempted to project it to the dreamer. The dreamer was immediately awakened when the REM episode had ceased and was asked to describe his dream. And this is what the dreamer dreamed: "...a feeling of memory... of New Mexico... a lot of mountains...Indians, Pueblos" (reviewed in Broughton, 1991; Ullman & Krippner, 1989).

Another "example of significant results, described by Krippner and colleagues

(1989) occurred on a night when the randomly selected art print was "School of the Dance" by Degas, depicting a dance class of several young women. The "receiver's" dream reports included such phrases as "I was in a class made up of maybe half a dozen people; it felt like a school." "There was one little girl that was trying to dance with me."

Quantum Physics, Time Travel and Accelerated Dream Consciousness

Dreams are the royal road to often inaccessible realms of the mind and to realities that lie just beyond conscious awareness; realities which may exist only as probabilities and alternate worlds as predicted by the Many Worlds interpretation of quantum physics.

Dream-consciousness and dream-time are associated with accelerated levels of brain activity. Relativity predicts that observers with an accelerated frame of reference experience time-contraction and a shrinking of time-space such that the future and the present come closer together relative to those with a different frame of reference. However, space-time includes other observers and their conscious minds. Minds are also part of the quantum continuum; and this continuum extends into the future and the past, such that all are entangled.

Every particle, every object, person, planet, star, galaxy, the entire universe has a wave function. As based on a coupling of the Copenhagen with the Many Worlds interpretations, dream-consciousness can become entangled with the space-time continuum which includes the wave functions of other dreams and conscious minds.

Dream-consciousness, and dream-time, because they are maintained by certain brain structures which, during sleep, are freed of inhibitory restraint, also have a sensitivity to the quantum continuum which differs from and is more wide-reaching than the mind during waking. Much of the human brain is subject to inhibition, which prevents sensory overload and allows for the focusing of attention. However, if that inhibition is removed from structures such as the amygdala, specific neurons and brain areas may greatly increase their activity and begin processing information which is normally filtered from the conscious mind; and this may include the future and the contents of other conscious minds. All of space-time, the quantum continuum, is entangled, everything is connected: stars, planets, dogs, cats, past, present, future, brains and minds. And all may interact through wave functions which intersect. Just as the ripples of a pond may strike distant shores, the space-time quantum continuum and wave function of minds, brains, the future and the past, may also effect distant shores and the eternal now of the event horizon of consciousness.

REFERENCES

Al-Khalili (2011). Black Holes, Wormholes and Time Machines, Taylor & Francis.

Almheiri, A. et al. (2013). Black Holes: Complementarity or Firewalls? J. High Energy Phys. 2, 062

Bethe, H. A., et al., (2003) Formation and Evolution of Black Holes in the Galaxy, World Scientific Publishing.

Bilaniuk, O.-M. P.; Sudarshan, E. C. G. (1969). "Particles beyond the Light Barrier". Physics Today 22 (5): 43–51.

Bo, L., Wen-Biao, L. (2010). Negative Temperature of Inner Horizon and Planck Absolute Entropy of a Kerr Newman Black Hole. Commun. Theor. Phys. 53, 83–86.

Bohr, N., (1913). "On the Constitution of Atoms and Molecules, Part I". Philosophical Magazine 26: 1–24.

Bohr, N., (1913). "On the Constitution of Atoms and Molecules, Part I". Philosophical Magazine 26: 1–24.

Bohr, N. (1934/1987), Atomic Theory and the Description of Nature, reprinted as The Philosophical Writings of Niels Bohr, Vol. I, Woodbridge: Ox Bow Press.

Bruno, N. R., (2001). Deformed boost transformations that saturate at the Planck scale. Physics Letters B, 522, 133-138.

Blandford, R.D. (1999). "Origin and evolution of massive black holes in galactic nuclei". Galaxy Dynamics, proceedings of a conference held at Rutgers University, 8–12 Aug 1998,ASP Conference Series vol. 182.

Carroll, S (2004). Spacetime and Geometry. Addison Wesley.

Casimir, H. B. G. (1948). "On the attraction between two perfectly conducting plates". Proc. Kon. Nederland. Akad. Wetensch. B51: 793.

Chodos, A. (1985). "The Neutrino as a Tachyon". Physics Letters B 150: 431.

Einstein, A. (1915a). Fundamental Ideas of the General Theory of Relativity and the Application of this Theory in Astronomy, Preussische Akademie der Wissenschaften, Sitzungsberichte, 1915 (part 1), 315.

Einstein, A. (1915b). On the General Theory of Relativity, Preussische Akademie der Wissenschaften, Sitzungsberichte, 1915 (part 2), 778–786.

Einstein A. (1939) A. Einstein, Ann. Math. 40, 922.

Einstein, A. (1961), Relativity: The Special and the General Theory, New York: Three Rivers Press.

Einstein, A. and Rosen, N. (1935). "The Particle Problem in the General Theory of Relativity". Physical Review 48: 73.

Einstein, A., Lorentz, H.A., Minkowski, H., and Weyl, H. (1923). Arnold Sommerfeld. ed. The Principle of Relativity. Dover Publications: Mineola, NY. pp. 38–49.

Einstein A, Podolsky B, Rosen N (1935). "Can Quantum-Mechanical Description of Physical Reality Be Considered Complete?". Phys. Rev. 47 (10):

777–780.

Eisberg, R., and Resnick. R. (1985). Quantum Physics of Atoms, Molecules, Solids, Nuclei, and Particles. Wily,

Everett, A., & Roman, T. (2012). TIme Travel and Warp Drives, University Chicago Press.

Feinberg, G. (1967). "Possibility of Faster-Than-Light Particles". Physical Review 159 (5): 1089–1105.

Fuller, Robert W. and Wheeler, John A. (1962). "Causality and Multiply-Connected Space-Time". Physical Review 128: 919.

Garay, L. J. (1995). Quantum gravity and minimum length Int.J.Mod.Phys. A10 (1995) 145-166

Ghez, A. M.; Salim, S.; Hornstein, S. D.; Tanner, A.; Lu, J. R.; Morris, M.; Becklin, E. E.; Duchene, G. (2005). "Stellar Orbits around the Galactic Center Black Hole". The Astrophysical Journal 620: 744.

Geiss, B., et al., (2010) The Effect of Stellar Collisions and Tidal Disruptions on Post-Main-Sequence Stars in the Galactic Nucleus. American Astronomical Society, AAS Meeting #215, #413.15; Bulletin of the American Astronomical Society, Vol. 41, p.252.

Giddings, S. (1995). The Black Hole Information Paradox," Proc. PASCOS symposium/Johns Hopkins Workshop, Baltimore, MD, 22-25 March, 1995, arXiv:hep-th/9508151v1.

Hawking, S. W., (1988) Wormholes in spacetime. Phys. Rev. D 37, 904–910.

Hawking, S., (1990). A Brief History of Time: From the Big Bang to Black Holes. Bantam.

Hawking, S. (2005). "Information loss in black holes". Physical Review D 72: 084013.

Hawking, S. W. (2014). Information Preservation and Weather Forecasting for Black Holes.

Heisenberg, W. (1927), "Über den anschaulichen Inhalt der quantentheoretischen Kinematik und Mechanik", Zeitschrift für Physik 43 (3–4): 172–198.

Heisenberg. W. (1930), Physikalische Prinzipien der Quantentheorie (Leipzig: Hirzel). English translation The Physical Principles of Quantum Theory, University of Chicago Press.

Heisenberg, W. (1955). The Development of the Interpretation of the Quantum Theory, in W. Pauli (ed), Niels Bohr and the Development of Physics, 35, London: Pergamon pp. 12-29.

Heisenberg, W. (1958), Physics and Philosophy: The Revolution in Modern Science, London: Goerge Allen & Unwin.

Jaffe, R. (2005). "Casimir effect and the quantum vacuum". Physical Review D 72 (2): 021301.

Joseph, R (2010a) The Infinite Cosmos vs the Myth of the Big Bang: Red Shifts, Black Holes, and the Accelerating Universe. Journal of Cosmology, 6,

1548-1615.

Joseph, R. (2010b). The Infinite Universe: Black Holes, Dark Matter, Gravity, Acceleration, Life. Journal of Cosmology, 6, 854-874.

Joseph, R. (2014a) Paradoxes of Time Travel: The Uncertainty Principle, Wave Function, Probability, Entanglement, and Multiple Worlds, Cosmology, 18, 282-302.

Joseph, R. (2014a) The Time Machine of Consciousness, Cosmology, 18, In press.

Kerr, R P. (1963). "Gravitational Field of a Spinning Mass as an Example of Algebraically Special Metrics". Physical Review Letters 11 (5): 237–238.

Lambrecht, A. (2002) The Casimir effect: a force from nothing, Physics World, September 2002

Lorentz, H. A. (1892), "The Relative Motion of the Earth and the Aether", Zittingsverlag Akad. V. Wet. 1: 74–79

McClintock, J. E. (2004). Black hole. World Book Online Reference Center. World Book, Inc.

Melia, F. (2003). The Edge of Infinity. Supermassive Black Holes in the Universe. Cambridge U Press. ISBN 978-0-521-81405-8.

Melia, F. (2007). The Galactic Supermassive Black Hole. Princeton University Press. pp. 255–256.

Merloni, A., and Heinz, S., (2008) A synthesis model for AGN evolution: supermassive black holes growth and feedback modes Monthly Notices of the Royal Astronomical Society, 388, 1011 1030.

Minchin, R. et al. (2005). "A Dark Hydrogen Cloud in the Virgo Cluster". The Astrophysical Journal 622: L21–L24.

Morris, M. S. and Thorne, K. S. (1988). "Wormholes in spacetime and their use for interstellar travel: A tool for teaching general relativity". American Journal of Physics 56 (5): 395–412.

O'Neill, B. (2014) The Geometry of Kerr Black Holes, Dover

Penrose, R. (1969) Rivista del Nuovo Cimento.

Preskill, J. (1994). Black holes and information: A crisis in quantum physics", Caltech Theory Seminar, 21 October. arXiv:hep-th/9209058v1.

Pollack, G. L., & Stump, D. R. (2001), Electromagnetism, Addison-Wesley.

Rindler, W. (2001). Relativity: Special, General and Cosmological. Oxford: Oxford University Press.

Rodriguez, A. W.; Capasso, F.; Johnson, Steven G. (2011). "The Casimir effect in microstructured geometries". Nature Photonics 5 (4): 211–221.

Ruffini, R., and Wheeler, J. A. (1971). Introducing the black hole. Physics Today: 30–41.

Russell, D. M., and Fender, R. P. (2010). Powerful jets from accreting black holes: Evidence from the Optical and the infrared. In Black Holes and Galaxy Formation. Nova Science Publishers. Inc.

Schrödinger, E. (1926). "An Undulatory Theory of the Mechanics of Atoms and Molecules". Physical Review 28 (6): 1049–1070. Bibcode:1926PhRv...28.1049S. doi:10.1103/PhysRev.28.1049.

Sen, A. (2002) Rolling tachyon," JHEP 0204, 048

Slater, J. C. & Frank, N. H. (2011) Electromagnetism, Dover.

Smolin, L. (2002). Three Roads to Quantum Gravity. Basic Books.

Taylor, E. F. & Wheeler, J. A. (2000) Exploring Black Holes, Addison Wesley.

Thorne, K. (1994) Black holes and time warps, W. W. Norton. NY.

Thorne, K. S. & Hawking, S. (1995). Black Holes and Time Warps: Einstein's Outrageous Legacy, W. W. Norton.

Thorne, K. et al. (1988). "Wormholes, Time Machines, and the Weak Energy Condition". Physical Review Letters 61 (13): 1446.

Wilf, M., et al. (2007) The Guidebook to Membrane Desalination Technology: Reverse Osmosis, Balaban Publishers.